Materials Science and Engineering in Food Product Development

Materials Science and Engineering in Food Product Development

Edited by

Wing-Fu Lai
The Hong Kong Polytechnic University
Hong Kong

WILEY

This edition first published 2023

Registered Offices
John Wiley & Sons, Inc., 111 River Street, Hoboken, NJ 07030, USA
John Wiley & Sons Ltd, The Atrium, Southern Gate, Chichester, West Sussex, PO19 8SQ, UK

For details of our global editorial offices, customer services, and more information about Wiley products visit us at www.wiley.com.

Wiley also publishes its books in a variety of electronic formats and by print-on-demand. Some content that appears in standard print versions of this book may not be available in other formats.

Library of Congress Cataloging-in-Publication Data
Names: Lai, Wing-Fu, editor. | John Wiley & Sons, publisher.
Title: Materials science and engineering in food product development / edited by Wing Fu Lai.
Description: Hoboken, NJ : John Wiley & Sons, 2023. | Includes bibliographical references and index.
Identifiers: LCCN 2023000273 (print) | LCCN 2023000274 (ebook) | ISBN 9781119860358 (Hardback) | ISBN 9781119860365 (pdf) | ISBN 9781119860587 (epub) | ISBN 9781119860594 (ebook)
Subjects: LCSH: Food industry and trade--Technological innovations | Food--Composition | Food--Analysis. | Food science.
Classification: LCC TP370 .M365 2023 (print) | LCC TP370 (ebook) | DDC 664/.07--dc23/eng/20230213
LC record available at https://lccn.loc.gov/2023000273
LC ebook record available at https://lccn.loc.gov/2023000274

Cover Image: © Sompong Sriphet/EyeEm/Getty Images
Cover Design: Wiley

Set in 9.5/12.5pt STIXTwoText by Integra Software Services Pvt. Ltd, Pondicherry, India

Contents

About the Editor *xiv*
List of Contributors *xv*
Preface *xix*
List of Abbreviations *xxi*

1 Overview of Different Materials Used in Food Production *1*
Nahed A. Abd El-Ghany and Mahmoud H. Abu Elella
1.1 Introduction *1*
1.2 Advanced Materials Engineering for Food Product Development *3*
1.2.1 Microstructured and Nanostructured Materials *3*
1.2.2 Preparation Methods *3*
1.2.2.1 Spray-Drying Technique *3*
1.2.2.2 Electrospinning Technique *3*
1.2.2.3 Coacervation Technique *4*
1.2.2.4 Emulsion Technique *5*
1.2.2.5 Ionic Gelation Technique *5*
1.2.2.6 Liposome Formulations *5*
1.3 Encapsulation of Food Ingredients for Food Product Development *6*
1.3.1 Encapsulation Based on Polysaccharides *8*
1.3.1.1 Chitosan *8*
1.3.1.2 Starch *8*
1.3.1.3 Alginate *9*
1.3.1.4 Carrageenans *9*
1.3.2 Encapsulation Based on Liposomes *9*
1.3.3 Encapsulation Based on Proteins *10*
1.4 Food Packaging Approach for Food Product Development *13*
1.4.1 Polysaccharide-Based Food Packaging *13*
1.4.1.1 Chitosan *14*
1.4.1.2 Alginate *15*
1.4.1.3 Starch *15*
1.4.2 Protein-Based Food Packaging *16*

1.4.2.1 Gelatin *16*
1.4.2.2 Casein *17*
1.5 Hydrogel Structures and Their Efficiency in Food Development *17*
1.6 Conclusion *18*
Glossary *19*
References *20*

2 Introduction to Food Properties and Techniques in Food Product Development *27*
Sara A. Al-Hafiry, Omar A. Alaboudi, Ghada A. Ahmed, and Abdulrahman M. Abdulrahman
2.1 Introduction *28*
2.2 Structural Impact on Properties *28*
2.2.1 Food Materials Science *28*
2.2.2 Food Matrix *29*
2.2.3 Food Characterization and Study Standardization *30*
2.3 Food Consumer Demands *31*
2.4 Food Properties to Be Improved *31*
2.4.1 Introduction to Food Properties *31*
2.4.2 Functional Ingredients and Health Properties *32*
2.4.3 Physiochemical and Sensory Properties *32*
2.4.4 Physical Properties *33*
2.4.5 Food Stability *34*
2.4.6 Kinetic Properties *34*
2.5 Food Materials Synthesis Techniques *34*
2.5.1 Food Formulation *34*
2.5.2 Food Processing *35*
2.5.3 3D Printing *36*
2.6 Concluding Remarks *37*
Glossary *37*
References *38*

3 Basic Concepts of Bulk Rheology in Food Emulsions *41*
Carlos Bengoechea, Estefanía Álvarez-Castillo, José Manuel Aguilar, and Antonio Guerrero
3.1 Introduction *41*
3.2 Emulsification Process *42*
3.3 Rheology of Continuous Phase *44*
3.4 Rheology of Emulsions *45*
3.5 Microstructure *48*
3.6 Destabilization Mechanisms *49*
3.7 Concluding Remarks *52*
Acknowledgments *52*
Glossary *52*
References *54*

4 Understanding Interfacial Rheology in Food Emulsions *57*
Cecilio Carrera, Manuel Felix, María Luisa López-Castejón, and Víctor Manuel Pizones
4.1 Introduction *57*
4.2 Interfacial Engineering of Food Emulsifiers *58*
4.3 Rheological Techniques for the Characterization of Interfacial Films *60*
4.3.1 Interfacial Dilatational Rheology *61*
4.3.1.1 Oscillating Droplet Techniques *62*
4.3.1.2 Trough Methods *62*
4.3.2 Interfacial Shear Rheology *63*
4.3.2.1 Du Noüy Ring Tool *65*
4.3.2.2 Bicone Tool *65*
4.3.2.3 DWR Tool *65*
4.3.2.4 Magnetic Rods Based Tools *66*
4.3.3 Comparison Between Interfacial Dilatational and Shear Rheology *66*
4.4 Concluding Remarks and Future Perspectives *68*
Acknowledgments *69*
Glossary *69*
References *69*

5 Overview of Types of Materials Used for Food Component Encapsulation *73*
Zahra Emam-Djomeh, Mohammad Ekrami, and Ali Ekrami
5.1 Introduction *73*
5.2 Major Techniques Used for Food Component Encapsulation *74*
5.3 Materials Used as Carrier Source for Encapsulation *77*
5.4 Protein-Based Carriers *78*
5.4.1 Soy Proteins *78*
5.4.2 Cereal Proteins *78*
5.4.3 Egg Proteins *79*
5.4.4 Pulse Proteins *79*
5.4.5 Silk Proteins *79*
5.4.6 Meat Proteins *79*
5.4.6.1 Milk Proteins *79*
5.5 Carbohydrate-Based Carriers *80*
5.5.1 Agar *81*
5.5.2 Starch *81*
5.5.3 Cellulose and Its Derivatives *81*
5.5.4 Pectin *81*
5.5.5 Gums *81*
5.5.6 Chitin and Chitosan *82*
5.5.7 Alginate *82*
5.5.8 Carrageenan *82*
5.5.9 Dextran *82*
5.5.10 Cyclodextrins *83*
5.6 Lipid-Based Carrier *83*

5.6.1 Emulsions *83*
5.6.2 Liposomes *83*
5.6.3 Solid Lipid Nanoparticles (SLN) *84*
5.6.4 Nanostructured Lipid Carriers (NLC) *84*
5.7 Roles Played by Materials in Food Component Encapsulation *86*
5.8 Improved Dispersibility *87*
5.9 Addition of Inhibitors *87*
5.10 Reducing the Interactions *87*
5.11 Control of Light Scattering and Absorption *87*
5.12 Increased Bioavailability *87*
5.13 Controlled or Targeted Release *88*
5.14 Conclusions *88*
Glossary *88*
Reference *89*

6 Design and Use of Microcarriers for the Delivery of Nutraceuticals *93*
Maxim V. Kiryukhin, Su Hui Lim, and Cheryl Yingxue Chia
6.1 Introduction *93*
6.2 Protection Against Environmental Conditions *96*
6.3 Controlled Release by Responsive Carrier Material *100*
6.4 Active Enhancement of M&Ns' Bioavailability Through Microencapsulation *107*
6.5 Conclusion *112*
Glossary *113*
Reference *114*

7 Design and Use of Lipid-Based Systems for Food Component Encapsulation *117*
Kalpani Y. Perera, Dileswar Pradhan, Shubham Sharma, Amit K. Jaiswal, and Swarna Jaiswal
7.1 Introduction *117*
7.2 Lipid-Based Nano Delivery Systems for Food Component Encapsulation *121*
7.2.1 Nanoemulsions *121*
7.2.2 Liposome *122*
7.2.3 Solid Lipid Nanoparticles *123*
7.2.4 Nanostructured Lipid Carriers *125*
7.3 Mechanism of Action of Encapsulated Food Components *127*
7.4 Encapsulation of Food Components in Lipid-Based Nano Delivery Systems *129*
7.4.1 Essential Oils *129*
7.4.2 Antioxidants *129*
7.4.3 Natural Colorants *130*
7.4.4 Flavors *131*
7.4.5 Vitamins *131*
7.4.6 Probiotics *132*
7.4.7 Fatty Acids *133*

7.5 Conclusion and Future Perspectives *134*
Glossary *134*
Reference *135*

8 Working Principles and Use of Gelatin for Food Component Encapsulation *139*
Youssef S. Abdelaziz, Rana Tarek, Donia G. Youssef, Mariam Khaled Abdel-Latif, Habiba Mohamad Ibrahim, Sohaila Mohammed Salah Saleh, and Heba M. Fahmy *139*
8.1 Introduction *139*
8.2 Why Use Gelatin in Encapsulation Technology? *140*
8.3 Techniques for Food Encapsulation Using Gelatin *142*
8.3.1 Coacervation *142*
8.3.2 Extrusion *144*
8.3.3 Fluid-Bed Coating System *144*
8.3.4 Spray Drying *145*
8.4 Microencapsulation Using Gelatin *145*
8.5 Nanoencapsulation of Food Components Using Gelatin *149*
8.6 Mechanisms of Release of Gelatin Encapsulation Systems for Food Components *153*
8.7 Conclusion *155*
Glossary *156*
Reference *156*

9 Working Principles and Use of Chitosan for Food Component Encapsulation *161*
Gastón Bravo-Arrepol, Plamen Dimitrov Katsarov, Bissera Asenova-Pilicheva, Paolina Kancheva- Lukova, Danilo Escobar-Avello, Hazel Peniche, Lorenzo García, Carlos Peniche-Covas, Philippe Michaud, Cédric Delattre, Liliam Becheran-Maron, Johanna Castaño, Maria Dolores Lopez, Oscar Valdes, Aleksandra Nesic, and Gustavo Cabrera-Barjas
9.1 Introduction *162*
9.2 Encapsulation Technologies *164*
9.3 Agent Encapsulation Using Chitosan as Polymeric Matrix *166*
9.3.1 Encapsulation of Vitamins and Omega Fatty Acids *166*
9.3.2 Encapsulation of Peptides and Proteins *184*
9.3.3 Encapsulation of Probiotics, Prebiotics, and Enzymes *185*
9.3.4 Encapsulation of Essential Oils *187*
9.3.5 Encapsulation of Phenolic Compounds *187*
9.3.6 Encapsulation of Carotenoids *189*
9.4 Potential Applications of Microencapsulated Materials in Food Packaging *191*
9.4.1 3D Food Printing *192*
9.4.2 Spinning Technology *197*
9.5 Market for Chitosan Uses in Food Application *199*
9.6 Concluding Remarks *200*
Glossary *201*
Reference *201*

10 Design and Use of Hydrogels for Food Component Encapsulation *209*
Zahra Emam-Djomeh, Mohammad Ekrami, and Ali Ekrami
10.1 Introduction *209*
10.2 Classification of Hydrogels *212*
10.3 Hydrogel Formation *214*
10.4 Recent Advances in Hydrogel Development *214*
10.4.1 3D Printed Hydrogels *214*
10.4.2 Superabsorbent Hybrid Hydrogels *215*
10.4.3 Conducting Polymer Hydrogels *215*
10.5 Retention and Release Properties *216*
10.6 Applications of Hydrogels in Food Production *217*
10.6.1 Encapsulation of Bioactive Compounds *218*
10.6.2 Encapsulation of Fats and Oils *219*
10.7 Conclusions *221*
Glossary *221*
References *222*

11 Optimization of Pasting and Textural Properties of Food Products *227*
Filopateer Nasser, Salma Hossam Mohamed, Mariam Ashraf Fouad Khalil, Omaima Ali Mostafa Mohammed, Radwa Magdy Mohamed, and Heba Mohamed Fahmy
11.1 Introduction *227*
11.2 Physical and Chemical Modification of Starch Structures *228*
11.3 Manipulation of Starch Properties Using Hydrocolloids *232*
11.4 Enzymatic Modification of Starch Properties *234*
11.5 Use of Starch Modification in Food Production *235*
11.6 Concluding Remarks *236*
Glossary *237*
References *237*

12 Phase Change Materials in Food Dryers *243*
Hasibul Hasan Himel, Sabit Hasan, Nufile Uddin Ahmed, and Mahadi Hasan Masud
12.1 Introduction *243*
12.2 Phase Change Materials and Their Properties *244*
12.2.1 Basic Mechanism *245*
12.2.2 Classification *245*
12.2.2.1 Based on Melting Temperature *245*
12.2.2.2 Based on the Mode of Phase Change *246*
12.2.2.3 Based on Chemical Composition *246*
12.2.3 Key Properties *247*
12.2.3.1 Thermal Properties *248*
12.2.3.2 Physical Properties *248*
12.2.3.3 Kinetic and Chemical Properties *248*
12.3 Potential of PCMs in Food Drying *250*

12.3.1 Energy Potential of the Upgrade *250*
12.3.2 Economy of PCM Integration *251*
12.3.3 Safety Concerns Related to Food and PCMs *251*
12.4 Current Status of Utilizing PCMs for Food Drying *252*
12.5 Recommendation for Optimization of PCM for Use in Solar Dryers *255*
12.5.1 Thermal Conductivity *255*
12.5.2 Long-Term Stability *255*
12.5.3 Environmental Impact *256*
12.6 Concluding Remarks and Future Perspectives *257*
Glossary *258*
References *258*

13 Multi-Functional Properties of Halloysite Nano-Clays in Food Safety and Security *261*
Satwik Majumder and Saji George
13.1 Overview *261*
13.2 Halloysite Nanotubes (HNT): A Versatile Natural Nanomaterial *263*
13.2.1 Halloysite as Nanofillers in Packaging Matrix *265*
13.2.2 Halloysite as Nano-Carrier in Active Packaging Systems *266*
13.2.3 Halloysite as a Colorimetric Indicator in Intelligent Packaging Systems *267*
13.2.4 Halloysite in Food Coatings, Capsules, and Fibrous Films *268*
13.2.5 Halloysite in Plant and Animal Agriculture *268*
13.3 Toxicity and Migration Associated with Halloysite *270*
13.4 Future Perspectives *271*
13.5 Conclusive Remarks *272*
Glossary *273*
References *273*

14 Electrospinning Technologies for Encapsulation of Probiotics *279*
Seethu, B.G., Aditya Sukumar P., Devikrishna P., Harshvardhan Kulkarni, Magdaline Eljeeva Emerald, Chandram Grover, and Heartwin A. Pushpadass
14.1 Introduction *279*
14.2 Major Methods for Encapsulation of Probiotics *281*
14.2.1 Freeze Drying *282*
14.2.2 Spray Drying *283*
14.2.3 Electrospinning *285*
14.2.4 Our Approach to Encapsulate *L. Rhamnosus* CRD11 by Electrospinning *287*
14.2.4.1 Determining the Viability and Properties of Encapsulated *L. Rhamnosus* *290*
14.2.4.2 Determining the Effect of Prebiotics on the Electrospinning of *L. Rhamnosus* *293*
14.2.5 Comparisons of Electrospinning with Other Encapsulation Methods *298*
14.3 Conclusions *298*
Glossary *299*
References *299*

15 Three-Dimensional Printing in Food Manufacturing and Mechanics *303*
Stefania Chirico Scheele, Martin Binks, and Paul F. Egan
15.1 Introduction *303*
15.2 Print Process *306*
15.3 Material Preparation *308*
15.4 Printing Parameters *309*
15.5 Food Mechanics *311*
15.6 Consumer Validation *314*
15.7 Concluding Remarks *315*
Glossary *316*
References *316*

16 Techniques for Characterization of Food-Packaging Materials *321*
Shubham Sharma, Kalpani Y. Perera, Dileswar Pradhan, Brendan Duffy, Amit K. Jaiswal, and Swarna Jaiswal
16.1 Introduction *321*
16.2 Characterization of Food-Packaging Material *323*
16.2.1 Microstructure of Food-Packaging Material *323*
16.2.1.1 Scanning Electron Microscope *323*
16.2.1.2 Transmission Electron Microscopy *323*
16.2.1.3 Atomic Force Microscopy *327*
16.2.2 Optical Properties of Food-Packaging Material *328*
16.2.2.1 Ultraviolet-Visible Spectroscopy *328*
16.2.2.2 ColorQuest XE Spectrophotometer *329*
16.2.2.3 Dynamic Light Scattering Spectroscopy *330*
16.2.2.4 Zeta Potential Spectroscopy *330*
16.2.3 Chemical Properties of Food-Packaging Material *331*
16.2.4 Thickness of Food-Packaging Material *331*
16.2.5 Oxygen Barrier Properties of Food-Packaging Material *332*
16.2.6 Mechanical Properties of Food-Packaging Material *332*
16.2.7 Thermal Properties of Food-Packaging Material *333*
16.2.7.1 Thermogravimetric Analysis *333*
16.2.7.2 Differential Scanning Calorimetry *334*
16.2.8 Permeation Properties of Food-Packaging Material *334*
16.2.8.1 Water Contact Angle *334*
16.2.8.2 Water Vapor Permeability *335*
16.2.9 Biodegradability of Food-Packaging Material *336*
16.3 Conclusion and Prospects *336*
Glossary *337*
References *337*

17 Development and Use of Edible Materials for Food Protection and Packaging *341*
Lin Lin, Mohamed Abdel-Shafi Abdel-Samie, Sherif M. Abed, and Haiying Cui
17.1 Introduction *341*
17.2 Antimicrobial and Antioxidant Active Agents Used in the Field of Food Packaging *343*

17.2.1 EOs Used as AMA and AOA in Food Packaging *343*
17.2.2 Peptides and Organic Acids Used as AMA and AOA in Food Packaging *344*
17.2.3 Natural Extracts and Other Agents Used as AMA and AOA in Food Packaging *344*
17.3 Carriers Applied in Food-Packaging Applications *345*
17.3.1 Starch and Chitosan *345*
17.3.2 Cellulose and Edible Bioplastics *347*
17.3.3 Other Materials Used as Carriers in Food Packaging *347*
17.4 Methods of Fabrication or the Enhancement Activity of Edible Packaging Films *349*
17.4.1 Coacervation *349*
17.4.2 Electrospinning *350*
17.4.3 Liposomes *352*
17.5 Controlled Release of the BACs from Encapsulation Materials *353*
17.6 Conclusion *353*
Glossary *354*
References *355*

18 Packaging Design as Part of a Holistic Food Quality Assurance Process *361*
Agnieszka S. Cholewa-Wójcik and Agnieszka K. Kawecka
18.1 Introduction *361*
18.2 Essence of Quality-Oriented Product Designing and Its Role in Quality Assurance *362*
18.3 Quality-Oriented Product-Designing Process *362*
18.4 Integrated Product Designing as the New Approach to Packaged Product Designing Process *365*
18.5 Methods to Aid Shaping of Quality of Products Being Designed *370*
18.6 Concluding Remarks and Future Perspectives *371*
Acknowledgments *372*
Glossary *372*
References *372*

19 Determinants of the Quality and Safety of Food Packaging *377*
Agnieszka K. Kawecka and Agnieszka S. Cholewa-Wójcik
19.1 Introduction *377*
19.2 Literature Review Concerning Food-Packaging Safety *378*
19.3 Packaging Safety Hazards *379*
19.4 Legal Requirements for the Safety of Food Packaging *381*
19.5 The Process of Ensuring Security – the Supply Chain *384*
19.6 Packaging Safety Features and Attributes of Food Packaging *386*
19.7 Concluding Remarks *388*
Acknowledgments 388
Glossary 388
References 389

Index *393*

About the Editor

Wing-Fu Lai

Department of Applied Biology and Chemical Technology, The Hong Kong Polytechnic University, Hong Kong

Wing-Fu Lai received his MSc degree in materials engineering and nanotechnology from the City University of Hong Kong, and earned his PhD in chemistry from the University of Hong Kong. He has been accredited as Shenzhen Municipal "Overseas High-caliber Personnel". Professionally, he is a registered nutritionist in the United Kingdom, and a Fellow of the UK Higher Education Academy. He has also received his certified food scientist credential from the International Food Science Certification Commission in the United States. His research focus is on the development and engineering of polymeric materials for food and pharmaceutical applications.

List of Contributors

Nahed A. Abd El-Ghany
Cairo University
Egypt

Youssef S. Abdelaziz
Cairo University
Egypt

Mariam Khaled Abdel-Latif
Cairo University
Egypt

Mohamed Abdel-Shafi Abdel-Samie
Arish University
Egypt

Abdulrahman M. Abdulrahman
Cairo University
Egypt

Sherif M. Abed
Arish University
Egypt

Mahmoud H. Abu Elella
Cairo University,
Egypt

Sukumar P. Aditya
ICAR-National Dairy Research Institute
India

José Manuel Aguilar
Universidad de Sevilla
Spain

Ghada A. Ahmed
Cairo University
Egypt

Nufile Uddin Ahmed
Rajshahi University of Engineering and Technology
Bangladesh

Omar A. Alaboudi
Cairo University
Egypt

Sara A. Al-Hafiry
Cairo University
Egypt

Estefanía Álvarez-Castillo
Universidad de Sevilla
Spain

Bissera Asenova-Pilicheva
Medical University of Plovdiv
Bulgaria

Liliam Becheran-Maron
University of Habana
Cuba

Carlos Bengoechea
Universidad de Sevilla
Spain

Martin Binks
Texas Tech University
USA

Gastón Bravo-Arrepol
Universidad de Concepción
Chile

Gustavo Cabrera-Barjas
Universidad de Concepción
Chile

Cecilio Carrera
Universidad de Sevilla
Spain

Johanna Castaño
Universidad San Sebastián
Chile

Cheryl Yingxue Chia
Agency for Science Technology and Research (A*STAR)
Singapore

Agnieszka S. Cholewa-Wójcik
Cracow University of Economics, Poland

Haiying Cui
Jiangsu University, China

Cédric Delattre
Université Clermont Auvergne
France

P. Devikrishna
ICAR-National Dairy Research Institute
India

Brendan Duffy
Technological University Dublin – City Campus
Ireland

Paul F. Egan
Texas Tech University
US

Ali Ekrami
University of Tehran
Iran

Mohammad Ekrami
University of Tehran
Iran

Zahra Emam-Djomeh
University of Tehran
Iran

Magdaline Eljeeva Emerald
ICAR-National Dairy Research Institute
India

Danilo Escobar-Avello
Universidad de Concepción
Chile

Heba Mohamed Fahmy
Badr University in Cairo
Egypt

Manuel Felix
Universidad de Sevilla
Spain

Lorenzo García
University of Havana
Cuba

Saji George
McGill University
Canada

Chandram Grover
ICAR-National Dairy Research Institute
India

Antonio Guerrero
Universidad de Sevilla
Spain

Sabit Hasan
Rajshahi University of Engineering and Technology
Bangladesh

Hasibul Hasan Himel
Rajshahi University of Engineering and Technology
Bangladesh

Habiba Mohamad Ibrahim
Cairo University
Egypt

Amit K. Jaiswal
Technological University Dublin-City Campus
Ireland

Swarna Jaiswal
Technological University Dublin-City Campus
Ireland

Plamen Dimitrov Katsarov
Medical University of Plovdiv
Bulgaria

Agnieszka K. Kawecka
Cracow University of Economics
Poland

Mariam Ashraf Fouad Khalil
Cairo University
Egypt

Maxim V. Kiryukhin
Singapore Institute of Food and Biotechnology Innovation
A*STAR
Singapore

Harshvardhan Kulkarni
ICAR-National Dairy Research Institute
India

Su Hui Lim
Agency for Science,
Technology and Research (A*STAR)
Singapore

Lin Lin
Jiangsu University
China

Maria Dolores Lopez
Universidad de Concepción
Chile

María Luisa López-Castejón
Universidad de Sevilla
Spain

Paolina Kancheva-Lukova
Medical University of Plovdiv
Bulgaria

Satwik Majumder
McGill University
Canada

Mahadi Hasan Masud
Rajshahi University of Engineering & Technology
Bangladesh

Philippe Michaud
Université Clermont Auvergne
France

Radwa Magdy Mohamed
Cairo University
Egypt

Salma Hossam Mohamed
Cairo University
Egypt

Omaima Ali Mostafa Mohammed
Cairo University
Egypt

Filopateer Nasser
Cairo University
Egypt

Aleksandra Nesic
Universidad de Concepción
Chile

Hazel Peniche
University of Havana
Cuba

Carlos Peniche-Covas
University of Havana
Cuba

Kalpani Y. Perera
Technological University Dublin-City Campus
Ireland

Víctor Manuel Pizones
Universidad de Sevilla
Spain

Dileswar Pradhan
Technological University Dublin-City Campus
Ireland

Heartwin A. Pushpadass
ICAR-National Dairy Research Institute
India

Sohaila Mohammed Salah Saleh
Cairo University
Egypt

Stefania Chirico Scheele
Texas Tech University
US

B. G. Seethu
ICAR-National Dairy Research Institute
India

Shubham Sharma
Technological University Dublin-City Campus
Ireland

Rana Tarek
Cairo University
Egypt

Oscar Valdes
Universidad Católica del Maule
Chile

Donia G. Youssef
Cairo University
Egypt

Preface

Food materials science is a discipline focusing on the development, characterization, processing, and engineering of materials used in food. Although it has practical significance in the food industry, it is an emerging area in the larger field of food science and little scholarly works are available in this field. Among those available works, most of them only focus on the characterization and properties of food materials *per se*. Efforts paid to explore how related advances can be translated into the development and improvement of a functional food product are almost absent. The objective of this book is to fill this gap by exploring and illustrating the roles played by materials science and engineering in the process of food production, particularly food microencapsulation and food packaging.

This book covers a wide range of topics in food materials science and engineering. Chapters 1–6 will present major concepts related to material properties in food science. An overview of different types of materials used in food applications will be provided. Concepts of bulk rheology and interfacial rheology in food emulsions will also be presented. This section lays a theoretical foundation for subsequent sections in this book. In Chapters 7–10, detailed discussions about the design and use of lipid-based and polymer-based materials in enrichment and protection of food components will be provided. Apart from chemical means, various engineering techniques (including 3D printing and electrospinning) can be applied to manipulate the properties of food materials or those used for food microencapsulation. Some of the major engineering techniques relevant to the process of food production will be discussed in Chapters 11–15. Chapters 16–19 will focus mainly on the design, characterization, and use of materials for food packaging. Various determinants of the quality and safety of food packaging, as well as concepts for the design of food packages, will be covered.

Contrary to existing books that largely focus on the chemical and physical principles of food materials science and hence are sometimes too theoretical to be directly adopted by food manufacturers in their professional practice, this book will approach the subject of food materials science with practical and industrial perspectives. Real-life examples will be provided to demonstrate how food products, especially functional foods, can benefit from the incorporation of materials science technologies. In addition, to benefit scholars, students, and a broader audience of interested readers, the book includes helpful glossary sections in each chapter. Important notes and tips to food manufacturers to translate the contents of the chapter from theory to real-life practice will also be provided in each chapter. This is the first book of its kind. It is not only a valuable reference book to researchers in the field, but can also serve as a guide for food manufacturers during the development of the food product.

Here I would like to express my gratitude to the contributors of different chapters of this book. Their support and efforts have made publication of this book possible. Thanks are extended to the staff in Wiley. The quality of this book, and its value to its readers, depends largely on the promptness with which submitted manuscripts are reviewed. I would like to thank the reviewers for putting their efforts to evaluate and select the best manuscripts for inclusion in this book, and for providing constructive suggestions to the contributors of those selected chapters. Haotian Zhang from the Chinese University of Hong Kong is also acknowledged for his editorial assistance throughout the process of this book's publication. I would like to express my appreciation in advance for every observation and suggestion toward further improvement of this material.

Wing-Fu Lai

List of Abbreviations

AFM	Atomic force microscopy
AMA	Antimicrobial agents
AMR	Antimicrobial resistance
AN	Anthocyanin nanoliposomes
AOA	Antioxidant agents
BSA	Bovine serum albumin
CAD	Computer-aided design
CEO	Clove essential oil
DD	Degree of deacetylation
DLS	Dynamic light scattering
DoE	Design of experiment
DP	Degree of polymerization
DQM	Design quality management
DS	Degree of substitution
DSC	Differential scanning calorimetry
DSD	Droplet size distribution
DWR	Double-wall ring
EAMPS	Edible antimicrobial packaging systems
EB	Elongation at break
EM	Electron microscopy
EMA	European Medicines Agency
EO	Essential oil
EVA	Ethylene vinyl acetate
F&D	Food and Drugs
FD&C	Food Drug and Cosmetic
FDA	Food and Drug Administration
FESEM	Field emission gun scanning electron microscope
FOS	Fructo-oligosaccharides
GA	Gum arabic
GK	Garcinia kola
GMP	Good manufacturing practice
GOS	Galacto-oligosaccharides
GRAS	Generally recognized as safe
GTE	Green tea extract

HIU	High-intensity ultrasound
HOSO	High-oleic sunflower oil
HSH	High-shear homogenization
HU	*Hunteria umbellata*
LA	Lactic acid
LH	Latent heat
LHS	Latent heat storage
LMWE	Low molecular weight emulsifiers
M&N	Micronutrients and nutraceuticals
MW	Molecular weight
NLC	Nanostructured lipid carrier
NMR	Nuclear magnetic resonance
OEO	Oregano essential oil
OP	Oxygen permeability
OSD	Open sun-drying
OTR	Oxygen transmission rate
PBAT	Polyadipate butylene terephthalate
PCM	Phase change materials
PPE	Pineapple peel extract
PV	Peak viscosity
QFD	Quality function deployment
RA	Rosmarinic acid
RB	Relative breakdown
RS	Resistant starch
SAOS	Small amplitude oscillatory shear
SDS	Sodium dodecyl sulfate
SEM	Scanning electron microscopy
SH	Sensible heat
SLN	Solid lipid nanoparticle
SPI	Soy protein isolate
SPM	Scanning probe microscopy
TC	Thermal conductivity
TEM	Transmission electron microscopy
TES	Thermal energy storage
TGA	Thermogravimetric analysis
TP	Tea polyphenol
TS	Tapioca starch
WPI	Whey protein isolate
WVP	Water vapor permeability
ZLO	*Zanthoxylum limonella* oil

1

Overview of Different Materials Used in Food Production

Nahed A. Abd El-Ghany and Mahmoud H. Abu Elella*

Chemistry Department, Faculty of Science, Cairo University, Giza, Egypt
** Corresponding author*

Highlights

- Materials science and engineering can be applied to different aspects of food science, ranging from encapsulation of food ingredients to food packaging.
- Materials engineering, depending on biopolymers, has gained extensive interest because polymers have shown outstanding properties, such as nontoxicity, ease of availability, biocompatibility, biodegradability, and low cost.
- Materials engineering can enhance food product quality, which is all about sensory features, such as taste, flavor, palatability, and semblance.
- Advances in materials science and engineering are expected to bring new opportunities to the food industry.

1.1 Introduction

Rapid growth of materials engineering science has provided a lot of functionalized materials for food product development application in the recent years. Three advanced types of functional materials that have been widely applied in food industry are nanostructured and microstructured materials, and three-dimensional hydrogels [1]. In general, materials engineering science normally represents a solid state of matter and is an integrated field comprising chemistry, physical attributes, and processing. Additionally, it involves the maintenance of the materials' properties, for example, chemical (structure and composition), physical (thermal and optical), dimensional (shape and size), and mechanical (toughness and strength). On the other hand, food product development has been gaining more interest among many industrial and academic researchers around the world to improve the quality of food products. Notably, the major components of food are carbohydrates, and proteins that are called biopolymers [2]. In general, nowadays, polymeric materials are considered as an important class of materials in a wide range of applications, thanks to their physicochemical properties [3]. They are macromolecules composed of repeating units that are known as monomers joined by covalent bonds. According to their origin, they are classified as either natural (if produced from natural sources, such as plants, animals, and

Materials Science and Engineering in Food Product Development, First Edition. Edited by Wing-Fu Lai.

microorganisms) or synthetic [4]. Recently, biopolymers have gained more attention from global researchers in food development applications since they have fabulous properties, such as biodegradability, biocompatibility, low cost, nontoxicity, and ease of availability [4a, 5]. They include naturally extracted polymers from animal and plant origins, for example, polysaccharides and proteins. Their repeated units include sugar or protein chains [6].

Polysaccharides are an example of natural biopolymers that are composed of carbohydrate chains with a large polymeric oligosaccharide formed through glycosidic linkages between multiple monosaccharides as repeating units [7]. Polysaccharides are the most abundant natural organic compounds. Additionally, they can be extracted from natural renewable resources, including plants (e.g. cellulose), animals (e.g. chitosan and alginate), and microorganisms (e.g. xanthan gum) [5a, 7a, 8]. Also, they are classified into two categories, for instance, homopolysaccharides and heteropolysaccharides. Homopolysaccharides are composed of the same monosaccharide-repeating unit as cellulose, whereas, heteropolysaccharides are composed of various repeating units including alginate [9]. Furthermore, polysaccharides have been used in various applications owing to their sustainable properties, such as ease of availability at less cost, ease of modifications and manufacturing, biocompatibility, biodegradability, nontoxicity, and bioactivity [5a, 7a, 10]. Conversely, proteins have polyamide chains, and they are one of the main constituents of the human body because they play both dynamic and diverse roles, such as catalyzing reactions, building cellular structures, and controlling cell fates. They have fabulous physicochemical properties, including isoelectric point (pI), chemical compositions, denaturation thermal temperature (Tm), and solubility [11].

Nanotechnology has revolutionized several scientific and industrial fields, including the food business. Food processing, food packaging, functional food development, food safety, detection of foodborne pathogens, and shelf-life extension of food and/or food products have emerged because of the growing need for nanoparticles in various fields of food science and food microbiology. On the other hand, hydrogels in the food science sector are efficient materials in the field of food quality improvement, nutrient-modification, sensory perception optimization, targeted nutrient delivery and protection, calorie control, risk monitoring for food safety, and food packaging. Although applications of hydrogels in the food industry are still limited, there are large areas to promote their use in food science. As a result, it is expected that the hydrogel structure's reasonable design will lead to more useful applications in order to keep up with the development of new foods [12]. In this chapter, we focus on shaping up the biopolymer-based nanostructured, microstructured, and hydrogel materials as shown

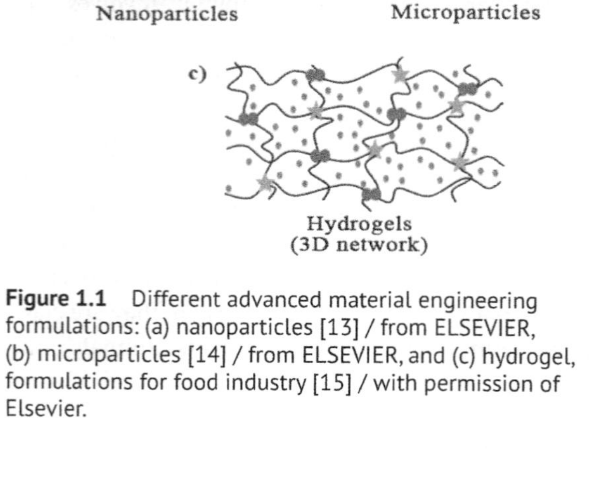

Figure 1.1 Different advanced material engineering formulations: (a) nanoparticles [13] / from ELSEVIER, (b) microparticles [14] / from ELSEVIER, and (c) hydrogel, formulations for food industry [15] / with permission of Elsevier.

in Figure 1.1, for encapsulation of different vital food ingredients in the food packaging field and explore their effect on food safety and quality that are essential for food development.

1.2 Advanced Materials Engineering for Food Product Development

1.2.1 Microstructured and Nanostructured Materials

Microstructured materials refer to the formulation of particle-sized compounds in the range of 1–1000 μm in diameter for different purposes, such as controlling and sustained bioactive compounds delivery, in addition to protecting the bioactive compounds from harsh environmental conditions. They have outstanding properties including a micro-size diameter and have the ability to encapsulate macromolecules with a high molecular weight [16]. For example, microcapsules based on the biopolymer mixture of chitosan and alginate have been reported in the literature [17], for encapsulating biologically active compounds, such as *Garcinia kola* (GK) and *Hunteria umbellata* (HU) seeds. The results showed that the extracted seeds have selective release patterns based on the pH of the medium. Also, a slower release of GK and HU from microcapsules was observed in an acidic medium (pH 1.2), but rose in a slightly neutral medium (pH 6.8). Nanostructured materials can be described as chemically and morphologically deposited matters in the range of 1–300 nm in diameter. All sorted materials used form the nanoscale and are classified from atoms to polymers. Moreover, nanostructured biopolymers are functional materials and controlling their architecture leads to achieved materials with amazing properties. For example, due to their nanometric dimension, which is less than the wavelength of light, they can display optical properties such as anti-reflectivity and structural colors [18].

1.2.2 Preparation Methods

1.2.2.1 Spray-Drying Technique

Spray-drying technique has been one of most widely used methods to design microparticle materials in the past decades due to its fabulous features, such as simplicity, speed, low cost, ease of scaling up, and flexibility [19]. It is also used to prepare microcapsule formulations for drug delivery applications in which the core material is dispersed in the solution of the shell material, such as water, after which, it is fed into the drying chamber while atomized under hot air coming from a pressure nozzle. Subsequently, the solvent is evaporated under the hot air stream, leaving a microparticle of solid. Additionally, this approach is a simple and flexible one to yield consistently distributed particle size in the range of 10–40 μm in diameter (Figure 1.2a) [20]. The spray-drying method allows a large-scale yield and high encapsulation efficiency in pharmaceutical applications, as well as excellent stability of the prepared product and ease of handling and maintenance of their properties [21].

1.2.2.2 Electrospinning Technique

Electrospinning technique is an effective method of fabricating micro- and nanoscale fibrous materials based on different biopolymers owing to their sustainable properties,

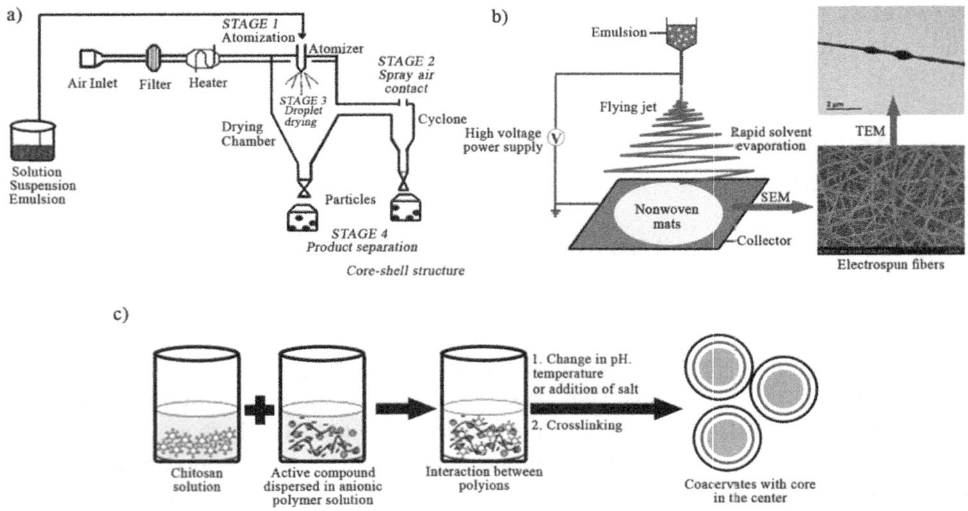

Figure 1.2 Schematic illustration of the (a) spray-drying technique Adapted from [20b], (b) electrospinning technique [22b] / with permission of Elsevier, and (c) coacervation technique [25b] / with permission of ScienceAsia.

including effectivity, low cost, and versatile technique. Also, it has been widely applied in recent years since it has many valuable advantages, such as high surface-to-volume ratio, high porosity, and ultrafine structures of the prepared fibers (Figure 1.2b). Figure 1.2b shows that it is non-mechanical technique and includes a high-voltage electrostatic field to charge droplets on a polymer solution surface, and then, induce the ejection of a liquid jet via a spinneret [22].

In this route, different natural polymers, such as biopolymers (proteins and polysaccharides), and biocompatible synthetic polymers, including polyvinyl alcohol and polycaprolactone, may be used individually or by mixing according to the specific type of usage of the food ingredients. Because these polymers are biodegradable, biocompatible, and nontoxic compounds, their micro- and nano-electrospun fibers have been applied for food and biomedical applications [22b, 23].

1.2.2.3 Coacervation Technique

Coacervation technique is widely used in food applications to prepare micro- and nanoparticle formulation. It involves the phase separation between the hydrocolloids phase from its starting solution using the change of ionic strength, temperature, solvent type, and pH. And then, subsequent deposition of the separated coacervate on the droplet core surface in the solution is noticed [24]. Generally, the coacervation technique involves many steps as shown in Figure 1.2c. The first step includes the dispersion of oil phase in the hydrocolloid solution to form oil/water emulsion. After that, the precipitation of hydrocolloid is put through different conditions, such as pH, temperature, ionic strength, and solvent polarity, to form a polyelectrolyte complex using the salting out method in the presence of salts, such as sodium sulfate, or the desolvation method in simple coacervation using a water miscible non-solvent [25]. However, in the complex coacervation method, there are polymer/polymer electrostatic interactions between two different and opposite charges that hydrocolloid. In addition, it contains other weak interaction bonds such as H-bonding and hydrophobic interactions. The obtained complex is stabilized through crosslinking interaction using tripolyphosphate, calcium chloride, and glutaraldehyde as crosslinkers. It is advantageous, thanks to high-encapsulated bioactive ingredients, up to 99% [25a, 26]. This technique is vastly applied in food industry, particularly, for encapsulating lipophilic ingredients, for example, essential oil, vegetable oil, and palm oil [27].

1.2.2.4 Emulsion Technique

Emulsion method based on the polymer hydrocolloid-delivery system is vastly applied in food and pharmaceutical applications to encapsulate, protect, and deliver bioactive ingredients. Based on the droplet size diameter, the emulsion product can be classified into three forms as shown in Figure 1.3a: nanoemulsion, miniemulsion, and macroemulsion formulations [28]. This technique basically depends on the mixing of two totally or partially immiscible liquids. Additionally, it involves amphiphilic surface-active surfactants that decrease the interfacial tension among both the liquids used to achieve good stability (Figure 1.3b). Generally, emulsion method can be found in two forms: oil-in-water (o/w) and water-in-oil (w/o) types that depend on oil dispersed as droplets in water or vice versa [28a, 29].

1.2.2.5 Ionic Gelation Technique

Ionic gelation method includes the reaction between polycation polymer as chitosan and polyanions, such as proteins, alginate, hyaluronic acid, etc., in the presence of crosslinking agents, including tripolyphosphate, aluminum chloride, calcium chloride, etc. (Figure 1.3c), to produce the desired nanoparticle formulation in different ranges of 84–600 nm. Ionic gelation technique has several outstanding advantages such as (i) a simple, easy, non-toxic, and mild technique, (ii) an organic solvent-free method; and (iii) prepared nanoparticles with excellent encapsulation efficiency. Additionally, it has main drawbacks as the prepared nanoparticles often appear with a broad size distribution and non-uniform composition [25b, 30].

1.2.2.6 Liposome Formulations

Liposome formulations are bilayer phospholipid vesicles with a definite diameter of 25 nm–10 μm. They could encapsulate polar materials inside their core and the hydrophobic materials through their lipid bilayer. They fabricate by the film hydration method (Figure 1.3d) with lipid and cholesterol and solvent as well. However, it has many instability issues referred to as aggregation, hydrolysis, and oxidation. So, to decrease its oxidation, an appropriate buffer is used, and the freeze-drying technique is also used to overcome the effect of temperature on liposomes [31]. Gomez et al. [32] reported the encapsulation efficiency of any liposome preparation based on the encapsulated active ingredient.

The aforementioned techniques of design or engineering of the materials science for food product development is governed by some factors summarized in Box 1.1.

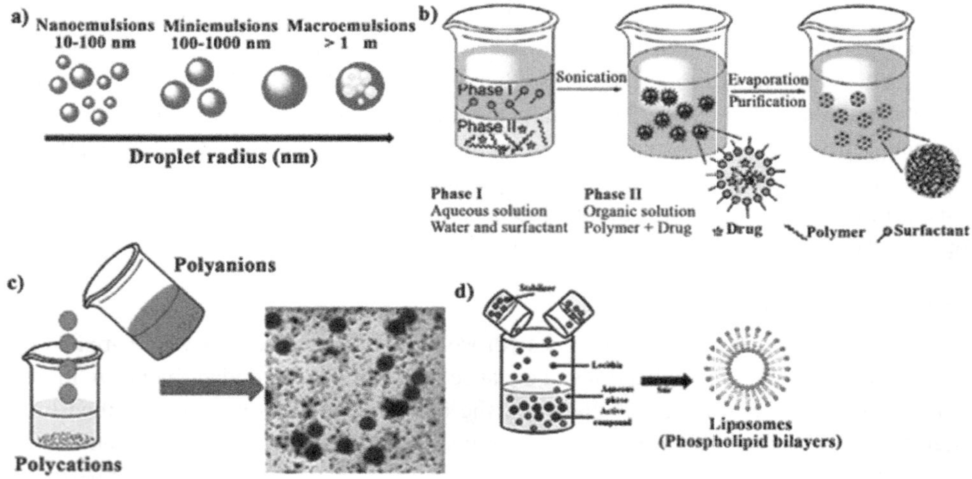

Figure 1.3 Schematic illustration of (a) emulsion fabricated products: nanoemulsion, miniemulsion, and macroemulsion formulations, (b) emulsion technique [28b] / with permission of Elsevier, (c) ionic gelation technique; and (d) liposomes formulation [25b] / with permission of ScienceAsia.

Box 1.1 Developing the product's concept

- A product is an amalgamation of hard values, or basic attributes, and soft values, or distinguishing qualities, such as aesthetic appearance and environmental friendliness that the consumer expects.
- Product design, or the process of creating a product, is a synthesis of consumer and market research and technological and engineering studies.
- The areas of engineering the product include: (i) determining what consumers want; (ii) creating a product brief for the target market; (iii) comparing with similar items; (iv) generating new ideas.

1.3 Encapsulation of Food Ingredients for Food Product Development

Encapsulation is a common technology in the food business for creating engineered products, especially in functional and specialized food industrialization, food processing, and product invention. Encasing a functionally active core material into an inert substance is what is required (Figure 1.4). The material that will be encapsulated is referred to as the core or active material. It is also known as the payload state, the fill state, or the internal state. The substance used to encapsulate the active component is known as the coating material, shell, matrix, membrane, wall, capsule, or carrier material [33]. As

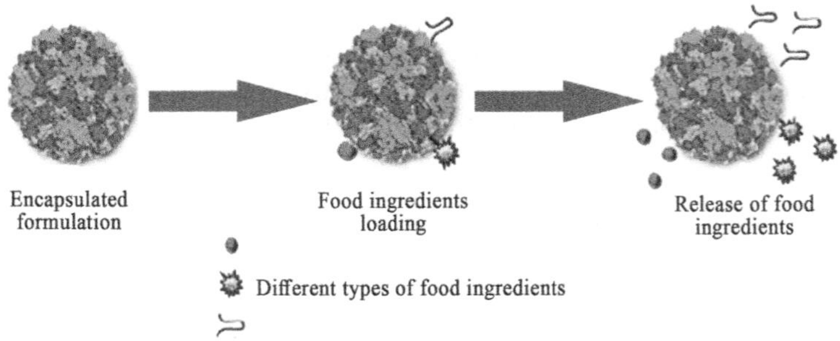

Figure 1.4 Schematic illustration of encapsulation of food ingredients.

the first to uncover the concept of cell encapsulation [34], pressure-sensitive coatings for carbonless copying paper were created using encapsulation technology for the first time in industry about two decades ago [35]. The main branches of nanotechnology, nanoencapsulation, and microencapsulation have been widely used in the food industry to protect bioactive food ingredients from processing and environmental stresses [36], as well as for controlled-release applications to solve the major problem of food ingredients that food industries face. They have received a lot of attention in the industrial world because of their capacity to safeguard unstable bioactive components, add new functional features into sophisticated food products, and release active material at a controlled rate. As a result, abundant encapsulation approaches have been studied for a long time.

The process selection is influenced by the nature of the active ingredient, the qualities of the shell material, and the wanted attributes of the final product based on the intended use. To improve shelf life and/or hide a disagreeable flavor or taste, food-grade proteins and polysaccharides are used to encapsulate sensitive and bioactive food constituents, such as highly unsaturated edible oils (e.g. fish oils), enzymes, vitamins, or diverse flavors [37]. However, recent research works have focused on enhancing the functionality and health benefits of processed foods, as well as enhancing the efficacy of probiotics and transportation of various enzymes or coenzymes, bioactive peptides, and so on [38]. Controlled and prolonged release and targeted delivery have been achieved by encapsulating artificial sweeteners, therapeutic proteins, and other bioactive ingredients [39]. Although the primary objectives of encapsulation research are to control the release of active ingredients with a desired rate, in the appropriate place at the appropriate time, and to protect bioactive food species from environmental factors (radiation, oxygen, light, moisture, and different pH states), recent developments have been made to improve product handling in terms of reduced toxicity, lowered cost, and reinforced nutrient bioavailability. The ultimate goal is to extend the shelf life of the designed product and promote its overall acceptability [40]. The physical and chemical properties of the resulting encapsulation are determined by the wall material chosen [33b]. As a result, the module that forms the wall is chosen based on the following criteria: (i) compatibility and degree of reactivity or inertness with the core and external environment; (ii) the material used as a wall should be certified as safe for food applications, i.e. "generally recognized as safe" by the Food and Drug Administration or the European Food Safety Authority (EFSA); (iii) cost-effectiveness; (iv) good encapsulation capability; (v) submissiveness to the specifications of the final product.

1.3.1 Encapsulation Based on Polysaccharides

1.3.1.1 Chitosan

Chitosan (CS), which has multiple amino and hydroxyl functional groups, is the second most prevalent natural biopolymer after cellulose [41]. Chitosan is the only commercially available cationic polymer so far due to the presence of the positive charge on its amino groups. Labile substances, such as tea polyphenols, essential oils, vitamins, enzymes, and others, are encapsulated with chitosan and its derivatives, because they are unstable and can lose their bioactivities during formulation, food processing, digestion, and delivery [42].

Nanoencapsulation allows for the regulated release and stability of bioactive food ingredients at the appropriate time and location. Vitamins' low stability and sensitivity to harsh environmental conditions, such as freezing, heating, and oxidation, limit their use in the food industry [43]. Vitamins have been encapsulated using chitosan-based nanoparticles. Ionotropic gelation was used to encapsulate vitamin C utilizing the chitosan nanostructure. Using the ionotropic polyelectrolyte pre-gelation approach, vitamin B2 was encapsulated by alginate/chitosan nanoparticles with a 59.6% encapsulation efficiency [36].

Several long-chain polyunsaturated omega-3 fatty acids extracted from fish oil were microencapsulated into chitosan shell to lower their susceptibility to atmospheric oxidation, using a spray-drying process. The results showed that these fish oil extracts had improved stability and storage time [44]. Chitosan encapsulation of fish oil was also achieved utilizing an ultrasonic atomizer and an emulsification process in a study. As a result, chitosan not only produced a stable emulsion, but it also improved its stability when combined with maltodextrin [45].

1.3.1.2 Starch

Starches are extracted from various sources, but maize is the most common source. They have been used to encapsulate a wide array of essential food compounds. Different modifications of starch, such as gelatinization, hydrogel, and acetylation have been mentioned. Different food types, including essential oils, flavors, lipids, vitamins, and probiotics, are encapsulated within starch modification using various preparation methods: electrospinning, extrusion, spray drying, physical trapping, freeze drying, and co-precipitation. Porous starch is a hollow particle that has high absorbency and can retain a wide range of chemicals [46]. The microencapsulated loading rate of olive oil with porous starch was higher than that of free olive oil, and the best porous starch adsorption capacity was obtained after a 12-hour reaction at a temperature of 45°C and a pH of 5.0 [46]. Nanoencapsulated flaxseed oil by high-amylose corn starch nanoparticles and impregnation into bread developed biofunctional food products with no impact on sensorial features. Beside the sensorial benefits, nanoencapsulation diminished the oxidation of lipids during the baking process [47]. Further, antioxidant compounds as curcumin yellow dye were microencapsulated in a mixture of modified starch with gum arabic and maltodextrin, which controlled the high loading rate after spray drying [48].

1.3.1.3 Alginate

Alginate is an unbranched anionic polysaccharide made up of glycosidic linkages between -D-mannuronic acid (M) and -L-guluronic acid (G). Because of its excellent thickening and film-forming ability, stability, chelation, and biocompatibility, alginate is widely used in microencapsulation applications [49]. Hydrogel, emulsion, emulsion-filled alginate hydrogel, nanoparticle, microparticle, core–shell particle, liposome, edible film, and aerogel are alginate-based delivery methods that have been utilized to encapsulate food ingredients and bioactive species [50]. The ionic gelation process was used to make nanoparticles that might be used as an oral iron delivery mechanism [51]. Iron-loaded alginate nanoparticles with negative surface charges were found to have excellent iron encapsulation (75%) and a prolonged iron release process, making them an appealing delivery system for traditional oral iron supplements [51]. Encapsulation of curcumin in zein/caseinate/sodium alginate nanoparticles improved the water solubility, photochemical stability, and antioxidant activity of curcumin. Important fatty acids, as linoleic and alpha-linolenic, as well as carotenoids, sterols, free and esterified triterpenoids, and isoprenoids, are abundant in sea buckthorn oil. All of these important chemicals are protected by microencapsulation in alginate, which also improves oil stability [52].

1.3.1.4 Carrageenans

Carrageenans are extensively utilized as culinary components because of their stabilizing, gelling, texturizing, and water-binding capacities. Carrageenan is increasingly being employed as a biomaterial in the pharmaceutical sector to improve medication formulations and ensure long-term drug release, and its benefits are currently being researched [53]. The three types of carrageenan identified by Liu et al. are kappa (κ)-carrageenan, lambda (λ)-carrageenan, and iota (ι)-carrageenan [54]. The presence of anhydrous bridges in carrageenan molecules influences their gelation process significantly, especially when utilizing κ and λ carrageenan. Carrageenan was used as a coating material for the microencapsulation of phycocyanin to enhance its application as a food colorant in the industry; the encapsulation efficiency was 68.66% [55]. Spray-drying technology was used to encapsulate several food waste compounds (collagen hydrolysate from squid tunics, polyphenolic extract from pomegranate peel and albedo, and shrimp lipid extract) utilizing iota-carrageenan as the wall material [56].

1.3.2 Encapsulation Based on Liposomes

Liposomes are made up of one or more lipid and/or phospholipid bilayers, and they can also contain other molecules like proteins and carbohydrates. The encapsulation of antioxidants in liposomes is used in the food domain to prevent nutrient oxidation (degradation) [57]. Incorporating α-tocopherol into liposomes [58] or nanoliposomes [57] can improve its efficacy as a dietary antioxidant. Liposomal and nanoliposomal carriers are viable alternatives for protecting enzymes, controlling their release, and enhancing their technological functions. Because of their hydrophilic/hydrophobic properties, enzymes can be entrapped in the core or membranes of liposomes, where they can be shielded from environmental stresses by a liposome layer. Enzymes are widely used in the food industry because of their ability to improve functional, sensory, and nutritional properties of products, such as improving bread quality, developing cheese flavor, producing lipolyzed milk fat, sweetening milk, extending the shelf life of food products, and improving food flavor [59].

Flavourzyme, derived from *Aspergillus oryzae*, is used in cheese production to speed up the ripening process, resulting in a bitter taste and enhanced flavor. Fish oil could be encapsulated using liposome technology to improve its sensory qualities. Vitamin C encapsulated in liposomes had shelf life extended from up to two months, particularly in the presence of prevalent dietary components that would normally enhance decomposition. Liposomes can also be employed to distribute encapsulated ingredients at a predetermined temperature [33b].

1.3.3 Encapsulation Based on Proteins

Natural proteins are biological macromolecules made up of strings of amino acids connected together by peptide bonds that have developed to provide vital biological functions, such as structure creation, signaling, and so on. Because nanoencapsulated antioxidant biopeptides are extremely permeable through the human intestines, where peptides are rapidly degraded and better absorbed into the bloodstream, their inclusion into food systems can result in a variety of health benefits [60]. Because of their unique nutritional and functional properties, encapsulation of bioactive compounds in protein-based delivery systems has received a lot of attention [61]. Encapsulation and delivery systems made of gelatin, which is produced from collagen, have been widely adopted [62]. Acid or alkaline hydrolysis is commonly used to extract gelatin from collagen; gelatin powder expands in cold water, and it becomes liquid in hot water. Tuna oil has been encapsulated using gelatin–sodium hexametaphosphate; D-limonene has been encapsulated using pectin–whey protein, and sulforaphane has been encapsulated using gelatin–gum arabic and gelatin–pectin complexes [63]. In this chapter, we summarize some reported works on the encapsulation of different food ingredients using various material engineering formulations (Table 1.1). In the food industry, the encapsulation of food ingredients always depends on some factors that help achieve the quality and safety of the final product as illustrated in Box 1.2.

Table 1.1 Encapsulation of various active food ingredients using different material engineering formulations.

Carrier formulation	Wall materials	Active core ingredients	Encapsulation method	Activity and applications	References
Nanoparticles	Chitosan	Clove essential oil (CEO)	Ionic gelation	Enhanced antifungal, antibacterial, and antioxidant properties of CEO due to the controlled release of CEO	[64]
Nanoparticles	Alginate/chitosan	Vitamin B2	Ionic gelation	Controlled release of Vitamin B	[25b]
Nanoparticles	Chitosan	Curcumin	Coacervation	Enhancing the antioxidant activity	[25b]
Nanoparticles	Alginate	Iron	Ionic gelation	Prolong the time of iron release	[65]
Nanoparticles	Zein/caseinate/alginate	Curcumin	Coacervation	Water solubility, photochemical stability, and antioxidant activity were all improved	[66]
Nanofibers	Chitosan/gelatin	Thyme (*Zataria multiflora Boiss*)	Electrospinning	Produce the best odor and taste sensory characteristics while avoiding color and texture differences	[67]
Nanofibers	Potato starch	Thyme (*Thymus vulgarius*)	Electrospinning	Enhances antioxidant action	[68]
Nanocapsules	Liposome	α-Tocopherol	Emulsion	Enhances antioxidant action	[69]
Nanocapsules	Liposome	Tea polyphenol	Film hydration	Bioavailability is being improved	[70]
Nanocapsules	Liposome	Vitamin C with medium chain fatty acids (MCFA)	Film hydration	Improved antioxidant activity and activating dietary supplements	[71]
Microcapsules	Liposome	Beta-carotene	Spray drying	Chemical stability and bioavailability are improved	[37a]
Microcapsules	Gelatin and gum arabic	f Black raspberry ANCs	Coacervation	Improving the thermal and storage stability	[72]
Microcapsules	Cashew gum and gelatin	Pequi oil	Coacervation	Protect pequi oil bioactive compounds to increase kinetic stability and functional activity	[73]
Microcapsules	Gelatin–sodium hexametaphosphate	Tuna oil	Coacervation	Decreasing environmental pollution and production costs	[66]

(Continued)

Table 1.1 (Continued)

Carrier formulation	Wall materials	Active core ingredients	Encapsulation method	Activity and applications	References
Microcapsules	k-Carrageenan	Phycocyanin	Spray drying	Enhance its application as a food colorant in the industry	[74]
Microcapsules	Chitosan /k-carrageenan	Pimenta dioica	Coacervation	Increasing a release rate with the chitosan content	[75]
Microcapsules	iota-Carrageenan	Polyphenolic and shrimp lipid extract	Spray drying	Enhancing the antioxidant properties	[76]
Microcapsule	Corn starch and maltodextrin	Curcumin	Spray drying	Increases antioxidant activity while maintaining a high retention rate	[77]
Microcapsule	Maltodextrin and modified starch	Iron	Spray drying	Increases iron bioavailability	[78]
Microcapsule	S. alginate	Sea buckthorn oil	Ionic gelation	Increases pressure resistance as well as antioxidant activity	[79]
Microspheres	Chitosan and carrageenan	Neem seed oil (NSO)	Coacervation	Pesticides, insecticides, and herbicides can all be utilized effectively	[42]
Microcapsule	Chitosan	Fish oil (long-chain polyunsaturated omega-3 fatty acids)	Spray drying	Reduces fish oil's susceptibility to environmental oxidation while also improving its stability and storage time	[80]
Hydrogel capsules	Gelatin and carrageenan	Gallic acid, catechin, chlorogenic acid, and tannic acid	Electrospinning	Hydrogels have the ability to release considerable amounts of tannic acid in different food simulants than others	[81]
Hydrogel beads	Alginate	Spirulina	Ionic gelation	Hydrogels are suitable for encapsulation of spirulina protein	[82]
Hydrogels	Chitosan and alginate	Plant extracts	Ionic gelation	Improved stability and controlled release	[83]

Box 1.2

- Industrial application of food encapsulation techniques depends partly on the kind of active food ingredient to be encapsulated, the physicochemical properties of the active ingredient, and the kind of instabilities the chosen ingredient exhibits, with respect to its surrounding environment and processing conditions.
- The kind of coating material that should be selected to entrap the chosen active ingredient is a factor that significantly affects the design and optimization of the encapsulation process.
- The encapsulation technique to be employed to convert the selected core and wall components to an encapsulated product must be considered during food microencapsulation.
- Optimization of the microencapsulation process should take the particle size, density, and stability requirements for the encapsulated ingredient into consideration.

1.4 Food Packaging Approach for Food Product Development

The major functions of food packaging are to protect food products and their effectiveness from chemical, physical, and biological reactions by deferring food deterioration, and preserving and extending the advantages of processing, and maintaining food quality and safety by extending shelf life [84]. Advances in food packaging and processing are critical to keeping the United States' food supply among the safest in the world. In the food industry, packaging is a critical component for durability and functionality of food products. Many customers are attracted to the design of food packaging; however, food packaging must be appropriate for the function of food packaging, such as protection, security, and consumer information [85].

Additionally, edible films and coatings are substances that can be directly consumed. They supply barrier and protection, while maintaining quality and safety of food products. One of the main benefits of using edible packaging materials, as compared with synthetic packaging, is that they are an integral part of the food product; they can be eaten, without having to unpack and dispose of the package. Moreover, these films usually are manufactured from renewable and edible components, displaying the significant advantage of being biodegradable and environmentally friendly [86], and they leave a thin layer on the product immediately after application [51]. They can help improve food preservation while lowering the cost and weight of traditional packaging. Biopolymers based on hydrocolloids, such as polysaccharides and cellulose, starch, alginates, chitosan, gums, pectin, and proteins, are used to create edible coatings and films. The use of composites or blends to manage the release of food additives and nutrients is a new inventive development in addition to the fundamental functional features of creating a barrier to gases and moisture.

1.4.1 Polysaccharide-Based Food Packaging

Polysaccharide-based films and coatings used to prolong the shelf life of vegetables, fruit, shellfish, or meat products usually have an extreme effect by minimizing, darkening of the food material, dehydration, and oxidative corruption. In edible packaging, some polysaccharides can act as carriers of certain organic acids that prevent the microbial contamination of food. These new edible materials are environmentally friendly that are considered better suppliers of vitamins and minerals and can improve the taste of food products [87]. Table 1.2 illustrates some of the reported studies on food packaging for some essential foods using polysaccharide-based edible films.

1.4.1.1 Chitosan

Chitosan can be utilized in food packaging applications as flexible packaging sheets or as coatings directly on food materials. Chitosan is a natural bioactive polymer with antibacterial properties, which has led to its use as a food preservation film or coating [88]. Chitosan films and coatings are nontoxic, renewable, biocompatible, and biodegradable [89]. Chitosan films can be either edible or inedible, unlike coatings, which are virtually always

Table 1.2 Reported edible films based on polysaccharides for food packaging of some essential foods.

Food	Edible coating films	Main findings	References
Fruits and vegetable	Alginate, glycerol, and oleic acid	These biofilms are designed to be used as primary packaging for the storage of postharvest fruit and vegetable products during cold chain storage to ensure food safety and nutritional quality.	[120]
Bread	Starch/chitosan/nanoclay	Chitosan/nanoclay bionanocomposite coatings showed antifungal activity in a preserved bread sample kept at 25°C for 20 days.	[121]
Fresh-cut pears, poultry, meat	Chitosan/rosemary oil	Chitosan films incorporated with rosemary extract were found to extend shelf life.	[122]
Fruits and vegetables	Chitosan/citric acid/glycerol	Films showed thermal stability, water resistance, and potent antioxidant activity, which were attributed to the presence of citric acid.	[90]
Chicken breast	Chitosan/0.25% oregano essential oil (OEO)	Shelf life of chicken fillets was extended by roughly 6 (with only OEO) and more than 15 days when they were coated with chitosan with or without OEO.	[123]
Guava fruit	Chitosan/pomegranate peel extract	The coated chicken samples were acceptable during the entire refrigerated storage period (21 days), with no negative effects on the chicken samples' flavor.	[124]
Vegetables and fruits	Animal proteins, meat proteins, gelatin	Excellent film-forming capacity, excellent barrier properties against oxygen, good mechanical resistance, and high elasticity, and a melting point near body temperature.	[107]
Food	Clove essential oil/graphene oxide nanosheets/PLA	The antimicrobial activity against food pathogens improved.	[125]
Red grapes	Chitosan/gelatin/AgNPs films	Composite film extended shelf life of red grapes by 14 days.	[126]

edible because they create a layer immediately on the top surface of foodstuff. Several additives have been added to each form of packaging to give it different qualities. To avoid endangering chitosan's food-grade designation, edible coatings contain just the bare minimum of additives, if any at all. Natural essential oils, which are extracted from renewable plant parts or agricultural wastes, are nontoxic and do not have any adverse effect on human health if consumed in small quantities [90]. They are proving to be good candidates as food film additives. Incorporation of essential oils in chitosan films imparts plasticizing and water-resisting effects on the film; they exhibit antimicrobial, antifungal, and antioxygenic activity against foodborne pathogens [91].

Nanomaterials included with chitosan matrix significantly improve the antibacterial, mechanical, and barrier properties of manufactured films and coatings, resulting in improved quality and shelf-life extension [60]. Silver nanoparticles, zinc oxides, and titanium dioxide have all been employed extensively in the manufacture of chitosan nanocomposite films. Due to their physical and chemical qualities, chitosan micro/nanofibers are also utilized for antibacterial packaging, preserving the food components' quality and safety through distribution and storage [92]. As a result, future usage of chitosan nanoparticles in food packaging materials and processing could potentially improve food production efficacy, sustainability, and storage capacity, as well as offer promise for more environmentally friendly food distribution [93].

1.4.1.2 Alginate

Alginates are mostly accessible commercially as sodium alginate, which is widely employed as a thickening agent in stabilizers, and as a stabilizer in foods, including desserts and sauces [94]. Alginates can combine to create a multifunctional food-packing system, which is the definitive objective of recent food packaging technology. A main disadvantage of alginate is its high water evaporation rate due to its high hydrophilicity. This drawback was solved by using cellulose nanoparticles or montmorillonite inorganic clay inside its chains [95].

Thus, when combined with another substance, the alginate activity in food packaging can be improved, for example, by adding (i) polylactic acid to boost oxygen barrier; (ii) glycerol and sorbitol to increase alginate flexibility; and (iii) calcium chloride to improve mechanical properties, water resistance, and water vapor transmission rate [58, 95]. Alginate-based edible films and coatings should reduce water evaporation, loss of edible odor and flavor volatiles, microbial growth, respiration suppression, and gas exchange [96]. Alginate films incorporated with graphene oxide and/or graphene oxide/zinc oxide nanofillers decrease the water vapor permeability and wettability of bionanocomposite films [97]. Copper-deposited graphitic carbon nitride nanoparticles were used as reinforcement and antibacterial agents in the fabrication of starch/sodium alginate films for food packaging applications [98].

1.4.1.3 Starch

Synthetic plastics are particularly detrimental to the environment since they are nonbiodegradable in the long run; nonetheless, these polymers are frequently used for food packaging, causing serious environmental concerns. To address this issue, food packaging materials should be replaced with environmentally friendly materials. Starch is a viable bio-source for packaging due to its great availability, low cost, renewability, environmental friendliness, and ease of processing [99]. Starch-derived edible food films have great potential as biodegradable food packaging materials because they minimize the overuse of traditional petroleum-based plastic [100]. Starch food packaging film that consisted of starch nanofibers prepared by using electrospinning technique is preferred to conventional starch-based composite films, whereas electrospinning can directly produce films of pure starch without plasticizers. Moreover, the mechanical properties of starch-based films can be improved by the presence of nanofibers [101]. Nanofibers are frequently impregnated into thermoplastic starch to improve its mechanical qualities for good packaging applications [102]. Further, starch has been successful in generating viable alternatives to petroleum-based plastics by blending it with other polymers or using it in composite materials [103]. Moreover, improvement of starch-based films by improving their mechanical, UV, and water inability properties can be obtained by integration of different nanoparticles, such as graphene oxide, titanium oxide, and zinc oxide nanoparticles. The mechanical properties, transmittance, and water vapor permeability of the starch/PVA/graphene oxide composite film significantly improved due to the presence of graphene nanosheets [104].

1.4.2 Protein-Based Food Packaging

Proteins are ideal materials for making edible films because of their intrinsic properties [105]. Furthermore, nitrogen sources of protein applied as fertilizers must be added through the disintegration of the film based on protein, which positive effect non-protein-based films lack [106]. On the other hand, antimicrobial compounds in protein-based products suppress the growth of germs on the surface of the product, extending their shelf life and improving their safety. Antimicrobial films and coatings can keep high concentrations of active chemicals on food surfaces, releasing them either through evaporation into the atmosphere or diffusing into the food in the case of nonvolatile additives [107].

1.4.2.1 Gelatin

Films made of gelatin are transparent, oxygen-impervious, and thermoreversible. Because gelatin's melting point temperature is similar to human body temperature, films based on gelatin are becoming more significant sources for the packaging of edible materials. Edible film based on gelatin has low oxygen permeability and can be made to have antioxidant and antibacterial effects by adding substances, such as citrus essential oils, carvacrol, and other compounds [108]. Gelatin films were vastly employed in fruit and vegetable preservation. The use of zein coatings on tomatoes causes color change, and softening can be postponed without the formation of ethanol [109]. When zein coatings are designed for apples and pears, the respiration rate is varied, with the apples having a lower rate and the pears having a higher rate [110]. Peanut fat degradation can be avoided by covering them with gluten [111]. Various vegetables and fruits, including carrots [112], cherry tomatoes [113], peppers [114], pineapple fruit [115], and blueberry fruit [116], are frequently coated with gelatin-based films to extend their shelf life.

1.4.2.2 Casein

Without further processing, casein can design films easily from aqueous solutions. Casein is biodegradable and nontoxic and has high thermal stability; in addition, it has the ability to interact with tiny molecules and ions, and the ability to form micelles, all of which make it an excellent material for biodegradable films [117].

Casein-based films demonstrated precise mechanical characteristics, indicating that they could be used in a variety of biocompatible and/or biodegradable coating applications. Furthermore, these films demonstrated intriguing antimicrobial characteristics. That is why it is critical to emphasize the necessity of using casein films in food packaging. More research is needed to see how successful these films are on certain food products, especially those that are highly microbially sensitive [118].

Edible casein films and coatings serve as a barrier between food components or as a protective layer on foods. Mechanical protection, mass transfer control, and sensory characteristics are all crucial features. To safeguard the integrity of packaging throughout distribution, it must have sufficient mechanical strength. Controlling mass transfers entails monitoring ingredient and additive migration in food systems, limiting food dehydration, and regulating gas microenvironments around foods. The sensory appeal of an edible film or coating is also a crucial determinant in its acceptance. Furthermore, because of their intricate intermolecular interactions, simple milk protein films are effective barriers to gas transmission [119]. Food packaging as a broad technique allows people to participate in the development of the food product and has a lot of industrial and social advantages as displayed in Box 1.3.

Box 1.3

- The industrial and societal benefits of packaging include the possibility of avoiding or lowering product damage and food spoilage, conserving energy and important nutrients, while also safeguarding the consumer's health.
- Proper use of packaging technologies can lower or eliminate the possibility of adulteration and tampering, and present food in a hygienic and often aesthetically attractive way.
- Food packaging can not only protect food products but can also convey crucial food information and assist customers in making informed purchase decisions.

1.5 Hydrogel Structures and Their Efficiency in Food Development

Hydrogels are hydrophilic three-dimensional polymeric networks that are prepared chemically or physically through the crosslinking method by using natural and synthetic polymers [127]. They have different hydrophilic functional groups, such as hydroxyl (–OH), carboxylic acid (–COOH), sulfonyl (–SO_3H), imide (–CONH–), and amide (–$CONH_2$) groups along polymeric chains, so they have the ability to capture both water and biological fluids through their structural compositions [128]. Also, hydrogels are known as hydrophilic gels and aqua gels [129]. In addition, their swelling ability can be controlled by various parameters, such as the presence of hydrophilic functional groups, nature of the crosslinking agent, and technique of polymerization [130]. The nature of the crosslinking agent for the formation of 3D polymeric networks is divided based on physically and chemically crosslinking properties. In chemically crosslinked hydrogels that are also known as either thermosetting hydrogels or permanent gels, the polymer chains are covalently bonded using a chemically crosslinking agent. While in physically crosslinked hydrogels, the polymers are connected via H-bonding or electrostatic interactions, the responsiveness of polymeric hydrogels is enhanced by different external factors, such as pH, electric field, light, temperature, or ionic strength [129c, 131]. Moreover, hydrogels demonstrate their dual nature pertaining to different properties on both the macroscopic and microscopic scales. They are found as solids with a definite shape on a macroscopic scale, while, on a microscopic scale, their behavior is similar to that of solutions. In addition, hydrogels can be used as a molecular filter in the biological sciences because they can trap water-soluble molecules. Therefore, they are essential in electrophoresis [129c, 132]. Hydrogels are classified into different categories depending on their physical structures, properties, electric charges, sources (origin), and nature of crosslinking agents [129c].

Because of the variety of hydrogel (natural or synthetic) structures, one can expect greater possibilities of functional regulation, making hydrogel applications in the field of food science more adjustable. High-strength hydrogels, topological hydrogels, and tetra-arm hydrogels, for example, have specific advantages in the design of food packaging materials, providing significant strength and toughness, while porous hydrogels may provide suitable permeability. Microgels, nanogels, core-shell-structured hydrogels, and emulsion hydrogels, on the other hand, are better suited as encapsulation and nutraceutical delivery platforms [118].

Hydrogels have also been widely used in various fields such as tissue engineering and scaffolds and cell culture, drug delivery, wastewater treatment in food industries, and antimicrobial fields [133–136].

1.6 Conclusion

Consumer demand for fresh, healthy, and nutritious food has prompted the food industry to seek out alternative technologies that can produce higher-quality food while also ensuring consumer safety at reasonable costs. Similarly, other new processing technologies could be employed to customize foods with improved functional and nutritional benefits. Application of nanotechnology supplies complete food solutions from food manufacturing to packaging. To address this concern, nanostructured and microstructured materials based on natural biopolymers (polysaccaharides and proteins) have been used in food encapsulation and the food packaging process to ensure preservation from spoilage and environmental conditions. Polysaccharides have been used in encapsulation applications owing to their sustainable properties, such as bioactivity, biocompatibility, biodegradability, and nontoxicity. They are also easily available at less cost and are easy to modify and manufacture. Several techniques have been used in the design of nano- and microparticles of the capsule's wall, of which, the spray-drying technique is one of the most widely used methods due to its favorable features, such as flexibility, simplicity, low cost, easy, and rapid scaling up. On the other hand, edible films and coatings from natural polymers have been used in coating food ingredients directly. They supply barrier properties and protection while maintaining the quality and safety of food. In addition to their industrial benefits to produce high-quality and safe products, these new technologies in food preservation largely reduce environmental pollution compared with traditional preservation tools. Future perspectives on the development of the food industry are summarized in Box 1.4.

Box 1.4

- The food industry takes a broad perspective of the word "quality" of the food product that includes several factors. One important factor is safety. This includes establishing toxicology and microbiology hazard criteria and methods and activities to verify that they are met.
- Nutrition is another factor that determines the quality of a food product. It is about keeping nutrient levels in food components and substituting foods with nutritional properties that contribute to consumer attention in healthful diets.
- Manufacturers can consider highlighting the use of organic ingredients and support for regional and local producers and the absence of additives, such as high-corn fructose, gluten, and genetically modulated organisms.

Glossary

Polymers Macromolecule, and composed of repeating units that are known as monomers that are joined by covalent bonds.

Polysaccharides Natural biopolymers composed of carbohydrate chains with a large polymeric oligosaccharide that are formed through glycosidic linkages between multiple monosaccharides as repeating units.

Proteins Polyamide chains and they are one of the main constituents of the human body.

Microstructure Formulation of particle-sized compounds in the range of 1–**1000** μm in diameter for different purposes such as controlling and sustained bioactive compounds delivery.

Nanostructure Chemical and morphological deposit matters in the range of 1–300 nm in diameter.

Coacervation Preparation technique includes separation of hydrocolloids phase from its starting solution by using the change in ionic strength, temperature, solvent type, and pH.

Liposome Bilayer phospholipid vesicles with definite diameter of 25 nm–10 μm.

Hydrogels Hydrophilic three-dimensional polymeric networks that are prepared chemically or physically by the crosslinking method using natural and synthetic polymers.

Chitosan Cationic polysaccharide that is composed of glycosidic linkage between glucosamine and acetylglucosamine.

Alginate It is an unbranched anionic polysaccharide made up of glycosidic linkages between -D-mannuronic acid (M) and -L-guluronic acid (G).

Food packaging Technique used to protect food products from food spoilage and in delaying food deterioration, retaining, and prolonging the advantages of processing, also ensuring quality and safety of food by increasing the shelf life.

References

1 a) L. Y. Maroufi, M. Tabibiazar, M. Ghorbani, A. Jahanban-Esfahlan, *Int. J. Biol. Macromol.* **2021**, *187*, 179; b) C. Demitri, V. M. De Benedictis, M. Madaghiele, C. E. Corcione, A. Maffezzoli, *Measurement* **2016**, *90*, 418; c) Y. Wu, F. Geng, P. R. Chang, J. Yu, X. Ma, *Carbohydr. Polym.* **2009**, *76*, 299.

2 J. M. Aguilera, P. J. Lillford, H. Watzke, in *Food Materials Science*, Springer, **2008**.

3 S. Kalia, M. W. Sabaa, S. Kango, in *Polysaccharide Based Graft Copolymers*, Springer, **2013**.

4 a) V. Bansal, P. K. Sharma, N. Sharma, O. P. Pal, R. Malviya, *Adv. Biclogical Res.* **2011**, *5*, 28; b) E. S. Goda, M. H. Abu Elella, H. Gamal, S. E. Hong, K. R. Yoon, in *Materials and Chemistry of Flame-Retardant Polyurethanes Volume 1: A Fundamental Approach*, ACS Publications, **2021**.

5 a) A. Kumar, K. M. Rao, S. S. Han, *Carbohydr. Polym.* **2017**; b) R. S. Singh, N. Kaur, V. Rana, J. F. Kennedy, *Carbohydr. Polym.* **2016**, *153*, 455; c) M. H. A. Elella, R. R. Mohamed, E. Abd ElHafeez, M. W. Sabaa, *Carbohydr. Polym.* **2017**, *173*, 305.

6 V. G. Muir, J. A. Burdick, *Chem. Rev.* **2020**.

7 a) T. A. Debele, S. L. Mekuria, H.-C. Tsai, *Mater. Sci. Eng. C* **2016**, *68*, 964; b) N. Zhang, P. R. Wardwell, R. A. Bader, *Pharmaceutics* **2013**, *5*, 329; c) R. R. Mohamed, M. H. A. Elella, M. W. Sabaa, *Int. J. Biol. Macromol.* **2015**, *80*, 149; d) R. R. Mohamed, M. H. A. Elella, M. W. Sabaa, *Int. J. Biol. Macromol.* **2017**, *98*, 302.

8 a) M. Leung, C. Liu, J. Koon, K. Fung, *Immunol. Lett.* 2006, *105*, 101; b) N. C. Ngwuluka, N. Y. Abu-Thabit, O. J. Uwaezuoke, J. O. Erebor, M. O. Ilomuanya, R. R. Mohamed, S. M. Soliman, M. H. A. Elella, N. A. Ebrahim, *Nano-and Microencapsulation: Techniques and Applications* **2021**, 55; c) E. S. Goda, M. H. A. Elella, M. Sohail, B. S. Singu, B. Pandit, A. El Shafey, A. M. Aboraia, H. Gamal, S. E. Hong, K. R. Yoon, *Int. J. Biol. Macromol.* **2021**, *182*, 680; d) M. H. A. Elella, E. S. Goda, K. R. Yoon, S. E. Hong, M. S. Morsy, R. A. Sadak, H. Gamal, *Compos. Commun.* **2021**, *24*, 100614; e) M. M. Abdel-Aziz, M. H. A. Elella, R. R. Mohamed, *Int. J. Biol. Macromol.* **2020**, *142*, 244.

9 a) T. A. Debele, S. L. Mekuria, H.-C. J. M. S. Tsai, E. C, **2016**, *68*, 964; b) S. Kumbar, U. Toti, M. Deng, R. James, C. Laurencin, A. Aravamudhan, M. Harmon, D. J. B. M. Ramos, **2011**, *6*, **0650**05; c) P. B. Malafaya, G. A. Silva, R. L. J. A. d. d. r. Reis, **2007**, *59*, 207.

10 a) S. Kumbar, U. Toti, M. Deng, R. James, C. Laurencin, A. Aravamudhan, M. Harmon, D. Ramos, *Biomed. Mater.* 2011, *6*, 065005; b) P. B. Malafaya, G. A. Silva, R. L. Reis, *Advanced Drug Delivery Reviews* **2007**, *59*, 207; c) N. B. Shelke, R. James, C. T. Laurencin, S. G. Kumbar, *Polymers for Advanced Technologies* **2014**, *25*, 448.

11 a) Y. Lu, W. Sun, Z. Gu, *J. Control. Release* **2014**, *194*, 1; b) I. El-Sherbiny, I. Khalil, I. Ali, M. Yacoub, *Drug Development and Industrial Pharmacy* **2017**, *43*, **1567**.

12 J. Li, X. Jia, L. J. F. R. I. Yin, **2021**, *37*, 313.

13 S. N. Van, H. D. Minh, D. N. J. B. Anh, *A. Biotechnol.* **2013**, *2*, 289.

14 M. Muhaimin, R. J. D. I. B. Bodmeier, **2020**, *30*, **1055**74.

15 N. Iglesias, E. Galbis, C. Valencia, M. J. Díaz-Blanco, B. Lacroix, M.-V. J. I. J. O. B. M. De-paz, **2020**, *165*, **2205**.

16 a) F. Fu, L. Lin, E. Xu, in *Advanced High Strength Natural Fibre Composites in Construction*, Elsevier, **2017**; b) M. Lengyel, N. Kállai-Szabó, V. Antal, A. J. Laki, I. Antal, *Scientia Pharmaceutica* **2019**, *87*, 20.

17 A. I. Matthew, **2016**.

18 W. Dudefoi, A. Villares, S. Peyron, C. Moreau, M.-H. Ropers, N. Gontard, B. Cathala, *Innov. Food Sci. Emerg. Technol.* **2018**, *46*, 107.

19 M. Chaumun, V. Goëlo, A. M. Ribeiro, F. Rocha, B. N. Estevinho, *Food Bioprod. Process.* **2020**, *122*, 124.

20 a) Z. L. Zhang, L. J. Li, D. Sun, M. Wang, J. R. Shi, D. Yang, L. H. Wang, S. C. Zou, *Food Sci. Nutr.* **2020**, *8*, **1933**; b) M. Dixit, P. Kulkarni, A. Kini, H. Shivakumar, *Int. J. Drug Form. Res.* **2010**, *1*, 1.

21 B. N. Estevinho, F. Rocha, in *Nanotechnology Applications in Food*, Elsevier, **2017**.

22 a) A. C. Mendes, K. Stephansen, I. S. Chronakis, *Food Hydrocoll.* **2017**, *68*, 53; b) C. Zhang, F. Feng, H. Zhang, *Trends in Food Science & Technology* **2018**, *80*, 175.

23 L. Tian, M. P. Prabhakaran, X. Ding, S. Ramakrishna, *J. Biomater. Sci. Polym. Ed.* **2013**, *24*, **1952**.

24 L. C. Corrêa-Filho, M. Moldão-Martins, V. D. Alves, *Appl. Sci.* **2019**, *9*, 571.

25 a) A. P. Capelezzo, L. C. Mohr, F. Dalcanton, J. M. M. de Mello, M. A. Fiori, *Cyclodext-Versatile Ingred* **2018**; b) A. Singh, A. Mittal, S. Benjakul, *Science Asia* **2021**, *47*, 1.

26 D. Ach, S. Briançon, V. Dugas, J. Pelletier, G. Broze, Y. Chevalier, *Coll. Surf. A Physicochem. Eng. Aspects* **2015**, *481*, 367.

27 J. K. Rutz, C. D. Borges, R. C. Zambiazi, M. M. Crizel-Cardozo, L. S. Kuck, C. P. Noreña, *Food Chem.* **2017**, *220*, 59.

28 a) T. G. Mason, J. N. Wilking, K. Meleson, C. B. Chang, S. M. Graves, *J. Phys. Condensed Matter* **2006**, *18*, R635; b) C. I. Crucho, M. T. Barros, *Mater. Sci. Eng. C* **2017**, *80*, 771.

29 a) D. J. McClements, *Soft Matter.* **2012**, *8*, **1719**; b) E. Gukelberger, C. Hitzel, R. Mancuso, F. Galiano, M. D. L. Bruno, R. Simonutti, B. Gabriele, A. Figoli, J. Hoinkis, *Membranes* **2020**, *10*, 246; c) M. S. Numin, K. Jumbri, A. Ramli, N. Borhan, *Processes* **2020**, *8*, 762.

30 a) M. R. Avadi, A. M. M. Sadeghi, N. Mohammadpour, S. Abedin, F. Atyabi, R. Dinarvand, M. Rafiee-Tehrani, *Nanomed. Nanotechnol. Biol. Med.* **2010**, *6*, 58; b) S. Vaezifar, S. Razavi, M. A. Golozar, S. Karbasi, M. Morshed, M. Kamali, *Journal of Cluster Science* **2013**, *24*, 891; c) H. Wang, H. Deng, M. Gao, W. Zhang, *Frontiers in Bioengineering and Biotechnology* **2021**, 9.

31 a) A. Rahman, V. Uahengo, D. Likius, *Open J. Pharmacol. Pharmacotherapeut.* **2018**, *3*, 005; b) N. C. Ngwuluka, N. Y. Abu-Thabit, O. J. Uwaezuoke, J. O. Erebor, M. O. Ilomuanya, R. R. Mohamed, S. M. Soliman, M. H. A. Elella, N. A. Ebrahim, in *Nano-and Microencapsulation-Techniques and Applications*, IntechOpen, **2020**.

32 R. Lap, **2021**.

33 a) N. C. Ngwuluka, N. Y. Abu-Thabit, O. J. Uwaezuoke, J. O. Erebor, M. O. Ilomuanya, R. R. Mohamed, S. Soliman, M. Elella, N. J. N. M. T. A. Ebrahim, **2021**; b) M. Hadidi, N. Majidiyan, A. Z. Jelyani, A. Moreno, Z. Hadian, A. J. F. Mousavi Khanegah, **2021**, *10*, **2215**.

34 a) B. Vafania, M. Fathi, S. J. F. Soleimanian-Zad, *B. Process.*, **2019**, *116*, 240; b) L. M. Fonseca, M. Radünz, H. C. dos Santos Hackbart, F. T. da Silva, T. M. Camargo, G. P. Bruni, J. L. Monks, E. da Rosa Zavareze, A. R. J. J. o. t. S. o. F. Dias, *Agriculture*, **2020**, *100*, **4263**.

35 M. R. Mozafari, K. Khosravi-Darani, G. G. Borazan, J. Cui, A. Pardakhty, S. J. I. J. O. F. P. Yurdugul, **2008**, *11*, 833.

36 T. Subramani, H. J. J. o. f. s. Ganapathyswamy, *Technology* **2020**, *57*, **3545**.

37 a) D. Hudiyanti, H. Fawrin, P. Siahaan, Presented at *IOP conference series: materials science and engineering*, **2018**; b) R. Shaddel, J. Hesari, S. Azadmard-Damirchi, H. Hamishehkar, B. Fathi-Achachlouei, Q. J. I. j. o. b. m. Huang, **2018**, *107*, **1800**.

38 H. Hadiyanto, M. Christwardana, M. Suzery, H. Sutanto, A. M. Nilamsari, A. J. I. J. O. F. E. Yunanda, **2019**, 15.

39 D. Cano-Higuita, C. Malacrida, V. J. J. O. F. P. Telis, *Preservation* **2015**, *39*, **2049**.

40 C. Gupta, P. Chawla, S. Arora, S. Tomar, A. J. F. H. Singh, **2015**, *43*, 622.

41 A. Dabija, I. J. A. F. S. Nechifor, *Technology* **2015**, *16*, 20.

42 Z. Shariatinia, in *Natural Polysaccharides in Drug Delivery and Biomedical Applications*, Elsevier, **2019**.

43 S. A. Strobel, K. Hudnall, B. Arbaugh, J. C. Cunniffe, H. B. Scher, T. J. F. Jeoh, *B. Technol.* **2020**, *13*, 275.

44 L. G. Gómez-Mascaraque, M. Martínez-Sanz, M. J. Fabra, A. J. F. H. López-Rubio, **2019**, *90*, 523.

45 D. Rajmohan, D. J. I. J. O. F. S. Bellmer, **2019**, **2019**.

46 A. Belščak-Cvitanović, R. Stojanović, V. Manojlović, D. Komes, I. J. Cindrić, V. Nedović, B. J. F. R. I. Bugarski, **2011**, *44*, **1094**.

47 H. Rostamabadi, S. R. Falsafi, S. M. Jafari, in *Biopolymer Nanostructures for Food Encapsulation Purposes*, Elsevier, **2019**.

48 K. Marsh, B. J. J. O. F. S. Bugusu, **2007**, *72*, R39.

49 E. Díaz-Montes, R. J. F. Castro-Muñoz, **2021**, *10*, 249.

50 M. A. Elella, M. M. Abdel-Aziz, N. A. J. C. C. T. S. Abd El-Ghany, **2021**, *55*, 75.

51 A. E. Pavlath, W. Orts, in *Edible Films and Coatings for Food Applications*, Springer, **2009**.

52 A. Nešić, G. Cabrera-Barjas, S. Dimitrijević-Branković, S. Davidović, N. Radovanović, C. J. M. Delattre, **2019**, *25*, 135.

53 R. A. Holley, D. J. F. M. Patel, **2005**, *22*, 273.

54 S. R. Kanatt, R. Chander, A. J. F. C. Sharma, **2008**, *107*, 845.

55 J. Liu, X. Zhan, J. Wan, Y. Wang, C. Wang, *Carbohydr. Polym.* **2015**, *121*, 27.

56 F. Şen, İ. Uzunsoy, E. Baştürk, M. V. J. C. P. Kahraman, **2017**, *170*, 264.

57 V. Jost, K. Kobsik, M. Schmid, K. J. C. P. Noller, **2014**, *110*, 309.

58 M. G. J. F. Kontominas, Vol. *9*, Multidisciplinary Digital Publishing Institute, **2020**.

59 T. Senturk Parreidt, K. Müller, M. J. F. Schmid, **2018**, *7*, 170.

60 J. Lawrence, R. J. I. J. O. M. S. Nho, **2018**, *19*, 778.

61 Y. Zhao, M. D. J. T. J. O. S. F. Saldaña, **2019**, *143*, 97.

62 J. Cai, D. Zhang, R. Zhou, R. Zhu, P. Fei, -Z.-Z. Zhu, S.-Y. Cheng, W.-P. J. J. O. A. Ding, *F. Chem.* **2021**, *69*, **5067**.

63 A. Nazrin, S. Sapuan, M. Zuhri, R. Ilyas, R. Syafiq, S. J. F. I. C. Sherwani, **2020**, *8*, 213.

64 N. Hasheminejad, F. Khodaiyan, M. Safari, *Food Chem.* **2019**, *275*, 113.

65 N. P. Katuwavila, A. Perera, D. Dahanayake, V. Karunaratne, G. A. Amaratunga, D. N. Karunaratne, *Int. J. Pharmaceut.* **2016**, *513*, 404.

66 M. Hadidi, N. Majidiyan, A. Z. Jelyani, A. Moreno, Z. Hadian, A. Mousavi Khanegah, *Foods* **2021**, *10*, **2215**.

67 B. Vafania, M. Fathi, S. Soleimanian-Zad, *Food Bioprod. Process.* **2019**, *116*, 240.

68 L. M. Fonseca, M. Radünz, H. C. Dos Santos Hackbart, F. T. da Silva, T. M. Camargo, G. P. Bruni, J. L. Monks, E. da Rosa Zavareze, A. R. Dias, *J. Sci. Food Agric.* **2020**, *100*, **4263**.

69 M. R. Mozafari, K. Khosravi-Darani, G. G. Borazan, J. Cui, A. Pardakhty, S. Yurdugul, *Int. J. Food Proper.* **2008**, *11*, 833.

70 T. Subramani, H. Ganapathyswamy, *J. Food Sci. Technol.* **2020**, *1*.

71 S. Yang, C. Liu, W. Liu, H. Yu, H. Zheng, W. Zhou, Y. Hu, *Int. J. Mol. Sci.* **2013**, *14*, **1976**3.

72 R. Shaddel, J. Hesari, S. Azadmard-Damirchi, H. Hamishehkar, B. Fathi-Achachlouei, Q. Huang, *Int. J. Biol. Macromol.* **2018**, *107*, **1800**.

73 J. D. B. Alexandre, T. L. C. T. Barroso, M. D. A. Oliveira, F. R. D. S. Mendes, J. M. C. D. Costa, R. D. A. Moreira, R. F. Furtado, *Ciência Rural* **2019**, *49*.

74 H. Hadiyanto, M. Christwardana, M. Suzery, H. Sutanto, A. M. Nilamsari, A. Yunanda, *Int. J. Food Eng.* **2019**, *15*.

75 C. Dima, M. Cotârlet, P. Alexe, S. Dima, *Innov. Food Sci. Emerg. Technol.* **2014**, *22*, 203.

76 D. Marín-Peñalver, A. Alemán, M. P. Montero, M. C. Gómez-Guillén, *Food Funct.* **2021**, *12*, **2137**.

77 D. Cano-Higuita, C. Malacrida, V. Telis, *J. Food Process. Preserv.* **2015**, *39*, **2049**.

78 C. Gupta, P. Chawla, S. Arora, S. Tomar, A. Singh, *Food Hydrocol.* **2015**, *43*, 622.

79 A. Dabija, I. Nechifor, *Ann. Food Sci. Technol.* **2015**, *16*, 20.

80 S. A. Strobel, K. Hudnall, B. Arbaugh, J. C. Cunniffe, H. B. Scher, T. Jeoh, *Food Bioprocess Technol.* **2020**, *13*, 275.

81 L. G. Gómez-Mascaraque, M. Martínez-Sanz, M. J. Fabra, A. López-Rubio, *Food Hydrocol.* **2019**, *90*, 523.

82 D. Rajmohan, D. Bellmer, *Int. J. Food Sci.* **2019**, **2019**.

83 A. Belščak-Cvitanović, R. Stojanović, V. Manojlović, D. Komes, I. J. Cindrić, V. Nedović, B. Bugarski, *Food Res. Int.* **2011**, *44*, **1094**.

84 K. Marsh, B. Bugusu, *J. Food Sci.* **2007**, *72*, R39.

85 A. Jiménez, R. Requena, M. Vargas, L. Atarés, A. Chiralt, *Role of Materials Science in Food Bioengineering* **2018**, 263.

86 E. Díaz-Montes, R. Castro-Muñoz, *Foods* **2021**, *10*, 249.

87 A. Nešić, G. Cabrera-Barjas, S. Dimitrijević-Branković, S. Davidović, N. Radovanović, C. Delattre, *Molecules* **2020**, *25*, 135.

88 V. G. Souza, J. R. Pires, C. Rodrigues, I. M. Coelhoso, A. L. Fernando, *Polymers* **2020**, *12*, 417.

89 M. H. A. Elella, E. Goda, M. Gab-Allah, S. E. Hong, Y. G. Lijalem, K. R. Yoon, in *Advances in Nanocomposite Materials for Environmental and Energy Harvesting Applications*, Springer, **2022**.

90 R. A. Holley, D. Patel, *Food Microbiol.* **2005**, *22*, 273.

91 S. R. Kanatt, R. Chander, A. Sharma, *Food Chem.* **2008**, *107*, 845.

92 L. Brannon-Peppas, N. A. J. C. E. S. Peppas, **1991**, *46*, 715.

93 S. Kumar, A. Mukherjee, J. Dutta, *Trends Food Sci. Technol.* **2020**, *97*, 196.
94 F. Şen, İ. Uzunsoy, E. Baştürk, M. V. Kahraman, *Carbohydr. Polym.* **2017**, *170*, 264.
95 V. Jost, K. Kobsik, M. Schmid, K. Noller, *Carbohydr. Polym.* **2014**, *110*, 309.
96 K. Müller, M. Schmid, *Foods (Basel, Switzerland)* **2018**, *7*.
97 J. Lawrence, R. Nho, *Int. J. Mol. Sci.* **2018**, *19*, 778.
98 S. N. Mousavi, H. Daneshvar, M. S. S. Dorraji, Z. Ghasempour, V. Panahi-Azar, A. Ehsani, *Mater. Chem. Phys.* **2021**, *267*, **1245**83.
99 Y. Zhao, M. D. Saldaña, *J. Supercrit. Fluids* **2019**, *143*, 97.
100 N. A. Abd El-Ghany, M. S. A. Aziz, M. M. Abdel-Aziz, Z. M. Mahmoud, **2019**, *134*, 912.
101 J. Cai, D. Zhang, R. Zhou, R. Zhu, P. Fei, -Z.-Z. Zhu, S.-Y. Cheng, W.-P. Ding, *J. Agricult. Food Chem.* **2021**, *69*, **5067**.
102 A. Nazrin, S. Sapuan, M. Zuhri, R. Ilyas, R. Syafiq, S. Sherwani, *Front. Chem.* **2020**, *8*, 213.
103 S. Jha, P. Rohilla, K. Singh
104 J. Jose, M. A. Al-Harthi, M. A. A. AlMa'adeed, J. Bhadra Dakua, S. K. J. J. O. A. P. S. De, **2015**, 132.
105 H. Chen, J. Wang, Y. Cheng, C. Wang, H. Liu, H. Bian, Y. Pan, J. Sun, W. Han, *Polymers* **2019**, *11*, **2039**.
106 E. D. R. Zavareze, S. L. M. E. Halal, R. Marques E Silva, A. R. G. Dias, C. Prentice-Hernández, *J. Food Process. Preserv.* **2014**, *38*, **1973**.
107 S. J. Calva-Estrada, M. Jiménez-Fernández, E. Lugo-Cervantes, *Food Eng. Rev.* **2019**, *11*, 78.
108 a) P. Tongnuanchan, S. Benjakul, T. Prodpran, *Food Chem.* **2012**, *134*, **1571**; b) M. Ahmad, S. Benjakul, T. Prodpran, T. W. Agustini, *Food Hydrocolloids* **2012**, *28*, 189.
109 E. Emragi, D. Kalita, S. S. Jayanty, *LWT* **2022**, *153*, **1125**80.
110 H. J. Park, J. W. Rhim, H. Y. Lee, *Food Sci. Biotechnol.* **1996**, *5*, 59.
111 S. K. Paul, **2020**.
112 X. Wang, D. Kong, Z. Ma, R. Zhao, *Irish J. Agricult. Food Res.* **2015**, *54*, 64.
113 B. Zhang, X. Feng, P. Han, X. Duan, *J. Chin. Inst. Food Sci. Technol.* **2016**, *16*, 159.
114 E. Poverenov, Y. Zaitsev, H. Arnon, R. Granit, S. Alkalai-Tuvia, Y. Perzelan, T. Weinberg, E. Fallik, *Postharvest Biol. Technol.* **2014**, *96*, 106.
115 M. Bizura Hasida, M. Nur Aida, M. Zaipun, M. Hairiyah, Presented at *VII International Postharvest Symposium* **1012**, **2012**.
116 D. Feng, W. Zhengguang, Z. Yimei, Z. Xiang, G. Meng, Y. Xu', Y. Bi, Presented at *II International Symposium on Discovery and Development of Innovative Strategies for Postharvest Disease Management* **1053**, **2013**.
117 Y. D. Livney, *Curr. Opin. Coll. Interf. Sci.* **2010**, *15*, 73.
118 S. Saez-Orviz, A. Laca, M. Rendueles, M. J. A. Díaz, **2017**, 74.
119 A. Shendurse, G. Gopikrishna, A. Patel, A. J. J. N. H. F. E. Pandya, **2018**, *8*, 219.
120 M. A. Castro-Yobal, A. Contreras-Oliva, V. Saucedo-Rivalcoba, J. L. Rivera-Armenta, G. Hernández-Ramírez, J. Salinas-Ruiz, A. Herrera-Corredor, *e-Polymers* **2021**, *21*, 82.
121 P. Jha, *J. Food Sci. Technol.* **2020**, *1*.
122 R. Priyadarshi, B. Kumar, Y. S. Negi, *Carbohydr. Polym.* **2018**, *195*, 329.
123 R. Priyadarshi, J.-W. Rhim, *Innov. Food Sci. Emerg. Technol.* **2020**, *62*, **1023**46.
124 S. Petrou, M. Tsiraki, V. Giatrakou, I. Savvaidis, *Int. J. Food Microbiol.* **2012**, *156*, 264.
125 Y. A. Arfat, J. Ahmed, M. Ejaz, M. Mullah, *Int. J. Biol. Macromol.* **2018**, *107*, 194.
126 S. Kumar, A. Shukla, P. P. Baul, A. Mitra, D. Halder, *Food Packag. Shelf Life* **2018**, *16*, 178.

127 T. Jayaramudu, H.-U. Ko, L. Zhai, Y. Li, J. Kim, *Soft Mater.* **2017**, *15*, 64.

128 N. A. Mohamed, N. A. Abd El-Ghany, M. M. Abdel-Aziz, *Int. J. Biol. Macromol.* **2021**, *181*, 956.

129 a) N. A. Peppas, A. R. Khare, *Adv. Drug Deliv. Rev.* **1993**, *11*, 1; b) N. Pekel, O. Güven, *Polym. Int.* **2002**, *51*, **1404**; c) G. Sharma, B. Thakur, M. Naushad, A. Kumar, F. J. Stadler, S. M. Alfadul, G. T. Mola, *Environ. Chem. Lett.* **2017**, 1; d) E. M. Ahmed, *J. Adv. Res.* **2015**, *6*, 105.

130 A. El-Ghany, A. Nahed, Z. M. Mahmoud, *Polym. Bull.* **2021**, *78*, **6161**.

131 a) L. Brannon-Peppas, N. A. Peppas, *Chem. Eng. Sci.* **1991**, *46*, 715; b) K.-H. Liu, T.-Y. Liu, S.-Y. Chen, D.-M. Liu, *Acta biomaterialia* **2008**, *4*, **1038**; c) D. Roy, J. N. Cambre, B. S. Sumerlin, *Progress in Polymer science* **2010**, *35*, 278; d) K. R. Kamath, K. Park, *Adv. Drug Deliv. Rev.* **1993**, *11*, 59; e) B. Manjula, A. B. Reddy, T. Jayaramudu, E. Sadiku, S. Owonubi, O. Agboola, T. Mokrani, *Handbook Composit. Renew. Mater. Nanocomposit. Sci. Fundament.* **2017**, *7*, 67.

132 Y. Tanaka, J. P. Gong, Y. Osada, *Progress Polym. Sci.* **2005**, *30*, 1.

133 A. Mellati, S. Dai, J. Bi, B. Jin, H. Zhang, *RSC Adv.* 2014, *4*, 63951.

134 A. M. Elgamal, M. H. Abu Elella, G. R. Saad and N. A. Abd El-Ghany, *Mater. Today Commun.* **2022**, *33*, **1042**09.

135 N. Y. Elmehbad, N. A. Mohamed and N. A. Abd El-Ghany, *Int. J. Biol. Macromol.* **2022**, *205*, 719.

136 Mahmoud H. Abu Elella, Heba M. Abdallah, Heba Gamal, Essam B. Moustafa, and Emad S. Goda, *Cellulose* 2022, *29*, 7317.

2

Introduction to Food Properties and Techniques in Food Product Development

Sara A. Al-Hafiry[1,*], *Omar A. Alaboudi*[2], *Ghada A. Ahmed*[2], *and Abdulrahman M. Abdulrahman*[3]

[1] *Department of Biophysics, Faculty of Science, Cairo University, Cairo, Egypt*
[2] *Chemistry and Zoology Department, Faculty of Science, Cairo University, Cairo, Egypt*
[3] *Chemistry/Biochemistry Department, Faculty of Science, Cairo University, Cairo, Egypt*
[*] *Corresponding author*

Highlights

- Food materials science offers the best tools to enhance food properties by understanding their underlying structure and composition and finding ways to modify them to produce a certain product with defined properties. This can be achieved by understanding the link between food matrix characteristics and its qualities and behaviors.
- Food industry, and hence the main focus of food materials science, is now driven by consumers' preferences, which are heading toward healthier food consumption.
- Among the many food properties, some have attracted great attention according to industry trends and consumers' preferences. For instance, efficient delivery of food functional ingredients is important to maximize their benefits and minimize their toxicity. In addition, textural properties are what is perceived first by consumers and making them likable is of great importance. Finally, food stability is what keeps other properties functional overtime, and is the key in food production and storage.
- Production techniques of food products should be compatible with the food design, aiming for a final product. It requires basic and applicable knowledge in food microstructure manipulation and the processes that affect it: for example, in food formulation. Food processing alters the chemical composition of food ingredients to enhance the product properties. Unlike earlier, 3D printing is not only used to make products of complicated geometrical design any longer, but also to acquire other properties like texture and nutritional bioavailability.

Materials Science and Engineering in Food Product Development, First Edition. Edited by Wing-Fu Lai.

2.1 Introduction

The development of a new product is to design it from scratch, produce it, and make it available in the market. Most of the new food products are developed according to precise procedures and models that depend mainly on experience [1–3]. A time-consuming process with a high cost is required to design new food products, as food products are characterized by their complexity and limited shelf lifetime. Therefore, accelerating the development of new food products by effective processes is critical [4]. Designing food products relies on many factors, such as the availability of raw materials, machines used in production, food supply chain, and consumers [5]. On finding that the consumer's expectation is one of the most important factors in food formulation and design, food engineering has shifted from focusing on procedures and unit operations to designing food products with high quality that promote convenience, health, and well-being [6].

Materials science involves the connections between the processing, characteristics, and structures of materials. It includes all the chemical and physical aspects of all the materials used besides their performance at atomic, molecular, and bulk levels [7]. As the sensory, physical, and transport properties of food are influenced by the food structure [6], materials science concepts must be incorporated into food formulation and design, to gain a better understanding of how the micro monomers of food are assembled to provide the food structure, the biomolecular interactions of food ingredients, and the food matrix. Incorporating materials science into the food product design field also helps in the design, development, formulation, and processing of new food products that are nutritious, safe, and have high functionality and an extended shelf life.

2.2 Structural Impact on Properties

2.2.1 Food Materials Science

A food material is a composite material, and its overall properties are different from its components' properties when separated from each other. Moreover, its components can be simultaneously present in different phases with different mobilities. The compatibility of these components and their ability to form bonds within the food material gives rise to the stability of this material [7]. Manipulation of this material to enhance it requires an understanding of its physical and chemical characteristics. On that note, materials science offers the best tools to achieve that, opening the way for an interdisciplinary food materials science. Materials science proposes that the properties of materials are defined by their composition and structure; hence, food materials properties can be modified by adjusting not only their composition, but also their structure [8]. Table 2.1 enlists some food structures and how those structures affect food properties. The addition of new components or the altering of the food microstructure by different processes can enhance certain properties in the food material, which can be utilized in industrial applications as in Industrial Corner 1 [7].

Table 2.1 Food structures and corresponding effects.

Food structure	Effect	Examples	Working on these structures as examples	References
Colloidal structure	Bioaccessibility and bioavailability in the GIT	Lipid emulsions	Adjusting emulsion concentration or oil to water ratio	[9]
Food particle size	Textural properties	Casein micelles	Shear stress, pressure, and homogenization can adjust particle size	[7, 9]
Crystalline structure	Mechanical properties and food stability	Water crystallization in frozen foods, lactose, and whey protein during dried milk processing	Creating dried milk powder, co-crystallization of sugar for active ingredient entrapment	[7]
Gel structure	Elasticity, viscosity	Jelly, yogurt, cheese, soy tofu	Stress and strain properties can be modified by pH and temperature manipulation	[7, 10]

Industrial Corner 1

- A food product's mechanical properties and shelf life can be improved by the addition of plasticizers.
- Glassy structures can be produced by rapid cooling or drying.
- Generally, elastic gels are composed of polymers, while viscous gels are composed of particles.
- Small particle gels can be used as encapsulation systems keeping an active ingredient trapped inside until it is delivered to the proper site.
- Amorphous powders can be produced from crystalline or semicrystalline materials by drying and extrusion.

2.2.2 Food Matrix

One of the most interesting concepts in food materials science is the food matrix that aids in understanding food as a material and explains the physical, chemical, and physiological interactions in the food. Aguilera [11] defines the food matrix as, "A physical domain that contains and/or interacts with specific constituents of a food (e.g. a nutrient) providing functionalities and behaviors which are different from those exhibited by the components in isolation or a free state." This definition revolves around the notion that a composite material has different qualities from its isolated constituents. It is important to understand the food matrix characteristics for their huge impact on food properties and the functionality of a food product. For instance, it was found that for better oil-soluble vitamins (ADEK) bioavailability, there should be lipid content in the food matrix; therefore, when food is broken down in the intestines, this lipid content can form missiles to contain the oil-soluble vitamins and improve their absorption through intestinal cell walls [12, 13]. Consequently, improving food bioavailability and other functional food qualities requires understanding the interactions of the food's active ingredients (i.e. flavors, vitamins, minerals, colorants) with the food matrix they are embedded in [8, 14]. Some types of food matrices are liquid matrices (i.e. milk, juices, and soft drinks), emulsion matrix (i.e. oil in water emulsions), gel matrix (i.e. a semisolid matrix that entraps water inside), cellular matrix, where micronutrients are held inside cell walls, fibrous extracellular matrix (e.g. collagen fibers), viscoelastic matrix (e.g. dough), dense matrix (i.e. crystalline, semicrystalline, and glassy material), porous material matrix or foams (i.e. cream and bread), and artificial matrix (i.e. capsules) [11]. Food gel formulation is discussed in Industrial Corner 2 [15, 16].

Industrial Corner 2

- Plastic fat, starch, globular proteins, casein, gelatin, carrageenan, and alginate are food grade polymers that can be used to form gels.
- Starch gels can be crystallized and get stiffer during storage.
- Mechanical fracture properties of the gels vary according to the processing conditions, temperature, pH, etc.
- When there is incompatibility in the gel thermodynamic system, phase separation may occur.

While focusing on health properties, some effects of the food matrix on the properties of many types of food are well defined, such as the reduction of bioaccessibility of starch in cooked pasta because of the entrapment of starch granules in the gluten network and their consequent gelatinization when cooked. This gelatinization of starch embedded in gluten makes it inaccessible by amylase, and such a conclusion in food materials science is acceptable because of a clear understanding of the structural interactions of pasta and digestion [11, 17, 18]. On the other hand, the exact effects of the food matrix on some other properties, such as the bioavailability of carotenoids or milk proteins, are still not fully understood, and inconsistency in results, combined with the inability to interpret them properly, shows the need for a more standardized form of studies that depends on materials science for study design and data interpretation [19–23]. We suggest, in that context, that food molecular interactions on different structural levels need to be further investigated, and consequently, how these interactions are affected in the different phases of *in vivo* digestion. As will be mentioned next, food characterization is a suitable tool to "see" those molecular interactions and to understand food from a materials science perspective.

2.2.3 Food Characterization and Study Standardization

Food materials characterization is important to find a link between food characteristics and properties and be able to manipulate them to produce a desired product with certain properties by different processes. Study standardization is important to compare results from different studies. For example, the optical properties of emulsions obtained by light-scattering experiments, confocal microscopy, and instrumental colorimetry during every stage of food fabrication show its droplet size and concentration, besides its degree of aggregation [24]. In addition, measuring the concentration of salt and sugar may be indicative of altered electrostatic interactions, osmotic effects, and surface tension [24, 25].

2.3 Food Consumer Demands

Understanding consumer's demands defines the direction of the efforts of food materials science [26]. Previously, food manufacturers used to produce foods based on the available food materials and production techniques, and consumers used to have the products available in the market without being a part of the decision of manufacturing a certain product. Currently, the direction of this decision-making process is somehow reversed where manufacturers focus on producing what is of value to the consumer's interests [8]. In addition, consumers' interests are always changing. For instance, they are currently focused on health issues [8]. People are looking for healthier ways of food consumption, either by avoiding unhealthy food or bad eating habits or by choosing beneficial foods and adopting better eating habits.

There are a few common attributes that different healthy meals share, such as meals with fruits and vegetables, as they attract great attention among consumers. Fruit extracts and products are of great potential to a wide area of applications [27]. Besides that, value-added products can be highly beneficial if their nutrients' bioavailability is optimized. Avoiding toxic and pathogenic effects is also essential [24]. To conclude, the core concept in current food product design is the focus on what the consumers want and how to deliver the best product that meets their demands, knowing that the best approach to achieve that is via materials science.

2.4 Food Properties to Be Improved

2.4.1 Introduction to Food Properties

A food property is a certain measure of the food behavior as matter or energy and its interaction with the senses or its effectiveness in improving human health and well-being. [28] Scientists and engineers understand food properties to solve food issues starting from preservation to consumption and processes thereafter, so a suitable classification of food properties is needed to avoid confusion and inaccuracy. Wherefore properties classification is needed to corroborate obtainable data, create a global database, and improve predictive correlations. A good categorization will promote a sound interdisciplinary approach to the research of food properties, and the measurement and application of food property data. As a result, the design process and characterization of food products are improved. The major four categories of food properties are physical, physicochemical, kinetic and sensory, and health properties [29].

2.4.2 Functional Ingredients and Health Properties

Customers taking an active role in preserving their health has resulted in a plethora of items on the market, including food ingredients, intended to improve their health [30]. Functional ingredients vary, and they are a series of substances that have a favorable impact on customer health. "Functional ingredient" is a term that refers to the chemicals that are intended to produce a favorable health effect through physiological activity in the body. They include medicinal, nutritional, and food additive components. Because of the physiological activity of the functional components, they can have a wide range of consequences in the body, from suboptimal physiological action and therapeutic influence to outright toxicity, depending on the quantity of intake. Understanding the mechanisms of pharmacological activity and the toxicological potential is required to foresee the effects of exposure on various dose levels [30].

Delivery systems differ from one another; each one has its unique advantages and limitations for the distribution and encapsulation of functional substances. One of the delivery systems of functional ingredients is microencapsulation; it has been utilized to help the active ingredients be induced into food, to keep up and manage their release into the gastrointestinal tract, and to improve their desired function in the body. Microencapsulation technology was created to protect these bioactive components throughout processing and storage and to target bioactive release into specific regions in the gastrointestinal tract [31]. As a result, the scientific community should create novel functional foods that have the potential to provide physiological benefits or reduce the long-term risk of disease development [32].

Phytochemicals play a critical role as functional components in food that link it with health. These components provide disease-fighting and health-promoting foods to prevent chronic degenerative diseases or support traditional medicines [33]. Not all the consumed food can be considered safe; foods may have negative or positive effects on health. So, health properties are linked to the potency of foods in promoting health [34]. Some of the health properties that link these functional components to health are further categorized into antioxidant and anti-cancerous properties [35]. Antioxidant properties, as those involving phytochemicals, such as phenol compounds, can protect the body against oxidative stress [36]. Some foods contain anti-cancerous properties as it is reported that some functional components of food have anti-proliferative effects on cancer cell lines and inhibition of DNA damage [37]. There are also many categorized health properties, such as antidiabetic, anti-inflammatory, cardiovascular, antimicrobial, immunomodulatory, anti-hypersensitive, anticoagulant, and foods that have protective effects on gastric organs, and many more. Therefore, consuming a healthy diet helps in preventing noncommunicable degenerative diseases due to the presence of these functional components, especially in plant-based foods.

2.4.3 Physiochemical and Sensory Properties

Most food is likely to be spoiled and can be changed easily by chemical, physical, and biological agents, because of the chemical composition and physicochemical properties [38], such as ash, nitrogen, viscosity, and degree of deacetylation, differ based on food products, and these properties show the interaction between them and the composition and structure of food components with the conditions under which these are linked or measured. Such functional characteristics are critical to determine the performance of a food product and have an impact on the sensory properties of the product [39, 40]. Sensory properties depend mainly on the primary senses of humans: sight, touch, smell, taste, and hearing. The five major sensory properties considered in the food industry are appearance, texture, taste, and aroma as well as whether the food can cause irritation [41]. To formulate food products with the desirable sensory properties, scientists use well-designed sensory tests with human experts that are well-trained to analyze, operate, and interpret reactions to food materials perceived by consumers' senses. Many other methods depend on devices present besides the regular sensory tests to analyze the sensory properties, for example, tests using gas and liquid chromatography are developed to analyze flavor compounds in roasted coffee beans [42].

2.4.4 Physical Properties

Physical properties, such as size, shape, volume, weight, and appearance, are assigned to the main characteristics of food. These physical properties are of great interest in food engineering as it is a well-known fact that nanomaterials have shown unique characteristics that can improve the properties of food biopolymers when incorporated into the food industry, and such nano-applications require in-depth research and a complete understanding of the physical structure. Another reason for such interest is that most of these physical properties can be measured easily, so a good knowledge of physical/engineering properties, such as density, thermal conductivity, and viscosity is critical for food process operation, rational design, and the prediction of food response for processing and preservation conditions. Also, food characteristics that are linked to the physical properties of food, such as texture and structure, are the most important aspects to define the stability and quality of the food product [43]. Texture has been historically used to describe the structure, feel, and appearance of fabric materials. Texture qualities and customer preferences differ dramatically from one type of food to the next. Food texture has become the sensory phrase of food description in the middle of the twentieth century when food production shifted to an industrial scale and the first food science and technology degree was offered. The broad idea that texture refers to the constitution, structure, or substance of everything in terms of its constituents and formative aspects has created an excess of definitions [33]. The types and amounts of structural carbohydrates (cellulose, starches, and pectic components), hydrocolloids, and proteins contained in foods determine the texture of the food. Consumers' judgments of "quality" are influenced by texture, and information regarding changes in texture is sent to the brain while chewing through mouth sensors, hearing, and memory [34].

For numerous physical forms of food, key texture traits must be specified at the category level: solid, soft solid, and fluid. Solid food contains specific geometric properties and shapes. Only a small amount of tension can be applied to this type of food before it cracks catastrophically. On the other hand, fluid food has a moderate amount of resistance to a small amount of stress but is unable to withstand gravity. Fluid food is distinguished by the absence of a geometrical shape, and instead, it is shaped after the container in which it is stored. Finally, soft solid foods (semisolid) are the materials that fall somewhere between liquid and solid materials. These meals are neither liquid nor solid [33].

2.4.5 Food Stability

Stability is utilized to explain how a certain parcel of food would retain its freshness, nutrition, and flavor under stressful environmental conditions, such as travel or storage in a home [44]. For example, protein stability refers to a protein's capacity to withstand diverse factors while maintaining its biological activity. It is required to ensure that protein molecules are in the form of relatively low free energy to preserve a specific shape. There are many factors affecting protein stability, such as their qualities and external conditions affecting them (temperature, modifier, or pH) [45]. By studying polyphenols, it has been discerned that their benefits depend on their bioavailability and not only on their consumption. The biotransformation of phenolics that reach the blood and tissues after absorption causes them to differ from those contained in food, making it difficult to determine their biological activity [46]. In order to enhance the bioavailability of polyphenols, they are incorporated into nanoparticles. Nanotechnology has recently exploded in popularity in the food and pharmaceutical industries, especially with the use of nanoencapsulation of bioactive substances to be utilized in biological systems [47]. Encapsulating phenolic compounds on the nanoscale protects them against a variety of stimuli that promote their chemical breakdown, hence expanding their potential applications [48].

2.4.6 Kinetic Properties

Kinetic properties are a group of several kinetic constants that characterize several rates of changes in foods. These constants can be classified into two categories. The first one includes kinetic constants that characterize rates of biological, biochemical, chemical, physicochemical, and chemical changes in food, such as respiratory constants, rate constant, half-life, and Arrhenius equation constants. The second category includes constants that characterize the growth, decline, and death rates of food microorganisms, such as specific growth rate, generation time, and decimal-reduction time. It is worth mentioning that such properties are not food properties, but the properties of the microorganisms that are moderated by the food they are present in [30].

2.5 Food Materials Synthesis Techniques

2.5.1 Food Formulation

Innovating and renovating food products can be considered as the main processes that outline the definition of food formulation. The major target of food formulation is to give the food some functionality such as desired nutritional supplementation besides its fundamental role and this can be achieved by creating or developing food products [49]. Food formulation is the direction that guides in reaching a certain product, so it is critical to list all the ingredients with their quantities and operational steps precisely. Whatever the site of production, industrial or not, based on science or observations, the targeting goal is to produce a good food product [50].

Food formulation is a complex methodical process with crucial importance that includes [49]:

1) Picking out the raw materials according to the end products required and their utility;
2) Understanding raw materials interactions at the molecular levels. The structure of food is made up of colloidal molecules of biopolymers and their complexes. These biopolymers are essential for giving the food its structure and physical, chemical, and sensory properties. Also, the rheology of foodstuff depends mainly on these biopolymers. Proteins are macromolecules that modify the structure of food, affecting its texture accordingly [51], and other molecules interact with each other to form essential complexes that identify the quality of food. Examples of these macromolecular interactions are protein–protein, protein–polysaccharides, protein–lipid, protein–water, polysaccharide–lipid, and protein–polysaccharide–lipid interactions;
3) Functioning technology according to ingredients, consumer demands, required quality, shelf lifetime, and cost.

To control the design of the food microstructure, one should consider the way food components are self-assembled, paving the way from raw materials to end products to be utilized by the consumer. Controlling the food microstructure will affect the stability of food in terms of chemistry, biochemistry, microbiology, mass transfer, and food texture. It will also impact the accessibility of nutrients and their transport [52, 53]. And by controlling food microstructure and structure, assembling strategies of new food components can be developed especially on a nanoscale, leading to maintaining a higher food structure and functionality, better look and taste, improved nutrients transport, with the capability of delivering to the consumer's needs. Consequently, developing new strategies and technologies will lead to better quality products that are safe, efficient, and cost-effective.

One of the fundamental strategies of developing food microstructures is to apply force to reduce size. The degree of reduction and refinement affects the properties and characteristics of the materials. It is important to decide carefully which technique will be used for the production process to produce the smallest functional size, which is balanced with functionality as well. When reducing the size, the surface area will be increased, which is required to improve flavor release and water absorption and increase the rates of catalysis. Smaller size distribution also allows reaching the best quality and the desired functionality control of the product. The three applied forces in the strategy used to reduce the size are compression, impact, and shear. In the case of food applications, such as forming nanoparticles, impact and shear are the forces used. Along with size reduction, other strategies are applied to food materials to control the food structure, such as dry milling, grinding, emulsification, and supercritical fluids [54, 55], whereas building supramolecular higher hierarchical structures depends mainly on the assembly control of the molecules under the effect of thermodynamics [55, 56].

2.5.2 Food Processing

Food processing can be defined as changing the characteristics of food substances to preserve and improve food quality and functionality. It includes a sequence of steps to attain desirable modifications with regard to the raw materials. Any food other than raw agricultural materials and that which contains modified/processed raw agricultural materials that have been subjected to milling, cooking, heating, freezing, canning, or dehydration is called processed food [57–59]. At this point, it is obvious that formulation is a recipe that can be modified according to the ingredients, available equipment and machines, and target consumers. So, for different procedures, there are varied formulations applied to control all parameters of the quality and food safety categories besides controlling the production yield [50].

2.5.3 3D Printing

Although it has been applied in the food industry only since 2007, 3D printing is not a very fresh technique [60]. Back then, it was only used to produce complex geometric visuals that were more advanced than the traditional structures produced [61, 62]. Nowadays, 3D printing is not only used in visual aspects but also with a higher focus in the control and customization of the nutritional properties of food [62–64]. Attractive visualization of food is not enough. Food should also have a healthy nutritional profile. Besides taste and appearance, texture and mouthfeel of the food product are major concerns by the consumer to decide whether the food is acceptable or not [65–67].

The 3D printing technique uses the application of layers, one by one, to build the desired structure. One of the most widely used techniques of 3D printing is extrusion 3D printing. On applying extrusion 3D printing, food materials are extruded through a dynamic nozzle by a hydraulic piston to attain layered structures [61]. According to the characteristics of these food materials, layers are then fused based on whether the materials solidify or form hydrogels on cooling [61, 68–70]. For creating surface decorations and fillings, low-viscosity materials are layered continuously or as droplets through a head on a support structure. This head can be thermal or piezoelectric [61, 70, 71].

Chocolate is a food material that possesses success in extrusion food 3D printing, as it solidifies after layering; it is a mixture of non-fat particles that are sugar, cocoa particles, and milk powders. This dense suspension is diffused in the continuous solidifying phase of cocoa butter. To characterize the cooling dynamics, one should study how the printing structure of chocolate is affected by printing velocity and environmental temperature. That type of study is achieved by an IR thermography. A critical print velocity depends on the environmental temperature and should not be exceeded in creating 3D edible structures. Printing velocity should be lower than 16 mm/s if the environment temperature is at 18°C and should be lower than 8 mm/s if the environmental temperature is at 20°C. Applying such conditions is critical to attaining the desired sufficient solidification of the cocoa butter phase, which, in turn, affects the print stability [72–74]. By this time, more innovations and developments would exist to widely help in manufacturing food substances with perfect nutritional values, highest quality, and safety, with a very attractive texture and appearance. One can expect that traditional meals would be replaced by 3D printed food, as the traditional meals are limited in terms of decorations and control and are also difficult to be prepared in any other way. By 3D printing technology, ready, healthy meals can be easily programmed to fit individual needs.

2.6 Concluding Remarks

It is clear that to improve food properties, a better understanding of the food structure and interaction of components on the molecular level is needed. Established classifications of food properties would help in aiming for a certain product that is of value to consumers. Product design is the discipline focused on optimizing the best approach toward creating that product and deciding the best techniques to achieve it. Many existing techniques have been discussed in this chapter and many are yet to be developed. Future work should be focused on understanding food properties in terms of materials science for better result interpretation. This can be done by finding the link between those properties and food matrix characteristics and their underlying interactions. Following that would be food product design and finding the best existing techniques to convert that design into a product or inventing new ones. In addition, the main goal of food product design and food materials science should always be focused on providing what is best for consumers' health and well-being. While food materials can be solid in nature, liquid foods play an important role in the food industry. Their properties are totally different from solid foods and are largely affected by rheological features. In the next two chapters, our colleagues will discuss in further detail about bulk rheology and interfacial rheology of liquid foods, particularly food emulsions.

Glossary

Bioavailability The extent at which a particular food ingredient enters plasma circulation.

Biopolymers Naturally existing polymers in living organisms usually produced by cells.

Emulsions Microscopical scattering of two immiscible liquids, one for the continuous phase and the other for the dispersion phase of the system.

Food stability The ability of food to keep its physical and chemical properties over time.

Food structure The order of food constituents in the space on different scales.

Gels Viscoelastic semisolid materials that entrap water inside.

IR thermography Infrared thermography is using thermal imaging for detecting radiations.

Micronutrients Molecules of nutritional value that are present in food, like minerals and vitamins.

Oil-soluble vitamins Vitamins that are insoluble in water and need a different medium to be solubilized and absorbed in the intestines.

Organoleptic qualities The aspects of food, water, or other substances that create an individual experience via the senses – including taste, sight, smell, and touch.

Phytochemicals Bioactive nutrient plant chemicals that provide desirable health benefits besides basic nutrition to reduce the risk of chronic diseases.

Polyphenols They are plant secondary metabolites with hydroxylated phenyl moieties that safeguard human health by reducing oxidative stress and thereby preventing many diseases.

Raw materials The primary input unprocessed goods of manufacturing used to reach an end processed product.

Sensory properties The human physiological–psychological perception of several physical and other properties of food and their interactions.

Supercritical fluids Fluids that are highly compressed at temperature and pressure higher than their critical ones combining the properties of both gas and liquid such as CO_2.

References

1 C. Pinna, F. Galati, M. Rossi, C. Saidy, R. Harik, S. Terzi, *Comput. Ind.* **2018**, *100*, 184.
2 G. Rossetti, F. Giraudo, P. Murer, L. Arcusin, *J. Ind. Eng.* **2014**, *2*, 5.
3 M. Rudolph, *Br. Food J.* **1995**, *97*, 3.
4 X. Zhang, T. Zhou, L. Zhang, K. Fung, K. Ng, *Ind. Eng. Chem. Res.* **2019**, *58*, 16743.
5 R. Guiné, S. Florença, M. Barroca, O. Anjos, *Food Prod. Dev.* **2020**, *9*, 1317.
6 J. Aguilera, *J. Sci. Food Agr.* **2006**, *86*, 1147.
7 B. Bhandari, Y. Roos, in *Food Materials Science and Engineering* (Eds: B. Bhandari, Y. Roos), Wiley-Blackwell, Oxford, **2012**, Ch.1.
8 J. Aguilera, P. Lillford, *Food Materials Science*, Springer, New York, **2008**.
9 R. Zhang, D. McClements, *Food Struct.* **2016**, *10*, 21.
10 S. Banerjee, S. Bhattacharya, *Crit. Rev. Food Sci. Nutr.* **2012**, *52*, 334.
11 J. Aguilera, *Crit. Rev. Food Sci. Nutr.* **2019**, *59*, 3612.
12 Y. Tan, D. McClements, *Food Chem.* **2021**, *348*, 129148.
13 B. Gleize, M. Hiolle, N. Meunier, B. Pereira, R. Richard, I. Savary-Auzeloux, C. Buffière, M. Peyron, C. Halimi, C. Caris-Veyrat, F. Nau, E. Reboul, *Mol. Nutr. Food Res.* **2020**, *64*, 2000228.
14 A. Fardet, D. Dupont, L. Rioux, S. Turgeon, *Crit. Rev. Food Sci. Nutr.* **2018**, *59*, 1987.
15 P. Walstra, in *Food Chemistry* (Ed: O. Fennema), Marcel Dekker, New York, **1996**, Ch.3.
16 P. Sukha, Ph.D. Thesis, The University of Queensland, **2005**.
17 E. H. J. Kim, J. R. Petrie, L. Motoi, M. P. Morgenstern, K. H. Sutton, S. Mishra, L. D. Simmons, *Food Biophys.* **2008**, *3*, 229.
18 M. Petitot, J. Abecassis, V. Micard, *Trends Food Sci. Technol.* **2009**, *20*, 521.
19 S. Van Buggenhout, L. Lemmens, *Trends Food Sci. Technol.* **2010**, *21*, 607.
20 R. M. Faulks, S. Southon, **2005**, *1740*, 95.
21 J. J. M. Castenmiller, C. E. West, J. P. H. Linssen, K. H. Van Hof, A. G. J. Voragen, **1999**, 349.
22 R. P. Ross, G. F. Fitzgerald, C. Stanton, **2013**, 1161.
23 T. K. Thorning, H. C. Bertram, J. Bonjour, L. De Groot, D. Dupont, E. Feeney, R. Ipsen, J. M. Lecerf, A. Mackie, M. C. Mckinley, U. Ris, C. Weaver, A. Astrup, I. Givens, **2017**, *105*, 1033.
24 Z. Zhang, R. Zhang, H. Xiao, K. Bhattacharya, D. Bitounis, P. Demokritou, D. McClements, *NanoImpact* **2019**, *13*, 13.

25 L. Figura, A. Teixeira, *Food Physics: Physical Properties-Measurement and Applications*, Springer Science & Business Media, **2007**.
26 J. Ubbink, in *Food Materials Science and Engineering* (Eds: B. Bhandari, Y. Roos), Wiley-Blackwell, Oxford, **2012**, Ch.7.
27 A. Oliveira, A. Amaro, M. Pintado, *Curr. Opin. Food Sci.* **2018**, *22*, 153.
28 O. McCarthy, *Physical Properties of Foods and Packaging Materials – an Introduction*, Palmerston North, New Zealand, **1997**.
29 M. Rahman, O. McCarthy, *Int. J. Food Prop.* **1999**, *2*, 93.
30 C. Kruger, S. Mann, *Food Chem. Toxicol.* **2003**, *41*, 793.
31 J. Smith, E. Charter, *Functional Food Product Development*, Wiley, United Kingdom, **2010**.
32 R. Elliott, T. J. Ong, *Bmj.* **2002**, *324*, 1438.
33 J. Chen, A. Rosenthal, *Modifying Food Texture: Novel Ingredients and Processing Techniques*, Woodhead Publishing, United Kingdom, **2015**.
34 P. Fellows, *Food Processing Technology*, Elsevier, Cambridge, UK, **2017**.
35 C. I. Abuajah, in *Bioactive Molecules in Food* (Eds: J. M. Mérillon, K. G. Ramawat), Springer, **2017**.
36 A. Chandrasekara, F. Shahidi, *J. Agri. Food Chem.* **2011**, *59*, 1, 428.
37 J. M. Awika, L. Yang, J. D. Browning, A. Faraj, *Food Sci. Tech.* **2009**, *42*, 1041.
38 S. Martínez, J. Carballo, *Foods.* **2021**, *10*, 2970.
39 S. Chandra, S. Singh, D. Kumari, *J. Food Sci. Tech.* **2015**, *52*, 3681.
40 Y. I. Cho, H. K. No, S. P. Meyers, *J. Agri. Food Chem.* **1998**, *46*, 3839.
41 Y. Lee, S. Y. Lee, S. J. Schmidt, in *Modern Magnetic Resonance* (Ed: G. A. Webb), Springer, Dordrecht, **2008**.
42 K. D. Deibler, J. Delwiche, *Handbook of Flavor Characterization*, Marecel Dekker, New York, **2003**.
43 Z. Berk, *Food Process Engineering and Technology*, Academic Press, **2018**.
44 J. Anderson, in *Encyclopedia of Food Security and Sustainability*, Vol. *2* (Eds: P. Ferranti, E. M. Berry, J. Anderson), Elsevier, **2018**, Ch.1.
45 X. Xiang, T. Zhou, L. Zhang, K. Fung, K. Ng, *ACS Pub.* **2019**, *58*, 16743.
46 G. Velderrain-Rodriguez, H. Palafox-Carlos, A. Wall-Medrano, J. Ayala-Zavala, C. Y. Chen, M. Robles-Sanchez, H. Astiazaran-Garc, E. Alvarez-Parrillab, G. Gonzalez-Aguilar, *R. Soc. Chem.* **2014**, *5*, 189.
47 R. Rahim, P. Jayusman, N. Muhammad, F. Ahmad, N. Mokhtar, I. Mohamed, A. Shuid, *Int. J. Environ. Res. Public Health.* **2019**, *16*, 4962.
48 C. Dicastillo, G. López-Carballo, R. Gavara, V. Galet, A. Guarda, M. Galotto, *J. Food Process. Preserv.* **2019**a, *43*, e14044.
49 S. Pathania, C. Bhatia, B. K. Tiwari, in *Food Formulation: Novel Ingredients and Processing Techniques* (Eds: S. Pathania, B. K. Tiwari), Wiley, Hoboken, **2021**, Ch.1.
50 R. Botelho, W. Araújo, L. Pineli, *Crit. Rev. Food Sci. Nutr.* **2016**, *58*, 4.
51 J. Lopes da Silva, M. Rao, in *Rheology of Fluid and Semisolid Foods*, Springer, NY, **2007**, Ch.6.
52 J. Aguilera, *Microstr. Food Tech.* **2000**, *54*, 56.
53 J. Aguilera, *J. Food Eng.* **2005**, *67*, 3.
54 J. Jung, M. Perrut, *J. Super. Fluids.* **2001**, *20*, 179.
55 P. Sanguansri, M. Augustin, *Trends Food Sci. Technol.* **2006**, *17*, 547.

56 S. Föster, M. Konrad, *J. Mat. Chem.* **2003**, *13*, 2671.

57 Infoods **2005**, Available at: http://www.fao.org/infoods/infoods/tables-and-databases/en.

58 A. S. Truswell, D. J. Bateson, K. C. Madafiglio, J. A. Pennington, W. M. Rand, J. C. Klensin, *J. Food Compos. Anal.* **2006**, *38*, 18.

59 P. Fellows, *Food Processing Technology: Principles and Practice*, Elsevier, **2009**.

60 D. Periard, N. Schaal, M. Schaal, E. Malone, H. Lipson, *International Solid Freeform Fabrication Symposium* **2007**, Austin, TX, USA.

61 F. Godoi, S. Prakash, B. Bhandari, *J. Food Eng.* **2016**, *179*, 44.

62 J. Sun, W. Zhou, L. Yan, D. Huang, L. Lin, *J. Food Eng.* **2018**, *220*, 1.

63 P. Krishnaraj, T. Anukiruthika, P. Choudhary, J. Moses, C. Anandharamakrishnan, *Food Bioprocess Tech.* **2019**, *12*, 1776.

64 A. Derossi, R. Caporizzi, D. Azzollini, C. Severini, *J. Food Eng.* **2018**, *220*, 65.

65 J. Stokes, M. Boehm, S. Baier, *Curr. Opin. Colloid Interface Sci.* **2013**, *18*, 349.

66 P. Sethupathy, J. Moses, C. Anandharamakrishnan, *Food Ris. Int.* **2021**, *37*, 538.

67 S. Huang, M. Zhang, B. Bhandari, *Food Bioprocess Tech.* **2019**, *12*, 1185.

68 A. Le-Bail, B. Maniglia, P. Le-Bail, *Curr. Opin. Colloid Interface Sci.* **2020**, *35*, 54.

69 R. Caporizzi, A. Derossi, C. Severini, in *Fundamentals of 3D Food Printing and Applications*, (Eds: F. C. Godoi, B. Bhandari, S. Prakash, M. Zhang), Academic Press, **2019**, Ch.4.

70 Z. Liu, M. Zhang, in *Fundamentals of 3D Food Printing and Applications*, (Eds: F. C. Godoi, B. Bhandari, S. Prakash, M. Zhang), Academic Press, **2019**, Ch.2.

71 T. Pereira, S. Barroso, M. Gil, *Foods* **2021**, *10*, 320.

72 C. Fernanda, P. Sangeeta, B. Bhandari, *J. Food Eng.* **2016**, *179*, 44.

73 S. Beckett, *Industrial Chocolate Manufacture and Use*, 4th ed., Wiley-Blackwell, **2009**.

74 P. Rando, M. Ramaioli, *J. Food Eng.* **2021**, *294*, 110415.

3

Basic Concepts of Bulk Rheology in Food Emulsions

Carlos Bengoechea, Estefanía Álvarez-Castillo, José Manuel Aguilar, and Antonio Guerrero*

Departamento de Ingeniería Química, Universidad de Sevilla, Escuela Politécnica Superior. C/ Virgen de África, 7, 41011 Sevilla Spain
** Corresponding author*

Highlights
• A great number of emulsion systems are found in the food industry, such as mayonnaises, mustards, sauces, butters, or creams. • Rheology provides information about emulsion systems, related to their microstructure, stability, or texture. • Some destabilization mechanisms can be predicted from viscoelasticity measurements.

3.1 Introduction

Emulsions are colloidal systems consisting of two immiscible liquids, usually an oily phase (O) and an aqueous phase (W). One of the phases is dispersed as small spherical droplets in the other. Therefore, the droplets make up the so-called dispersed, discontinuous, or internal phase, whereas the surrounding liquid phase is referred to as the continuous or external phase. Conventionally, simple emulsions are denoted as O/W or W/O emulsions when the oily phase is dispersed in the aqueous phase or vice versa. Mayonnaises, dressings, or dips are common O/W emulsions manufactured industrially; butter or margarine are W/O emulsions. There are also multiple emulsions, also known as double emulsions, which can be briefly defined as an emulsion of an emulsion [1]. They are typically denoted as O/W/O or W/O/W, and include small droplets of the first mentioned internal phase dispersed in the second one as a primary emulsion. Then this primary emulsion is subsequently dispersed in a third liquid that constitutes the main continuous phase. These emulsions are commonly used in the manufacture of food products with a reduced fat content or for encapsulating specific nutrients that will be eventually delivered during food digestion [1].

Most food emulsions exhibit droplet diameters within a 0.1–100 μm range, although there is an increasing interest in developing emulsion systems in the nanometric scale (nanoemulsions) with applications for the delivery of specific ingredients (e.g. nutraceuticals, antimicrobials) in the food industry, or even for developing biodegradable coating and packaging films to promote the quality, functional properties, nutritional value, and shelf

Materials Science and Engineering in Food Product Development, First Edition. Edited by Wing-Fu Lai.

life of food products [2]. Apart from immiscible liquids, a stable food emulsion requires the presence of an emulsifier in its formulation. An emulsifier is a surface-active agent that adsorbs onto the interface between the two immiscible liquid phases, allowing the dispersion of one of them into the other. Moreover, emulsifiers modify the texture of the food product, altering its stickiness or controlling crystallization. The most commonly used food emulsifiers include mono- and diglycerides (representing around 75% of the total emulsifier production), stearoyl lactylates, sorbitan esters, polyglycerol esters, sucrose esters, and lecithin [3]. There are some emulsions that use solid particles as emulsifiers (Pickering emulsions), being more resistant to changes in the external environment (e.g. pH, temperature). However, these emulsions lack great stability and are expensive to produce when prepared from food-grade materials, as it is difficult to produce at a large scale [4].

In food science, rheology is an important subject discipline. It is the science of flow and deformation of materials and involves the study of the relationship between the stress acting on a given material and the resulting deformation, or the other way around. Rheological properties, such as viscoelastic moduli or viscosity, are determined by measuring force and deformation as a function of time. Emulsions are thermodynamically unstable systems that can achieve kinetic stability through the presence of suitable emulsifiers that are adsorbed on the interface. Thus, the interface between the immiscible phases play a critical role that justifies the importance of interfacial rheology for dispersed systems like emulsions or foams. However, this chapter focuses on bulk rheology dealing with macroscopic rheological properties that are commonly required for optimizing both the processing and the formulation of emulsions in the industry. Rheological properties can be related to certain structural features, organoleptic properties, or stability of many foods. The purpose of this chapter is to give an overview of food emulsions basics and the importance of bulk rheology for the unexperienced. Main emulsification processes (i.e. based on mixing, pressure, or ultrasonication devices) are described. Additionally, a focus on the importance of the rheology of the continuous phase is made, followed by a report of the most common rheological behavior of final emulsions. Finally, main destabilization mechanisms are briefed, interrelating the rheology and the microstructure of the emulsion system. Extra information is listed in Industrial Corner 1–3 (i.e. master curves, other rheological tests, ways to avoid destabilization), as well as a specific glossary.

3.2 Emulsification Process

Emulsification can be achieved through different continuous techniques with a large-throughput for a wide range of materials. All these techniques are based on the deformation and possible breakup of droplets, while surfactant molecules are transported to and adsorbed onto the interface of the two immiscible fluids to produce a physically stable emulsion. The main emulsification devices are the following:

a) Mixing devices
   Mixing devices can be of two types. One is for high-shear homogenization (HSH), in which emulsions are usually produced by first using a low shear device for obtaining a coarser emulsion that, secondly, is processed through devices that impart a greater mechanical energy, such as a colloid mill (for high-viscosity emulsions) or a

high-pressure homogenizer (for low-viscosity emulsions). High-shear homogenizers are typically based on a rotor-stator principle operating at atmospheric pressure. Turbulent rotor-stator mixers may operate in a batch or in continuous mode, and they all are composed by a certain number of rotors operating at high speed (1000–25 000 rpm) in a number of static stator screens. Rotors and stators are commonly separated by a short distance. The fluid is accelerated by the rotor tangentially and is redirected radially through the stator holes or slots. Thus, a narrow region of high-intensity hydrodynamics stresses is created in the stator slots that enhances the mixing and dispersing efficiency [5]. Apart from HSH, colloid mills can be used. The widespread version of a colloid mill is based on a conical rotor rotating inside a static stator. In this case, droplet disruption is achieved when the emulsion is passed once or several times through the narrow rotor-stator gap that produced intense shearing of the sample.

b) Pressure vessels
Pressure vessels can be of two types. One is for high-pressure valve homogenization. This type of emulsification device is basically composed by a high-pressure pump and a homogenizing nozzle. Droplets are disrupted when the pressurized crude emulsion is depressurized in the nozzle. It is habitually employed for low-viscosity finely dispersed emulsions with a disperse phase volume fraction that is not too high. In standard high-pressure homogenizers, pressures up to 500 bar can be achieved, resulting commonly in a reduction of the droplet size, sometimes even helping to disperse formed aggregates, which finally can enhance the physical stability by inhibiting the creaming process [6]. An alternative to high-pressure valve homogenization is microfluidization. Higher pressures (2,500 bar) can be achieved in a microfluidizer (Microfluidics Corp., Newton, USA), thanks to a nozzle where two jets of a primary emulsion from two opposite channels collide with one another. The throughput largely will depend on the cross section of the nozzle employed.

c) Ultrasonication
Ultrasound, mechanical waves at frequencies higher than 20 kHz, can be of high intensity (10–1000 $W \cdot cm^{-2}$) or low intensity (<1 $W \cdot cm^{-2}$). High-intensity ultrasound (HIU) can be used to prepare stable emulsions as droplets are disrupted by capillary surface waves, acoustic cavitation, and liquid microjets [7]. They may operate continuously or discontinuously. Higher frequencies are normally used for de-emulsification processes, as cavitation is limited, while low frequencies, like 20 kHz, are typically used for producing stable emulsions [7].

An increase in mechanical energy applied to the emulsion system during its processing typically produces higher values for the viscoelastic moduli of the resulting products, showing a tendency toward smaller sizes in a narrower droplet size distribution (DSD) [8, 9]. The effect of energy input can be described in terms of an increase in specific interface area, which can also lead to enhancement of network structures in closely packed O/W emulsions. For shear-mixing devices, mechanical energy can be determined from the power input into the dispersing volume and the volume flow rate, while in high-pressure homogenizers, it corresponds to the pressure difference operating at the nozzle [10]. The increase of sonication power can also help to reduce droplet diameters, but it may increase flocculation processes [11]. All emulsification processes can be carried out continuously, which is the preferred type of throughput in the industry. The selection of one emulsification device or another can be done on the basis of the viscosity and/or droplet diameter of the pursued food emulsion: the colloid mill is preferable for systems with medium-to-high viscosities and minimum diameters of 1 μm. For emulsions with a lower viscosity, the greater the mechanical energy, the lower the droplet diameter. Thus, HSH can achieve diameters as small as 2 μm, while smaller diameters can be reached by ultrasound (0.1–1 μm), high-pressure (0.1 μm), or micro-fluidization (< 1 μm) [12].

3.3 Rheology of Continuous Phase

Most destabilization processes of emulsion systems, specially coalescence of droplets, are slowed down on increasing the viscosity of the continuous phase to a certain extent. Thus, the rheology of the continuous phase is an essential aspect to consider when designing a kinetically stable emulsion. However, a "too thick" or "too creamy" emulsion cannot be selected just based on the viscosity of its continuous phase. Instead, formulation of food emulsions can be thought of as a balance between the nutritional, organoleptic properties required for a specific food product, either legally or by the consumer's preferences, and the rheology needed to guarantee physical stability during its shelf life. Moreover, viscosity of the continuous phase greatly affects the type of emulsification process that is recommended for an emulsion. Therefore, an emulsification process solely based on an ultrasound (HIU) technique is not typically used on mayonnaise-like emulsions, although low-frequency ultrasound can be used in the emulsification of emulsions with a lower oil content. When comparing the stability of emulsions produced by ultrasonic processing with that generated by a conventional rotor-stator device at similar specific mechanical energies, it has been found that the high-shear technique requires longer processing times to produce stable emulsions, which suggests a greater efficiency of the ultrasound emulsification process [13].

An increase in viscosity has proven to retard the coalescence of droplets, although surfactant adsorption is also slowed down. The increase in the viscosity and viscoelastic properties of the continuous phase may be the result of the well-known thickening effect of biopolymers added to the formulation. This effect may be related to different factors, like the increase in the solids total content or the existence of oligo and polysaccharide interactions [14–16]. The addition of different polysaccharides, such as xanthan gum, as structuring/thickening/gelling agents of aqueous solutions is of common use in the food industry to retard or even inhibit creaming through modification of the continuous phase rheology [17–22]. Likewise, the effect of the pH of the continuous phase should not be neglected as it also clearly affects the rheology and stability due to the change of interactions within the emulsion system [23–25].

When the viscosity of the continuous phase is increased, longer times are required for the coalescence of droplets. This is due to the lower motion of the liquid occupying the space between two approaching droplets, which retards the destabilization mechanism. This is the main reason why the coalescence frequency decreases as thickening agents are included in the formulation, as long as they do not cause any destabilizing effect, and especially when poorly stabilizing surfactants are used [26]. Some authors have already reported that increasing the viscosity of the continuous phase is a way of increasing the shelf life for protein-stabilized oil-in-water emulsions [27].

3.4 Rheology of Emulsions

Rheology provides highly valuable information about the microstructure and physical stability of emulsions. There are several factors that determine the rheology of emulsion systems, such as the continuous phase rheology, the features of the particles included (e.g. concentration, size distribution, deformability, internal viscosity), and the nature of particle–particle interactions [28]. Therefore, rheological techniques should be an essential part in any research and development department of any important food industry producing emulsion systems. Among the great variety of tests that can be performed for the bulk system, dynamic and rotational shear tests take a predominant place. Most typical dynamic tests involve a small amplitude oscillatory shear (SAOS) performed within the linear viscoelastic region of the studied system. From a practical point of view, SAOS tests are ideally performed on a macroscopically unaltered system that keeps its main structural features during the characterization tests. These tests give a different information from large amplitude oscillatory shear tests, which apply large stresses and deformations to the studied systems, altering their structure mostly in an unrecoverable manner [29, 30]. SAOS tests provide information about the viscoelasticity of the system, through the elastic or storage modulus (G') and the viscous or loss modulus (G''). The former is in phase with the oscillatory signal applied to the sample, while the latter is 90° out of phase with the same signal. The G''/G' ratio is the loss tangent ($\tan \delta$), which is lower than one for solid-like systems (e.g. gel emulsions) and higher than unity for fluid-like systems (e.g. low oil content O/W emulsions). Frequency sweep tests performed isothermally at a constant stress or strain within the linear viscoelastic range are typically plotted in a double log scale as G' and G'' (both in Pa) versus frequency, ω (rad/s). This plot is the so-called mechanical spectra of the emulsion. A typical mechanical spectrum of structured fluids appears in Figure 3.1a. Different regions can be distinguished, although commonly only one or two of those regions are presented under a given experimental range of frequencies:

i) *Viscous or terminal region ($G'' > G'$).*
ii) *Lower transition region ($G' - G''$ crossover).*
iii) *Rubbery or plateau region ($G' > G''$).*
iv) *Leathery or higher transition crossover region (G'-G'' crossover).*
v) *Glassy region ($G'' > G'$).*

The first region requires very low frequencies and is often undetected for structured emulsions, although it is quite common in low-viscosity systems. In the second region, the inverse of the crossover frequency provides a relaxation time τ, which can be valuable for modeling. In the plateau region, G'' is much lower than G' and displays a minimum, while G' keeps a rather stable value with frequency. A plateau modulus can be defined as the G' value at the frequency where G'' displays the minimum value. At higher frequencies, another crossover is found at a frequency that also provides a characteristic time. The transition from a predominantly elastic behavior to a prominently viscous one is explained

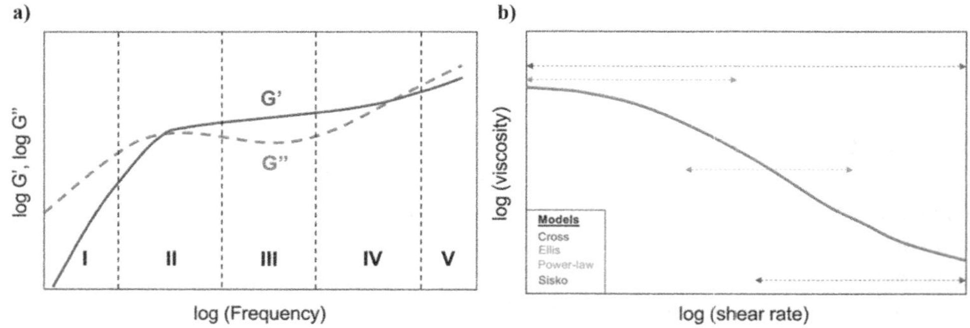

Figure 3.1 (a) General viscoelastic behavior of non-Newtonian fluids; (b) general viscosity curve of shear-thinning emulsions.

on the basis of high-frequency relaxation and dissipation mechanisms. Nevertheless, regions IV and V are uncommon in emulsions.

There are different models that correlate viscoelastic and microstructural parameters for concentrated emulsion [31–33]. These models establish that the elastic shear modulus is inversely proportional to the drop diameter, as long as the viscous forces can be ignored (i.e. deformation is sufficiently low) [34]. Also, many studies have pointed out that higher viscoelastic moduli values typically result in greater physical stability [27, 35]. SAOS tests are a sensitive experimental indicator of flocculation in moderately concentrated systems, giving a characteristic response of an entanglement gel network (region III) with strong bonds that tend to restore the structure after large strains have been applied [8, 36, 37]. A higher temperature or specific mechanical energy supplied during processing, a higher disperse phase volume fraction or a higher emulsifier concentration generally results in emulsions with lower droplet sizes and higher viscoelastic moduli [8].

Industrial Corner 1. How to obtain master curves for mechanical spectra or viscosity curves

- In some cases, a qualitatively similar rheological response is obtained for different systems, but displaced onto lower or greater shear rates or frequencies of the flow curve or mechanical spectrum, respectively. In those cases, all similar experimental data can be overlapped into a single master curve by multiplying the shear rate or frequency range by a shift factor (a_c), displacing the parameters to lower ($a_c < 1$) or higher ($a_c > 1$) shear rates or frequencies, respectively. This procedure is called factorization and results in a displacement on the *x*-axis.
- In some other cases, a qualitatively similar rheological response is obtained for different systems, but displaced onto lower or greater values of the corresponding parameter (viscosity or viscoelastic moduli), at the same shear rates or frequencies, respectively. In those cases, every set of the experimental data can be divided by a single factor that makes all of them match in a master curve. Typical factors used are the low shear rate viscosity (η_0) or the elastic modulus measured at 1 Hz (G'_1) of every

Industrial Corner 1. (Continued)

flow curve or mechanical spectrum, respectively. This procedure is called normalization and results in a displacement on the *y*-axis.

- Commonly, bot factorization and normalization should be carried out to obtain a single master curve. The knowledge of the parameters used to normalize and/or factorize the curves can give us relevant information to predict the behavior of other systems with similar compositions or processed at different conditions (e.g. time, temperature).

Shear viscosity of emulsions can be measured through rotational steady shear rate–shear stress tests. A typical response for a shear thinning or pseudoplastic emulsion appears in Figure 3.1b. At low shear rates, viscosity is constant (η_0) up to a point where a critical shear rate for the onset of the power law region is surpassed and it begins to decrease with a constant slope (power law region), displaying what is known as shear-thinning behavior. Eventually, if a high enough shear rate is reached, a second constant viscosity region is found (η_∞). When the volume fraction of the disperse phase increases, viscosity increases and the onset of the power law region appears at lower shear rates. There is a point when a further increase of the volume fraction takes place, a yield stress can be defined and the Newtonian behavior totally disappears [38]. Results should be analyzed cautiously as experimental artifacts may lead to the detection of an erroneous η_0 in those cases where the steady state has not been achieved [39]. Wall slip should also be avoided or minimized as they interfere with the accuracy of the measurement [40].

The frame of detection of this viscosity curve depends on the shear rates applied and the nature of the emulsion system studied. Several models have been used extensively for predicting the flow behavior of shear-thinning emulsions (e.g. Cross, Power law, Sisko) [41,–43]. Viscosity of emulsions can be directly related to the volume fraction of the dispersed phase and, at relatively low oil volume fraction (i.e. lower than 0.6), viscosity curves of emulsions with different drop diameters can be horizontally shifted onto a single master curve. On the other hand, relatively high oil volume fractions (higher than 0.7), typically require a vertical shift for superimposing viscosity curves with different droplet sizes. Critical volume fraction corresponds to close packing of undeformed drops [34], which acquire a polygonal shape for higher-volume fractions [44].

Industrial Corner 2. Other Rheological Tests

- Time-dependent shear rate–shear stress. These tests are useful when the emulsion system exhibits thixotropic behavior, that is, when there is a continuous decrease in viscosity with time when a constant shear rate is applied to a sample, and the subsequent viscosity recovery in time when the flow is finished [45]. Thixotropy can be measured by performing a shear rate ramp test from low to high values and then backward on the same sample. Thixotropic behavior is displayed as a hysteresis loop and can be found in emulsions where aggregates are broken as a constant shear rate is applied, although they may be formed when stopped.

(Continued)

Industrial Corner 2. (Continued)

- Transient tests. Transient flow behavior is studied through stress growth and decay at a constant shear rate. Initially, the sample is held at rest and suddenly a shear rate is applied ("start-up flow"). They are typically plotted in a double logarithmic scale stress versus time graph, where a sudden stress overshoot at shorter times is followed by a decay onto a steady-state value at longer times. This response is due to the viscoelastic behavior of the sample and its analysis can provide characteristic times and stresses at certain shear rates for every formulation.
- Creep tests. Viscoelastic behavior can also be studied through tests that involve stress application (creep) and recovery after stress removal. In these tests, the deformation of a sample is measured during the application of a constant stress amplitude, which is kept for a certain period of time. Subsequently, the stress is removed, while still measuring the deformation for some time. The deformation to stress ratio is called compliance (J), which may be correlated with some sensory attributes. Creep test results commonly are represented in a graph of J as a function of time.

3.5 Microstructure

An emulsion can be considered as a dynamic system, where even at equilibrium after its formation, a continuous migration of molecules between the different regions takes place. This displacement depends on the mass transport of the molecules through the various phases of the system and results in a modification of their properties over time. Thus, emulsion design should be based on the need to develop a relatively stable structure that can withstand thermal or mechanical stresses and biological activity during the shelf life of the product. However, different factors, such as the actual availability of the raw materials or economic issues, should be taken into account when creating a desired formulation [46].

In terms of microstructure, a simple O/W emulsion consists of three distinct regions: an oily liquid substance that makes up the emulsion droplets, the aqueous medium where droplets are dispersed, and the interfacial material between lipid and aqueous phases. The lipid may be partially or totally crystallized, depending on temperature. The aqueous phase can contain ions and biomolecules, such as polysaccharides, which may either stabilize or destabilize the emulsion. The interfacial region can be composed of surface-active molecules, such as proteins or emulsifier molecules of low molecular weight (e.g. fatty acids, monoglycerides, esters, or phospholipids), or mixtures of a number of these components. Thus, any molecule that is part of an emulsion will tend to distribute between these three regions according to their concentration and polarity. Polar molecules will be typically found in the aqueous phase as nonpolar substances will situate in the oil phase, and amphiphilic compounds will place at the interface. Moreover, more complex emulsion-based systems can be used in foods, such as multilayer emulsions, colloidosomes, multiple emulsions, filled hydrogel particles, filled solid particles, or micro clusters [47].

Understanding the structural properties of emulsions requires a precise knowledge of the behaviors of these three phases, both individually and cooperatively. Thus, different techniques are available to provide information about the structure, dimensions, and organization of the different components. Structural images can be obtained through different microscopy techniques (optical, laser scanning confocal microscopy, electron microscopy, atomic force microscopy) [20, 35, 48–50]. DSD can be analyzed by dynamic light scattering, diffusing wave spectroscopy, electrical pulse counting, sedimentation, ultrasonic spectrometry, nuclear magnetic resonance (NMR), or neutron scattering [9, 51–54]. Disperse phase volume fraction can also be estimated by proximate analysis, density measurements, or electrical conductivity. Droplet crystallinity can be estimated by dilatometry, or NMR, and droplet charge by particle electrophoresis or electro-acoustic signals [54].

3.6 Destabilization Mechanisms

Considering that an emulsion is a dispersed system containing two immiscible phases that is thermodynamically incompatible, the study of destabilization mechanisms is of great importance when defining the shelf life of food emulsion products. The stability of an emulsion relies on different parameters such as DSD, concentration, and composition of both phases (oleic or aqueous) [55]. This section briefly highlights the main destabilization mechanisms affecting emulsions and also numbers some techniques to identify them, which are depicted in Figure 3.2.

a) **Creaming**
   Destabilization might be promoted by the creaming phenomenon, which is caused by the buoyant effect of the droplets that tend to rise to the top of the system [56]. An acceleration of the separation of the two immiscible phases can take place based on differences in their densities [57]. Some studies have highlighted the importance of controlling the rheology of the continuous phase to minimize this undesired phenomenon [58, 59]. In order to monitor creaming mechanisms, visual observations, turbidity, and ultrasound could be used [59]. A vertical scan analyzer is usually employed to identify this type of destabilization mechanism. This device optically scans the system that is vertically placed inside a glass tube. Transmittance is measured along the whole length of the tube and differences in the backscattering profile during

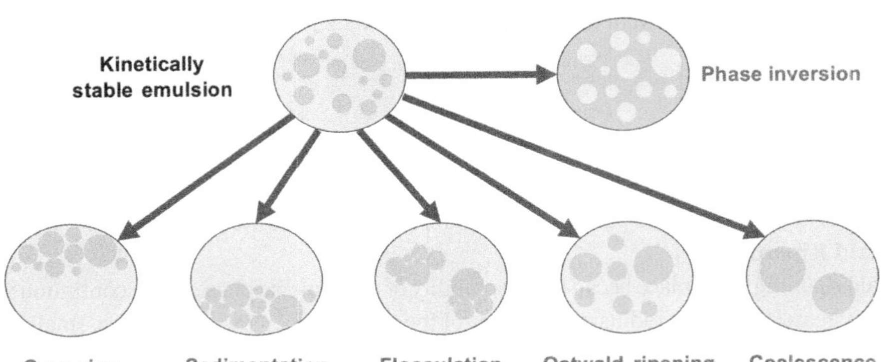

Figure 3.2 Overview of destabilization mechanisms for food emulsions.

storage provide an idea of the nature of the destabilization mechanisms that are taking place in the emulsion [49]. Eventually, a global destabilization percentage, D (%), can be estimated using Equation 3.1.

$$D(\%) = \frac{BS_{avmax} - BS_{avt}}{BS_{avmax}} \tag{3.1}$$

where $BS_{av\,max}$ is the maximum backscattering value corresponding to the middle zone of creamed phase, and $BS_{av\,t}$ is the mean value at a certain time of the study.

b) Sedimentation

The sedimentation rate of the dispersed phase under static conditions represents a key parameter in determining the overall stability characteristics of an emulsion. A high sedimentation rate, would be detrimental for the stability of an emulsion, affecting the shelf life of the product [60, 61]. During the sedimentation procedure, the action of the gravity produces a concentration gradient as droplets tend to progressively accumulate at the bottom of the emulsion container [62]. Sedimentation, in the same way as creaming, not only affects uniformity but also enhances thermodynamic instabilities [63]. However, a high rate of sedimentation can be useful in some industrial processes where the separation of phases is pursued [60]. As creaming, this mechanism could also be recognized through the vertical scan of a tube containing the emulsion as the droplets placed at the bottom would produce changes in the backscattering or transmittance profile during storage time.

c) Flocculation and Aggregation

Droplets from the disperse phase of an emulsion can interact with each other due to the existence of inter-drop forces, forming aggregates or floccules that promote extensive droplet flocculation [9, 64]. This mechanism results in the reduction of the specific surface area of the emulsions, finally leading to their destabilization, as it leads to phase separation [65]. This is due to the higher diameter of the floccules when compared with those measured for the unaggregated droplets. Moreover, an important flocculation degree may alter the functionality of a food emulsion, such as a modification in the digestion rate of fats [65]. To determine if flocculation or aggregation is taking place in a dispersed system, the droplet size can be measured in the presence and in the absence of a low molecular weight surfactant (i.e. sodium dodecyl sulfate [SDS]). Then, a flocculation index can be calculated employing Equation 3.2 [19, 20, 22].

$$FI = \frac{|D_{3,2} - D_{3,2SDS}|}{D_{3,2}} \tag{3.2}$$

where $D_{3.2}$ is the Sauter mean droplet diameter from DSD estimated in the absence of SDS and $D_{3.2\,SDS}$ is the same parameter measured after the addition of SDS.

d) Ostwald Ripening

Ostwald ripening takes place because the solubility of the disperse phase in the continuous phase increases when the radius of curvature of the droplets decreases. Therefore, smaller droplets can dissolve into the continuous phase and then diffuse to larger ones, increasing their overall size as they redeposit upon them. It is also known as isothermal distillation or molecular diffusion [66]. This destabilization mechanism can be also studied through the evolution of DSD during aging, as the cube of the mean droplet radius of the emulsions increases linearly with time when Ostwald ripening takes place, with a slope that depends on the physical characteristics of the oil and aqueous phases [67].

e) **Coalescence**
The mechanism of coalescence is caused by the rupture of the films covering the droplets in an emulsion [68]. This involves the destabilization of the system as it leads to the formation of droplets of greater size from smaller ones. It is uneasy to characterize the coalescence phenomenon due to its rapidness and because it is linked to several other mechanisms, such as gravity effects, leading to the vertical motion of drops/bubbles and ripening [68]. Several factors affect the coalescence of a dispersed system like concentration, density, viscosity, ion strength, pH, or the duration and frequency of the collisions between droplets [69, 70]. To identify the droplet coalescence within an emulsion, it is necessary to measure the droplet size along with time [71, 72]. It is important to distinguish if the size increase might be due to coalescence or flocculation. For that purpose, a DSD analysis can be performed with and without an anti-flocculant reagent (e.g. SDS), where coalescence would not be affected in any case. The droplet coalescence could be reduced while enhancing the steric barrier [73], which is commonly achieved by the use of a surfactant reagent [61]. Some authors have identified a shift from bimodal to multimodal DSD with aging time as a consequence of coalescence, while a shift from bimodal to monomodal has been reported when Ostwald ripening takes place [74].

f) **Phase Inversion**
Phase inversion is the phenomenon that takes place when a O/W emulsion reverts to a W/O emulsion, or vice versa [75]. Alternatively, it can also separate into its component phases, contributing to the instability of the emulsion [70]. At the point when inversion occurs (balance point), W/O regions coexist with O/W in the system. This mechanism might be promoted by several factors (e.g. a change in temperature, the election, and the nature of the ingredients). Shinoda et al. [76] introduced the phase inversion temperature concept at which the hydrophilic/lipophilic character of a nonionic shift promotes the inversion [77]. Furthermore, one type of destabilization mechanism can accelerate another, since an extensive droplet coalescence, flocculation, and disruption can also cause a phase inversion [70]. Moreover, some studies have highlighted that phase inversion can also be considered a low-energy method to prepare emulsions, as it imparts the mechanical strength required to produce good texture and stability in a newly formed emulsion [70, 78].

Industrial Corner 3. Ways to Avoid Destabilization

- Destabilization is one of the most important issues in the industrial production of emulsions. It is mainly caused by the absence of an interfacial barrier preventing the coalescence of droplets, which is among the principal causes of destabilization, the differences between the densities of the continuous and dispersed phases, the low continuous phase viscosity, and a heterogeneous DSD.
- In the main, the addition of some emulsifying agents (i.e. lecithin, mono and diglycerides of fatty acids, or some proteins) might be highly recommended to achieve desirable kinetic stability.

(Continued)

Industrial Corner 3. (Continued)

- Some destabilization mechanisms, such as sedimentation, might be slowed down by increasing the viscosity of the continuous phase. The addition of some thickeners, such as carboxymethyl cellulose, xanthan gum, guar gum, or other polysaccharides, might contribute to achieve this goal.
- Storing at low temperature might also help to increase the viscosity of the continuous phase.
- DSD is intimately related to the kinetic stability of the emulsion, as unimodal droplet size distributions show better stability than those with a bimodal one. Thus, an appropriate selection of the homogenizer type and of the operation conditions should be set to obtain a desirable size range in which the emulsion is stable.

3.7 Concluding Remarks

This chapter summarizes some basic concepts related to food emulsions, together with the importance of rheology as a tool to characterize those systems. In the industry, rheological devices can be used to predict the stability of a system mostly finding a correlation to other techniques that assess destabilization mechanisms during the storage of the food product. Controlling of the viscosity of the continuous phase through the presence of food-grade surfactants and thickeners or through the management of the storage temperature is of paramount importance to retard those destabilization mechanisms to permit an acceptable shelf life of the food product. Novel food emulsions, such as Pickering emulsions, have gained attention due to the many advantages they provide (low emulsifiers content, high biocompatibility, and safety). The need for food-grade solid particles and optimization of their industrial production is increasingly studied by scientific researchers. Moreover, there is a growing focus on naturalness, health, and sustainability of food emulsions, which promotes the introduction of more sustainable ingredients, like those from plant sources. Rheology will for sure be an essential tool to guarantee the desired organoleptic properties and stability of these new food emulsion systems in the market.

Acknowledgments

This work was funded by the Junta de Andalucía (Consejería de Economía, Conocimiento, Empresas y Universidad) through the P20_01046 project (BionanoWOW).

Glossary

Coalescence Merging of droplets into a single larger droplet, which eventually can lead to phase separation.

Creaming Appearance of an oil layer on top of an O/W emulsion due to the lower density of the oil phase.

Emulsion Disperse system containing two different immiscible phases where a disperse phase is found as droplets in a continuous phase. Kinetic stability is achieved by the presence of a surface-active agent that is adsorbed on the interface, reducing surface tension.

Emulsification Process by means of which one liquid is dispersed as small droplets in another immiscible liquid. The high energy required for the process is typically supplied by mechanical devices, such as high-speed blenders, high-pressure valve homogenizers, or colloid mills.

Flocculation Aggregation of small droplets to form floccules in which the droplets retain their individual integrity. Eventually, it can lead to the coalescence of the droplets.

Flow index Measure of the degree of shear thinning (<1) or shear thickening (>1) character of an emulsion. Fitting parameter of different rheological models employed for describing the flow behavior of emulsion systems.

High and low shear rate Newtonian viscosities Expressed as η_0 and η_∞, respectively, they are the viscosity values displayed at the beginning (η_0) and at the end (η_∞) of a whole viscosity curve. Both of them are independent of the shear rate applied.

Linear viscoelastic range (LVR) Region where the strain or stress values applied to a system are low enough for not irreversibly altering its structure. It may be determined through amplitude (strain or stress) sweep tests, being characterized by a constant value of the viscoelastic moduli until critical values of strain and stress are surpassed.

Phase inversion Conversion of the dispersed phase into the continuous one. For instance, an O/W emulsion suffering phase inversion would shift to W/O, or vice versa.

Rheology Science that studies the deformation and flow of materials when submitted to external forces. This term comes from Greek *rheos*, which means to flow.

Sedimentation Downward movement of droplets due to gravitational forces, in case they are denser than the continuous phase.

Thickener Substance that increases the viscosity of a solution or dispersion. It is usually added in the aqueous phase of commercial emulsions to improve their stability by interfering with the movement of droplets and/or to obtain a desired texture in the mouth.

Viscoelastic moduli Parameters indicating the viscoelastic behavior of a sample obtained by dynamic tests. The storage or elastic modulus (G′) gives purely elastic information and is related to the energy stored in a deformed material, while the loss or viscous modulus (G″) measures the viscous component of the viscoelastic behavior that arises from the internal friction in the sample. The ratio of the loss modulus (G″) to the storage modulus (G′) is named loss tangent (tan δ, where δ is the phase angle between stress and strain).

Viscosity Property of materials measuring the resistance to the flow of fluid materials.

Yield stress Stress threshold above which systems starts to flow sharply. At lower stresses, the system shows a solid-like behavior, with a relatively high-shear viscosity value. Yield stress reflects the evolution of the physical structure of emulsions in time. Its existence remains controversial since its value may depend on the chosen procedure of estimation and experimental data fitting.

References

1 E. Dickinson, *Food Biophys.* **2011**, *6*, 1.
2 J. B. Aswathanarayan, R. R. Vittal, *Front. Sustain. Food Syst.* **2019**, *3*, DOI 10.3389/fsufs.2019.00095.
3 R. Miller, in (Eds: B. Caballero, P.M. Finglas, F. Toldrá), Academic Press, Oxford, **2016**, pp. 498–502.
4 L. Chen, F. Ao, X. Ge, W. Shen, *Molecules* **2020**, *25*, 3202.
5 A. Håkansson, *Process* **2018**, *6*, DOI 10.3390/pr6040032.
6 J. Floury, A. Desrumaux, J. Lardières, *Innov. Food Sci. Emerg. Technol.* **2000**, *1*, 127.
7 A. Taha, E. Ahmed, A. Ismaiel, M. Ashokkumar, X. Xu, S. Pan, H. Hu, *Trends Food Sci. Technol.* **2020**, *105*, 363.
8 C. Bengoechea, F. Cordobés, A. Guerrero, *Rheol. Acta.* **2006**, *46*, 13.
9 J. Li, L. Xu, Y. Su, C. Chang, Y. Yang, L. Gu, *Food Res. Int.* **2020**, *132*, 108990.
10 M. Stang, H. Schuchmann, H. Schubert, *Eng. Life Sci.* **2001**, *1*, 151.
11 K. C. G. Silva, A. C. K. Sato, *Ultrason. Sonochem.* **2019**, *52*, 286.
12 D. J. McClements, *Food Emulsions: Principles, Practices, and Techniques*, CRC Press, Boca Raton, USA, **2015**.
13 A. Shanmugam, M. Ashokkumar, *Food Hydrocoll.* **2014**, *39*, 151.
14 R. K. Robinson, *Br. Food J.* **1995**, *97*, 30.
15 A. Syrbe, W. J. Bauer, H. Klostermeyer, *Int. Dairy J.* **1998**, *8*, 179.
16 E. Fernandez-Garcia, J. U. McGregor, S. Traylor, *J. Dairy Sci.* **1998**, *81*, 655.
17 L. De Vuyst, B. Degeest, *FEMS Microbiol. Rev.* **1999**, *23*, 153.
18 E. Dickinson, *Food Hydrocoll.* **2009**, *23*, 1473.
19 M. Felix, A. Romero, A. Guerrero, *Food Hydrocoll.* **2017**, *72*, 372.
20 M. Felix, A. Romero, A. Guerrero, *Food Hydrocoll.* **2017**, *64*, 9.
21 J. Piermaría, C. Bengoechea, A. G. A. G. Abraham, A. Guerrero, *Carbohydr. Polymer.* **2016**, *152*, 97.
22 M. Felix, I. Martínez, A. Sayago, M. A. Fernández Recamales, *Innov. Food Sci. Emerge. Technol.* **2021**, *74*, 102810.
23 C. Bengoechea, A. Romero, J. M. Aguilar, F. Cordobés, A. Guerrero, *Food Hydrocoll.* **2010**, *24*, 783.
24 M. Felix, N. Isurralde, A. Romero, A. Guerrero, *Food Sci. Technol. Int.* **2018**, *24*, 555.
25 A. Romero, F. Cordobés, A. Guerrero, *Food Hydrocoll.* **2009**, *23*, 244.
26 S. Tesch, H. Schubert, *J. Food Eng.* **2002**, *52*, 305.
27 M. L. López-Castejón, C. Bengoechea, S. Espinosa, C. Carrera, *Food Hydrocoll.* **2019**, *87*, 382.
28 H. A. Barnes, *Colloids Surf. A Physicochem. Eng. Asp.* **1994**, *91*, 89.
29 Q. Li, M. Xu, J. Xie, E. Su, Z. Wan, L. M. C. Sagis, X. Yang, *Food Res. Int.* **2021**, *140*, 110076.
30 V. Tanna, C. Wetzel, H. Henning Winter, *Rheol. Acta* **2017**, *56*, 527.
31 T. G. Mason, M.-D.-D. Lacasse, G. S. Grest, D. Levine, J. Bibette, D. A. Weitz, *Phys. Rev. E* **1997**, *56*, 3150.
32 H. M. Princen, A. D. Kiss, *J. Colloid Interface Sci.* **1986**, *112*, 427.
33 H. M. Princen, *J. Colloid Interface Sci.* **1983**, *91*, 160.
34 Y. Otsubo, R. K. Prud'homme, *Rheol. Acta* **1994**, *33*, 29.

35 A. Romero, F. Cordobés, M. C. Puppo, A. Guerrero, C. Bengoechea, *Food Hydrocoll.* **2008**, *22*, DOI 10.1016/j.foodhyd.2007.05.019.

36 M. C. Sánchez, M. Berjano, A. Guerrero, C. Gallegos, *Langmuir* **2001**, *17*, 5410.

37 E. Dickinson, *Curr. Opin. Colloid Interface Sci.* **1998**, *3*, 633.

38 A. J. Poslinski, M. E. Ryan, R. K. Gupta, S. G. Seshadri, F. J. Frechette, *J. Rheol. (N. Y. N. Y).* **1988**, *32*, 703.

39 C. E. Wagner, A. C. Barbati, J. Engmann, A. S. Burbidge, G. H. McKinley, *Appl. Rheol.* **2016**, *26*, 1.

40 M. C. Sánchez, C. Valencia, J. M. Franco, C. Gallegos, *J. Colloid Interface Sci.* **2001**, *241*, 226.

41 M. M. Cross, *Rheol. Acta* **1979**, *18*, 609.

42 A. W. Sisko, *Ind. Eng. Chem.* **1958**, *50*, 1789.

43 J. W. Gooch, in *Encycl. Dict. Polym.*, Springer New York, New York, NY, **2007**, pp. 781–781.

44 I. Masalova, A. Y. Malkin, *Appl. Rheol.* **2007**, *17*, 42250.

45 J. Mewis, N. J. Wagner, *Adv. Colloid Interface Sci.* **2009**, *147–148*, 214.

46 D. G. Dalgleish, *Food Hydrocoll.* **2006**, *20*, 415.

47 D. J. McClements, *Curr. Opin. Colloid Interface Sci.* **2012**, *17*, 235.

48 J. Yang, I. Thielen, C. C. Berton-Carabin, E. van der Linden, L. M. C. Sagis, *Food Hydrocoll.* **2020**, DOI 10.1016/j.foodhyd.2019.105466.

49 S. A. Fioramonti, M. J. Martinez, A. M. R. Pilosof, A. C. Rubiolo, L. G. Santiago, *Food Hydrocoll.* **2015**, *43*, 8.

50 Y. Baimark, Y. Srisuwan, *Int. J. Polym. Sci.* **2013**, DOI 10.1155/2013/761870.

51 D. C. Thorn, H. Ecroyd, J. A. Carver, in *Bio-Nanoimaging* (Eds.: V.N. Uversky, Y.L.B.T.-B. Lyubchenko), Elsevier, Boston, **2014**, pp. 323–331.

52 C. Bengoechea, M. L. López, F. Cordobés, A. Guerrero, M. L. Lopez, F. Cordobes, A. Guerrero, *Food Sci. Technol. Int.* **2009**, *15*, 367.

53 F. Scheffold, J. Dispers, *Sci. Technol.* **2002**, *23*, 591.

54 D. J. McClements, *Curr. Protoc. Food Anal. Chem.* **2001**, DOI 10.1002/0471142913.fad0303s00.

55 C. Griffith, H. Daigle, *J. Colloid Interface Sci.* **2019**, *547*, 117.

56 M. M. Robins, *Curr. Opin. Colloid Interface Sci.* **2000**, *5*, 265.

57 R. Murakami, H. Moriyama, T. Noguchi, M. Yamamoto, B. P. Binks, *Langmuir* **2014**, *30*, 496.

58 R. Chanamai, D. J. McClements, *Colloids Surf. A Physicochem. Eng. Asp.* **2000**, *172*, 79.

59 M. M. Robins, A. D. Watson, P. J. Wilde, *Curr. Opin. Colloid Interface Sci.* **2002**, *7*, 419.

60 N. Kishore, R. P. Chhabra, V. Eswaran, *Chem. Eng. Res. Des.* **2006**, *84*, 1180.

61 R. Pal, *Fluids* **2019**, *4*, DOI 10.3390/fluids4040186.

62 I. Aranberri, B. P. Binks, J. H. Clint, P. D. I. Fletcher, others, *Rev. Iberoam. Polímeros* **2006**, *7*, 211.

63 S. Simon, S. Theiler, A. Knudsen, G. Øye, J. Sjöblom, J. Dispers, *Sci. Technol.* **2010**, *31*, 632.

64 E. Dickinson, *Food Hydrocoll.* **2019**, DOI 10.1016/j.foodhyd.2019.05.021.

65 X. Zhu, Q. Wang, Y. Leng, F. Chen, F. Wu, G. Mu, X. Wu, *Food Chem.* **2021**, *346*, 128918.

66 P. Taylor, *Adv. Colloid Interface Sci.* **1998**, *75*, 107.

67 J. Weiss, D. J. McClements, *Langmuir* **2000**, *16*, 2145.

68 D. Langevin, *Curr. Opin. Colloid Interface Sci.* **2019**, *44*, 23.

69 T. M. Ho, A. Razzaghi, A. Ramachandran, K. S. Mikkonen, *Adv. Colloid Interface Sci.* **2022**, *299*, 102541.
70 D. J. Mcclements, *Crit. Rev. Food Sci. Nutr.* **2007**, *47*, 611.
71 I. Vinckier, P. Moldenaers, A. M. Terracciano, N. Grizzuti, *AIChE J.* **1998**, *44*, 951.
72 A. Bhardwaj, S. Hartland, *J. Dispers. Sci. Technol.* **1994**, *15*, 133.
73 B. P. Binks, *Curr. Opin. Colloid Interface Sci.* **2002**, *7*, 21.
74 J. Santos, N. Calero, L. A. Trujillo-Cayado, M. C. Garcia, J. Muñoz, *Colloids Surf. B Biointerfaces* **2017**, *159*, 405.
75 F. M. Fernandes Barros, C. Chassenieux, M. M. de Souza Lima, L. Benyahia, *Colloids Surf. A Physicochem. Eng. Asp.* **2020**, *593*, 124630.
76 K. Shinoda, H. Arai, *J. Phys. Chem.* **1964**, *68*, 3485.
77 J. F. Ontiveros, C. Pierlot, M. Catté, V. Molinier, J.-L. Salager, J.-M. Aubry, *Colloids Surf. A Physicochem. Eng. Asp.* **2014**, *458*, 32.
78 X. Hua, P. Ding, M. Wang, K. Chi, R. Yang, Y. Cao, *Int. J. Biol. Macromol.* **2019**, *128*, 167.

4

Understanding Interfacial Rheology in Food Emulsions

Cecilio Carrera, Manuel Felix, María Luisa López-Castejón, and Víctor Manuel Pizones*

Departamento de Ingeniería Química, Universidad de Sevilla, Escuela Politécnica Superior. C/ Virgen de África, 7, 41011 Sevilla Spain
** Corresponding author*

Highlights

- Many food products are found in the form of dispersed systems (emulsions).
- Interfacial rheology plays a key role in characterizing the amphiphilic molecules adsorbed at food-grade interfaces.
- Dilatational and interfacial shear rheology are complementary techniques that help predict the stability of final food products.

4.1 Introduction

Many of the products that we consume in our life are in the form of an emulsion or foam (e.g. mayonnaise, coffee cream, salad dressing, cream ...). They represent a thermodynamically unstable dispersion of two immiscible fluids, which possess nonpolar and polar nature, respectively. Typically, one of them forms small droplets dispersed in the other phase. The former is called the dispersed phase and the latter is the continuous or external phase, being essential for the stability and organoleptic properties that this structure remains stable along time. This is one of the main concerns of the food industry to establish these structures and to maintain them with certain stability, which is achieved by processing conditions, and formulation [1]. In any case, the presence of a third component, known as an emulsifier (which has a surfactant nature) or stabilizer, is essential. In this sense, low molecular weight emulsifiers (LMWEs) and/or biopolymers (proteins, polysaccharides...) are clear examples that are widely used in this type of products that can also provide nutritional and functional benefits (mainly when proteins are used). The role of these emulsifiers is to create stable spherical structures between both constituent fluids, due to their adsorption at the oil/water (O/W) interface. In this sense, outside the oil droplets, there are hydrophilic heads with a high affinity for the aqueous phase and, on the other hand, there are lipophilic tails with a high affinity for the hydrophobic fluid [2].

The formation and stabilization of emulsions largely depend on the behavior of these interfaces during processing and later aging, and in the course of their mutual interaction.

Materials Science and Engineering in Food Product Development, First Edition. Edited by Wing-Fu Lai.

The aspects of importance during the formation and stabilization of emulsions can be divided into those related to the physical properties of fluids and those related to the composition and structure of the adsorbed material, surface-active species, at the interface between the fluids, which, in turn will determine the properties of the dispersions formulated. Therefore, these properties must be studied extensively. Among the different techniques, interfacial rheology has shown an enormous influence on emulsion stability, particularly droplet coalescence and Ostwald ripening [3]. Furthermore, this technique has shown a high sensitivity to detect the interactions between the components that form the interfacial film [4].

The purpose of this chapter is to show some insights related to the role of surface-active agents on the stability of emulsified systems, as well as to give a related overview of the importance of controlling the interfacial properties of the formed film. The latter would allow the controlled release of bioactive ingredients encapsulated in an emulsion. Thus, this chapter begins by highlighting the importance of interfacial engineering and how interfacial rheology can help it. In the following sections, both types of interfacial rheology (dilatational and shear) are defined, commenting on the different techniques used to measure them. A comparative analysis concludes this section. Lastly, some future applications of tailored interfaces in the food industry are discussed.

4.2 Interfacial Engineering of Food Emulsifiers

Emulsions are present in many food products. In all the definitions of emulsions that can be found in the bibliography, the existence of two immiscible phases is always highlighted. However, the existence of another region of the emulsion that is of vital importance is seldom mentioned. This region is the interfacial region that separates both aqueous and oily media, and although it represents only a small fraction of the total volume, it has a decisive influence on the final properties of food emulsions (e.g. stability, rheology, taste). In fact, emulsions are unstable systems because of their large interfacial area. The kinetic stability of these systems is generally achieved through a protective interfacial layer around the droplets [1]. The properties of this interfacial layer are governed by the composition and structure of the adsorbed material (emulsifiers) and, in turn, determine the properties of the dispersion. Therefore, from an industrial point of view, the correct choice of an emulsifier that stabilizes an emulsion is a key point.

In general, an emulsifier should have several desirable characteristics to be used in the food industry. Some of these characteristics are related to the physicochemical properties of the emulsifier, whereas others are related to more practical and economic factors [5]. Thus, a food emulsifier should have the following characteristics during the two stages of formation and stability of emulsion-based products:

- emulsion formation: it must be rapidly adsorbed and reduce interfacial tension, facilitating droplet breakup.
- emulsion stability: it must generate strong repulsive forces (steric and/or electrostatic), forming a resistant interfacial layer that prevents droplet coalescence.

The optimum use of emulsifiers depends on our knowledge of their interfacial physicochemical characteristics, such as surface activity, amount adsorbed, structure, thickness, topography, ability to desorb, lateral mobility, interactions between adsorbed molecules, ability to change conformation, interfacial rheological properties, etc. The key component of this nanoscopic approach is molecular self-assembly. Thus, an important topic is how the macroscopic characteristics of dispersions can be improved by controlling the nanostructure formation at fluid interfaces, i.e. by means of molecular or interfacial engineering (Figure 4.1) [4]. Thus, at present, there is great interest in seeking strategies and techniques (engineering) for the production of tailored interfaces. The latter can be achieved using simple and mixed emulsifiers, such as proteins [6], LMWEs [7], and/or polysaccharides [8]. Other very interesting strategies are those that seek to modulate the interfacial properties by altering the structure of the emulsifier, before or after being adsorbed at the interface. To achieve this, different physical, chemical, and/or enzymatic strategies can be used [4, 9, 10].

Traditional instruments are used for the characterization of interfacial films (tensiometry, film balance, surface rheology, etc.), as well as new techniques belonging to microscopic, nanoscopic, and spectroscopic analysis (e.g. atomic force microscopy, AFM [11], Brewster angle microscopy, BAM [12], infrared reflection absorption spectroscopy (IRRAS) [13], imaging ellipsometry (IE) [14], surface plasmon resonance (SPR) [15], etc.). Among all these techniques, interfacial rheology is presented as a very interesting tool for the study of interfaces, due to its high capacity to detect the interactions between the components that form the interfacial film. Interfacial rheology deals with studying the relationship

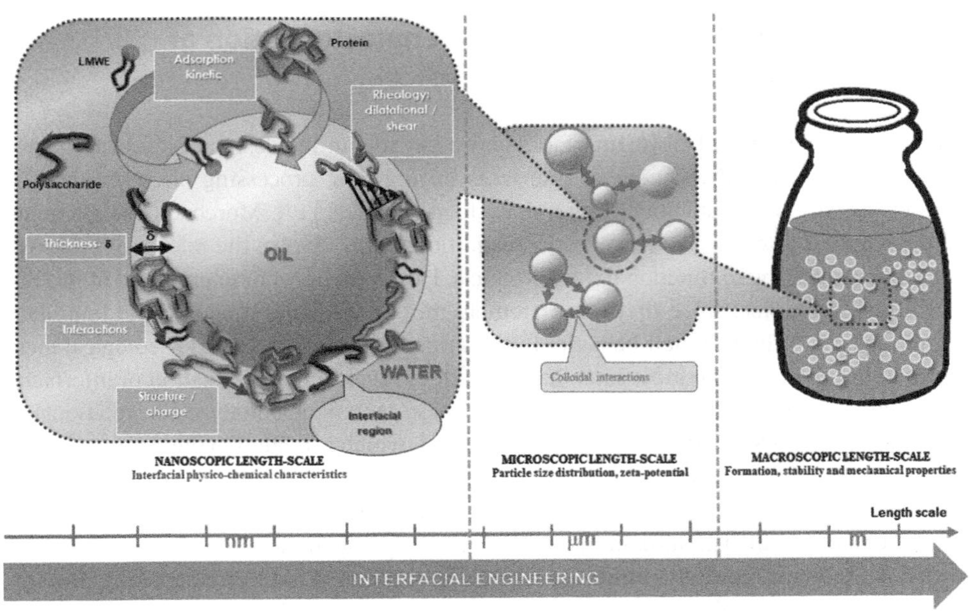

Figure 4.1 Graphic representation of three levels in emulsion analysis.

between stress and strain magnitudes associated with the deformation obtained or applied to an interfacial layer, which may cause significant changes in its microstructure [16]. Due to the variety in deformations of the interfacial layer (shear and expansions or compressions), the field of interfacial rheology is divided into the subcategories of shear and dilatational rheology. The characterization of both types is of great practical interest since shear rheology is primarily linked to the long-term stability of dispersions, and dilatational rheology provides information regarding short-term stability. Thus, the rheological characterization of interfaces is key for the formulation and stability of emulsions. The next section covers the main techniques applied for this purpose.

Industrial Corner 1

- Interfacial science plays a key role when the food industry aims to stabilize emulsions. Some examples of food products where interfacial science significantly contributes to this are margarine, ice cream, chocolates, beverages, and whipped toppings.
- Surfactants are amphiphilic molecules that are adsorbed at interfaces, and the interfacial films formed confer the kinetic stability of these emulsions.
- The rheological characterization of these interfacial films gives food makers general information related to their microstructure, which eventually helps to predict the stability (i.e. viability) of the final food products developed.

4.3 Rheological Techniques for the Characterization of Interfacial Films

As has been mentioned before, interfaces play a key role in the design of food emulsions, since the stability and behavior of interfaces during food processing often control the further formation and stability of two-phase food products [17]. Moreover, the adequate control of their integrity (under digestion conditions) may involve the release of bioactive compounds, leading to functional food products. However, the characterization of interfacial films is complex, since they are extremely thin at the same time and the boundary conditions can influence their behavior [18]. In this sense, the interfacial rheological measurements are among the most extended techniques for the characterization of interfacial films with applications in food products [3]. These techniques analyze the relationship between the stress applied and the strain associated, which is related to the microstructure of the interfacial film developed [19, 20]. In any case, the study of the rheological behavior of interfacial films can be mainly addressed by interfacial dilatational and interfacial shear measurements, as the results obtained are key for analyzing the stability of food emulsions, which is the main requirement for industrial applications of food products [18]. The following subsections will cover the characterization of interfacial films by dilatational and shear measurements. Note that whereas bulk rheology comes from three-dimensional measurements, interfacial shear rheology comes from a two-dimensional measurements, where the dimensions change from $\left[M\right]\left[L^{-1}\right]\left[T^{-2}\right]$ to $\left[M\right]\left[T^{-2}\right]$.

4.3.1 Interfacial Dilatational Rheology

Interfacial dilatational rheology is an interesting tool to analyze the equilibrium and dynamic properties of simple and complex interfacial layers, formed by LMWEs, amphiphilic biopolymers (protein, polysaccharides), and their mixtures. It consists of deforming the interfacial film by increasing and decreasing the interfacial area, through expansion and compression oscillations at defined deformation amplitude (dA/A) and frequency, while conserving the shape (constant shape but change in area) [21]. These modifications generate gradients in the interfacial tension (γ). Thus, it is possible to determine the interfacial dilatational parameters, namely, the elastic and viscous moduli, from the relationship between the interfacial tension and interfacial area modifications.

The interfacial dilatational modulus (E), defined by Gibbs, can be calculated from the following ratio (Equation 4.1):

$$E = \frac{d\gamma}{dA / A} = \frac{d\gamma}{dlnA} \tag{4.1}$$

where E is a complex parameter that consists of a real and an imaginary component (E' and E'', respectively). The real component (elastic or storage modulus) is the surface dilatational elasticity, defined as Equation (4.2), that represents the recoverable energy stored in the interface. The imaginary part (viscous or loss modulus) is the surface dilatational viscosity, defined as Equation (4.3), that represents the energy loss through relaxation processes [22, 23].

$$E' = \left[\mathrm{E}\right]\cos\delta \tag{4.2}$$

$$E'' = \left[\mathrm{E}\right]\sin\delta = \omega\eta_d \tag{4.3}$$

where ω is the frequency of the oscillatory changes in the droplet area and η_d is the interfacial dilatational viscosity. These components can be related by the phase angle (δ) tangent according to Equation (4.4) that can be employed as an indication of the viscoelasticity of the interfacial film. In this sense, the stress and strain are in phase ($\delta = 0$) for a purely elastic interfacial film, and E_v is zero, while in the case of a perfectly viscous interfacial film, $\delta = 90°$ and the real part is zero.

$$\tan\delta = \frac{E''}{E'} \tag{4.4}$$

The interfacial dilatational rheology is sensitive to the interactions between the emulsifiers adsorbed at fluid interfaces [4]. Once the surface-active components have been adsorbed at the O/W interface, they tend to undergo rearrangement and aggregation processes to form the interfacial film. In this sense, the evolution of E with time is associated with these stages occurring during the formation of the interfacial film, related to the interface coverage, as well as with the molecular structure of the constituents, and the oscillation frequency. It is also important to highlight that, depending on the amphiphilic molecule adsorbed at the O/W interface, different behaviors could be obtained. For example, globular proteins tend to produce more elastic interfacial films than those formed with random-coil proteins, due to the difference in the intermolecular interactions between the adsorbed components, as well as the different packing densities. However, an interfacial film formed by LMWEs is more fluid, with lower viscoelasticity than proteins. This can infer significant differences in the physical properties of emulsions [3, 18, 24]. Moreover, the scientific literature attributes the short-term stability of emulsified systems to interfacial dilatational measurements, particularly droplet coalescence and Ostwald ripening. Interfaces with high interfacial dilatational viscoelasticity facilitate the inhibition of flocculating droplets, coalescence, and help to resist the deformation of the liquid film because of mechanical disturbance [3, 4, 25].

Various techniques exist to measure interfacial dilatational rheology that induces a change in the interfacial area while simultaneously measuring the interfacial tension. They can be mainly divided into droplets- and trough-based methods.

4.3.1.1 Oscillating Droplet Techniques

Two main types of droplet techniques exist to determine the interfacial tension when a sinusoidal expansion/compression deformation is applied to an axisymmetric droplet [21].

4.3.1.1.1 Droplet Profile Tensiometer

It is the technique most frequently used to assess the dilatational rheology of interfacial layers, becoming a suitable method for routine studies. In this tensiometer, sufficiently large droplets can be formed, where their shape is determined by a combination of surface tension (makes drops spherical) and gravity (makes drops vertically elongated) effects. The volume of the droplet is modified and from the variation of its shape (Young–Laplace equation) is determined the interfacial tension response, and by applying Fourier analysis, the dilatational rheology parameters can be extracted. Small amounts of samples are required, and according to its working principle, an appreciable density difference between the fluids explored and the interface is required, which must not be far from the mechanical equilibrium. As a drawback, this technique is not suitable to work at high frequencies.

4.3.1.1.2 Capillary Pressure Tensiometry

This is a versatile technique that consists in measuring the pressure difference between two phases separated by the spherical interface of a sub-millimetric droplet, practically not deformed by the gravity field, that allows the acquirement of capillary pressure values, which are typical of a hundred pascal (Pa) order. The device employed consists of two chambers that are connected by a capillary tube. One of the chambers is closed and contains both the pressure sensor and a piezoelectric rod that is used to control the volume of the droplet formed at the tip of a glass capillary. The other cell is open to atmospheric pressure or, if closed, contains another pressure sensor. The changes of the pressure variations, resulting from the compression/expansion of the liquid, allow the extraction of the rheological parameters.

4.3.1.2 Trough Methods

The Langmuir trough is also another widely used equipment for evaluating the dilatational properties of interfaces. It consists of a Wilhelmy type film balance that comprises two oscillating barriers that compress and expand the interfacial film, applying a series of sinusoidal waves (with different amplitude or oscillation frequencies of the barriers), parallel to the interface. The response to these oscillations is reflected by the variation of the interfacial pressure, that is directly measured with a roughened platinum plate that is placed between the barriers. The measurement of the interfacial dilatational viscoelasticity during these oscillations is then used to investigate the resistance of the interfacial film to a mechanical disturbance [26, 27].

There are many experimental studies in the literature concerning the analysis of dilatational rheology of surfactant adsorption layers by using these experimental devices. In this sense, numerous contributions concerning the interfacial and emulsifying characteristics of plant-based proteins have appeared, reflecting the enormous interest in this topic. Dilatational properties of legume proteins (chickpea or fava bean) have been analyzed using pendant droplet measurements, as they represent an interesting alternative to model proteins, with the aim to avoid the use of animal sources. These vegetal proteins formed an interfacial film with a gel-like behavior at the O/W interface and the results from the stability of the emulsion seemed to agree with the interfacial rheology [28].

In many experimental studies, the frequency dependence of the dilatational viscoelasticity is also used to access dynamic adsorption mechanisms, such as diffusion, molecular reorientation, and surface aggregation, which in turn depend on the molecular characteristics of the surfactant molecules. LMWEs, in contrast to proteins, form a mobile interfacial layer with low viscoelasticity, which is more effective in reducing the dynamic interfacial tension. The addition of LMWEs to protein-stabilized emulsions can lead to a decrease in the stability of the emulsions due to conflicts between the two different stabilization mechanisms of these emulsifiers. The LMWEs competitively displace proteins from the O/W interface by the proposed "orogenic mechanism" [4, 29].

Due to the increasing interest by consumers for healthier natural foods, interest in the analysis of interfacial properties of mixtures of natural food emulsifiers has also grown up to produce low-fat or fat-free formulations, including fat replacers. In this sense, the use of polysaccharides is common due to their several functions: some polysaccharides possess interfacial activity properties, such as thickening and gelling properties. In this sense, emulsions including a prebiotic like inulin have great potential in the food industry as they increase emulsions stabilized by a milk protein because of its thickening effect [30]. Furthermore, the use of surface-active polysaccharides has also been demonstrated to affect the interfacial properties of pure proteins [4, 31, 32].

4.3.2 Interfacial Shear Rheology

Contrary to dilatational measurements, the interfacial shear rheology applies a shearing motion to an interfacial film, and analyses the deformation generated. These experiments are performed generally at a constant area (avoiding undesired adsorption/desorption effects described for dilatational experiments). Contrary to droplet experiments, the interface can be in thermodynamic equilibrium with the bulk solution [4]. Unlike dispersed food products, the interface formed in these experiments is planar, although this morphological difference is important only if the droplet diameter is below 1 μm [33, 34]. The application of this technique has been mainly limited to small amplitude tests, within the linear viscoelastic range, where the linear viscoelastic moduli can be obtained as follows:

$$G' = \frac{\sigma}{\gamma}\cos(\delta) \tag{4.5}$$

$$G'' = \frac{\sigma}{\gamma}\sin(\delta) \tag{4.6}$$

where the G' and G'' are the in-phase and the out-of-phase moduli (also known as the elastic and viscous moduli). σ is the stress applied and δ is the phase angle.

The phase angle also has implications in interfacial shear rheology, which can be defined as follows (Equation 4.7):

$$\tan\delta = \frac{G'}{G''} \tag{4.7}$$

where a purely elastic response leads to an in-phase stress/strain ($\delta = 0$), a purely viscous response is given by $\delta = 90°$.

The interfacial shear characterization of O/W has been used to analyze the stability of protein-adsorbed interfacial films, and their relationship with the stability of dispersed systems such as emulsions [16, 35]. Although the transient response over protein adsorption has been analyzed, most of these studies have been focused on linear viscoelastic interfacial properties once the pseudo-equilibrium state has been reached. Interestingly, the complex interfaces developed to stabilize food systems must involve molecular interactions, leading to a viscoelastic response, which can be analyzed through interfacial shear experiments [36].

One of the main challenges of this technique is to decouple the effects of the adjoining phases. In this sense, different measurement devices have been developed to highlight the influence of the interfacial film over the bulk contribution. The Boussinesq number (*Bo*) accounts for the influence of each response, and it is calculated for the specific response when measured with a specific geometry (Equation 4.8) [37]:

$$Bo = \frac{\eta_s}{\eta_b \cdot G} \tag{4.8}$$

where η_s and η_b correspond to the interfacial shear viscosity and the average bulk viscosity (related to water and oil), respectively, whereas G depends on the dimensions of the measuring tool used ($G = A_S / P_S$, A_S is the contact area between the tool and the adjoining subphase and P_s is the perimeter of the geometry in contact with the O/W interface).

The initial definition of Bo number was for a continuous shear flow. However, it can also be used for interfacial small amplitude oscillatory shear (i-SAOS) tests substituting the apparent viscosity by the complex viscosity [38–40]. The response obtained is attributed to the interfacial films developed when $|\mathrm{Bo}| \gg 1$ [38]. The main differences between interfacial shear accessories are related to the contact area, and, consequently, to the Bo number.

Thus, the sensitivity of the measurement increases when the fluid dynamics of the tool used is considered [16], isolating the interfacial response from the bulk contribution. These are the most important tools used for interfacial shear characterization with application in food products:

4.3.2.1 Du Noüy Ring Tool

The Du Noüy ring fitted to interfacial rheometers has been used to perform dynamic surface shear measurements for many years [41]. However, the sensitivity of these interfacial rheometers is limited to a narrow torque range. Unfortunately, this limitation becomes important when the interfacial dynamic moduli of protein-based interfacial layers may change by six or more orders of magnitude as the protein migrates to the surface [42]. Unfortunately, commercial rheometers exhibit lower sensibility and higher inertia than required for interfacial measurements. Thus, the Du Noüy ring fitted to conventional rheometers shows limitations such as non-isotropic stress and low sensitivity associated with the use of small contact areas [4]. Some attempts have been carried out by using highly sensitive and hybrid magnetic bearing-based rheometers [42]; however, the developments of Du Noüy geometry have evolved toward the double-wall ring (DWR) tool, which considers the flow dynamics and maximizes the surface contact area to obtain a suitable interfacial response [38]. This measurement tool will be described later.

4.3.2.2 Bicone Tool

This measurement tool has been widely used for the characterization of protein-based interfacial films with applications in food products [43]. As an advantage, the bicone can be fitted to conventional rheometers. However, one of the main limitations of this geometry is its large contact surface area, which leads to a potential noticeable contribution of the adjoining subphases (note the high G values for Equation 4.8). The specific flow field for this geometry was initially described by Oh and Slattery [44] in the absence of inertia. However, Erni et al. [45] updated the instrumental setup, improving the calculation of the velocity profile distribution for this measurement tool. In any case, this geometry requires a high interfacial rheological response to neglect the contribution of the adjoining phases. Despite this limitation, this solution is among the three most widely used methods for measuring the interfacial shear response [4].

4.3.2.3 DWR Tool

The DWR is an evolution of the Du Noüy ring, which has been designed to minimize the characteristic length scale (G, Equation 4.8), being 48-fold smaller than the standard bicone geometry. Moreover, the inertia effect (due to its higher size compared with the Du Noüy ring) was also reduced by the alloy selected, whereas the section of the ring is rhombohedral instead of cylindrical [38]. One limitation of devices coupled to a rheometer (either standard or interfacial) is their ability to measure the rheological response at the same time as the interfacial tension. However, some improvements have been introduced in some configurations simultaneously for both responses over compression-expansion cycles [46, 47]. Moreover, a unit for subphase renovation has also been adapted to this measuring tool, increasing the versatility of this tool.

4.3.2.4 Magnetic Rods Based Tools

This technique consists of the application of a magnetic field to a magnetic rod, which is deposited on the interface formed. The oscillation of the needle will depend on the magnetic field applied and the viscoelastic response of the interface [48]. Since the characteristic length scale (G) comes from the needle, the contribution of the subphases is minimized [38]. Thus, the magnetic interfacial needle rheometer is considered the most sensitive interfacial shear rheometer, where the response is dominated by the flow properties of the interfacial films formed [49]. Recently, the advantages of this device have been maximized after using magnetic microwires that have characteristic lengths one order of magnitude lower, leading to lower Bo number and inertia effects. Consequently, the frequency window available for this device has extended when analyzing the rheological properties of interfacial films [50].

All these measurement tools have been used for the characterization of protein-based interfacial films. Thus, bicone has been used with many proteins adsorbed at the interface such as β-lactoglobulin, β-casein, and ovalbumin, among others [51–53]. Moreover, proteins adsorbed at the interfacial region has also been studied by the DWR geometry. These proteins also can come from animal sources, such as crayfish, whey, βLG, lysozyme [38, 54–58], or from plant sources, such as chickpea, fava bean, rice, or potato [40, 59–61]. Although these are the two techniques that are more extended, protein-based interfaces have also been analyzed by shear using magnetic rods for amphiphilic molecules, such as albumin, crayfish protein isolate, and eicosanol [50, 58, 62]. Less important has been the findings based on the Du Noüy ring, although some are found in the scientific literature [42, 63]. The applications of interfacial shear rheology are mostly related to the prediction of long-term stability of the emulsions formed later [3, 64]. This prediction typically comes from the analysis of frequency sweep tests after protein adsorption, although the viscoelastic moduli are also obtained during protein adsorption at constant frequency [59]. Less common has been the determination of interfacial rheological properties by means of continuous deformation tests, such as flow curves, stress-relaxation tests, and creep tests [39, 40].

4.3.3 Comparison Between Interfacial Dilatational and Shear Rheology

Dilatational rheology is related to the response of interfacial films when they are subjected to a reduction/increase in the interfacial area. In contrast, interfacial shear rheology is related to the interfacial response at a constant area. This different approach leads to a mechanical response where different mechanisms are involved. In this sense, the interfacial shear measurements have been related to the inter- and intra-molecular interactions that occur at a high surface coverage (which is typically found in food products) [65]. Note that the stresses found include deviatoric stresses, which can be determined directly by shear measurements. However, the dilatational measurements involve changes in the surface area, which may lead to changes in protein concentration, and consequently, surface tension also is affected [20].

Moreover, the adsorption/desorption kinetics can influence the interfacial tension when changes in the surface area occur. Food products contain both, low- and high molecular weight surfactants, which can behave in a very different way. In this sense, proteins tend to be irreversibly adsorbed at the interface (leading to excess concentration), whereas phospholipids tend to be adsorbed and desorbed [66, 67]. Thus, the primary advantage of the interfacial shear rheology is its response in the absence of changes in the interfacial area, and consequently, in the film composition [68]. Despite the advantages and drawbacks that each method may have during the measurement, it should be borne in mind that dilatational measurements have been previously related to short-term stability and emulsion formation, whereas interfacial shear rheology has been related to the long-term stability of emulsions [64]. Thus, a combination of interfacial dilatational and shear measurements (Figure 4.2) would be required to extract the untangled response of the complex fluid–fluid interfaces [19, 69].

Industrial Corner 2

- Dilatational and interfacial shear techniques are complementary, where the dilatational results are related to emulsion formation and short-term stability, and interfacial shear rheology is related to long-term stability.
- The elastic prevalence, together with a low dependence of the viscoelastic moduli are always the advantages of using stable interfacial films. In this sense, both the typical tests are performed at interfaces:
  - Evolution of viscoelastic moduli during protein adsorption, at a constant frequency.
  - Frequency sweep tests, typically from 0.075 to 0.1 Hz (lower frequency values take a long experimentation time, while higher frequency values may not be suitable for droplet experiments or create inertia problems in interfacial shear rheology).
- The interfacial response in complex interfaces (i.e. in food products) depends on many factors; consequently, it is not possible to define a simple response model. In any case, a comparison between a suitable interfacial film (for a specific formulation) and a new formulation can always be useful for the food industry.

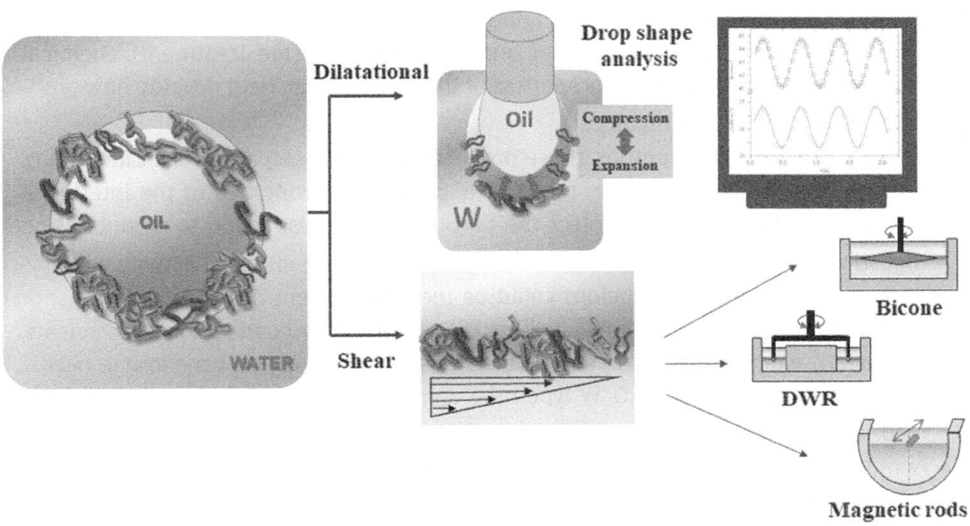

Figure 4.2 Graphic representation comparing interfacial dilatational and shear rheology.

4.4 Concluding Remarks and Future Perspectives

The latest advances in interfacial rheology techniques have shown the great potential of this approach to characterize the interactions between the emulsifiers adsorbed on interfacial films. As has been commented, the characterization of interfacial films is of great importance since the main macroscopic properties of dispersed systems (e.g. formation, stability, and texture) depend on it. For all these reasons, if the properties of interfacial films are controlled, interesting properties could be investigated, such as the controlled release of components with bioactive properties (design of functional foods) and the control of the oxidative stability of emulsified oils.

In emulsions, the bioavailability of fats is ultimately controlled by the accessibility of lipase to the fat contained within the emulsion droplets and hence to the interfacial properties of the emulsions. On the contrary, for other nutrients, such as polyphenols, vitamins, bioactive peptides, etc., it could be interesting to increase their absorption or protect their bioavailability during food digestion. To do this, these nutrients must be encapsulated to protect them during storage and after digestion and then released at the appropriate site of action within the human body [4, 70]. That is to say, what is involved is to know which colloidal transformations occur during the digestion of a dispersion. Since these colloidal transformations are closely related to the characteristics of the interface, it is essential to know the role of interfacial engineering in biophysical processes, which in turn can be analyzed by the rheological characterization of the interfacial film formed [71, 72]. Thus, it is necessary to adequately design the properties of the film that protects the particles of a dispersion to modulate the digestion of fats and the bioavailability of nutrients with biological activity, and in all of them, interfacial rheology has proven to be a tool with great potential.

Industrial Corner 3

- The interfacial engineering of emulsions has attracted big interest due to their great potential for application in the formulation of dispersed systems.
- The control over the physical (heat and ultrasound-assisted treatment), chemical (cross-linking), and enzymatic properties of the interface can help develop new products with specific functionality.
- Based on their final properties, these products may have specific applications. Thus, for this purpose, it is necessary to properly design the properties of the interfacial film, such as being able to protect the particles of the dispersion to modulate the digestion of fats and the bioavailability of nutrients with biological activity.
- Another use of these formulations could be the development of protein-based nanovehicles for the encapsulation and delivery of hydrophobic or poorly soluble ingredients, due to its great potential use in formulations of functional products beneficial to health. Double emulsions of the W/O/W type have applications in the formulation of low-fat foods and as a vehicle for the encapsulation and release of nutrients during the digestion.
- Another way of presenting these formulations could be as nanoemulsions, which are emulsions whose droplet size is very small (less than 200 nm); their small droplet size provides greater stability to these emulsions that they can be used in food and beverage formulations. For this reason, the interfaces can be modified by altering their physical, chemical, or enzymatic properties, thus allowing the desired characteristics to be achieved.

Acknowledgments

This work was funded by the Junta de Andalucía (Consejería de Economía, Conocimiento, Empresas y Universidad) through the P20_01046 project (BionanoWOW).

Glossary

Emulsifier It is a surface-active agent that acts as a border between two immiscible liquids. These molecules work by having a hydrophilic end (water-loving) and a hydrophobic end (water-hating). Each of these ends is adsorbed on the corresponding phase so that the emulsifier remains adsorbed on the oil/water interface.

Interfacial elastic modulus It refers to the recoverable deformation when a viscoelastic interfacial film is subjected to stress (or strain).

Interfacial dilatational rheology It is an interesting tool to analyze the equilibrium and dynamic properties of interfacial films, formed by emulsifiers. It consists of deforming the interfacial film by increasing and decreasing the interfacial area (expansion and compression oscillations), while conserving the shape (constant shape but change in area). Interfacial tension follows the change in area of the droplet when emulsifiers are present.

Interfacial film It is the layer located in the interfacial region that is formed by the adsorption of molecules with surface activity.

Interfacial rheology It is a branch of rheology that analyzes the response of the adsorbed interfacial film to the deformation caused by external stimuli. The response depends on the layer composition. By selecting the appropriate layer composition, one can form a more stable emulsion and thus improve the quality of the product.

Interfacial shear rheology It is an interesting tool that applies a shearing motion on interfacial films and analyzes the deformation generated (constant interfacial area but change in shape). These experiments are performed generally at a constant area (avoiding the undesired adsorption/desorption effects described for dilatational experiments).

Interfacial tension It is the force of attraction between the molecules at the interface of two immiscible fluids. The SI unit is mN/m.

Interfacial viscous modulus It refers to the non-recoverable deformation when a viscoelastic interfacial film is subjected to stress (or strain).

LMWEs Low molecular weight emulsifiers (e.g. phospholipids, mono-diglycerides ...).

Phase angle In an oscillatory test, the phase angle is the delay between the sinusoidal stress applied and the strain experienced by the interfacial film.

Viscoelasticity A property of materials that exhibit both viscous and elastic characteristics when they undergo deformation.

References

1 D. J. McClements, *Food Emulsions: Principles, Practices, and Techniques*, CRC Press, USA, Boca Raton, **2015**.

2 Y. Zhang, F. Zhou, P. Shen, Q. Zhao, M. Zhao, *Food Hydrocoll.* **2020**, *103*, 105646.

3 J. Maldonado-Valderrama, J. M. R. Patino, *Curr. Opin. Colloid Interface Sci.* **2010**, *15*, 271.

4 C. Carrera Sánchez, J. M. Rodríguez Patino, *Food Hydrocoll.* **2021**, *119*, 106838.
5 D. J. McClements, L. Bai, C. Chung, *Annu. Rev. Food Sci. Technol.* **2017**, *8*, 205.
6 C. Chang, S. Tu, S. Ghosh, M. T. Nickerson, *Food Res. Int.* **2015**, *77*, *Part 3*, 360.
7 D. J. McClements, S. M. Jafari, *Adv. Colloid Interface Sci.* **2018**, *251*, 55.
8 J. M. R. Patino, A. M. R. R. Pilosof, J. M. Rodriguez Patino, A. M. R. R. Pilosof, *Food Hydrocoll.* **2011**, *25*, 1925.
9 M. Felix, J. Yang, A. Guerrero, L. M. C. Sagis, *Food Hydrocoll.* **2019**, *97*, 105235.
10 H. Chen, D. J. McClements, E. Chen, S. Liu, B. Li, Y. Li, *Langmuir* **2017**, *33*, 14608.
11 J. Yang, I. Thielen, C. C. Berton-Carabin, E. van der Linden, L. M. C. Sagis, *Food Hydrocoll.* **2020**, 10.1016/j.foodhyd.2019.105466.
12 D. Vollhardt, *Curr. Opin. Colloid Interface Sci.* **2014**, *19*, 183.
13 M. Krajewska, K. Dopierała, P. Wydro, M. Broniatowski, K. Prochaska, *J. Mol. Liq.* **2020**, *319*, 114089.
14 W. Ogieglo, H. Wormeester, K.-J. Eichhorn, M. Wessling, N. E. Benes, *Prog. Polym. Sci.* **2015**, *42*, 42.
15 X. Du, Y. Wang, *J. Phys. Chem. B* **2007**, *111*, 2347.
16 N. Jaensson, J. Vermant, *Curr. Opin. Colloid Interface Sci.* **2018**, *37*, 136.
17 P. Wilde, *Curr. Opin. Colloid Interface Sci.* **2000**, *5*, 176.
18 B. S. Murray, *Curr. Opin. Colloid Interface Sci.* **2011**, *16*, 27.
19 P. Wilde, A. Mackie, F. Husband, P. Gunning, V. Morris, *Adv. Colloid Interface Sci.* **2004**, *108–109*, 63.
20 L. M. C. Sagis, E. Scholten, *Trends Food Sci. Technol.* **2014**, *37*, 59.
21 R. Miller, J. K. Ferri, A. Javadi, J. Krägel, N. Mucic, R. Wüstneck, *Colloid Polym. Sci.* **2010**, *288*, 937.
22 V. Pizones Ruiz-Henestrosa, C. Carrera-Sanchez, J. J. Pedroche, F. Millan, J. M. Rodriguez Patino, *Food Hydrocoll.* **2009**, *23*, 377.
23 B. Zhou, J. T. Tobin, S. Drusch, S. A. Hogan, *Adv. Colloid Interface Sci.* **2021**, *295*, 102347.
24 P. J. Wilde, M. J. Ridout, A. R. Mackie, M. S. J. Wickham, R. M. Faulks, *ACS Symp. Ser.* **2009**, 1007, 67.
25 M. Félix, C. Carrera-Sanchez, A. Romero, C. Bengoechea, A. Guerrero, *Food Hydrocoll.* **2020**, *104*, 105719.
26 A. Lucero Caro, M. R. Rodríguez Niño, J. M. Rodríguez Patino, *Colloid. Surface. A* **2008**, *327*, 79.
27 C. C. Sánchez, M. Rosario Rodríguez, J. M. R. P. Niño, *Food Hydrocoll.* **2005**, *19*, 395.
28 M. Félix, A. Romero, C. Carrera-Sanchez, A. Guerrero, *Fluids* **2019**, *4*, doi:10.3390/fluids4020065.
29 J. Pelipenko, J. Kristl, R. Rošic, S. Baumgartner, P. Kocbek, *Acta Pharm.* **2012**, *62*, 123.
30 M. L. López-Castejón, C. Bengoechea, S. Espinosa, C. Carrera, *Food Hydrocoll.* **2019**, *87*, 382.
31 Y. Cai, L. Huang, X. Tao, J. Su, B. Chen, F. Zhou, M. Zhao, Q. Zhao, P. Van der Meeren, *Food Hydrocoll.* **2021**, *113*, 106529.
32 N. A. Camino, C. C. Sánchez, J. M. Rodríguez Patino, A. M. R. Pilosof, *Food Hydrocoll.* **2011**, *25*, 1.
33 S. Mun, D. J. McClements, *Langmuir* **2006**, *22*, 1551.
34 D. Langevin, *Adv. Colloid Interface Sci.* **2000**, *88*, 209.

35 E. H. Lucassen-Reynders, J. Benjamins, V. B. Fainerman, *Curr. Opin. Colloid Interface Sci.* **2010**, *15*, 264.

36 D. L. Leiske, S. R. Raju, H. A. Ketelson, T. J. Millar, G. G. Fuller, *Exp. Eye Res.* **2010**, *90*, 598.

37 D. A. Edwards, H. Brenner, D. T. Wasan, A. M. Kraynik, *Phys. Today* **1993**, *46*, 63.

38 S. Vandebril, A. Franck, G. G. Fuller, P. Moldenaers, J. Vermant, *Rheol. Acta* **2010**, *49*, 131.

39 M. Felix, A. Romero, J. Vermant, A. Guerrero, *Colloid. Surface. A* **2016**, *499*, 10.

40 M. Felix, A. Romero, C. C. Sanchez, A. Guerrero, *Appl. Surf. Sci.* **2019**, *469*, 792.

41 B. Warburton, Interfacial rheology, *in Rheological Measurement* (Eds: A. A. Collyer, D. W. Clegg), Springer Netherlands, Dordrecht, **1998**, pp. 723–754.

42 B. Costello, P. Hodder, A. Franck, *Annu. Trans. Rheol. Soc.* **2006**, *14*, 209.

43 R. A. Mohammed, A. I. Bailey, P. F. Luckham, S. E. Taylor, *Colloid. Surface. A* **1994**, *91*, 129.

44 O. H. Soo-Gun, J. C. Slattery, *J. Colloid Interface Sci.* **1978**, *67*, 516.

45 P. Erni, P. Fischer, E. J. Windhab, V. Kusnezov, H. Stettin, J. Lauger, *Rev. Sci. Instrum.* **2003**, *74*, 4916.

46 E. Hermans, J. Vermant, *Soft Matter.* **2014**, *10*, 175.

47 P. J. Beltramo, M. Gupta, A. Alicke, I. Liascukiene, D. Z. Gunes, C. N. Baroud, J. Vermant, *Proc. Natl. Acad. Sci.* **2017**, *114*, 10373.

48 C. O. Klein, A. Theodoratou, P. A. Rühs, U. Jonas, B. Loppinet, M. Wilhelm, P. Fischer, J. Vermant, D. Vlassopoulos, *Rheol. Acta* **2019**, doi:10.1007/s00397-018-01122-y.

49 S. Reynaert, C. F. Brooks, P. Moldenaers, J. Vermant, G. G. Fuller, *J. Rheol. (N. Y. N. Y).* **2008**, *52*, 261.

50 J. Tajuelo, J. M. Pastor, F. Martínez-Pedrero, M. Vázquez, F. Ortega, R. G. Rubio, M. A. Rubio, *Langmuir* **2015**, *31*, 1410.

51 L. Salvia-Trujillo, R. Soliva-Fortuny, M. A. Rojas-Graü, D. J. McClements, O. Martín-Belloso, *Annu. Rev. Food Sci. Technol.* **2017**, *8*, 439.

52 E. Assadpour, S.-M. Jafari, Y. Maghsoudlou, *Int. J. Biol. Macromol.* **2017**, *95*, 238.

53 N. Dasgupta, S. Ranjan, S. Mundra, C. Ramalingam, A. Kumar, *Int. J. Food Prop.* **2016**, *19*, 700.

54 L. M. C. Sagis, K. N. P. Humblet-Hua, S. van Kempen, *J. Phys.-Condens. Matter.* **2014**, *26*, 9.

55 L. M. C. Sagis, B. Liu, Y. Li, J. Essers, J. Yang, A. Moghimikheirabadi, E. Hinderink, C. Berton-Carabin, K. Schroen, *Sci. Rep.* **2019**, *9*, 2938.

56 S. G. Baldursdottir, M. S. Fullerton, S. H. Nielsen, L. Jorgensen, *Colloid. Surface. B* **2010**, *79*, 41.

57 M. Felix, A. Romero, A. Guerrero, *Food Hydrocoll.* **2017**, *64*, 9.

58 A. Romero, T. Verwijlen, A. Guerrero, J. Vermant, *Food Hydrocoll.* **2013**, *30*, 470.

59 M. Felix, A. Romero, C. Carrera-Sanchez, A. Guerrero, *Food Hydrocoll.* **2019**, *90*, 353.

60 A. Romero, V. Beaumal, E. David-Briand, F. Cordobes, A. Guerrero, M. Anton, *Food Hydrocoll.* **2012**, *29*, 1.

61 A. Romero, V. Beaumal, E. David-Briand, F. Cordobes, A. Guerrero, M. Anton, *J. Agric. Food Chem.* **2011**, *59*, 9466.

62 P. Dhar, Y. Cao, T. M. Fischer, J. A. Zasadzinski, *Phys. Rev. Lett.* **2010**, *104*, 16001.

63 W. Li, Y. Wang, H. Zhao, Z. He, M. Zeng, F. Qin, J. Chen, *Food Hydrocoll.* **2016**, *60*, 453.

64 E. H. Lucassen-Reynders, J. Lucassen, D. Giles, *J. Colloid Interface Sci.* **1981**, *81*, 150.

65 J. Krägel, S. R. Derkatch, *Curr. Opin. Colloid Interface Sci.* **2010**, *15*, 246.
66 Y. J. Choi, S. Hur, B.-D. Choi, K. Konno, J. W. Park, *J. Food Sci.* **2009**, *74*, C17.
67 P. A. Rühs, C. Affolter, E. J. Windhab, P. Fischer, *J. Rheol. (N. Y. N. Y).* **2013**, *57*, 1003.
68 B. S. Murray, B. S. Murray, *Curr. Opin. Colloid Interface Sci.* **2002**, *7*, 426.
69 L. M. C. Sagis, *Eur. Phys. J.-Special Top.* **2013**, *222*, 39.
70 D. J. McClements, *Adv. Colloid Interface Sci.* **2018**, *253*, 1.
71 E. Dickinson, *Curr. Opin. Food Sci.* **2020**, *33*, 52.
72 A. Sarkar, S. Zhang, M. Holmes, R. Ettelaie, *Adv. Colloid Interface Sci.* **2019**, *263*, 195.

5

Overview of Types of Materials Used for Food Component Encapsulation

Zahra Emam-Djomeh[1,3,4,*], *Mohammad Ekrami*[1], *and Ali Ekrami*[2]

[1] *Transfer Phenomena Laboratory (TPL), Department of Food Science, Technology and Engineering, University College of Agriculture & Natural Resources, University of Tehran, Karaj Campus, Karaj, Iran*
[2] *Department of Nursing, University of Social Welfare and Rehabilitation Sciences, Tehran, Iran*
[3] *Functional Food Research Core (FFRC), University of Tehran, Tehran, Iran*
[4] *Center of Excellence in Biothermodynamics, University of Tehran, Tehran, Iran*
[*] *Corresponding author*

Highlights

- Recent trends suggest biodegradable polymers as the most promising solution for synthetic encapsulation systems.
- Safety, efficiency, biocompatibility, and biodegradability are some of the properties that biodegradable systems in encapsulation can provide for a broad spectrum of applications.
- The ease, desirability, and speed of the controlled release of encapsulated food components depend on the techniques used for encapsulation.
- For each polymer, depending on the carrier source, the relevant properties, applications, and existing industrial samples are different.

5.1 Introduction

In recent years, there has been a growing interest in improving life. Consumers are becoming more health-aware, focusing on healthy lifestyles and diets to avoid chronic illnesses and enhance health and well-being. Food products and industries have benefited from developing food matrices that conserve and release bioactive substances, such as food components, phytochemicals, and probiotics [1, 2]. However, bioactive compounds are susceptible to processing and storage conditions and have reduced bioavailability when exposed to upper gastrointestinal (GI) disorders. Thus, incorporating bioactive chemicals while retaining their qualities has been difficult for the industry.

Encapsulation is a popular way to secure and distribute dangerous active components (core material) to specific sites while extending product shelf life [3]. Increased efficiency is achieved by using bioactive compounds in encapsulation, allowing precise quantity determination. Pharmaceuticals' bioavailability, preservation, and stability can be enhanced by

Materials Science and Engineering in Food Product Development, First Edition. Edited by Wing-Fu Lai.

protecting them from deterioration during food processing and digestion. In this way, the active components are protected from harmful elements like light and air. It may also enhance the product's organoleptic properties by concealing flavor without modifying taste, aroma, color, or texture [4, 5]. This chapter explores colloidal particles' production by encapsulating these food components in natural polymers also known as biopolymers.

5.2 Major Techniques Used for Food Component Encapsulation

Numerous strategies for encapsulating bioactive compounds for food and pharmaceutical applications have been explored. By applying physical and chemical modifications, conventional techniques develop capsule-gel structures. Although these procedures are affordable, possess high load encapsulation, and are straightforward to use on an industrial scale, high temperatures throughout the process may degrade bioactive components. Consequently, developing electrohydrodynamic (EHD) technologies, including electrospinning, electrospray, and 3D-printing, are effective encapsulating techniques. Table 5.1 shows several methods and fundamental concepts of encapsulation techniques for food components.

Electrospinning can produce fibers as small as submicron or nanometers by using an electric field. Excellent porosity and surface-to-volume ratio may be achieved with this

Table 5.1 Principles of encapsulation techniques for the food component.

Method	Principle	Advantage
Spray drying	The food component is atomized in a drying chamber after being disseminated in an encapsulant solution	• Acceptable encapsulation efficiency • Stable encapsulated product • Cost-effective • Applicable at the industrial level • Easy to operate
Spray chilling	The carrier material is solidified by spraying a coating solution with a food component distributed in it and then letting it cool	• Useful for temperature-sensitive nutrients • Applicable on a large scale
Co-extrusion	To solidify the particle, the emulsion dispersion containing the food component was pushed through a die at high temperature and pressure	• Prolong the shelf life of products • Useful for temperature-sensitive ingredients • The shape of the extruded products can be easily controlled
Fluidized bed coating	The food component is suspended in air; then, a coating is applied to the particles	• Useful for hydrophilic and hydrophobic nutrients • Easy process control • High production efficiency

Table 5.1 (Continued)

Method	Principle	Advantage
Inclusion complexation	Cyclodextrin with food components constitutes an inclusion complex	• Stable products under oxidation conditions • Can operate at a low temperature • Useful for unstable nutrients in aqueous media
Complex coacervation	Biopolymers that have opposing charges may develop coacervates when they come into contact and then phase-separate	• Useful for temperature-sensitive nutrients • Organic solvent usage • Low cost
Electrohydrodynamic	Electrohydrodynamic processes dictate the assembly of fibers and particles from a polymer solution in a voltage-driven process	• Beneficial for sensitive, hydrophilic, and hydrophobic nutrients • Easy to control the particle size • High encapsulation efficiency
3D-printing	A biopolymer-based method for manufacturing three-dimensional solid structures from digital files	• High adaptability • Ability to use a variety of biopolymers

approach. Fibers are affected by changes in the fluid's polymer composition, molecular weight, and viscosity [6]. An electric charge builds up on the polymer solution's surface due to the voltage applied. The polymer droplets' surface tension is overcome by a critical voltage, resulting in the jet's release. Due to electrospun micro-nanofiber's biodegradability and biocompatibility, breakthrough products in the food, pharmaceutical, and biomedical industries have been challenging to develop [7]. Figure 5.1a depicts an electrospinning system schematic.

Electrospraying, also known as electrohydrodynamic atomization (EHDA), is a potential method for creating bioactive polymeric nanoparticles or fibers. Conventional methods have scalability, particle polydispersity, and encapsulation efficiency limitations, which this method eliminates. Using food technology to improve product quality is both easy and effective. The standard electrospraying arrangement may be shown in Figure 5.1b. This is done by spraying polymer solution with high voltage.

As a function of 3D-food printing, a wider variety of raw materials may create ready-to-eat and ready-to-process items [8]. In comparison with previous techniques of manufacturing objects, 3D-printing offers a wide range of possibilities. Complex interior structures with varied infill densities and patterns are possible with 3D-printing [9]. Compared with standard encapsulation technologies, process times may be decreased, which benefits food components and nutrients. Post-processing techniques, materials, and other variables all influence the quality and accuracy of printed products. Figure 5.2 displays two prevalent 3D-printing methods in the food industry.

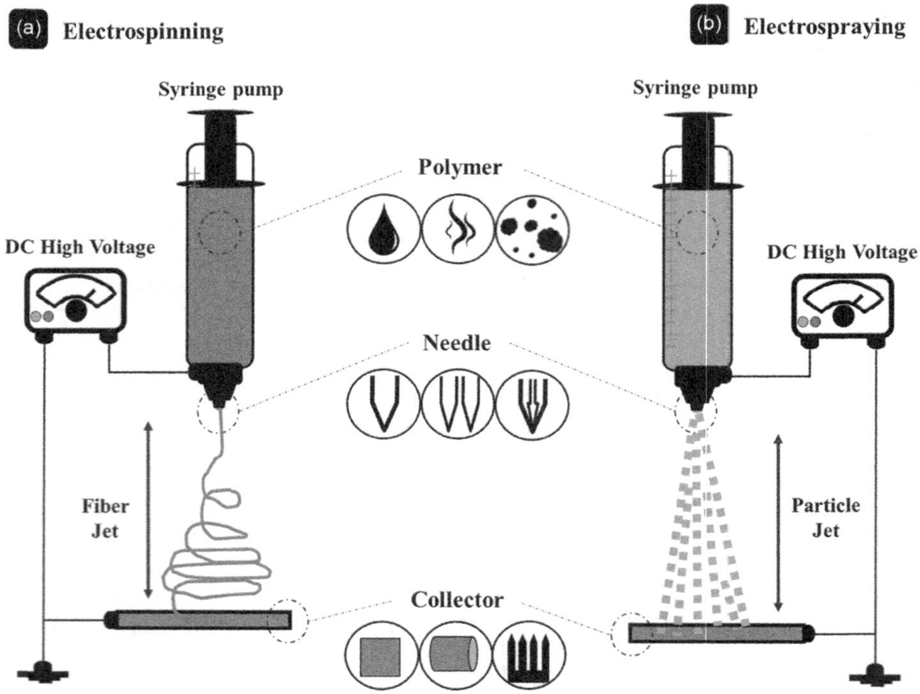

Figure 5.1 Schematic illustration of (a) electrospinning setup; (b) electrospraying setup.

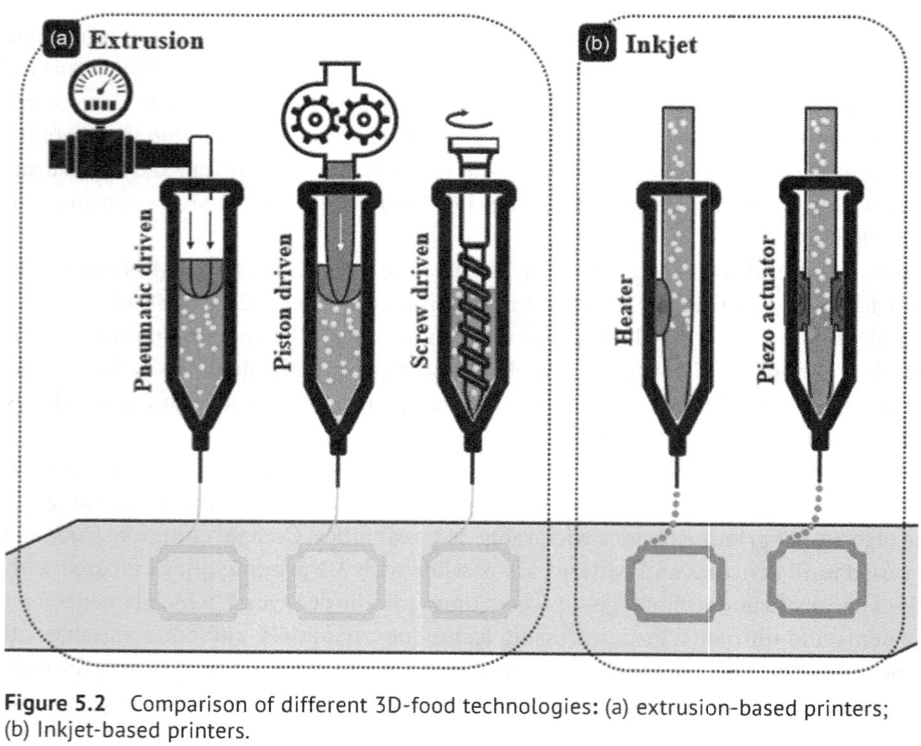

Figure 5.2 Comparison of different 3D-food technologies: (a) extrusion-based printers; (b) Inkjet-based printers.

Industrial Corner
• Proper selection of encapsulation methods can affect the effectiveness of food component encapsulation. • Electrospinning is an effective strategy to produce fibers as small as submicron or nanometers for food component encapsulation purposes. • 3D-printing extends the versatility and flexibility in food component encapsulation.

5.3 Materials Used as Carrier Source for Encapsulation

Proteins, carbohydrates, and lipids are some of the major types of materials used for food component encapsulation. Figure 5.3 shows encapsulant materials for proteins, carbohydrates, and lipids that may be employed alone or in combination to improve efficiency. Film-forming, non-hygroscopic materials are often used as encapsulants. Depending on the processing needs of the encapsulated component and when it is integrated into the food product, aqueous solubility or phase transition capabilities, such as melting or gelling, might be desirable. Other ingredients, including emulsifiers, plasticizers, and defoaming agents, are sometimes added to customize the final product. To obtain the appropriate

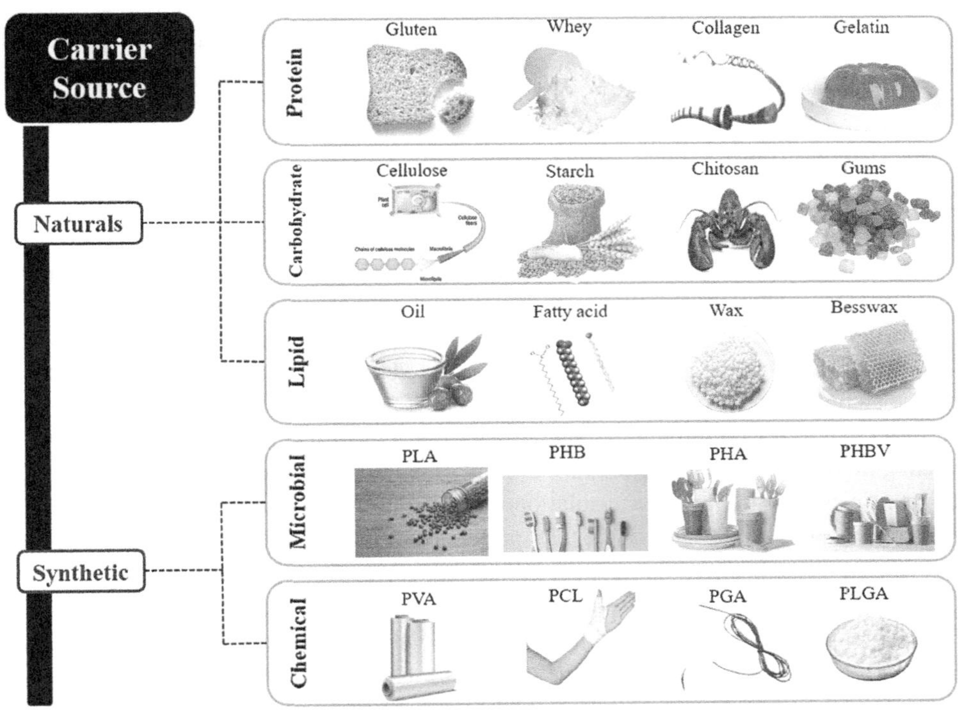

Figure 5.3 Carrier source for food component encapsulation. *Source:* Adobe stock photos/ Dotdash Media, Inc.

encapsulation matrix functionality, the encapsulant materials may also be physically or chemically changed. The choice of encapsulant material is consequently influenced by its physical and chemical qualities, compatibility with the intended food application, and sensory and aesthetic features.

5.4 Protein-Based Carriers

The most significant protein-based nanoparticles are safe and biodegradable [10]. The results of targeted strategies and preparatory procedures were also addressed. Hybrid nanoparticles and other components improve carriers of protein-based bioactive compounds. Proteins and carbohydrates are typically mixed with encasing food components. Molecular weight, amino acid composition, and quantity and type of amino acid residues dictate their orientation in the protein–polymer chain, directly impacting the carrier's quality derived from spherical proteins [1]. The type and sequencing of amino acid residues alter the chain-to-chain contact that creates cohesive carriers. A polymer chain with polar, hydrophobic, and/or thiol groups increases the possibility of interactions [11, 12].

5.4.1 Soy Proteins

A dilute neutral salt solution is required to dissolve most of the protein in soybeans since it is insoluble in water. As a result, soy protein is classified as a globulin. Asparagine and glutamine are found in significant concentrations in soy protein [13]. Nesterenko et al. illustrated that ascorbic acid and α-tocopherol encapsulation efficiency could be improved by the proper modification of soy protein isolate [14].

5.4.2 Cereal Proteins

Cereal proteins originating from maize, such as zein (also known as prolamine), are necessary for various biological processes. Cereal proteins that can form distinct nanostructures in different solvents owing to their superb self-assembly are not water-soluble, biodegradable, and biocompatible. Chen et al. showed it to be an effective vehicle for the co-delivery of synergistic curcumin and resveratrol by the zein–chitosan nano complex [1].

Water-insoluble proteins in wheat flour are referred to as "gluten," and they are made up of various polypeptide molecules called "globular proteins." Gliadin and glutenin are the common names for the prolamine and glutelin portions of wheat flour proteins found in wheat gluten. Wang et al. reported that deamidation improved the stability of gliadin-base nanoparticles for curcumin encapsulation [15]. Barley's storing protein is called hordein. Hordeins make about 35–50% of the total protein content in seeds. For a variety of reasons, they are crucial in the brewing process. Glutamate and proline, two essential amino acids, are found in hordenine proteins. He et al. found that hordein had a suitable structure for encapsulation of (−)-epigallocatechin-gallate [16]. Patatin, a protein found in potato, has a vigorous antioxidant activity and great emulsifying and foaming capabilities. It is also low-cost, GRAS-certified, and devoid of allergens. Edelman et al. highlight the great potential of patatin insolubilization, protection, and bioavailability improvement of lipophilic bioactive compounds, such as astaxanthin, for food and dietary supplements [17].

5.4.3 Egg Proteins

Egg whites and yolks include a variety of proteins with qualitative features that make them appropriate for controlled encapsulation and release mechanisms [18]. Wan et al. fabricated ovalbumin-derived biopolymer nanoparticles to encapsulate naringin, hesperidin, and anthocyanin [19].

5.4.4 Pulse Proteins

The main protein components of pulses, including peas, chickpeas, lentils, beans, and lupins, are globulins (water-soluble) and albumins (salt-soluble), with prolamins and glutelins as minor proteins. Pulse proteins have excellent foaming, emulsifying, and trapping properties [20]. Xu et al. found the characteristics of the chickpea protein isolate/stevioside/lutein complex to be of small average particle size, high emulsifying and encapsulation efficiency, and good stability [10].

5.4.5 Silk Proteins

Silk is a natural fiber generated in the glands of arthropods, such as silkworms, scorpions, bees, mites, and spiders (*Nephila clavipes*), having structural and functional similarities and distinctions [21]. The most well-known, characterized, and used silk is *Bombyx mori* silkworm silk (Bombycidae family), commonly known as mulberry silk; on the other hand, Saturniidae (*Antheraea mylitta*) makes non-mulberry silk. Fibroin and sericin are the two main proteins found in silk [22]. Silk proteins' chemical variety allows for site-specific functionalization through the ion addition of the chemical species [23]. He et al. found an alternative to RNA lyophilization based on aqueous fibroin to fabricate storage matrices [24].

5.4.6 Meat Proteins

Invertebrates and other species rely on collagen as a structurally essential protein. Collagen is widely accepted as a safe drug ingredient because of its biocompatibility, biodegradability, and low immunogenicity [23]. It is derived from the fibrous, insoluble protein collagen, which is present in abundance in the human body's skin, bones, and connective tissues. On cooling, gelatin aqueous solutions in the sol state produce physical, thermoreversible gels. Aytac et al. fabricated fast-dissolving electrospun gelatin nanofibers as the carrier of ciprofloxacin [7]. Also, Hadidi et al. suggested that the encapsulation of *L. acidophilus* in alginate/fish gelatin capsule had great potential to improve probiotic bacteria's survival during baking and storage and serve as an effective bread enhancer [3].

5.4.6.1 Milk Proteins

To synthesize nanoparticles from milk proteins, many manufacturing methods have been established. To create nanoparticles, the casein micelle structure may be reassembled from single casein molecules [25]. β-casein exhibited the most promising characteristics for developing encapsulation and hybrid encapsulation systems among the various casein fractions. Cheng et al. reported β-casein can bind *trans*- and/or *cis*-resveratrol to form β-casein–ligand complexes [11].

Whey proteins, unlike caseins, are thermostable and suffer thermal denaturation when heated beyond 70°C. Whole whey proteins and their fractions, such as β-lactoglobulin (β-LG) and α-lactalbumin (α-LA), may be used to make encapsulation systems. de Araújo Etchepare et al. suggested that multilayer microparticles based on calcium alginate/whey protein present an effective method to improve the probiotic viability under storage, thermal treatments, and gastrointestinal conditions [26]. In another study, Chavoshpour-Natanzia and Sahihi encapsulated quercetin-loaded β-lactoglobulin for drug delivery application [12].

5.5 Carbohydrate-Based Carriers

Carbohydrates have been employed extensively in food systems to distribute delicate components [4]. Carbohydrate acid modification has been found to increase the encapsulation capabilities of dietary components relative to the native carbohydrates. The physical alteration of carbohydrate characteristics by applying heat, shear, and pressure has also been investigated [27]. Modified carbohydrates have been combined with proteins to encapsulate nutritional components [9]. Carbohydrates have been used to encapsulate food components because they create hydrogels (Figure 5.4). Cross-linked polymeric networks known as hydrogels can absorb a significant quantity of water in a short period and retain moisture even under adverse conditions, such as high temperatures and pressures [28].

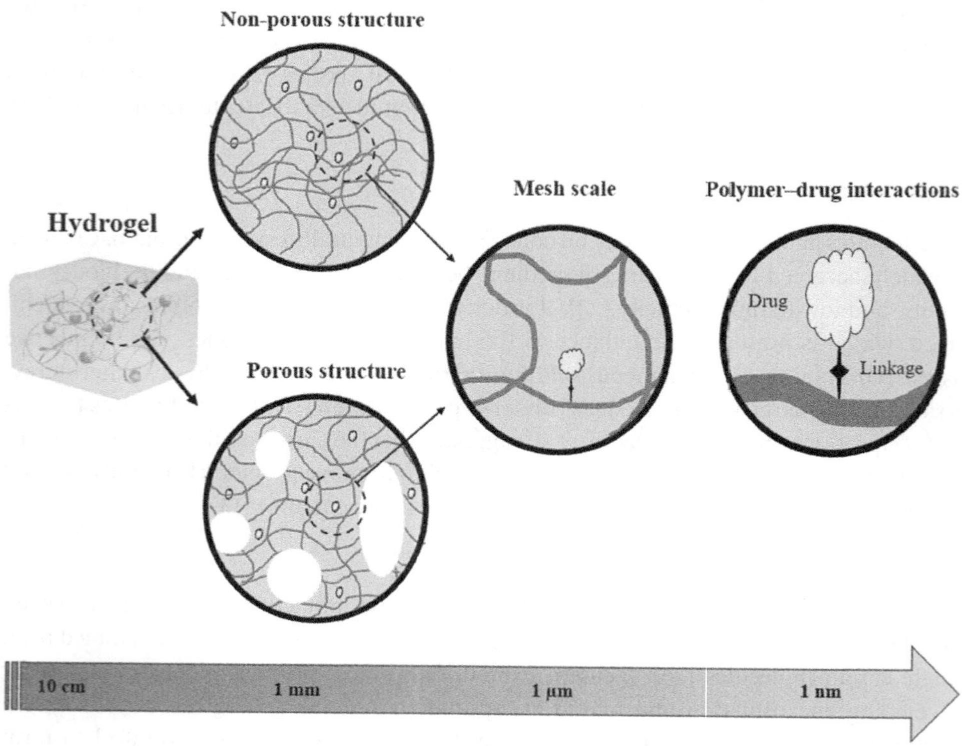

Figure 5.4 Multiscale properties of hydrogels.

5.5.1 Agar

It is possible to make a reversible gel out of agar by chilling a hot aqueous solution. When heated, agar gel melts and returns to its original state when cooled. Gelled agar chains may stabilize carrier-forming emulsions to form a macro network [29]. This macro network enhances barrier performance by trapping flattened lipid particles by enhancing tortuosity. Wu et al. designed an injectable hydrogel by dissolving MoS_2/Bi_2S_3-PEG, doxorubicin, and agar into the water for the concurrent tumor photothermal therapy and chemotherapy [30].

5.5.2 Starch

Anhydrous glucose units are the building blocks of starch, which is a polymeric carbohydrate. The straight chain molecule amylase and the branched glucose polymer amylopectin make up polysaccharides known as starch [31]. Another method for encapsulating food and flavor components is to use modified starches containing octenyl succinic anhydride. Li et al. demonstrate that porous maize starch granules allow for a high probiotic loading efficiency and provide enhanced protection to various stressful conditions compared with free cells [32].

5.5.3 Cellulose and Its Derivatives

The ß-1,4 glycosidic linkages connect D-glucose repeating units in the cellulose molecule. The hydroxymethyl groups of anhydroglucose residues may be found either above or below the polymer backbone in the structure of cellulose [33]. Oxygen and moisture transmission are moderately affected by the use of cellulose and its derivatives; carriers are generally tasteless and odorless, transparent, resistant to oil and fats, water-soluble, moderately resistant to moisture and oxygen transmission, and naturally odorless and tasteless [34]. Winuprasith et al.'s results indicated that mangosteen fiber can stabilize oil-in-water emulsions and only has a minor effect on lipid digestion and vitamin bioaccessibility when used at relatively low levels [35].

5.5.4 Pectin

Pectin is a linear anionic polysaccharide of α-D-(1→4) galacturonic acid residues that are linearly interconnected. Pectin's function as an emulsifier and encapsulation wall material has recently received greater attention. Ca^{2+} and CO_3^{2-} ions and glutaraldehyde were used to cross-link pectin nanospheres in the aqueous phase [36]. The highly porous nature of calcium pectinate carriers may result in poor entrapment efficiency and a rapid release of absorbed nutrients, particularly for hydrophilic, low molecular weight molecules [37]. Vancauwenberghe et al. developed a 3D-printing method based on the extrusion of bio-inks composed of a low-methoxylated pectin gel and embedded lettuce leaf cells [9].

5.5.5 Gums

It is possible to disperse a gum in cold or hot water and develop gels, very viscous mixtures, or solutions, depending on the molecular structure of the gum [38]. They are used in a range of food items to provide the desired stability and viscosity. The need for excellent quality and fat reduction has boosted the business of gums in the food industry [39].

Natural gums may be enhanced to provide bespoke goods for encapsulation systems, emulating the synthetic additives already on the market. Wissam et al. studied the encapsulation of flaxseed oil extract in the alginate–salep system by ionic gelation [40].

5.5.6 Chitin and Chitosan

In addition to crustaceans, fungi, and other organisms, chitin is the second most common biopolymer found in nature. Because it is structurally similar to cellulose, except for substituting an acryl group for the second carbon atom of the hexose repeat unit, poly ß-(1–4)-2-acetamide-D-glucose accounts for the vast majority of the substance's composition [41]. Chitosan can form semi-permeable coverings on fruits and vegetables, which may alter the interior environment to delay ripening and reduce transpiration rates [42]. They are transparent, substantial, flexible, and excellent oxygen barriers in aqueous chitosan. Hasheminejad et al. improved the antifungal activity of clove essential oil encapsulated by chitosan nanoparticles [43].

5.5.7 Alginate

Polysaccharides isolated from the cell walls of several brown algae species are known as alginates. β-D-mannuronic acid (M blocks) and β-L-guluronic acid (G blocks) residues are connected in varied ratios and sequential sequences to form this biopolymer [44]. Sarker et al. reported that alginate hydrogels covalently cross-linked with gelatin show the highest degree of cell adhesion, spreading, migration, proliferation, and a faster degradation rate and are a particularly suitable material for encapsulation [45].

5.5.8 Carrageenan

As a carrier-forming substance, carrageenans are water-soluble polymers made of partly sulfated galactans. When drying at moderate temperatures, the polysaccharide double helices of carrageenan form a three-dimensional network, and the solvent evaporates; the solid structure known as a hydrogel is developed [46]. Zhang et al. suggested that carrageenan hydrogel beads may be helpful to encapsulation systems, but further work is needed to inhibit lactase (β-galactosidase) enzyme leakage [47].

5.5.9 Dextran

Dextran is a bacterial glucan polysaccharide comprising glucose chains of various lengths linked mainly by α-(1→6) and a minor proportion of α-(1→3) connections. Dextran, a natural-derived polysaccharide made chiefly of glucose, has received much interest as a good biomacromolecule for making semi-synthetic hydrogels because of its biocompatibility and ease of chemical modification [48]. Alibolandi et al. reported the synthesis and preparation of biodegradable hybrid dextran hydrogel incorporated with biodegradable curcumin nanomicelles for full-thickness wound healing [4].

5.5.10 Cyclodextrins

Cyclodextrins (CDs) are cyclic oligosaccharides made from starch via enzymatic modification. The most popular ones are made up of 6–8 glucopyranose units and are referred to as α, β, and γ CD, respectively, with β-cyclodextrin (βCD) being the most researched and extensively used [49]. As a result, various CDs (αCD, βCD, and γCD) have varied entrapment capacities for the same molecule. CDs help encase dietary components that are poorly soluble, temperature-sensitive, or chemically labile [50]. Azzi et al. evaluated quercetin encapsulation in cyclodextrin conventional liposomes [51].

5.6 Lipid-Based Carrier

The compatibility of lipid particle matrices with the loaded food components must be addressed in many initiatives to build food industry lipid-based carriers [52]. Lipid-based carriers offer the benefits of industrial-scale biocompatible components, customizable physicochemical characteristics, and a wide range of compound entrapment capabilities. Aside from that, digestible fats are utilized to aid in the gastrointestinal absorption of interest compounds [53]. It is possible to classify the emulsions depending on emulsion, size, stability, and formulation processes [54, 55].

5.6.1 Emulsions

Emulsions have become a popular method of loading bioactive components into the food system since they are simple carriers. The emulsion may have a wide range of physical and chemical characteristics depending on the application. There are isotropic, clear colloidal dispersions in the emulsions. Nonpolar phase, polar phase, and surfactant are all required to make them [56]. As a cosurfactant or solvent, these extra components may be necessary for certain instances. Most micro- and nanoemulsions are more resistant to aggregation and gravity separation in large volumes than traditional emulsions [57]. Emulsions may be divided into O/W and W/O emulsions based on their arrangement, as seen in Figure 5.5. Although emulsions offer several advantages, they are not without certain drawbacks, the most significant one being the lack of formulation stability [58].

5.6.2 Liposomes

The pharmaceutical, food, agricultural, and cosmetics industries have found nano-sized delivery vehicles particularly enticing from their initial creation [59]. Liposomes are made by combining surfactants and lipids with water under low shear force. Surfactant sheets placed in a bilayer membrane present some water inside an emulsion-like vesicle of phospholipid [60].

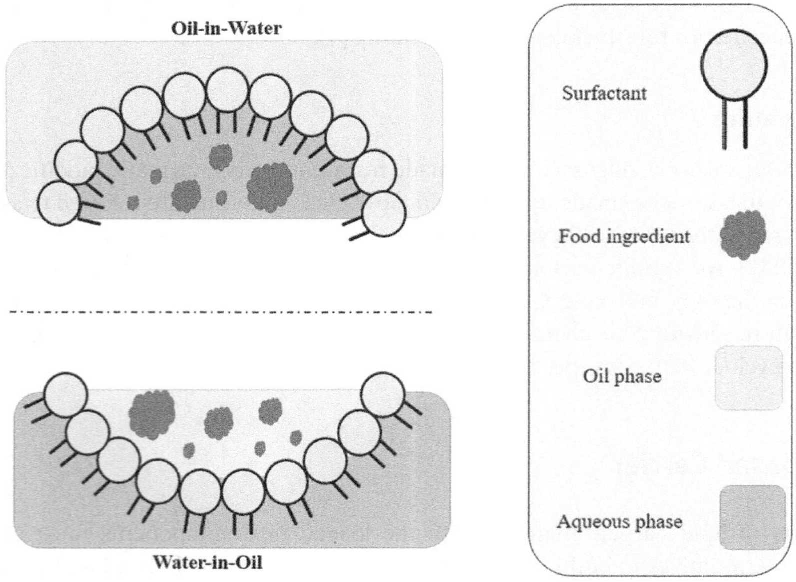

Figure 5.5 Schematic diagram of oil-in-water and water-in-oil emulsions.

Many studies have established the qualities of persistent antimicrobial-containing liposome vesicles. However, there are just a few studies on using bacteriocin-loaded liposomes in food. Liposomes can only be used as food component carriers to a limited extent because of the chemical and physical instability of aqueous dispersions, particularly over long periods [61].

5.6.3 Solid Lipid Nanoparticles (SLN)

It is possible to disintegrate a emulsion system if its composition and temperature vary. Lipophilic molecules are difficult to integrate into the particle structure of certain materials, which further complicates matters [62]. Liposomal drawbacks include a high price, poor payload, bioactive leakage, quick release, and stability issues during storage. To address the problems and limits of the existing lipid-based carriers, the SLN was designed as an alternate carrier system [63]. Particles in SLN typically have a size between 40 and 1000 nm. There are several benefits to employing physiological lipids in the formulation of SLN, including enhanced bioavailability, high physicochemical stability, and prolonged release [64]. SLN provides a longer half-life for molecules with lower diffusion rates. Large-scale manufacturing, biocompatibility, improved loaded bioactivity stability, and superior formulation stability is all made possible by the SLN [65].

5.6.4 Nanostructured Lipid Carriers (NLC)

To overcome the inherent limitations of SLN and other lipid-based carriers, NLC was specifically designed to accommodate the incorporation of bioactive compounds [66]. To prevent the burst release of active chemicals during storage, NLC has a lower water content than SLN [67]. The NLC is a mixture of liquid and solid lipids, while the SLN is solely a solid lipid. The substance placed into the NLC has a greater loading capacity and is released in a more controlled manner [68]. As a result of the disordered structure of liquid and solid lipids, the loaded substance can fit into the lipids [69]. Low crystallinity index and delayed polymorphic transition are the results of NLC's lipid combination.

Consequently, the NLC was shown to be a beneficial choice for increasing the controlled release and bioavailability of functional food components in addition to boosting encapsulation production and physical stability [66, 70]. While SLN contains just a solid lipid, NLC has a mixture of both liquid and solid lipids (Figure 5.6). Lipid-based carriers have been utilized to transfer food components effectively in the past. It is still necessary to do more studies on novel delivery methods that can meet the ever-changing needs of the food sector. For the most part, emulsions are transparent and stable because they are thermodynamically stable mixtures. The SLN and NLC's properties improve the colloidal stability of particle dispersions. Furthermore, the various lipid-based carriers for the delivery of food components are thoroughly explained. Table 5.2 lists several lipid-based encapsulating food components that have been authorized for multiple uses, including nutraceutical development, food supplements, and food preservation [8, 28, 52–55, 71–75].

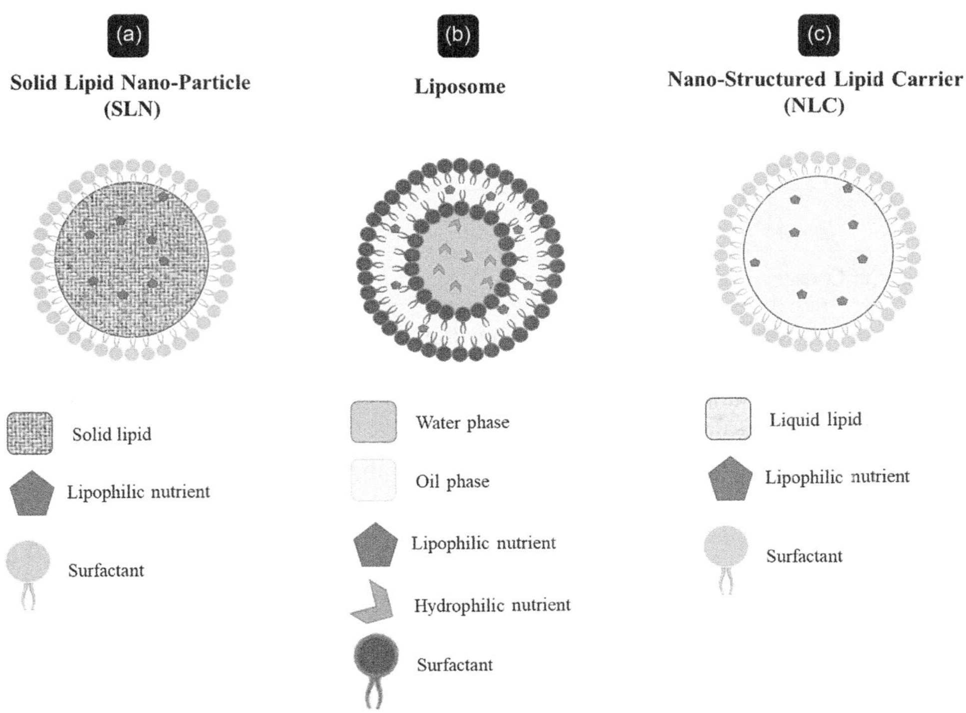

Figure 5.6 Structural features of (a) solid lipid nanoparticles; (b) liposomes; (c) nanostructured lipid carriers.

Table 5.2 Lipid-based encapsulating food components and their applications.

Type of carrier	Food component	Characterization	References
Nanoemulsion	Gallic acid	Nanoemulsion increased the in vitro transport of gallic acid	[71]
Nanoemulsion	Astaxanthin	Nanoemulsion enhanced astaxanthin bioavailability in functional foods	[72]
Microemulsion	Resveratrol and coenzyme Q10	Coenzyme Q10 protected by resveratrol-loaded nanoparticles	[52]
NLC	Eugenol	Lower toxicity of NLC on human cultured cells	[53]
NLC	Vitamin D3	NLCs protected vitamin D3 under simulated gastric conditions	[73]
NLC	β-carotene	NLC developed an effective inhibition zone against bacteria growth	[74]
SLN	Echium oil	Lauric acid was more effective in protecting Echium oil against oxidation than both palmitic and stearic acids	[55]
SLN	β-lactoglobulin	The presence of α-tocopherol at the SLN surface enhances protein adsorption	[75]
SLN	Vulgaris essential oil	SLN shows cytotoxicity effect on breast cancer cell line (MCF7)	[54]
Liposome	Linseed oil and quercetin	Chemical stability of bioactive compounds was improved by loading liposome into hydrogel bead	[28]
Liposome	Curcumin	3D-printed scaffold eradicated the osteosarcoma cells and promoted osteoblast proliferation	[8]

Industrial Corner

- Natural polymers, such as proteins, carbohydrates, and lipids, can be used alone or in combination to improve encapsulation efficiency.
- The appropriate industry varies for each encapsulated structure depending on the carrier polymer source.
- By using polymers to enclose food components, many industrial applications can be designed for the created structure.
- Numerous strategies for enclosing bioactive compounds in polymers for food and pharmaceutical applications have been investigated.

5.7 Roles Played by Materials in Food Component Encapsulation

As presented in the section above, different types of materials can be used to encapsulate food components. In this section, we will talk about how various practical goals can be achieved by using carrier materials in the food production process.

5.8 Improved Dispersibility

Many lipophilic bioactive compounds cannot be integrated into water-based food items, which is a severe drawback. Examples include polyunsaturated fatty acid-rich oil-based foods that cannot be readily dissolved in water because of their immiscibility and subsequent separation [76]. However, it is possible to scatter the lipid droplets into an aqueous solution if they are first transformed into an oil-in-water emulsion of nanoemulsion. Nanoscale delivery technologies, such as bioactive-fortified particles dispersed in water, are viable options [77].

5.9 Addition of Inhibitors

The various reactive species must be near one other such species during a chemical reaction, like lipophilic bioactive, whose breakdown is mediated by a mixture of nonpolar and polar reactants [78]. The degradation rate of citral lowers when oil droplet concentration, solid fat content, or surfactant micelle concentration is adjusted to enhance citral's concentration in the oil phase [79]. It is possible to reduce the activity of many reactions by adding specific components that interact with the reactants or intermediates as encapsulant systems and reduce their activity [80].

5.10 Reducing the Interactions

Creating impermeable barriers inside a nanoscale delivery system is another option for limiting the interactions between chemical reactants. Some nanoparticles may be covered with a material that blocks the diffusion of one or more reactants into or out of the particles [81]. The kind of reactants involved will determine the type of material utilized. Embedding lipid droplets in a protein-rich microgel has been found to lower the rate of lipid oxidation in oil-in-water emulsions [82].

5.11 Control of Light Scattering and Absorption

Many beneficial compounds degrade chemically when exposed to UV or visible light. Incorporating nanoparticles that absorb or scatter light may help increase the stability of these compounds by lowering the quantity of light that reaches bioactivity [83]. Recent investigations have revealed that nanoemulsions may increase the UV stability of resveratrol. The oil–surfactant combination had to be titrated into water [77].

5.12 Increased Bioavailability

It is possible to build efficient delivery systems by optimizing droplet composition, size, and interfacial qualities, promoting bioavailability. According to simulated GIT investigations, digestible emulsified oils have better bioavailability of coenzyme Q10 than indigestible ones

[84]. The size of the particles encasing lipophilic bioactivity affects their bioavailability in nanoscale delivery systems. According to these findings, emulsions with smaller lipid droplets are digested more quickly and completely [85].

5.13 Controlled or Targeted Release

Nanoscale delivery methods might also administer lipophilic bioactive components in a regulated or targeted manner [86]. Many factors may cause a release to occur, including, but not limited to, time, dilution, pH, ionic strength, temperature, mechanical forces, and enzyme activities. Nanoscale delivery methods may be engineered to release lipophilic active compounds in the mouth, stomach, small intestine, or colon for GIT applications [83]. In particular situations, delivery systems should hold the bioactive chemicals and then release them when the conditions are right. Due to their ease and simplicity, many possible delivery systems may be quickly screened using in vitro techniques; only the most promising options are investigated in vivo due to ethical, time, and economic considerations [86].

5.14 Conclusions

In the food industry, nutrient encapsulation systems have become increasingly popular in recent years due to the rise in consumer demand for more nutritious and healthier foods. Food-grade materials have been used to achieve several vital breakthroughs in the encapsulation of food components. Materials that are safe for human consumption are being developed and used across the food supply chain as effective encapsulation methods for nutrients. With the emergence of innovative approaches such as 3D-printers and electrohydrodynamic techniques to fabricate the encapsulation systems needed to achieve the desired features, we expect that food materials science will play an increasingly important role in food applications.

Glossary

3D-printing Variety of processes in which material is deposited, joined, or solidified under computer control to create a three-dimensional object.

Bioavailability The rate and extent to which the active constituent or active moiety of a drug is absorbed from a drug product and reaches circulation.

Biodegradable Able to decay naturally and in a way that is not harmful.

Curcumin (1,7-bis(4-hydroxy-3-methoxyphenyl)-1,6-heptadiene-3,5-dione), also called diferuloylmethane, is the primary natural polyphenol found in the rhizome of *Curcuma* spp.

Drug delivery Approaches, formulations, manufacturing techniques, storage systems, and technologies involved in transporting a pharmaceutical compound to its target site to achieve a desired therapeutic effect.

Electrohydrodynamics The study of the dynamics of electrically charged fluids.
Emulsion Mixture of two or more liquids in which one is present as droplets, of microscopic or ultramicroscopic size, distributed throughout the other.
Encapsulation Process of stabilization of active compounds through the structuring of systems capable of preservation of their properties.
Essential oil Concentrated hydrophobic liquid containing volatile chemical compounds extracted from plants.
Food component Essential nutrients like carbohydrates, proteins, fats, vitamins, and minerals.
Gum A viscous secretion of some trees and shrubs that hardens on drying but is soluble in water, and from which adhesives and other products are made.
Hydrogel Three-dimensional (3D) network structures able to imbibe large amounts of water.
Polysaccharide Polymers consisting of chains of monosaccharide or disaccharide units joined by glycosidic bonds.
Polyunsaturated fatty acid Fatty acids that contain more than one double bond in their backbone.
Surfactant Materials that lower the surface tension (or interfacial tension) between two liquids or between a liquid and a solid.
UV Ultraviolet is a form of electromagnetic radiation with wavelength from 10 to 400 nm, shorter than that of visible light.

References

1 S. Chen, Y. Han, L. Jian, W. Liao, Y. Zhang, Y. Gao, *Carbohydr. Polym.* **2020**, *236*, 116090.
2 M. Shakouri, M. Salami, L.-T. Lim, M. Ekrami, M. Mohammadian, G. Askari, Z. Emam-Djomeh, D. J. McClements, *J. Food Meas. Charact.* **2022**, 1.
3 M. Mirzakhani, M. Ekrami, S. Moini, *JFBE* **2018**, *1*, 8.
4 M. Ekrami, Z. Emam-Djomeh, P. Joolaei-Ahranjani, S. Mahmoodi and S. Khaleghi, *Int. J. Biol. Macromol.* **2021**, *168*, 591.
5 A. Bratovcic, J. Suljagic, *Croat. J. Food Sci. Technol.* **2019**, *11*, 113.
6 S. C. Coelho, B. N. Estevinho, F. Rocha, *Food Chem.* **2021**, *339*, 127850.
7 Z. Aytac, S. Ipek, I. Erol, E. Durgun, T. Uyar, *Colloids Surf. B: Biointerfaces* **2019**, *178*, 129.
8 N. Sarkar, S. Bose, *ACS Applied Mat. Inter.* **2019**, *11*, 17184.
9 V. Vancauwenberghe, V. B. M. Mbong, E. Vanstreels, P. Verboven, J. Lammertyn, B. Nicolai, *J. Food Eng.* **2019**, *263*, 454.
10 Y. Xu, J. Song, Z. Dai, L. Niu, L. Dajing, C. Wu, *J. Sci. Food Agric.* **2021**, *102*, 1872.
11 H. Cheng, H. Dong, L. Liang, *Food Chem.* **2020**, *330*, 127209.
12 Z. Chavoshpour-Natanzi, M. Sahihi, *Food Hydrocoll.* **2019**, *96*, 493.
13 Z. Teng, Y. Luo, Q. Wang, *J. Agri. Food Chem.* **2012**, *60*, 2712.
14 A. Nesterenko, I. Alric, F. Silvestre, V. Durrieu, *Food Hydrocoll.* **2014**, *38*, 172.
15 Y. Wang, W. Yan, R. Li, X. Jia, Y. Cheng, *J. Food Eng.* **2019**, *260*, 30.
16 A. He, X. Guan, H. Song, S. Li, K. Huang, *Food Biosci.* **2020**, *37*, 100727.

17 R. Edelman, S. Engelberg, L. Fahoum, E. G. Meyron-Holtz, Y. D. Livney, *Food Hydrocoll.* **2019**, *96*, 72.

18 M. Mohammadian, M. I. Waly, M. Moghadam, Z. Emam-Djomeh, M. Salami, A. A. Moosavi-Movahedi, *Food Sci. Hum. Wellness* **2020**, *9*, 199.

19 M. Wan, Z. Huang, X. Yang, Q. Chen, L. Chen, S. Liang, Q. Zeng, R. Zhang, L. Dong, D. Su, *Int. J. Food Sci. Technol.* **2022**, *57*, 643.

20 K. Shevkani, N. Singh, Y. Chen, A. Kaur, L. Yu, *J. Food Sci. Technol.* **2019**, *56*, 2787.

21 S. Rahaiee, M. Zare, S. M. Jafari, *Biopolymer Nanostructures for Food Encapsulation Purposes, Elsevier*, **2019**, 305.

22 R. Elia, J. Guo, S. Budijono, V. Normand, D. Benczédi, F. Omenetto, D. L. Kaplan, *J. Coat. Technol. Res.* **2015**, *12*, 793.

23 M. Fathi, F. Donsi, D. J. McClements, *Compr. Rev. Food Sci. Food Saf.* **2018**, *17*, 920.

24 J. He, B. Yavuz, J. A. Kluge, A. B. Li, F. G. Omenetto, D. L. Kaplan, *ACS Biomater. Sci. Eng.* **2018**, *4*, 1708.

25 G. M. Tavares, T. Croguennec, A. F. Carvalho, S. Bouhallab, *Trends in Food Sci. Technol.* **2014**, *37*, 5.

26 M. de Araujo Etchepare, G. L. Nunes, B. R. Nicoloso, J. S. Barin, E. M. M. Flores, R. de Oliveira Mello, C. R. de Menezes, *LWT* **2020**, *117*, 108601.

27 M. Fathi, A. Martin, D. J. McClements, *Trends in Food Sci. Technol.* **2014**, *39*, 18.

28 J. Huang, Q. Wang, L. Chu, Q. Xia, *LWT* **2020**, *117*, 108615.

29 D. Zhang, M. Zhang, X. Gu, *Bioactive Seaweeds for Food Applications*, Elsevier, **2018**, 153.

30 C. Wu, J. Zhao, F. Hu, Y. Zheng, H. Yang, S. Pan, S. Shi, X. Chen, S. Wang, *Carbohydrate Polym.* **2018**, *180*, 112.

31 F. Zhu, *Food Chem.* **2017**, *229*, 542.

32 H. Li, M. S. Turner, S. Dhital, *Lwt* **2016**, *74*, 542.

33 Y. Liu, S. Ahmed, D. E. Sameen, Y. Wang, R. Lu, J. Dai, S. Li, W. Qin, *Trends in Food Sci. Technol.* **2021**, *112*, 532.

34 X. He, W. Lu, C. Sun, H. Khalesi, A. Mata, R. Andaleeb, Y. Fang, *Carbohydrate Polym.* **2021**, *255*, 117334.

35 T. Winuprasith, P. Khomein, W. Mitbumrung, M. Suphantharika, A. Nitithamyong, D. J. McClements, *Food Hydrocoll.* **2018**, *83*, 153.

36 P. Ishwarya, S. P. Nisha, *Cri. Rev. Food Sci. Nut.* **2021**, *1*.

37 A. Rehman, T. Ahmad, R. M. Aadil, M. J. Spotti, A. M. Bakry, I. M. Khan, L. Zhao, T. Riaz, Q. Tong, *Trends in Food Sci. Technol.* **2019**, *90*, 35.

38 M. Ekrami, Z. Emam-Djomeh, *J. Food Process. Preserv.* **2014**, *38*, 1812.

39 M. Ekrami, Z. Emam-Djomeh, S. A. Ghoreishy, Z. Najari, N. Shakoury, *Int. J. Biol. Macromol.* **2019**, *133*, 529.

40 M. Ekrami, N. Roshani-Dehlaghi, A. Ekrami, M. Shakouri, Z. Emam-Djomeh, *Chem.* **2022**, *4*, 1360.

41 S. Akbari-Alavijeh, R. Shaddel, S. M. Jafari, *Biopolymer Nanostructures for Food Encapsulation Purposes*, Elsevier, **2019**, Vol. *381*.

42 V. L. Kabanov, L. V. Novinyuk, *Food Sys.* **2020**, *3*, 10.

43 N. Hasheminejad, F. Khodaiyan, M. Safari, *Food Chem.* **2019**, *275*, 113.

44 E. Martins, D. Poncelet, R. C. Rodrigues, D. Renard, *J. Microencapsul.* **2017**, *34*, 754.

45 B. Sarker, J. Rompf, R. Silva, N. Lang, R. Detsch, J. Kaschta, B. Fabry, A. R. Boccaccini, *Int. J. Biol. Macromol.* **2015**, *78*, 72.

46 S. Chakraborty, *J. Carbohydr. Chem.* **2017**, *36*, 1.

47 Z. Zhang, R. Zhang, L. Chen, D. J. McClements, *Food Chem.* **2016**, *200*, 69.

48 M. Fathi, M. N. Nasrabadi, J. Varshosaz, *Int. J. Food Prop.* **2017**, *20*, 2665.

49 L. Szente, J. Szejtli, *Trends in Food Sci. Technol.* **2004**, *15*, 137.

50 C. Dos Santos, P. Buera, F. Mazzobre, *Curr. Opin. Food Sci.* **2017**, *16*, 106.

51 J. Azzi, A. Jraij, L. Auezova, S. Fourmentin, H. Greige-Gerges, *Food Hydrocoll.* **2018**, *81*, 328.

52 Y. Wei, Z. Yu, K. Lin, S. Yang, K. Tai, J. Liu, L. Mao, F. Yuan, Y. Gao, *J. Agr. Food Chem.* **2020**, *68*, 1405.

53 R. Cortesi, G. Valacchi, X. M. Muresan, M. Drechsler, C. Contado, E. Esposito, A. Grandini, A. Guerrini, G. Forlani, G. Sacchetti, *J. Microencapsulation* **2017**, *34*, 63.

54 M. Teimouri, M. Odoumizadeh, *Archives of Adv. Biosci.* **2021**, *12*, 11.

55 M. Azizi, A. Kierulf, M. C. Lee, A. Abbaspourrad, *Food chemistry* **2018**, 246, 448.

56 W. W. Mwangi, H. P. Lim, L. E. Low, B. T. Tey, E. S. Chan, *Trends in Food Sci. Technol.* **2020**, *100*, 320.

57 M. S. Nejat, M. Ekrami, Z. Emam-Djomeh, *Biopolymers in Nutraceuticals and Functional Foods, RSC* **2022**, 232.

58 R. Zhang, T. Belwal, L. Li, X. Lin, Y. Xu, Z. Luo, *Carbohydr. Polym.* **2020**, *242*, 116388.

59 M. Ekrami, A. Ekrami, M. A. Hosseini, Z. Emam-Djomeh, *Mol.* **2022**, *27*, 7032.

60 N. Karimi, B. Ghanbarzadeh, H. Hamishehkar, F. Keyvani, A. Pezeshki, M. M. Gholian, *Appl. Food Biotechnol.* **2015**, *2*, 17.

61 M. Ekrami, A. Ekrami, R. H. Moghadam, P. Joolaei-Ahranjani, Z. Emam-Djomeh, *Biopolymers in Nutraceuticals and Functional Foods, RSC* **2022**, 488.

62 H. Salminen, C. Gommel, B. H. Leuenberger, J. Weiss, *Food Chem.* **2016**, *190*, 928.

63 M. Ekrami, A. Ekrami, R. Esmaeily, Z. Emam-Djomeh, *Biopolymers in Nutraceuticals and Functional Foods, RSC* **2022**, 545.

64 H. Salminen, J. Ankenbrand, B. Zeeb, G. B. Bonisch, C. Schafer, R. Kohlus, J. Weiss, *Food Res. Int.* **2019**, *119*, 741.

65 K. Oehlke, D. Behsnilian, E. Mayer-Miebach, P. G. Weidler, R. Greiner, *PloS one* **2017**, *12*, e0171662.

66 F. Tamjidi, M. Shahedi, J. Varshosaz, A. Nasirpour, *Innovative Food Sci. Emerging Technol.* **2013**, *19*, 29.

67 M. Mohammadi, E. Assadpour, S. M. Jafari, *Lipid-based Nanostructures for Food Encapsulation Purposes*, Elsevier, **2019**, 217.

68 O. P. Almeida, M. B. de Freitas Marques, J. P. de Oliveira, J. M. G. da Costa, A. E. Rodrigues, M. I. Yoshida, W. d. N. Mussel, G. Carneiro, *J. Food Sci. Technol.* **2022**, *59*, 805.

69 F. A. N. Azar, A. Pezeshki, B. Ghanbarzadeh, H. Hamishehkar, M. Mohammadi, *J. Agri. Food Res.* **2020**, *2*, 100084.

70 J. Huang, Q. Wang, T. Li, N. Xia, Q. Xia, *J. Food Engineering* **2017**, *215*, 1.

71 E. Mitsou, V. Pletsa, G. T. Sotiroudis, P. Panine, M. Zoumpanioti, A. Xenakis, *Coll. Sur. B: Biointer.* **2020**, *190*, 110974.

72 X. Liu, R. Zhang, D. J. McClements, F. Li, H. Liu, Y. Cao, H. Xiao, *Food Biophys.* **2018**, *13*, 412.

73 S. J. Park, C. V. Garcia, G. H. Shin, J. T. Kim, *Food Chem.* **2017**, *225*, 213.

74 I. Lacatusu, N. Badea, O. Ovidiu, D. Bojin, A. Meghea, *J. Nanoparticle Res.* **2012**, *14*, 1.

75 K. Oehlke, J. K. Keppler, J. Milsmann, E. Mayer-Miebach, R. Greiner, A. Steffen-Heins, *J. Food Engineer.* **2019**, *247*, 144.

76 C. Bao, P. Jiang, J. Chai, Y. Jiang, D. Li, W. Bao, B. Liu, B. Liu, W. Norde, Y. Li, *Food Res. Int.* **2019**, *120*, 130.

77 D. J. McClements, *J. Food Sci.* **2015**, *80*, N1602.

78 Q. Liu, H. Huang, H. Chen, J. Lin, Q. Wang, *Molecules* **2019**, *24*, 4242.

79 M. Zanetti, T. K. Carniel, F. Dalcanton, R. S. Dos Anjos, H. G. Riella, P. H. de Araujo, D. de Oliveira, M. A. Fiori, *Trends Food Sci. Technol.* **2018**, *81*, 51.

80 I. Simionato, F. C. Domingues, C. Nerin, F. Silva, *Food and Chem. Toxico.* **2019**, *132*, 110647.

81 A. Mohan, S. R. Rajendran, Q. S. He, L. Bazinet, C. C. Udenigwe, *Rsc Adv.* **2015**, *5*, 79270.

82 N. Eghbal, R. Choudhary, *Lwt* **2018**, *90*, 254.

83 I. Katouzian, S. M. Jafari, *Trends Food Sci. Technol.* **2016**, *53*, 34.

84 J. Grgić, G. Šelo, M. Planinić, M. Tišma, A. Bucić-Kojić, *Antioxidants* **2020**, *9*, 923.

85 C. Soukoulis, T. Bohn, *Cri. Rev. Food Sci. and Nutrition* **2018**, *58*, 1.

86 A. Ekrami, M. Ghadermazi, M. Ekrami, M. A. Hosseini, Z. Emam-Djomeh, R. Hamidi-Moghadam, *J. Drug Deliv. Sci. Technol.* **2022**, *69*, 103148.

6

Design and Use of Microcarriers for the Delivery of Nutraceuticals

Maxim V. Kiryukhin, Su Hui Lim, and Cheryl Yingxue Chia*

*Singapore Institute of Food and Biotechnology Innovation, Agency for Science, Technology and Research (A*STAR), 31 Biopolis Way, #01-02, Nanos, Singapore 138669*
** Corresponding author*

Highlights

- Vitamin D_3 encapsulated in casein micelles does not degrade upon heat treatment, only 10% is lost after 28 days of storage, and 85% survives digestion. It was used to fortify low fat milk and demonstrated high D_3 bioavailability in a randomized double-blinded placebo-controlled clinical trial with 87 human volunteers.
- A functional whey protein, lactoferrin, is protected from gastric digestion via encapsulation in protein-polyphenol multilayer microcapsules. Shells erosion by α-chymotrypsin triggers lactoferrin release in the small intestine, providing 6.5 times higher local concentration compared with a control group dosed with the same amount of unprotected lactoferrin.
- Complexation of vitamin B_{12} with recombinant human intrinsic factor enables its active absorption through the natural receptor, cubam, for people with pernicious anemia. A clinical trial demonstrated a significantly higher level of holo-transcobalamin in serum a day after the intake of encapsulated B_{12} complex among nine patients with evident signs of B_{12} deficiency.

6.1 Introduction

Microencapsulation stands for the encapsulation of ingredients in micron- or submicron-sized carriers. In the food industry, the intended benefits of microencapsulation include masking the odor or unpleasant taste of certain *micronutrients* (e.g. vitamins or minerals), making them more appealing to customers; protection of active ingredients from degradation when exposed to harsh environmental conditions upon processing or storage (toward extended shelf life), and retention of highly volatile flavors. Reducing the size of

Materials Science and Engineering in Food Product Development, First Edition. Edited by Wing-Fu Lai.

carriers toward the submicron range helps to improve the transparency of fortified food products or hinder the unwanted color while maintaining the same weight percentage of an ingredient.

The two main microencapsulation approaches, named core-shell and matrix, are overviewed in Table 6.1. In the core-shell approach, an ingredient (core) is encased in a continuous shell made of an encapsulating material and is typically completely isolated from contact with the surrounding environment. The approach allows high loading

Table 6.1 General overview of core-shell and matrix encapsulation.

	Core-shell encapsulation	Matrix encapsulation
Schematic	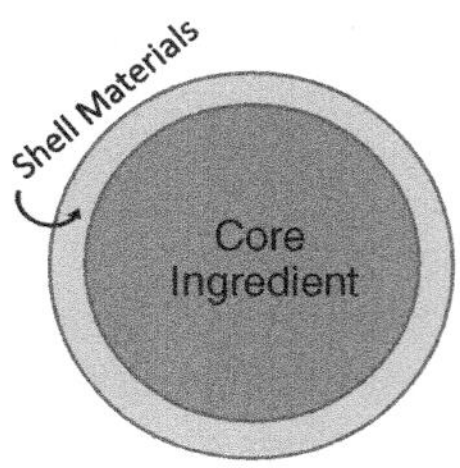	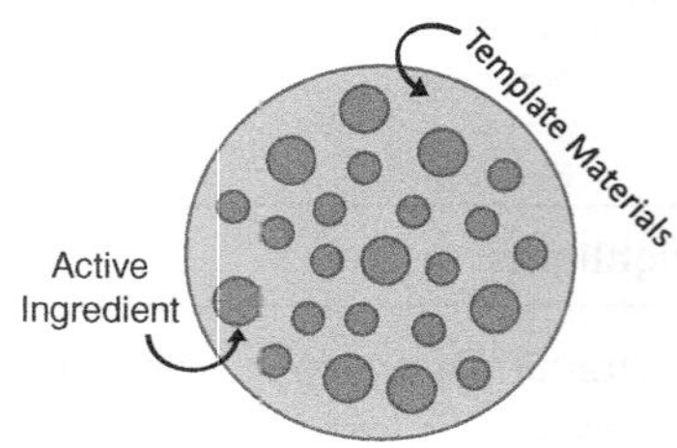
Principle differences	Active ingredients (core) are encased in a continuous shell made of encapsulating material and are isolated from contact with the surrounding environment The shell erodes under specific environmental conditions (pH, temperature, enzymes, etc.), triggering the release of active ingredients	Active ingredients are homogeneously dispersed in a template material that makes the carrier particle The matrix undergoes bulk erosion releasing active ingredients from the exposed surface
Advantages	High loading capacities as coating-to-core volume ratios are typically low A wide range of shell materials can be customized, for various release triggers and target receptors, providing better control over the release site Release mechanism is ideal for burst release applications	More direct methodology for continuous production at a scale Typically, they are more durable under mechanical stress Release mechanism is ideal for sustained release applications
Disadvantages	Time-consuming, often multistep production methodology Typically, less durable under mechanical stress posing higher risk of entire payload leakage	Low loading capacity as active ingredient makes just a fraction of the template matrix Leakage or degradation of active ingredients exposed at the surface of matrix May have issues with homogeneous distribution
Examples in food industry	Complex coacervation Interfacial polymerization in emulsions Hot melt or top-spray fluidized bed coating	Coprecipitation Ionic or thermal gelation Spray drying

capacities as the ingredient occupies the whole inner volume of microcapsules and coating-to-core volume ratios are typically low. Some examples of such microencapsulation techniques are complex coacervation, interfacial polymerization in oil-in-oil or oil-in-water emulsions, some fluidized bed technologies, such as hot melt or top-spray fluidized bed coating. In the matrix approach, an ingredient is dispersed in a template material that makes the carrier particle. Here, some portion of the ingredient may appear at the carriers' surface being in direct contact with the surrounding media while the rest is hidden underneath. The loading capacities are typically lower as the active ingredient makes just a fraction of the template matrix. Absorption of active ingredients in the templates with high open porosity, ionic or thermal gelation, spray drying or coprecipitation, e.g. in supercritical carbon dioxide, are some examples resulting in such microcapsules.

A new and emerging application of microencapsulation is aimed at enhancing the *bioavailability* of active ingredients. Considerable progress was achieved over the past decade in the development of smart micron- and submicron-sized carriers for precision vaccine and drug delivery. The most sophisticated systems combine the protective, targeting, and controlled release functions. Some examples include block-copolymer *micelles*, dendrimers, liposomes, and layer-by-layer assembled systems, as summarized in several recent reviews [1–3]. By 2020, about 21 of them had been approved by the Food and Drug Administration (FDA) or European Medicines Agency (EMA), while hundreds are undergoing clinical trials and thousands are under development [4]. More recently, a number of research groups attempted to translate the developed technologies toward an emerging area of functional foods that benefits go beyond basic nutritional function. Interestingly, a recent survey has shown that consumers regard it justifiable to consume functional foods aimed at boosting a weakened immune system, alleviating high cholesterol levels, or preventing digestive problems [5].

In this chapter, we will discuss the latest achievements in the area of micron- and submicron-sized carriers for the oral delivery of essential M&Ns that were summarized in a number of beautiful reviews [6–13], largely focusing on the works that clearly demonstrate how carriers enhance the bioavailability of M&Ns. Not surprisingly, the most common materials for the carriers aimed at food applications are carbohydrates [14–16], proteins, [17–19], and lipids [20–23]. First, we will discuss the protective function of the carriers preventing degradation of M&Ns upon processing, storage, and premature digestion in the upper sections of the gastrointestinal tract (GIT). Second, we will touch on the ways of providing control over the release of M&Ns from the carriers when they reach a part of the GIT most beneficial for their bioavailability. These two approaches toward enhanced bioavailability of M&Ns may be called "passive." Third, we will overview the approaches, where the carriers play an active role in the enhancement of M&Ns' bioavailability through an increased residence time of M&Ns at the right part of the GIT (*mucoadhesive* carriers), improved permeation through *epithelial* barrier often enabling novel transport pathways (permeation enhancers), selective binding to the identified receptors of choice, or even overcoming issues with impaired absorption of M&Ns, arising from a variety of physiological factors. Finally, we will give our vision of the future trends in the area of personalized functional foods, overcoming malabsorption of M&Ns, the trigger and adverse outcome of many chronic diseases and frailty.

Industrial Corner – Market Overview

- The major food product categories fortified with microencapsulated ingredients are dairy products, infant formula, confectionary, breakfast cereals, bakery, and vegetable oils.
- In 2019, the global revenue of microencapsulated products in the food additives industry was US$802M, compared with US$6.2B in the pharmaceutical industry. The main barriers to penetration include:
  1) Lack of technologies developed for targeted oral delivery (compared with those developed for subcutaneous injections).
  2) High fixed costs of microencapsulation.
- However, the five-year compound annual growth rate of microencapsulated products revenue in food additives was 14.5% (!) even outpacing the CAGR of 11.7% in the pharmaceutical industry. The growth is driven by the following value proposition:
  1) superior product claims (due to higher efficacy)
  2) lower cost of materials (due to improved bioavailability of high-value active ingredients that allows reduced dosages).

6.2 Protection Against Environmental Conditions

Many M&Ns are not stable at specific environmental conditions. For example, they may be subjected to oxidation promoted by change in pH and exposure to light, heat, or moisture. When consumed, they may be susceptible to premature digestive degradation in the upper parts of the GIT. Therefore, they require protection to retain bioavailability and preserve their function. Encapsulation in micron- and submicron-sized carriers often serves this purpose. Table 6.2 lists a few examples illustrating this approach.

Polyunsaturated fatty acids (PUFA) play an important role in human physiology and can only be obtained through diet. However, they are vulnerable to oxidation when exposed to air. Encapsulation in 100 nm liposomes was shown to inhibit oxidation of shrimp oil rich in eicosapentaenoic acid and docosahexaenoic acid [24]. PUFA content in that sample decreased by 16% upon storage over a period of 8 weeks at 30°C (compared with a 67% decrease in the control sample of unencapsulated oil). High efficiency of the inhibition was achieved through oil localization at the core of the liposomes rather than between the bilayers. Anthocyanins are food pigments of plant origin that also possess antioxidant and immunomodulatory functions. Chemically, they are glycosides of flavylium cations prone to a nucleophilic attack that makes them unstable in aqueous media, particularly under the slightly alkaline conditions of the small intestine. For example, only ~50% of anthocyanins survive 28 days of storage in the dark (pH 5, 25°C), and less than 30%, when exposed to light [25]. Encapsulation in 100 nm liposomes made of lecithin and cholesterol improves survival to ~70% and 45%, respectively. Nearly 100% of encapsulated anthocyanins survive 28 days of storage at pH 3 and 4°C. Importantly, encapsulation in liposomal carriers increases anthocyanins survival upon digestion. In vitro simulation demonstrated that 89% of anthocyanins survived 2 hours of gastric digestion and 52% survived 2 hours of subsequent intestinal digestion. When encapsulated in liposomes, the survival rate improved to 99.8%

Table 6.2 Examples of encapsulated M&Ns providing protection against environmental conditions.

M&Ns	Environmental factor	Encapsulation system	Type of loading	Performance	References
Polyunsaturated fatty acids	Oxidation	Liposomes	Core-shell	Improved survivability upon storage: the content decreases by 16% over 8 weeks at 30°C (vs. 67% in a control)	[24]
Anthocyanins	Alkaline condition, light	Liposomes	Core-shell	Improved survivability upon storage: 70%/45% survives 28th day of storage at pH 5, 25°C in the dark/light condition (vs. 50%/30% in a control) Increased survivability during gastrointestinal digestion (in vitro): ~99.8%/73% survives 2 hours of gastric/intestinal digestion (vs. ~89%/52% in a control)	[25]
Astaxanthin	Oxidation, heat, light	Liposomes	Core-shell	Increased survivability during heating (processing): 80% survives 30 minutes of heating at 50°C (vs. 50% in a control) Improved storage survivability: 20% degrades on the 8th day of storage in the dark at 4°C (vs. 80% in a control)	[26]
Vitamin D_3	Oxidation, heat, acidic condition	Casein micelles	Core-shell	Improved storage survivability: 90% survives 28 days of storage at 4°C with oxygen penetration (vs. 70% in a control) Survivability during gastrointestinal digestion (in vitro, clinical trials): 85% survives digestion (vs. 15% in a control)	[27, 28]
Vitamin D_3	Oxidation, heat, acidic condition	Potato protein nanoparticles	Core-shell	Increased survivability during heating (processing): 90% survives pasteurization at 88°C for 10 seconds (vs 80% in a control) Improved storage survivability: no detectable degradation on the 21st day of storage in the dark at 4°C (vs 70% in a control)	[29]
Lutein	Oxidation	Casein micelles	Core-shell	Improved storage survivability: 85% survives on the 16th day of storage at room temperature (vs. 56% in a control)	[30, 31]

(Continued)

Table 6.2 (Continued)

M&Ns	Environmental factor	Encapsulation system	Type of loading	Performance	References
Resveratrol	Oxidation	Casein micelles	Core-shell	Improved storage survivability: ~70% survives on the 10th day of storage at 25°C (vs. 60% in a control), and 20% survives on the 80th day of storage at 25°C (vs. 5% in a control)	[31]
Curcumin	Oxidation, light, acidic condition	Casein micelles coated with soy polysaccharides	Core-shell	Improved storage survivability: 97% survives on the 5th day of storage at 25°C in the dark at pH 4.0 (vs. 6% in a control) Increased survivability during gastrointestinal digestion (in vitro): 10.2% released in gastric and 52.3% released in the intestinal stage Increased bioavailability (in vivo): ~3.5-fold higher plasma concentration in mice	[32]
Catechin	Alkaline condition	Starch carrier	Matrix	Increased activity after gastrointestinal digestion (in vitro): 77% inhibitory activity toward pancreatic lipase (vs 44% in a control); 88% inhibitory activity toward cholesterol esterase (vs. 36% in a control)	[33]

and 73%, respectively. In a similar work, liposomes were used to improve the oxidative stability of astaxanthin, a keto-carotenoid widely used as lipo-soluble pigment and antioxidant [26]. Nearly 50% of free astaxanthin does not survive 30 minutes of heating at 50°C and 80% degrade upon 8 days of storage in the dark at 4°C. However, astaxanthin retention improves to 80% in both scenarios upon its encapsulation in liposomes.

Reconstituted ~100 nm casein micelles were used to protect vitamin D_3 upon processing, storage, and gastrointestinal digestion [27, 28]. Encapsulated vitamin did not exhibit detectable degradation upon heat treatment (80°C, 1 min) compared with 13% and 14% degradation in the emulsion stabilized by Tween-80 or free vitamin D_3 suspension, respectively [27]. Only 10% of encapsulated vitamin was lost after 28 days of storage at 4°C in the conditions allowing oxygen penetration, compared with 50% loss in the emulsion and 70% loss in the suspension. About 85% of encapsulated vitamin D_3 survived the simulated digestion process compared with only 15% in the free vitamin D_3 suspension [28]. As a result, low fat (1%) milk fortified with vitamin D_3 encapsulated in reconstituted casein micelles demonstrated as high bioavailability of the vitamin as Tween-80-emulsified supplement, as shown in a randomized double-blinded placebo-controlled clinical trial with 87 human volunteers [27]. Similarly, vitamin D_3 was solubilized into hydrophobic patches of potato protein isolate forming ~30 nm carriers [29]. Encapsulation provided vitamin D_3 protection from degradation upon pasteurization and storage. For example, pasteurization at 88°C for 10 seconds caused ~10% of encapsulated vitamin D_3 to degrade (compared with 20% of unprotected vitamin), while no detectable degradation of encapsulated vitamin D_3 was detected after 21 days of storage at 4°C and in the absence of light (compared with degradation of 70% of unprotected vitamin). An additional benefit of encapsulation here was the absence of turbidity characteristic to the beverages fortified with suspensions of unprotected vitamin D_3. Micelles formed by milk proteins, casein and whey, increased stability of lutein, a type of carotenoid, against oxidation upon storage at room temperature [30]. Casein micelles showed better protection than whey, as survival rates for lutein after 16 days of storage were 79% vs 73% and 85% vs 75%, at 0.25 and 0.50 mg/mL protein concentration, respectively (compared with only 56% lutein left in the unencapsulated control). Casein micelles improved the stability of resveratrol, a plant polyphenol with antioxidant and anti-inflammatory properties, upon storage at 25°C. About ~70% of resveratrol survived 10 days and 20% survived 80 days of storage, compared with 60% and 5% survival rates in the absence of casein [31]. Interestingly, molecular complexes of resveratrol with casein formed at the concentration of the latter well below critical micelle concentration provide significantly lower encapsulation capacity (~5% vs 70% for the micelles) but better resveratrol protection over long-term storage. The resveratrol survival rate was 80% and 40% after 10 and 80 days of storage as a molecular complex, respectively. Possible reason could be protein oxidation upon long-term storage at high concentration.

Curcumin is another potent anti-inflammatory and antioxidant but highly unstable plant polyphenol. About ~94% of curcumin degrades within 5 days of storage as a suspension at pH 4.0 and 25°C when exposed to natural light. However, only 3% of curcumin degrades after 30 days of storage if encapsulated in ~360 nm carriers formed by casein and soluble soy polysaccharides at a loading content of 48.7% [32]. While casein provides stability over storage, polysaccharide component enhances curcumin stability under gastric conditions where casein degrades under the action of a gastric protease, pepsin. As a result, only 10.2% of curcumin was released after 2 hours of digestion in the simulated gastric fluid. The following 2 hours of simulated intestinal digestion releases 52.3% of the curcumin into the mixed micelles phase. In vivo studies in mice demonstrated 3.4–3.5-fold higher plasma concentration of curcumin when ingested as encapsulated in the casein–polysaccharide carriers compared with a curcumin suspension stabilized with Tween-20. Encapsulation of catechin, also a plant polyphenol, in starch submicron-sized carriers protected it from degradation under intestinal conditions where elevated pH and the presence of reactive oxygen species provide favorable conditions for its auto-oxidative reactions [33]. As a result, encapsulated catechin retained its inhibitory activity toward pancreatic lipase and cholesterol esterase, providing important anti-obesity health benefits. For example, catechin encapsulated in 560 nm carriers made of starch from water chestnut retained 77% and 88% of inhibitory activity toward pancreatic lipase and cholesterol esterase, respectively (compared with 44% and 36%, respectively, for unencapsulated catechin).

6.3 Controlled Release by Responsive Carrier Material

Once the carriers reach a required site along the gastrointestinal tract, responsive material they are made from starts to degrade releasing the encapsulated payload. In the pharmaceutical industry, different types of materials that respond to external triggers of the release are used widely, such as light, ultrasound, magnetic field, etc. [34]. All of them are largely inapplicable in food products where the only reliable release trigger is the environmental change along the GIT. The majority of the controlled oral delivery systems developed so far rely on the pH changes as the release trigger [10, 11, 35]. This trigger works well for food supplements. To fortify food products with encapsulated M&Ns, the carriers should be stable in a food matrix, while pH levels vary among different food products in a wide range, from 2–3 (e.g. vinegar, cider chili, and many jams and juices) to 7.5–8 (camembert cheese, egg white, some dressings). Therefore, more GIT-specific triggers are highly beneficial. Figure 6.1 shows different sections of the GIT and lists specific triggers available in each section, while Table 6.3 lists a few examples of responsive carrier materials and their performance.

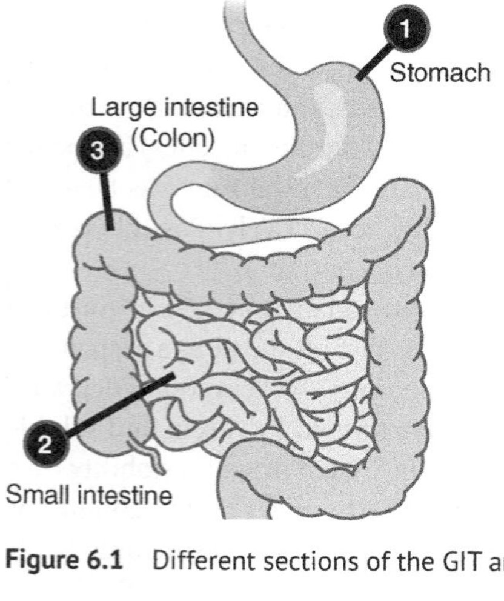

1. Stomach
   - The pH in stomach is about 1–3 and the gastric juice contains pepsin,lipases,etc. Transit time: 15 minutes to 3 hours.
2. Small intestine
   - The pH in small intestine is 6.8 and the main components are bile salt, amylase, trypsin, chymotrypsin, lipases, etc. Transit time: 2–5 hours.
3. Colon
   - The pH is 7.5 and it contains the digestive enzymes of gut microbiota. Transit time: 12–24 hours.

Figure 6.1 Different sections of the GIT and specific release triggers available in each section.

Table 6.3 Examples of responsive carriers for controlled release of M&Ns.

M&Ns	Microencapsulation system	Type of loading	Release trigger	Performance	References
Coenzyme Q10	Starch emulsifier	Matrix	Bile	Increased bioavailability during gastrointestinal digestion (in vivo): 1.8-fold increase of coenzyme Q10 uptake in the form of emulsion and 2.8-fold increase of coenzyme Q10 uptake in the form of emulsion-containing protein beverage	[36]
Polyphenols from olive leaves	Liposomes coated with chitosan	Core-shell	Bile	Increased bioaccessibility during gastrointestinal digestion (in vitro): 80% conversion to mixed micelles (vs. ~30% in a control)	[27, 37]
Calcium	Liposomes	Core-shell	Bile	Increased bioaccessibility during gastrointestinal digestion (in vitro): 79% conversion to mixed micelles (vs. 51% in a control)	[38]
β-carotene	Solid lipid nanoparticles	Core-shell	Lipase	Increased bioaccessibility during gastrointestinal digestion (in vitro): 36% conversion to mixed micelles (vs. 28%)	[39]
Curcumin/quercetin/resveratrol	Nano emulsion	Core-shell	Lipase	Increased bioaccessibility during gastrointestinal digestion (in vitro): 92% released	[40, 41]
Curcumin/resveratrol	Zein cores and polyethylene glycol shell	Core-shell	Protease	Increased bioaccessibility during gastrointestinal digestion (ex vivo): 2.22-fold increase for resveratrol and 3.51-fold for curcumin	[42]
Curcumin	Nanogels with Pickering emulsion	Matrix	Protease	Increased bioaccessibility during gastrointestinal digestion (in vitro): ~80% of curcumin was transferred to mixed micelles (vs. 47% in a control)	[43]
5-Demethylnobiletin	Pickering emulsion	Core-shell	Protease	Increased bioaccessibility during gastrointestinal digestion (in vitro): 18.3% conversion to mixed micelles (vs. 9.2% in a control)	[44]

(Continued)

Table 6.3 (Continued)

M&Ns	Microencapsulation system	Type of loading	Release trigger	Performance	References
Lactoferrin	Layer-by-layer assembled protein-polyphenol microcapsules	Core-shell	Chymotrypsin	Increased bioaccessibility during gastrointestinal digestion (in vitro, ex vivo): 65% survived gastric digestion, 6.5-fold higher concentration in the small intestine compared with a control	[45, 46]
Curcumin	Octenylsuccinate oat β-glucan micelles	Core-shell	Enzymes of colonic microbiota	Protection during gastric digestion (in vitro): 97.3% survives Sustained release during intestinal digestion (in vitro): ~9% per hour Increased bioaccessibility in the colon (in vivo): release of ~50% in the first 10 minutes, followed by sustained release of ~1–3% of per hour over 24 hours	[47]
Resveratrol	Pectin–chitosan polyelectrolyte complex	Core-shell	Pectinase of colonic microbiota	Protection during gastrointestinal digestion (in vitro): ~79% survives Target release in the colon: complete release in the presence of pectinase Improved storage performance: ~13% decrement in resveratrol activity on 30th day storage at 4°C (vs. 26% in a control)	[48]
Carp skin gelatin hydrolysate	Furcellaran encapsulation	Core-shell	Enzymes of colonic microbiota	Protection during gastrointestinal digestion (in vivo) Target release in the colon	[49]
Curcumin	Solid lipid nanoparticles	Matrix	Enzymes of colonic microbiota	Protection during gastrointestinal digestion (in vitro) Sustained release in the colon over 12 hours, significant inhibition of TNF-α Target release to colon (in vivo): by the eighth day, significant reduction of histological score of inflammation, TNF-α level, and myeloperoxidase activity	[50]

In the small intestine, a widely used release trigger comprises bile salts that distort the structure of many lipid-based micron- and submicron-sized carriers of hydrophobic M&Ns simultaneously solubilizing them into mixed micelles for subsequent adsorption through passive diffusion mechanism. In one example, ~200 nm droplets of soybean oil containing 20 wt.% of a model hydrophobic bioactive ingredient, coenzyme Q10, were stabilized with a food-grade emulsifier and an octenyl succinic anhydride-modified starch [36]. The emulsion was found to be fairly stable over 2 hours of simulated gastric and subsequent 4 hours of simulated intestinal digestion avoiding massive aggregation typical of many other emulsifiers. Bile salts induce conversion of the emulsion into even smaller, 140–150 nm, mixed micelles. Up to 73% of coenzyme Q10 turns into those mixed micelles, and it is generally considered that this fraction is available for intestinal absorption. However, an in vitro model of the trans-epithelial transfer using IEC-6 cells confirmed translocation of only 2.3% of the mixed micelles from the apical to the basolateral side following 2 hours of incubation that corresponds to the permeability coefficient P_{app} of 4.79×10^{-6} cm/s, while another 36.4% of the mixed micelles was taken up by or was bound to the cell monolayer. In vivo tests in rats demonstrated 1.8- and 2.8-fold increase in bioavailability of coenzyme Q10 encapsulated in the emulsion, and an emulsion-containing protein beverage, respectively, if compared with the same amount of coenzyme Q10 dissolved in oil.

About ~70 nm liposomes made of lecithin and cholesterol were used to encapsulate polyphenols extracted from olive leaves [37]. Liposomes were additionally coated with chitosan to improve their stability under the conditions of the small intestine and achieve sustained release. Resulting chitosomes were found to be relatively stable in simulated gastric conditions releasing ~30% of the payload within the first hour of digestion, mainly due to the Fickian diffusion of polyphenols through the liposomes. In the simulated medium of the small intestine, ~60% of the payload were released upon 40 min of digestion due to erosion of the liposomes. Similarly, astaxanthin-containing liposomes were additionally modified with a surface layer of positively charged biopolymer, chitosan hydrochloride, or lactoferrin [27]. Under the action of bile salts upon intestinal digestion, liposomes were converted into mixed micelles improving *bioaccessibility* (defined as a percentage of astaxanthin transition into the micellar phase) from ~30% (for unencapsulated astaxanthin) to 65–80% depending on the liposomal formulation. Encapsulation in liposomes improved bioaccessibility of calcium ions chelated with peptides from egg white [38]. The free peptide–calcium complex partially degrades during gastrointestinal digestion, providing only 51% of solubilized calcium; in liposomes, the content of calcium in the micellar phase was 76–79% at the end of gastrointestinal digestion.

Digestive enzymes of the intestinal juice provide another release trigger. Lipases were used to initiate release from lipid nanoparticles. The digestion of ~100 nm liquid and solid lipid nanoparticles with encapsulated β-carotene was studied depending on the composition of the lipid phase [39]. Lipid nanoparticles were stabilized with Tween 80, a nonionic surfactant that is not sensitive to pH changes providing high protection against gastric digestion. In the small intestine, lipases convert triacylglycerols into free fatty acids initiating the formation of mixed micelles. For example, solid lipid nanoparticles made of hydrogenated palm oil form mixed micelles containing large amounts of oleic acid. Being monounsaturated and long chain, it provides high solubilization capacity toward β-carotene and better transition into the micellar phase, increasing its bioaccessibility from 28% (for nanoparticles made of medium chain triacylglycerol, MCT) to 36% at the end of intestinal digestion. Similarly, quercetin, curcumin, and resveratrol have been encapsulated in plant-based oil droplets of submicron size [40, 41]. After simulated digestion, quercetin exhibits a lower release of 44% while curcumin is at 92% and resveratrol is at 100% [40]. This is due to the lower oil–water partition coefficient (logD) of quercetin (2.17) than curcumin (4.12) and resveratrol (3.39). When the molecule is more hydrophilic, it will transfer into the digestive fluid to a larger extent, and thus will be more chemically degraded or precipitated. However, even for quercetin, the bioaccessibility was relatively high (>70%) after digestion of the oil. Another study shows that the triacylglycerol oil type based on medium and long chain fatty acids affects the GI stability and bioaccessibility of quercetin, which highlights the importance of selecting an appropriate oil phase [41]. This type of system is therefore suitable to increase the bioavailability or bioaccessibility of encapsulated hydrophobic bioactive ingredients in GIT.

Intestinal proteases (trypsin and chymotrypsin) facilitate the erosion of proteinaceous carriers triggering the release. Submicron-sized carriers made of zein, a prolamine-rich storage protein from maize, are widely used to improve bioaccessibility of hydrophobic phytochemicals. For example, ~300 nm core-shell carriers having zein cores and *polyethylene glycol* (PEG) shell were loaded with 3.4 wt.% of curcumin [42]. Under physiological conditions, curcumin demonstrated sustained release mainly governed by Fickian diffusion through the carrier material, achieving the release of ~45% of the payload in 10 hours. However, an erosion of carriers upon two hours of the intestinal digestion resulted in the transition of more than 80% of encapsulated curcumin into the mixed micelles fraction (compared with ~20% bioaccessibility for a curcumin suspension and ~50% bioaccessibility for curcumin encapsulated in zein nanoparticles without PEG shell). The impact of encapsulation on bioaccessibility of a more water-soluble phytochemical, resveratrol, was less prominent: 80% when in PEG-coated zein nanoparticles vs 50% for resveratrol suspension. The data were further confirmed upon ex vivo analysis of the curcumin and resveratrol intestinal permeability using chicken small intestine. The permeability coefficient of the suspensions was found to be 0.94×10^{-7} cm/s and 1.93×10^{-7} cm/s for curcumin and resveratrol, respectively. It increased 2.22-fold and 3.51-fold for resveratrol and curcumin encapsulated in zein-PEG core-shell carriers, respectively. Protein nanogel particles are also used to stabilize Pickering emulsions. For example, nanogels made of whey protein isolate were applied to stabilize an aqueous emulsion of an MCT oil containing 2 wt.% of curcumin [43]. In the small intestine, whey protein particles at the surface of oil droplets were hydrolyzed by trypsin making them incapable of stabilizing the emulsion, while lipases caused triacylglycerol hydrolysis into free fatty acids. At this point, ~80% of encapsulated curcumin was transferred into the micellar fraction, a significant enhancement when compared with 47% of curcumin bioaccessibility in the non-emulsified oil. In contrast, the 18.3% bioaccessibility was achieved for 5-demethylnobiletin, a polymethoxyflavone from aged citrus peel (chenpi) exhibiting various bioactivities, when encapsulated in a Pickering emulsion of corn oil stabilized with 250 nm particles made of peanut protein isolate [44]. Still, it was ~two times higher than the 9.2% bioaccessibility of that in the bulk oil. The apparent permeability coefficient for the formed mixed micelles measured in caco2 cells model was 26.9×10^{-6} cm s^{-1} and the efflux ratio was 1.46, an indication of passive transport mechanism.

Layer-by-layer assembled protein-polyphenol shells were suggested for the controlled release of bioactive ingredients in the small intestine triggered by proteolysis of the protein component of the shells by an intestinal protease, α-chymotrypsin [45, 46, 51]. For hydrophobic M&Ns, the protein-polyphenol shell was assembled on the surface of the emulsion stabilized with a protein emulsifier [51]. Treatment with α-chymotrypsin caused complete degradation of the shell and coalescence of the emulsion droplets. Water-soluble M&Ns were adsorbed onto the porous inorganic microparticles that served as sacrificial templates for the shell assembly followed by the templates' extraction with a trisodium salt of ethylenediaminetetraacetic acid or hydrochloric acid. The approach was used to encapsulate a functional whey protein with multiple health-beneficial functions, lactoferrin [45, 46]. When the templates were made of 2–4 μm vaterite particles, a polymorph of calcium carbonate, the loading capacity of the carriers was ~2.2 wt.% of lactoferrin. A more porous ~8 or ~20 μm templates made of surface-reacted calcium carbonate containing up to 50% of hydroxyapatite allowed much higher loading capacity of 57 wt.%. The final carriers protected 65% of encapsulated lactoferrin from simulated gastric digestion and eroded in the simulated intestinal fluid. Ex vivo tests demonstrated 6.5 times higher concentration of lactoferrin in the small intestine of mice dosed with encapsulated lactoferrin than in a control group dosed with the same amount of the unprotected one. Moreover, significant amount of lactoferrin released from the microcarriers was absorbed into the bloodstream and later detected in the liver.

For many nutraceuticals, their most beneficial site of action is not plasma but distal parts of the alimentary canal. In this case, an enhanced residential time of the microcarriers in the gut on a par with the highly localized sustained release of the active compounds are more beneficial than increased permeability of microcarriers through the epithelium and absorption into systemic circulation. Enzymes produced by the colonic microbiota is a natural and highly specific trigger of the colonic release. Some polysaccharides (e.g. pectin, chitosan, alginate, resistant starch, etc.) are resistant to gastric and intestinal enzymes but digestible by the enzymes of colonic microbiota. One approach relies on designing a carrier with a chemically bound corona made of those resistant polysaccharides. For example, ~200 nm micelles made of self-assembled octenylsuccinate oat β-glucan were used to solubilize curcumin [47]. An in vitro digestion model demonstrated that 97.3% of capsules remained stable under gastric conditions. A slow but sustained release of curcumin (~ 9% per hour) happened under intestinal conditions probably due to a significant (~ 2 times) swelling of the carriers. In vivo tests with rats demonstrated the highest plasma concentration of curcumin was achieved 60 mins after oral administration, indicating some intestinal adsorption. However, in the colonic conditions, a burst release of ~50% of encapsulated curcumin was recorded within 10 minutes followed by a slower (~1–3% per hour) sustained release of the remaining curcumin over a period of 24 hours. The authors speculate that the oat β-glucan consists of several fractions. Some are rapidly cleaved by the enzymes of gut microbiota, while others are cleaved slower, and the rest are resistant.

Resveratrol was encapsulated in ~700 nm carriers made of the pectin–chitosan polyelectrolyte complex [48]. In vitro digestion of the carriers with simulated gastric and intestinal juices induced the cumulative release of ~21% of resveratrol. Another 22% was released at pH 7.4 of the colonic environment. However, a complete release of the encapsulated resveratrol was achieved in the presence of pectinase simulating pectin-digesting enzymes of the colonic microbiota. As an additional benefit, pectin–chitosan carriers provide antioxidant activity of encapsulated resveratrol upon storage. There was just a marginal decrease in the resveratrol activity (by 13%) after 30 days of storage at 4°C compared with a twofold decrease in the activity of the unprotected one. In another example, furcellaran, a negatively charged polysaccharide, was used to encapsulate antioxidant peptides of carp skin gelatin hydrolysate [49]. Both encapsulated and non-encapsulated peptides were included in a diet of male Wistar rats for a period of 35 days followed by the analysis of the total antioxidant level of serum. It was assumed that the furcellaran shell of the microcapsule did not undergo digestion in the small intestine of the tested animals, preventing peptides adsorption into the bloodstream but were fermented in the colon. Solid submicron carriers (280 nm in size) made of biocompatible and biodegradable lipids (glyceryl palmitostearate, Miglyol 812N/F) demonstrate suppressed intestinal absorption and increased colon residence time. They were suggested as carriers for curcumin in inflammatory bowel disease treatment [50]. The carriers were localized within caco2 cell monolayer but did not penetrate the layer. Their sustained presence in the colon was observed over 12 hours. A solid lipid core provides an additional advantage protecting encapsulated curcumin from degradation. In vitro tests demonstrated significant inhibition of *tumor necrosis factor alpha* (TNF-α) secretion by activated J774 macrophages. Encapsulated curcumin significantly reduced the histological score of inflammation, TNF-α level, and myeloperoxidase activity at day 8 of the in vivo study of a murine dextran sulfate-induced colitis, a model of human inflammatory bowel disease.

Industrial Corner – Technological Challenges

- Sensory qualities of food cannot be compromised by the addition of encapsulated components
- Stability of the carriers in food matrices and upon transition through the gastrointestinal tract
- Availability of food-grade carrier shell material that fits within requirements of religion, food habits, and clean labeling. For example, it may not be suitable to use casein or lactose as shell material for vegan consumers
- The four main thrusts of technological innovation to facilitate microencapsulation adoption by the food industry are:
  1) To increase effectiveness of microencapsulation through system design and process needs
  2) To push, nurture, and facilitate different microencapsulation systems and provide solutions to scale up operation with simple and economic processes
  3) To focus on end-user applications, in accordance with carriers' stability and safety in food, and in compliance with regulations
  4) To provide scientifically substantiated health benefit claims and achieve consumer acceptance of products.

6.4 Active Enhancement of M&Ns' Bioavailability Through Microencapsulation

In all examples above, the carriers played a passive role in the enhancement of M&Ns bioavailability improving miscibility with a food matrix, providing a barrier protection function, and then eroding at the right section of the GIT. However, the examples shown in Table 6.4 demonstrate that the carriers may play a more important and active role in fixing the impaired or providing new routes for their absorption. Figure 6.2 shows the most important pathways of M&Ns through the epithelial wall.

Natural absorption of essential micronutrients may be impaired due to a variety of physiological factors that may become a trigger or adverse outcome of chronic diseases and frailty. Microencapsulation may provide a means to overcome malabsorption and enhance M&Ns' bioavailability for the affected cohort. For example, vitamin B_{12} is primarily absorbed as a complex with its *intrinsic factor* (IF) via active transport through the cubam receptor in the ileal [59]. Only a small fraction (1–2%) is absorbed through passive diffusion but this route becomes of outmost importance for people with pernicious anemia caused by impaired secretion of IF. The condition becomes more common with age, with a prevalence of 2–3% among the people of 60+ years of age, and 23–35% among the people of 80+ years of age [60]. Submicron carriers formed upon crosslinking of a *soy protein isolate* (SPI) with Ca^{2+} can encapsulate B_{12} through hydrophobic interactions [52]. Loading capacity of B_{12} in SPI carriers was reported to be 0.02–0.03 wt.%. Active transport of encapsulated B_{12} through jejunum epithelial was demonstrated in an ex vivo rat model. Clathrin-mediated transcytosis and micropinocytosis were suggested as the primary mechanisms of 30 and 100 nm carriers' transport across caco2 cell monolayer, which was followed by the degradation of carriers by acid hydrolase enzymes and B_{12} release into the cytoplasm. A minor (~20%) fraction of the carriers was transported by

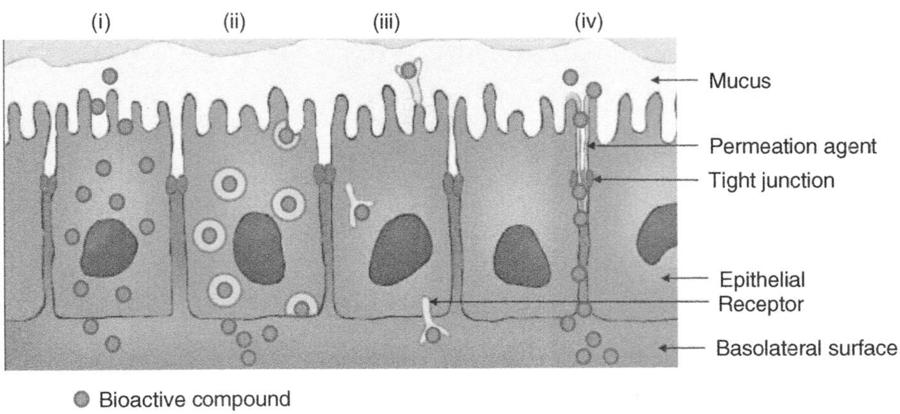

Figure 6.2 Different transport pathways of M&Ns through the epithelial wall: (a) transcellular methods: (i) facilitated passive diffusion; (ii) endocytosis; (iii) receptor-mediated endocytosis. (b) paracellular method: (iv) tight junction opening paracellular transport.

Table 6.4 Examples of encapsulation systems that actively enhance absorption of M&Ns.

M&Ns	Microencapsulation system	Type of loading	Mechanism of action	Performance	References
Vitamin B_{12}	Soy protein isolate – Ca^{2+} nanogels	Matrix	Enable absorption through clathrin-mediated transcytosis and micropinocytosis	Active transport through jejunum epithelial (ex vivo);	[52]
Curcumin	Solid lipid nanoparticles with PEG corona	Matrix	Facilitated passive diffusion	Protection during gastric digestion and improved intestinal absorption (in vitro) Increased bioavailability (in vivo): 6 times higher peak concentration in plasma and at a later time (30–45 minutes vs 15 minutes) compared with control	[53]
Daidzein	Liposomes	Core-shell	Facilitated passive diffusion	Protection during gastrointestinal digestion (in vitro): ~18% was released in 1 hour of gastric digestion and ~3% in 1 hour of intestinal digestion Target release in the colon (in vitro): the remaining ~79% possess prolonged mean resident time in systemic circulation Increased bioavailability (in vivo): ~2.5-fold higher peak concentration in plasma, ~1.8-fold longer residence time compared with control	[54]
Peptide RLSFNP (β-lactoglobulin F (148–153))	Liposomes	Core-shell	Facilitated passive diffusion	Improved bioavailability: sustained release in the small intestine and twofold higher permeability through intestinal epithelium compared with a control	[55]

Curcumin	Layer-by-layer assembled whey protein isolate – chitosan microcapsules	Core-shell	Enable paracellular transport through tight junction opening	Protection during gastric digestion Improved bioaccessibility during intestinal digestion: ~20% reach jejunum and further 15% reach ileum (vs. 0.15% in a control) 1.55-fold higher permeation rate	[56]
Lactoferrin	Layer-by-layer assembled protein-polyphenol microcapsules with incorporated immunoglobulin G	Core-shell	Mucoadhesive material	Increased residential time in the small intestine: 2.7-fold higher adhesion toward intestinal epithelial than in the control	[57]
Lactoferrin hydrolysate	Conjugates with galacto-oligosaccharides	Matrix	Binding toward bacteria receptors, phosphotransferase system, or lactose permease	Improved bioaccessibility: twofold higher proliferation of *Lactobacillus casei* on a minimal growth medium	[58]

passive diffusion through Ca^{2+}/Mg^{2+} chelating-induced opening of the tight junctions. Those carriers may avoid intracellular degradation and enter systemic circulation. A better approach could be an encapsulation of B_{12} complex with IF, e.g. recombinant human IF expressed in the transgenic plant *Arabidopsis thaliana* that may target the natural B_{12} receptor, cubam. A clinical trial demonstrated significantly higher increase in a biochemical marker of B_{12} level in serum, *holo-transcobalamin* (holoTC), one day after the intake of 9 μg of B_{12}-IF complex among nine patients with evident signs of B_{12} deficiency [61]. Notably, this dosage is not enough to achieve an increased level of holoTC through passive B_{12} adsorption, meaning active B_{12} adsorption through cubam has been achieved successively.

Unlike vitamin B_{12}, the main adsorption route for many hydrophobic M&Ns is passive diffusion. For some, e.g. isoflavones, there are no active transporters. For others, e.g. flavonoid glucosides, the efficacy of their active adsorption by sodium-dependent glucose transporter 1 is suppressed to a large extent by efflux back to lumen by the apical transporter multidrug resistance-associated protein 2. Achieving the size of the carriers below 200 nm is the key factor facilitating passive diffusion through the mucus and entering systemic circulation (that determines the bioavailability of nutraceuticals). Solid lipid nanoparticles modification with PEG allowed longer systemic circulation. For example, ~150 nm lipid carriers made of tristearoylglycerol were used for curcumin encapsulation (5 wt.% of the lipid phase) by the ultrasound-assisted oil-in-water emulsion technique upon cooling from 85°C to 45°C in the presence of a nonionic surfactant, polyoxyethylene(100) stearyl ether [53]. Carriers did not undergo aggregation upon in vitro digestion simulation, i.e. subsequent treatment with simulated saliva (5 mins), gastric (2 hours), and intestinal (2 hours) fluids. In the small intestine, pancreatic lipase hydrolyzes tristearoylglycerol to monostearoylglycerol and two stearic acids. Lipolysate makes mixed micelles with bile solubilizing most (92–95%) of the loaded curcumin. Its intestinal adsorption via passive diffusion was simulated in vitro using caco-2 cell monolayer covered with an artificial mucus layer. Permeability coefficient of $(2–3) \times 10^{-6}$ cm s^{-1} was measured. An in vivo *pharmacokinetic* study following curcumin concentration in plasma isolated from the rat tail vein demonstrated six times higher peak concentration (~100 ng/mL) and at a later time (30–45 min instead of 15 min) when compared with the control group administered with curcumin solution in ethanol. Encapsulation in solid lipid nanoparticles allowed 12–13 times higher curcumin bioavailability compared with control. In contrast, solid lipid nanoparticles stabilized with shorter polyoxyethylene(10) stearyl ether had relatively low bioavailability. Submicron-sized carriers with the terminal polypeptide layer were shown to further enhance penetration through intestinal membrane. For example, carriers made of a lipid core and a surrounding protamine corona were suggested for curcumin encapsulation [50]. The permeability coefficient of ~200 nm carriers across caco-2 cell monolayer was found to be 3.54×10^{-5} cm/s, while the permeability coefficient of lipid carriers of comparable size but without polypeptide corona could not be measured because the levels of curcumin were below the detection limit.

About ~150 nm liposomes were used for the encapsulation of daidzein, an isoflavone found in legumes and estrogenic agonist, with a loading capacity of 8.2% [54]. In vitro digestion simulation studies have shown that ~18% of encapsulated daidzein was released in 1 hour of gastric digestion while only ~3% was released in 1 hour of intestinal digestion.

Poorer stability of liposomes under acidic conditions was suggested as a possible explanation. Nevertheless, the remaining liposomes promoted the absorption and prolonged mean resident time of daidzein in the systemic circulation. In vivo pharmacokinetic studies in rats demonstrated ~2.5 higher dosage of daidzein in plasma and ~1.8 times longer residence time compared with unencapsulated control. Liposomes were also shown to enhance permeability of peptides. For example, peptide RLSFNP (β-lactoglobulin F (148 – 153)), an effective inhibitor of the angiotensin-converting enzyme, is primarily absorbed by energy-dependent transcytosis. However, the efficiency of this process is low (RLSFNP permeability coefficient is $<10^{-6}$ cm/s). Encapsulation in ~200 nm liposomes enhances peptide permeability across the enterocyte more than two times (permeability coefficient was 1.72×10^{-6} cm/s), shifting it from poorly absorbed to moderately absorbed level [55]. Moreover, liposomes demonstrated a sustained release of RLSFPN in systemic circulation, maintaining its concentration within the therapeutic range for a longer time than unencapsulated peptide.

Curcumin was encapsulated in 190 nm droplets of an MCT oil stabilized with whey protein isolate and then additionally coated with a layer of chitosan [56]. The chitosan terminal layer protected underlying whey protein from gastric digestion and delayed digestion in the jejunum. As a result, curcumin bioaccessibility in the jejunum was ~20% for both emulsions (compared with 0.15% bioaccessibility of free curcumin in a non-stabilized MCT emulsion), but the chitosan layer significantly enhanced curcumin bioaccessibility in the ileum, from 9% to 15% (while curcumin was not detected in the ileal filtrate of digested non-stabilized MCT emulsion). Moreover, the caco-2 cells-based in vitro permeability assay demonstrated that chitosan disrupted the tight junctions enabling paracellular transport of curcumin, while the droplets stabilized with the single whey protein layer permeated the cells through transcellular pathway only. As a result, the permeation rate for encapsulated curcumin with terminal chitosan layer was 1.93×10^{-6} cm/s, 1.55-fold higher than without chitosan. Besides, the use of chitosan significantly increased the cellular uptake of curcumin by endocytosis, from 4.44% to 10.50%. In another example, nanogel particles made of whey protein isolate were used to make Pickering emulsions of MCT oil containing curcumin, and whey protein was conjugated with, or complexed to dextran [43]. Interestingly, dextran did not improve emulsion stability under the intestinal conditions, as curcumin bioavailability was not statistically different from the one obtained in the emulsion stabilized by non-modified whey protein isolate. However, dextran significantly increased curcumin uptake by caco-2 cells, from 6.3% to 7.4–7.6%, as dextran increased the mucosal permeability.

Incorporation of non-specific mucoadhesive materials, e.g. non-toxic lectins, glycoproteins, or some of milk proteins [62], into the carriers is aimed to achieve their anchoring at certain sites along the alimentary canal dramatically increasing residential time at the respective site and, consequently, bioaccessibility of encapsulated M&Ns. Formulations enhancing retention at the oral mucosa were excluded from this review. Transmucosal delivery of M&Ns in the oral cavity has some advantages as it bypasses transit through gastrointestinal digestion and hepatic first-pass metabolism. Readers interested in oral mucoadhesive formulations may refer to a very nice recent review [63]. Here, we focus on more traditional ways of M&Ns adsorption from foods. In one example, incorporation of a whey protein, *immunoglobulin G* (IgG), into the layer-by-layer assembled shell made of bovine serum albumin (BSA) and tannic acid dramatically increased adhesion of the carrier toward intestinal epithelial [57]. Interestingly, the highest adhesion was achieved when IgG was assembled as the terminal layer, 6.3-fold higher than in a control sample made of BSA-tannic acid shells, as demonstrated in the in vitro epithelial adherence test using the caco-2 cell model. However, the terminal IgG layer was completely digested by gastric fluids, as shown using in vitro gastrointestinal digestion model, which does not leave the carrier a chance to perform its mucoadhesive function in the small intestine. However, further coating the IgG layer with a couple of BSA-tannic acid layers effectively protects it from gastric digestion. And the resulted carriers still possess 2.7-fold higher adhesion toward intestinal epithelial than the control.

More precise targeting of the carriers to specific receptors in the small intestine or colon is achieved through their conjugation with targeting ligands, e.g. vitamins, peptides, proteins, aptamers, or antibodies. These ligands ensure selective binding to chosen receptors, as summarized in a recent review [64]. The cost of antibodies and many of the special proteins (e.g. TRAIL) is prohibitively high to be used for food products. However, aptamers and peptides (e.g. iRPG, SP94) provide comparably high avidity and are nowadays synthesized at large scales and low costs. Folate moiety is another promising ligand for targeted oral delivery. Folate receptors are expressed mainly in the jejunum but also in the colon [65]. Some nutraceuticals are supposed to be delivered to colonic microbiota to facilitate the proliferation of probiotic bacteria (prebiotics) or to be fermented by bacteria into favorable metabolites (postbiotics). In this case, precise targeting of the carriers toward probiotic bacteria is highly beneficial. This mechanism was suggested as a possible explanation of the twofold higher proliferation of *Lactobacillus casei* on a minimal growth medium containing conjugates of lactoferrin hydrolysate and galacto-oligosaccharides, when compared with unconjugated ingredients [58]. In the small intestine, the conjugates self-assemble into 370 nm aggregates having indigestible oligosaccharide corona protecting the proteinaceous core from intestinal proteases. It was hypothesized that the conjugates may interact with the receptors on the surface of lactobacilli involved in the uptake of galacto-oligosaccharides, either phosphotransferase system or lactose permease. This triggers release of the conjugated protein fraction in close proximity to the peptides' transporters enhancing uptake of this important nitrogen source that is so scarce in the colonic environment.

6.5 Conclusion

While the application of micron- and submicron-sized carriers for protection and precise delivery of pharmaceuticals had been largely established, their use to enhance bioavailability of M&Ns started to be uncovered. Unlike pharmaceuticals, the benefits of foods fortified with M&Ns in the prevention of diseases and improving the quality of life are more long-term and less fit for an unambiguous randomized double-blinded placebo-controlled clinical trial – a gold standard in the pharmaceutical industry. Moreover, for many bioactive phytochemicals or functional proteins, their true efficiency for humans, or certain cohorts of humans (e.g. immunocompromised, elderly, infants, etc.) is still under investigation and research often brings controversial outcomes. To a large extent, the reason for ambiguity could arise from poor bioavailability of those nutraceuticals, leading to a big variance in dosage. Therefore, the vast majority of research in the area stops at in vitro trials, demonstrating improved bioaccessibility of encapsulated M&Ns. A few go into a detailed study of the mechanisms behind the facilitated absorption of encapsulated M&Ns in the small intestine or by the gut microbiota, or follow the improved bioavailability in vivo using animal models. The importance of taking into account the carriers' stability and performance in a food matrix is being realized and investigated. Having said that the potential of smart carriers for functional foods is already clear to be enormous. Ability to tune the carriers' functionality addressing particular issues with nutrients malabsorption opens an avenue for a true personalized nutrition designed to maximize nutritional value and health benefits of foods for a particular person. Continuous development of the encapsulation and biochemical technologies to bring down the production costs could make such foods affordable for the general public.

Glossary

Bioavailability The proportion of M&Ns that enter the systemic circulation when introduced into the body.

Bioaccessibility The fraction of the total amount of a M&Ns that are potentially available for absorption or other function.

CAGR Compound annual growth rate is a business and investing specific term for the geometric progression ratio that provides a constant rate of return over the period.

EMA European Medicines Agency is in charge of evaluation and supervision of medical products.

Epithelial Layers of cells that form the lining of the outer surface of the body, including hollow organs and glands.

FDA Food and Drug Administration is a US federal agency responsible for protecting public health by ensuring safety, efficacy, and security of human and veterinary drugs, biological products, medical devices, food products, cosmetics, and devices that emit radiation.

GIT Gastrointestinal tract is the passageway of the digestive system where food taken in through the mouth is digested to extract nutrients and absorb energy, and waste is expelled at anus as feces.

holoTC Holo-transcobalamin, a marker of vitamin B_{12} status in serum.

IF Intrinsic factor is a glycoprotein produced by the parietal cells in the stomach of humans necessary for the absorption of vitamin B_{12}.

IgG Immunoglobulin G – one of the whey proteins, an antibody secreted in milk that possesses enhanced mucoadhesive properties, among others.

MCT Medium chain triacylglycerol with two or three fatty acids having an aliphatic tail of 6–12 carbon atoms, commonly extracted from palm kernel oil or coconut oil.

Micelle An aggregate of surfactant molecules dispersed in aqueous media that has a hydrophobic core and polar surface, making up a colloidal suspension.

Micronutrients Essential dietary elements such as vitamins and minerals required by organisms in varying quantities throughout life to orchestrate a range of physiological functions to maintain health.

Mucoadhesive The adherence ability of a biomaterial to the mucous membrane of the epithelial surfaces, which can be found in the gastrointestinal tract, eyes, lungs, etc.

Nutraceuticals Any substance that is a food or part of a food and provides medical or health benefits, including the prevention and treatment of disease.

PEG Polyethylene glycol – a highly hydrophilic synthetic polymer commonly used to tailor "stealth" functionality to the microcarriers reducing the rate of clearance from the blood.

Pharmacokinetic The study of the administration time of drug in four stages: absorption, distribution, metabolism, and excretion.

TNF-α Tumor necrosis factor alpha is a cytokine, a small protein commonly used as an inflammation marker. If macrophages detect an infection, they release TNF-α to alert other immune system cells as part of an inflammatory response.

PUFA Polyunsaturated fatty acids. Many PUFA are important regulators of lipid metabolism; have anti-inflammatory and anticancer effects.

SPI Soy protein isolate; proteins isolated from soybeans that have been dehulled and defatted.

References

1 Y. Qiao, J. Wan, L. Zhou, W. Ma, Y. Yang, W. Luo, Z. Yu, H. Wang, *Wiley Interdiscip. Rev. Nanomed. Nanobiotechnol.* **2019**, *11*, 1527.

2 W. Cui, J. Li, G. Decher, *Adv. Mater.* **2016**, *28*, 1302.

3 S. Correa, E. C. Dreaden, L. Gu, P. T. Hammond, *J. Controll. Release* **2016**, *240*, 364.

4 P. L. Chariou, O. A. Ortega-Rivera, N. F. Steinmetz, *ACS Nano* **2020**, *14*, 2678.

5 B. Plasek, Z. Lakner, G. Kasza, A. Temesi, *Nutrients* **2020**, *12*, 69.

6 J. P. Gleeson, S. M. Ryan, D. J. Brayden, *Trends Food Sci. Technol.* **2016**, *53*, 90.

7 M. R. I. Shishir, L. Xie, C. Sun, X. Zheng, W. Chen, *Trends Food Sci. Technol.* **2018**, *78*, 34.

8 R. F. S. Goncalves, J. T. Martins, C. M. M. Duarte, A. A. Vicente, A. C. Pinheiro, *Trends Food Sci. Technol.* **2018**, *78*, 270.

9 E. Nowak, Y. D. Livney, Z. Niu, H. Singh, *Trends Food Sci. Technol.* **2019**, *91*, 557.

10 K. Feng, Y. Wei, T. Hu, R. J. Linhardt, M. Zong, H. Wu, *Trends Food Sci. Technol.* **2020**, *102*, 203.

11 S. H. Lee, R. Bajracharya, J. Y. Min, J. W. Han, B. J. Park, H. K. Han, *Pharmaceutics* **2020**, *12*, 68.

12 M. Yang, Z. Liang, L. Wang, M. Qi, Z. Luo, L. Li, *Adv. Polym. Technol.* **2020**, 2020, 7531810.

13 M. Pateiro, B. Gomez, P. E. S. Munekata, F. J. Barba, P. Putnik, D. B. Kovacevic, J. M. Lorenzo, *Molecules* **2021**, *26*, 1547.

14 M. Manzoor, J. Singh, J. D. Bandral, A. Gani, R. Shams, *Int. J. Biol. Macromol.* **2020**, *165*, 554.

15 A. Rehman, T. Ahmad, R. M. Aadil, M. J. Spotti, A. M. Bakry, I. M. Khan, L. Zhao, T. Riaz, Q. Tong, *Trends Food Sci. Technol.* **2019**, *90*, 35.

16 M. Nooshkam, A. Babazadeh, H. Jooyandeh, *Trends Food Sci. Technol.* **2018**, *80*, 23.

17 A. Ashfaq, K. Jahan, R. U. Islam, K. Younis, *LWT – Food Sci. Technol.* **2022**, *154*, 112667.

18 A. Abaee, M. Mohammadian, S. M. Jafari, *Trends Food Sci. Technol.* **2017**, *70*, 69.

19 U. Sadiq, H. Gill, J. Chandrapala, *Foods* **2021**, *10*, 1965.

20 O. Torres, B. Murray, A. Sarkar, *Trends Food Sci. Technol.* **2016**, *55*, 98.
21 D. J. McClements, *Food Funct.* **2018**, *9*, 22.
22 S. F. Hosseini, L. Ramezanzade, D. J. McClements, *Trends Food Sci. Technol.* **2021**, *109*, 322.
23 L. Barroso, C. Viegas, J. Vieira, C. Ferreira-Pego, J. Costa, P. Fonte, *J. Food Eng.* **2021**, *295*, 110451.
24 S. Gulzar, S. Benjakul, *Food Chem.* **2020**, *310*, 125916.
25 J. Chi, J. Ge, X. Yue, J. Liang, Y. Sun, X. Gao, P. Yue, *LWT - Food Sci. Technol.* **2019**, *109*, 101.
26 M. Qiang, X. Pang, D. Ma, C. Ma, F. Liu, *Molecules* **2020**, *25*, 610.
27 M. Haham, S. Ish-Shalom, M. Nodelman, I. Duek, E. Segal, M. Kustanovich, Y. D. Livney, *Food Funct.* **2012**, *3*, 737.
28 Y. Cohen, M. Levi, U. Lesmes, M. Margier, E. Reboul, Y. D. Livney, *Food Funct.* **2017**, *8*, 2133.
29 S. David, Y. D. Livney, *Food Hydrocoll.* **2016**, *57*, 229.
30 J. Yi, Y. Fan, W. Yokoyama, Y. Zhang, L. Zhao, *Food Chem.* **2016**, *200*, 91.
31 H. Cheng, H. Dong, L. L. Wusigale, *Food Chem.* **2020**, *330*, 127209.
32 G. Xu, L. Li, X. Bao, P. Yao, *Food Biosci.* **2020**, *35*, 100569.
33 M. Ahmad, P. Mudgil, A. Gani, F. Hamed, F. A. Masoodi, S. Maqsood, *Food Chem.* **2019**, *270*, 95.
34 M. V. Kiryukhin, *Curr. Opin. Pharmacol.* **2014**, *18*, 69.
35 C. Y. Wong, H. Al-Salami, C. R. Dass, *Int. J. Pharm.* **2018**, *537*, 223.
36 Z. Niu, A. Acevedo-Fani, A. McDowell, A. Barnett, S. M. Loveday, H. Singh, *J. Controll. Release* **2020**, *327*, 444.
37 I. Katouzian, R. A. Taheri, *J. Food Sci. Technol.* **2021**, *58*, 3430.
38 P. Zhang, Z. Bao, P. Jiang, S. Zhang, X. Zhang, S. Lin, N. Sun, *J. Food Sci.* **2021**, *86*, 1418.
39 H. H. de Abreu-martins, M. Artiga-Artigas, R. H. Piccoli, O. Martin-Belloso, L. Salvia-Trujillo, *Food Chem.* **2020**, *311*, 126024.
40 H. L. Zhou, B. J. Zheng, D. J. McClements, *J. Agric. Food Chem.* **2021**, *69*, 3340.
41 H. L. Zhou, B. J. Zheng, D. J. McClements, *Food Funct.* **2021**, *12*, 3420.
42 M. Maria Leena, T. Anukiruthika, J. A. Moses, C. Anandharamakrishnan, *Food Hydrocoll.* **2022**, *124*, 107200.
43 A. Araiza-Calahorra, Y. Wang, C. Boesch, Y. Zhao, A. Sarkar, *Curr. Res. Food Sci.* **2020**, *3*, 178.
44 F. Ning, X. Wang, H. Zheng, K. Zhang, C. Bai, H. Peng, Q. Huang, H. Xiong, *J. Funct. Foods* **2019**, *55*, 76.
45 E. Kilic, M. V. Novoselova, S. H. Lim, N. A. Pyataev, S. I. Pinyaev, O. A. Kulikov, O. A. Sindeeva, O. A. Mayorova, R. Murney, M. N. Antipina, B. Haigh, G. B. Sukhorukov, M. V. Kiryukhin, *Sci. Rep.* **2017**, *7*, 44159.
46 M. V. Kiryukhin, S. H. Lim, H. H. Lau, M. Antipina, Y. W. Khin, C. Y. Chia, P. Harris, M. Weeks, C. Berry, D. Hurford, O. Wallace, M. Broadhurst, C. J. Ridgway, J. Schoelkopf, *J. Colloid Interface Sci.* **2021**, *594*, 362.
47 J. Liu, L. Lei, F. Ye, Y. Zhou, H. G. R. Younis, G. Zhao, *Food Funct.* **2018**, *9*, 491.
48 H. Andishmand, H. Hamishehkar, A. Babazadeh, A. Taghvimi, M. A. Mohammadifar, M. Tabibiazar, *Pharmaceut. Sci.* **2017**, *23*, 42.
49 J. Tkaczewska, E. Jamroz, E. Piatkowska, B. Borczak, J. Kapusta-Duch, M. Morawska, *Nutrients* **2019**, *11*, 2502.
50 A. Beloqui, P. B. Memvanga, R. Coco, S. Reimondez-Troitino, M. Alhouayek, G. G. Muccioli, M. J. Alonso, N. Csaba, M. de la Fuente, V. Preat, *Colloids Surf., B* **2016**, *143*, 327.

51 M. V. Lomova, A. I. Brichkina, M. V. Kiryukhin, E. N. Vasina, A. M. Pavlov, D. A. Gorin, G. B. Sukhorukov, M. N. Antipina, *ACS Appl. Mater. Interfaces* **2015**, *7*, 11732.

52 J. Zhang, C. J. Field, D. Vine, L. Chen, *Pharm. Res.* **2015**, *32*, 1288.

53 C. Ban, M. Jo, Y. H. Park, J. H. Kim, J. Y. Han, K. W. Lee, D. H. Kweon, Y. J. Choi, *Food Chem.* **2020**, *302*, 125328.

54 Q. Wang, W. Liu, J. Wang, H. Liu, Y. Chen, *Nanoscale Res. Lett.* **2019**, *14*, 321.

55 T. Zhang, M. Su, X. Jiang, Y. Xue, J. Zhang, X. Zeng, Z. Wu, Y. Guo, D. Pan, *J. Agric. Food Chem.* **2019**, *67*, 5544.

56 H. D. Silva, E. Beldikova, J. Poejo, L. Abrunhosa, A. T. Serra, C. M. M. Duarte, T. Branyik, M. A. Cerqueira, A. C. Pinheiro, A. A. Vicente, *J. Food Eng.* **2019**, *243*, 89.

57 H. H. Lau, R. Murney, N. L. Yakovlev, M. V. Novoselova, S. H. Lim, N. Roy, H. Singh, G. B. Sukhorukov, B. Haigh, M. V. Kiryukhin, *J. Colloid Interface Sci.* **2017**, *505*, 332.

58 A. Seifert, S. Freilich, Y. Kashi, Y. D. Livney, *Polym. Adv. Technol.* **2019**, *30*, 2577.

59 A. Brito, E. Habeych, I. Silva-Zolezzi, N. Galaffu, L. H. Allen, *Nutr. Rev.* **2018**, *76*, 778.

60 R. Green, L. H. Allen, A. L. Bjorke-Monsen, A. Brito, J. L. Gueant, J. W. Miller, A. M. Molloy, E. Nexo, S. Y. Stabler, B. H. Toh, P. M. Ueland, C. Yajnik, *Nat. Rev. Dis. Pri.* **2017**, *3*, 17040.

61 A. M. Hvas, T. Buhl, N. B. Laursen, B. Hesse, L. Berglund, E. Nexxo, *Haematol./Hematol. J.* **2006**, *91*, 805.

62 C. Schmidmeier, *Ph.D. thesis*, Massey University **2016**.

63 P. Subramanian, *Foods* **2021**, *10*, 1362.

64 M. Srinivasarao, P. S. Low, *Chem. Rev.* **2017**, *117*, 12133.

65 M. Kohlmeier, *Nutrient Metabolism: Structures, Functions, and Genes*, 2nd ed., Elsevier, **2015**.

7

Design and Use of Lipid-Based Systems for Food Component Encapsulation

Kalpani Y. Perera[1,2$], *Dileswar Pradhan*[1,2$], *Shubham Sharma*[1,2$], *Amit K. Jaiswal*[1,2], and *Swarna Jaiswal*[1,2*]

[1] *School of Food Science and Environmental Health, Faculty of Science, Technological University Dublin-City Campus, Central Quad, Grangegorman, Dublin D07 ADY7, Ireland*
[2] *Environmental Sustainability and Health Institute (ESHI), Technological University Dublin-City Campus, Grangegorman, Dublin D07 H6K8, Ireland*
* *Corresponding author*

Highlights

- Food encapsulation allows the targeted, prolonged, and controlled release of food components.
- Food component encapsulation improves the characteristics and quality of food products by increasing the solubility, stability, and bioactivity of the associated ingredients.
- Lipid-based nano delivery systems for food component encapsulation are nanoemulsions, liposomes, solid lipid nanoparticles, and nanostructure lipid carriers.
- Essential oils, antioxidants, natural colorants, flavors, vitamins, probiotics, and fatty acids are encapsulated in lipid-based nano delivery systems.

7.1 Introduction

Food component encapsulation has been developed as a beneficial approach in the food industry for improving the characteristics and quality of food products. Encapsulation is able to increase the solubility, stability, and bioactivity of the food ingredients. In addition, encapsulation can improve the functioning of food ingredients in a variety of processing and storage settings while lowering degradation. This technique also enables to improve the adherence capacity and residence time of components at the target areas, allowing a prolonged, controlled release and enhanced oxidative stability [1]. The increased stability of the food ingredients results in prolonged shelf life and effectiveness. Encapsulation masks the taste, flavor, and odor of the food ingredients until they are released, and most importantly, it prevents their interactions with other components. During encapsulation, the encapsulation compound (i.e. active ingredient, base material, fill, internal phase, core,

$ *Authors contributed equally*

Materials Science and Engineering in Food Product Development, First Edition. Edited by Wing-Fu Lai.

or payload phase) is surrounded by other carrier materials (i.e. shell, wall, matrix, or coating), which are of nano, micro, and millimeter scale [2–4].

The carriers used in encapsulation can be either protein, carbohydrate, or lipid. However, lipid-based carriers have gained popularity, owing to their capacity to carry a variety of hydrophobic and hydrophilic molecules. Further, the demerits of food components, such as physicochemical instabilities, high volatility, light sensitivity, and low aqueous solubility, can be overcome by using a lipid-based delivery system. These lipid-based delivery systems are nontoxic and they have the ability of ease of production on a large scale. They have excellent biocompatibility, targeted delivery, and the ability for controlled release [1].

The different types of lipid-based systems used to carry food components are nanoemulsions, liposomes, solid lipid nanoparticles (SLNs), lipid nanospheres, lipid nanotubes, and nanostructured lipid carriers (NLCs) [3] as depicted in Figure 7.1. The lipid-based systems are usually produced by electrospinning, gelation, layer-by-layer deposition, extrusion, co-precipitation, coacervation, freeze-drying, emulsification, and high-pressure homogenization. Lipid-based carriers are usually formed from emulsifiers, lipids, bilayer lipids, and non-bilayer lipids, and the compatibility of these carriers for the food component is a very crucial factor [1, 3].

Industrial Corner

- Encapsulation is employed in more than 40% of the food industry. Many food additives that use lipid-based carriers are now on the market or are in the process of being developed [1].
- The food industry and pharmaceutical industries have increased food and nutrient encapsulation due to the growing concern of food waste, climate change, aging populations, and increased health concerns [2].
- According to the Markets and Markets market research report, the worldwide food encapsulation market was worth USD9.9 billion in 2020 and is estimated to be worth USD14.1 billion by 2025 [5].

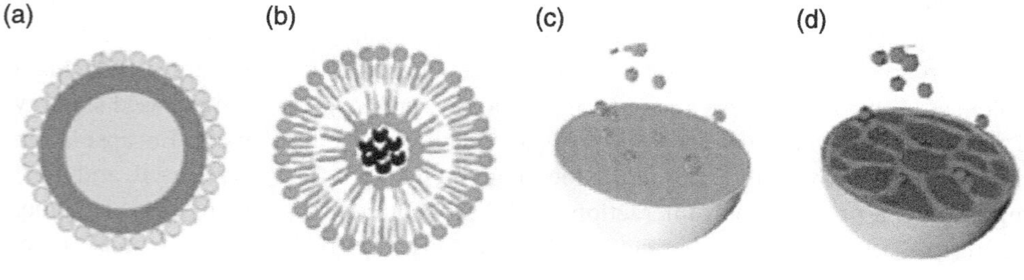

Figure 7.1 Lipid-based delivery systems: (a) nanoemulsions; (b) liposomes; (c) solid lipid nanoparticles (SLN); and (d) nanostructured lipid carriers (NLC). *Source:* Nahum and Domb [4]/ MDPI/ CC BY 4.0.

There is a variety of applications in the food industry in using lipid-based systems for food component encapsulations, such as encapsulation of vitamins, minerals, food ingredients, macronutrients, antioxidants, phenolic compounds, prebiotics, probiotics, food colorants, antimicrobial agents, and enzymes [6]. For example, gallic acid is encapsulated by monoglycerides of vegetable fatty acids using nanoemulsions [7], α-tocopherol is encapsulated by lemon oil using nanoemulsions [8], resveratrol and coenzyme Q10 is encapsulated by medium-chain triglyceride using microemulsion [9], vitamin D is encapsulated by stearic acid and oleic acids using NLCs [10], β-carotene is encapsulated by phospholipid S-100 using SLNs [11], and fish oil is encapsulated by palm oil using solid lipid microparticles [12]. The applications of lipid-based systems encapsulated food components are described in detail in a later section and also in Table 7.1.

Table 7.1 Recent studies on active food component incorporated lipid-based nano delivery system.

Active food components	Lipid-based nano delivery system	Results	References
Turmeric extract	Nanostructured lipid carriers	• Turmeric extract-NLC (T-NLC) had high antioxidant activity • T-NLC had high antibacterial activity against *Escherichia coli*	[50]
α-Tocopherol	Solid lipid nanoparticles	• Higher antioxidant capacity • Decrease in crystallinity	[31]
Carvacrol and astaxanthin	Solid lipid nanoparticles	• Enhanced stability and radical scavenging • Remove 92.87% and 84.25% and kill 94.81% and 83.26% of total *Pseudonymonas aeruginosa* and *Staphylococcus aureus* biofilms	[69]
Beeswax	Solid lipid nanoparticles	• Increase the shelf life of strawberries stored	[70]
Citral essential oil	Nanoemulsions	• The interaction of citral with microbial cell membranes was enhanced	[71]
Nisin	Nanoliposomes	• Nisin was encapsulated with lysozyme to prevent *Listeria monocytogenes* and *Salmonella* from growing in skim milk	[72]
Vitamin D	Nanostructured lipid carriers	• Showed positive zeta potential (+30 mV) • Remain stable for 60 days at 25°C	[10]
Vitamin D3	Nanostructured lipid carriers	• High encapsulation efficiency (85.6%) • Stable under acidic condition • No significant changes in stability while stored at 25°C for 0–20 days • Controlled release under simulated gastrointestinal conditions • More than 90% release after 8 h of digestion in simulated intestinal fluid	[73]

(Continued)

Table 7.1 (Continued)

Active food components	Lipid-based nano delivery system	Results	References
Quercetin	Nanoliposomes	• High encapsulation efficiency (71.14%) • 9% degradation after 14 days of storage at 4°C • 35.83% degradation after 14 days of storage at 37°C • Improvement in antioxidant activity • Inhibitory effect on HepG2 cells	[74]
β-Carotene	Solid lipid nanoparticles	• Improved anticancer activity • Increase in bioavailability of β-carotene on oral administration • Retention of antioxidant activity • Good physical stability during three months of storage	[11]
Zataria multiflora Bioss. essential oil	Nanoliposomes	• Essential oil can affect the size of nanoliposomes • 18.36% encapsulation efficiency • Proper antibacterial activity against *E. coli* O157:H7	[75]
Zataria multiflora essential oil	Solid lipid nanoparticles	• Good physical stability (zeta potential: −37.8 ± 0.8 mV) • Encapsulation efficacy of 84 ± 0.92% • 79% inhibition on the growth of fungal pathogens	[76]
Anthocyanin	Liposomes	• Highest encapsulation efficiency of 78.2 ± 1.4% with 40% anthocyanin addition • Highest bioactive loading of 23.8 ± 0.2% • Slow-release (≤35.9%) in the simulated gastric fluid • Rapid release in the simulated intestinal fluid	[77]
Curcumin	Nanoliposomes	• The entrapment efficiency of 57.1% • Good stability against alkaline pH and metal ions • Good storage stability at 4°C • Equal cellular antioxidant activity when compared with free curcumin • The cellular uptake rate of 69.6% ± 5.4% after 4 h of incubation	[78]
Omega-3 fish oil and α-tocopherol	Nanostructured lipid carriers	• Sustained oil release at a physiological pH, and long-term stability after 75 days of storage • More successful in inhibiting the formation of the primary oxidation products	[79]

Table 7.1 (Continued)

Active food components	Lipid-based nano delivery system	Results	References
Linalool	Nanostructured lipid carriers	• Encapsulation efficiency of 79.563% • Drug loading of 7.555% • 15.564% release of linalool at 180 min • Relative bioavailability around 393.34%	[80]
Anthocyanin	Solid lipid nanoparticles	• Entrapment efficiency about 89.2 ± 0.3% • Percent of release at pH = 7.4 was more than at pH = 5 and pH = 3 • Most appropriate condition for anthocyanin storage is at low pH (pH = 3) and temperatures ($T < 25°C$)	[81]

7.2 Lipid-Based Nano Delivery Systems for Food Component Encapsulation

7.2.1 Nanoemulsions

Nanoemulsions comprise two immiscible phases of water and oil that are thermodynamically stable, which can be stabilized by an interfacial layer of surfactants and cosurfactants [13]. Water-in-oil (W/O) and oil-in-water (O/W) are the two common types of nanoemulsions that are distinguishable by their structural characteristics as shown in Figure 7.2. W/O nanoemulsions are prepared by dispersing small water droplets in oil, whereas O/W

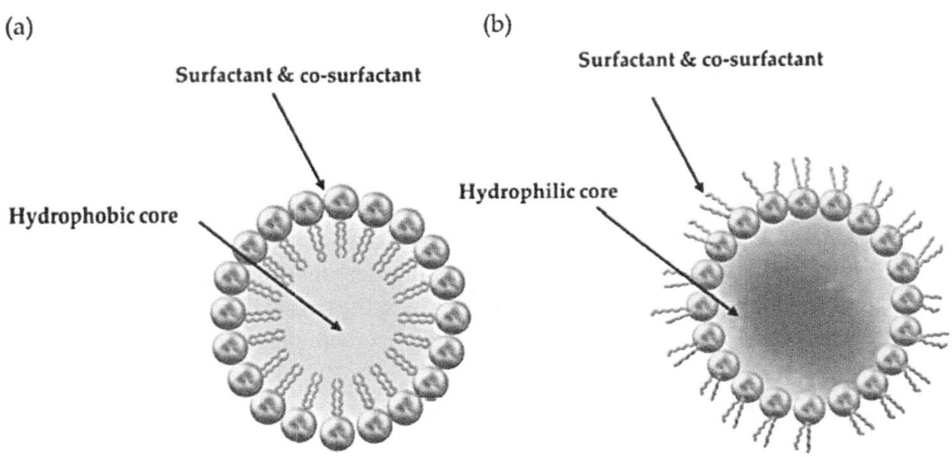

Figure 7.2 Structure of an (a) O/W nanoemulsion and (b) W/O nanoemulsion. *Source:* Adapted from de Oca-ávalos et al. [16].

nanoemulsions are prepared by dispersing small oil droplets in water. More complex nanoemulsions, such as O/W/O or W/O/W, are also possible, although they are uncommon [14]. Because nanoemulsions are essentially unaffected by chemical and physical changes, they have demonstrated promising potential in a variety of sectors, including food component encapsulation and drug delivery [15].

The production of nanoemulsions can be classified into two types such as low-energy and high-energy emulsification. The low-energy emulsification techniques work by causing small oil droplets to form spontaneously when the temperature or composition of a system, including surfactant, water, and oil, is altered in a specific manner. Emulsion inversion, phase inversion temperature, and spontaneous emulsification methods are included in this category [14]. The high-energy emulsification techniques work by producing huge pressure differences within mechanical devices or ultrasound waves to create droplet disruption. Ultrasonic homogenization, high-pressure homogenization, and microfluidization are the most common techniques included in this category. Due to their capability of producing fine emulsions from a range of materials and the flexibility of controlling emulsion droplet size distribution, these approaches are preferred in industrial operations [16].

Today, there is an increasing interest in the food-processing industry to develop nanoemulsions using label-friendly substances so that novel food items can find appropriate applications. Various cuisines and food ingredients are available in emulsion form, including sauces, soups, desserts, dressings, butter, and beverages, with the majority of them falling into the O/W emulsion category [17]. In addition, because of their capability to enhance bioactive solubilization and the possibility of improving absorption in the gastrointestinal (GI) tract due to surfactant-induced permeability changes, nanoemulsions are an excellent option for delivering poorly water-soluble food components, such as lipophilic vitamins and fish oil. After ingestion, droplets easily disperse in the stomach to form tiny nanoemulsion droplets, allowing for widespread diffusion of the encapsulated bioactive compounds throughout the GI tract [18].

Nanoemulsions have several advantages, including potential application for health care and personal products, thermodynamic stability, and high kinetic stability against creaming and flocculation. They can also be transported easily through the cell membrane, increasing their biocompatibility and bioavailability [19]. However, nanoemulsions have drawbacks like particle attraction during storage, creaming, gelling, membrane permeability, and further release of bioactive, flocculation, and Ostwald ripening. However, these challenges can be minimized by using a suitable surfactant and controlling the surface charges. Besides, the production of nanoemulsions is thought to be unprofitable in the food industry. However, with the introduction of new equipment for the fabrication of nanocarriers and increased competition among food companies, the cost of nanoemulsion preparation will likely reduce and reach the price of normal macroemulsions [17].

7.2.2 Liposome

Liposomes were discovered by Alec D Bangham in the 1960s [20]. Liposomes are closed bilayer structures formed autonomously by hydrated phospholipids that are widely used as effective delivery systems because of their ability to encapsulate bioactive amphipathic hydrophilic and lipophilic molecules into the inner water phase or within lipid leaflets [20]. Because of their amphipathic character, phospholipids have a strong inclination to form membranes when dispersed in aqueous solutions. Their polar heads prefer to interact with the aqueous environment, whereas their long apolar aliphatic chains encourage contact among themselves. These dual features facilitate the production of two lipid layers in an aqueous solution. Each layer's hydrophobic chains face each other, forming a lipophilic inner compartment that works as an interior and outward permeability barrier [21]. These lipid bilayers are formed by hydrophobic contacts, while van der Waals forces maintain the long hydrocarbon tails together, strengthening the architecture. Finally, this arrangement is stabilized by hydrogen bonds and polar contacts between water molecules in the aqueous environment and the polar heads of lipids as shown in Figure 7.3. The composition, concentration, temperature, and geometric form of lipids all influence their eventual organization [21, 22].

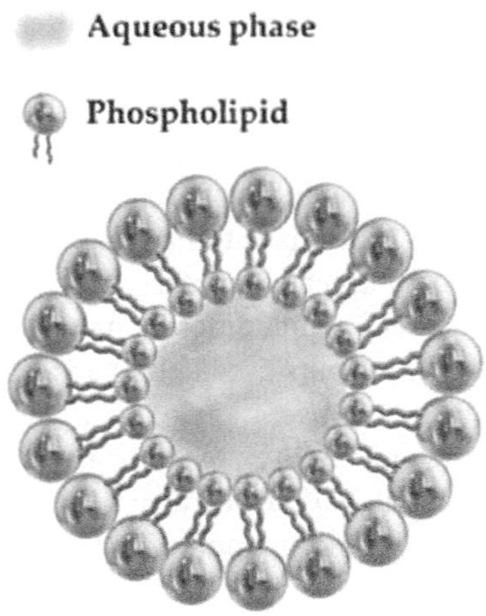

Figure 7.3 The structure of a liposome. *Source:* Adapted from de Oca-ávalos et al. [16].

Liposomes can be classified into two types: standard and liposomes with modified surface characteristics. Traditional liposomes are first-generation products with a variety of lipid compositions and physicochemical features. Because of their reported functional properties, such as their physicochemical properties, kinetic stability, efficient encapsulation capacity, biocompatibility with food constituents, low cost of raw materials used in manufacturing, and feasibility as nutraceuticals, antimicrobials, and food-flavoring agents, interest in liposome nanoparticles has grown [23]. Liposome nanoparticles' key advantage is their natural compositions, which can remove or even lessen concerns associated with their use in various food systems [23].

Food fortification could also benefit from liposome nanoparticles. Because many countries have low fish consumption, dietary supplementation with fish oil capsules looks to be a simple way to increase omega-3 (ω-3) long-chain polyunsaturated fatty acid intake [23, 24]. Fish oil that is highly refined, odorless, or microencapsulated may be able to alleviate unpleasant sensory qualities. Iafelice et al. have nano-encapsulated fish oil with liposomes for yogurt fortification for these reasons. In addition, another study has also fortified yogurt with liposome-encapsulated fish oil and found that the sensory features were similar to those of the control yogurt compared with yogurt fortified with free fish oil [25]. Amjadi et al. studied the improvement in betanin stability by liposomal nanocarriers. They have also studied its application in gummy candy. According to their findings, the betanin concentration and antioxidant activity of samples containing betanin-nanoliposomes were at least double that of samples containing free betanin. The sensory metrics of panelists for gummy candies fortified with betanin-loaded nanoliposomes and those supplemented with betanin alone reveal no differences [26].

7.2.3 Solid Lipid Nanoparticles

Lipid nanoparticles comprise lipid matrix, bioactive compound, water, and emulsifiers. They are mainly divided into two types: SLNs and NLCs [27]. SLNs are the first-generation nanocarriers with a highly ordered crystalline structure and a solid oil/water emulsion (Figure 7.4). Numerous lipid components, emulsifiers, and co-emulsifiers are generally recognized as safe (GRAS) – or have approved regulatory conditions for the preparation of bioactive component nanocarriers among the materials used to develop lipid nanoparticles [28]. The mobility of embedded substances is regulated by the physical lipid matrix structure in SLNs made from a single solid lipid species or a combination of lipid classes. Different lipid modifications, such as α-, β-, and β'-modification or supercooled melt, can coexist in SLN. As a result, different nanoparticle forms emerge, such as spheres or platelets [29].

Components used to make SLNs, such as lipids, surfactants, cosurfactants, and cryoprotectants, impact their status in terms of encapsulation and loading effectiveness, stability, and targetability. As a result, selecting appropriate and suitable active compounds is essential. For the effective integration of produced SLNs into food products, stability is crucial. Considering antimicrobial agents have lower mobility in solid lipids than in liquid lipids, the release of antimicrobial agents from SLNs is relatively modest. Different strategies, such as homogeneous matrices, bioactive-enriched shells, or bioactive-enriched core models, are used to include antimicrobial agents into SLNs. These nanoparticles are susceptible to disintegration by coagulation, agglomeration, flocculation, and sedimentation due to the complex nature of the food matrix (temperature, pH, ionic strength, etc.) and sterilizing methods used to keep foods. As a result, fabricating SLNs for food applications requires more precision than fabricating SLNs for pharmaceutical applications.

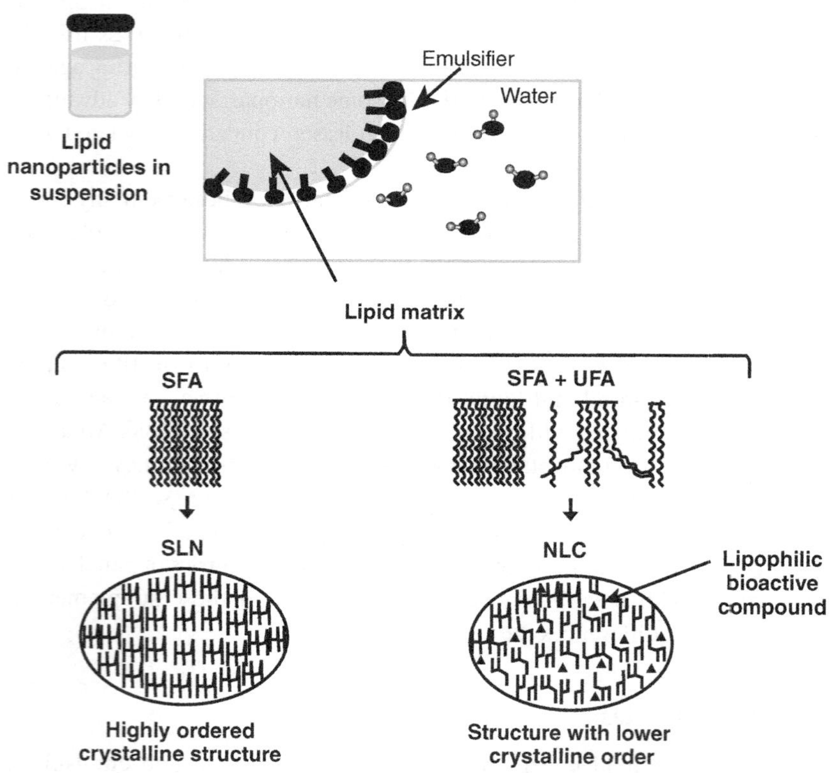

Figure 7.4 Structure of a lipid nanoparticle. *Source:* Adapted from Aditya and Ko [28].

In addition, the crystallinity of the lipid is crucial in influencing the nanoparticles' stability. Fabrication of SLNs containing lipids with no structural distortion resulted in active discharge from the nanoparticle matrix due to polymorphic transformation of unstable subcells into stable β-subcells. The space in the matrix of the SLNs that provides accommodation for bioactive is obliterated when a subcell is converted into β-subcells. As a result, optimizing the mixture of mono-, di-, and triglycerides is critical [28]. De Carvalho et al. have produced polymeric films containing 30%, 50%, and 70% SLNs encapsulated in tocopherol based on poly (vinyl alcohol) (PVA). They observed incorporation of nanoparticles affected the structure of the films, resulting in a decrease in crystallinity, according to thermal studies. Furthermore, films containing-tocopherol entrapped in SLNs had greater antioxidant potential and the tocopherol-controlled release validated the viability of its usage as active packaging for food preservation [30]. In addition, Tonyali et al. produced an antimicrobial packaging system using pullulan-SLNs loaded with thymol, cinnamaldehyde, and eugenol, which allows for the regulated release of the active chemicals [31].

7.2.4 Nanostructured Lipid Carriers

NLC is a type of O/W emulsion system, comprising both solid lipids with melting points above the room temperature and liquid lipids with melting points far below the ambient temperature within the inner phase along with appropriate surfactant in the outer phase to achieve a fine dispersion of these lipids in the aqueous medium [32]. However, the choice of these elements and their ratios might significantly impact the final characteristics and functionalities of the developed formulation [33]. Solid lipids are often combined with liquid lipids in a 70:30 to 99.9:0.1 ratio, with surfactant levels ranging from 1.5% to 5% (w/v) [34].

Depending on the polymeric nanoparticle production technique, NLCs can be formulated using various methods [35]. These methods are membrane contractor, high-pressure homogenization, solvent injection/solvent displacement method, solvent diffusion, solvent emulsification–evaporation, emulsification sonification, phase inversion, and microemulsion [36]. However, the high-pressure homogenization technique is preferred over the others since it requires no solvents during the production and can be readily scaled up from a laboratory scale (batch size 2 l) to a large-scale production (batch size > 50 l) [34].

Three distinct types of NLCs are produced based on the composition of the lipid blends and techniques used for their formulation [37]. The NLCs type 1 (imperfect type) are made up of spatially different lipids, which increases their capability to incorporate bioactive components [38]. The NLCs type II (structureless type) is generated when the lipid solidifies in a solid but in the amorphous state, which decreases the expulsion of encapsulated bioactive compounds owing to the crystallization process [39]. In NLCs type III (multiple types), bioactive components' solubility in oil is greater than that of solid lipids, and the solubilized components in oil are surrounded by solid lipids during production [32]. Subsequently, fixation of the oily nano compartments occurs as the solid lipid enveloping these droplets solidifies [40]. As a result, a high bioactive loading may be attained while avoiding bioactive leakage until the solid lipid is degraded [32].

Because of the particles' tiny size, the characterization of different types of NLCs can be difficult due to the system's complexity, which also includes dynamic phenomena. However, characterization of the NLCs may be a necessary precondition for product quality management [41]. The primary characterization that needs to be carried out for the description of NLCs is the biological fate of nanoparticles, release manner, crystalline status, drug loading, encapsulation efficiency, chemical stability during processing and storing (alone or as a food additive), particle morphology, zeta potential, size, and size distribution [42]. The advantages and the disadvantages of the lipid-based nano delivery systems are depicted in Table 7.2.

Table 7.2 Pros and cons of lipid-based nano delivery systems.

Lipid-based nano delivery systems	Pros	Cons	References
Nanoemulsions	• Unaffected by chemical and physical changes • Enhanced bioactive solubilization • Improved absorption in the gastrointestinal tract • Thermodynamic stability • High kinetic stability • Transport easily through cell membrane	• Article attraction during storage • Creaming • Gelling • Membrane permeability • Further release of bioactive • Flocculation • Ostwald ripening • Expensive to prepare	[15, 18–20]
Liposome	• Physicochemical properties • Kinetic stability • Efficient encapsulation capacity[23] • Low cost of raw materials used in manufacturing • Encapsulate both hydrophilic and hydrophobic components • Biocompatibility • Low toxicity • Biodegradability • Protect drug from environmental conditions	• Poor stability • Requirement of special storage conditions • Limited penetration	[23]
Solid lipid nanoparticle (SLN)	• Ease of production on a large scale • Biocompatibility • Low toxicity • Biodegradability • Low cost • Long-term stability	• Susceptible to disintegration by coagulation, agglomeration, flocculation, and sedimentation • Low drug loading capability • Risk of gelation	[28]
Nanostructured lipid carrier (NLC)	• Ease in producing on a large scale • Increased capability to incorporate bioactive components • High bioactive loading • Avoid bioactive leakage until the solid lipid is degraded • Low toxicity • Biodegradability • Low cost	• Higher dilution of dispersion • Not suitable for thermolabile drugs	[39]

7.3 Mechanism of Action of Encapsulated Food Components

The microencapsulation method, as well as the physicochemical parameters of the core and shell materials, determine the resulting release properties of the lipid-based nano delivery systems. In reservoir systems, such as liposomes, the shell can operate as a rate-limiting barrier, inhibiting the release of bioactive substances that have been encapsulated. The release of the active component is triggered when the microencapsulated systems come in contact with the external liquid. The first step is the influx of the active component through the lipid shell. This is followed by the dissolution of the encapsulated ingredients in the core. Finally, effluxion of the dissolved bioactive components takes place. Liquid solutions develop inside the core, coexisting with the encapsulated contents. The liquid inside the core grows increasingly concentrated over time. When the concentration of the main constituents is compared with the concentration in the liquid from the outside, a positive gradient is formed. The concentration gradient is the driving force for bioactive chemicals to be released into the external environment. As the concentration of the bioactive substance in the core diminishes and the concentration gradient decreases, the release rate will begin to slow. The factors affecting the release of the bioactive components from the core are the chemical nature of the shell material, such as the type of lipid and the thickness. In addition, the processing conditions and parameters also affect the release of the bioactive components [43].

The release mechanism of the bioactive components from the matrix-type lipid systems such as SLN and NLC are more complex as they entail structural changes to encapsulated bioactive substances. Dissolution and erosion of the matrix can control the release of the bioactive components from the lipid-based nanocarriers. Due to the elongation of the diffusion channel, the diffusion release rate reduces over time. This impact is overcompensated by increased permeability due to aging after a given amount of time that leads to gradual polymer erosion. As a result, the bioactive ingredient's release rate is generally diminished at first, but it then grows with time. On the other hand, for poorly soluble bioactive compounds, dissolution and diffusion are almost nonexistent. The release of soluble or non-soluble bioactive substances is generally mediated by erosion [43]. Factors such as the particle sizes of the bioactive components determine the release rate of the components. Faster release will be facilitated by smaller particle sizes and irregular forms of pellets.

The most common type of bioactive compound release method is solvent-activated release. Here, lipid-based systems can release enzymes in response to a change in pH. Melt activation is another method of release, where melting destroys the membrane on the wall or the wall itself, which comprises lipids or waxes, releasing salts, leavening, flavorings, and nutrients. Spray chilling is a common technique used to form encapsulation components for this method. The above technique is limited to water-soluble flavorings because only hydrophobic flavors will pass through the wall material. Another method used for the release of bioactive components is the alteration of surface properties. This is done so that the lipid carriers will accumulate at the target site [44].

The food components encapsulated in the lipid-based systems provide site-specific targeting, controlled release of food components, and lead to the increased shelf life of the food component. Most of the encapsulated food components, such as essential oils (EOs), fatty acids, and vitamins, are hydrophobic, thus having poor solubility characteristics, leading to the gradual reduction of the bioavailability and sustainability of these components. Lipid-based carriers aid in the favorable approach of these food components to the GI tract. Lipid-based systems increase the intestinal absorption of the water-insoluble food components. These lipid-based systems improve intestinal absorption by: (i) increasing solubilization in the GI tract by delivering bioactive substances in a soluble form and reducing precipitation in the intestinal environment; (ii) endogenous and exogenous lipid transport, as well as intestinal lymphatic transport of bioactive substances; or (iii) interacting with apical membrane lipid transporters. The movement of lipid-based system-encapsulated food components is through the intestinal membrane, involving a variety of transfer methods, including paracellular absorption, M-cell mediated transport, and endocytosis. The transport of molecules across adjacent epithelial cells is referred to as paracellular absorption, although the rate-limiting stage in this absorption process is the transport across the tight junction. The nutrition absorption from the colon and transfer of intestinal microbial antigens to gut-associated lymphoid tissue for mucosal and systemic immune responses are both facilitated by M-cell mediated transport. Clathrin-mediated endocytosis, caveolae, micropinocytosis, and phagocytosis are the four methods by which intestinal cells take up nutrients from the external environment during endocytosis as demonstrated in Figure 7.5. These delivery systems are designed to be released at a specific target at the GI tract by the trigger of environmental factors, such as ionic strength, pH, or enzyme activity [41, 45].

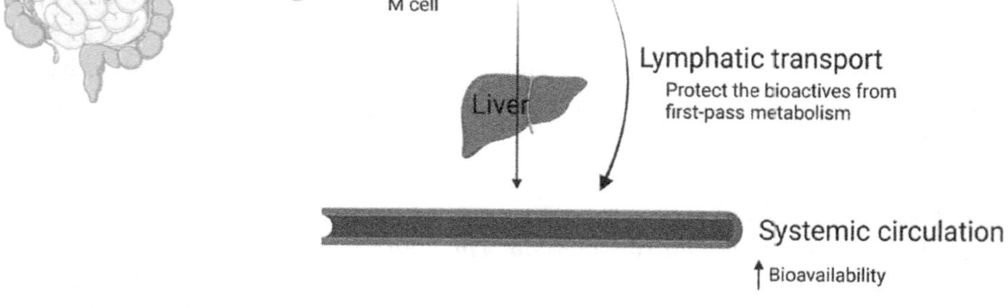

Figure 7.5 Intestinal transport mechanism of lipid-based systems for food component encapsulation. *Source:* Iqbal et al. [43].

7.4 Encapsulation of Food Components in Lipid-Based Nano Delivery Systems

7.4.1 Essential Oils

Microencapsulation technology has been employed in food technology to protect and enhance essential materials, such as EOs. In the food industries, lipid-based colloidal delivery systems for encapsulating bioactive components, such as SLNs, emulsions, NLCs, and vehicular carriers (e.g. liposome, phytosome, etc.) are used. These techniques can be used to convert liquid and volatile molecules to solids, alter the solubility of compounds, and protect and modulate their release and functionality.

Baldim et al. had developed a nanostructured lipid system loaded with *Lippia sidoides* EO. They investigated the effect of the system's composition on its properties, such as stability, retention of an active marker compound, and antifungal activity, against *Candida albicans*. Smaller droplets, stronger zeta potential, and thymol retention were seen in oleic acid-containing formulations, although stability was decreased. Both the EO and lipid formulations of *L. sidoides* were found to have potent antifungal action against *Candida albicans* [46]. Pabast et al. formulated a chitosan-based coating loaded with nanoliposomes encapsulated with *Satureja khuzestanica* EO specifically for application on lamb meat to enhance shelf-life properties. The coating successfully delayed the microbial and chemical spoilage on the meat while the encapsulation process decelerated the release of the EO from the matrix that extended the antimicrobial and antioxidant activities [47]. Thus, as per this study, encapsulation of EO in lipid-based delivery systems allows the controlled release of EO making them act as antimicrobial and antioxidant agents for a longer time period.

7.4.2 Antioxidants

Antioxidants are also known as "free-radical scavengers." Antioxidants can come from both natural and synthetic sources. Antioxidants are known to be abundant in certain plant-based foods. Plant-based antioxidants are a type of phytonutrient or nutrient that comes from plants. Because of their auto-oxidation qualities, antioxidants, such as tocopherols, carotenoids, and ubiquinones, have been employed to preserve foods against deterioration, lipid oxidation, and discoloration, which are also relevant in dietary supplements. Encapsulation using a NLC is recommended to compensate for the absence of unsaturated fatty acids in foods. In comparison with other particulate systems, lipid nanoparticles have a number of advantages that include the convenience of large-scale production, biodegradable and biocompatible materials, low toxicity ability, the ability to control and modify active agent release, solubility enhancement, and the ability to incorporate both hydrophilic and lipophilic agents [48].

In recent years, using nanocarriers for food fortification has been a good method that can be effective in the transport and stability of hydrophobic chemicals. Nanocarriers also offer improved bioavailability and suppress the development of off-flavor and off-color offood during digestion [48, 49]. Karimi et al. used a nanostructured lipid carrier to encapsulate turmeric extract as a water-insoluble bioactive compound (NLC). The high shear homogenization process was used to make turmeric extract loaded NLC (T-NLC). The study observed T-NLC had significantly higher antioxidant activity than free turmeric extract, and it can therefore be used in food products with high functional effects, according to its findings [50]. Huang et al. have used a high-pressure homogenization process to encapsulate quercetin and linseed oil into NLC. The antioxidant properties of quercetin-loaded NLCs were enhanced with the addition of linseed oil [51]. The NLC has superior drug loading, entrapment efficiency, and physical stability due to its structure. Further, NLC can also improve the solubility and chemical stability of lipophilic compounds, making them suitable carriers for antioxidant substances.

7.4.3 Natural Colorants

Natural colorants are increasingly utilized in food and beverage industries instead of synthetic colorants due to the rising demand for healthy food products [4]. However, the challenges that remain in their utilization are their higher costs than synthetic colorants, inherent molecular instability, a narrower range of hues, and the requirement of higher concentrations to attain equivalent color intensity [52]. In addition, most natural colorants are oxygen, light, and heat-sensitive, which can lead to shifting or loss of color. Besides, they are also sensitive to various characteristics of food products, such as the presence of some organic compounds, metal ions, proteins, and pH [53]. Therefore, the colorants' stability throughout industrial processing and in the food matrix is a critical parameter to consider during the development of food products [54].

Encapsulation techniques have appeared as a viable method for improving natural colorants' stability, bioaccessibility, and bioavailability. Lipid-based carriers enable the delivery of a wide variety of food components. Lipid-based encapsulation has proven to be a good alternative and remarkable approach for delivering bioactive compounds, with high chemical, physical stability, higher bioavailability, and prolonged release. The ability to protect embedded food ingredients from external influences could also extend shelf life, which has immediate advantages for the food industry and tends to minimize food waste. To date, a number of lipid-based carriers have been developed, addressing a variety of difficulties in the food industry and assisting in the creation of higher-quality foods with greater health benefits. Nanoemulsions, liposomes, SLNs, lipid nanospheres, lipid nanotubes, and NLCs are some of the lipid-based systems utilized to transport food components. Encapsulation of vitamins, minerals, food ingredients, macronutrients, antioxidants, phenolic compounds, prebiotics, probiotics, food colorants, antimicrobial agents, and enzymes are a few of the possibilities for lipid-based food component encapsulation in the food industry. Lipid-based delivery methods are currently employed in the food industry; however, they must be improved in terms of reproducibility, affordability, and scalability. The primary goal of this field in the future is to produce more affordable, high-quality lipid-based methods for food component encapsulation [54]. Several studies have recently demonstrated the encapsulation of natural colorants utilizing a lipid-based delivery system. In a study, Oliveira et al. developed tristearin/high oleic sunflower oil (HOSO) NLC and tristearin SLN for the encapsulation of β-carotene. Compared with SLN, the loading capacity of β-carotene in NLC was higher, and the loading capacity increased with the increase in HOSO concentration. Even though overall β-carotene deterioration was equivalent across all delivery systems, the color analysis revealed that encapsulated β-carotene deterioration was less for the HOSO content [39]. In another study, Chi et al. fabricated anthocyanin nanoliposomes (AN) to improve the bioavailability and stability of anthocyanins (ACNs). For preparing optimal AN, 0.15 g ACNs, lecithin to cholesterol ratio of 5.98, and lecithin concentration of 10.75 g/L were used. After 28 days of storage at 4°C under dark conditions, AN (encapsulated form) presented a high retention rate of 83.34%, while AS (nonencapsulated form) had a 76.63% retention rate. Further, the ACNs from the AN exhibited a prolonged release profile and higher stability in vitro digestion. Liposomes can minimize degradation while assisting the in-target delivery and prolonged release of colorants [55].

7.4.4 Flavors

Flavor compounds are utilized as food additives primarily to improve the gustatory and olfactory sensations of the product. However, their chemical instability and volatility are the major challenges regarding their utilization. Besides, most flavor compounds are oxygen, heat, or light-sensitive; thus, they can degrade during production, and their shelf life can be reduced or lost through consumer manipulation and storage [56]. In addition, flavor compounds are very vulnerable to changes in enzymes, salt, acid, moisture, and pH, which can result in off-flavors. Encapsulation is the most efficient and widely used method for increasing the availability of flavor components and mitigating their deterioration. Flavors are protected from thermal and oxidative damage by encapsulation, which also ensures their efficient processing. On the other hand, the type of flavor encapsulation technology to be used is determined by the type of food product it will be used in and the processing conditions it will be subjected to [57].

In recent times, various lipid-based delivery systems have been reported by researchers for the encapsulation of flavor compounds. For example, Rodenak-Kladniew et al. developed SLN of cetyl palmitate, cetyl esters, and myristyl myristate for the encapsulation of linalool (LN). More than 80% of encapsulation efficiency of LN was observed in all formulations of SLN [58]. Further, the SLN (MM)-LN formulation exhibited in vitro antiproliferative effects on lung adenocarcinoma (A549) and hepatocarcinoma (HepG2) cell lines in a dose-dependent response. Here SLN provides a LN encapsulation by providing good physical stability, controlled drug release, and reduced degradation of LN. In another study, Eltayeb et al. prepared core-shell lipid (stearic acid) nanoparticles to encapsulate ethylvanillin. Increasing the stearic acid concentration increased ethylvanillin's encapsulation efficiency, and maximum efficiency of 71 ± 3% was achieved at 4 wt%. Furthermore, after 60 minutes, the 4 wt% lipid nanoparticles released around 5% of ethylvanillin, whereas the 1, 2, and 3 wt% lipid nanoparticles released the same amount in 12, 20, and 48 minutes. These findings suggested that stearic acid at the core's interface works as a shell, protecting the ethylvanillin inside the nanoparticles [58].

7.4.5 Vitamins

Vitamins are essential micronutrients that perform a variety of biochemical operations, including the improvement of the immune system and the treatment of neurological problems. Vitamins may also aid in the reduction of cholesterol levels and the prevention of cardiovascular disease, as well as the improvement of physical endurance and the maintenance of a standard metabolic rate [35]. However, vitamins need protection from various physicochemical factors, such as oxidants, light, and temperature, as vitamins are sensitive molecules because of their unique structure. Encapsulation of vitamins within different nanocarriers is one of the innovative strategies for their protection [59]. Besides, the physicochemical properties of vitamins are substantially influenced by the nanoencapsulation method and delivery mechanism used. As a result, the requisite size, physicochemical characteristics, nature of the encapsulated vitamin, and wall material must all be considered when selecting a nanoencapsulation technique [60].

In recent years, various authors have reported using a lipid-based nano delivery system to encapsulate vitamins. In a study, folic acid was encapsulated in liposomes (containing histidine) prepared using soybean lecithin, SPAN 20, and cholesterol. The liposomes exhibited an 84% encapsulation efficiency and a drug loading of 0.123 mg/g. Further, a reduction of about 30% folic acid content in the acceptor compartment was observed after 300 minutes of incubation at pH 7, which was due to the liposomal barrier properties. On the other hand, the presence of both SPAN 20 and cholesterol in the liposomes led to a quicker release of folic acid when compared with plain liposomes. Encapsulation of folic acid overcame its poor solubility issue in addition to its added advantage of controlled release [61]. In another research, Couto et al. used the supercritical CO_2 method for Vitamin B2 encapsulation in SLNs. They obtained an encapsulation efficiency of $12 \pm 0.03\%$ and a bioactive load of 0.54 ± 0.05 mg/g using optimized parameters, such as 15 MPa of pressure, 2% w/w of Vitamin B2 concentration, 35 000 Da molecular weight of stabilizer (polyethylene glycol), and 5% w/w of surfactant (sodium lauryl sulfate) concentration. SLN was able to prolong the shelf life of vitamins by reducing the exposure to microorganisms and reducing the degradation caused due to heat, oxygen, and acid [62].

7.4.6 Probiotics

Probiotics are living microorganisms that provide several health benefits when administered in appropriate quantity. However, the consummation of a product containing probiotics concentration of 10^6–10^8 CFU/g of food and the safe arrival of microorganisms in the intestine are the most critical factors in achieving health benefits [63]. Nevertheless, the microorganisms' growth in the food matrix depends on intrinsic characteristics of food, such as buffering capacity, nutrient availability, pH, and water activity, and various environmental factors, such as temperature and relative humidity. Thus, selecting an appropriate food matrix is critical in developing probiotic functional food products as it should help the inoculated microbes regulate colonization and survival in the GI system [64]. Using encapsulation techniques, the viability, survivability, and functionality of probiotics in food products can be promoted [65]. In this regard, various lipid-based encapsulation systems have emerged as promising solutions in recent years.

In a study, Hosseini et al. developed nanoliposomes coated with chitosan (CH)–gelatin (GE) polyelectrolytes for encapsulation of *Lactobacillus rhamnosus* probiotics. The encapsulation efficiency was around $81.8 \pm 0.6\%$ to $87.5 \pm 0.8\%$ for uncoated and CH–GE-coated liposomes. Further, a substantial improvement in the viability of cells in simulated GI fluids was observed in the CH–GE-coated liposomes. After administration, encapsulation technologies aid in effectively loading the probiotic cells and shielding them from the acidic environment of the stomach. However, the manner in which the encapsulation is carried out is critical, because high temperatures destroy microorganisms [66]. Similarly, Matos-Jr. et al. encapsulated *L. rhamnosus* 64 and *L. paracasei* BGP1 in solid lipid microparticles covered by a complex of gum arabic and gelatin. The stability of the probiotics in GI conditions and the presence of salt improved due to the encapsulation. In addition, during the storage period, the encapsulated microbes retained their viability. Further, the dosage of cytokines (IL-2, IL-6, IL-10, and TNF-α) and S-IgA in mice revealed that the encapsulated probiotics preserved their functionality. Microencapsulation has the power to enhance probiotic viability under extreme conditions without diminishing their immunomodulatory activity.

7.4.7 Fatty Acids

The food industry is constantly seeking to develop novel food products incorporated with bioactives in response to the increased demand for healthier products. Among these, ω-3 fatty acids, particularly docosahexaenoic acid (DHA) and eicosapentaenoic acid (EPA) have piqued the interest of researchers owing to their distinct health advantages. To obtain the beneficial characteristics of DHA and EPA, the North Atlantic Treaty Organization and the World Health Organization have recommended daily consumption of 0.3–0.5 g DHA and EPA [4]. Developing [66] food products rich in ω-3 fatty acids can help increase the intake of DHA and EPA. However, the intrinsic properties of ω-3 fatty acids, such as limited bioavailability, sensitivity to oxidation, and poor water solubility, present significant hurdles in developing an efficient composition. In this aspect, nano delivery systems have gained the attention of several researchers as a possible medium for the effective encapsulation of ω-3 fatty acids and their targeted release [67].

Several authors have reported the utilization of lipid-based nanocarriers for the encapsulation of fatty acids. Cassano et al. reported α-tocopheryl linolenate based SLNs that had a 77% efficiency in encapsulating ω-3 α-linolenic acid. Notably, the developed SLNs are more efficient than both α-tocopherol and α-linolenic acid in inhibiting human melanoma cell viability in vitro. The findings suggested that the SLNs could be utilized to deliver and protect unstable compounds, such as α-linolenic acid, from deterioration caused by oxidative stress pathways [68]. In another research, Inapurapu et al. developed nanoemulsions using sunflower oil, isopropanol, and Tween 85 for ω-3 fatty acids encapsulation. After 4 weeks of storage at either 4°C or 25°C, around a 15% decrease in fatty acid content was recorded in the nanoemulsions; however, storage at 40°C resulted in a reduction of around 50% within 2 weeks. Besides, the ω-3 fatty acids bioaccessibility was around 53% in the developed nanoemulsions. In addition, nanoemulsions can be employed in food-based approaches and supportive therapies for enhancing ω-3 fatty acid nutrition. These nanoemulsion-loaded fatty acids are kinetically stable, protected from oxidative damage, and has improved bioavailability [68].

Industrial Corner

The growing demand for encapsulation of taste and colorants has boosted the encapsulation market. The consumer demand for fortified and functional meals has resulted in an increase in food additive encapsulation [3].

Food industry application of encapsulation involves encapsulation of food products, such as phenolic compounds (e.g. quercetin, kaempferol), antioxidants (e.g. ferulic acid, curcumin), vitamins (e.g. vitamin D, C, B), antimicrobial agents (e.g. cinnamon oil, clove oil), natural food colorants (e.g. astaxanthin, β-carotene), mineral ions and salts (e.g. iron), and other miscellaneous food components (e.g. DHA/EPA, honey) [6].

7.5 Conclusion and Future Perspectives

Lipid-based carriers offer possibilities for delivering a wide range of food ingredients. Due to its ability to deliver bioactive substances, lipid-based encapsulation has proven to be a good alternative and outstanding method, with high chemical and physical stability, enhanced bioavailability, and sustained release. The capacity to protect incorporated food ingredients from environmental influences may also improve shelf life, which has clear benefits for the food industry and contributes to reduced food waste. Various lipid-based carriers have been created to date, solving various issues in the food industry and assisting in the production of higher-quality food and improved health benefits.

Despite all of the research demonstrating the use of lipid-based carriers in food, much more effort is required to bring such systems to market with demonstrable cost-effectiveness and good food feature enhancement properties. Toxicological safety, impact on the human body, and commercialization are the primary hurdles in the development of lipid-based nanocarriers. Furthermore, combining food-grade materials and healthy lipids to construct lipid-based nanocarriers is critical for the food industry.

Reproducibility, affordability, and scalability are essential variables that might forecast the deployment of novel microencapsulation technologies that are constantly being created to solve diverse challenges faced by the food industry. Customers play an important role in the adoption of new items on the market. Lipid-based delivery systems are expected to grow dramatically as consumers become more concerned about food quality and technological advancement. Despite its broad applicability, encapsulation has struggled to find a foothold in the food business due to its high cost. In addition, given the cost of implementation and the necessity for training, industries frequently reject the adoption of new technology. Therefore, the main future perspective of this field is to develop cheaper, high-quality lipid-based systems for food component encapsulation.

Glossary

- Encapsulation: Entrapment of the active food component in a polymeric material while providing a controlled release of the substance.
- Active ingredient: The food product/nutrient incorporated into the polymer material. Also known as the active ingredient, base material, fill, internal phase, core, or payload phase.

- Carrier material: A polymeric material that encapsulates the active ingredient. Also known as the shell, wall, matrix, or coating.
- Nanoemulsions: Thermodynamically stable immiscible phases of water and oil that can be stabilized by an interfacial layer of surfactants and cosurfactants.
- Liposome: Closed bilayer structures formed autonomously by hydrated phospholipids that are widely used as effective delivery systems because of their ability to encapsulate bioactive amphipathic hydrophilic, and lipophilic molecules into the inner water phase or within lipid leaflets.
- Solid lipid nanoparticles: First-generation nanocarriers with a highly ordered crystalline structure and a solid oil/water emulsion.
- Nanostructure lipid carriers: A type of oil-in-water emulsion system that comprises both solid lipids having melting points above room temperature and liquid lipids having melting points far below the ambient temperature within the inner phase along with appropriate surfactant in the outer phase to achieve a fine dispersion of these lipids in the aqueous medium.

References

1 L. Barroso, C. Viegas, J. Vieira, C. Ferreira-Pego, J. Costa, P. Fonte, *J. Food Eng.* **2021**, *295*, 110451.

2 M. Arenas-Jal, J. M. Sune-Negre, E. Garcia-Montoya, *Eur. Food Res. Technol.* **2020**, *246*, 1371.

3 I. Alemzadeh, M. Hajiabbas, H. Pakzad, S. S. Dehkordi, A. Vossoughi, *Int. J. Eng.* **2020**, *33*, 1,400.

4 V. Nahum, A. J. Domb, *Foods* **2021**, *10*.

5 M. R. P. Ltd, in Report Code: FB 2207, 2021, MarketsandMarkets Research Private Ltd, https://www.marketsandmarkets.com/Market-Reports/food-encapsulation-advanced-technologies-and-global-market-68.html, **2020**.

6 A. Tahir, R. S. Ahmad, M. Imran, M. H. Ahmad, M. K. Khan, N. Muhammad, M. U. Nisa, M. T. Nadeem, A. Yasmin, H. S. Tahir, A. Zulifqar, M. Javed, *Int. J. Food Prop.* **2021**, *24*, 1074.

7 E. Mitsou, V. Pletsa, G. T. Sotiroudis, P. Panine, M. Zoumpanioti, A. Xenakis, *Coll. Surf. B-Biointerfaces* **2020**, *190*, 110974.

8 M. A. Aboudzadeh, E. Mehravar, M. Fernandez, L. Lezama, R. Tomovska, *ACS Omega.* **2018**, *3*, 10999.

9 Y. Wei, Z. P. Yu, K. S. Lin, S. F. Yang, K. D. Tai, J. F. Liu, L. K. Mao, F. Yuan, Y. X. Gao, *J. Agri. Food Chem.* **2020**, *68*, 1405.

10 R. S. Rabelo, I. F. Oliveira, V. M. da Silva, A. S. Prata, M. D. Hubinger, *Int. J. Biol. Macromol.* **2018**, *119*, 902.

11 A. Jain, D. Thakur, G. Ghoshal, O. P. Katare, U. S. Shivhare, *Int. J. Biol. Macromol.* **2016**, *87*, 101.

12 A. L. Fadini, I. D. Alvim, K. B. D. Paganotti, L. da Silva, M. B. Queiroz, A. Miguel, R. A. F. Rodrigues, *Food Sci. Technol. Int.* **2019**, *25*, 359.

13 A. Naseema, L. Kovooru, A. K. Behera, K. P. P. Kumar, P. Srivastava, *Adv. Coll. Interface Sci.* **2021**, *287*, 102318.

14 D. J. McClements, *Prog. Lipid Res.* **2021**, *81*, 101081.

15 N. Kumar, A. Verma, A. Mandal, *J. Petroleum Sci. Eng.* **2021**, *206*, 109042.
16 J. M. M. de Oca-ávalos, R. J. Candal, M. L. Herrera, *Curr. Opin. Food Sci.* **2017**, *16*, 1.
17 S. Akhavan, E. Assadpour, I. Katouzian, S. M. Jafari, *Trends in Food Sci. Technol.* **2018**, *74*, 132.
18 M. Fathi, M. R. Mozafari, M. Mohebbi, *Trends in Food Sci. Technol.* **2012**, *23*, 13.
19 L. Rashidi, *Food Biosc.* **2021**, *43*, 101258.
20 R. Nisini, N. Poerio, S. Mariotti, F. De Santis, M. Fraziano, *Front. Immunol.* **2018**, *9*, 155.
21 G. Bozzuto, A. Molinari, *Int. J. Nanomed.* **2015**, *10*, 975.
22 V. A. Frolov, A. V. Shnyrova, J. Zimmerberg, *Cold Spring Harbor Perspect. Biol.* **2011**, *3*, a004747.
23 S. Shukla, Y. Haldorai, S. K. Hwang, V. K. Bajpai, Y. S. Huh, Y. K. Han, *Front. Microbiol.* **2017**, *8*, 02398.
24 G. Iafelice, M. F. Caboni, R. Cubadda, T. Di Criscio, M. C. Trivisonno, E. Marconi, *Cereal Chem.* **2008**, *85*, 146.
25 T. Ghorbanzade, S. M. Jafari, S. Akhavan, R. Hadavi, *Food Chem.* **2017**, *216*, 146.
26 S. Amjadi, M. Ghorbani, H. Hamishehkar, L. Roufegarinejad, *Food Chem.* **2018**, *256*, 156.
27 V. D. Santos, A. P. B. Ribeiro, M. H. A. Santana, *Food Res. Int.* **2019**, *122*, 610.
28 N. P. Aditya, S. Ko, *RSC Adv.* **2015**, *5*, 30902.
29 W. Mehnert, K. Mader, *Adv. Drug Del. Rev.* **2012**, *64*, 83, V. Wiedenmann, K. Oehlke, U. van der Schaaf, H. M. Koivula, K. S. Mikkonen, H. P. Karbstein, *J. Food Sci.* **2019**, *84*, 3642.
30 S. M. de Carvalho, C. M. Noronha, C. G. da Rosa, W. G. Sganzerla, I. C. Bellettini, M. R. Nunes, F. C. Bertoldi, P. L. M. Barreto, *Coll. And Surfaces a-Physicochem. Eng. Aspects* **2019**, *581*, 123793.
31 B. Tonyali, A. McDaniel, J. Amamcharla, V. Trinetta, U. Yucel, *Food Packag. Shelf Life* **2020**, *24*, 100484.
32 M. Mohammadi, E. Assadpour, S. M. Jafari, in *Lipid-Based Nanostructures for Food Encapsulation Purposes* (Ed: S. M. Jafari), Academic Press, **2019**, *217*, 114–137.
33 M. Elmowafy, M. M. Al-Sanea, *Saudi Pharmaceutical J.* **2021**, *29*, 999.
34 A. Beloqui, M. A. Solinis, A. Rodriguez-Gascon, A. J. Almeida, V. Preat, *Nanomed.-Nanotechnol. Biol. Med.* **2016**, *12*, 143.
35 S. S. Panigrahi, I. Syed, S. Sivabalan, P. Sarkar, *Chem. Papers* **2019**, *73*, 1.
36 F. A. N. Azar, A. Pezeshki, B. Ghanbarzadeh, H. Hamishehkar, M. Mohammadi, *J. Agri. Food Res.* **2020**, *2*, 100084.
37 P. Jaiswal, B. Gidwani, A. Vyas, *Artif. Cells Nanom. Biotechnol.* **2016**, *44*, 27.
38 M. Yousefi, A. Ehsani, S. M. Jafari, *Adv. Coll. Interface Sci.* **2019**, *270*, 263.
39 D. R. B. Oliveira, M. Michelon, G. D. Furtado, R. Sinigaglia-Coimbra, R. L. Cunha, *Food Res. Int.* **2016**, *90*, 139.
40 B. Gaba, M. Fazil, A. Ali, S. Baboota, J. K. Sahni, J. Ali, *Drug Del.* **2015**, *22*, 691.
41 M. A. Iqbal, S. Md, J. K. Sahni, S. Baboota, S. Dang, J. Ali, *J. Drug Target.* **2012**, *20*, 813.
42 F. Tamjidi, M. Shahedi, J. Varshosaz, A. Nasirpour, *Innovative Food Sci. Emerging Technol.* **2013**, *19*, 29.
43 S. S. Kuang, J. C. Oliveira, A. M. Crean, *Crit. Rev. Food Sci. Nutrition* **2010**, *50*, 951.
44 B. F. Gibbs, S. Kermasha, I. Alli, C. N. Mulligan, *Int. J. Food Sci. Nutrition* **1999**, *50*, 213.
45 S. L. Perry, D. J. McClements, *Mol.* **2020**, *25*, 1161.
46 I. Baldim, L. Tonani, M. R. V. Kress, W. P. Oliveira, *Ind. Crops and Prod.* **2019**, *127*, 73.
47 M. Pabast, N. Shariatifar, S. Beikzadeh, G. Jahed, *Food Control* **2018**, *91*, 185.

48 F. S. Hashemi, F. Farzadnia, A. Aghajani, F. A. NobariAzar, A. Pezeshki, *Food Sci. Nutrition* **2020**, *8*, 4185.

49 T. Mehmood, *Food Chem.* **2015**, *183*, 1.

50 N. Karimi, B. Ghanbarzadeh, H. Hamishehkar, B. Mehramuz, H. S. Kafil, *Coll. Interface Sci. Comm.* **2018**, *22*, 18.

51 J. Huang, Q. Wang, T. Li, N. Xia, Q. Xia, *J. Food Eng.* **2017**, *215*, 1.

52 S. Juric, M. Juric, Z. Krol-Kilinska, K. Vlahovicek-Kahlina, M. Vincekovic, V. Dragovic-Uzelac, F. Donsi, Food reviews international.

53 F. Y. de Boer, A. Imhof, K. P. Velikov, *Coloration Technol.* **2019**, *135*, 183.

54 S. Ghosh, T. Sarkar, A. Das, R. Chakraborty, *Appl. Biochem. Biotechnol.* **2021**, *193*, 3787.

55 J. P. Chi, J. Ge, X. Y. Yue, J. Liang, Y. Sun, X. L. Gao, P. X. Yue, *Lwt-Food Sci. Technol.* **2019**, *109*, 101.

56 D. R. Perinelli, G. F. Palmieri, M. Cespi, G. Bonacucina, *Mol.* **2020**, *25*, 5878.

57 M. Saifullah, M. R. I. Shishir, R. Ferdowsi, M. R. T. Rahman, Q. V. Vuong, *Trends in Food Sci. Technol.* **2019**, *86*, 230.

58 B. Rodenak-Kladniew, G. A. Islan, M. G. de Bravo, N. Duran, G. R. Castro, *Coll. Surf. B-Biointerfaces* **2017**, *154*, 123.

59 C. Dima, E. Assadpour, S. Dima, S. M. Jafari, *Crit. Rev. Food Sci. Nutrition* **2021**, *61*, 3031.

60 I. Katouzian, S. M. Jafari, *Trends in Food Sci. Technol.* **2016**, *53*, 34.

61 P. M. Batinic, V. B. Dordevic, S. I. Stevanovic, B. D. Balanc, S. B. Markovic, N. D. Lukovic, D. Z. Mijin, B. M. Bugarski, *J. Drug Delivery Sci. Technol.* **2020**, *59*, 101920.

62 R. Couto, V. Alvarez, F. Temelli, *J Supercrit. Fluids* **2017**, *120*, 432.

63 F. E. de Matos, M. P. da Silva, M. G. C. Kasemodel, T. T. Santos, P. Burns, J. Reinheimer, G. Vinderola, C. S. Favaro-Trindade, *J. Funct. Foods* **2019**, *54*, 98.

64 D. H. L. Kumar, P. Sarkar, *Environ. Chem. Lett.* **2018**, *16*, 59.

65 P. M. Reque, A. Brandelli, *Trends in Food Sci. Technol.* **2021**, *114*, 1.

66 S. F. Hosseini, B. Ansari, A. Gharsallaoui, *Food Chem.* **2022**, *372*, 131358.

67 S. P. Inapurapu, A. Ibrahim, S. R. Kona, S. C. Pawar, S. Bodiga, V. L. Bodiga, *Colloids Surf. A-Physicochem. Eng. Asp.* **2020**, *606*, 125515.

68 R. Cassano, S. Mellace, M. Marrelli, F. Conforti, S. Trombino, *Colloids Surf. B-Biointerfaces* **2017**, *151*, 128.

69 M. Shakeri, S. H. Razavi, S. Shakeri, *Lwt-Food Sci. Techno.* **2019**, *107*, 280.

70 M. L. Zambrano-Zaragoza, D. Quintanar-Guerrero, A. Del Real, R. M. Gonzalez-Reza, M. A. Cornejo-Villegas, E. Gutierrez-Cortez, *Coatings* **2020**, *10*, 253.

71 W. C. Lu, D. W. Huang, C. C. R. Wang, C. H. Yeh, J. C. Tsai, Y. T. Huang, P. H. Li, *J. Food Drug Anal.* **2018**, *26*, 82.

72 N. A. Lopes, C. M. B. Pinilla, A. Brandelli, *Food Hydrocoll.* **2019**, *93*, 1.

73 S. J. Park, C. V. Garcia, G. H. Shin, J. T. Kim, *Food Chem.* **2017**, *225*, 213.

74 J. P. Hao, B. Guo, S. X. Yu, W. T. Zhang, D. H. Zhang, J. L. Wang, Y. R. Wang, *Lwt-Food Sci. Technol.* **2017**, *85*, 37.

75 S. A. Khatibi, A. Misaghi, M. H. Moosavy, A. A. Basti, M. K. Koohi, P. Khosravi, F. Haghirosadat, *J. Food Process. Preserv.* **2017**, *41*, e12955.

76 M. Nasseri, S. Golmohammadzadeh, H. Arouiee, M. R. Jaafari, H. Neamati, *Iran. J. Basic Med. Sci.* **2016**, *19*, 1231.

77 L. S. Zhao, F. Temelli, L. Y. Chen, *J. Funct. Foods* **2017**, *34*, 159.

78 X. Chen, L. Q. Zou, J. Niu, W. Liu, S. F. Peng, C. M. Liu, *Molecules* **2015**, *20*, 14293.

79 Y. Shahparast, M. Eskandani, A. Rajaei, A. Y. Khosroushahim, *Adv. Pharm. Bull.* **2019**, *9*, 393.

80 F. Shi, Y. Y. Zhao, C. K. Firempong, X. M. Xu, *Pharm. Biol.* **2016**, *54*, 2320.

81 R. Ravanfar, A. M. Tamaddon, M. Niakousari, M. R. Moein, *Food Chem.* **2016**, *199*, 573.

8

Working Principles and Use of Gelatin for Food Component Encapsulation

Youssef S. Abdelaziz[1], Rana Tarek[2], Donia G. Youssef[3], Mariam Khaled Abdel-Latif[2], Habiba Mohamad Ibrahim[2], Sohaila Mohammed Salah Saleh[2], and Heba M. Fahmy[3,]*

[1] *Chemistry & Microbiology Department, Faculty of Science, Cairo University, 12613, Egypt*
[2] *Biotechnology & Biomolecular Chemistry Department, Faculty of Science, Cairo University, 12613 Egypt*
[3] *School of Biotechnology and Science Academy, Badr University in Cairo, Badr City, Cairo 11829, Egypt*
* *Corresponding author*

Highlights

- Gelatin has a variety of sources for production, such as pigs, poultry, fish, and cows. Gelatin can blend with different biopolymers to enhance its mechanical and encapsulation efficiencies.
- Microencapsulation stabilizes the core material, provides controlled release, and masks the unpleasant odor or flavor.
- The nanoencapsulation process increases the stability and solubility and enhances the bioavailability of the core substance, and it has various advantages. The techniques for fabricating the nanocapsules are divided into two categories: top-down and bottom-up techniques.
- The release mechanisms are diffusion, degradation, dissolution, melting, swelling, osmosis, fracturation, erosion, and partitioning or fragmentation.
- The release process depends on the chemical and physical capsule, coating properties, and experimental parameters. The release of the food ingredient at a suitable time with a controlled rate is the main point during the release process.

8.1 Introduction

Encapsulation is a food-processing technique that protects compounds in their solid, liquid, or gaseous state in a membrane against any dangerous factors from the environment, such as moisture, light, and free radicals [1, 2]. Structural systems of encapsulation are characterized by the core and wall [3]. Microencapsulation is a method of entrapping small particles in a coat or incorporating them into matrices that can be homogeneous or heterogeneous to produce small capsules with various valuable properties [4, 5]. The ideal coating material will disperse or emulsify the active ingredient, stabilize the emulsion, protect the active agent from environmental circumstances, and not react with the material encapsulated during the procedure or prolonged storage [6–8]. The reactions of the

Materials Science and Engineering in Food Product Development, First Edition. Edited by Wing-Fu Lai.

implicated coating material and the rheological and physical properties of the prepared emulsion are taken into account to effectively perform the encapsulation process [9]. The coating material's composition impacts the encapsulated material's stability [3, 10]. As a result, the coating material must have good emulsifying properties [8]. However, providing all of the criteria mentioned above with a single coating material is difficult. A combination of coating materials is frequently preferred for a proper encapsulation process, favored powder properties, and high microencapsulation efficiency [11].

Gelatin is considered an essential component of the food industry in modern cuisine due to its higher gelling potential. It is used in the food industry to manufacture gourmet desserts because of its texture, clarity, and food structure stability [12]. Moreover, it can be used as an edible film and coating material because of its outstanding film ability and nutritional benefits [13]. Chemical encapsulation can be accomplished through various methods, including ionic gelation, coacervation, co-crystallization, molecular inclusion, and electrostatic interactions. Spray drying, freeze/cold drying, extrusion, and a fluidized bed are mechanical processes. In this chapter, the food encapsulation process using gelatin biopolymer or in other blended forms will be covered from different points of view, including microencapsulation, nanoencapsulation, mechanisms of release, and encapsulation techniques, such as coacervation, extrusion, fluidized-bed coating, or spray drying.

8.2 Why Use Gelatin in Encapsulation Technology?

Gelatin is a biodegradable, biocompatible protein biopolymer formed due to the partial hydrolysis of natural collagens [14]. Collagens are extracted from animal sources, such as pigs, poultry, and cows [15, 16], insects [17], and fish [18, 19]. About 50% of gelatin production depends on pigs' skin extraction [16]. Gelatin is still popular in commercials coating wall materials in food encapsulation due to its excellent water solubility, emulsifying capacity, stabilizing activity, thickening ability, high crosslinking activity via its primary amino group, generally recognized as safe (GRAS), etc. [20, 21]. It has attractive properties, such as great mechanical properties, including barrier capabilities for carbon dioxide, oxygen, and volatile compounds. Gelatin can also blend effectively with other biopolymers [22]. Besides the numerous advantages of gelatin, certain drawbacks prevent its expansion into the food encapsulation market. Gelatin extracted from animal resources has not been accepted by a large group of consumers, depending on their dietary preferences and religious obligations [23]. It also suffers from the lack of barrier capacity to moisture due to its hydrophilicity [22]. In addition, gelatin structure changes according to the medium's conditions such as temperature, pH, and ionic strength, affecting gelatin encapsulation efficiency. Accordingly, researchers looked through blending gelatin with other materials, especially carbohydrate-based biopolymers, which decreased the surrounding medium effect on gelatin encapsulation [23]. Some of the strengths and weaknesses of some coating materials are summarized in Table 8.1.

Polymer mixtures have attracted researchers' concern in food encapsulation applications due to their excellent capabilities and low cost. Also, from a chemical point of view, making mixtures of gelatin with other oppositely charged polymers leads to the formation of a complex of lower solubility, facilitating the phase separation process [20]. Mixtures have been made between gelatin and other biopolymers, such as chitosan [24], carboxymethyl cellulose [25], and alginates [26]. Chitosan/gelatin mixtures provide a compact structure via the

Table 8.1 Summary of properties and sources of some coating materials.

Coating material	Origin	Characteristics
Carbohydrates	Starch, dextran, sucrose	Relatively higher tensile strength than gums, hydrocolloidal
Proteins	Gelatin, albumin	Gelation, water-binding capacity, emulsification, foaming
Chitosan	Shells of crustaceans	Good barrier to gases and water vapor
Lipids	Phospholipids, stearic acid, bees wax	Good barrier to gases and water vapor, plasticizing properties
Cellulose and its derivatives	Plant cells	Good filming capability and surface activity, hydrophilic properties
Gums	Sodium alginate, gum arabic, carrageenan	Elastic gels, weak tensile strength, soft elastic gels, hydrocolloidal properties

Source: Adapted from Choudhury et al. [30].

formation of electrostatic interaction and hydrogen bonding between both biopolymers. This results in forming an enhanced flexible and permeable edible film [26]. In the case of carboxymethyl cellulose and gelatin blend, the compact matrix of the prepared film showed better permeability to water vapor and oxygen and enhanced tensile strength. The Fourier-transform infrared spectroscopy results showed intramolecular and intermolecular hydrogen bond formation between the two biopolymers [27]. Gelatin can form an athletic complex with sodium alginates through strong intermolecular ionic interaction and hydrogen bonds [28]. Sodium alginates and gelatin complex were observed to encapsulate probiotics and increase their survival [26]. The complex was also tested for the encapsulation of black pepper (*Piper nigrum L.*) essential oil with an efficiency of 49.13–82.36% [29]. A summary of critical information regarding the usage of gelatin and its blends in industry is mentioned in Box 8.1. The capabilities of applying such combinations in food encapsulation will be discussed later.

8.1 Industrial corner

- Gelatin is a cheap encapsulating material due to its variety of natural resources, making it an ideal coating material of interest for starting small businesses in the food industry.
- In order to enhance the mechanical and encapsulation capabilities of gelatin, polymer mixtures can be used.
- Certain aspects should be considered while preparing a biopolymer blend, such as optimizing the used polymer concentrations, pH, and temperature of the blending procedure.
- Before applying a certain polymer blend in the food encapsulation process, the type of bonds formed between the target polymers should be studied, and the solubility of the resulting complex is critical in the phase separation process.

8.3 Techniques for Food Encapsulation Using Gelatin

The encapsulation process is effective in covering active components, especially in the food industry [31]. Recently, it has helped in producing healthy and desired food products. Widely, encapsulation is used in the food industry in flavoring and by covering active components with essential oils and in other food applications [31]. Many techniques are used to encapsulate the different types of food ingredients [32]. The major types of encapsulations are micro/nanoencapsulation, which are discussed later in the chapter, especially with gelatin as the coating material [33]. This process depends mainly on the coating or core material, which should be GRAS [32]. The studies showed impressive results of using gelatin as the coating material, improving the stability and consistency of food products [31]. The proficiency of gelatin appears mainly in its characteristics, such as the formation of collagen-like triple helices and thermally reversible hydrogels [31]. Generally, the most commonly used techniques in gelatin food encapsulation are coacervation and centrifugal extrusion: fluid-bed coating and spray drying [34]. As a result of these characteristics, gelatin is assumed to be the only protein used on a commercial scale effectively.

8.3.1 Coacervation

Phase separation coacervation is considered one of the oldest and most widely used technologies in the food encapsulation industry [33]. It depends primarily on separating the liquid phase of the coating material and then coating it with the core substance as a uniform layer [35]. The principal concept of the coacervation method is the formation of an emulsion, the subsequent precipitation of the continuous phase around the discontinued phase. In this process, we use three main phases of the solvent: system containing/manufacturing visible material, material that is encapsulated, and coating material [33]. This technique is used extensively in gelatin and gelatin–acacia microcapsules preparation and many products with cellulose derivatives and synthetic polymer techniques [31]. The coacervation method is complex and straightforward. This part of the method implies the phase separation of hydrocolloids from a layer of polymeric solution around the core material is suspended in the same reaction [35]. The simple coacervation method is based on a single coating material (e.g., gelatin only) as the gelatin spreads in water to form a core solution, then on adding the core material (e.g., hydrophobic citrus oil), the blend is agitated. With experiments, notice that the solubility of gelatin decreases when the temperature is lowered or sodium sulfate is added [34], so the process depends on both external factors, such as pH and temperature, and internal factors, such as the structure of the macromolecule and ionic strength. The most commonly used polymers in simple coacervation are non-solvent and water-soluble polymers. The polymer competes for the solubility of the gelatin protein solution through hydrophobic interaction [35]. The complex coacervation mostly depends on the ionic interaction between two differently charged polymers. This method is called polymer-to-polymer interaction. It depends mainly on the pH concentration of each polymer [35]. The coacervation process begins when the base material is suspended in, for example, a gelatin or gum arabic solution. Then, a gelatin or gum arabic solution (whichever is not used to suspend the core material) is added to the system, and the system is cooled to 5°C with a 3.8–4.3 pH range. Both simple and complex conservation follow the same steps, except for the separation of used phases, which will be discussed in Table 8.2 [31].

Table 8.2 A summary of some techniques used in food encapsulation using gelatin.

Technique	Major technique step	Advantage	Disadvantage	Product size/ encapsulated form	References
Coacervation	• Preparing the core material • Preparing coating material solution • Coextrusion of both coating and core material using a nozzle • Uses of organic solvents	• Used effectively with heat-sensitive ingredients • Sensitive to environmental conditions • Expose excellent storage stability • Have high loading capacity • Decrease in thermal degradation • Low agitation utilization	• Using some toxic chemical agent • Product complex is highly unstable • Small size range of pH and other conditions • Expensive and sometimes complex method	About 10–800 mm paste powder capsules	[36–41]
Extrusion	• Prepare molten coating material • Dispersion of core material into the polymer • Cooling or passing of core–coat mixture through dehydrating liquid	• Low temperature entrapping method • Suited for encapsulating highly sensitive citrus flavors	• It necessitates additional techniques for the drying process. • Scaling up is difficult.	About 150–8000 mm powder granule	[36, 37, 41, 42, 138, 139]
Spray drying	• Establishing a stable emulsion • Homogenizing the dispersion • Atomizing the emulsion • Dehydrating the atomized particle	• Rapid, adaptable, and reproducible method • In terms of industrial perspective, it can generate large quantities of encapsulates	• Requirement of high temperature to be operated • Limitations of shell or matrix materials	Spherical droplets with diameters ranging from 10 to 100 μm	[43, 44]
Fluid-bed coating	• Coating solution preparation • Fluidizing of core particles • Core particles coating with the required coating solution	• Because high temperatures are not required, this method is effective in terms of saving energy and time	• Consume high energy, complex method to be performed, and submicron processing particles is limited	Results in a product size with diameter in the range 100–5000 μm	[44–46]

8.3.2 Extrusion

Centrifugal extrusion is one of the unique procedures extensively used in gelatin encapsulation. It is based on liquid coextrusion that has been used widely to produce a capsule size of 500–1000 μm, which is of unstable flavors and vitamins [38, 39]. The central concept for applying this technique is extrusion, which applies high pressure on the mixing of wall and core materials, producing a high-density and low-porosity capsule [39]. The rotating cylinder consists of coextensive orifice nozzles found on the outer circumference, and the head cylinder consists of the concentric tube, core, and wall material [39]. The significant steps and advantages of this process with the size of an encapsulated product will be shown in Table 8.2 [31].

8.3.3 Fluid-Bed Coating System

Fluidized-bed coating provides increasingly diverse encapsulated versions of food ingredients and additives to the food industry [47]. Fluid-bed coating is commonly used to coat solid particles [48]. Coated particles are synthesized by spraying an encapsulating agent onto a fluidized powder bed in the fluidized-bed coating process [49]. It coats dry particles with thin layers of coating material that are of extremely low density and/or small [50]. The coating solution and suspension are sprayed onto the fluidized powder surface via nozzles to form a shell-like structure in a fluidized coating process. It can be used in both batches and continuous processes. [51]. The formation mechanism of microparticles using this technique can be divided into three stages: nucleation, transition, and ball growth. During the initial encapsulation process, particles are suspended in the coating chamber. Droplets of the polymer solution are then sprayed, increasing the probability of impacting particle droplets and spreading on the particle's surface. As a result of this process, the droplets are evaporated and formed into microparticles [52]. This manufacturing technique is ideal for the food industry due to its simplicity, versatility, and relatively large batch size [53].

Bottom spray, top spray, and tangential spray are the three types of fluid-bed coating, and those types are used to encapsulate liquids or solids absorbed into granules. Fluid-bed coating involves processing variables, such as solid circulating rate and nozzle atomization pressure [54], humidity [45], coating feed rate, and temperature [53]. Those variables are important because they influence particle agglomeration and film formation, affecting coating efficiency. In a fluidized-bed system, menthol powder was coated with different materials, such as gelatin, an oil–gelatin emulsion, modified starch, and Aquacoat® ECD; the gelatin emulsion was found to be a promising coating material for controlling menthol release and high microencapsulation. These findings suggest that the coating's solubility and integrity are critical to the fluidized-bed coating process's success in controlling flavor release from this degradation-activated release system [55]. Also, the properties of gelatin and its emulsion as coating agents on the release characteristics of mint flavor powder had been studied, and the results showed that the emulsion of 10% gelatin with 20% oil content based on gelatin mass achieves a higher quality of the coated granules that had 90.44% encapsulation efficiency and reached a 60% release rate in 11 minutes. The above results showed that a fluidized-bed coating operation could improve the property of sustained release by coating the gelatin emulsion [56].

8.3.4 Spray Drying

In the food industry, spray drying is considered to be the oldest and most extensively used technology in the microencapsulation process because of its wide range of wall materials, capacity to dry even heat-sensitive components, allowing for production rates at relatively cheap operational costs, as well as a stable and scalable output [57–60]. Spray drying is a microencapsulation procedure used to convert a solution, suspension, or slurry into a powdered form with a hot gas (usually air) [61]. Spray drying is a physical dehydration method that produces microparticles with variable sizes and morphology that changes quickly, continuously, and scalably. The selection of the wall material is a critical stage for the performance and stability of the encapsulation procedure using the spray-drying technique. The wall material serves as a barrier and keeps the active components safe from harmful environmental conditions. Gelatin is a natural, biocompatible, economical, and easily accessible biopolymer. The products generated from the biodegradation of gelatin are nontoxic and easily excreted, making it the core interest in the food industry. Gelatin has been employed as a coating material to encapsulate various food ingredients using various processes because of its film formation, emulsification, and capability of network formation on drying [62–64]. Gelatin is the most appropriate wall material for encapsulation during the spray-drying procedure [57]. The drying rate of the gelatin-drying curve drops rapidly at the beginning of the drying process, resulting in a tight, thick film formation as well as the protection of the active ingredient, showing the excellent effect of gelatin as a wall material for encapsulation during the drying process [65]. Some fundamental industrial aspects of food encapsulation techniques are summarized in Box 8.2.

8.2 Industrial Corner

- Coacervation process aids in extending the lifespan of the sensitive food ingredients in addition to increasing its loading efficiency. It can also ease the handling techniques by improving the physical parameters of the active ingredients via transforming liquids into dry solids.
- Spray drying can be a cost-effective method in a variety of ways because of the limited number of moving parts; it also necessitates minimal labor because its design allows for continuous production, enabling the setup of larger capacity units at a lower cost.
- Fluidized-bed coating is considered to be an inexpensive method in addition to its simple setup and high moisture reduction rate.
- Extrusion method can protect oil volatilization to expand the shelf life of active ingredients. This technique costs more than the spray-drying procedure.

8.4 Microencapsulation Using Gelatin

Microencapsulation is defined as using protective coating material to produce small particles of solid particles, liquid droplets, or gaseous components that give small capsules [66–69]. This capsule isolates the core compound from the surrounding environment [68]. Capsule contents are released in their targeted site according to stimuli, such as enzyme action, shear, or pH [67]. The microencapsulation technique for the food-processing industry was introduced in the middle of the twentieth century despite its discovery by the Dutch chemist H.G. Bungenberg de Jong in the 1930s [68]. The microcapsules are prepared by various methods and divided into physicochemical, physical, and chemical categories [70]. Microencapsulation has various benefits to be utilized in the food industry [32, 66, 70–72], such as the following: (i) it preserves the core material and prevents its degradation as it lowers the "core material – external environment" reactivity; moreover, it helps separate the mixture's components to avoid their reaction with each other; (ii) reduction of the core material's loss rate to the surrounding environment, thereby reducing the volatility of various food products; (iii) protection of the bioactive components by providing a barrier against the various environmental conditions; (iv) provide easier handling for the original material through physical characteristics modifications, such as the possibility to convert food material from a liquid form to a solid or free-flowing powder or vice versa; (v) control the release mechanism of the core material and mask the undesirable taste or smell of the core material; (vi) help stabilize various food products against the surrounding conditions such as their exposure to light, change in pH, or oxidation; (vii) effective process that helps deliver bioactive compounds and living cells into food. It also helps in extending the shelf life of the encapsulated material. When it comes to the selection of the wall material for food microencapsulation, it is chosen to be of the following qualities: (i) edible; (ii) food grade; (iii) nonreactive with the core material; (iv) tasteless; (v) strengthens the capsule; (vi) impermeable; (vii) GRAS; also some physicochemical parameters, such as its molecular weight, solubility, and the release of the core material should be considered. As summarized in Table 8.3, gelatin was generally used in the encapsulation of different active compounds as well as in the microencapsulation of different active compounds, as mentioned in Table 8.4.

Table 8.3 Various active compounds encapsulated using gelatin.

Core material	Coating material	Technique	References
Omega-3 fatty acids	Gelatin	Spray drying (fish oil) or freeze drying (flaxseed oil)	[66, 73]
	Gelatin–gum arabic with transglutaminase (TG) as crosslinking agent	Complex coacervation	
Polyphenols (yerba mate extract, EGCG, black currant extract, pimento oil)	Gelatin (type A)	Coacervation	[32, 43, 66]
Vitamins and minerals (Vitamin A)	Gelatin and acacia	Coacervation	[66, 74]
Microorganisms (*Lactobacillus lactis*)	Gelatin and gum	Dual process by combining double emulsification, followed by complex coacervation	[66, 75]
Protein hydrolysate and peptide (casein hydrolysate, whey protein hydrolysate, rapeseed peptide, chicken hydrolysate)	Gelatin	Spray drying	[66]

Table 8.4 Benefits of gelatin in microencapsulation for different active compounds.

Core material	Coating material	Technique	Results	References
Anthocyanins of barberry extract	Gum arabic/ gelatin	Spray drying	Stabilization of active ingredient	[76]
Phenolic compounds of plum	Maltodextrin/ gelatin	Spray drying	(i) Storage stability (ii) Plum phenolics might be used commercially as nutraceuticals	[76, 77]
Phenolic compounds of echium oil	Gelatin	Spray drying	Stabilization of active ingredient	[76]
Phenolic compounds of broccoli	Gum arabic/ gelatin	Complex coacervation	The microencapsulation process allows maintaining chlorophyll content and masking broccoli off odor	[76, 78]
Proanthocyanidin of cinnamon	Arabic gum/ gelatin	Complex coacervation	(i) High encapsulation yield and high efficiency, but the extract of the active material has high solubility with water and can mix with polymers; (ii) maintains the core material bioavailability and masks its undesired sensorial characteristics	[76, 79]
Probiotic bacteria of *Bifidobacterium*	Arabic gum/ gelatin	Spray drying	In vitro simulated gastrointestinal digestion release of probiotics from microcapsules; storage stability	[76, 80]
Xylitol	Arabic gum/ gelatin	Spray drying	Stabilization of active ingredient; controlled release of bioactive material from microcapsules	[76, 81]
L. casei from skim milk	Gelatin	Spray drying	Providing a survival rate above 85% in the gastrointestinal tract	[16–22, 76, 82]
Oregano essential oil	Gelatin/ sucrose	Spray drying	(i) High antioxidant and antimicrobial activity (ii) High diffusion coefficient about (10 – 15 m^2/s)	[71, 83, 84]
Peumus boldus Mol. (boldo) essential oil (EO)	Arabic gum/ gelatin	Complex coacervation	(i) High-efficiency percentage (~94%) (ii) Maintaining antifungal capacity against *Aspergillus* and *Penicillium*	[85, 86]

(*Continued*)

Table 8.4 (Continued)

Core material	Coating material	Technique	Results	References
Lactobacillus plantarum and DHA-rich oil	Alginate, pectin, and gelatin complexes	Freeze drying	(i) Increase smoothness and compactness of the particle surface (ii) Increase thermal stability (iii) Higher survivability of encapsulated probiotics	[87, 88]
Zanthoxylum limonella oil (ZLO)	Genipin-crosslinked modified chitosan–gelatin	Coacervation	(i) Could be used in the intestine-targeted delivery of hydrophobic compounds used in food and pharmaceutical products (ii) The desired delivery performance could be tailored by simply varying the coacervation pH	[71]

Moreover, the source of the coating material selected must be considered to fulfill food standards in manufacturing and exporting countries, knowing that the allowed quantity, purity, and use requirements differ among countries [89–91]. The wall material of the microcapsule acts as the major constituent of the final product, so the cost of the wall-forming material is an important parameter to be considered, especially in the microencapsulation of flavors. Also, the type of the wall material affects the tendency of the microcapsule to absorb moisture from the surrounding environment in case their exposure to a relatively high humidity environment [72, 89–91]. Various factors should be considered in coating the core material, such as (i) molecular structure; (ii) physical state; (iii) biological structure; (iv) solubility and surface activity; (v) optical properties; and (vi) chemical stability, also taking into account that the concentration of the core material affects solubility. The releasing of the core material can be optimized concerning application parameters, such as dissolution, pH, temperature, and pressure [89, 91, 92]. The microcapsule's morphology depends on the core, wall material, and the microencapsulation technique producing various types of particles: mononuclear, poly/multinuclear, matrix, multi-wall, and irregular particles [35]. The size of microcapsules ranges from about 1 to 1000 μm [71, 72]. The microcapsule morphology is greatly important as its impact on the microcapsule controls its performance, protection, and release mechanism. The nature of the starting material and the manufacturing process parameters produce a group of different intermediate structures. In addition, the morphology can affect the loading and release of the core material. For example, microcapsules accomplish the higher load of the core material and the burst release mechanism. In contrast, microspheres possess fewer core materials and a slow rate of release mechanism [89, 93]. Recommendations and tips have been included in Box 8.3 for further concerns regarding food microencapsulation.

8.3 Industrial Corner

- It is recommended that the manufacturers use spray-drying technique by using gelatin or its mixture with another substance as a coating material in microencapsulation, as the microcapsules produced effectively stabilize and control the release of the active ingredient.
- Manufacturers should consider the type of wall material used in food microencapsulation is edible and nonreactive with the core material because it is also important to estimate the cost of the wall material as it is a major constituent of the microcapsule.
- Manufacturers can easily control the loading and the release of the microcapsule by controlling the morphology of the microcapsule. Also, the release of the core material can be optimized using other parameters.
- The economics of the microencapsulation process should be considered as the complexity of the system chosen and the production scale surely affects the cost of the produced microcapsule.
- The complex coacervation method is recommended to be used in industries especially those that require masking off undesired odor, for example, as it should a successful masking off of undesired sensorial characteristics and the maintenance of the core material in the microencapsulation process using gelatin with gum arabic as the coating material.

8.5 Nanoencapsulation of Food Components Using Gelatin

Nanoencapsulation is described as the trapping of substances in the nanoscale as it carries the active ingredient (core material) to the required site in the body [94, 95]. The coating material encapsulates the substance on the nanometer scale, creating a protective coat on the core material [94]. There are various advantages to nanoencapsulation, such as: (i) the capsule absorption in the human body increases in addition to its bioavailability [96–99]; (ii) enhancing the solubility and stability [97, 98]; (iii) the core material will be maintained from the surrounding environment and protected through the coating material [96, 97, 99]; (iv) increasing the activity levels [97]; (v) the unfavorable smell and taste will be masked [96, 99]; (vi) the capability of the delivery of the active material will be enhanced [97, 99, 100]; (vii) these capsules will be in powder form; so, the cost of the storage and transport will be reduced [99].

The assembly techniques of nanoencapsulation are classified into two main mechanisms: the first one is top-down technique that involves emulsification [94, 96, 97], solvent evaporation [94, 96], high pressure homogenization [94], and electrospinning [96, 101], and the second mechanism is bottom-up technique that involves self-assembly and self-organization of molecules [96, 97], supercritical fluids [94, 96], conservation [94, 96, 97], and nanoprecipitation [94, 96]. However, this classification is no longer in general use [96, 100, 102] because, according to the point of view of some scientists, the classification of the nanoencapsulation techniques should be broader [102]. Moreover, more categories should be made based on the mechanism or the ingredient used to form the nanocapsules, such as lipid formulations, natural nanocarriers, specialized equipment, functional biopolymeric NPs, and disparate techniques [102]. In addition to the classification of the assembly techniques, the nanoencapsulation systems' classification is liquid–liquid, solid–liquid, and solid–solid [96]. The nanocapsules in the food industry are better to be in powder form. Consequently, a solvent media is used, eradicating the solvent to produce powdered nanocapsules [94]. This is used for spray drying, lyophilization, freeze drying, and nano-spray drying techniques [94]. Although various techniques are followed to prepare the nanocapsules, some main factors should be taken into account: (i) the desired size of the nanocapsule; (ii) different properties and characteristics of different bioactive materials (core substance) to be entrapped in gelatin nanocapsule; and (iii) the mechanism of the drug release [103]. Therefore, the manufactured nanocapsules can be tested in five different phases: (i) when they are prepared; (ii) when they are used in food products; (iii) when they are in the food network (the relationships between different food products [104]); (iv) during measuring the toxicity; and (v) when they are delivered in the human body [105]. Gelatin is used as the coating material for nanocapsules as it is one of the biopolymers used in food nanoencapsulation [96, 99, 106] due to its low cost and its presence in large quantities [97, 107]. A summary of the use of gelatin, whether it is free or combined with another material to form a complex for coating different food ingredients, is shown in Table 8.5.

Unfortunately, nanoencapsulation may cause many problems to the human body as these nanocapsules are of nano size and very reactive; they may accumulate in several biological organs and tissues inside the body since they can cross cell membranes easily [122–125]. Therefore, they can destroy the cells, cause tissue inflammation, and change the redox balance [122–125]. Gelatin nanoparticles may lead to many immune system responses; for example, the content of the antibodies in the human body will increase [105]. Therefore, the safety of the nanocapsules on human health and the environment cannot be neglected. The nanocapsules are GRAS, but immunological assays, toxicological assays, and safety experiments must be done to ensure that these products are not dangerous to humans [105, 122–124]. As a result, the United States Food and Drug Administration (FDA) has done some guidelines called "Draft Guidance for Industry" for manufacturers to take into account, for example: (i) the changes in the physicochemical properties and impurities of the food components should be observed; (ii) an assessment should be submitted to the FDA; (iii) the safety of the manufactured food products after their encapsulation should be estimated [105, 126, 127]. Nanoencapsulation usage in industrial field should be managed with more special care than other techniques. There are certain factors covered in Box 8.4 regarding nanoencapsulation applications in industry.

Table 8.5 Summary of the most recent studies in nanoencapsulation using gelatin.

Core material	Coating material	Technique	Results	References
Carotenoids extracted from tomato peels	Gelatin	Electrospinning technique	1) The stability and solubility of carotenoids in water increased, whereas solubility of carotenoids in water increased from $0.89 \pm 0.017\%$ to $99.08 \pm 0.20\%$ 2) They became more thermally stable and had higher antioxidant activity by increasing them 10 times 3) The encapsulation efficiency was more than 90% 4) The preservation of lycopene, a type of carotenoid that gives the red to pink color in tomatoes, was improved	[95, 108, 109]
Carotenoids extracted from cantaloupe melon (Cucumis melo L.)	Gelatin	Emulsification O/W	1) The nanoencapsulation increased its stability for 60 days when added to yogurt, unlike its crude extract 2) It was proved that melon carotenoids are used in food as natural dyes 3) Its solubility in water improved	[95, 110]
Carotenoids extracted from cantaloupe melon (Cucumis melo L.)	Gelatin	Oil in water (o/w) emulsification technique	1) The antioxidant potential of carotenoids increased by 57%, and the stability of beta-carotene was enhanced 2) The nanoencapsulation maintained the physical structure and concentration of beta-carotene 3) It will be beneficial to human health	[111]
Omega 3 (ω3) fatty acids	Gelatin and gum arabic complex	The soluble technique (electrostatic attraction)	1) Improvement of food products and human health 2) Increase its solubility in water and oxidation susceptibility 3) The unfavorable scent and taste disappeared	[112]
Jasmine essential oil	Gelatin and gum arabic complex	Complex coacervation (the technique used depends on the electrostatic interaction between gelatin and gum arabic)	1) The resultant nanocapsules were resistant to heat and thus could bear a water bath (80^0C) for seven hours	[113]

(Continued)

Table 8.5 (Continued)

Core material	Coating material	Technique	Results	References
Zeaxanthin	Gelatin and carboxymethylcellulose (CMC) complex	Complex coacervation	2) The stability of water solubility, and protection from external conditions were enhanced 3) These nanocapsules have a great impact on improving the food industry and economics 4) Increase in the thermal stability of zeaxanthin 5) The nanocapsule showed controlled release of zeaxanthin as, during digestion and ingestion, zeaxanthin nanocapsules were absorbed and processed in the human body	[114]
Thyme essential oil	Gelatin and chitosan complex	Nozzle-less electrospinning	1) The encapsulation efficiency was 98.9%, and the loading capacity was 39.5% 2) The stability of the core substance increased, their water solubility increased, and it had a better taste and smell than nonencapsulated ones 3) According to the results of this study, these nanocapsules can be an alternative to nitrite in meat products, as they decreased the nitrite concentration in sausages	[115, 116]
Stigmasterol	Gelatin and chitosan complex	Electrospinning method	1) The encapsulation efficiency was $87 \pm 5\%$ 2) The antioxidant ability and stability increased, and it was characterized by having a controlled release	[117]
Pork slices	Gelatin and chitosan complex containing free or encapsulated tarragon essential oil (TEO)	Ionic gelation method (for encapsulating TEO)	1) Microbial growth, lipid oxidation, and the pork slices' decay were inhibited 2) Tarragon essential oil was released sustainably 3) The preservation and the packaging of the meat were improved	[28]
Oregano essential oil	Gelatin and chia mucilage complex	Complex coacervation	1) The stability and bioactivity of oregano essential oil increased 2) The nanoencapsulation protected the core material from damage due to unfavorable external conditions 3) These nanocapsules could have numerous applications in the food industry	[118–120]
Capsaicin	Gelatin and acacia	Complex coacervation	The efficiency of the encapsulation is increased by 81.17%. Moreover, the mean drug-loading content is 20.81%, representing a high result. These results are associated with good dispersion	[121]

8.4 Industrial Corner

- Gelatin complexes are preferred as the coating material for nanoencapsulation. The compounds used to make complexes with gelatin are gum arabic, carboxymethylcellulose (CMC), chitosan, chia mucilage, and acacia. These complexes increase the stability, solubility, encapsulation efficiency, and antioxidant potential. Hence, it is preferred to use gelatin with another compound to make a complex for the coating substance, other than using it alone.
- Using gelatin complexes for nanoencapsulation will protect the core material from external conditions, enhance the thermal stability and oxidation susceptibility, and, finally, mask the unpleasant odor.
- The most common nanoencapsulation techniques used are electrospinning techniques and complex coacervation; thus, manufacturers prefer to use these two techniques to obtain the best results.
- The manufacture of the powdered form of nanocapsules is better than the liquid form, using suitable techniques for nanoencapsulation, such as spray drying, lyophilization, freeze drying, and nano-spray drying techniques.
- The nanoencapsulation technique to make the nanocapsule should be chosen wisely. The procedure used should be fast, easy, available, cheap, and give the best results. Manufacturers should be updated about the best and new techniques that scientists discover. They should check for the FDA rules and guidelines before preparing any nanocapsule keeping in mind the safety of human beings.

8.6 Mechanisms of Release of Gelatin Encapsulation Systems for Food Components

Release of the core at a suitable time with a controlled rate with high encapsulation efficiency is the main point to consider during the release process and can be determined as shown in Table 8.6 [128]. Generally, the available release rates of a single microcapsule are in the order of zero-, half-, or first-order. Zero-order is established when a pure active core is liberated through the mononuclear microcapsule (core-shell) wall in the form of pure material. Half-order generally implies the release of matrix particles. When the core material is a solution, first-order release occurs. The solute-material release occurs from the capsule, and the desired solute concentration is attained [129]. Different release rates (zero-, half-, or first-order) may occur by a mixture of microcapsules, including the distribution of capsules varying in wall thickness and size due to the microcapsule assembly. Therefore, it is crucial to carefully examine the release rate from an ensemble of microcapsules to determine the deviation from theory due to the distribution in wall thickness and size [130]. The release process depends on the chemical and physical capsules, coating properties, and experimental parameters. Coating properties involve orientation, crystallinity, plasticizer level, crosslinking, density, solubility, and pretreatments. Capsule properties exhibited configuration, conformity, wall thickness, size, coating layers, and post-treatment effects. Chemical composition, the roughness of the surface, porosity, and

Table 8.6 Equations used in studying food encapsulation.

Equation		References
Release (%)	$\frac{\text{The released amount of TPP at time t}}{\text{The total amount of TPP entrapped in nanoparticles}} \times 100$	[134]
Encapsulation efficiency (%)	$\frac{Wi}{Wt} \times 100$ W_i is the actual ingredient amount; W_t is the theoretical amount of ingredient	[137]

the carrier and core degradation rate mighty have been affected. Experimental parameters include temperature, solvent, partial pressure differential (outside and inside of coating), mechanical action, pH, and moisture [131]. A variety of release mechanisms have been suggested for the food encapsulation process.

The main pathways of the release include (i) mechanical means/shear force being used to break the capsule; (ii) the wall of the shell being melted/dissolved away from around the core material; and (iii) the ingredient or core begins to diffuse via the shell wall [129]. These mechanisms are diffusion, degradation, dissolution, melting, and swelling. Diffusion is when the individual core molecules are transferred out of the shell or wall due to differences in a concentration gradient, or interchain attractive forces may be hydrogen bonding and van der Waals' interactions. Activated diffusion may be used without pinholes, other flaws, or cracks [128, 130]. Degradation occurs when the capsule is present in a degradative media/environment. Dissolution transfers the core material from its solid phase into the surrounding environment (water). The melting process involves the capsule wall melting when heated above its melting point to release the desired ingredient. A swelling release mechanism may occur as gelatin exhibits hydrophilic behavior. When polymer matrix contacts water/fluid, both physical and chemical bonds occur with an expansion of the polymeric network and the stored ingredient is released. The osmosis mechanism is termed the solvent flow through a semipermeable membrane that separates two fluid spaces toward high solute concentration to equalize the solute concentration on both sides of the membrane by osmotic pressure [132, 133]. The use occurs in fracturation places by using external forces, such as shearing, ultrasonics, pressure, or internal forces, in microcapsules having a permeation-selective coating [130]. Erosion occurs by chemical reactions resulting in the release of the wanted ingredient. Also, partitioning or fragmentation may occur in addition to releasing the core material [128], using PBS as a release medium. Tea polyphenols (TPP) are encapsulated in surface-modified gelatin nanoparticles (GNP) with polyelectrolyte layers (TPP-GNP). Fast TPP release (burst release in the first hour with > 20% release) was observed in TPP-GNPs without surface modification by the polyelectrolyte layer, while the release rate has been considerably slowed as the number of polyelectrolyte layers has increased due to its protection. Burst release in non-surface-modified GNPs was established due to the hydrophilicity of gelatin. Release of TPP with 99–100% in TPP-GNP with six layers of polyelectrolyte took between 120 and 150 hours through erosion mechanism, whereas TPP-GNP with no polyelectrolyte layer took 24 hours through a diffusion and erosion mechanism coupling [134]. The diffusion mechanism is based on the geometry and size of the carrier, the core's solubility, and its diffusion coefficient, porosity, and temperature. The erosion rate depends on the release medium, the biopolymer size, molecular mass (the smaller the size, the lower the molecular mass, the higher the erosion), and physicochemical stability [135].

They used gelatin-encapsulated vitamin C as the wall material for feeding larval shrimp (*Penaeus japonicus*). Flavor is released by melting the capsule wall as the capsules are kept well below the melting point during storage, then heated above this temperature during cooking or preparation to release the active material [132]. The release is based upon protein-hydrolyzing enzymes in the larvae, and it was thus easy to digest and break the capsule with protein walls. The leaching rates of feed particles were affected by pH, temperature, solute concentration, and particle diameter [136].

Electrosprayed gelatin sub-microparticles were used as edible carriers to encapsulate an antioxidant molecule, (−)-epigallocatechin gallate (EGCG). Ethanol, 2-(N134 morpholino) ethane sulfonic acid, hemisodium salt MES and phosphate-buffered saline system PBS were chosen as release media. It is observed that an initial burst release was recorded as the swelling consequences of the gelatin matrix and then followed by a slower sustained release, which was more significant in the ethanolic suspension due to the antioxidant molecule encapsulation, which will therefore have an impact on the degradation kinetics in the solution [137].

8.7 Conclusion

Gelatin is known for its excellent drying and film-forming properties. The encapsulation and the phase separation processes can be enhanced by forming a complex from gelatin and oppositely charged polymers. Various techniques, such as coacervation, encapsulate food ingredients, depend on external and internal factors. Centrifugal extrusion is another technique that is used for the encapsulation of unstable flavors and vitamins. Fluid-bed coating is controlled by variables that affect coating efficiency, such as temperature and humidity. Nanoencapsulation has various benefits, including the enhancement of the active compound delivery. It also helps reduce storage and transport costs as the capsules will be in powder form. The release of the core material from the capsule depends on various factors, such as the chemical and physical capsule, coating properties, and experimental parameters. To sum up, gelatin is widely used as a commercial coating material, but it is restricted in markets of certain regions due to consumers' dietary preferences and religious obligations. Also, gelatin's tendency to absorb moisture is considered another limitation as it needs specific conditions for its storage. Gelatin can be blended with other materials for enhanced encapsulation to overcome this problem. Further studies are expected to show the effect of various materials combined with gelatin to enhance encapsulation efficiency.

Glossary

Bottom-up technique This technique depends on combining nanoscale molecules or parts to form large-sized molecules.

Co-crystallization An encapsulation technique that refers to incorporating flavorings in the carbohydrate surface between lipophilic flavorings and the water phase.

FTIR Fourier-transform infrared spectroscopy is a technique in which liquid, solid, and gas samples are analyzed to produce an infrared spectrum of absorption, photoconductivity, and emission. It is also implied to define different functional chemical groups of the analyte.

Ionic gelation A simple food encapsulation technique where the bioactive ingredient is injected into another solution via a biopolymer solution or emulsion under circumstances that favor biopolymer gelation. It is a competitive alternative to the most-used encapsulation techniques, such as spray drying, freeze drying, and complex coacervation.

Molecular inclusion The encapsulation technique that occurs at the molecular level using α-cyclodextrin as the medium of encapsulation.

Redox balance The balance between oxidants (or pro-oxidants) and antioxidants. Oxidants such as free radicals and other reactive species are constantly created.

Top-down technique It depends on breaking down a large-sized complex material into its small parts (nanoscale). It is easier than the bottom-up technique.

References

1 P. Balasubramani, P. T. Palaniswamy, R. Visvanathan, V. Thirupathi, A. Subbarayan, J. Prakash Maran, *Int. J. Biol. Macromol.* **2015**, *72*, 210.

2 H. Rajabi, M. Ghorbani, S. M. Jafari, A. Sadeghi Mahoonak, G. Rajabzadeh, *Food Hydrocoll.* **2015**, *51*, 327.

3 S. Anandaraman, G. A. Reineccius, *Food Technol.* **1987**, *40*, 88.

4 S. Gouin, *Trends Food Sci. Technol.* **2004**, *15*, 330.

5 T. Kondo, *J. Oleo Sci.* **2001**, *50*, 143.

6 H. C. F. Carneiro, R. V. Tonon, C. R. F. Grosso, M. D. Hubinger, *J. Food Eng.* **2013**, *115*, 443.

7 R. V. D. B. Fernandes, S. V. Borges, D. A. Botrel, *Carbohydr. Polym.* **2014**, *101*, 524.

8 Y. Kagami, S. Sugimura, N. Fujishima, K. Matsuda, T. Kometani, Y. Matsumura, *J. Food Sci.* **2003**, *68*, 2248.

9 K. G. H. Desai, H. J. Park, *Dry. Technol.* **2007**, *23*, 1361.

10 Y. Beatus, A. Raziel, M. Rosenberg, I. J. Kopelman, *Leb. Technol.* **1985**, *18*, 28.

11 W. Kolanowski, G. Laufenberg, B. Kunz, *Int. J. Food Sci. Nutr.* **2009**, *55*, 333.

12 J. Calvarro, T. Perez-Palacios, J. Ruiz, *Int. J. Gastron. Food Sci.* **2016**, *5–6*, 27.

13 M. Ramos, A. Valdés, A. Beltrán, M. C. Garrigós, *Coatings* **2016**, *6*, 41.

14 L. Liu, L. Tao, J. Chen, T. Zhang, J. Xu, M. Ding, X. Wang, J. Zhong, *Lot* **2021**, *146*, 111500.

15 X. M. Sha, L. J. Zhang, Z. C. Tu, L. Z. Zhang, Z. Z. Hu, Z. Li, X. Li, T. Huang, H. Wang, L. Zhang, H. Xiao, *LWT* **2018**, *89*, 74.

16 Z. K. Farahani, *Human, Heal. Halal Metrics.* **2021**, *2*, 12.

17 A. A. Mariod, *African Edible Insects As Altern. Source Food, Oil, Protein Bioact. Components*, Springer, Cham, **2020**, 149–158.

18 L. C. Sow, J. M. Nicole Chong, Q. X. Liao, H. Yang, *J. Food Eng.* **2018**, *239*, 92.

19 L. C. Sow, S. J. Tan, H. Yang, *Food Hydrocoll.* **2019**, *90*, 9.

20 R. Shaddel, J. Hesari, S. Azadmard-Damirchi, H. Hamishehkar, B. Fathi-Achachlouei, Q. Huang, *Int. J. Biol. Macromol.* **2018**, *107*, 1800.

21 R. M. Hathout, M. K. Omran, *Pharm. Dev. Technol.* **2015**, *21*, 379.

22 L. J. Pérez-Córdoba, I. T. Norton, H. K. Batchelor, K. Gkatzionis, F. Spyropoulos, P. J. A. Sobral, *Food Hydrocoll.* **2018**, *79*, 544.

23 Y. P. Timilsena, M. A. Haque, B. Adhikari, Y. P. Timilsena, M. A. Haque, B. Adhikari, *Food Nutr. Sci.* **2020**, *11*, 481.

24 H. Zhang, Y. Liang, X. Li, H. Kang, *Meat Sci.* **2020**, *166*, 108137.

25 Y. Shahbazi, N. Shavisi, N. Karami, R. Lorestani, F. Dabirian, *Lwt* **2021**, *152*, 112322.

26 R. C. N. da Conceição, R. D. Batista, F. M. D. A. L. Zimmer, I. K. M. Trindade, A. F. de Almeida, C. C. A. D. A. Santos, *Brazilian J. Microbiol.* **2021**, *52*, 1503.

27 M. S. Samsi, A. Kamari, S. M. Din, G. Lazar, *J. Food Sci. Technol.* **2019**, *56*, 3099.

28 C. Xiao, H. Liu, Y. Lu, L. Zhang, *J. Macromol. Sci. Part A Pure Appl. Chem.* **2007**, *38 A*, 317.

29 L. P. Heckert Bastos, J. Vicente, C. H. Corrêa Dos Santos, M. Geraldo de Carvalho, E. E. Garcia-Rojas, *Food Hydrocoll.* **2020**, *102*, 105605.

30 N. Choudhury, M. Meghwal, K. Das, *Food Frontiers.* **2021**, *2*, 426.

31 P. Kumar, *Crit. Rev. Ther. Drug Carr. Syst.* **2010**, *27*, 509.

32 V. Nedovic, A. Kalusevic, V. Manojlovic, S. Levic, B. Bugarski, *Procedia. Food Sci.* **2011**, *1*, 1806.

33 Y. P. Timilsena, M. A. Haque, B. Adhikari, *Food Nutr. Sci.* **2020**, *11*, 481.

34 G. V. Barbosa-Cánovas, E. Ortega-Rivas, P. Juliano, H. Yan, in *Food Powders. Food Eng. Ser.*, (Ed: Gustavo V. Barbosa), Springer, Boston, MA, **2005**, 199–219.

35 M. Peanparkdee, S. Iwamoto, R. Yamauchi, *Rev. Agric. Sci.* **2016**, *4*, 56.

36 A. E. Abd El Kader, H. M. A. Hashish, *Egypt. J. Chem.* **2020**, *63*, 1881.

37 V. Suganya, V. Anuradha, *Int. J. Pharm. Clin. Res.* **2017**, *9*, 233.

38 P. Maroulas, S. Kripotou, P. Pissis, A. Fainleib, I. Bei, V. Bershtein, Y. Gomza, *J. Compos. Mater.* **2009**, *43*, 943.

39 A. K. K., A. K. K., M. V. A., L. R. G. Kumar, *Recent TRENDS Harvest POST-HARVEST Technol. Fish.*, Director, CIFT, Cochin-29, **2017**, 230–242.

40 B. Balanc, K. Trifkovic, *Food Eng. Rev.* **2015**, *7*, 452.

41 G. Ozkan, P. Franco, I. De Marco, J. Xiao, E. Capanoglu, *Food Chem.* **2018**, *272*, 494.

42 K. G. H. Desai, H. J. Park, *Recent Developments in Microencapsulation of Food Ingredients*, 2005.

43 A. M. Bakry, S. Abbas, B. Ali, H. Majeed, M. Y. Abouelwafa, A. Mousa, L. Liang, *Compr. Rev. Food Sci. Food Saf.* **2016**, *15*, 143.

44 V. Đorđević, A. Paraskevopoulou, F. Mantzouridou, S. Lalou, M. Pantić, B. Bugarski, V. Nedović, V. Đorđević, B. Bugarski, A. Paraskevopoulou, F. Mantzouridou, S. Lalou, M. Pantić, V. Nedović, *Encapsulation Technologies for Food Industry*, Springer, Cham, **2016**.

45 K. Dewettinck, A. Huyghebaert, *Trends Food Sci. Technol.* **1999**, *10*, 163.

46 A. K. Anal, H. Singh, *Trends Food Sci. Technol.* **2007**, *18*, 240.

47 S. Ray, U. Raychaudhuri, R. Chakraborty, *Food Biosci.* **2016**, *13*, 76.

48 F. Ronsse, J. G. Pieters, K. Dewettinck, *J. Food Eng.* **2007**, *78*, 296.

49 M. Hemati, R. Cherif, K. Saleh, V. Pont, *Powder Technol.* **2003**, *130*, 18.

50 D. Solís-Morales, C. M. Sáenz-Hernández, E. Ortega-Rivas, *J. Food Eng.* **2009**, *93*, 236.

51 F. Debeaufort, A. Voilley, *Edible Film. Coatings Food Appl., Springer*, **2009**, 135.

52 V. Đorđević et al., Encapsulation technologies for food industry. in *Emerging and Traditional Technologies for Safe, Healthy and Quality Food. Food Engineering Series.* (Eds: V. Nedović, P. Raspor, J. Lević, V. Tumbas Šaponjac, G. Barbosa-Cánovas), Springer, Cham, **2016**, https://doi.org/10.1007/978-3-319-24040-4_18.

53 K. Dewettinck, A. Huyghebaert, Fluidized bed coating in food technology. *Trends Food Sci. Technol.* **1999**, 10, 163–168. https://doi.org/10.1016/S0924-2244(99)00041-2.

54 H. Kage, R. Abe, R. Hattanda, T. Zhou, H. Ogura, Y. Matsuno, *Powder Technol.* **2003**, *130*, 203.

55 P. Sun, M. Zeng, Z. He, F. Qin, J. Chen, *Dry. Technol.* **2013**, *31*, 1619.

56 P. Sun, M. Zeng Z. HE, ... F. Q.-S. and T. of, undefined **2013**, *Sci. Technol. Food Ind*, **2013**, 15.

57 A. Gharsallaoui, G. Roudaut, O. Chambin, A. Voilley, R. Saurel, *Food Res. Int.* **2007**, *40*, 1107.

58 D. T. Santos, M. Angela, A. Meireles, *Open Chem. Eng. J.* **2010**, *4*, 42.

59 A. Verma, S. V. Singh, *Crit. Rev. Food Sci. Nutr.* **2014**, *55*, 701.

60 E. Ortega-Rivas, P. Juliano, H. Yan, *Food Powders: Physical Properties, Processing, and Functionality*, Springer, Berlin, **2006**.

61 A. L. R. Rattes, W. P. Oliveira, *Powder Technol.* **2007**, *171*, 7.

62 A. A. Karim, R. Bhat, *Trends Food Sci. Technol.* **2008**, *19*, 644.

63 Y. Wang, Z. Lu, F. Lv, X. Bie, *Eur. Food Res. Technol.* **2009**, *229*, 391.

64 Y. Wang, Z. Lu, X. Bie, F. Lv, *Eur. Food Res. Technol.* **2010**, *231*, 189.

65 R. Matsuno, S. Adachi, *Trends Food Sci. Technol.* **1993**, *4*, 256.

66 A. Jeyakumari, A. Zynudheen, U. Parvathy, **2016**, *2*.

67 M. A. Augustin, Y. Hemar, *Chem. Soc. Rev.* **2009**, *38*, 902.

68 N. Chikkamalapura, *Int. J. Chem. Stud.* **2020**, *8*, 2560.

69 G. Ozkan, P. Franco, I. De Marco, J. Xiao, E. Capanoglu, *Food Chem.* **2019**, *272*, 494.

70 P. Henrique Rodrigues Do Amaral, P. Lopes Andrade, L. Costa de Conto, in *Microencapsul. – Process. Technol. Ind. Appl*, (Ed: F. Salaün), IntechOpen, 2019.

71 A. Trojanowska, M. Giamberini, I. Tsibranska, M. Nowak, Ł. Marciniak, R. Jatrzab, B. Tylkowski, *J. Membr. Sci. Res.* **2017**, *3*, 265.

72 S. P. Dhakal, J. He, *Food Res. Int.* **2020**, *137*, 109326.

73 P. Kaushik, K. Dowling, C. J. Barrow, B. Adhikari, *J. Funct. Foods* **2015**, *19*, 868.

74 A. Masuang, P. Finot, S. Reyna, *ASEAN Food J* **2007**, *14*, 1.

75 D. de Almeida. Paula, E. M. F. Martins, N. D. A. Costa, P. M. de Oliveira, E. B. de Oliveira, A. M. Ramos, *Int. J. Biol. Macromol.* **2019**, *133*, 722.

76 L. C. Corrêa-Filho, M. Moldão-Martins, V. D. Alves, *Appl. Sci* **2019**, *9, 571.*

77 L. Yinbin, L. Wu, M. Weng, B. Tang, P. Lai, J. Chen, *Powder. Technol.* **2018**, *340*, 459.

78 F. M. Sánchez, F. García, P. Calvo, M. J. Bernalte, D. González-Gómez, *Innov. Food Sci. Emerg. Technol.* **2016**, *34*, 243.

79 V. B. de Souza, M. Thomazini, M. A. Echalar Barrientos, C. M. Nalin, R. Ferro-Furtado, M. I. Genovese, C. S. Favaro-Trindade, *Food Hydrocoll* **2018**, *77*, 297.

80 T. Marques da Silva, E. J. Lopes, C. F. Codevilla, A. J. Cichoski, É. M. de M. Flores, M. H. Motta, C. D. B. da Silva, C. R. F. Grosso, C. R. de Menezes, *LWT* **2018**, *90*, 412.

81 M. G. Santos, F. T. Bozza, M. Thomazini, C. S. Favaro-Trindade, *Food Chem* **2015**, *171*, 32.

82 L.-K. Liao, X.-Y. Wei, X. Gong, J.-H. Li, T. Huang, T. Xiong, *LWT – Food Sci. Technol.* **2017**, *82*, 82.

83 S. Beirão-da-Costa, C. Duarte, A. I. Bourbon, A. C. Pinheiro, M. I. N. Januário, A. A. Vicente, M. L. Beirão-da-Costa, I. Delgadillo, *Food Hydrocoll* **2013**, *33*, 199.

84 S. Beirão da Costa, C. Duarte, A. I. Bourbon, A. C. Pinheiro, A. T. Serra, M. M. Martins, M. I. N. Januário, A. A. Vicente, I. Delgadillo, C. Duarte, M. L. Beirão da Costa, *J. Food Eng.* **2012**, *110*, 190.

85 J. Castro-Rosas, C. R. Ferreira-Grosso, C. A. Gómez-Aldapa, E. Rangel-Vargas, M. L. Rodríguez-Marín, F. A. Guzmán-Ortiz, R. N. Falfan-Cortes, *Food Res. Int.* **2017**, *102*, 575.

86 N. S. Girardi, D. García, S. N. Robledo, M. A. Passone, A. Nesci, M. Etcheverry, *Ind. Crops Prod.* **2016**, *92*, 93.

87 M. Yang, Z. Liang, L. Wang, M. Qi, Z. Luo, L. Li, *Adv. Polym. Technol* **2020**, *2020*.

88 A. S. Vaziri, I. Alemzadeh, M. Vossoughi, A. C. Khorasani, *Carbohydr. Polym.* **2018**, *199*, 266.

89 R. Arshady, *J. Microencapsul.* **1993**, *10*, 413.

90 M. A. Jal, J. M. S. Negre, E. G. Montoya, *Eur. Food Res. Technol.* **2020**, *246*, 1371.

91 N. Choudhury, M. Meghwal, K. Das, *Food Front*, **2021**, 1.

92 Q. Ye, N. Georges, C. Selomulya, *Trends Food Sci. Technol.* **2018**, *78*, 167.

93 M. Arenas-Jal, J. M. Suñé-Negre, E. García-Montoya, *Eur. Food Res. Technol.* **2020**, *246*, 1371.

94 A. J. Paredes, C. M. Asencio, L. J. Manuel, D. Alberto, **2016**.

95 A. Bratovcic, J. Suljagic, *Croat. J. Food Sci. Technol.* **2019**, *11*, 113.

96 M. Pateiro, B. Gómez, P. E. S. Munekata, F. J. Barba, P. Putnik, D. B. Kovačević, J. M. Lorenzo, *Molecules* **2021**, *26*, 1547.

97 O. M. Ali, Y. Hashem, A. A. Bekhit, S. N. Khattab, K. A. Elkhodairy, M. S. Freag, M. Teleb, A. O. Elzoghby, in *Biopolym. Nanostructures Food Encapsulation Purp.*, (Ed: S. Mahdi Jafari), Academic Press, **2019**, 189–216.

98 V. J. Morris, *Encycl. Food Saf.*, Academic Press, **2014**, 208–210.

99 K. N. C. Murthy, P. Monika, G. K. Jayaprakasha, B. S. Patil, *ACS Symp. Ser.*, **2018**, 383–405.

100 A. M. Pisoschi, A. Pop, C. Cimpeanu, V. Turcuş, G. Predoi, F. Iordache, *Eur. J. Med. Chem.* **2018**, *157*, 1326.

101 J. Anu Bhushani, C. Anandharamakrishnan, *Trends Food Sci. Technol.* **2014**, *38*, 21.

102 S. M. Jafari, in *Nanoencapsulation Technol. Food Nutraceutical Ind.*, (Ed: S. Mahdi Jafari), Academic Press, **2017**, 1–34.

103 F. Akbar, I. Jaswir, P. Jamal, Octavianti, *Int. Food Res. J.* **2017**, *24*, 255.

104 M. F. Fontefrancesco, H. Sidsaph, **2019**, 1.

105 I. Katouzian, S. M. Jafari, *Trends Food Sci. Technol.* **2016**, *53*, 34.

106 K. Khoshakhlagh, M. Mohebbi, A. Koocheki, A. Allafchian, *Bioact. Carbohydrates Diet. Fibre* **2018**, *16*, 43.

107 C. S. S. R. Kumar, *Biofunctionalization of Nanomaterials*, Wiley-VCH, **2005**.

108 T. İnanç Horuz, K. B. Belibağlı, *Food Chem* **2018**, *268*, 86.

109 E. N. Story, R. E. Kopec, S. J. Schwartz, G. K. Harris, *Annu. Rev. Food Sci. Technol.* **2010**, *1*, 189.

110 A. K. D. O. C. Medeiros, C. de C. Gomes, M. L. Q. de A. Amaral, L. D. G. de Medeiros, I. Medeiros, D. L. Porto, C. F. S. Aragão, B. L. L. Maciel, A. H. de A. Morais, T. S. Passos, *Food Chem.* **2019**, *270*, 562.

111 G. L. R. de Oliveira, I. Medeiros, S. S. da C. Nascimento, R. L. S. Viana, D. L. Porto, H. A. O. Rocha, C. F. S. Aragão, B. L. L. Maciel, C. F. de Assis, A. H. de A. Morais, T. S. Passos, *Food Chem.* **2021**, *348*, 129055.

112 R. Esfahani, S. Mahdi, A. Jafarpour, D. Dehnad, *Food Hydrocoll.* **2019**, *90*, 291.

113 Y. Lv, F. Yang, X. Li, X. Zhang, S. Abbas, *Food Hydrocoll.* **2014**, *35*, 305.

114 J. Zhang, G. Jia, Z. Wanbin, J. Minghao, Y. Wei, J. Hao, X. Liu, Z. Gan, A. Sun, *Food Hydrocoll.* **2021**, *112*, 106280.

115 B. Vafania, M. Fathi, S. Soleimanian-Zad, *Food Bioprod. Process.* **2019**, *116*, 240.

116 T. Kt, M. NZ, D. Vb, K. Mt, K.-J. Zd, N. VA, B. BM, *Carbohydr. Polym.* **2014**, *111*, 901.

117 M. M. Mousavi, M. Torbati, P. Farshi, H. Hosseini, M. A. Mohammadi, S. M. Hosseini, S. Hosseinzadeh, *Adv. Pharm. Bull.* **2021**, *11*, 514.

118 R. Hernández-Nava, A. López-Malo, E. Palou, N. Ramírez-Corona, M. T. Jiménez-Munguía, *Food Hydrocoll.* **2020**, *109*.

119 F. Donsì, M. Annunziata, M. Sessa, G. Ferrari, *LWT – Food Sci. Technol.* **2011**, *44*, 1908.

120 Z. Xiao, W. Li, G. Zhu, R. Zhou, Y. Niu, *Flavour Fragr. J.* **2016**, *31*, 283.

121 F. Xing, G. Cheng, K. Yi, L. Ma, *J. Appl. Polym. Sci.* **2005**, *96*, 2225.

122 S. M. Jafari, I. Katouzian, S. Akhavan, in *Nanoencapsulation Technol. Food Nutraceutical Ind.*, (Ed: S. Mahdi Jafari), Elsevier Inc., **2017**, 545–590.

123 T. Singh, S. Shukla, P. Kumar, V. Wahla, V. K. Bajpai, *Front. Microbiol.* **2017**, *8*, 1501.

124 K. Savolainen, L. Pylkkänen, H. Norppa, G. Falck, H. Lindberg, T. Tuomi, M. Vippola, H. Alenius, K. Hämeri, J. Koivisto, D. Brouwer, D. Mark, D. Bard, M. Berges, E. Jankowska, M. Posniak, P. Farmer, R. Singh, F. Krombach, P. Bihari, G. Kasper, M. Seipenbusch, *Saf. Sci.* **2010**, *48*, 957.

125 S. Arora, J. M. Rajwade, K. M. Paknikar, *Toxicol. Appl. Pharmacol.* **2012**, *258*, 151.

126 M. N. Duvall, *FDA Regulation of Nanotechnology, Beveridge and Diamond*, PG, Washington, DC, USA., **2012**.

127 S. M. Jafari, I. Katouzian, S. Akhavan, *Nanoencapsulation Technol. Food Nutraceutical Ind.*, **2017**, 545.

128 M. Bruschi, *Strategies to Modify the Drug Release from Pharmaceutical Systems*, Woodhead Publishing, **2015**.

129 M. Mishra, *Handbook of Encapsulation and Controlled Release*, CRC, 2015.

130 F. Shahidi, X. Q. Han, F. Shahidi, X. Han, **1993**, 37.

131 J. M. Unagolla, A. C. Jayasuriya, *Eur. J. Pharm. Sci.* **2018**, *114*, 199.

132 A. Madene, M. Jacquot, J. Scher, S. Desobry, *Int. J. Food Sci. Technol.* **2006**, *41*, 1.

133 J. Siepmann, R. A. Siegel, M. J. Rathbone, *Fundamentals and Applications of Controlled Release Drug Delivery*, Springer US, **2012**.

134 K. Kulandaivelu, A. K. Azad Mandal, *ieeexplore.ieee.org* **2017**, *11*, 469.

135 S. Chirico, A. Dalmoro, G. Lamberti, G. Russo, G. Titomanlio, *J. Control. Release* **2007**, *122*, 181.

136 Z. Xie, F. Wang, H. Liu, S. Guo, A. Zhu, H. Niu, *Aquac. Res.* **2010**, *42*, 65.

137 L. G. Gómez-Mascaraque, J. M. Lagarón, A. López-Rubio, *Food Hydrocoll.* **2015**, *49*, 42.

138 C. S. Favaro-Trindade, R. J. B. Heinemann, D. L. Pedroso, *CAB Rev. Perspect. Agric. Vet. Sci. Nutr. Nat. Resour.* **2011**, *6*.

139 M. P. Silva, F. L. Tulini, E. Martins, M. Penning, C. S. Fávaro-Trindade, D. Poncelet, *LWT* **2018**, *89*, 392.

9

Working Principles and Use of Chitosan for Food Component Encapsulation

Gastón Bravo-Arrepol[1], *Plamen Dimitrov Katsarov*[2,3], *Bissera Asenova-Pilicheva*[2,3], *Paolina Kancheva- Lukova*[4], *Danilo Escobar-Avello*[1,5], *Hazel Peniche*[6], *Lorenzo García*[6], *Carlos Peniche-Covas*[6], *Philippe Michaud*[7], *Cédric Delattre*[7,8], *Liliam Becheran-Maron*[9], *Johanna Castaño*[10], *Maria Dolores Lopez*[11], *Oscar Valdes*[12], *Aleksandra Nesic*[1], and *Gustavo Cabrera-Barjas*[1,5,*]

[1] *Universidad de Concepción, Unidad de Desarrollo Tecnológico (UDT), Av. Cordillera 2634, Parque Industrial Coronel, BioBio, Coronel, Chile*
[2] *Department of Pharmaceutical Sciences, Faculty of Pharmacy, Medical University of Plovdiv, Plovdiv, Bulgaria*
[3] *Research Institute at Medical University of Plovdiv (RIMU), Plovdiv, Bulgary*
[4] *Department of Pharmacognosy and Pharmaceutical Chemistry, Faculty of Pharmacy, Medical University of Plovdiv, Plovdiv, Bulgary*
[5] *Centro Nacional de la Madera (CENAMAD), Av Libertador Bernardo O'Higgins 340, Santiago, Región Metropolitana, Chile*
[6] *Center of Biomaterials, University of Havana Ave. Universidad s/n entre G y Ronda, Vedado, La Habana, Cuba*
[7] *Université Clermont Auvergne, Clermont Auvergne INP, CNRS, Institut Pascal, Clermont-Ferrand, France*
[8] *Institut Universitaire de France (IUF), 1 Rue Descartes, Paris, France*
[9] *Materials Science and Technology Institute, University of Habana, La Habana, Cuba*
[10] *Facultad de Ingeniería y Tecnología, Universidad San Sebastián, Lientur, Concepción Chile*
[11] *Facultad de Agronomía, Universidad de Concepción, Campus Concepción*
[12] *Centro de Investigación de Estudios Avanzados del Maule (CIEAM), Vicerrectoría de Investigación y Postgrado, Universidad Católica del Maule, Talca Chile*
* *Corresponding author*

Highlights

- Chitosan is a natural product (GRAS) that can be isolated from different animal and plant sources, with application in food industry.
- This biopolymeric material with antimicrobial properties can be used to encapsulate high-value food components.
- Different nano and microencapsulation techniques are shown using chitosan as a polymeric matrix.
- Chitosan can be used in food-packaging applications directly or as an additive, encapsulating the active ingredient.
- Three-dimensional (3D) bioprinting and electrospinning offer an opportunity to use chitosan to encapsulate food components.
- Chitosan's market had decreased during the pandemic, but it is expected to rise again, driven by food applications.

Materials Science and Engineering in Food Product Development, First Edition. Edited by Wing-Fu Lai.

9.1 Introduction

Chitin and chitosan are a family of copolymers varying in the ratio of β-(1→4) linked N-acetylglucosamine and glucosamine sugar residues. Chitin is a widespread polysaccharide in nature and can be obtained from various species of crustaceans, insects, algae, and fungi (Figure 9.1) [1]. Chitin's chemical structure is similar to cellulose, but acetyl amine groups substitute the hydroxyl groups at the C2 position. Unlike other abundant polysaccharides containing carbon, hydrogen, and oxygen, chitin also contains nitrogen (6.89%) in its structure. Although extremely abundant and affordable, chitin has limited application, mainly due to its low solubility in water. Conversely, high content of glucosamine units allows chitosan to be soluble in an aqueous acidic medium [2]. The most common chitin

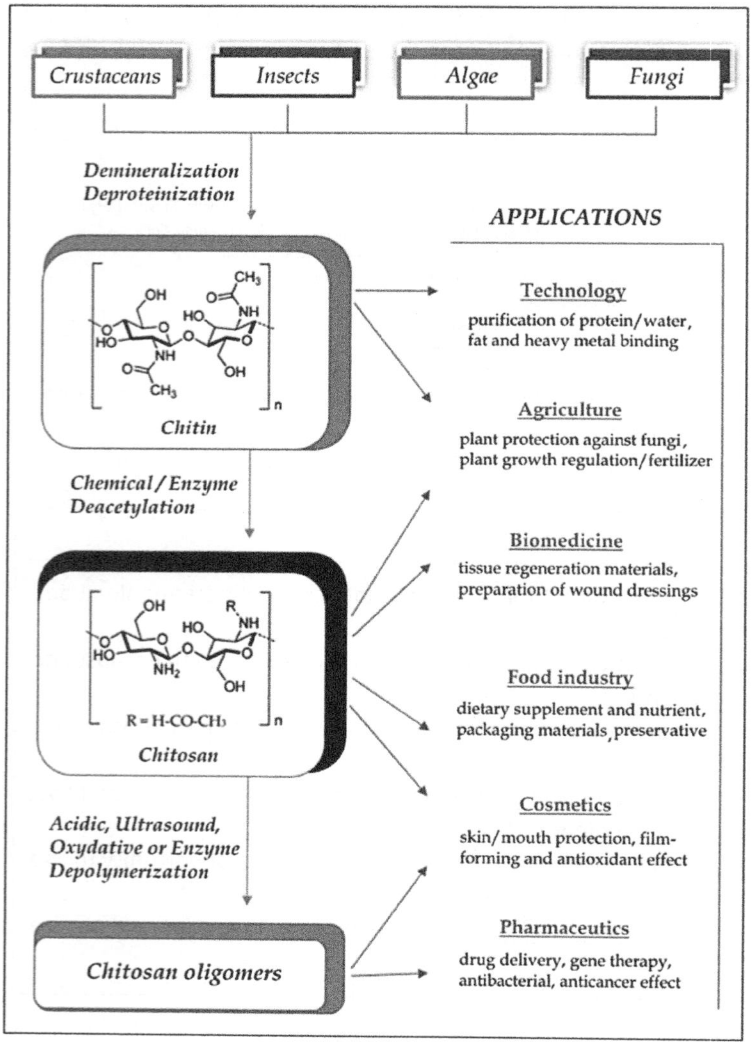

Figure 9.1 Sources, chemical structure, and some applications of chitin and chitosan.

and chitosan industrial production sources are shrimps, crabs, lobsters, and krills. In the shells of crustaceans, chitin forms a complex network with proteins and calcium carbonate deposits. Therefore, the isolation of chitin (15–30 wt%) from shells requires the removal of proteins (20–30 wt%) by deproteinization and inorganic calcium carbonate (~50 wt%) following demineralization [3]. To ensure the high quality of the final polysaccharides it is strongly recommended that shells of the same species be used. The shells of different marine sources have different compositions, implying a wide variety of production process parameters. Nowadays, about half of the total mass of the crustaceans used in the food sector is disposed of as waste. Only seafood waste can provide thousands of tons (~10^6) of chitin a year. Therefore, sources of chitosan are available and inexpensive. Moreover, the valorization and upcycling of seafood waste is important from an environmental and economic point of view. Usually, seafood waste is burned, landfilled, dumped into the sea, or left to spoil. If not processed properly, it can negatively impact human health, biodiversity, and the environment [4].

When chitin is deacetylated, and the repeating units in the polymer are predominantly free of acetyl functional groups such as *β*-1,4-*D*-glucosamine, the polymer is known as chitosan. The deacetylation process is carried out in hot alkali to varying degrees, depending on target applications. Then, numerous products with different degrees of deacetylation and molecular weights can be obtained. The mole fraction of the *N*-acetylated repeating units is defined as the degree of acetylation. In contrast, the *β*-1,4-*D*-glucosamine repeating units percentage in the polysaccharides is defined as the degree of deacetylation (DD). The resulting polymer is chitosan when the polysaccharide structure has a DD ~ 50%, but must be fully soluble in acid media. The DD is one of the essential characteristics of chitosan, which determines many of its properties and applications. The polymer can be classified based on its DD value as chitosan with a high DD (70–99%) or with a low DD (50–70%). The properties of chitin and chitosan also depend on the degree of polymerization (DP), which represents the length of the polymer chain and correlates with the molecular weight (MW). The DP is the number of monomer units in the polymer or oligomer structure. The chain length of most chitosans cannot be determined accurately and the average MW in kDa is often used for its characterization [1]. The MW of chitosan usually varies in the range 50–1000 kDa, depending on the starting material used for its production. The polymer can be classified according to its MW as low molecular weight chitosan (LMW) (<300 kDa) or high molecular weight chitosan (HMW) (>300 kDa) [2]. MW affects vital polymer properties, such as solubility, viscosity, adsorption on solids, strength, elasticity, and bioactivity.

Regarding its biocidal activity, chitosan antibacterial, antifungal, and antiviral properties are well-known, making it an interesting candidate for food applications. In fact, chitosan has been widely studied as a wall material for bioactive substances to be applied in agriculture, pharmacy, biomedicine, food, and food-packaging sectors, due to its natural origin and biodegradable, biocompatible, nontoxic, and biological functions. This chapter reviews the role of chitosan as a material used for food product encapsulation. Also, it provides an overview of using microencapsulated products to prepare functional foods and food-packaging products. Two emerging technologies, like 3D bioprinting and electrospinning, are discussed, and the food chitosan market outlook is presented.

Industrial Corner

- Chitosan, polyglucosamine polymer, is a GRAS material whose application in food has long data.
- Chitosan is a well-known emulsifier, coagulant, wine thinner, and rheology modifier.
- Special attention has been paid during the past few years to chitosan obtained from plants, e.g. mushrooms and fungal cell walls. These are an alternative for including additives on foods for vegan and Kosher consumers.
- Chitosan is widely used as fining in the wine and beer industry because it can bind yeast, other fruit particles, and debris with mild electrostatic attraction when added to wine or beer.
- The addition of chitosan speeds up the sedimentation (flocculation) process, allowing the beverage to be clarified more rapidly and effectively.
- Because chitosan does not require acidic tannins (found in red wines) to flocculate, it is becoming a popular fining agent for white wines when paired with colloidal silica.

9.2 Encapsulation Technologies

Encapsulation is the mechanical or physicochemical process in which valuable nutraceutical substances are confined or coated inside a material (wall material or carrier) to give small capsules or particles. Encapsulation permits the protection of sensitive active ingredients from the environment, ensuring the preservation of their physical, chemical, and biological properties besides their release or delivery under determined conditions. In terms of particle size, capsules are classified as nanocapsules (100–1000 nm), microcapsules (1–1000 μm), and millimeter capsules (>1 mm). During the past few years, nanoencapsulation technologies have been gaining attention due to the many advantages of using nanoparticles. The methods used to prepare food component nanoparticles are based on different strategies (Figure 9.2).

In the food industry, encapsulation is performed by various physical processes (e.g. spray drying, emulsification) and chemical processes (e.g. coacervation, ionotropic gelation). The selection of a particular process for a given application greatly depends on the properties of the active principle and the carrier [5]. In a typical spray-drying procedure, the carrier is dissolved in a solvent (preferably water), and the active ingredient is emulsified, dissolved, or dispersed in the carrier solution, usually with a surfactant. This dispersion is pumped to an atomizer, and the generated droplets are expelled into a hot drying gas chamber where moisture is evaporated, leading to particle formation. Particles ranging from 10 to 40 μm in size are then separated and collected as a dry powder. The yield can be affected by losses of material that remains attached to the drying chamber walls and the particle collector [6].

Coacervation is when a hydrocolloid solution is separated into two coexisting liquid phases: a concentrated colloidal phase (coacervate) and a highly dilute, almost colloidal free phase. It can be subdivided into simple and complex coacervation. In simple coacervation, the colloid is deprived of the solvent by adding a nonsolvent, an electrolyte, or a pH or

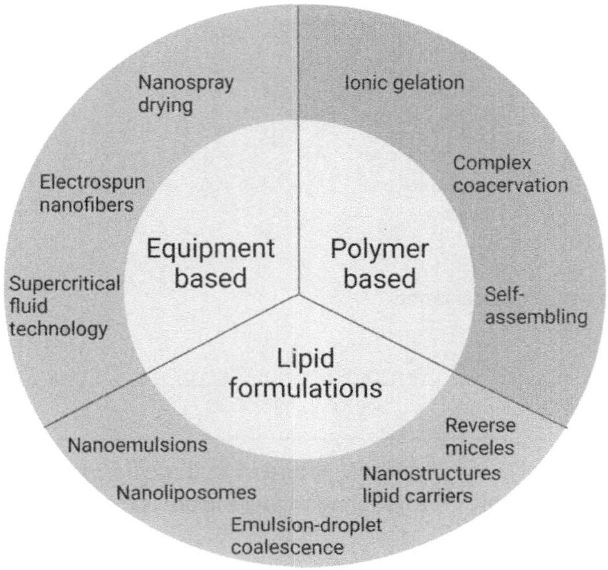

Figure 9.2 Nanoencapsulation technologies applied in the food industry.

temperature change. Complex coacervation is achieved by mixing two oppositely charged polyelectrolytes or colloids; see Figure 9.3 [7]. The coacervate formed from their electrostatic interaction is called polyelectrolyte complex. A typical polyelectrolyte complex is obtained by reacting polycationic chitosan with polyanionic polysaccharides, such as pectin, alginate, carrageenan, and hyaluronic acid. Ionotropic gelation is the crosslinking of a polyelectrolyte by the addition of a di- or multivalent oppositely charged ion. It is frequently used to prepare microcapsules and nanoparticles of polycationic chitosan crosslinked with sodium tripolyphosphate or polyanionic alginate crosslinked with calcium chloride. As a result of crosslinking, a precipitate of polymer particles containing the encapsulated substance is formed.

An emulsion is a system of at least two phases in which a liquid is dispersed in another liquid (continuous phase) with the aid of a surfactant or emulsifier, forming small droplets (discontinuous phase). In oil-in-water emulsions (O/W), the continuous phase is the aqueous component. To prevent degradation during storage, these emulsions are appropriate for encapsulating lipophilic active ingredients, such as fat-soluble vitamins and carotenoids. In water-in-oil (W/O) emulsions, the continuous phase is the lipophilic component, and the aqueous phase contains the hydrophilic components of interest. For some applications, double emulsions of W/O/W and O/W/O types can also be prepared. In recent years, an emerging technology to encapsulate food products is spinning, which prepares fibers used to encapsulate the desired products [8–10]. Fibers can be made from non-thermoplastic carbohydrate polymers (polysaccharides) like chitosan or cellulose found in some cell walls. Four technologies produced polysaccharide fibers, e.g. wet spinning, dry-jet wet spinning, blow spinning, and electrospinning, which will be explained in another topic of this chapter [11–15].

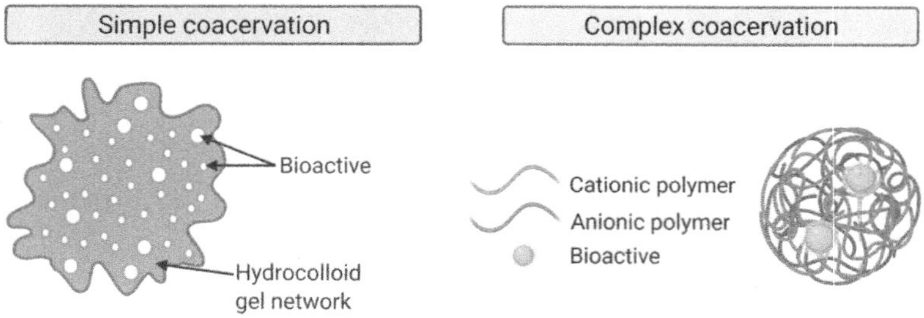

Figure 9.3 Simple coacervation and complex coacervation.

Industrial Corner

- Chitosan is the only polysaccharide with a polycationic nature in acid media, making it very attractive for polyelectrolyte complex formation with anionic polysaccharides, proteins, and nucleic acids.
- Polyelectrolyte complexation allows entrapping different molecules inside the chitosan matrix. Thus, it is possible to encapsulate materials and prepare capsules and particles of various sizes (nano or micrometer).
- Novel applications are obtained when used to protect high-value food components (vitamins, antioxidants, antimicrobials, prebiotics, etc.). Those nano and microparticles can be used as antimicrobials, food additives, or nutraceuticals to obtain new functionality or to improve food shelf life and human health.

9.3 Agent Encapsulation Using Chitosan as Polymeric Matrix

Table 9.1 summarizes the main nutraceutical compounds encapsulated into the chitosan matrix by different technologies and their efficiency. Some examples will be discussed below according to the active ingredient nature.

9.3.1 Encapsulation of Vitamins and Omega Fatty Acids

Vitamins are classified as micronutrients that play an important role in human development [70]. Most of them cannot be synthesized by humans or are synthesized in small quantities. Therefore, they must be ingested through the diet with foods rich in them, fortified, or supplements [71]. Despite these advantages, the vitamins are sensitive to degradation and must be protected by encapsulation. Emulsions, lipid nanoparticles, and polymers, such as chitosan, are widely used in the industry for this purpose [72]. In recent years, micro and nanoparticles based on chitosan for supplying vitamins used in food science (e.g. vitamin C, B complex, vitamin D, and vitamin K) were studied. In fact, there are several studies of micro and nanoparticles preparation using chitosan as carriers for the controlled release of certain vitamins, presenting better characteristics of biodegradability,

Table 9.1 Composition of chitosan-based formulations used to encapsulate nutraceutical compounds.

Food compound	Composition	Preparation	Chitosan properties	Application	References
Omega fatty acids					
Fish tuna oil	Chitosan (0.5% w/v), maltodextrin (10% w/v) and whey protein isolate (1% w/v).	Emulsification, ultrasonic atomization, and freeze drying	Chitosan with a degree of deacetylation (DD) = 80	Nutraceutical compounds by encapsulation and stabilization of fish tuna oil containing ultra-pure omega-3 fatty acid by preserving the fatty acids, EPA and DHA, at a level of 24 g/100 g	[16]
Ultra-pure omega-3 fatty acids	Chitosan (1.5%, w/v)–whey protein isolate (5% w/v), glycerol (2% w/v), Tween 80 (3 times weight of FO), and oregano oil (0.5% w/v)	Emulsification	Shrimp derived chitosan from Primex ehf (149 kDa MW, 9 cPs viscosity of a 1% (w/w) and DD 97%)	Carry fish oil for up to 37% in the film with higher oxidative stability	[17]
Powder of tuna fish oil	Chitosan: maltodextrin (1:1 ratio). Concentration of wall material was secured at 30% (w/w) while core material concentration used was 10%, 20%, 30%, and 40% (w/w)	Emulsification, spray drying Spray dryer operating factors were set as following: inlet temperature 180 ± 2°C, outlet temperature 85 ± 2°C, emulsion feed rate 0.7 L/h and air flow rate 250 kg h − 1		The fortification of food product by micropasules with improving fishy odor and oxidation	[18]

(Continued)

Table 9.1 (Continued)

Food compound	Composition	Preparation	Chitosan properties	Application	References
Omega-3 fish oil	β-lactoglobulin (β-LG) fibril variants, chitosan, and maltodextrin	Spray drying The inlet air temperature was set at 160°C, 170°C, and 180°C, with a fixed flow rate of 15 mL/min	Chitosan (100–300 kDa) and maltodextrin (dextrose equivalent 10)	High MW chitosan coating provided a physical barrier against high thermal condition and aided in overcoming the poor heat stability of β-LG fibrils	[19]
Omega-3 rich oils and curcumin	Chitosan, alginate, and curcumin NaOH or $CaCl_2$ as the crosslinking agents	Nanoemulsion and emulsification	SC with a DD and average MW of 85% and 100.000 Da, respectively	Incorporation of curcumin into fish oil–CS, fish oil–AL has improved the overall activity against Gram negative bacteria but reduced the activity toward Gram positive strains	[20]
Essential oils					
Thyme essential oil	Ethanol (50% v/v), acetic acid (45% v/v) and deionized water (5% v/v); chitosan (2% w/v) and gelatin (9% w/v at 40°C)	Solutions were mixed with different ratios (chitosan to gelatin = 1:10, 1:8 and 1:6) and then thyme essence was added with the ratios of 20% and 40% (V/W of solid biopolymer powdcr). Disk-shaped nozzle-less electrospinning was used to fabricate nanofibers	Molecular mass of chitosan used was 60 000–120 000	To reduce nitrite in sausages	[21]

Lime essential oil	Chitosan (0.05% w/v) in acetic acid. 2.5 mL of the solvent phase was added to the nonsolvent phase composed of 20% lime and methanol (40 mL)	30 mg of lecithin were dissolved in 0.5 mL of ethanol. 0.125 mL of lime essential oil and 9.5 mL of acetone were added. The organic phase was incorporated into 20 mL of chitosan (0.5% w/v) under constant agitation for 10 min	Medium MW chitosan (deacetylation degree 75–85%) was used	Antibacterial activity against food-borne pathogens	[22]
Cuminun cyminum	Chitosan (0.5 % w/v) in acid (1% v/v) and 85 mL of methanol and mixed with ethylene dichloride (668 uL) and 221 mg of caffeic acid. pH was adjusted at 8.5–9 using sodium hydroxide (1 M) to precipitate the nanogel	*C. cyminum* essential oils (5000 mg) were dissolved in ethanol (1:1), and the solution was then added to nanogels (5000 mg). The mixture was diluted with distilled water to a volume of 1 L followed by encapsulation through sonication (5 min, 70 Hz)	Chitosan purchased from Sigma-Aldrich (not specified)	Antimicrobial activity against *Aspergillus flavus*	[23]
Mentha piperita	Chitosan (0.5 g) was dissolved in acetic acid (1% v/v) solution (100 mL) to the final concentration of 0.5% (w/v). Then it was diluted with 85 mL of methanol under continuous stirring at 250 rpm for 20 min at room temperature. Stock solution was tuned to pH 3.5–4 and sonicated for 5 min	*M. piperita* essential oils (6000 ppm) was added to the nanogel (0.5% w/v), and sonication was performed (5 min, 70 Hz)	Low MW Chitosan	Antimicrobial activity against *Aspergillus flavus*	[24]

(Continued)

Table 9.1 (Continued)

Food compound	Composition	Preparation	Chitosan properties	Application	References
Carum copticum	A chitosan solution (1% w/v) was prepared in acetic acid (1% v/v). Tween 80 (0.56 g) was added to the 50 mL of chitosan solution with constant stirring at 50°C for 1.5 h	Different amounts of essential oils (0.125, 0.25, 0.375, 0.5 and 0.625 g) were dissolved in CH_2Cl_2 (5 mL) and oil phase was added dropwise into the chitosan solution (50 mL) with different weight ratios of chitosan to essential oil of 1:0, 1:0.25, 1:0.50, 1:0.75, 1:1, 1:1.25, respectively. Pantasodium tripolyphosphate aqueous solution at various concentrations (0.25, 0.5, 0.75, 1% (w/v), 50 mL) was gradually dropped into the o/w emulsion	Medium MW chitosan (75–85% DD)	Several biological activities	[25]
Zataria multiflora	Chitosan (0.3% w/v, 50 mL) was dispersed in acetic acid (1%, v/v)	Zataria essential oil was dropped into the stirring mixture of chitosan solution for 30 min to obtain an oil-in-water emulsion. Various contents (0, 375, 750, 1500, and 3000 ppm) were used to obtain different weight ratios (chitosan-essential oil) of 1:0, 1:0.25, 1:0.50, 1:0.75 and 1:1.00	Chitosan derived from crab shell with a MW of 684 kDa and deacetylation degree of ~ 85%	Antifungal activity for controlling *Botrytis cinerea*	[26]

Cinnamon essential oil	Chitosan (3.2 mg/mL) was dissolved in acetic acid solution. pH of the solutions adjusted to 4.0. Tween-80 (80 mg) was added as a surfactant to the solution (50 mL) and stirred at 60°C for 30 min	The cinnamon essential oil incorporated chitosan nanoparticles were prepared by a two-step process: oil-in-water (o/w) emulsification and ionic gelation	Chitosan (75–85% DD and Mw = 760 kDa)	Additive for the controlled release and shelf-life extension of beef patties	[27]
Coriandrum sativum	Chitosan solution (4% w/v) in acetic acid 1% (v/v), sodium alginate solution (4% w/v), inulin solution (15% w/v) These emulsions contained 20% (w/w) solid wall material, 5% (w/w) CEO and 75% (w/w) water	Solutions of biopolymers were mixed with the coriander essential oil in a weight ratio of solid content: essential oil 4:1 (w/w), in the presence of soy lecithin (0.1% w/w from the content of the emulsion). The spray-drying process was carried out 6 h after the preparation of the emulsions	The chitosan (CAS 9012–76-4) was characterized by a deacetylation degree of (DDA) 75% and the average MW $M_w = 387$ kg/mol, determined through the viscosimetric method using the following equation [η]25°C = 1.81 · 10-3MW0.93 (0.1 M CH_3COONa)	To ensure the maintenance of the nutritive and sensorial characteristics of foods and prolongs the period of storage	[28]
Copaiba / lemon / geranium	Chitosan solution (1% w/v) in acetic acid (0.5% v/v)	Prepared in the proportions of 1:10 and 1:5 (oil mass / polymer mass), Tween-80 surfactant 1% in relation to the polymer mass	Chitosan of low MW (DD 75–85% and 50–190 kDa)	Anticholinesterase activity (Alzheimer's disease)	[29]
Phenolics compounds					
Curcumin	Zein-egg white derived peptides-chitosan (Z-EWDP-CS) ternary nanoparticles (NPs)	coacervate		The thermal, photochemical, and salt stability and antioxidant activity were considerably improved in the NPs after the addition of wall material	[30]

(Continued)

Table 9.1 (Continued)

Food compound	Composition	Preparation	Chitosan properties	Application	References
α-tocopherol	Whey protein isolate/chitosan particles	Emulsificate	Medium molecular weight	The α-tocopherol-loaded emulsion particle structure remained unchanged after gastric digestion for 120 min	[31]
Phlorotannins	Chitosan-tripolyphosphate	Freeze-dried/dispersion	Raw crab shells were obtained from Taprobane Seafoods (Pvt) Ltd	Encapsulation did not improve the storage stability but significantly improved processing stability at high temperatures	[32]
Cobalamin and ascorbic		Spray-drying under 120°C on inlet and 65°C on outlet	Chitosan of medium molecular 127 weight (Cat. No. 448877)	Higher antioxidant effect for encapsulated vitamins	[33]
Vitamin B12 and vitamin C	Chitosan, modified chitosan, and sodium alginate	Spray drying air pressure and inlet temperature were set at 4 ml/min (15%), 32 m3/h (80%), 6.0 bar and 120°C, respectively	Chitosan of medium molecular 127 weight (Cat. No. 448877)	The release time of the total amount of vitamins was around 120 min for microparticles made with chitosan, 15 min 357 min for microparticles made with alginate and 10 min for the microparticles made with 358 modified chitosan	[29]
Vitamin B1	Chitosan, carrageenan, arabic gum, and modified chitosan were 1% (w/V) content of encapsulating agent. 1% (w/V) and 0.125% (w/V) of vitamin B1	Spray drying solution flow rate: 4 mL/min (15%), air flow rate: 32 m3/h (80%), air pressure: 6 bar, and inlet temperature: 120°C	Chitosan (Cat. No. 448877)	Higher encapsulation efficiency of vitamin (values of 94–96%) arabic gum, modified chitosan, and sodium alginate show slow release, while chitosan and pectin adequate results were obtained for vitamins microencapsulation	[34]

Quercetin (flavonols)	CS + 2% GAA s + pH 4.7 by (20% (w/v) NaOH) + 3 mg/ml concentration in 1:1 ratio of PEG 400: DW	Ionic gelation method, stirring, centrifugation, and lyophilized	Yarrow Chem Products, India	Antitumor effect, mammalian cells, and tumor xenografts in mice	[35]
Myricetin (flavonols)	Chitosan (1 mg/mL) + acetic acid (1% v/v) (pH ~ 4.8), myricetin (5 mg/mL), TPP (0.5 mg/mL), 2:1 (chitosan, w/w)	Ionic gelation methods, stirred under room temperature, centrifuged, dried under vacuum	Sigma, MW 50–190 kDa	Antioxidant activity	[36]
Kaempferol (flavonols)	Chitosan (0.2%, w/v) in acetic acid (0.1%, v/v) + pH 4.8 (NaOH) solution (20%, w/v) + TPP (0.2%, w/v)	Ionic gelation of chitosan with TPP anions, stirring, centrifuged, sonicated to avoid the aggregation of particles	Sigma-Aldrich, MW Low	Anti-biofilm agents, antioxidant activity-quorum sensing inhibition	[37]
Rutin (flavonols)	CS (3 mg/mL) in acetic acid (0.5%) + TPP (1.0 mg/mL) + 100 mg of rutin	Ionic gelation process, stirring, centrifugation, lyophilized	No data	Cytotoxicity – apoptosis, breast cancer cells	[38]
Naringenin (flavanone)	CS to TPP ratios of 2:1 and 5:1 (w/w) + naringenin (0.1, 0.3, and 0.5 mg/ml) + pH 4.8 + 1 ml trehalose (1.7 mg/ml)	Ionic gelation technique, stirred for 1 h, centrifuged, lyophilized	Sigma-Aldrich	Therapeutic drug delivery, lung cancer cells	[39]
Naringin (flavanone)	Chitosan-TPP concentration (0.5–0.12%). 0.5% of chitosan (w/v), 1% (v/v) acetic acid. Naringin (100 μM)	Ionic gelation technique, stirring	Chitosan from shrimp shells	Bone regeneration, anti-inflammatory – anticoagulant – antioxidant – antiproliferative – osteoblast differentiation	[40]
Hesperetin (flavanone)	Hesperetin (1 mg) + 2 ml of DMSO: PBS (1:8) + TPP (12 mg) + 2 ml of PBS solution	Chitosan folate stirred and lyophilized	Himedia, Chennai, Tamil Nadu, India	Treatment of colorectal cancer, cellular uptake – antitumoral – colony formation – apoptosis – gene expression	[41]

(Continued)

Table 9.1 (Continued)

Food compound	Composition	Preparation	Chitosan properties	Application	References
Epigallocatechin-3-gallate (EGCG) (flavanol)	45 ml of a solution of EGCG (10 mg/ml in deionized water) + 5 ml of 1% water-soluble chitosan + 1 ml of a TPP (10 mg/ml in deionized water)	Chitosan nanoparticles solutions, stirring, sonicator, lyophilized	Polysciences (Warrington, PA)	Tumor xenograft in mice release studies – antiproliferative	[42]
Anthocyanin (anthocyanins)	CS + 1 vol% aqueous acetic acid (1.5–3.5 mg mL − 1) + TPP distilled water (0.5–1.5 mg mL − 1) + final pH (4.6–4.8 using 0.5 mol L − 1 NaOH) + anthocyanin (3–9 mg)	Ionic gelation method, stirring, dialyzed, vacuum freeze-dried	Sigma-Aldrich; medium MW CS (deacetylation degree: 75–85%)	Food additive – antioxidant (DPPH, FRAP) – simulated gastrointestinal digestion	[43]
Genistein (isoflavones)	Chitosan powder in a 10 ml + 2.0 ml acetic acid 10% (v/v) + 8.0 ml of 70% (v/v) + ethanol (containing 1 mg/ml of genistein) + 1.0 M sodium hydroxide (pH 4.5) + TPP (0.075% w/v) (6 ml, 2 mg/ml)	Chitosan to tripolyphosphate particles, sonication, separated by centrifugation	Sigma-Aldrich, MW low, 2% deacetylated	Treatment of colon cancer-antiproliferative – anti-angiogenic colorectal adenocarcinoma cells – chick embryos	[44]
Chlorogenic acid (phenolic acids)	3 ml of TPP + 5 ml of chitosan solutions, final concentration of chitosan and TPP in nanoparticle suspensions were 1.5, 2, 2.5, and 3 mg/ml and 0.5, 0.6, and 0.7 mg/ml	Ionic gelation of chitosan with TPP anions, stirring, centrifugation, lyophilization	Sigma-Aldrich, MW LMW 448869	Food additive, antioxidant (ABTS)	[45]
Ellagic acid (phenolic acids)	Chitosan solution 0.175% (w/v) + 1% (v/v) + TPP 0.1% (w/v) + PEG-400 1% (v/v) + pH 5.2 (1 M NaOH) + ellagic acid (0.3 mg/mL)	Ionotropic gelation method, centrifuged, ultrasonication, freeze-dried	Sigma – Aldrich, (>75% deacetylated)	Nutraceuticals – drug delivery, antioxidant – antiapoptotic	[46]

Resveratrol (stilbenes)	CS (1% w/v) + acetic acid solution (1% v/v) + TPP (2.5% w/w)	Ionic gelation method, stirring, ultrasonic processor	Prawns, INTI-Mar del Plata (Argentine), 50–190 kDa	Nutraceuticals – drug delivery (ocular)	[47]
Carotenoids					
Astaxanthin (carotenoids)	1 mL of astaxanthin + acetone + alginate aqueous solution (20 mL, 0.6 mg/mL dissolved in Milli-Q® water) + Tween™ 80 + + $CaCl_2$ (4 mL, 0.67 mg/mL dissolved in Milli-Q water)	Ionotropic gelation, stirring, sonication, dispersion in an aqueous solution	Kono Chem Co., Ltd. (Xi'an, China), Mw 1500–1600 Da	Nutraceuticals, antioxidant activity, in vitro bio accessibility, bioavailability	[48]
Lutein (carotenoids)	CS (1 mg/mL) + 0.1% acetic acid + distilled water at 0.5 mg/mL + 0.1–0.2% tween 80	Ionic gelation method, stirring, sonicated, homogenization	Sigma-Aldrich chitosan (DD ≥ 75%)	Food and pharma, bioavailability of lutein, no toxicity	[49]
Lycopene (carotenoids)	10 g/kg of sodium alginate + without 200 g/kg trehalose + 2.5 g/kg of pectin/arabic gum	Emulsification and ionotropic gelation	Sigma-Aldrich medium MW from Ultra-Turrax	Enhance the stability toward isomerization and control the release	[50]
Probiotics and prebiotics					
Lactobacillus reuteri	Chitosan, alginate $CaCl_2$ as the crosslinking agents	Spray drying	Chitosan with a degree of deacetylation (DD) ≥ 75%	Microencapsulated probiotic cells are more resistant to heat stress and simulated gastrointestinal conditions compared with free cells	[51, 52]
Lactobacillus casei, Bifidobacterium bifidum	Maize starch, alginate, inulin, chitosan NaOH and $CaCl_2$ as the crosslinking agents; vegetable oil containing 0.2% tween 80 as the emulsifying agents	Emulsification	Chitosan with a DD ≥ 75%	The survival of microencapsulated probiotic increased significantly in simulated gastrointestinal condition; excellent bead forming	[53–55]

(Continued)

Table 9.1 (Continued)

Food compound	Composition	Preparation	Chitosan properties	Application	References
Saccharomyces boulardii	Chitosan, alginate, yeast $CaCl_2$ as the crosslinking agents	Ionic gelation	Chitosan with a DD ≥ 75%	The encapsulated yeast was resistant to the simulated gastrointestinal conditions	[56]
Lactobacillus acidophilus, Lactobacillus rhamnosus	Chitosan, Eudragit S100 Tween 80 as the surfactant NaOH and $CaCl_2$ as the crosslinking agents	Nanoemulsion and emulsification polyelectrolyte complex Ionic gelation	Chitosan with a DD ≥ 75%	The strength of the beads and the viability rate of the probiotic bacteria were increased when using Eudragit nanoparticles as the outer coating layer in comparison with single-coated beads	[57]
Lactobacillus acidophilus	. Chitosan, carboxymethyl cellulose	Polyelectrolyte complex (layer by layer)	chitosan (Mw = 70 kDa)	A coating layer by layer was produced on the surface of the bacteria, the survival of encapsulated cells in simulated gastric, and intestinal fluids, was significant when compared with nonencapsulated cells, the nanolayers are stable at low pH, higher pH, and osmotic stresses, the nanodimensions of layers exclude larger enzyme molecules (pepsin and pancreatin), but allowing smaller nutrient molecules	[58]

Lactobacillus pentosus	Chitosan, sodium phytate	Polyelectrolyte complex (layer by layer)	Chitosan with a DD ≅ 90%	The new probiotic encapsulation strategy via layer by layer produced higher survival rates in SGF and bile salts than the plain-probiotic; moreover, effectively improved the survival rate during storage and heat treatment	[59]
Lactobacillus plantarum, Lactobacillus rhamnosus	Alginate, skim milk–alginate, chitosan; $CaCl_2$ as the crosslinking agents	Extrusion polyelectrolyte complex Ionic gelation	Chitosan with a degree of deacetylation (DD) ≥ 75%	The chitosan-coated alginate-skim milk microspheres protected probiotics in simulated gastrocolic fluid, 99% of the probiotic cells being released at 120 min in simulated intestinal fluid, skim milk to be a better matrix for encapsulation and coating, making the microspheres stronger compared with uncoated microspheres, the storage stability of the microspheres at −4°C for one month showed that viable counts of encapsulated probiotics were much higher than free cells	[60]

(Continued)

Table 9.1 (Continued)

Food compound	Composition	Preparation	Chitosan properties	Application	References
***Lactobacillus rhamnosus* GG**	Alginate, chitosan, inulin NaOH and $CaCl_2$ as the crosslinking agents	Extrusion Polyelectrolyte complex Ionic gelation	Chitosan with a degree of deacetylation (DD) ≥ 75%	The encapsulated *L. rhamnosus* experienced greater survival rate, which was 4.5 times higher than that of free bacteria at day 90, microencapsulated bacteria exhibited higher survival rate during gastrointestinal treatments compared with the free bacteria, inulin fortification improved viability of encapsulated bacteria during product storage as well as during GI transition assay, the apple juice may be used as an appropriate food product for delivering encapsulated probiotic bacteria to humans	[61]
***Lactobacillus casei* 01**	Alginate, κ-carrageenan, gums, chitosan locust bean gum, gellan gum, xanthan gum	Extrusion Polyelectrolyte complex Ionic gelation	Chitosan with a degree of deacetylation (DD) ≥ 75%	The morphology of beads varied depending on gels used, the capsules containing prebiotics exceptionally resistant to starch, gcncrally provide better protection against low pH and bile salt conditions, the incorporation of probiotic-loaded gel capsules into the food products with no alteration of sensory properties, such as texture or flavor	[62]

Pullulanase	Chitosan, Fe3O4, glutaraldehyde as activating agent	Emulsion covalent bonding	Chitosan with a degree of deacetylation (DD) ≥ 75%	The pullulanase immobilized by sol–gel encapsulated or covalent bound onto magnetic chitosan/Fe_3O_4 nanoparticles increase the resistant starch (RS) formation, optimum conditions were found to obtain the highest content of RS (35.1%), for RS preparation pullulanase immobilized by sol–gel encapsulation offered major advantages over pullulanase immobilized by crosslinking	[63]
***Lactococcus lactis*, β-galactosidase**	Alginate, chitosan sodium triphosphate and $CaCl_2$ as the crosslinking agents	Extrusion Polyelectrolyte complex Ionic gelation	Chitosan with a degree of deacetylation (DD) ≥ 75–85%	Chitosan-based immobilized hyperthermostable β-galactosidase enzyme was developed and this enabled efficient galacto-oligosaccharides (GOS) production for food-grade use; chitosan was found to be a superior immobilization material compared with alginate, the final GOS-enriched product was free of any recombinant material	[64]

(Continued)

Table 9.1 (Continued)

Food compound	Composition	Preparation	Chitosan properties	Application	References
β-galactosidase	Chitosan, glutaraldehyde as activating agent	Covalent bonding	Chitosan with a degree of deacetylation (DD) ≥ 75%	Immobilized β-galactosidase had their thermal stability improved compared with the free enzyme, the immobilization increased the range of pH and, GOS yield (40% for the milk whey and 42% for the milk whey permeate), the main factor that affected the formation was the concentration of lactose present in the reaction, the biocatalyst could be reused for 17 repeated cycles in the lactose hydrolysis without loss of enzyme activity and 10 repeated cycles in the GOS synthesis with 26% loss of relative activity	[65]
Fructosyltransferase	Chitosan, alginate Sodium triphosphate and CaCl2 as the crosslinking agents	Polyelectrolyte complex Ionic gelation	Chitosan with a degree of deacetylation (DD) ≥ 75%	Chitosan and alginate served as good matrix to entrap mycelial fructosyltransferase (FTase) for the successful production of fructooligosaccharides (FOS), the optimum concentration of sodium alginate and CaCl2 prevented enzyme leakage from beads thus stabilizing entrapment efficiency, the recycling efficiency of alginate beads were more successful as compared with chitosan beads in FOS production, the thermal stability was improved in alginate beads than free biocatalyst	[66]

β-fructofuranosidase, β-fructosyltransferase	Chitosan; NaOH as the crosslinking agents; glutaraldehyde as activating agent e	Covalent bonding	CS with a degree of deacetylation (DD) ≥ 75%	Is possible to modulate the FOS composition by varying the flow rates in packed-bed and fluidized-bed reactors used in the sucrose hydrolysis and FOS synthesis, respectively, packed-bed reactors are the best choice for the synthesis of booth products	[67]
β-fructofuranosidase	Chitosan, Fe_3O_4 Tripolyphosphate as the crosslinking agents	Coprecipitation Covalent bonding	Chitosan with a degree of deacetylation (DD) ≥ 75%	The enzyme can be immobilized onto the surface of nanoparticles without using a covalent bonding reagent (i.e. glutaraldehyde), the composition and yield of FOS were similar between the immobilized and the free enzyme systems, the immobilized b-fructofuranosidase showed higher activity at a wider range of temperatures and pHs than the free form, the immobilized enzyme can easily be recovered by applying a magnetic field, and reused for FOS production	[68]

(Continued)

Table 9.1 (Continued)

Food compound	Composition	Preparation	Chitosan properties	Application	References
***Lactobacillus reuteri* 121 inulosucrase**	Chitosan NaOH as the crosslinking agents Glutaraldehyde as activating agent	Covalent bonding	Chitosan with a degree of deacetylation (DD) $\geq$ 80%	Chitosan beads in core-shell format were superior to traditional hydrogel beads used for the bacterial inulosucrase immobilization, the inulosucrase immobilized on chitosan beads in core-shell manifested resistance to thermal denaturation and operational stability for batch and continuous production of inulin-type fructooligosaccharides	[69]

biocompatibility, and bioactivity [73]. For instance, vitamin C, which helps build biological systems through various physiological functions, such as growth and repair of the skin and connective tissues, was successfully encapsulated in chitosan [74]. Other authors [75] demonstrated that the encapsulation of vitamin B2 in alginate/chitosan nanoparticles might be achieved using an ionotropic pre-gelation method. The alginate/chitosan nanoparticles loaded with vitamin B2 were stable for five months. Those studies demonstrated that it is possible to encapsulate both water-soluble and nonsoluble vitamins using chitosan as a polymeric matrix.

On the other side, omega fatty acids have proven to help prevent coronary artery disease, hypertension, inflammation, and autoimmune disorders and improve health if used in a balanced ratio. The preferred ratios are usually 2:1:1 of fatty acids 3–6–9, respectively [76]. Recently, several studies have focused on the regular consumption of lipids rich in omega-3 fatty acids, demonstrating a wide spectrum of nutritional and health benefits, such as the inhibition of inflammation, cardiovascular disease, diabetes, arthritis, and ulcerative colitis [20, 77]. However, they have at least one double bond placed on the carbon chain, leading to susceptibility to oxidation conditions and chemical instability [19]. Therefore, to get the maximum health benefit and minimize or prevent oxidation, omega-rich oils must be protected against atmospheric oxygen. Thus, microencapsulation technology has been used as a viable method to maintain and improve the biological and functional characteristics of the oils against different processes, such as oxidation, evaporation, or uncontrolled release. Hence the great interest in studying and developing encapsulation methodologies that allow greater efficiency in the delivery and release of bioactive substances, such as omega fatty acids [77, 78].

A capsule containing omega fatty acids should be designed with a high capacity to encapsulate large amounts of active ingredients per unit mass of food-grade wall materials and to protect the oils against chemical degradation (e.g. oxidation and hydrolysis). Also, to release the bioactive components at a particular site at a controlled rate in response to environmental factors (e.g. pH, ionic strength, enzyme activity, and temperature). Further, it must be compatible with the surrounding food matrix without adverse changes in the final food product's appearance, taste, texture, and acceptability [79]. Proper selection of wall materials is an important factor in designing microcapsules containing oils rich in omega fatty acids, as wall materials influence their stability, capture efficiency, and degree of protection. Several carbohydrates have been studied as wall materials, such as chitosan, pectin, maltodextrin, and sodium alginate. The main advantages of these carbohydrates are their excellent water solubility and low viscosity at high concentrations [80]. However, carbohydrates are mostly combined with proteins (gelatin, casein, and soy protein isolate), which offer good emulsifying and integrity properties of the wall materials to develop microcapsules [81, 82].

Chitosan is used as a delivery system or wall material for bioactive substances by encapsulating the active ingredient to be administered. Since chitosan is polycationic, it interacts with the negatively charged mucous membrane adhering to it, thus increasing the contact time for the active ingredient to penetrate through it and thereby increasing the release efficiency. Some authors investigated how to encapsulate and stabilize fish oil containing ultra-pure omega-3 fatty acids by emulsifying using chitosan–whey protein isolate (WPI) as wall materials [17]. The chitosan–WPI films developed demonstrated an effective fish oil load for up to 37%. In addition, the incorporation of oregano essential oil in the films significantly improved the oxidative stability of the encapsulated fish oil. To protect fish oil from oxidation and achieve maximum health benefits, Kanwal et al. studied the concentration of fish oil and its impact on microencapsulation using the spray-drying technique and chitosan as a wall material [18]. The results showed that the microcapsules prepared with 10% and 20% fish oil loading showed the highest oxidative stability during the storage period [17, 18]. Fishy odor and oxidation of the oil were prevented by encapsulation of the oil, and the microcapsules are used to fortify food products.

Because of the low cost and easy implementation, spray-drying technology has become one of the most widely employed in the food industry [83]. However, the disadvantage of this technology is that the high temperature used during the drying process accelerates the oxidation process and negatively affects the stability of encapsulated oil, damaging the wall structure of microcapsules (pores, shrinkages, and invaginations). Then, emulsification, ultrasonic atomization, and freeze-drying encapsulation technologies were used in combination, to increase the fish oil's encapsulation efficiency, and in this case, chitosan, maltodextrin, and whey protein were used as wall materials [16, 84]. Chitosan and maltodextrin blends led to the smallest particle size, and they showed the highest emulsion stability with EPA and DHA content (240 mg/g) of the encapsulated powder, which was slightly higher than the commercial specification (100 mg/g) for those products. The incorporation of sodium triphosphate as a chitosan crosslinker during oil encapsulation considerably improved the stability of the capsules. Degradability tests have shown a rapid loss of chitosan capsule shape after 24 hours, which may facilitate their application in products that require high degradation levels [85].

9.3.2 Encapsulation of Peptides and Proteins

Bioactive compounds, such as proteins, vitamins, minerals, phytosterols, polyphenols, anthocyanins, and bioactive peptides, currently have a trending market worldwide due to their diverse applications and health benefits. Specifically, incorporating these bioactive compounds in the development of functional foods has attracted tremendous attention for their diverse biological activities, antioxidant capacity, and improved consumer health [30]. Food bioactive proteins and/or peptides are labile compounds that are sensitive to light, oxygen, and temperature, capable of interacting with other components when they are dispersed in the food matrix, and susceptible to the harsh conditions of the gastrointestinal tract [86]. On the other hand, the direct incorporation of bioactive agents into functional foods is minimal due to the unpleasant organoleptic attributes they impart to foods and also due to their low bioavailability and loss of activity after food processing and storage. Encapsulation has been able to overcome these limitations and prolong shelf life, control delivery, mask the bitter taste (improve their sensory and physicochemical properties), and reduce harmful effects during passage through the gastrointestinal tract [87, 88]. Therefore, similarly to other nutrients and food bioactivity, the incorporation of peptides into carriers has been demonstrated to increase their stability and bioavailability [86]. Microencapsulation based on proteins and polysaccharides has been increasingly studied due to the high payload, high encapsulation efficiency, high thermal stability, and a complex structure that allows the sustained release of core materials. Animal–plant proteins and polysaccharides are widely used to produce complex coacervates with excellent physicochemical properties due to a positive protein charge below the isoelectric point, while anionic polysaccharides are negatively charged above their pKa, leading to the formation of soluble or insoluble electrostatic complexes [89]. Vegetable proteins are preferred because there are fewer allergenic products in the encapsulation of bioactive compounds by spray drying for human consumption [33].

Li et al. explored the effects of pH and polysaccharide type on the rheological properties of soy protein isolate–chitosan (SPI–CS) and soy protein isolate–carboxymethyl cellulose (SPI–CMC) prepared by coacervates [90]. They found SPI–CS complexes occurred at neutral pH, while SPI–CMC coacervates occurred at acidic pH. Hence, biopolymer ratio, pH, ionic strength, charge density, and MW of biopolymers are paramount factors in achieving maximum conditions for complex coacervation. Chitosan is a recognized promising candidate to be applied in the development of bioactive protein and peptides-based functional foods and nutraceuticals due to its antibacterial activity, biodegradability, and biocompatibility [91]. Recent studies conducted by Liu et al. developed a ternary system of self-assembled NPs based on zein-egg white derived peptides-chitosan (Z-EWDP-CS) to encapsulate, protect, and improve the stability of curcumin. Their results demonstrate that the presence of EWDP in the NPs is effective in enhancing the stability and antioxidant properties of curcumin, which provided new insight for the development of more potent NPs for hydrophobic nutraceuticals in food applications.

Chitosan has been used to encapsulate by spray drying hydrophilic vitamins, such as cobalamin and ascorbic acid, achieving a high yield (55% and 45%, respectively), and preventing their oxidation. Also, it was successfully used for the encapsulation of glutathione to incorporate it as an antioxidant in wine. Furthermore, chitosan was used to encapsulate tea polyphenols to improve their bioavailability in the gastrointestinal tract, and the authors obtained improved capsule stability against oxidation [33]. The complex based on whey protein isolate/chitosan particles was used for prepared emulsions (O/W) under acidic conditions ($pH = 5.7$), loaded with α-tocopherol with a high α-tocopherol encapsulation rate; it showed good water solubility and excellent stability [31, 32]. In conclusion, there is potential to exploit the maximum bioactivity of bioactive compounds by encapsulation in chitosan as support in functional food and nutraceutical formulations.

9.3.3 Encapsulation of Probiotics, Prebiotics, and Enzymes

Recently, the variety of functional foods has been extended through the years to include food products with several health-promoting properties. This kind of food could contain probiotics, prebiotics, and enzymes. It has attracted consumers' attention due to its health benefits compared with traditional products [92–94]. According to the FAO/WHO, probiotics were defined as living microorganisms that provide health benefits to the host when administered in a certain amount [95]. The health benefits of probiotics include preventing or even treating acute diarrhea symptoms and gastrointestinal infections caused by pathogens, such as *Helicobacter pylori*, which causes type B gastritis, peptic ulcers, and gastric cancer [96]. Moreover, they could benefit against carcinogenesis, allergies, antibiotic consumption, and cardiovascular or urogenital disorders [94]. These benefits can be attributed to probiotics improving the microbial balance, particularly in the gastrointestinal tract.

On the other hand, prebiotics were selectively fermented substrates that result in specific changes in the composition and/or activity of the gastrointestinal microbiota, thus conferring benefit(s) upon host health [97]. Prebiotics are specialized plant fibers and, they act like fertilizers that stimulate the growth of healthy bacteria in the gut. These substances come from many fruits and vegetables, especially those containing complex carbohydrates (mostly fiber) that humans cannot digest. Basically, these carbohydrates pass through the digestive system to become food for the bacteria and other microbes. The health benefits of prebiotics include the improvement and/or stabilization of gut microbiota composition, intestinal functions (stool bulking, stool regularity, stool consistency), mineral absorption, and improvement of bone health (bone calcium content, bone mineral density). In addition, they modulate peptides production that assists in gastrointestinal functions, energy metabolism, and satiety. Finally, their use reduces the risk of intestinal infections, obesity, type 2 diabetes, metabolic syndrome, colon cancer, intestinal inflammation, etc. [98]. The list of prebiotic foods is long, from asparagus to yams. An internet search will yield dozens of examples of foods consumed in occidental countries that contain fructooligosaccharides, galactooligosaccharides, and lactulose. In contrast, a more extensive list of prebiotics exists in Japan, i.e. soyoligosaccharides, xylooligosaccharides, isomaltooligosaccharides, gentiooligosaccharides, lactosucrose, and glucooligosaccharides [99].

Finally, enzymes are primarily proteins and peptides that act as biological catalysts (biocatalysts) and can be used in different biochemical processes. These molecules exhibit several features that make their use advantageous compared with conventional chemical catalysts, such as biodegradability, high level of catalytic efficiency, high degree of specificity, higher reaction yields, and mild conditions of the reaction [100]. These characteristics make enzymes used as catalysts that have been successfully introduced to the industry to produce certain foodstuffs, pharmaceuticals, and agrochemicals. It is known that probiotic microorganisms, prebiotics, and enzymes are susceptible to adverse environmental conditions in which they are used. Probiotics are vulnerable to pH, oxygen, and temperature, significantly affecting the viability and survival rate during processing, storage, and passage through the gastrointestinal tract [51]. Prebiotics has the same problem, and they require protection to go through the digestive tract and arrive in the colon intact without being absorbed into other sites or degraded [101]. Finally, the enzymes have undoubted advantages, but they present several practical problems to use. Some of the issues mentioned are the high cost of isolation and purification, the instability of their structures once isolated from their natural environments, and their sensitivity to inhibitors present in the active medium [102].

Taking into account the aforementioned is of extreme importance in protecting them, making it possible to add them to a food matrix that can result in positive benefits to the human body. Encapsulation represents the leading modern solution to this problem. Based on the literature available, chitosan appears to be one of the most promising coating materials, among the most common polymers used for microencapsulation. It improves the efficiency of maintaining viability, protection, and intestinal delivery of probiotics, prebiotics, and enzymes [64, 103, 104]. As previously discussed, chitosan has a significant protective role against external damages in food products. The antimicrobial capacity of chitosan and the little or no sensory influence on food make it an ideal biopolymer to be applied in the food industry [105, 106]. Table 9.1 shows the main results obtained from some reports on the encapsulation of probiotic bacteria, prebiotics, and enzymes, in which chitosan is present as a coating material.

9.3.4 Encapsulation of Essential Oils

Essential oils (EOs), also known as volatile oils, are hydrophobic substances soluble in most organic solvents [107]. EOs are obtained from various natural plant tissues such as leaves, roots, flowers, and fruits by steam distillation or extraction [108]. EOs and terpenes emerge as natural antifungal, antimicrobial, and antioxidant agents, and they are being studied and developed for the agri-food industry [109–111]. Different methods have been used to formulate EOs by using chitosan as a carrier. For instance, thyme oil, *Cuminum cyminum*, or *Mentha piperita* have been encapsulated through nanoprecipitation procedure [22–24] whereas *Carum copticum*, *Zataria multiflora*, or cinnamon oil by ionic gelation [25, 27, 112], and *Coriandrum sativum* has been encapsulated through spray drying [28]. A method widely used to carry out chitosan-EOs release systems has been ionic gelation due to its slight experimental conditions, nonorganic solvents use, innocuous properties, and controllable nature. Nanoemulsions could be developed through ionic gelation and profit from emulsion particle size and stability advantages. Nanoemulsions have shown to be an effective strategy to enhance the bioactivity of EOs due to their higher dispersibility, stability, and penetration [25, 113, 114]. A study on basil (*Ocimum basilicum* L.) essential oil emulsification with chitosan observed that the addition of surfactant before ionic gelation improved the formulation since the lipophilic components could be rapidly integrated into the aqueous solution [115]. Furthermore, the process depends on many factors, such as pH, concentration, relative ratio, and degree of mixing, which directly reflect other properties such as morphology, retention rate, and antimicrobial activities.

9.3.5 Encapsulation of Phenolic Compounds

Phenolic compounds are plant metabolites found in human and animal diets and are classified into two major groups: flavonoids and non-flavonoids. Flavonoids are polyphenolic compounds having 15 carbons with two aromatic rings connected by a three-carbon bridge (C6–C3–C6). According to the C-ring hydroxylation pattern as well as the degree of saturation, oxidation, and degradation, flavonoids can be further subclassified as flavonols, flavones, flavanones, flavanols, isoflavones, chalcones, and anthocyanins [116–118]. Non-flavonoid compounds have a single aromatic ring in their most basic configuration and include (i) phenolic acids, which can be subdivided into two main subclasses: benzoic acid and cinnamic acid; (ii) stilbenes, such as resveratrol and viniferin; (iii) lignans, such as secoisolariciresinol, pinoresinol, and lariciresinol; and (iv) coumarins, which are naturally occurring benzopyrone derivatives [116, 118].

Phenolic compounds are interesting natural products used in industry in both biomaterials and food product development for humans and animals. They have a wide variety of biological effects on human health, including antioxidant, anti-inflammatory, antibacterial, and antiviral properties [116, 119–121]. However, their low environmental stability, photosensitivity, low water solubility, low bioavailability, rapid metabolism, and excretion are major limitations for their use, which could be overcome by encapsulating phenolic compounds into chitosan [122]. Chitosan is one of the most extensively used encapsulating agents for phenolic compounds, and some reviews have lately been published on this topic [122–126]. On the other hand, the US Food and Drug Administration has awarded CS the generally recognized as safe (GRAS) category for use as a dietary supplement [122, 127].

Several encapsulation strategies have been developed to encapsulate phenolic compounds in chitosan to improve their biological performance, such as (i) molecular self-assembly; (ii) desolvation; (iii) complex coacervation; (iv) lyophilization; (v) nano spray-drying technology; (vi) nanoemulsions; (vii) ionic gelation, among others [122].

Ionic gelation has been widely used to encapsulate flavonoids and non-flavonoids into chitosan (CS)-tripolyphosphate (TPP)-based nanoparticles (NPs) [122]. The flavonols, quercetin [35], myricetin [36], kaempferol [37], and rutin [38, 128], stand out among the flavonoids encapsulated using this method. The antitumor activity of quercetin-loaded chitosan nanoparticles (QCT-CS-NPs) has been studied in vitro, and in animal models [35], the study produced biocompatible QCT-CS-NPs with a small size (200 nm) and reported effective encapsulation of QCT in CS-NPs to target the tumor microenvironment as well as increased QCT-CS-NPs efficacy in cancer treatment. Chrysin, a flavone documented for its anticancer, antioxidant, and antibacterial properties, has been encapsulated into CS-NPs using TPP as a linker. Because the NPs were anti-biofilm against *Staphylococcus aureus*, nanocomposites might be employed to build anti-biofilm coatings [129]. Flavanones, such as naringenin, naringin, hesperetin, and hesperidin, were encapsulated in CS-TPP-NP due to their anticancer properties. Kumar et al. [39] used an ionic gelation process with tripolyphosphate as a crosslinker to make chitosan-encapsulated naringenin (NAR) nanoparticles (CS-NPs/NAR). The findings pointed to a practical approach for administering NAR with antioxidant and anticancer properties since their cytotoxic effects on lung cancer cells (A549) were seen.

Similarly, Mary Lazer et al. [41] developed a chitosan folate hesperetin nanoparticle by covalently conjugating folic acid with chitosan molecules, which was a good carrier for hesperetin delivery to colorectal cancer cells. The most frequent encapsulated flavanol in CS-NPs is (–)-epigallocatechin-3-gallate (EGCG), which has proven utility as a food additive and medication delivery and improved oral administration therapeutic application for a variety of disorders [42, 130–134]. Three studies in cancer therapy described EGCG encapsulation in CS-TPP NP to increase therapeutic effectiveness. In preclinical animal models, oral treatment of naturally occurring chitosan-based nanoformulation with EGCG inhibited prostate cancer cell proliferation [42]. Other investigations found that EGCG encapsulated in CS-NPs had good anti-proliferative and pro-apoptotic effects on human melanoma cell development in vitro and in vivo models [133]. Similarly, NPs loaded with EGCG showed inhibitory effects on human breast cancer MCF-7 cells [134]. On the other hand, attempts have been made to increase the stability of anthocyanins when encapsulated in CS-TPP-NP [43, 135–137]. In fact, nanoencapsulation in CS-TPP might be an excellent way to improve the color and antioxidant stability of black soybean anthocyanin [136]. Another study found that chitosan nanoencapsulation improved anthocyanins' physical and oxidative stability in black chokeberry (*Aronia melanocarpa*) [43]. On the other hand, in human oral cancer cells, phloretin, a natural dihydrochalcone, loaded with CS-NPs, enhances pH-dependent mitochondrial-mediated intrinsic apoptosis [43]. Meanwhile, chitosan-encapsulated genistein, an isoflavone, reduced the development and multiplication of human colorectal cancer cells (HT-29), suggesting that it might be a potential carrier for genistein delivery to the colon [44].

The group of non-flavonoids comprises numerous polyphenol subclasses such as phenolic acids, stilbenes, and lignans, that were encapsulated into CS-NPs. Phenolic acids, such as chlorogenic [45, 138], rosmarinic [139, 140], ellagic [46, 141, 142], gallic [143], ferulic [144, 145], protocatechuic [146], and sinapic [147], have been encapsulated with CS agents. Nallamuthu et al. [45] used ionic gelation to encapsulate chlorogenic acid. The encapsulation efficiency was determined to be up to 59%, and the particles had a size of 210 nm and zeta potential of 33 mV, respectively. The CS-NPs have been employed to create a rosmarinic acid (RA) delivery method that is both reliable and efficient. Da Silva et al. [139] used ionic gelation with sodium tripolyphosphate to make CS-NPs. Individual and tiny-sized NPs, about 300 nm, were found in the findings. According to the authors, the encapsulated extracts are potential vehicles for RA drug delivery in CS nanocarriers. Furthermore, CS-NPs-RA prepared by ionic gelation with TPP was shown to be safe in testing (in vitro) for developing drug delivery systems for ocular use in oxidative eye diseases [140]. In another study, ionotropic gelation was used to encapsulate ellagic acid, a hydroxybenzoic acid. When ellagic acid-encapsulated CS was compared with free ellagic acid at concentrations as low as 0.1 mg/mL, a decrease of 34% in clot time and 16.4% in retraction time was seen, demonstrating the promise of the ellagic acid–CS system as an effective anti-hemorrhagic system [142].

Ferulic acid has also been encapsulated in CS-NPs and used as an agent against *Candida albicans*, an opportunistic fungal pathogen that forms a superficial biofilm and infects medical equipment [148]. Ferulic acid-loaded CS-NPs showed anticancer effects against human cervical cancer cell lines (ME-180 and HeLa) as well as cytotoxicity against melanoma tumor cells (B16-F10) [145, 149]. The pharmacological and biological applications of encapsulated stilbenes have primarily been reported for resveratrol [47, 150–152], which has a wide range of applications in the food and pharmaceutical industries. Wu et al. [152] developed resveratrol-ionically crosslinked CS-TPP nanoparticles with a size range of 172–217 nm that was reloaded. The resveratrol-loaded CS-TPP nanoparticles presented long-term storage stability and UV light stability and retained resveratrol's antioxidant and anticancer properties. According to another study, encapsulating resveratrol in high MW chitosan-based nanogels can be employed in eye therapeutics since the NPs created are biocompatible with human cells [47]. Another non-flavonoid compound is podophyllotoxin (PPT), a significant lignan that has a wide range of anticancer properties. Chen et al. [153] modified the pH-sensitive interactions of podophyllotoxin and chitosan for more controlled drug release, concluding that PPT-CS-NPs provide a safe and effective drug delivery system. Chitosan-based nanostructures seem to be promising systems for exploiting a variety of phenolic compounds since the latter's sensitivity in adverse environments restricts their efficacy and bioavailability. Most biological activity evaluations are done in vitro, and more efforts should be made to investigate in vivo antioxidant and biological efficacy for eventual commercial application.

9.3.6 Encapsulation of Carotenoids

Carotenoids are natural pigments that are classified into two families: carotenes and xanthophylls, based on the presence or lack of oxygen functional groups in their chemical structure [154]. Plants and some fungi and bacteria produce these chemicals, which are responsible for the red, yellow, or orange coloring of fruits, vegetables, flowers, fungus, birds, insects, and crustaceans, among other things [155]. Food scientists and nutritionists have long been intrigued by carotenoids, initially due to the pro-vitamin A activity of some of them, particularly β-carotene, and more recently due to their link to a variety of health benefits such as antioxidant activity, antitumor effects, and immune stimulation [156]. Despite the numerous benefits described for carotenoids, their use as antioxidant additives in functional foods is limited by two major drawbacks: (i) they are easily oxidized due to their double bond structure; (ii) they have low solubility in aqueous solutions. Although carotenes have been encapsulated ideally with cyclodextrins to minimize these issues [157], recently, some publications on the encapsulation of carotenes with chitosan have been published as well to improve their biological activities.

Carotenoids, such as astaxanthin [48, 158–160], fucoxanthin [161], lutein [49, 162], lycopene [50, 163], β-carotene [164], crocin [165, 166], and urucum (bixin and norbixin) [167], have been encapsulated by chitosan. Recently, oil-in-water emulsion was used to make astaxanthin-loaded chitosan oligosaccharide/alginate nanoparticles, which were then ionotropically gelated. Astaxanthin nanoparticles encapsulated in chitosan oligosaccharide/alginate were shown to be an excellent carrier for astaxanthin, enhancing stability, in vitro oral bioaccessibility, and bioavailability [48]. In another study, Koo et al. [161] developed fucoxanthin (FX)-loaded casein nanoparticles (FX-CN) and chitosan (CS)-coated FX-CN. FX-CS-CN can enable the application of FX, with improved bioavailability and water dispersibility, in the food industry. In vitro simulated digestion demonstrated better FX bioaccessibility from the nanoparticles versus *Phaeodactylum tricornutum* powder. For retinal administration, lutein was likewise encapsulated with a chitosan–sodium alginate–oleic acid hybrid nanocarrier system (NCs). The improved bioavailability of lutein (L) from LNCs, together with the lack of toxicity, offers potential in food and pharmaceuticals [49].

Another carotenoid, lycopene, can protect against loss of bone mass, chronic diseases, skin cancer, prostate cancer, and cardiovascular disease. High instability and low oral bioavailability limit further clinical development of lycopene. Organic green tea catechin derivative, oligomerized (−)-epigallocatechin-3-O-gallate (OEGCG) as a carrier for oral delivery. Lycopene-loaded OEGCG nanoparticles (NPs) have been prepared by the nanoprecipitation method, followed by coating with chitosan to form a shell. OEGCG's antioxidant property may help in oral drug delivery in various indications [163]. In another study, β-carotene and palm oil were encapsulated in chitosan/TPP by ionic gelation. Encapsulation efficiency was greater than 95%, and the yield of microparticles coated with chitosan/sodium tripolyphosphorus was approximately 55%. The release profile was evaluated by simulating gastric and intestinal conditions. The microparticles were applied to bread and yogurt. When applied to foods, the release of carotenoids was increased [164]. Crocin is a water-soluble carotenoid and has shown potential to improve neurodegenerative diseases, such as Alzheimer's disease. Crocin-loaded hollow zein NFs were produced by a layer-by-layer self-assembly. In vitro simulated gastrointestinal digestion and antioxidant activity assays indicated that the NPs exhibited better-controlled release behavior and stronger antioxidant activity than uncoated crocin-loaded zein NPs dispersed in water and could be a promising delivery system for an effective therapeutic approach to Alzheimer's disease [166]. Various studies have shown that encapsulating carotenoids improves stability, bioaccessibility, and bioavailability. However, most biological activity assessments are carried out in vitro, and greater efforts should be made to research in vivo for eventual commercialization in the pharmaceutical and food sectors.

9.4 Potential Applications of Microencapsulated Materials in Food Packaging

Chitosan has been extensively used in food coating as edible films in combination that include active substances, such as antioxidants and antimicrobial agents, generally of natural origin. It is also mixed with pigments, whose appearance may change due to their interaction with food, thus indicating the freshness of the product. In general, depending on the properties of the packages, they are classified as active or intelligent food-packaging materials. Active food-packaging materials show better properties than conventional packaging materials and generally inhibit microbial growth or retard oxidation reactions, thereby extending food freshness over time [168]. For example, Nguyen et al. prepared edible packaging films from chitosan and different quantities of *Sonneratia caseolaris* (L.) Engl. (SCE) leaf extract and the antimicrobial activity of the films was verified. Bananas coated with chitosan–SCE extract films indicated longer shelf life than chitosan-coated or uncoated bananas, showing that the combination of SCE leaf extract in the chitosan matrix could be favorable for the preservation of food in active edible packaging [169].

On the other hand, intelligent packaging materials are intended to respond in a particular way when a change occurs in the food and must be able to facilitate its detection [168]. Ezati et al. designed a pH-responsive film based on chitosan with the addition of alizarin. During characterization, it was observed that the antioxidant activity of the film increased when adding alizarin. Films were placed inside fish packages, following the color changes for 24–48 hours, which was associated with the food variation of pH due to quality losses [170]. However, in many cases, these packaging systems are not efficient for food preservation. In this sense, the encapsulation of the active substances allows better release control of these agents under specific conditions, contributing to better results. For instance, some authors incorporated chitosan microcapsules containing sorbic acid in ethylene vinyl alcohol copolymer (EVOH) films, to achieve a slow release system of the antimicrobial agent, improving the packaging-system properties. These films were tested on snakehead fish fillets. The authors found that films containing encapsulated sorbic acid preserved food properties longer than those containing free antimicrobial agents. They also found that the shelf life of fish fillets could be extended by 4 days at 4ºC, compared with the EVOH films [171].

In other research, Zhang et al. used nanoencapsulation to enhance EOs' antimicrobial and antioxidant activities from *Paulownia tomentosa* (PTEO) in a chitosan coating to increase the shelf life of ready-to-cook stored pork chops at 4°C for 16 days. The results indicated that all treatments effectively decreased microbial growth, during storage, in the pork chops without coating. The coating containing the PTEO-loaded nanoparticles led to less discoloration, lipid oxidation, and microbial growth of the product. In addition, sensory properties for the stored product were better regarding samples with free PTEO or free PTEO chitosan coating [172].

Recently, Hasheminejad et al. incorporated clove essential oil (CEO) in a chitosan film for coating pomegranate arils during storage at 5ºC, and the results of shelf-life extension were compared with samples coated with chitosan, free CEO, and chitosan with CEO-loaded nanoparticles. Among the tested samples, coatings with CEO encapsulated in nanoparticles extended the shelf life of the food by 54 days, while the uncoated arils became useless on day 18, due to the influence of fungi. At the end of the study, in general, coatings with CEO encapsulated in nanoparticles allowed to maintain the best conservation conditions of the arils, turning out to be the most effective coating to extend the shelf life and control undesirable alterations of these fruits [173].

In smart packaging materials, Ma et al. developed a system based on gelatin and chitosan nanoparticles with blueberry anthocyanins as pH indicators to assess the freshness of milk. This system showed color variation at different pH values caused by milk spoilage. There were no substantial color differences at normal milk pH in the product. With increasing storage time, as the pH of the milk decreased, the nanoparticles released anthocyanins, which gave a red color to the milk sample. This color change indicated the deterioration of the milk solution. Thus, this system could be beneficial for monitoring changes in milk quality [174].

Recent advances have been reported on diverse systems that use chitosan to encapsulate EOs and other substances with antimicrobial and/or antioxidant properties, which show promising results for its future application in the food-packaging industry. Some of these investigations are summarized in Table 9.2.

9.4.1 3D Food Printing

Additive manufacturing or 3D food printing assists in upgrading the emerging industrial technology to prepare personalized 3D food production items and design food texture [189, 190]. Recently, 3D food printing was developed according to the expectations of modern consumers who are in constant search of novel, fresh and tasty, textural, and technological food that offers a new mouthfeel [189, 190]. This innovative technology is generally based on soft paste extrusion technology through nozzles (melting extrusion, soft material extrusion) that consider the viscosity of ingredients formulation, such as cake frosting, chocolate, mashed potatoes, puree, dough, meat paste, cheese, or other specific alternative food ingredient material, which are printed by layer-by-layer deposition to generate the design of 3D food constructs. Figure 9.4 summarizes the main steps investigated during the 3D food-printing strategies [65], that is to say: (i) the specific choice of printable/nonprintable/alternative food ingredient material to be used for the formulation of 3D food ink; (ii) the computer-assisted modeling and slicing of the 3D food structure (3D model building); (iii) the 3D food-printing technology (3D objects printing); (iv) the applied post-processing method for the final 3D food constructs (post-treatment), which could be baked, cooked, or just trimmed on demand [190].

One of the prevalent challenges wanted in 3D food-printing technology are the quality of the 3D shape stability/accuracy and the repeatability of the 3D construction for market commercialization. Consequently, the printing parameters, such as the slicing, shape of 3D model, flow rate, viscosity of 3D ink ingredients, nozzle/needle diameters, sizes and movement, and printing temperature are crucial and have to be rigorously selected and adjusted according to the properties of the food ingredients (thermal/chemical/rheological) [189, 190].

In this context, as recently described by Gholamipour-Shirazi et al. [191] in a very interesting review on 3D food printing as additive manufacturing, the main appropriate materials for 3D food printing are unquestionably proteins, fats, and polysaccharides [192]. As

Table 9.2 Packaging system, food application, and main results of research studies using chitosan as encapsulating material in food-packaging applications.

Packaging system	Preparation method	Additive functions	Food application	Results	References
Pomegranate peel extract in chitosan and poly (ethylene oxide) nanofibers	Electrospinning	Antibacterial agent	Beef meat	The growth of *E. coli* O157:H7 was effectively inhibited in beef stored at 4°C after 7 days and at 25°C after 10 days relative to the control sample	[175]
Nisin in poly-g-glutamic acid/chitosan nanoparticles embedded in poly (ethylene oxide) nanofibers	Self-assembly	Antibacterial agent	Cheese	The system showed satisfactory antibacterial effects on *L. monocytogenes* and little influence on the sensory quality of cheese	[176]
Green tea extract (GTE) in chitosan/sodium triphosphate microparticles (GTE-CHI-TPP) dispersed in chitosan films	Ionic gelation	Antioxidant agent	Hamburger patties	Samples coated with chitosan, chitosan and GTE, as well as GTE-CHI-TPP were more resistant to lipid oxidation during storage	[178]
Tea polyphenol in chitosan/sulfobutyl ether-β-cyclodextrin nanoparticles dispersed in gelatin films	Ionic gelation	Antioxidant agent	Sunflower oil	When films are used in contact with the oil, the antioxidant activity can be modified according to the proportion between free and encapsulated polyphenol in the film	[179]

(Continued)

Table 9.2 (Continued)

Packaging system	Preparation method	Additive functions	Food application	Results	References
Tea polyphenol in chitosan/sodium triphosphate nanoparticles (TE-CHI-TPP) dispersed in gelatin films	Ionic gelation	Antioxidant agent	Fish oil	The oxidation of fish oil was effectively retarded throughout the incubation period when packaged with films containing TE-CHI-TPP	[180]
Coating with *Paulownia Tomentosa* essential oil (PTEO)/chitosan nanoparticles dispersed in chitosan solution	Ionic gelation	Antibacterial/ antioxidant agent	Ready-to-cook pork chops	Chitosan coating containing PTEO/chitosan nanoparticles showed a better antimicrobial effect and best lipid oxidation stability on pork chops, compared with coating containing free PTEO	[172]
Cinnamon essential oil/ chitosan nanoparticles sprayed in low-density polyethylene films (LDPF)	Ionic gelation	Antibacterial/ antioxidant agent	Chilled pork meat	Films with nanoparticles of bigger size preserved the best properties of the pork during 15 days of refrigerated storage	[181]
Clove oil/ chitosan–sodium triphosphate nanoparticles in gelatin nanofibers	Oil-in-water emulsion/ionic gelation	Antibacterial agent	Cucumber	The system showed significant eradication effect on *E. coli* O157:H7 biofilms in vitro and satisfactory anti-biofilm effect on cucumber during 4 days of storage	[182]
Chrysanthemum essential oil (CHEO) in chitosan/ poly (ethylene oxide) nanofibers	Electrospinning	Antibacterial agent	Beef	The system showed long-term anti-*L. monocytogenes* effect during seven-day storage. The slow release of CHEO from nanofibers prolong the shelf life of beef and therefore could be very useful in the food-packaging industry	[183]

Moringa oleifera oil in chitosan/sodium triphosphate nanoparticles dispersed in gelatin nanofibers	Oil-in-water emulsion/ionic gelation	Antibacterial agent	Cheese	The nanofibers possessed high antibacterial effect against *L. monocytogenes* and *S. aureus* on cheese at 4°C and 25°C	[184]
Oregano essential oil in chitosan microcapsules added to low-density polyethylene (LDPE)/ ethylene vinyl acetate (EVA) copolymer films	Spray drying	Carbon dioxide and oxygen barrier	Strawberry	The carbon dioxide and oxygen barrier properties of the LDPE/ EVA films were improved by the addition of the loaded microcapsules. The shelf life of strawberries was effectively prolonged	[185]
Clove essential oil (CEO) in chitosan/sodium triphosphate nanoparticles	Oil-in-water emulsion/ionic gelation	Antimicrobial/ antioxidant agent	Pomegranate arils	The coating dispersion containing nanoparticles loaded with CEO was the most effective, which extended the shelf life of pomegranate arils for 54 days and protected them against undesirable microbial, physicochemical, and sensory changes, probably induced by the sustained release of CEO from the nanoparticles	[173]
Layer-by-layer coating with alginate solution and cinnamon essential oil in chitosan/sodium triphosphate microcapsules dispersed in chitosan solution	Oil-in-water emulsion/ionic gelation/complex coacervation	Antimicrobial/ antioxidant agent	Mangoes	Mangoes were coated with 3, 5, and 7 layers, and after 14 days of storage, mangoes coated with 5 layers still maintained their food and commercial value	[186]

(Continued)

Table 9.2 (Continued)

Packaging system	Preparation method	Additive functions	Food application	Results	References
Barberry anthocyanins in chitosan nanofibers/ methyl cellulose halochromic films	Color indicator film	pH sensor/ colorimetric change	Lamb meat	The pH-responsive color indicator film was successfully used to indicate the freshness of lamb meat	[187]
Saffron petal anthocyanin in chitosan nanofibers/ methyl cellulose halochromic films	Casting	Ammonia gas and pH sensor/colorimetric change	Lamb meat	Smart halochromic films changed their color from violet to green after 72 h of storage of lamb meat, showing potential as smart packaging materials to monitor meat freshness in real time	[188]
Blueberry anthocyanins in chitosan/gelatin nanoparticles	Complex coacervation	pH sensor/ colorimetric change	Milk	When adding anthocyanins loaded nanoparticles to spoiled milk, there was a significant color change within 60 s. With this method, it is possible to detect the difference between fresh and deteriorated milk	[174]

largely described in 3D bioprinting literature, many hydrogels have been developed, including food-grade polysaccharides, such as pectin, starch xanthan, methylcellulose, carrageen agar, alginate, and chitosan. All these well-described hydrocolloids are mostly used to fine-tune the properties of 3D materials and, more specifically, to improve and increase the printability of the 3D constructs [193].

In the past decade, 3D bioprinting of chitosan biomaterial has involved extensive concern because of their excellent zero toxicity, biodegradability, antibacterial activities, and biocompatibility. In this consideration, 3D food printing could be developed using chitosan as the polysaccharide and the main ingredient for intelligent food packaging, techno-functional food printing, and 3D food constructions and food-preserving agent.

9.4.2 Spinning Technology

Non-thermoplastic carbohydrate polymers (polysaccharides) such as chitosan or cellulose, occurring in fibrous materials of some cell walls can be shaped into fibers. Four technologies can be used for obtaining fibers from polysaccharides, namely electrospinning, blow spinning, dry-jet wet spinning, and wet spinning. Electrospinning and blow spinning are applied to produce submicron-sized chitosan fibers and chitosan micro/nanofibers. The electrospinning process aims to form a liquid jet induced by an electric field. After solvent evaporation, the fiber is collected. Blow spinning is a more recent technique. The polymer solution passes through the inner nozzle of a spinneret, and high-pressure gas is emitted from the outer nozzle forcing the solvent evaporation. Despite these assays and dry-jet spinning attempts applied to chitosan, wet spinning remains the largest method used to generate chitosan fibers. In wet spinning, the dissolved polysaccharide is forced and injected using a syringe pump through a spinneret submerged in a coagulation bath containing a nonsolvent for the polysaccharide but a solvent miscible with the spinning one [8].

In this technology, the polysaccharide spinnability is notably affected by the shear flow properties of the solutions and the rate of coagulation in the bath. Considering the fibers obtained from chitosan by wet-spinning, aqueous acidic solutions (e.g. acetic acid, lactic acid, glycine chloride, or formic acid) can dissolve chitosan, which will be extruded in an alkaline solution for coagulation [9]. Alternative coagulation systems like a mixture of ethanol/methanol/NaOH, sodium sulfate/NaOH solution, and calcium acetate/water/methanol have also succeeded. These solutions are very common, cheap, and green; even if numerous other solvent/solvent systems are available to dissolve chitin and chitosan, some of them have been previously used for wet spinning. Indeed, the solubility of chitosan in water or other solvents can be improved by chemical modification or other processes (xanthate process, formulation with an organic cosolvent, N-acylation, the introduction of bulkyl groups, preparation of carboxy methyl derivatives, covalent attachment of a hydrophilic sugar moiety, enzymatic hydrolysis or grafting, and lithium complexation) [10]. The efficiency of the wet-spinning process depends strongly on fiber quality, meaning not only their mechanical properties but also their ability to separate each other.

For that, chitosan fibers obtained after spinning have to be well separated, leading to the development of two strategies. In the first one, multifilament obtained after spinning is washed in an organic solvent to avoid the adhesion of fibers among themselves [11]. The second one aims to use monofilament spinning [194]. Chitosan fiber, microfibers, and

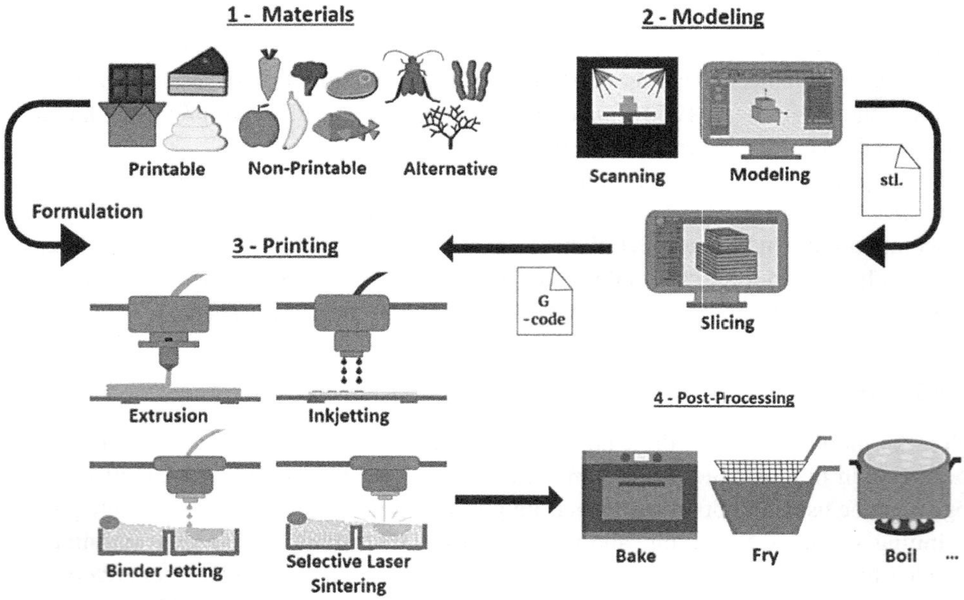

Figure 9.4 Schematic representation of the main steps investigated for 3D food-printing strategy. Source: Adapted from Hackenhaar et al. [65].

nanofibers have been proposed in recent literature for applications as active food packaging, enzyme immobilization, and nanofood carrier [13]. Chitosan fibers have been abundantly studied for biomedical applications, considering their nontoxicity, biocompatibility, and antimicrobial properties of chitosan. Moreover, chitosan has been reported as a green alternative to petroleum resources for food packaging. This biopolymer has interesting mechanical properties; it is biodegradable and has antimicrobial activities. Arkoun et al. [14] reported the potential chitosan/poly(ethylene oxide) nanofibers obtained after electrospinning and having bacteriostatic activities as food packaging for fresh meat preservation. Chitosan-based nanofibers also revealed high potential as a food carrier for the delivery of spices, pigments, and nutraceuticals [13]. Chitosan nanofibers have recently emerged for enzyme immobilization in food processing. In this field, a beta-galactosidase has been successfully immobilized on chitosan/poly(vinyl alcohol) nanofibers to hydrolyze lactose. This immobilization improved the stability of the enzyme [15].

Industrial Corner

- One way to improve the properties of chitosan-based coatings is to mix chitosan with organic molecules (peptide or polyphenolic compounds), inorganic salts, or antimicrobial agents (e.g. plant extracts, EOs).
- A second way consists of chitosan coating combined with noncoating treatments, e.g. heat treatment, hypobaric therapy, gas fumigation, and changed atmosphere packing.

(Continued)

Compared with a single chitosan coating, the preservation effects of the enhanced chitosan-based coating were better in most circumstances.

- As a food-packaging material, chitosan can be made into thin sheet material, film, or aerogel, which has many advantages, such as stability, antimicrobial properties, and durability (thermal and chemical stability).
- Due to its polysaccharidic nature, the hydrophilicity limitation can be improved by chemical modification, crosslinking, or physical mixture with hydrophobic materials (e.g. natural or synthetic waxes, fatty acid).
- The chitosan barrier and mechanical properties can be improved by mixing with nanoclays, nanofibers, nanocrystals, and nanocarbon materials. The antimicrobial properties to be used in food can be enhanced by combining with other antimicrobial compounds, both natural and synthetic ones.

9.5 Market for Chitosan Uses in Food Application

From the commercial point of view, there are some marketed nano-based products in the food industry including 24Hr Microactive® CoQ10 (nanosized coenzyme Q10 and β-cyclodextrin matrix) (Genceutic Naturals, United States), Nanoceuticals™ Slim Shake Chocolate from RBC Life Sciences in the United States (pure cocoa infused "Nanoclusters"), NanoCurcuminoids™ from Life Enhancement in the United States (curcuminoids encapsulated with solid–lipid nanospheres) [195], as well as Aquanova NovaSol® (nanoemulsions fabricated by ω-3 fatty acids and liposoluble vitamins), Asia Food Beverage® in Thailand (manufacturing food-grade nanoliposomes) and Curcosome® (quercetin-bearing nanoliposomes) [196]. The novel approach to utilizing nanoencapsulation in food-manufacturing sectors is using food-grade natural biodegradable polymers as shell material [197–201]. Polysaccharides' unique characteristics, including high modifiability, solubility, and binding ability via their functional groups, make them suitable for this purpose [202–204].

The production of chitosan in the industrial sector has expanded in recent decades. However, it has been hampered by several factors, such as the lack of constant supply of raw materials, irregular quality, the presence of toxic elements, ash, and other foreign materials, and high production costs [205]. These problems are becoming progressively less with new innovations from chitosan research that have boosted its use in various fields and have driven the development of new products, better processes for obtaining them, and more potential applications underway or yet to be discovered. Of the patent search that uses chitosan as the main ingredient or part of its components, a summary of the period 2016–2021 is shown in Figure 9.5. The most extensive application of chitosan is its use as food packaging, followed by direct use in food or feed. The third most important category is pharmaceuticals, including patents for use in food supplements and nutraceuticals. The graph also shows an upward trend of patents until **2018**, and then a significant drop due to the effect of the COVID-19 pandemic, which caused a decrease in patent applications. Consistent with the effect of the pandemic COVID-19, it is expected to harm the chitosan market. A nationwide lockdown has harmed the overall market in several economies and lowered the demand for cosmetics and toiletries.

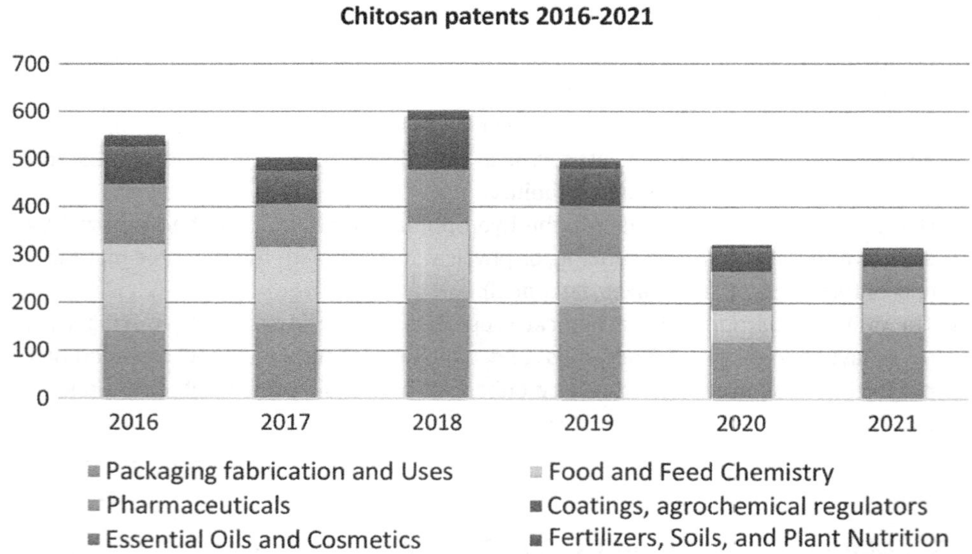

Figure 9.5 Patents from 2016 to 2021 with chitosan as an ingredient. Categories according to their application.

It is noted that 56% of the patents for the 2016–2021 period in the category "Packaging fabrication" correspond to inventions of chitosan in combination with plastics, such as polyethylene, polyacrylamide, PTFE, and other components to form fully or partially biodegradable materials. Patent applications with formulations for plastic composites with chitosan were the most abundant during **2021**, showing an increase over the previous year. About 85% of chitosan patent filings are made in Asia (China). Of the total number of patents, 59% are assigned by the private sector, compared with 41% assigned by universities or academic institutes. Due to product usage in a broad range of sectors, such as water treatment, health care, food and beverages, cosmetics, and agriculture, the global chitosan market was valued at USD1.83 billion in 2020. It is expected to increase at a compound annual growth rate of nearly 14.8% from 2021 to 2027 (USD4.35 billion). Chitosan-based packaging in the food and beverage sector will be aided by rapid growth in the sector and a growing need for sophisticated packaging materials with eco-friendly properties [206]. Besides, chitosan is a fibrillar polysaccharide used in dietary supplements for weight loss and nutraceuticals, as it binds the fats and lipids in the intestine and prevents them from saturating in the human body. It is also used in food preservatives, edible films, and food additives.

9.6 Concluding Remarks

Chitosan is a nontoxic, biodegradable polycationic material that can be isolated from different sources and is currently used as a food supplement and nutraceutical. Its water-soluble properties, broad compatibility with other food gums, easy processability, and biocidal activity strongly support its role as a food additive. This chapter shows that it can be exploited as a matrix to microencapsulate high-value food products to protect them from the environment and undesired reactions. The most common nano and microencapsulation techniques available at an industrial scale to encapsulate food components were shown through examples, using chitosan alone or combined with other macromolecules as a polymer matrix. Also, emerging technologies like 3D bioprinting and wet spinning can be an alternative to exploring shortly with chitosan as an encapsulation matrix in fiber form. We also presented recent developments on this biomaterial application in food-packaging applications, where it is required to act as a polymeric matrix and biocide compound. According to the patent number during the past five years and the market assessment, the food-packaging sector shows the highest market opportunity for chitosan-encapsulated materials in the future. All those examples demonstrated that natural polymers could be used to replace oil-derived plastics that contribute to environmental pollution (e.g. microplastics). Succeeding in finding the commercial opportunity to use chitosan as a material in the food sector depends on several factors like identifying market necessity, developing a good product (price–performance), and good marketing. This versatile material could contribute to achieving a more sustainable world.

Glossary

Chitosan A product of the natural polysaccharide chitin following deacetylation, with biodegradability, biocompatibility, nontoxicity, antibacterial, anticancer, lipid-lowering, immune enhancement, and other physiological functions.

Cellulose A large polysaccharide composed of glucose, being insoluble in water and general organic solvents and serving as the main component of plant cell walls.

Nutraceutical Food claimed to have health-care functions or for the purpose of supplementing vitamins, minerals, and other nutrients.

Nanoencapsulation A technology to embed powder particles or liquid droplets into nanocapsules generated by the wall material.

Coacervation The aggregation of hydrophilic colloidal particles into a thick sol, which is separated as microscopic droplets or visible phases.

Maltodextrin A low conversion product between starch and starch sugar.

Peptides A chain of α-amino acids linked together by peptide bonds and are intermediate products of protein hydrolysis.

Carotenoids A group of natural pigments found in the yellow, orange-red, and red pigments of animals, higher plants, fungi, and algae.

References

1 C. Schmitz, L. G. Auza, D. Koberidze, S. Rasche, R. Fischer, L. Bortesi, *Mar. Drugs* **2019**, *17(8)*, 452.

2 S. M. Joseph, S. Krishnamoorthy, R. Paranthaman, J. A. Moses, C. Anandharamakrishnan, Carbohydr, *Polym. Technol. Appl.* **2021**, *2*, 100036.

3 I. Younes, M. Rinaudo, D. Harding, H. Sashiwa, *Mar. Drugs* **2015**, *13*, 1133–1174.

4 M. Yadav, P. Goswami, K. Paritosh, M. Kumar, N. Pareek, V. Vivekanand, *Bioresour. Bioprocess* **2019**, *6(1)*, 1–201-20.

5 Anandharamakrishnan, C., and S. Padma Ishwarya. Encapsulation of bioactive ingredients by spray drying. *Spray Drying Techniques for Food Ingredient Encapsulation*; Anandharamakrishnan, C., Ishwarya, SP, Eds. **2015**, 156–179.

6 C. I. Piñón-Balderrama, C. Leyva-Porras, Y. Terán-Figueroa, V. Espinosa-Solís, C. Álvarez-Salas, M. Z. Saavedra-Leos, *Process* **2020**, *8*, 889.

7 Y. P. Timilsena, T. O. Akanbi, N. Khalid, B. Adhikari, C. J. Barrow, *Int. J. Biol. Macromol.* **2019**, *121*, 1276.

8 A. Ziabicki, *Fundamentals of Fibre Formation : The Science of Fibre Spinning and Drawing*, Wiley, London; New York, 1976.

9 O. Nechyporchuk, T. Yang Nilsson, H. Ulmefors, T. Köhnke, *ACS Appl. Polym. Mater.* **2020**, *2*, 3867.

10 C. K. S. Pillai, W. Paul, C. P. Sharma, *Prog. Polym. Sci.* **2009**, *34*, 641.

11 G. Toskas, R. Brünler, H. Hund, R. D. Hund, M. Hild, D. Aibibu, C. Cherif, *Autex Res. J.* **2013**, *13*, 134.

12 A. A. Enache, L. David, J. P. Puaux, I. Banu, G. Bozga, U. P. B. Sci, *Bull. Ser. B Chem. Mater. Sci.* **2018**, *80*, 13.

13 B. S. de Farias, T. R. Sant'Anna Cadaval, L. A. D. A. P. Junior, *Int. J. Biol. Macromol.* **2019**, *123*, 210.

14 M. Arkoun, F. Daigle, M. C. Heuzey, A. Ajji, *Molecules* **2017**, *22(4)*, 585.

15 S. Haghju, M. R. Bari, M. A. Khaled-Abad, *Carbohydr. Polym.* **2018**, *200*, 137.

16 W. Klaypradit, Y. W. Huang, *LWT - Food Sci. Technol.* **2008**, *41*, 1133.

17 J. Duan, Y. Jiang, Y. Zhao, *J. Food Sci.* **2011**, *76(1)*, C133–C141.

18 S. Kanwal, M. Hafeez-ur-rehman, A. Hussain, M. Nadeem, F. Abbas, M. Akram, M. Inayat, F. Sughra, K. Ali, *Brazilian J. Biol.* **2024**, *84*, doi:10.1590/1519-6984.254010.

19 C. Chang, M. T. Nickerson, *J. Food Sci. Technol.* **2018**, *55*, 2850.

20 S. F. Hamed, A. F. Hashim, H. A. Hay, K. A. Abd-Elsalam, I. M. El-Sherbiny, *Egypt. J. Chem.* **2022**, *65*, 623.

21 B. Vafania, M. Fathi, S. Soleimanian-Zad, *Food Bioprod. Process.* **2019**, *116*, 240.

22 M. E. Sotelo-Boyás, Z. N. Correa-Pacheco, S. Bautista-Baños, M. L. Corona-Rangel, *LWT - Food Sci. Technol.* **2017**, *77*, 15.

23 S. Zhaveh, A. Mohsenifar, M. Beiki, S. T. Khalili, A. Abdollahi, T. Rahmani-Cherati, M. Tabatabaei, *Ind. Crops Prod.* **2015**, *69*, 251.

24 M. Beyki, S. Zhaveh, S. T. Khalili, T. Rahmani-Cherati, A. Abollahi, M. Bayat, M. Tabatabaei, A. Mohsenifar, *Ind. Crops Prod.* **2014**, *54*, 310.

25 A. Esmaeili, A. Asgari, *Int. J. Biol. Macromol.* **2015**, *81*, 283.

26 A. Mohammadi, M. Hashemi, S. M. Hosseini, *Innov. Food Sci. Emerg. Technol.* **2015**, *28*, 73.

27 M. Ghaderi-Ghahfarokhi, M. Barzegar, M. A. Sahari, H. Ahmadi Gavlighi, F. Gardini, *Int. J. Biol. Macromol.* **2017**, *102*, 19.

28 C. Dima, L. Pətraşcu, A. Cantaragiu, P. Alexe, Ş. Dima, *Food Chem.* **2016**, *195*, 39.

29 B. N. Estevinho, I. Carlan, A. Blaga, F. Rocha, *Powder Technol.* **2016**, *289*, 71.

30 J. Liu, Y. Li, H. Zhang, S. Liu, M. Yang, M. Cui, T. Zhang, Y. Yu, H. Xiao, Z. Du, *Food Chem.* **2022**, *372*, 131266.

31 W. Xu, K. Lv, W. Mu, S. Zhou, Y. Yang, *LWT* **2021**, *148*, 111724.

32 K. G. D. Kaushalya, K. D. P. P. Gunathilake, *Food Chem.* **2022**, *377*, 132012.

33 P. P. S. Coimbra, F. D. S. N. Cardoso, É. C. B. D. A. Gonçalves, *Crit. Rev. Food Sci. Nutr.* **2021**, *61*, 2809.

34 I. C. Carlan, B. N. Estevinho, F. Rocha, *Can. J. Chem. Eng.* **2020**, *98*, 1682.

35 R. Baksi, D. P. Singh, S. P. Borse, R. Rana, V. Sharma, M. Nivsarkar, *Biomed. Pharmacother* **2018**, *106*, 1513.

36 P. Roy, S. Parveen, P. Ghosh, K. Ghatak, S. Dasgupta, *Biochimie* **2019**, *162*, 185.

37 S. Ilk, N. Sağlam, M. Özgen, F. Korkusuz, *Int. J. Biol. Macromol.* **2017**, *94*, 653.

38 C. Chang, L. Zhang, Y. Miao, B. Fang, Z. Yang, *J. Clust. Sci.* **2021**, *32*, 331.

39 S. P. Kumar, K. Birundha, K. Kaveri, K. T. R. Devi, *Int. J. Biol. Macromol.* **2015**, *78*, 87.

40 S. Malathy, P. R. Iyer, *J. Nanomed. Nanotechnol.* **2018**, *9(4)*, 4172.

41 L. Mary Lazer, B. Sadhasivam, K. Palaniyandi, T. Muthuswamy, I. Ramachandran, A. Balakrishnan, S. Pathak, S. Narayan, S. Ramalingam, *Int. J. Biol. Macromol.* **2018**, *107*, 1988.

42 N. Khan, D. J. Bharali, V. M. Adhami, I. A. Siddiqui, H. Cui, S. M. Shabana, S. A. Mousa, H. Mukhtar, *Carcinogenesis* **2014**, *35*, 415.

43 M. Wang, L. Li, M. Wan, Y. Lin, Y. Tong, Y. Cui, H. Deng, C. Tan, Y. Kong, X. Meng, *RSC Adv.* **2020**, *11*, 210.

44 F. Rahmani, E. Karimi, E. Oskoueian, *J. Microencapsul.* **2020**, *37*, 305.

45 I. Nallamuthu, A. Devi, F. Khanum, *Asian J. Pharm. Sci.* **2015**, *10*, 203.

46 J. Ahlawat, R. Neupane, E. Deemer, S. T. Sreenivasan, M. Narayan, *ACS Appl. Mater. Interfaces* **2020**, *12*, 18964.

47 F. S. Buosi, A. Alaimo, M. C. Di Santo, F. Elías, G. García Liñares, S. L. Acebedo, M. A. Castañeda Cataña, C. C. Spagnuolo, L. Lizarraga, K. D. Martínez, K. D. Martínez, O. E. Pérez, *Int. J. Biol. Macromol.* **2020**, *165*, 804.

48 F. N. Sorasitthiyanukarn, C. Muangnoi, P. Rojsitthisak, P. Rojsitthisak, *Food Hydrocoll.* **2022**, 124, doi: 10.1016/j.foodhyd.2021.107246.

49 V. Toragall, N. Jayapala, SS.SS. Muthukumar P. B. Vallikanan, *Food Chem.* **2020**, *330*, 127195.

50 T. R. Aguirre Calvo, P. R. Santagapita, *Chem. Biol. Technol. Agric.* **2017**, *4*, 4.

51 L. F. Calinoiu, B. E. Ştefanescu, I. D. Pop, L. Muntean, D. C. Vodnar, *Coatings* **2019**, *9(3)*, 194.

52 A. F. Groboillot, C. P. Champagne, G. D. Darling, D. Poncelet, R. J. Neufeld, Biotechnol, *Bioeng* **1993**, *42*, 1157.

53 A. Mortazavian, S. H. Razavi, M. R. Ehsani, S. Sohrabvandi, *Iran. J. Biotechnol.* **2007**, *5*, 1.

54 M. A. Khosravi Zanjani, B. G. Tarzi, A. Sharifan, N. Mohammadi, *Iran. J. Pharm. Res.* **2014**, *13*, 843.

55 N. Varankovich, M. F. Martinez, M. T. Nickerson, D. R. Korber, *Food Sci. Biotechnol.* **2017**, *26*, 189.

56 M. A. S. Santos, M. T. C. Machado, *Int. J. Food Sci. Technol.* **2021**, *56*, 2122.

57 F. Ansari, H. Pourjafar, V. Jodat, J. Sahebi, A. Ataei, *AMB Express* **2017**, *7(1)*, 1-8.

58 A. J. Priya, S. P. Vijayalakshmi, A. M. Raichur, *J. Agric. Food Chem.* **2011**, *59*, 11838.

59 M. Wang, J. Yang, M. Li, Y. Wang, H. Wu, L. Xiong, Q. Sun, *Food Chem.* **2019**, *285*, 260.

60 V. Padhmavathi, R. Shruthy, R. Preetha. *J. Food Science Technol.*, **2021**, 1-7.

61 H. Gandomi, S. Abbaszadeh, A. Misaghi, S. Bokaie, N. Noori, *LWT - Food Sci. Technol.* **2016**, *69*, 365.

62 L. P. Ta, E. Bujna, O. Antal, M. Ladányi, R. Juhász, A. Szécsi, S. Kun, S. Sudheer, V. K. Gupta, Q. D. Nguyen, *Int. J. Biol. Macromol.* **2021**, *183*, 1136.

63 J. Long, B. Zhang, X. Li, X. Zhan, X. Xu, Z. Xie, Z. Jin, *Food Chem.* **2018**, *239*, 276.

64 L. Yu, D. J. O'Sullivan, *J. Dairy Sci.* **2018**, *101*, 2974.

65 C. R. Hackenhaar, L. S. Spolidoro, E. E. E. Flores, M. P. Klein, P. F. Hertz, *Biocatal. Agric. Biotechnol.* **2021**, *36*, 102136.

66 M. A. Ganaie, H. K. Rawat, O. A. Wani, U. S. Gupta, N. Kango, *Process Biochem.* **2014**, *49*, 840.

67 A. S. G. Lorenzoni, L. F. Aydos, M. P. Klein, M. A. Z. Ayub, R. C. Rodrigues, P. F. Hertz, J. Mol, *Catal. B Enzym.* **2015**, *111*, 51.

68 S. C. Chen, D. C. Sheu, K. J. Duan, *J. Taiwan Inst. Chem. Eng.* **2014**, *45*, 1105.

69 T. Charoenwongpaiboon, K. Wangpaiboon, R. Pichyangkura, M. H. Prousoontorn, *RSC Adv.* **2018**, *8*, 17008.

70 K. Biesalski Hans, T. Jana, *NFS J.* **2018**, *11*, 1.

71 M. Granger, P. Eck, *Adv. Food Nutr. Res.* **2018**, *83*, 281.

72 Z. Jiao, X. Wang, Y. Yin, J. Xia, Y. Mei, *J. Microencapsul.* **2018**, *35*, 272.

73 G. H. Shin, S. K. Chung, J. T. Kim, H. J. Joung, H. J. Park, J. Agric, *Food Chem.* **2013**, *61*, 11119.

74 S. Chambial, S. Dwivedi, K. K. Shukla, P. J. John, P. Sharma, *Indian J. Clin. Biochem.* **2013**, *28*, 314.

75 J. T. Pinto, J. Zempleni, *Adv. Nutr.* **2016**, *7*, 973.

76 S. Vidyashri, A. Laksminarayanan, S. Rajeshkumar, T. Lakshmi, *Plant Cell Biotechnol. Mol. Biol.* **2020**, *21*, 69.

77 V. K. Venugopalan, L. R. Gopakumar, A. K. Kumaran, N. S. Chatterjee, V. Soman, S. Peeralil, S. Mathew, D. J. McClements, R. C. Nagarajarao, *Foods* **2021**, 10(7), 1566.

78 D. M. Hariyadi, A. Fitri, T. Purwanti, T. Erawati, *Rasayan J. Chem.* **2021**, *14*, 1680.

79 D. J. McClements, E. A. Decker, J. Weiss, *J. Food Sci.* **2007**, *72*, doi:10.1111/J.1750-3841.2007.00507.X.

80 A. Nesterenko, I. Alric, F. Silvestre, V. Durrieu, *Ind. Crops Prod.* **2013**, *42*, 469.

81 C. Chang, N. Varankovich, M. T. Nickerson, *Food Chem.* **2016**, *212*, 264.

82 T. Furuta, T. L. Neoh, *Dry. Technol.* **2021**, *39*, 1800.

83 A. M. Bakry, S. Abbas, B. Ali, H. Majeed, M. Y. Abouelwafa, A. Mousa, L. Liang, *Compr. Rev. Food Sci. Food Saf.* **2016**, *15*, 143.

84 N. Raak, H. Rohm, D. Jaros, *Colloids Surf. A: Physicochem. Eng. Asp.* **2021**, *619*, 126468.

85 C. Lucía, F. Marcela, L. Ainhoa, C. Lucía, F. Marcela, L. Ainhoa, *J. Encapsul. Adsorpt. Sci.* **2017**, *7*, 67.

86 I. J. Joye, G. Davidov-Pardo, D. J. McClements, *Trends Food Sci. Technol.* **2014**, *40*, 168.

87 X. Sun, O. D. Okagu, C. C. Udenigwe, In *Biol. Act. Pept., Academic Press.* **2021**, 331–356.

88 Z. Fang, B. Bhandari, *Trends Food Sci. Technol.* **2010**, *21*, 510.

89 B. Muhoza, B. Qi, J. D. Harindintwali, M. Y. Farag Koko, S. Zhang, Y. Li, *Food Hydrocoll.* **2022**, *124*, 107239.

90 G. Y. Li, Q. H. Chen, C. R. Su, H. Wang, S. He, J. Liu, A. Nag, Y. Yuan, *Innov. Food Sci. Emerg. Technol.* **2021**, *68*, 102612.

91 L. Amigo, B. Hernández-Ledesma, *Molecules* **2020**, *25(19)*, 4479.

92 M. Peng, Z. Tabashsum, M. Anderson, A. Truong, A. K. Houser, J. Padilla, A. Akmel, J. Bhatti, S. O. Rahaman, D. Biswas, *Compr. Rev. Food Sci. Food Saf.* **2020**, *19*, 1908.

93 M. R. Corbo, M. Albenzio, M. De Angelis, A. Sevi, M. Gobbetti, *J. Dairy Sci.* **2001**, *84*, 551.

94 G. Frakolaki, V. Giannou, D. Kekos, C. Tzia, *Crit. Rev. Food Sci. Nutr.* **2021**, *61*, 1515.

95 C. Hill, F. Guarner, G. Reid, G. R. Gibson, D. J. Merenstein, B. Pot, L. Morelli, R. B. Canani, H. J. Flint, S. Salminen, P. C. Calder, M. E. Sanders, *Nat. Rev. Gastroenterol. Hepatol.* **2014**, *11*, 506.

96 D. Boltin, B. Pract, *Res. Clin. Gastroenterol.* **2016**, *30*, 99.

97 ISAPP, ISAPP holds 6th Annual meeting November 9-11, 2008 - International Scientific Association for Probiotics and Prebiotics (ISAPP), London, Ontario, can be found under https://isappscience.org/isapp-holds-6th-annual-meeting-november-9-11-2008, 2008. (Accessed 25-06-2022)

98 M. Roberfroid, G. R. Gibson, L. Hoyles, A. L. McCartney, R. Rastall, I. Rowland, D. Wolvers, B. Watzl, H. Szajewska, B. Stahl, F. Guarner, F. Respondek, K. Whelan, V. Coxam, M. J. Davicco, L. Léotoing, Y. Wittrant, N. M. Delzenne, P. D. Cani, A. M. Neyrinck, A. Meheust, *Br. J. Nutr.* **2010**, *104 S2*, S1-S63.

99 G. Reid, M. E. Sanders, H. R. Gaskins, G. R. Gibson, A. Mercenier, R. Rastall, M. Roberfroid, I. Rowland, C. Cherbut, T. R. Klaenhammer, *J. Clin. Gastroenterol.* **2003**, *37*, 105.

100 B. Krajewska, *Enzyme Microb. Technol.* **2004**, *35*, 126.

101 A. G. Tarone, E. K. Silva, C. B. Betim Cazarin, M. R. Marostica Junior, *Food Hydrocoll.* **2021**, *111*, 106387.

102 M. Bilal, H. M. N. Iqbal, *Int. J. Biol. Macromol.* **2019**, *130*, 462.

103 F. Pavli, C. Tassou, G. J. E. Nychas, N. Chorianopoulos, *Int. J. Mol. Sci.* **2018**, *19(1)*, 150.

104 A. D. S. Pereira, M. M. Diniz, G. De Jong, H. S. Gama Filho, M. J. Dos Anjos, P. V. Finotelli, G. C. Fontes-Sant'Ana, P. F. F. Amaral, *Int. J. Biol. Macromol.* **2019**, *139*, 621.

105 M. Ye, H. Neetoo, H. Chen, *Food Microbiol.* **2008**, *25*, 260.

106 D. C. Vodnar, *Chem. Cent. J.* **2012**, *6(1)*, 1–6.

107 W. Dhifi, S. Bellili, S. Jazi, N. Bahloul, W. Mnif, L. Nahar, N. Basar, S. D. Sarker, *Med.* **2016**, *3*, Page 25, **2016**, 3, 25,.

108 J. Sharifi-Rad, A. Sureda, G. C. Tenore, M. Daglia, M. Sharifi-Rad, M. Valussi, R. Tundis, M. Sharifi-Rad, M. R. Loizzo, A. Oluwaseun Ademiluyi, R. Sharifi-Rad, S. A. Ayatollahi, M. Iriti, *Molecules* **2017**, *22(1)*, 70.

109 M. D'agostino, N. Tesse, J. P. Frippiat, M. Machouart, A. Debourgogne, *Molecules* **2019**, *24(20)*, 3713.

110 N. A. Mahizan, S. K. Yang, C. L. Moo, A. A. L. Song, C. M. Chong, C. W. Chong, A. Abushelaibi, S. H. Erin Lim, K. S. Lai, *Molecules* **2019**, *24(14)*, 2631.

111 M. Valdivieso-Ugarte, C. Gomez-Llorente, J. Plaza-Díaz, Á. Gil, *Nutrients* **2019**, *11(11)*, 2786.

112 M. R. Mozafari, A. Omri, *J. Pharm. Sci.* **2007**, *96*, 1955.

113 Z. Lou, J. Chen, F. Yu, H. Wang, X. Kou, C. Ma, S. Zhu, *LWT - Food Sci. Technol.* **2017**, *80*, 371.

114 E. Anastasiou, K. O. Lorentz, G. J. Stein, P. D. Mitchell, *Lancet Infect. Dis.* **2014**, *14*, 553.

115 M. Cai, Y. Wang, R. Wang, M. Li, W. Zhang, J. Yu, R. Hua, *Int. J. Biol. Macromol.* **2022**, *202*, 122.

116 D. Del Rio, A. Rodriguez-Mateos, J. P. E. Spencer, M. Tognolini, G. Borges, A. Crozier, *Antioxidants Redox Signal.* **2013**, *18*, 1818.

117 A. N. Panche, A. D. Diwan, S. R. Chandra, *J. Nutr. Sci.* **2016**, *5*, 41.

118 R. Tsao, *Nutrients* **2010**, *2*, 1231.

119 J. M. Lorenzo, P. E. Munekata, P. Putnik, D. B. Kovačević, V. Muchenje, F. J. Barba, *Stud. Nat. Prod. Chem.*, 2019, *60*, 189–221.

120 M. F. Montenegro-Landívar, P. Tapia-Quirós, X. Vecino, M. Reig, C. Valderrama, M. Granados, J. L. Cortina, J. Saurina, *Sci. Total Environ.* **2021**, *801*, 149719.

121 A. Scalbert, C. Manach, C. Morand, C. Rémésy, L. Jiménez, *Crit. Rev. Food Sci. Nutr.* **2005**, *45*, 287.

122 M. C. Di Santo, C. L. D' Antoni, A. P. Domínguez Rubio, A. Alaimo, O. E. Pérez, *Biomed. Pharmacother* **2021**, *142*, 111970.

123 A. Detsi, E. Kavetsou, I. Kostopoulou, I. Pitterou, A. R. N. Pontillo, A. Tzani, P. Christodoulou, A. Siliachli, P. Zoumpoulakis, *Pharmaceutics* **2020**, *12*, 1.

124 Q. Hu, Y. Luo, *Carbohydr. Polym.* **2016**, *151*, 624.

125 J. Liu, H. Pu, S. Liu, J. Kan, C. Jin, *Carbohydr. Polym.* **2017**, *174*, 999.

126 A. Singh, A. Mittal, S. Benjakul, *Food Rev. Int.* **2021**, *00*, 1.

127 M. A. Mohammed, J. T. M. Syeda, K. M. Wasan, E. K. Wasan, *Pharmaceutics* **2017**, *9(4)*, 53.

128 K. Konecsni, N. H. Low, M. T. Nickerson, *Food Chem.* **2012**, *134*, 1775.

129 B. Siddhardha, U. Pandey, K. Kaviyarasu, R. Pala, A. Syed, A. H. Bahkali, A. M. Elgorban, *Pathogens* **2020**, *9*, 1.

130 A. Dube, K. Ng, J. A. Nicolazzo, I. Larson, *Food Chem.* **2010**, *122*, 662.

131 A. Dube, J. A. Nicolazzo, I. Larson, *Eur. J. Pharm. Sci.* **2010**, *41*, 219.

132 A.-M. Safer, S. Leporatti, J. Jose, M. S. Soliman, *Int. J. Nanomed.* **2019**, *14*, 8033.

133 I. A. Siddiqui, D. J. Bharali, M. Nihal, V. M. Adhami, N. Khan, J. C. Chamcheu, M. I. Khan, S. Shabana, S. A. Mousa, H. Mukhtar, Nanomedicine Nanotechnology, *Biol. Med.* **2014**, *10*, 1619.

134 L. Zeng, J. Yan, L. Luo, M. Ma, H. Zhu, *Sci. Rep.* **2017**, *7(1)*, 1–15.

135 V. A. S. Augustia, M. Musdzalifah, D. A. Lestari, A. Chafidz, in AIP Conf. Proc., 2018.

136 A. Ko, J.-S. Lee, H. Sop Nam, H. Gyu Lee, *J. Food Biochem.* **2017**, *41(2)*, e12316.

137 W. Wang, J. Jung, Y. Zhao, *Carbohydr. Polym.* **2017**, *157*, 1246.

138 R. K. Rajan, M. Z. Hussein, S. Fakurazi, K. Yusoff, M. J. Masarudin, *Int. J. Mol. Sci.* **2019**, *20(19)*, 4667.

139 S. B. Da Silva, M. Amorim, P. Fonte, R. Madureira, D. Ferreira, M. Pintado, B. Sarmento, *Pharm. Biol.* **2015**, *53*, 642.

140 S. B. da Silva, D. Ferreira, M. Pintado, B. Sarmento, *Int. J. Biol. Macromol.* **2016**, *84*, 112.

141 V. Arulmozhi, K. Pandian, S. Mirunalini, *Colloids Surf. B Biointerf.* **2013**, *110*, 313.

142 L. Gopalakrishnan, L. N. Ramana, S. Sethuraman, U. M. Krishnan, *Carbohydr. Polym.* **2014**, *111*, 215.

143 J. Lamarra, S. Rivero, A. Pinotti, *Mater. Sci. Eng. C* **2016**, *67*, 717.

144 R. Panwar, N. Raghuwanshi, A. K. Srivastava, A. K. Sharma, V. Pruthi, *Mater. Sci. Eng. C* **2018**, *92*, 381.

145 R. Panwar, A. K. Sharma, M. Kaloti, D. Dutt, V. Pruthi, *Appl. Nanosci.* **2016**, *6*, 803.

146 T. T. Pham, T. H. Nguyen, T. V. Thi, -T.-T. Nguyen, T. D. Le, D. M. H. Vo, D. H. Nguyen, C. K. Nguyen, D. C. Nguyen, T. T. Nguyen, T. T. Nguyen, L. G. Bach, *Polymers (Basel)* **2019**, *11(1)*, 177.

147 K. Balagangadharan, R. Trivedi, M. Vairamani, N. Selvamurugan, *Carbohydr. Polym.* **2019**, *216*, 1.

148 R. Panwar, S. C. Pemmaraju, A. K. Sharma, V. Pruthi, *Microb. Pathog.* **2016**, *95*, 21.

149 I. A. D. Lima, N. M. Khalil, T. T. Tominaga, A. Lechanteur, B. Sarmento, R. M. Mainardes, *Artif. Cells, Nanomed. Biotechnol.* **2018**, *46*, 993.

150 A. R. Cho, Y. G. Chun, B. K. Kim, D. J. Park, *J. Food Sci.* **2014**, *79(4)*, E568-E576.

151 M. Y. Kim, H.-K. Ha, I. L. Ayu, K.-S. Han, W.-J. Lee, M.-R. Lee, *Food Sci. Anim. Resour.* **2019**, *39*, 831.

152 J. Wu, Y. Wang, H. Yang, X. Liu, Z. Lu, *Carbohydr. Polym.* **2017**, *175*, 170.

153 C. Chen, Y. Yu, X. Wang, P. Shi, Y. Wang, P. Wang, *Int. J. Biol. Macromol.* **2017**, *95*, 451.

154 C. A. Fuenmayor, O. G. Baron-Cangrejo, P. A. Salgado-Rivera, *Polysaccharides* **2021**, *2*, 454.

155 F. Delgado-Vargas, A. R. Jiménez, O. Paredes-López, F. J. Francis. *Critical Reviews Food Sci. Nutr.* 2000, *40(3)*, 173–289.

156 G. Britton, *Biochim. Biophys. Acta - Mol. Cell Biol. Lipids* **2020**, **1865**, 158699.

157 J. M. López-Nicolás, P. Rodríguez-Bonilla, F. García-Carmona, *Crit. Rev. Food Sci. Nutr.* **2014**, *54*, 251.

158 C. Liu, Z. Liu, X. Sun, S. Zhang, S. Wang, F. Feng, D. Wang, Y. Xu, J. Agric, *Food Chem.* **2018**, *66*, 6717.

159 C. Liu, Y. Tan, Y. Xu, D. J. McCleiments, D. Wang, *Int. J. Biol. Macromol.* **2019**, *140*, 985.

160 A. Tachaprutinun, T. Udomsup, C. Luadthong, S. Wanichwecharungruang, *Int. J. Pharm.* **2009**, *374*, 119.

161 S. Y. Koo, I.-K. Mok, C.-H. Pan, S. M. Kim, J. Agric, *Food Chem.* **2016**, *64*, 9428.

162 I. Bhat, U. G. Yathisha, I. Karunasagar, B. S. Mamatha, *Nutr. Rev.* **2020**, *78*, 709.

163 W. Li, M. Yalcin, Q. Lin, M.-S. M. Ardawi, S. A. Mousa, *J. Control. Release* **2017**, *248*, 117.

164 J. K. Rutz, C. D. Borges, R. C. Zambiazi, C. G. Da Rosa, M. M. Da Silva, *Food Chem.* **2016**, *202*, 324.

165 S. Rahaiee, M. Hashemi, S. A. Shojaosadati, S. Moini, S. H. Razavi, *Int. J. Biol. Macromol.* **2017**, *99*, 401.

166 G. Song, J. Liu, Q. Wang, D. Wang, B. Chu, L. Li, G. Xiao, J. Gong, F. Zheng, *Food Hydrocoll.* **2022**, *125*, 107420.

167 T. C. R. de Souza, A. L. Parize, I. M. C. Brighente, V. T. Fávere, M. C. M. Laranjeira, *J. Microencapsul.* **2005**, *22*, 511.

168 M. A. Sani, M. Azizi-Lalabadi, M. Tavassoli, K. Mohammadi, *D. J. McClements Nanomater.* **2021**, *11(5)*, 1331.

169 T. T. Nguyen, U. T. Thi Dao, Q. P. Thi Bui, G. L. Bach, C. N. Ha Thuc, H. Ha Thuc, *Prog. Org. Coat.* **2020**, 105487.

170 P. Ezati, J. W. Rhim, *Food Hydrocoll.* **2020**, *102*, 105629.

171 S. Hu, J. Yu, Z. Wang, L. Li, Y. Du, L. Wang, Y. Liu, *J. Food Sci.* **2017**, *82*, 1451.

172 H. Zhang, X. Li, H. Kang, *LWT* **2019**, *116*, 108580.

173 N. Hasheminejad, F. Khodaiyan, *Food Chem.* **2020**, *309*, 125520.

174 Y. Ma, S. Li, T. Ji, W. Wu, D. E. Sameen, S. Ahmed, W. Qin, J. Dai, Y. Liu, *Carbohydr. Polym.* **2020**, *247*, 116738.

175 D. Surendhiran, C. Li, H. Cui, L. Lin, *Food Packag. Shelf Life* **2020**, *23*, 100439.

176 H. Cui, J. Wu, C. Li, L. Lin, *LWT - Food Sci. Technol.* **2017**, *81*, 233.

177 G. A. Ojeda, S. C. Sgroppo, O. Martín-Belloso, R. Soliva-Fortuny, *Postharvest Biol. Technol.* **2019**, *156*, 110934.

178 E. B. Özvural, Q. Huang, M. L. Chikindas, *LWT* **2016**, *68*, 385.

179 F. Liu, J. Antoniou, Y. Li, J. Yi, W. Yokoyama, J. Ma, F. Zhong, *J. Agric. Food Chem.* **2015**, *63*, 3987.

180 S. Bao, S. Xu, Z. Wang, *J. Sci. Food Agric.* **2009**, *89*, 2692.

181 J. Hu, X. Wang, Z. Xiao, W. Bi, *LWT* **2015**, *63*, 519.

182 H. Cui, M. Bai, M. M. A. Rashed, L. Lin, *Int. J. Food Microbiol.* **2018**, *266*, 69.

183 L. Lin, X. Mao, Y. Sun, G. Rajivgandhi, H. Cui, *Int. J. Food Microbiol.* **2019**, *292*, 21.

184 L. Lin, Y. Gu, H. Cui, *Food Packag. Shelf Life* **2019**, *19*, 86.

185 L. Li, W. Song, C. Shen, Q. Dong, Y. Wang, S. Zuo, *J. Food Process. Preserv.* **2020**, *44*, doi:10.1111/JFPP.14799.

186 C. Yin, C. Huang, J. Wang, Y. Liu, P. Lu, L. Huang, *Materials (Basel)* **2019**, *12*, doi:10.3390/MA12132039.

187 M. Alizadeh-Sani, M. Tavassoli, E. Mohammadian, A. Ehsani, G. J. Khaniki, R. Priyadarshi, J. W. Rhim, *Int. J. Biol. Macromol.* **2021**, *166*, 741.

188 M. Alizadeh-Sani, M. Tavassoli, D. J. McClements, H. Hamishehkar, *Food Hydrocoll.* **2021**, *111*, 106237.

189 Z. Liu, M. Zhang, B. Bhandari, Y. Wang, *Trends Food Sci. Technol.* **2017**, *69*, 83.

190 T. Pereira, S. Barroso, M. M. Gil, *Foods* **2021**, *10*, Page 320, **2021**, 10, 320.

191 A. Gholamipour-Shirazi, M. A. Kamlow, I. T. Norton, T. Mills, *Foods* **2020**, *9*, 497.

192 H. Jiang, L. Zheng, Y. Zou, Z. Tong, S. Han, S. Wang, **2018**, *59(14)*, 2335-2347.

193 J. Izdebska, Z. Zołek-Tryznowska, *Agro Food Ind. Hi. Tech.* **2016**, *27*, 33.

194 A. Enache, L. David, I. Banu, 2018.

195 N. Dasgupta, S. Ranjan, **2018**, DOI 10.1007/978-981-10-6986-4.

196 S. Akhavan, E. Assadpour, I. Katouzian, S. M. Jafari, *Trends Food Sci. Technol.* **2018**, *74*, 132.

197 S. M. Jafari, M. Fathi, I. Mandala, *Food Waste Recover.* **2021**, 257–275.

198 I. Katouzian, S. M. Jafari, *J. Control. Release* **2019**, *303*, 302.

199 S. Mokhtari, S. M. Jafari, E. Assadpour, *Food Chem.* **2017**, *229*, 286.

200 H. Rostamabadi, S. R. Falsafi, S. M. Jafari, *Trends Food Sci. Technol.* **2019**, *88*, 397.

201 A. Taheri, S. M. Jafari, *Adv. Colloid Interface Sci.* **2019**, *269*, 277.

202 S. Akbari-Alavijeh, S. Soleimanian-Zad, M. Sheikh-Zeinoddin, S. Hashmi, *Int. J. Biol. Macromol.* **2018**, *107*, 808.

203 E. Assadpour, S. Mahdi Jafari. *Crit. Rev. Food Sci. Nutr.* **2019**, *59(19)*, 3129–3151.

204 M. R. Rostami, M. Yousefi, A. Khezerlou, M. Aman Mohammadi, S. M. Jafari, *Food Hydrocoll.* **2019**, *97*, 105170.

205 G. M. Oyatogun, T. A. Esan, E. I. Akpan, S. O. Adeosun, A. P. I. Popoola, B. I. Imasogie, W. O. Soboyejo, A. A. Afonja, S. A. Ibitoye, V. D. Abere, A. O. Oyatogun, K. M. Oluwasegun, I. E. Akinwole, K. J. Akinluwade, *Handb. Chitin Chitosan Vol. 1 Prep. Prop.* **2020**, 335.

206 "Chitosan Market Size and Share | industry Statistics - **2027**," can be found under https://www.gminsights.com/industry-analysis/chitosan-market, 2020. Accesed 26-06-2022.

10

Design and Use of Hydrogels for Food Component Encapsulation

Zahra Emam-Djomeh[1,3,4*], *Mohammad Ekrami*[1], *and Ali Ekrami*[2]

[1] *Transfer Phenomena Laboratory (TPL), Department of Food Science, Technology and Engineering, University College of Agriculture & Natural Resources, University of Tehran, Karaj Campus, Karaj, Iran*
[2] *Department of Nursing, University of Social Welfare and Rehabilitation Sciences, Tehran, Iran*
[3] *Functional Food Research Core (FFRC), University of Tehran, Tehran, Iran*
[4] *Center of Excellence in Biothermodynamics, University of Tehran, Tehran, Iran*
* *Corresponding author*

Highlights

- Hydrogels are new materials that have many applications due to their high water content.
- The classification of hydrogels is based on the structure and composition.
- The physicochemical properties of hydrogels determine their recent development potential.
- Hydrogels must be biocompatible and biodegradable for food component encapsulation applications.

10.1 Introduction

Materials made of cross-linked polymer chains with a strong attraction for water are known as polymeric hydrogels. Their mechanical qualities are acceptable for many technological applications because of their cross-linked nature [1]. Hydrogels can be designed to hold large amounts of water while maintaining their shape. Hydrogels that contain more than 99% water are not uncommon. Hence, they became highly biocompatible in a wide range of encapsulation applications [2]. Hydrogels are an excellent alternative for encapsulating biological molecules because of their capacity to replicate natural water conditions. For these applications, hydrogels have inherent advantages: (i) maintaining the biological function of the encapsulated material; (ii) providing a diffusion barrier capable of allowing molecules of a given size threshold to pass through while excluding larger molecules; and (iii) the size threshold can be tailored according to the degree of crosslinking of the hydrogel [3].

Materials Science and Engineering in Food Product Development, First Edition. Edited by Wing-Fu Lai.

Traditional hydrogel synthesis techniques have little control over their exact structure. The ability to encapsulate food components and create hybrid hydrogels with synthetic and natural chemicals has greatly expanded hydrogel's application potential. These methods allow the incorporation of biorecognition moieties into macromolecules, influencing their self-assembly into predetermined three-dimensional (3D) structures [4]. Consequently, hydrogels' potential applications have moved beyond food and medication delivery systems to microfluidics and nanotechnology. Hydrogels made of natural and artificial polymers have remained popular for food component encapsulation [5]. Table 10.1 lists commercially relevant gelling proteins and polysaccharides from diverse sources [6–18]. Because of their cross-linked structure, hydrogels preserve biological molecules encapsulated in food from degradation and/or immunological rejection. Entanglement of molecules or hydrophobic/ionic areas in physical hydrogels prevents homogeneity [19]. Polyelectrolytes generate ionotropic hydrogels when they combine with multivalent ions. According to Flory-Huggins, a hydrogel's ability to absorb water depends on temperature and water–polymer interaction [20]. Conditions like gel formation and food characteristics are affected by variables. Hydrogel production involves one or more of these intermolecular interactions [21]. Several kinds of junction zones may be created (Figure 10.1). In practice, small molecules and solutes may readily pass through the hydrogel layer in an enclosed system. Encapsulation of biological agents may prevent bigger molecules from interfering with them. Consequently, hydrogels may be employed in food industry applications where encapsulated polymers create sensitive chemicals. Encapsulated biological agents are protected from enzyme breakdown by hydrogels' size-exclusion properties. Hydrogels are well-suited for the controlled release of biomolecules because of the latter property [22].

Hydrogels with environmental sensitivities may be used in a wide range of applications (Figure 10.2). Release target environments contain certain adaptive factors, such as low pH

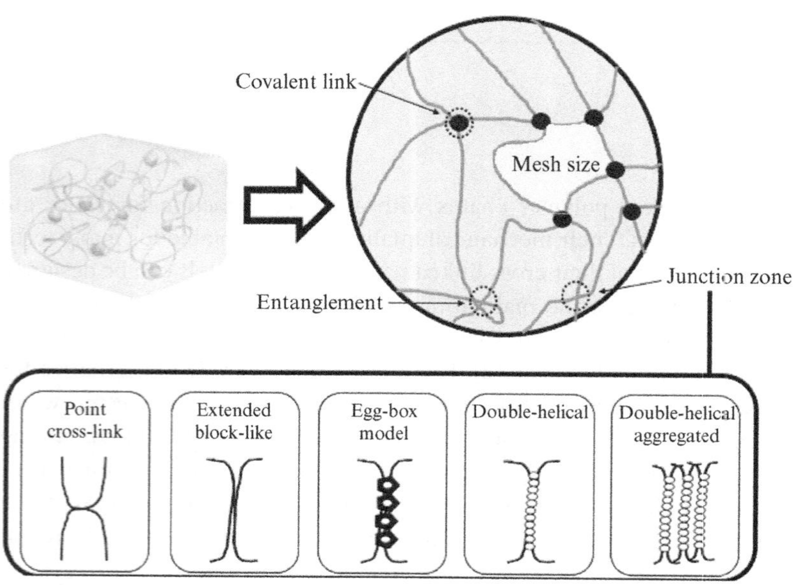

Figure 10.1 Structural chemistry of a hydrogel and idealized junction zones.

Table 10.1 Sources of hydrogel-structure proteins and polysaccharides.

Type	Category	Example	Reference
Protein	Animal	Silk, meat, milk, egg	[6–8]
	Plants	Soy, cereal, pulse	[9]
Polysaccharide	Plants	Starch, pectin, cellulose	[10]
	Tree gum exudates	Arabic, karaya, ghatti, tragacanth	[11]
	Seeds	Guar, locust bean, tara, tamarind seed xyloglucan	[12, 13]
	Tubers	Konjac glucomannan	[14]
	Red seaweeds	Agar (agarose), carrageenan	[15]
	Brown seaweeds	Alginate	[16]
	Microbial	Xanthan, gellan, curdlan, dextran, bacterial cellulose	[17]
	Animal	Chitosan, hyaluronan (hyaluronic acid)	[18]

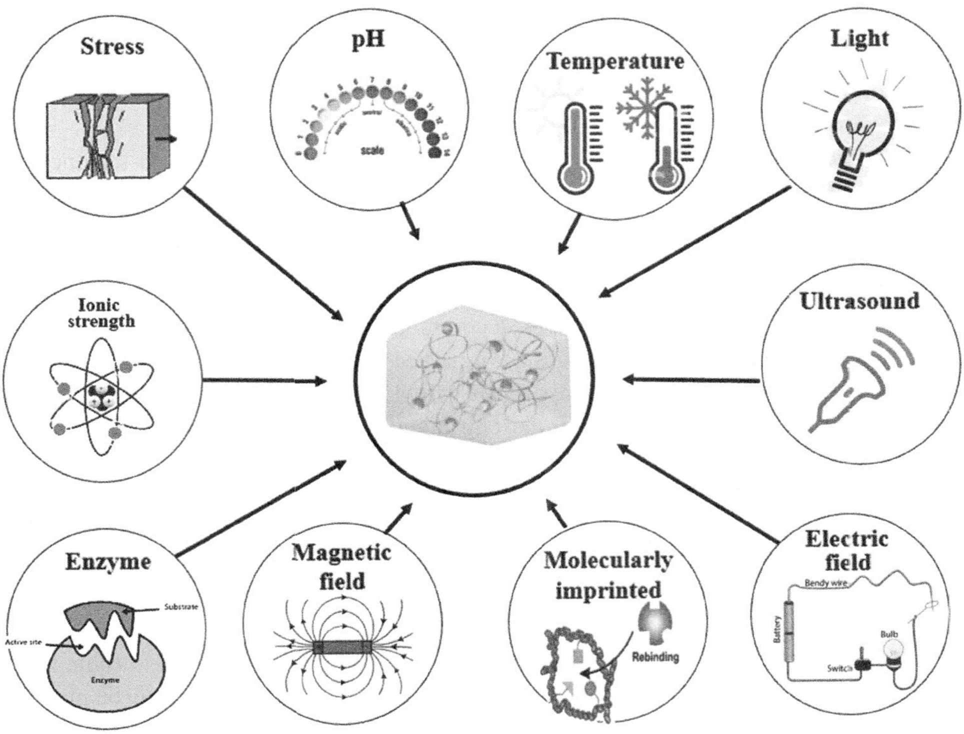

Figure 10.2 Environmental stimuli sensitive to hydrogels.

and high temperatures. Hydrogels that are pH-sensitive and/or temperature-sensitive may be employed to transport food components to specified locations. Specific hydrogels, such as glucose or antigens, may serve as both biosensors and medication delivery devices. Light-sensitive, pressure-sensitive, and electro-sensitive hydrogels may also be employed for food component delivery and bioseparation [23]. These environment-sensitive hydrogels need major improvements in their hydrogel characteristics for practical applications. The major flaw in all of these external stimuli-sensitive hydrogels is that their reaction time is excessively long. Consequently, fast-acting hydrogels must be created, and the simplest method is to build smaller and thinner hydrogels [24]. A lack of mechanical strength is a common problem in hydrogel systems. Biocompatibility is required for environmentally sensitive hydrogels used in food component delivery applications. Novel polymers and crosslinkers with greater biocompatibility and greater biodegradability are needed for effective applications. It is a huge undertaking to create hydrogels that are both ecologically friendly and functional. Responsive hydrogels with a broad range of desired qualities may be made possible if previous successes can be projected into the future [22].

10.2 Classification of Hydrogels

According to their physical and chemical properties and their rate of biodegradation and crosslinking, water-insoluble hydrogels are classed [8]. There are two basic types of hydrogels [25]:

When covalently cross-linked, the chemical gel is called "permanent" gels (networks replacing hydrogen bonds with stronger and stable covalent bonds). They attain an equilibrium swelling state that depends on the polymer–water interaction parameter and the cross-link density.

Physical gel: they are called "reversible" gels when the networks are held together by molecular entanglements and/or secondary forces, including ionic, hydrogen bonding, or hydrophobic interactions. Physical interactions between polymer chains limit gel breakdown in physically cross-linked gels. Changes in physical circumstances or the application of stress may disrupt any of these interactions, which are all reversible [26]. Figure 10.3 depicts an overview of a few of the most common hydrogel classifications.

Swelling and diffusion characteristics have gained a lot of attention in food industry applications owing to the creation of hydrogels with important physical properties. Three-dimensional microstructure plays an important function in various sectors because it stabilizes non-extracellular and extracellular matrixes. Soft tissue-like elastic qualities with stimulus sensitivity are found in polymeric hydrogels, nontoxic, biodegradable, and biocomparable. The hydrogel's composition significantly impacts these qualities [27]. Hydrogels are categorized into three primary categories based on their physical attributes for food industry applications. Hydrogels come in a variety of forms, including solid, semi-solid, and liquid.

Cross-linked hydrogels feature ionic or covalent crosslinkers and are solid at room temperature but may expand when exposed to water or buffer solutions. They may be utilized to make hydrogels for biomedical, environmental, and ecological purposes because of their unique features [28]. An increasing number of researchers are using inorganic metals to enhance their applications in advanced (ethnological and medical) fields. Because of their

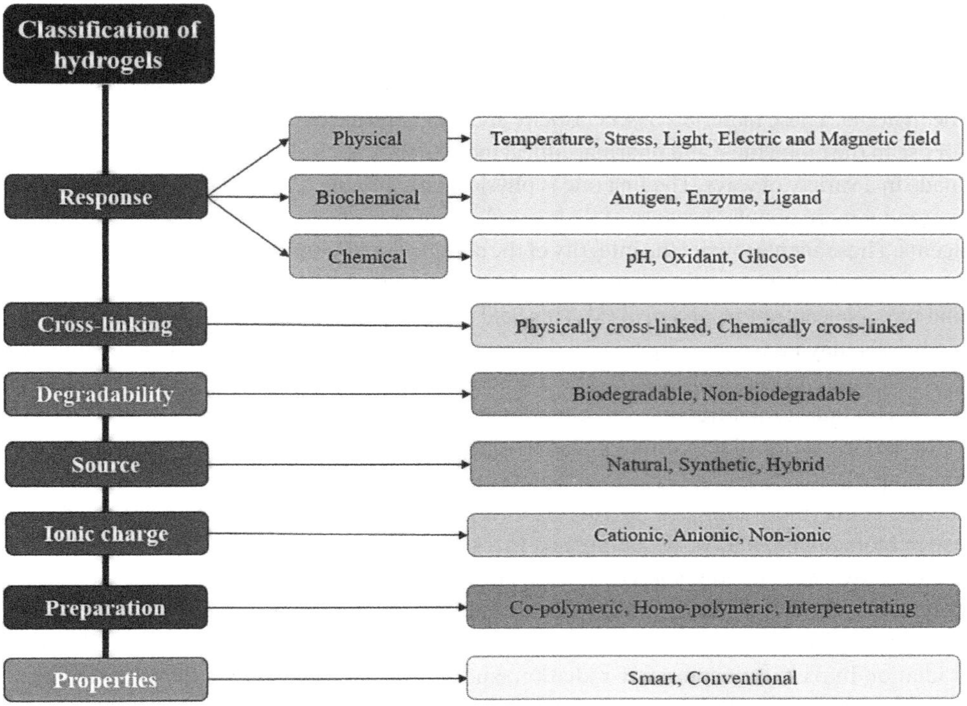

Figure 10.3 Classification of hydrogels by characteristics.

optical, electrical, and magnetic capabilities, inorganic materials are particularly well-suited for biomedical applications because of their easy functionalization with biomaterials [29]. Interfacial (van der Waals, hydrogen bonds, and electrostatic) forces and soft tissue networks exhibit significant adhesive interactions with semisolid hydrogels [30]. In biomedical applications, including buccal, ophthalmic, rectal, vaginal, and sublingual routes, this property is particularly important for the long-term administration and efficient dosing of drugs. Biological hydrogels may be described as such because of their bioadhesive characteristics. Hydrogel complexes are bioadhesive because of the biological components in the hydrogel complex [31].

Concerning the ambient temperature and the liquid phase of hydrogels, they are both liquid and elastic simultaneously. Future biological applications may greatly benefit from this self-assembling characteristic. While liquid hydrogels' unrivaled biocompatibility and functional ability and their simplicity and capacity to self-adjust their network (i.e. the size of the pores) make them very desirable, they are not without drawbacks [32]. These liquid hydrogels have several advantages: (i) they can be readily integrated into hydrogels and (ii) they may be injected into biological systems without any surgical procedures, as they have hydrophilic qualities [33]. Food components and regenerative medicine, as well as biomedical applications, benefit from these characteristics. Biodegradability, biocompatibility, and excellent porosity for hydration/dehydration are required for these hydrogels to be useful in future biomedical applications. These hydrogels should be biodegradable, biocompatible, and have good porosity for hydration/dehydration [34].

10.3 Hydrogel Formation

Crosslinking, grafting polymerization, and radiation crosslinking are methods used to prepare the material. These materials' viscoelasticity and mechanical characteristics may be enhanced for use in the biomedical and pharmaceutical industries by such alterations. Hydrogels may be made in a variety of ways. The first one is physical crosslinking. Physical or reversible gels are becoming more popular because of their simplicity of manufacturing and lack of crosslinking agents. These agents impact the integrity of the chemicals to be entrapped and the requirement to remove them before application. Gel textures may be created by selecting the right hydrocolloid type, concentration, and pH [35]. This field is garnering a lot of interest in the food sector. Hydrogels may be made using several different techniques, including heating and chilling a polymer solution, intermolecular ion interactions, sophisticated coacervation techniques, such as HF, H-bonding, maturation techniques, such as freeze–thawing, and many more [36].

Radiation-mediated crosslinking also has gained increasing interest in recent years. The biocompatibility of the biopolymer is maintained since radiation crosslinking does not need the use of chemical additives. Modifying and sterilizing biopolymers with a specified end-use in biomedicine may be accomplished in a single step, saving time and money [37]. This method uses high-energy sources like electron beams or X-rays to make free radicals in polymers. Whether direct or indirect, radiation action will be determined by the polymer's surroundings and composition (i.e. dilute solution, concentrated solution, solid state). Radiation in paste, aqueous state radiation, and solid-state radiation are all ways that have been described in the literature for obtaining radiationally cross-linked hydrogels [38].

Apart from the crosslinking methods mentioned above, chemical crosslinking can be applied. Here a crosslinking agent is used to bind two polymer chains together, or a monomer is grafted onto the polymer's backbone. Functional groups (such as OH, COOH, and N_2H_2) in natural and synthetic polymers may be cross-linked using crosslinkers like aldehydes (e.g. glutaraldehyde, adipic acid dihydrazide) [35]. There are a variety of ways to create chemically cross-linked hydrogels that are long-term stable. The IPN (polymerization of a monomer within another solid polymer to form an interpenetrating network structure) and hydrophobic interactions (hydrolysis or oxidation of a polar hydrophilic group followed by covalent crosslinking) methods are also used to obtain chemically cross-linked permanent hydrogels [39]. Chemical crosslinkers, chemical grafting, radiation grafting, and radiation crosslinking are some of the techniques mentioned in the literature for obtaining chemically cross-linked hydrogels.

10.4 Recent Advances in Hydrogel Development

Some scientists and engineers are always working to find new ways to improve hydrogels, which results in a wide range of breakthroughs (Figure 10.3). The following hydrogels, among others, have recently been developed to meet urgent societal demands.

10.4.1 3D-printed Hydrogels

Three-dimensional (3D) printing is a computer-aided method for building 3D objects layer by layer. Configuring the code is the initial step in 3D-food printing; as soon as the codes are loaded and the 3D structure is chosen, printing starts. Printed raw materials such as natural polymer hydrogels are popular for use in 3D-printing. Hydrocolloid polymers are very suitable as a substrate for printed meals [40]. For the elderly who have difficulty swallowing, 3D hydrogel technology has been improved to produce soft and ready-to-eat delicacies. The hydrogel's viscosity impacts the food's processability and ultimate texture. Printing food may also be enhanced by mixing proteins and polysaccharides with different degrees of esterification. 3D-printing may fabricate food component encapsulation [41].

10.4.2 Superabsorbent Hybrid Hydrogels

Hydrogels, which are superabsorbent polymers, can absorb and retain massive volumes of water. In comparison, the absorption capabilities of ordinary hydrogels are no more than 100% (1 g/g) [42], which can only absorb 100% of deionized water using these hydrogels. Biocompatible, flexible, superabsorbent hydrogels with tunable characteristics, synthesized by the one-pot synthesis (photopolymerization) approach, were described by Seraphim et al. [43]. Superabsorbent polymers suitable for agricultural use were described by Yang et al. [44]. By using point source irrigation and adding superabsorbent hydrogels to the intermediate layer of the soil root zone, they looked at the impact of hydrogels on moisture distribution during infiltration and evaporation [45].

10.4.3 Conducting Polymer Hydrogels

An entirely new family of materials called conducting polymer hydrogels combines the advantages of both hydrogels and organic conductors. Electrode and ionic carrying phases (electrolyte) may be connected via polymer hydrogels, which are good interfaces between natural and synthetic biological systems. A wide variety of contemporary applications, from energy storage devices like biofuel cells and supercapacitors to molecular and bioelectronics and medical electrodes, have shown encouraging results for conducting polymeric hydrogels [46]. Because of its outstanding processability and ease of casting into thin films and any required forms during their gelation stage, conducting polymer hydrogels have been extensively studied. Micro-patterns may be created by printing conducting polymer hydrogels using inkjet or screen printers. There are several advantages to using hydrogels made from conducting polymers, such as the inherent 3D structured leading frameworks that aid in the movement of ions and molecules [47].

Industrial Corner

- Hydrogels with diverse properties and functionalities have been generated over the years for food component encapsulation and other industrial applications.
- Fine manipulation of the microstructures of hydrogels have been made possible by 3D-printing.
- Micro-patterns may be created by printing conducting polymer hydrogels using inkjet or screen printers.

10.5 Retention and Release Properties

Structures of biopolymer molecules encapsulate themselves inside the gel network. The internal structure of food-grade hydrogels can be dense with small holes or open with large pores [13]. Because smaller molecules or particles can easily be dispersed through the hydrogel matrix, the pores in the granules of the hydrogel have an impact on the environment of food components (Figure 10.4). Depending on the nutraceutical to be encapsulated, it may be necessary to carefully manage the hydrogel beads' interior structure [20]. A change in pore size may also allow an encapsulated material to be released after it has reached a particular environment. An encapsulated nutraceutical may be released from a bead when it swells under appropriate pH and ionic strength or temperature conditions [48].

The retention and release characteristics of hydrogel beads are critical. The nutraceutical should be kept in the beads until discharged into the specific release sites. An encapsulated component is released when a hydrogel bead dissociates owing to physical, chemical, or enzymatic erosion or disintegration [49]. Absorbent and repulsive interactions between food components and the biopolymer molecules that make up the hydrogel network in the beads may also be affected by the electrostatic interaction. (Figure 10.5). This fact implies that proteins may be positively charged if the pH is low enough and vice versa. Using this technology, pH protein release systems were developed [6]. The electrostatic interaction

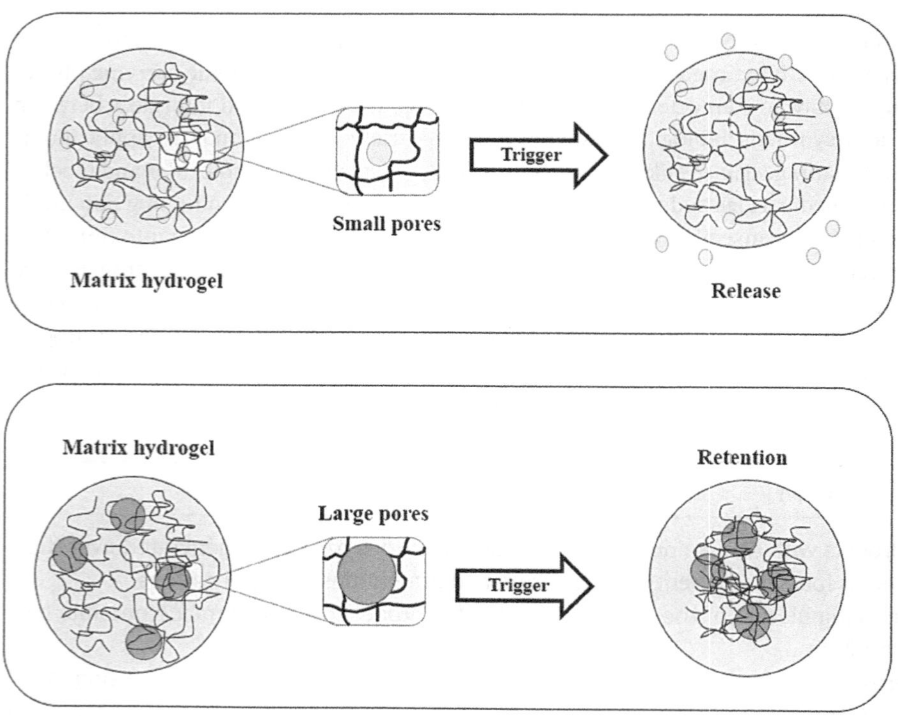

Figure 10.4 Release and retention of nutrients in the hydrogel structure based on size.

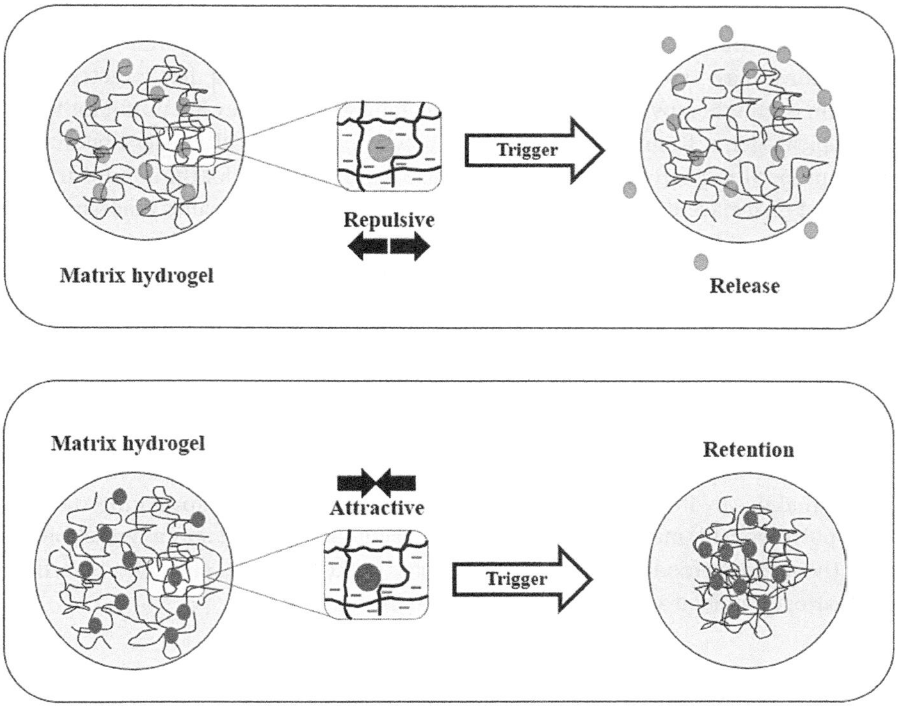

Figure 10.5 Release and retention of nutrients in the hydrogel structure based on electrostatic interactions.

between cationic proteins and anionic alginate molecules in hydrogels keeps hydrogel protein concentrations at pH 3. Specific GIT stimulation may cause the release of nutrients from hydrogel granules. Hydrogel granules may disintegrate under simulated oral circumstances if the pH or temperature of the granules is altered [49].

10.6 Applications of Hydrogels in Food Production

The biocompatibility and high water content of hydrogels make them desirable materials in medical applications. It is possible to add biological molecules into these systems to control the required biological reactions. Hydrogels have been cited as the first biomaterials rationally created for medicinal use [50]. As a result of their ability to be altered independently, polymeric chains and cross-links may be used in various applications. As time goes on, the cross-links in hydrogels may be designed to display qualities that allow the hydrogels to disintegrate. Hydrogel cross-links may be engineered to break down by hydrolysis, enzymatic activity, or a combination of both [51]. Hydrogels are ideal for minimally intrusive applications because they may be generated under moderate pH and temperature settings in aquatic environments. Water-soluble monomers and crosslinkers may be used as hydrogel precursors and photopolymerized in situ as hydrogel precursors under physiological pH and temperature conditions or macromers can self-assemble to create hydrogels.

As an alternative, biomolecules or even organisms may be included in hydrogels that have been mildly cross-linked using stimuli-sensitive polymers [52]. Temperature-sensitive molecular chains may transform a gel-like substance into a solid for targeted drug administration or cell therapy when they reach body temperature. They are useful for transporting biological agents, from basic biomolecules to whole cells and cell clusters, because of their unique properties. Therefore, it is possible to use these hydrogels for the administration of encapsulated pharmaceuticals and the transport of cells and/or molecules that have been enclosed inside the hydrogel environment [53]. When making hydrogels in vivo, the equilibrium swelling of the material has to be taken into consideration. The hydrogel may not always swell evenly and uniformly in different environments and times. The hydrogel's initial volume may fluctuate over time in certain scenarios. When hydrogels are employed in reconstructive surgery, this might have major ramifications (e.g. hydrogel materials containing cells and molecules intended to fill a tissue defect) [54]. While hydrogels have been used extensively in biomedical settings, they have also been widely exploited in food applications. Hydrogels for food component encapsulation can serve several functions: (i) preventing degradation; (ii) making hydrophobic substances soluble in water by encapsulating them in hydrophilic polymers; (iii) masking the strong flavor of compounds that may adversely affect food flavor; (iv) avoiding food component interactions. In the following parts of this section, the use of hydrogels in food applications will be presented.

10.6.1 Encapsulation of Bioactive Compounds

Due to their poor water solubility and chemical instability, carotenoids, such as β-carotene, lutein, lycopene, zeaxanthin, and astaxanthin, have struggled to achieve adequate oral bioavailability for usage as nutrients in meals. Recent research shows that encasing carotenoid-rich lipid droplets in calcium alginate beads improves their chemical stability [49]. Delivery methods are required to preserve carotenoids in meals while releasing them into the gastrointestinal tract (GIT). The chemical stability of encapsulated beta-carotene inside the GIT was most excellent in hydrogel beads, with 1% beads providing superior protection against degradation than 0.5% beads, according to Zhang et al. [56]. The ability of hydrogels to prevent excessive droplet flocculation in the stomach and small intestine, allowing lipase to very easily access the lipid droplet surfaces, was recently shown in some investigations [57]. Another research confined carotenoid-rich lipid droplets within starch-trimetaphosphate hydrogel beads. The hydrogel beads kept the carotenoids in the mouth and stomach, releasing them in the small intestine. Because specific carotenoids degrade quickly at low pH, this property may help protect carotenoids from chemical degradation in the stomach's severely acidic environment [58]. Hamdi et al. used a new blue crab chitosan–protein composite hydrogel enhanced with carotenoids [59]. Bera et al. suggested that future drug delivery applications might benefit from incorporating bacterial canthaxanthin into a polymeric hydrogel [49].

Curcumin is another example that can demonstrate the use of hydrogels in encapsulating bioactive compounds. It is a bioactive compound present in turmeric that has been demonstrated to have antioxidant, antibacterial, anti-inflammatory, and anti-cancer properties. Curcumin's low water solubility and poor chemical stability restrict its usage as a nutraceutical additive. According to Zhang et al., lipase must digest lipid droplets to release encapsulated curcumin and create mixed micelles capable of solubilizing released curcumin [4]. In another study, Yan et al. proposed that scallop (*Patinopecten yessoensis*) male gonad hydrolysates and carrageenan hydrogels with biocompatibility and acid tolerance might be trustworthy and desired soft materials for embedding and delivering curcumin to the colon [19]. Curcumin's limited water solubility may be addressed by encapsulating it in lipid-based carriers. Gunathilake et al. think that nonionic surfactant integrated chitosan/nanocellulose hydrogel may address curcumin delivery issues [48]. The same research found that the curcumin release from the beads was connected to their capacity to influence lipid digestion [4]. Curcumin-loaded lipid droplets could also be kept within calcium alginate beads, and the rate at which they released curcumin was influenced by the hydrogel hole size and the alginate and calcium concentrations [60]. The rate of lipid droplet release increased when calcium, alginate, and droplet size decreased. Xu et al. found curcumin released in a diffusion-controlled model [20].

10.6.2 Encapsulation of Fats and Oils

Lipid oxidation readily degrades polyunsaturated fatty acids (PUFAs), and results in rancid smells and potentially hazardous reaction products. Consequently, the potential of hydrogel beads to protect PUFA-rich lipid droplets against oxidation has received much attention. Table 10.2 covers numerous hydrogel-based polyunsaturated fatty acid carriers [5, 21, 61–73]. Lipid droplets encapsulated in caseinate–pectin hydrogels showed superior oxidation resistance to nonencapsulated lipid droplets due to the high concentration of hydrogels around the fat droplets, increasing their oxidative stability. Recent research found that these hydrogel beads might be utilized to add PUFAs to meat products.

Besides being used to encapsulate fats, hydrogels can be used to encapsulate essential oils or even herbal extracts. Herbal goods include herbs used as food or food additives, dietary supplements, and traditional treatments. Encapsulation might benefit herbal goods by controlling distribution, prolonging shelf life, isolating incompatible chemicals, and enhancing final product quality [74]. Encapsulation may also improve product stability by protecting active chemicals from the air, moisture, and incompatible substances. In the case of liquid herbal concentrates, absorption might be achieved by employing pre-prepared hydrogel beads [75]. Encapsulating essential oils in the hydrogel is a great way to solve issues associated with their usage. They were encouraging the bioactivity of antibacterial drugs by targeting them where microorganisms preferentially reside. Essential oils used as food preservatives were encapsulated in various ways. Here are some examples of plant extracts and essential oils encapsulated in hydrogel [75]. Table 10.2 lists numerous hydrogel-based herbal extracts and essential oil carriers.

Industrial Corner

- Encapsulation of food components is the most exciting area of hydrogel research.
- Hydrogels have been used to encapsulate diverse types of food components, including, but not limited, to essential oils and bioactive compounds.

The biocompatibility and high water content of hydrogels make them desirable materials in food industrial applications.

Table 10.2 Applications of hydrogel-based carriers on polyunsaturated fatty acids, herbal extracts, and essential oils encapsulation.

Polymer	Nutrient compound	Characterization	References
Polyacrylic acid	ω-3/ ω-6 polyunsaturated fatty acids	Good particle size stability and good chemical stability of the actives and excipients	[61]
Low methoxy pectin and caseinate	Fish oil	Protect fish oil from oxidation during storage	[62]
Pectin and sodium caseinate	Fish oil	Good oxidative stability over the storage time	[63]
Cress seed mucilage and chitosan	Fish oil	Postpone fish oil oxidation	[64]
Gelatin	Eicosapentaenoic acid	Prevent osteoarthritis progression in vivo	[65]
κ-Carrageenan	Grapefruit seed extract	Antibacterial solid activity and excellent wound healing effect	[5]
κ-Carrageenan and β-cyclodextrin	Honey bee propolis extract	High potential in the wound healing fabrics containing hydrogels	[66]
Polyvinyl alcohol, hydroxypropyl chitosan, and carbomer	*Periplaneta americana* extract	Accelerated wound healing process in vivo	[67]
Icodextrin and dextran	*Turkish galls* extract	Gradual degradation and absorption	[68]
Methacrylated chitosan	*Punica granatum* peel extract	Acceptable cytotoxicity against human fibroblast cells	[69]
Methylcellulose	*Melissa officinalis* essential oil	Inhibited *Candida albicans* growth as evident	[70]
Gelatin and poly(vinyl alcohol)	*Zataria multiflora* essential oil	Increased the antioxidant and antibacterial activities of the dispersions	[21]
Cellulose	*Clove* essential oil	Reasonable cell viability rates (74–101%) to human gingival fibroblast cells	[71]
Methyl-β-cyclodextrin and soy soluble polysaccharide	*Satureja montana L.* essential oil	Reduce the microbial count	[72]
Gelatin	*Eupatorium adenophorum* essential oil	Inhibited the growth of the tested pathogens	[73]

10.7 Conclusions

Hydrogels have been classified based on their physical and chemical characteristics throughout the past 20 years. They can be made in various ways, but the preparation method and the design process have a significant impact on the final result. Polymers that react to diverse stimuli (physical, chemical, and biological) must be discovered in the future. In a short span of time, hydrogels absorb a lot of water or aqueous fluids. Hydrogels with improved durability, mechanical characteristics, and biocompatibility must be miniaturized in this era of nanofabrication for novel applications. For the next several decades, the key objective will be to meet the practical needs while minimizing the complexity of the hydrogel composition. To address the special needs of modern, high-tech delivery systems, greater emphasis should be paid to future research. When referring to hydrogel systems, "green" refers to their processing methods, including nontoxic crosslinkers or low- or no-solvent solutions. Research into new hydrogels is ongoing, but a new conceptualization of hydrogel production may lead to specialized features translating into unique applications in a wide range of disciplines. The most pressing requirement will be to find a cost-effective solution to increase the effectiveness of the hydrogel system in food applications.

Glossary

3D-printing A variety of processes in which a material is deposited, joined, or solidified under computer control to create a three-dimensional object.

Biodegradable Able to decay naturally and in a way that is not harmful.

Carotenoid A class of naturally occurring pigments synthesized by plants, algae, and photosynthetic bacteria.

Conducting polymer A class of organic materials with unique electrical and optical properties similar to those of inorganic semiconductors and metals.

Cross-link A bond or a short sequence of bonds that link one polymer chain to another.

Curcumin (1,7-Bis(4-hydroxy-3-methoxyphenyl)-1,6-heptadiene-3,5-dione), also called diferuloylmethane, is the primary natural polyphenol found in the rhizome of *Curcuma* spp.

Drug delivery Approaches, formulations, manufacturing techniques, storage systems, and technologies involved in transporting a pharmaceutical compound to its target site to achieve a desired therapeutic effect.

Encapsulation Process of stabilization of active compounds through the structuring of systems capable of preservation of their properties.

Essential oil Concentrated hydrophobic liquid containing volatile chemical compounds extracted from plants.

Food component Essential nutrients like carbohydrates, proteins, fats, vitamins, and minerals.

Hydrogel Three-dimensional (3D) network structures that can imbibe large amounts of water.

Polysaccharide Polymers consisting of chains of monosaccharide or disaccharide units joined by glycosidic bonds.
Polyunsaturated fatty acid Fatty acids that contain more than one double bond in their backbone.
Stimuli State of responsiveness to sensory stimulation or excitability.
Swelling Enlargement of a structure due to water absorption and retention.
Superabsorbent Substances that can absorb up to several times their weight in aqueous fluids.

References

1 S. Mondal, S. Das, A. K. Nandi, *Soft Matter* **2020**, *16*, 1404.
2 M. C. Catoira, L. Fusaro, D. Di Francesco, M. Ramella, F. Boccafoschi, *J. Mater Sci. Mater Med.* **2019**, *30*, 1.
3 S. Aswathy, U. Narendrakumar, I. Manjubala, *Heliyon* **2020**, *6*, e03719.
4 Z. Zhang, R. Zhang, L. Zou, L. Chen, Y. Ahmed, W. Al Bishri, K. Balamash, D. J. McClements, *Food Hydrocoll.* **2016**, *58*, 160.
5 M. Ekrami, A. Ekrami, M. A. Hosseini, Z. Emam-Djomeh, *Mol.* **2022**, *27*, 7032.
6 L. G. L. Nascimento, F. Casanova, N. F. N. Silva, A. V. N. de Carvalho Teixeira, A. F. de Carvalho, *Food Chem.* **2020**, *314*, 126063.
7 M. Antman-Passig, S. Levy, C. Gartenberg, H. Schori, O. Shefi, *Tissue Eng. Part A* **2017**, *23*, 403.
8 M. Shakouri, M. Salami, L.-T. Lim, M. Ekrami, M. Mohammadian, G. Askari, Z. Emam-Djomeh, D. J. McClements, *J. Food Meas. Charact.* **2022**, 1.
9 N. Ni, M.-J. Dumont, *Waste Biomass Valorization* **2017**, *8*, 285.
10 B. Biduski, W. M. F. da Silva, R. Colussi, S. L. D. M. El Halal, L.-T. Lim, A. R. G. Dias, E. da Rosa Zavareze, Int. *J. Biol. Macromol.* **2018**, *113*, 443.
11 M. Ekrami, Z. Emam-Djomeh, P. Joolaei-Ahranjani, S. Mahmoodi and S. Khaleghi, *Int. J. Biol. Macromol.* **2021**, *168*, 591.
12 M. Ekrami, Z. Emam-Djomeh, *J. Food Process. Preserv.* **2014**, *38*, 1812.
13 A. Ajovalasit, M. A. Sabatino, S. Todaro, S. Alessi, D. Giacomazza, P. Picone, M. Di Carlo, C. Dispenza, *Carbohydr. Polym.* **2018**, *179*, 262.
14 R.-J. Mu, Y. Yuan, L. Wang, Y. Ni, M. Li, H. Chen, J. Pang, *Food Hydrocoll.* **2018**, *76*, 42.
15 C. Wu, J. Zhao, F. Hu, Y. Zheng, H. Yang, S. Pan, S. Shi, X. Chen, S. Wang, *Carbohydr. Polym.* **2018**, *180*, 112.
16 M. Zhang, X. Zhao, *Int. J. Biol. Macromol.* **2020**, *1*, 1414.
17 D. Hua, S. Gao, M. Zhang, W. Ma, C. Huang, *Carbohydr. Polym.* **2020**, *247*, 116743.
18 T. T. Demirtaş, G. Irmak, M. Gumuşderelioğlu, *Biofabrication* **2017**, *9*, 035003.
19 M. Ekrami, A. Ekrami, R. H. Moghadam, P. Joolaei-Ahranjani, Z. Emam-Djomeh, *Biopolymers in Nutraceuticals and Functional Foods, RSC* **2022**, 488.
20 W. Xu, L. Huang, W. Jin, P. Ge, B. R. Shah, D. Zhu, J. Jing, *Int. J. Biol. Macromol.* **2019**, *134*, 210.
21 G. Kavoosi, Z. Bordbar, S. M. Dadfar, S. M. M. Dadfar, *J. Appl. Polym. Sci.* **2017**, *134*, 45351.

22 S. Bashir, M. Hina, J. Iqbal, A. Rajpar, M. Mujtaba, N. Alghamdi, S. Wageh, K. Ramesh, S. Ramesh, *Polymers* **2020**, *12*, 2702.

23 M. S. Nejat, M. Ekrami, Z. Emam-Djomeh, *Biopolymers in Nutraceuticals and Functional Foods, RSC* **2022**, 232.

24 M. Ekrami, A. Ekrami, R. Esmaeily, Z. Emam-Djomeh, *Biopolymers in Nutraceuticals and Functional Foods, RSC* **2022**, 545.

25 J. Yang, R. Bai, B. Chen, Z. Suo, *Adv. Funct. Mater.* **2020**, *30*, 1901693.

26 N. A. Peppas, A. S. Hoffman, *Biomaterials Science*, Elsevier, 2020, 153.

27 A. S. Hoffman, *Adv. Drug Deliv. Rev.* **2012**, *64*, 18.

28 C. Yan, A. Altunbas, T. Yucel, R. P. Nagarkar, J. P. Schneider, D. J. Pochan, *Soft Matter* 2010, *6*, 5143.

29 R. Takahashi, K. Shimano, H. Okazaki, T. Kurokawa, T. Nakajima, T. Nonoyama, D. R. King, J. P. Gong, *Adv. Mater. Interfaces*. **2018**, *5*, 1801018.

30 S. Hamd-Ghadareh, A. Salimi, S. Parsa, S. J. Mowla, *Biosens. Bioelectric.* **2022**, *199*, 113895.

31 M. Zignani, C. Tabatabay, R. Gurny, *Adv. Drug Deliv. Rev.* **1995**, *16*, 51.

32 T. Zhou, X. Gao, B. Dong, N. Sun, L. Zheng, *J. Mater. Chem. A* 2016, *4*, 1112.

33 Z. Gao, L. Kong, R. Jin, X. Liu, W. Hu, G. Gao, *J. Mater. Chem. C* 2020, *8*, 11119.

34 S. Gallagher, L. Florea, K. J. Fraser, D. Diamond, *Int. J. Mol. Sci.* **2014**, *15*, 5337.

35 J. Maitra, V. K. Shukla, *Am. J. Polym. Sci* 2014, *4*, 25.

36 L. Rebers, R. Reichsöllner, S. Regett, G. E. Tovar, K. Borchers, S. Baudis, A. Southan, *Sci. Rep.* **2021**, *11*, 1.

37 L. Lu, S. Yuan, J. Wang, Y. Shen, S. Deng, L. Xie, Q. Yang, *Curr. Stem. Cell Res. Thera.* **2018**, *13*, 490.

38 M. Ekrami, N. Roshani-Dehlaghi, A. Ekrami, M. Shakouri, Z. Emam-Djomeh, *Chem.* **2022**, *4*, 1360.

39 C. Echalier, L. Valot, J. Martinez, A. Mehdi, G. Subra, *Mater. Today Communed.* **2019**, *20*, 100536.

40 J. Li, C. Wu, P. K. Chu, M. Gelinsky, *Mater. Sci. Eng. R Rep.* **2020**, *140*, 100543.

41 S. M. Park, H. W. Kim, H. J. Park, *J. Food Eng.* **2020**, *271*, 109781.

42 I. Călina, M. Demeter, A. Scărișoreanu, M. Micutz, *Gels* 2021, *7*, 189.

43 S. Sharma, S. Tiwari, *Int. J. Biol. Macromol.* **2020**, *162*, 737.

44 M. Yang, J. Wu, G. M. Graham, J. Lin, M. Huang, *Front. Chem.* **2021**, 9.

45 S. Behera, P. A. Mahanwar, *Polym.-Plast. Technol. Mater.* **2020**, *59*, 341.

46 J. Stejskal, *Chem. Pap.* **2017**, *71*, 269.

47 Z. Ma, W. Shi, K. Yan, L. Pan, G. Yu, *Chem. Sci.* **2019**, *10*, 6232.

48 T. M. S. U. Gunathilake, Y. C. Ching, C. H. Chuah, H. A. Illias, K. Y. Ching, R. Singh, L. Nai-Shang, *Int. J. Biol. Macromol.* **2018**, *118*, 1055.

49 M. Mirzakhani, M. Ekrami, S. Moini, *JFBE* **2018**, *1*, 8.

50 A. Ekrami, M. Ghadermazi, M. Ekrami, M. A. Hosseini, Z. Emam-Djomeh, R. Hamidi-Moghadam, *J. Drug Deliv. Sci. Technol.* **2022**, *69*, 103148.

51 S. J. Bryant, F. J. Vernerey, *Adv. Healthcare Mater.* **2018**, *7*, 1700605.

52 C. M. Madl, S. C. Heilshorn, *Chem. Mater.* **2019**, *31*, 8035.

53 O. Hasturk, K. E. Jordan, J. Choi, D. L. Kaplan, *Biomaterials* **2020**, *232*, 119720.

54 A. L. Lee, S. Venkataraman, C. H. Fox, D. J. Coady, C. W. Frank, J. L. Hedrick, Y. Y. Yang, *J. Mater. Chem. B* **2015**, *3*, 6953.

55 S. Afzal, M. Maswal, A. A. Dar, *Colloids Surf B: Biointerfaces* 2018, *169*,99.
56 Z. Zhang, R. Zhang, D. J. McClements, *Food Hydrocoll.* **2016**, *61*, 1.
57 S. Mun, Y.-R. Kim, D. J. McClements, *Food Chem.* **2015**, *173*, 454.
58 S. Wang, X. Chen, M. Shi, L. Zhao, W. Li, Y. Chen, M. Lu, J. Wu, Q. Yuan, Y. Li, *Int. Food Res. J.* **2015**, *67*, 315.
59 M. Hamdi, A. Feki, S. Bardaa, S. Li, S. Nagarajan, M. Mellouli, T. Boudawara, Z. Sahnoun, M. Nasri, R. Nasri, *Mater. Sci. Eng. C* 2020, *113*, 110978.
60 B. Zeeb, A. H. Saberi, J. Weiss, D. J. McClements, *Food Hydrocoll.* **2015**, *50*, 27.
61 S. Mangiafico, D. Aleo, M. Saita, M. Cro, *Invest. Ophthalmol. Vis. Sci.* **2010**, *51*, 1982.
62 Z. Zhang, E. A. Decker, D. J. McClements, *Int. Food Res. J.* **2014**, *64*, 520.
63 L. Salcedo-Sandoval, S. Cofrades, C. Ruiz-Capillas, A. Matalanis, D. J. McClements, E. A. Decker, F. Jiménez-Colmenero, *Food Chem.* **2015**, *184*, 207.
64 H. R. Kavousi, M. Fathi, S. A. Goli, *Int. J. Food Prop.* **2017**, *20*, 1890.
65 M. Tsubosaka, S. Kihara, S. Hayashi, J. Nagata, T. Kuwahara, M. Fujita, K. Kikuchi, Y. Takashima, T. Kamenaga, Y. Kuroda, *J. Orthop. Res.* **2020**, *38*, 2157.
66 S. Sharaf, M. E. El-Naggar, *Int. J. Biol. Macromol.* **2019**, *133*, 583.
67 Z. Chen, Y. Hu, J. Li, C. Zhang, F. Gao, X. Ma, J. Zhang, C. Fu, F. Geng, *Int. J. Pharm.* **2019**, *571*, 118707.
68 X. Li, B. Zou, N. Zhao, C. Wang, Y. Du, L. Mei, Y. Wang, S. Ma, X. Tian, J. He, *ACS Appl. Mater. Interfaces* **2018**, *10*, 24469.
69 T. I. Ragab, A. A. Nada, E. A. Ali, A. A. Soliman, M. Emam, M. A. El Raey, *Int. J. Biol. Macromol.* **2019**, *135*, 407.
70 E. Serra, F. Saubade, C. Ligorio, K. Whitehead, A. Sloan, D. W. Williams, A. Hidalgo-Bastida, J. Verran, S. Malic, *Microorganisms* **2020**, *8*, 215.
71 R. R. Huerta, E. K. Silva, T. El-Bialy, M. D. Saldaña, *Ultrason Sonochem.* **2020**, *64*, 104845.
72 H. Cui, Y. Wang, C. Li, X. Chen, L. Lin, *Lwt* **2021**, *152*, 112427.
73 P. Chuysinuan, N. Chimnoi, N. Reuk-Ngam, P. Khlaychan, A. Makarasen, N. Wetprasit, D. Dechtrirat, P. Supaphol, S. Techasakul, *Polym. Bull.* **2019**, *76*, 701.
74 M. Ekrami, Z. Emam-Djomeh, S. A. Ghoreishy, Z. Najari, N. Shakoury, *Int. J. Biol. Macromol.* **2019**, *133*, 529.
75 E. Yahya, M. A. Abdulsamad, *J. Appl. Sci. Process Eng.* **2020**, *7*, 564.

11

Optimization of Pasting and Textural Properties of Food Products

*Filopateer Nasser, Salma Hossam Mohamed, Mariam Ashraf Fouad Khalil, Omaima Ali Mostafa Mohammed, Radwa Magdy Mohamed, and Heba Mohamed Fahmy**

Faculty of Science, Cairo University, Egypt
** Corresponding author*

Highlights

- Pasting and textural properties significantly affect sensory attributes of the generated food product.
- Properties of starch can be manipulated using multiple approaches, including physical means, chemical treatment, and enzymatic treatment.
- Gum arabic can be used as a stabilizer and thickness control specialist in the food industry.
- Acetylation process reduces starches' pasting temperature, which helps in cooking.
- Modified starch may show changes in the pasting temperature, peak and final viscosity, and peak time when compared with native starch.

11.1 Introduction

Pasting properties comprise pasting temperature and peak viscosity, changes in the food due to water with heat. At the same time, the hardness determines textural properties (swell of granules of samples), cohesiveness (stand of sample to deformations), springiness (flexibility of food product), and chewiness (amount of energy required to break up and chew semi-solid foods) [1–3]. The modification in pasting and texture properties in some products produces and develops new features in these products, which help the industry. For example, starch is the most important element penetrating different food products. The starch of wheat and its products, such as flour, oils, and fats, must be considered; sometimes, some starches are resistant to enzymatic hydrolysis, which results in a lack of functional properties.

Materials Science and Engineering in Food Product Development, First Edition. Edited by Wing-Fu Lai.

In addition, some starch granules are inert and insoluble in water at room temperature [1–3]. So, the modification process needs to be applied to native starches to develop certain properties, such as solubility, texture, adhesion, and tolerance to the heating temperatures used in industrial processes [4]; wheat, with its types and its product, contains starch with different percentages, so the addition of different fats and oils affects hard wheat's hardness and adhesiveness more than soft wheat related to the high percentage of amylose content for hard wheat [5]. High amylose content leads to the hard texture because the conformation of amylose swells less than amylopectin, making the texture of the final food product hard [5], while gumminess, springiness, cohesiveness, and chewiness of soft wheat are affected more by the addition of fats and oils [5]. For example, addition of groundnut oil reduces the hardness of hard wheat flour more than that of soft wheat, which can happen due to the addition of fats and oils, which leads to the formation of an amylose–lipid complex, which is supposed to make an insoluble film on the outside of granules that leads to distribution of amylose chains so the strength will decrease [5]. Also, chewiness and gumminess will decrease after adding oils and fats, while springiness will increase [5].

When fats and oils are added, especially coconut oil and hydrogenated fat, to wheat flour that contains starch, it generally improves pasting temperature, which is the temperature when viscosity begins to rise as indicated [5]. Increasing pasting temperature indicates an interaction between the starch–lipid complex and amylose chains with fats and oils, which prevents starch's hydration and swelling, leading to a delay in viscosity [5]. Higher pasting temperature is required for good cookies because it improves gelatinization capacity, binds to water, and has less swelling properties [5, 6]. In the following parts of this chapter, we will discuss different ways to manipulate the pasting and textural properties of starch and the recent applications of these ways on different examples of starch.

11.2 Physical and Chemical Modification of Starch Structures

There are different ways to modify the structure and properties of starch (Figure 11.1 and Table 11.1) [7–19]. For example, Zheng et al. applied the ultrasound treatment on sweet potato starch and reported a reduction in its viscosity [11]. They found that the peak

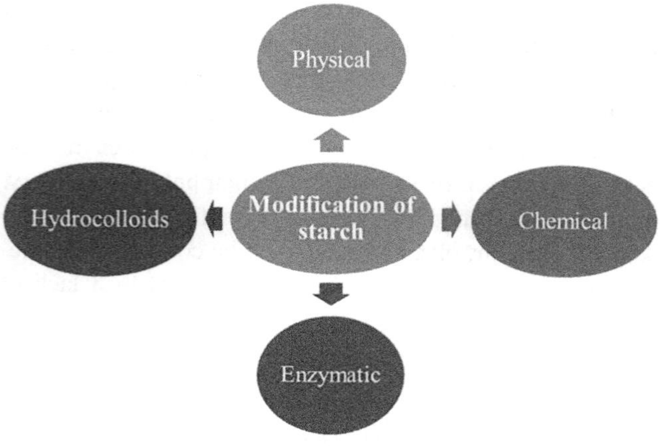

Figure 11.1 A scheme shows different methods to modify starch.

Table 11.1 Shows the different physical methods for starch modification.

Method	Description
Ultrasound treatment	The acoustic energy released by ultrasound causes cavitation forces, which induce the collapse of starch granules, making them more permeable to water during the heating step [7–9]. The effect of ultrasound treatment on the pasting properties of starch depends on treatment conditions as treatment duration [8, 10, 11].
Heat-moisture treatment	Hydrothermal treatment. It depends on changing the physicochemical properties of starch while saving its granular structure [12–15]. It is applied by heating starch at a temperature above 94^{0}C above its glass transition temperature and moisture contents (10–40%, w/w) [16, 17].
Annealing	Hydrothermal treatment. It is applied in excess water (60%) with a temperature between 5°C above its glass transitions temperature and below 15°C its gelatinization temperature [12, 15].
High-pressure treatment	Starch characteristics (structure, retrogradation, and chemical reactions) differ, depending on the treatment pressure, pressurization temperature, and pressurization time [10, 13].
Pulsed electric field treatment	Nonthermal method. The samples are exposed to high-intensity electric pulses for a short duration, as the duration needs to be short to prevent unwanted electrolytic reactions [18, 19].

viscosities of starch treated with ultrasound frequencies of 25 and 80 kHz were 6.94% and 9.84%, which is lower than that of native starch. That is due to the occurrence of cleavage of glycosidic linkages that leads to a weakness of the polymer network [4, 7, 12, 20, 21]. While Chan et al. and Carmona-Garcia et al. studied, they focused on the effect of the ultrasound treatment time on the starch [22, 23]. They concluded that the longer the ultrasound treatment time, the more the increase in the peak viscosities of starch, as the starch granules decompose and absorb more water, causing more swelling of granules and more viscous starch [17, 23, 24]. In addition, Subroto et al. applied the HMT technique to potato starch and studied its effect on the pasting characters of the starch. They heated the starch at 80°C, 90°C, 100°C, 110°C, and 120°C for 16 hours [25]. They found that the higher temperature of HMT decreased peak viscosity and hold viscosity but increased pasting point and setback viscosity [25]. They concluded that the HMT potato starch could be suitable for sterilization products as it was resistant to heating [25]. However, in Chen et al.'s study, they used the PEF technique. They investigated the pasting properties of potato starch after exposing it to an esterification process by using octenyl succinic anhydride OSA with an assistant of PEF treatment [26]. When their modified starch was compared with native starch, it showed induced lower pasting temperature and paste viscosity. Their experiment has approved that PEF can be used to produce starch products with desired pasting properties [26]. Recently, Yu et al. studied the effect of annealing on the textural properties of corn starch [27]. When they annealed corn starch at 45°C, the textural properties increased slightly, whereas when annealed at 55°C, those properties decreased [27]. They also studied the effect of additives, like sucrose, glucose, and sodium chloride, on annealed corn starch gels, as the strength of the annealed corn starch at 45°C increased with using 10% sucrose or 2% sodium chloride, while the strength of the annealed corn starch at 55°C increased with all [27].

Industrial Corner

- The effect of ultrasound treatment on the pasting properties of starch depends on treatment conditions as treatment duration.
- Applied annealing treatment caused the reduction of amylose leaching and granular swelling depending on the starch source.
- The modified potato starch, which assists in pulsed electric field treatment, showed lower pasting temperature and paste viscosity than the native starch.

Besides using physical methods, chemical modification of starch can be achieved [28–30]. This is accomplished by adding a new functional group into the polymer molecule of the starch granule in its natural state [28]. It produces reflective modification in the pasting of native starch granules, and it helps stabilize inter- and intramolecular bonds at different locations and positions [28]. There are many techniques to carry out chemical modification, such as acetylation, crosslinking, oxidation, cationization, and acid hydrolysis, but some of these methods may be limited due to the safety of both consumers and the environment (Table 11.2); this leads to having an increasing demand for mixing several types of chemical methods to produce new types of modification [28]. Acetylation is one of the commonest approaches to modifying starch. The acetylation process reduces the pasting temperature of starch, which helps in cooking as adding products with sensitivity to high temperatures to the acetylated starch gives more stable products [31]. It also gives starch more resistance versus heating and shearing [32]. Also, because of steric hindrance, acetylation of starches causes a structural reconfiguration. This results in starch molecules repelling each other; thus, increasing water percolation inside amorphous zones of granules, which needs an increase in swelling capacity, is made easier [33]. After the substitution, starch can be divided into three categories: starches with a low degree of substitution (DS) about 0.01–0.2 can act as adhesion, thickening, texturizing, film-forming, stabilizing, and binding agents as they have been used in frozen foods, baked goods, baby foods, sauces,

Table 11.2 Major chemical methods for starch modification.

Methods	Description
Acetylation	Replacing part of the hydroxyl groups of glucose molecules with acetyl groups by using acetic anhydride with an alkaline catalyst like sodium hydroxide [29].
Cationization	Replacing the hydroxyl group of the native starches by using functional groups like ammonium, amino, or sulfonium to introduce a positive ionic charge in one of the three ways: wet, dry, or intermediate process with 2,3-epoxypropyltrimethylammonium chloride ETA as a cationic monomer [30].
Crosslinking	Different crosslinking agents, such as sodium trimetaphosphate, sodium tripolyphosphate, epichlorohydrin, and phosphoryl chloride, chemically modify native starches. For food uses, sodium trimetaphosphate, sodium tripolyphosphate, and phosphoryl chloride are adipic acid and acetic anhydride mixtures [28].

snack foods, and salad dressings [28]. At the same time, starches with intermediate DS (0.2–1.5) and high DS (1.5–3.0) are used as thermoplastic materials, as they have high solubility in chloroform and acetone [34].

In the cationization process, the wet technique is the most widely employed among the three techniques used, as it is safe for the environment [30]. However, there is a possibility that a by-product will be formed alongside the targeted product, which may be costly to separate [30]. The dry cationization method produces more dense starch, while the wet approach produces many higher molar masses [35]. The cationization of starch reduced the pasting temperature and showed higher peak viscosities [36]. On the other hand, the crosslinking process helps the starch break down less during agitation, cooking, or exposure to low pH, which makes it useful in thickening and stabilizing the food system in the manufacturing of sauces, soups, gravies, infant meals, deep-fried dishes, fruit fillings, and pudding [28]. Among the chemicals used in that process, sodium trimetaphosphate, which is one of the essential food additives with low toxicity, is an excellent crosslinking agent [37].

Starch can be modified dually. In other words, it is modified by an approach involving a combination of chemical modifications and different types of modifications, like a combination of physical and chemical modifications, or it may involve two chemical modifications [4]. Dual modification optimizes modified starch functionality to be used industrially [4]. Starches can be modified by two chemical processes, such as emulsifiers, agglutinants, and thickeners, which are widely employed in the food industry, while they can also be used as heavy metal adsorbents in the nonfood industry [38]. One example of dual modification is chemical alteration, which involves oxidation and alkali treatment. Sodium hypochlorite was used as an oxidizing agent during the modification process, while sodium hydroxide was used as an alkaline agent [39]. It was carried out on commercial maize starch, causing various surface damage to the granules, making them more sensitive to enzymatic hydrolysis [39]. Another example is dual chemical modifications, which involve acid hydrolysis and hydroxypropylation. In Javadian et al.'s study [40], they investigated the effects of the dual modification on different properties of tapioca starch (TS). At first, they hydrolyzed TS using HCl and then hydroxypropylated it by using propylene oxide [40]. The results of their study showed lower pasting temperature, peak, and final viscosity, and peak time when compared with native starch. Even if their study had shown lower pasting parameters, it showed the possibility of using this dually modified starch in applications like coating and dip molding [40].

Industrial Corner

- Starch modified by substitution can be categorized with low, intermediate, and high degrees of substitution.
- The cationization of starch reduced the pasting temperature and showed higher peak viscosities.
- Starch can be modified chemically using crosslinking agents, such as sodium trimetaphosphate.
- Dual modification is physical and chemical or may involve two chemical modifications.

11.3 Manipulation of Starch Properties Using Hydrocolloids

Besides the physical and chemical modification, hydrocolloids can be used as an alternative way to enhance the characteristics of native starch, as they can interact with starch effectively. This method of modification is eco-friendly [41–43]. Improvement of the properties of starch by using this method can be exemplified by gum arabic (GA)-modified TS. TS is normally utilized as a condensing and gelling specialist in numerous diet items [44]. TS frames a reasonable, tacky, and strong glue with tasteless character and is utilized in various food and industrial applications [44–47]. Overall, local starches do not have goods that make them unmistakably appropriate for usage in food items. Normally, starch is adjusted via derivatization to work on its functionalities before being used in handled food forms [44]. Also, it is remarkable that expansion of gum/hydrocolloids' labors on starch's advantageousness via altering its rheological/key and gelatinization properties [48, 49]. Moreover, gum can enrich starch's dampness, water-sustaining limit, and generally point to quality devoid of much greasy worth [50–53]. It is very cost-effective to use custard starch than starch from other sources [44]. Due to its unique adhesiveness, which has an unappealing flavor and a relatively thick consistency, in its gelatinized structure, TS has been widely employed in the food industry, including bread, cakes, ejected oats, beverages, sauces, and pie fillings [45]. Local (unmodified or unadulterated) starches have numerous unfortunate properties, for example, low water-sustaining limit, retrogradation, and syneresis [54, 55]. The local starch's unfortunate properties can be righted both via adjustment or by enlarging gum (food hydrocolloids) [44].

Gum has benefits over the adjustment cycle because it is fatless, just as it has better surface retainment and steadiness throughout handling [56]. These benefits have provoked the inescapable starch and gum combinations in numerous diet plans [57–59]. The impact of thickener expansion on TS expansion, gelatinization, and freeze–defrost stability was studied by Chaisawang and Suphantharika [45]. They discovered that TS particles were tightly coated in thickeners [45]. This resulted in their delayed expansion and gelatinization hindrance. In an unusually late evaluation, GA expansion into a TS suspension exhibited a decline in expanding power. According to the study, the dissolvability file and pinnacle consistency can be used as a stabilizer and thickness control specialist in the food industry [60–62]. Although altering these food macromolecules through physical (warm/nonthermal handling) and material approaches can also achieve some of the aforementioned objectives, these strategies are not savvy [63–67]. GA is frequently used as a stabilizer in the food industry [44]. It is a weak polyelectrolyte that transports carboxyl groups and can be used as a dietary fiber source [68]. The ability of GA to expand to potato starch has been demonstrated to reduce the pinnacle consistency [63].

A starch gluing cell connection on the TA rheometer determined the sticking capabilities of TS and TS–GA mixes [44]. The mixtures were created by scattering GA in water at 0.1, 0.3, 0.6, and 1.0 percent (w/w) with attractive mixing and adding starch (5%) to the gum arrangements [44]. TS was gradually introduced to the combinations when mixing to avoid bump arrangement. Various boundaries were estimated to think about peak viscosity (PV), hot paste viscosity, cold paste viscosity, relative breakdown (RB), and total setback (TSB) [44]. Numerous agents have observed that the development of hydrocolloids alters starch-adhering characteristics [44]. According to this study, the growth of GA causes a crucial (P0.05) drop in PV, RB, and TSB esteems. The GA screened a large fragment of the TS granules and advanced granules affiliation, which confined enlargement of the granules and limited the increment inconsistency throughout gluing [44]. The introduction of GA into TS negatively lowered the PV of the biphasic TS–GA framework. According to a comparison pattern, PV, RB, and TSB's unlimited supply of thickener was reduced to custard starch [45].

When the negatively charged starch arrangement was warmed with anionic gum, one of the feasible causes for the delay in gluing and decreased PV was the impediment of granule annihilation and draining of amylose [56]. They likewise simplified that such impacts were due to collaborations flanked by drained amylose atoms and utilization of gum [44]. The GA employed in this study had a molar mass of (1.86 × 105 g mol^{-1}), nearly as low as some polysaccharides [69]. The majority of hydrocolloids were thought to increase thickness; however, hydrocolloids with lower atomic mass, such as corn fiber gum and guar gum, were thought to cause a decrease in thickness [49, 53], which might be helpful as far as controlling the consistency of emulsion/combination/scatterings. According to Funami et al., the basic reason for the decreased consistency was the lower molar mass of guar gum (4.7 × 105 g mol^{-1}) [53]. They discovered that the lower subatomic mass of guar gum created a smooth stream and prevented starch granules from scraping an area free of amylose or amylopectin from draining [44, 53]. When a TS–GA framework is heated, the starch granules get gelatinized, increasing the overall thickness of the framework. The primary rationale for this behavior is because the GA is in the constant stage, and its focus expands as the volume of the starch stage open to GA decreases due to gelatinization enlargement [70]. This increase in thickness is mostly due to bigger granules, which, when shear powers are applied, results in a lack of granule trustworthiness and further interruption, causing the glue consistency to expand. This latter characteristic is known as breakdown consistency, and it is observed in the TS–GA relationships in the current review, which is consistent with other examiners' findings for starch and gum combinations [49, 66, 71].

Apart from the case of GA-modified TS, the possibility of improving starch properties by using hydrocolloids has been demonstrated in the case of hydrocolloid gum-modified chickpea starch. Because of its pasting and gelatinization capabilities, chickpea starch is significant in the food sector [72]. The influence of hydrocolloids on chickpea starch's pasting and textural qualities was confirmed by Shahzad et al. who utilized arabic, xanthan, cress seed, fenugreek, flaxseed, and okra gum [72]. Their study showed the possibility of using chickpea starch blended with fenugreek and xanthan gum in gluten-free bread formulations, while others can be used as a protective coating in food applications [72]. In the case of pasting properties, the modified starch with fenugreek and xanthan gum showed an increase in PV, while other gums showed a decrease in PV [72]. After overnight storage at room temperature, the textural profile analysis of gels from starch/gum mixes revealed an increase in gel hardness in the case of blending with fenugreek, while modified starch with xanthan, flaxseed, and okra gum showed a decrease in hardness [72].

Industrial Corner

- Modified starch with fenugreek and xanthan gum showed an increase in PV, while other gum mixtures decreased it.
- The textural profile examination of gels from gum blends revealed a rise in gel hardness when combining with fenugreek but a decrease in hardness when blending with modified starch with xanthan and okra gum.
- Starch can be modified with enzyme properties as enzymes can change their amylose–amylopectin content.
- In kofter products, adding molasses concentration up to 50% increases pasting viscosity. However, the pasting viscosity decreases when the concentration increases above 50%.

11.4 Enzymatic Modification of Starch Properties

Another alternative for physical and chemical modification is enzymatic modifications. When starch is modified with enzyme, it usually produces starches with modified physicochemical properties as enzymes can change their amylose–amylopectin content [73]. In Hui Li et al.'s experiment, they modified wheat starch to enhance its viscoelastic properties [74]. They used wheat starch, as the main component of wheat flour, used in many food products. Nevertheless, wheat starch, when heated, becomes a soft and viscous paste, leading to reduced quality of the food product due to the formation of an undesirable gel during cooling [74–76]. They used β-amylase, transglucosidase, then pullulanase. This enzymatic modification increased the paste viscosity due to the reduction in the amylose content, which improved starch granules' swelling [74]. Their experiment has approved the ability of using these enzymes as suitable additives to modify foods based on wheat starch, such as desserts and soft candy [74].

Recently, Raphael et al. investigated the effect of hydrolyzing the native red rice starch SN with two different enzymes by adding jabuticaba powder [77]. They produced MS1 (hydrolyzed starch with α-amylase) and MS2 (hydrolyzed starch with α-amylase and amyloglucosidase), then mixed these samples with jabuticaba peel powder [77]. With regard to pasting properties, MS1 showed the highest PV and minimum viscosity, while the samples with jabuticaba peel powder showed lower peak viscosities with no difference in minimum viscosity. MS2 mixed with jabuticaba peel powder showed the higher paste temperature when compared with other samples [77]. MS2 showed lower textural properties and firmness than MS1 and SN [77], whereas the samples mixed with jabuticaba powder showed higher firmness and gumminess, which support the paste when gelatinized [77]. The enzymatic modification of red rice starch had lowered the cohesiveness compared with SN; on the other hand, adding jabuticaba peel powder to SN increased its cohesiveness [77]. Their study showed that adding jabuticaba peel powder with the enzymatic modification of native starch had a positive effect as their produced mixtures can be used in food applications [77].

11.5 Use of Starch Modification in Food Production

The use of starch modification methods in food production can be illustrated by the case of kofter, which is a traditional dessert and is manufactured from grape molasses, water, and starch. Yildiz et al. investigated the effect of different water and molasses on starch samples like kofter [78]. They aimed to find the best quality characteristics of kofter by determining the pasting and textural properties of different samples [78]. Concentration of water affects pasting properties significantly. It is noticed that the value of PV is related to the degree of swelling of starch granules during heating, which means that high-value pasting viscosity of starch means high swelling starch capacity [78]. The increasing ability of starch to swell is related to the presence of sugar; its concentration reaches 25% [78, 79]. In a kofter product, the addition of molasses concentration up to 50% leads to increasing pasting viscosity because the swelling power of starch increases with the presence of sugar, as indicated [78]. However, the addition of molasses, with a concentration higher than 50%, decreases the pasting viscosity of kofter [78]. Decreasing swelling may have resulted from the breakthrough of sugar molecules into starch granules during the gelatinization process [80]. When water concentration decreases, the trough viscosity increases, associated with the swelling of starch granules. In the end, molasses contains a lot of sugar content; so, if the molasses concentration increases up to 50%, the pasting viscosity and trough viscosity will increase because the interaction between starch granules increases [78]. It is noticed that the starch sample that contains an equal amount of water and molasses can bear more heating and stress, which is better for the quality of the product [78].

Final viscosity is the viscosity of the starch blend after cooling related to handling the product and stabilizing the colloid system [78]. The magnitude of the final viscosity of starch into kofter increases when the concentration of molasses increases up to 60%, but concentrations above 60% decrease the final viscosity [78]. It is noticed that the peak time required for swelling starch granules until rapture increases with increasing molasses concentration up to 60%. Pasting temperature also increases with the addition of molasses [81]. The reason for increasing pasting temperature is related to the ability of sugar to reduce the availability of water [82]. Types of sugar and starch affect the relationship between sugar and pasting temperature [78]. In general, the pasting temperature increases by adding disaccharides more than monosaccharides [83, 84]. Molasses contains monosaccharides as glucose and fructose contain more disaccharides [78]. Sugar can also inhibit the hydration of starch granules [85].

Characterizing starch's pasting and textural properties is critical in food production since they are a necessary component in many dishes cooked with water, leading to great changes in their structure. Pasting properties are characterized by pasting temperature and PV, while textural properties are determined by hardness, cohesiveness, springiness, and chewiness. Starch properties can be modified chemically, physically, enzymatically, and using additives, such as hydrocolloids. The simple change in the modification condition may produce new products from the same compound with different tastes, pasting, and texture properties. This is determined when the frequency intervals of ultrasound waves are changed. The physical characteristics of the compound will be changed, which opens the way to develop and produce new products from the same compound. Also, a change in HMT and chemical reaction conditions, for example (pressure or temperature of the reaction, the concentration of chemicals, and the pasting properties as the degree of starch gelatinization), may all lead to changes in pasting and textural properties. Heat-moisture treatment is used to modify the physicochemical properties of starch while saving its granular structure. The starch's ability to swell is related to the presence of sugar. Enzymatic modification increases the paste viscosity in the modified starch. Subsequent changes may produce new products.

11.6 Concluding Remarks

Wheat and starch are the most important elements of different food products. The importance of characterizing the pasting and textural properties of wheat and starch lies in the fact that they are an essential component in many foods cooked with water, which leads to great changes in their structure. Various methods have to be used to modify and optimize those properties as chemical, physical, and enzymatic modification. As a result, the physical modification is dependent on changing the physicochemical properties of starch: (i) by utilizing sonic energy released by cavitation forces caused by ultrasound, which induces a collapse of the starch granules, making them more permeable to water during the heating step; (ii) by using HMT, which has been shown to increase the gelatinization temperature, thermal stability, and crystallinity as the reduction of amylose leaching leads to diminished granular swelling without depending on the source of the starch; (iii) by using annealing treatment that can cause a reduction in amylose leaching and granular swelling, the induction of high gelatinization temperatures, and high granular stability and crystalline perfection; (iv) high-pressure treatment of starch where pasting properties are affected as the degree of starch gelatinization increases with higher pressurization treatment temperature (between 25°C and the heat gelatinization temperature) and higher pressurization treatment time (up to 60 minutes); or (v) by exposing the samples to high-intensity electric pulses for a short duration, as the duration needs to be short, to prevent unwanted electrolytic reactions. Chemical modification produces reflective modification in the pasting of native starch granules. In addition, it helps in the stabilization of inter- and intramolecular bonds at different locations and positions. There are many techniques for chemical modification, such as acetylation, crosslinking, oxidation, cationization, and acid hydrolysis, but some of these methods may be limited due to the safety of both consumers and the environment. This led to an increasing demand for mixing different chemical techniques to create new kinds of modification. The unfortunate properties of local starch, such as low water-holding limit, retrogradation, and syneresis, can be redressed by expanding the gum. Also, the enzymatic modification increases the paste viscosity in the modified starch due to the reduction in the amylose content, improving the swelling of starch granules.

Glossary

Annealing treatment Method of physical modification of starch depends on using water and temperature to decrease the swelling of granules of starch.
Cold paste viscosity Viscosity value at the end of the isothermal period at 30°C.
Heat-moisture treatment Method of physical modification of starch depends on changing physicochemical properties of starch without changing the granular structure.
Pulsed electric field treatment Method of physical modification of starch depends on using an electric pulse to get starch changed.
Peak viscosity The highest viscosity recorded throughout heating.
Native starch Pure form of starch.

References

1 O. B. Ocheme, O. E. Adedeji, C. E. Chinma, C. M. Yakubu, U. H. Ajibo, Proximate composition, functional, and pasting properties of wheat and groundnut protein concentrate flour blends, *Food Sci. Nutr.* **2018**, *6*(5), 1173–1178. doi:10.1002/FSN3.670.
2 A. Farahnaky, R. Azizi, M. Majzoobi, G. Mesbahi, N. Maftoonazad, Using power ultrasound for cold gelation of kappa-carrageenan in presence of sodium ions, *Innov. Food Sci. Emerg. Technol.* **2013**, *20*, 173–181. doi:10.1016/J.IFSET.2013.06.002.
3 S. Hedayati, M. Niakousari, Microstructure, pasting and textural properties of wheat starch-corn starch citrate composites, *Food Hydrocoll.* **2018**, *81*, 1–5. doi:10.1016/J.FOODHYD.2018.02.024.
4 S. C. Alcázar-Alay, M. A. A. Meireles, Physicochemical properties, modifications and applications of starches from different botanical sources, *Food Sci. Technol.* **2015**, *35*(2), 215–236. doi:10.1590/1678-457X.6749.
5 A. Devi, R. Sindhu, B. S. Khatkar, Effect of fats and oils on pasting and textural properties of wheat flour, *J. Food Sci. Technol.* **2020**, *57*(10), 3836–3842. doi:10.1007/S13197-020-04415-4/TABLES/3.
6 A. Devi, B. S. Khatkar, Effects of fatty acids composition and microstructure properties of fats and oils on textural properties of dough and cookie quality, *J. Food Sci. Technol.* **2018**, *55*(1), 321–330. doi:10.1007/S13197-017-2942-8/TABLES/4.
7 K. M. Albano, C. M. L. Franco, V. R. N. Telis, Rheological behavior of Peruvian carrot starch gels as affected by temperature and concentration, *Food Hydrocoll.* **2014**, *40*, 30–43. doi:10.1016/J.FOODHYD.2014.02.003.
8 S. Park, Y. R. Kim, Clean label starch: production, physicochemical characteristics, and industrial applications, *Food Sci. Biotechnol.* **2021**, *30*(1), 1–17. doi:10.1007/s10068-020-00834-3.
9 I. L. Herceg, A. R. Jambrak, D. Šubarić, et al., Texture and pasting properties of ultrasonically treated corn starch, *Czech J. Food Sci.* **2010**, *28*(2), 83–93. doi:10.17221/50/2009-cjfs.
10 F. Zhu, Impact of ultrasound on structure, physicochemical properties, modifications, and applications of starch, *Trends Food Sci. Technol.* **2015**, *43*(1), 1–17. doi:10.1016/J.TIFS.2014.12.008.

11 J. Zheng, Q. Li, A. Hu et al., Dual-frequency ultrasound effect on structure and properties of sweet potato starch, *Starch/Staerke.* **2013**, *65*(7–8), 621–627. doi:10.1002/star.201200197.

12 M. Sujka, J. Jamroz, Ultrasound-treated starch: SEM and TEM imaging, and functional behaviour, *Food Hydrocoll.* **2013**, *31*(2), 413–419. doi:10.1016/j.foodhyd.2012.11.027.

13 K. O. Adebowale, B. I. Olu-Owolabi, O. O. Olayinka, O. S. Lawal, Effect of heat moisture treatment and annealing on physicochemical properties of red sorghum starch, *African J. Biotechnol.* **2005**, *4*(9), 928–933. doi:10.5897/AJB2005.000-3197.

14 H. Jacobs, J. A. Delcour, Hydrothermal Modifications of Granular Starch, with Retention of the Granular structure: a review, *J. Agric. Food Chem.* **1998**, *46*, 8. doi:10.1021/jf980169k.

15 E. D. R. Zavareze, A. R. G. Dias, Impact of heat-moisture treatment and annealing in starches: a review, *Carbohydr. Polymer.* **2011**, *83*(2), 317–328. doi:10.1016/j.carbpol.2010.08.064.

16 H. J. Chung, Q. Liu, R. Hoover, Impact of annealing and heat-moisture treatment on rapidly digestible, slowly digestible and resistant starch levels in native and gelatinized corn, pea and lentil starches, *Carbohydr. Polym.* **2009**, *75*(3), 436–447. doi:10.1016/j.carbpol.2008.08.006.

17 H. T. Chan, R. Bhat, A. A. Karim, Effects of sodium dodecyl sulphate and sonication treatment on physicochemical properties of starch, *Food Chem.* **2010**, *120*(3), 703–709. doi:10.1016/j.foodchem.2009.10.066.

18 Z. Han, X. A. Zeng, N. Fu, S. J. Yu, X. D. Chen, J. F. Kennedy, Effects of pulsed electric field treatments on some properties of tapioca starch, *Carbohydr. Polym.* **2012**, *89*(4), 1012–1017. doi:10.1016/j.carbpol.2012.02.053.

19 F. Zhu, Modifications of starch by electric field based techniques, *Trends Food Sci. Technol.* **2018**, *75*, 158–169. doi:10.1016/j.tifs.2018.03.011.

20 J. Zheng, Q. Li, A. Hu et al., Dual-frequency ultrasound effect on structure and properties of sweet potato starch, *Starch – Stärke.* **2013**, *65*(7–8), 621–627. doi:10.1002/STAR.201200197.

21 A. R. Jambrak, Z. Herceg, D. Šubarić et al., Ultrasound effect on physical properties of corn starch, *Carbohydr. Polym.* **2010**, *79*(1), 91–100. doi:10.1016/j.carbpol.2009.07.051.

22 H. T. Chan, R. Bhat, A. A. Karim, Effects of sodium dodecyl sulphate and sonication treatment on physicochemical properties of starch, *Food Chem.* **2010**, *120*(3), 703–709. doi:10.1016/J.FOODCHEM.2009.10.066.

23 R. Carmona-García, L. A. Bello-Pérez, A. Aguirre-Cruz, A. Aparicio-Saguilán, J. Hernández-Torres, J. Alvarez-Ramirez, Effect of ultrasonic treatment on the morphological, physicochemical, functional, and rheological properties of starches with different granule size, *Starch/Staerke* **2016**, *68*(9–10), 972–979. doi:10.1002/star.201600019.

24 N. Sit, S. Misra, S. C. Deka, Yield and functional properties of Taro Starch as affected by ultrasound, *Food Bioprocess Technol.* **2014**, *7*(7), 1950–1958. doi:10.1007/s11947-013-1192-7.

25 E. Subroto, R. Indiarto, H. Marta, S. Shalihah, Effect of heat - Moisture treatment on functional and pasting properties of Potato, *Food Res.* **2019**, *3*(October), 469–476.

26 B. R. Chen, Q. H. Wen, X. A. Zeng, R. Abdul, U. Roobab, F. Y. Xu, Pulsed electric field assisted modification of octenyl succinylated potato starch and its influence on pasting properties, *Carbohydr. Polym.* **2021**, *254*, 117294. doi:10.1016/J.CARBPOL.2020.117294.

27 K. Yu, Y. Wang, Y. Wang, L. Guo, X. Du, Effects of annealing and additives on the gelatinization, structure, and textural characteristics of Corn Starch, *Int. J. Food Prop.* **2016**, *19*(6), 1272–1281. doi:10.1080/10942912.2015.1071842.

28 Zia-ud-Din, H. Xiong, P. Fei Physical and chemical modification of starches: a review, **2017**, *57*(12), 2691–2705. https://doi.org/101080/1040839820151087379. doi:10.1080/10408398.2015.1087379.

29 L. A. Bello-Pérez, E. Agama-Acevedo, P. B. Zamudio-Flores, G. Mendez-Montealvo, S. L. Rodriguez-Ambriz, Effect of low and high acetylation degree in the morphological, physicochemical and structural characteristics of barley starch, *LWT - Food Sci. Technol.* **2010**, *43*(9), 1434–1440. doi:10.1016/J.LWT.2010.04.003.

30 S. Radosta, W. Vorwerg, A. Ebert, A. H. Begli, D. Grülc, M. Wastyn, Properties of low-substituted cationic starch derivatives prepared by different derivatisation processes, *Starch – Stärke* **2004**, *56*(7), 277–287. doi:10.1002/STAR.200300210.

31 R. Colussi, V. Z. Pinto, E. H. Slm et al., Structural, morphological, and physicochemical properties of acetylated high-, medium-, and low-amylose rice starches, *Carbohydr. Polym.* **2014**, *103*(1), 405–413. doi:10.1016/J.CARBPOL.2013.12.070.

32 H. S. L. M. El, R. Colussi, V. Z. Pinto et al., Structure, morphology and functionality of acetylated and oxidised barley starches, *Food Chem.* **2015**, *168*, 247–256. doi:10.1016/J.FOODCHEM.2014.07.046.

33 O. S. Lawal, Composition, physicochemical properties and retrogradation characteristics of native, oxidised, acetylated and acid-thinned new cocoyam (Xanthosoma sagittifolium) starch, *Food Chem.* **2004**, *87*(2), 205–218. doi:10.1016/J.FOODCHEM.2003.11.013.

34 Z. G. Luo, Y. C. Shi, Preparation of acetylated waxy, normal, and high-amylose maize starches with intermediate degrees of substitution in aqueous solution and their properties, *J. Agric. Food Chem.* **2012**, *60*(37), 9468–9475. doi:10.1021/JF301178C/ASSET/IMAGES/MEDIUM/JF-2012-01178C_0006.GIF.

35 S. Santacruz, Characterisation of cationic potato starch by asymmetrical flow field-flow fractionation. Influence of ionic strength and degree of substitution, *Carbohydr. Polym.* **2014**, *106*(1), 166–171. doi:10.1016/J.CARBPOL.2014.02.002.

36 C. L. Siau, A. A. Karim, M. H. Norziah, W. D. Wan Rosli, Effects of cationization on DSC thermal profiles, pasting and emulsifying properties of sago starch, *J. Sci. Food Agric.* **2004**, *84*(13), 1722–1730. doi:10.1002/JSFA.1871.

37 M. Gui-Jie, W. Peng, M. Xiang-Sheng, Z. Xing, Z. Tong, Crosslinking of corn starch with sodium trimetaphosphate in solid state by microwave irradiation, *J. Appl. Polym. Sci.* **2006**, *102*(6), 5854–5860. doi:10.1002/APP.24947.

38 A. O. Ashogbon, E. T. Akintayo, Recent trend in the physical and chemical modification of starches from different botanical sources: a review, *Starch – Stärke* **2014**, *66*(1–2), 41–57. doi:10.1002/STAR.201300106.

39 F. Spier, E. D. R. Zavareze, R. M. Silva, M. C. Elias, A. R. G. Dias, Effect of alkali and oxidative treatments on the physicochemical, pasting, thermal and morphological properties of corn starch, *J. Sci. Food Agric.* **2013**, *93*(9), 2331–2337. doi:10.1002/JSFA.6049.

40 N. Javadian, A. Mohammadi Nafchi, M. Bolandi, The effects of dual modification on functional, microstructural, and thermal properties of tapioca starch, *Food Sci. Nutr.* **2021**, *9*(10), 5467–5476. doi:10.1002/FSN3.2506.

41 B. Zhang, B. Bai, Y. Pan, X. M. Li, J. S. Cheng, H. Q. Chen, Effects of pectin with different molecular weight on gelatinization behavior, textural properties, retrogradation and in vitro digestibility of corn starch, *Food Chem.* **2018**, *264*, 58–63. doi:10.1016/J.FOODCHEM.2018.05.011.

42 J. Babić, D. Šubarić, D. Ackar, V. Piližota, M. Kopjar, N. N. Tiban, Effects of pectin and carrageenan on thermophysical and rheological properties of tapioca starch, *Czech J. Food Sci.* **2006**, *24*(6), 275–282. doi:10.17221/3325-CJFS.

43 N. K. O. Ojijo, E. Shimon, Influence of xanthan gum and tapioca starch on the retrogradation and gelation of finger millet (Eleusine coracana L. Gaertner) starch pastes, *J. Texture Stud.* **2007**, *38*(1), 100–115. doi:10.1111/J.1745-4603.2007.00088.X.

44 A. Singh, D. J. Geveke, M. P. Yadav, Improvement of rheological, thermal and functional properties of tapioca starch by using gum arabic, *LWT* **2017**, *80*, 155–162. doi:10.1016/J.LWT.2016.07.059.

45 M. Chaisawang, M. Suphantharika, Pasting and rheological properties of native and anionic tapioca starches as modified by guar gum and xanthan gum, *Food Hydrocoll.* **2006**, *20*(5), 641–649. doi:10.1016/J.FOODHYD.2005.06.003.

46 Y. J. Chang, H. W. Choi, H. S. Kim et al., Physicochemical properties of granular and non-granular cationic starches prepared under ultra high pressure, *Carbohydr. Polym.* **2014**, *99*, 385–393. doi:10.1016/J.CARBPOL.2013.09.010.

47 R. Pongsawatmanit, T. Temsiripong, T. Suwonsichon, Thermal and rheological properties of tapioca starch and xyloglucan mixtures in the presence of sucrose, *Food Res. Int.* **2007**, *40*(2), 239–248. doi:10.1016/J.FOODRES.2006.10.013.

48 M. G. Lomelí-Ramírez, S. G. Kestur, R. Manríquez-González, S. Iwakiri, G. B. De Muniz, T. S. Flores-Sahagun, Bio-composites of cassava starch-green coconut fiber: part II—Structure and properties, *Carbohydr. Polym.* **2014**, *102*(1), 576–583. doi:10.1016/J.CARBPOL.2013.11.020.

49 S. Qiu, M. P. Yadav, H. Chen, Y. Liu, E. Tatsumi, L. Yin, Effects of corn fiber gum (CFG) on the pasting and thermal behaviors of maize starch, *Carbohydr. Polym.* **2015**, *115*, 246–252. doi:10.1016/J.CARBPOL.2014.08.071.

50 C. Arismendi, S. Chillo, A. Conte, M. A. Del Nobile, S. Flores, L. N. Gerschenson, Optimization of physical properties of xanthan gum/tapioca starch edible matrices containing potassium sorbate and evaluation of its antimicrobial effectiveness, *LWT - Food Sci. Technol.* **2013**, *53*(1), 290–296. doi:10.1016/J.LWT.2013.01.022.

51 H. M. Choi, B. Yoo, Steady and dynamic shear rheology of sweet potato starch–xanthan gum mixtures, *Food Chem.* **2009**, *116*(3), 638–643. doi:10.1016/J.FOODCHEM.2009.02.076.

52 G. Fadavi, M. A. Mohammadifar, A. Zargarran, A. M. Mortazavian, R. Komeili, Composition and physicochemical properties of Zedo gum exudates from Amygdalus scoparia, *Carbohydr. Polym.* **2014**, *101*(1), 1074–1080. doi:10.1016/J.CARBPOL.2013.09.095.

53 T. Funami, Y. Kataoka, T. Omoto, Y. Goto, I. Asai, K. Nishinari, Food hydrocolloids control the gelatinization and retrogradation behavior of starch. 2a. Functions of guar gums with different molecular weights on the gelatinization behavior of corn starch, *Food Hydrocoll.* **2005**, *19*(1), 15–24. doi:10.1016/J.FOODHYD.2004.04.008.

54 Y. Wang, L. J. Wang, D. Li, N. Özkan, X. D. Chen, Z. H. Mao, Effect of flaxseed gum addition on rheological properties of native maize starch, *J. Food Eng.* **2008**, *89*(1), 87–92. doi:10.1016/J.JFOODENG.2008.04.005.

55 F. Zhu, Composition, structure, physicochemical properties, and modifications of cassava starch, *Carbohydr. Polym.* **2015**, *122*, 456–480. doi:10.1016/J.CARBPOL.2014.10.063.

56 X. Shi, J. N. BeMiller, Effects of food gums on viscosities of starch suspensions during pasting, *Carbohydr. Polym.* **2002**, *50*(1), 7–18. doi:10.1016/S0144-8617(01.

57 K. Gul, C. S. Riar, A. Bala, M. S. Sibian, Effect of ionic gums and dry heating on physicochemical, morphological, thermal and pasting properties of water chestnut starch, *LWT - Food Sci. Technol.* **2014**, *59*(1), 348–355. doi:10.1016/J.LWT.2014.04.060.

58 A. Kayacier, M. Dogan, Rheological properties of some gums-salep mixed solutions, *J. Food Eng.* **2006**, *72*(3), 261–265. doi:10.1016/J.JFOODENG.2004.12.005.

59 Z. S. Ladjevardi, S. M. T. Gharibzahedi, M. Mousavi, Development of a stable low-fat yogurt gel using functionality of psyllium (Plantago ovata Forsk) husk gum, *Carbohydr. Polym.* **2015**, *125*, 272–280. doi:10.1016/J.CARBPOL.2015.02.051.

60 P. Chivero, S. Gohtani, H. Yoshii, A. Nakamura, Assessment of soy soluble polysaccharide, gum arabic and OSA-Starch as emulsifiers for mayonnaise-like emulsions, *LWT - Food Sci. Technol.* **2016**, *69*, 59–66. doi:10.1016/J.LWT.2015.12.064.

61 C. Chranioti, A. Nikoloudaki, C. Tzia, Saffron and beetroot extracts encapsulated in maltodextrin, gum Arabic, modified starch and chitosan: incorporation in a chewing gum system, *Carbohydr. Polym.* **2015**, *127*, 252–263. doi:10.1016/J.CARBPOL.2015.03.049.

62 T. Feng, R. Ye, H. Zhuang, Z. Fang, H. Chen, Thermal behavior and gelling interactions of Mesona Blumes gum and rice starch mixture, *Carbohydr. Polym.* **2012**, *90*(1), 667–674. doi:10.1016/J.CARBPOL.2012.05.094.

63 M. H. Lee, M. H. Baek, D. S. Cha, H. J. Park, S. T. Lim, Freeze–thaw stabilization of sweet potato starch gel by polysaccharide gums, *Food Hydrocoll.* **2002**, *16*(4), 345–352. doi:10.1016/S0268-005X(01)00107-2.

64 N. Singh, J. Singh, L. Kaur, N. S. Sodhi, B. S. Gill, Morphological, thermal and rheological properties of starches from different botanical sources, *Food Chem.* **2003**, *81*(2), 219–231. doi:10.1016/S0308-8146(02)00416-8.

65 S. K. Flores, D. Costa, F. Yamashita, L. N. Gerschenson, M. V. Grossmann, Mixture design for evaluation of potassium sorbate and xanthan gum effect on properties of tapioca starch films obtained by extrusion, *Mater Sci. Eng. C.* **2010**, *30*(1), 196–202. doi:10.1016/J.MSEC.2009.10.001.

66 J. Ahmed, A. Singh, H. S. Ramaswamy, P. K. Pandey, G. S. V. Raghavan, Effect of high-pressure on calorimetric, rheological and dielectric properties of selected starch dispersions, *Carbohydr. Polym.* **2014**, *103*(1), 12–21. doi:10.1016/J.CARBPOL.2013.12.014.

67 A. Singh, M. Sharma, H. S. Ramaswamy Effect of high pressure treatment on rheological characteristics of egg components. **2014**, *18*(3), 558–571. doi:10.1080/10942912.2013.837063.

68 O. Nasir, F. Artunc, A. Saeed et al., Effects of Gum Arabic (Acacia senegal) on water and electrolyte balance in healthy mice, *J. Ren. Nutr.* **2008**, *18*(2), 230–238. doi:10.1053/J.JRN.2007.08.004.

69 D. Renard, C. Garnier, A. Lapp, C. Schmitt, C. Sanchez, Structure of arabinogalactan-protein from Acacia gum: from porous ellipsoids to supramolecular architectures, *Carbohydr. Polym.* **2012**, *90*(1), 322–332. doi:10.1016/J.CARBPOL.2012.05.046.

70 M A, J L, G L, JL D, A rheological characterization of cereal starch-galactomannan mixtures, *Cereal Chem.* **1989**, *66*(2), 90–93, https://europepmc.org/article/AGR/IND89033777. Accessed November 27, 2021.

71 D. D. Christianson, J. E. Hodge, D. Osborne, R. W. Detroy, *Gelatinization of Wheat Starch as Modified by Xanthan Gum, Guar Gum, and Cellulose Gum*. Published online 1981.

72 S. A. Shahzad, S. Hussain, A. A. Mohamed, M. S. Alamri, M. A. Ibraheem, A. A. Abdo Qasem, Effect of hydrocolloid gums on the pasting, thermal, rheological and textural properties of chickpea starch, *Foods* **2019**, *8*, 12. doi:10.3390/foods8120687.

73 S. H. Park, Y. Na, J. Kim, S. D. Kang, K. H. Park, Properties and applications of starch modifying enzymes for use in the baking industry, *Food Sci. Biotechnol.* **2018**, *27*(2), 299–312. doi:10.1007/S10068-017-0261-5/FIGURES/5.

74 H. Li, J. Li, L. Guo, Rheological and pasting characteristics of wheat starch modified with sequential triple enzymes, *Carbohydr. Polym.* **2020**, *230*, 115667. doi:10.1016/J.CARBPOL.2019.115667.

75 N. Mohd Adzahan, Modification of wheat, sago and tapioca starches by irradiation and its effect on the physical properties of fish crackers (Keropok) - Universiti Putra Malaysia Institutional Repository. Master's thesis, Universiti Putra Malaysia. Published 2002. http://psasir.upm.edu.my/id/eprint/8466. Accessed May 17, 2022.

76 R. Ziobro, J. Korus, M. Witczak, L. Juszczak, Influence of modified starches on properties of gluten-free dough and bread. Part II: quality and staling of gluten-free bread, *Food Hydrocoll.* **2012**, *29*(1), 68–74. doi:10.1016/J.FOODHYD.2012.02.009.

77 R. L. J. Almeida, T. Dos Santos Pereira, R. D. Almeida et al., Rheological and technological characterization of red rice modified starch and jaboticaba peel powder mixtures, *Sci. Repor. 2021 111* **2021**, *11*(1), 1–11. doi:10.1038/s41598-021-88627-4.

78 O. Yildiz, B. Yurt, O. S. Toker, M. M. Ceylan, M. T. Yilmaz, A. Bastürk, Pasting, textural and sensory characteristics of the kofter, a fruit-based dessert: effect of molasses and water concentration, *Int. J. Food Eng.* **2015**, *11*(3), 349–358. doi:10.1515/IJFE-2014-0313/MACHINEREADABLECITATION/RIS.

79 F. B. Ahmad, P. A. Williams, Effect of sugars on the thermal and rheological properties of sago starch, *Biopolymers* **1999**, *50*(4), 401–412. doi:10.1002/(SICI)1097-0282(19991005)50:4<401::AID-BIP6>3.0.CO;2-V.

80 Q. Li, H. Li, Q. Gao, The influence of different sugars on corn starch gelatinization process with digital image analysis method, *Food Hydrocoll.* **2015**, *43*, 803–811. doi:10.1016/J.FOODHYD.2014.08.012.

81 M. H. Baek, B. Yoo, S. T. Lim, Effects of sugars and sugar alcohols on thermal transition and cold stability of corn starch gel, *Food Hydrocoll.* **2004**, *18*(1), 133–142. doi:10.1016/S0268-005X(03.

82 P. Chantaro, R. Pongsawatmanit, Influence of sucrose on thermal and pasting properties of tapioca starch and xanthan gum mixtures, *J. Food Eng.* **2010**, *98*(1), 44–50. doi:10.1016/J.JFOODENG.2009.12.006.

83 R. Sharma, D. P. S. Oberoi, D. S. Sogi, B. S. Gill, Effect of sugar and gums on the pasting properties of cassava starch, *J. Food Process Preserv.* **2009**, *33*(3), 401–414. doi:10.1111/J.1745-4549.2008.00283.X.

84 A. Eroglu, K. Bayrambas, Z. Eroglu et al., Steady, dynamic, creep/ recovery, and textural properties of yogurt/molasses blends: temperature sweep tests and applicability of Cox-Merz rule, *Food Sci. Technol. Int.* **2016**, *22*(1), 31–46. doi:10.1177/1082013214566478.

85 M. Yoshimura, T. Takaya, K. Nishinari, Effects of Konjac-Glucomannan on the gelatinization and retrogradation of corn starch as determined by rheology and differential scanning calorimetry, *J. Agric. Food Chem.* **1996**, *44*(10), 2970–2976. doi:10.1021/JF960221H.

12

Phase Change Materials in Food Dryers

Prospects and Challenges

Hasibul Hasan Himel[2], *Sabit Hasan*[2], *Nufile Uddin Ahmed*[2], *and Mahadi Hasan Masud*[1,2,*]

[1] *School of Engineering, RMIT University, Bundoora Campus, Melbourne, VIC-3083, Australia*
[2] *Department of Mechanical Engineering, Rajshahi University of Engineering and Technology, Rajshahi, Bangladesh*
* *Corresponding author*

Highlights

1. PCM increases the reliability of the solar food dryer by extending the drying period.
2. PCM integrated with a solar food dryer ensures uniform food drying.
3. The drawbacks of PCM regarding thermal conductivity can be improved by adding carbon-based nanoparticles and using proper encapsulation.
4. PCM increases the thermal efficiency and effectiveness of solar food dryers.

12.1 Introduction

One of the ancient methods of food preservation with origins dating as far back as 12 000 BC, which is still in use today, is drying – the process of dehydrating food for future consumption [1, 2]. Some developed countries still practice direct open sun-drying, which is time-intensive and causes the dried object to become susceptible to nutrition loss and contamination [3, 4]. However, scientific measures have taken this process into an advanced state [5]. Developed and developing countries currently employ drying systems that utilize convective heating and electricity [6]. These newer systems eliminate the limiting time and contamination factors at the cost of energy consumption [7]. For instance, a study by Kjaer et al. estimated that even though evaporation of free water requires merely 2.3 MJ/kg, post-harvest grain drying consumes at least 4.5 MJ/kg [8]. This makes drying one of the most energy-consuming tasks in the food-processing industry [9], taking up to 30% of direct energy input for purposes such as the cultivation of cereal [10]. Another study reveals that in First World countries, 10–15% of the total energy consumed in industries is expended on drying [6, 11]. As energy has monetary value, Menon et al. reported that energy used in drying accounts for tentatively 15% of manufacturing-related expenditures [12].

To strike a balance between prolonging the shelf life of food and minimizing energy input, researchers have implemented a myriad of hybrid solutions to complement the energy needs of food dryers [13]. They integrated conventional dryers with microwave, infrared radiation, or ultrasound [10]. The efficacy of such hybrid systems was demonstrated by Ye et al. in a study where a hybrid infrared hot air dryer consumed approximately

Materials Science and Engineering in Food Product Development, First Edition. Edited by Wing-Fu Lai.

41–63% less energy than conventional hot air dryers [14]. Another approach that is of interest to us is the investigation into the replacement of fossil fuels that acts as a primary source of energy in drying. In these arrangements, heat would be supplied using solar energy supplemented by geothermal or biomass energy [15]. In some cases, solar power and electricity can complement each other to maintain continuous drying operations. Catorze et al. reported that using solar power for a drying system led up to a 35% decrease in electrical energy usage under the right conditions [16]. Although solar power lessens the energy requirement by 35%, it is not accessible 24 hours a day [17], which presents a significant problem in bridging the gap between energy demand and supply [18]. Moreover, the availability of solar energy is season based and determined by the location's weather conditions [19]. The most significant impediment to widespread solar energy use is unreliability [20]. One of the best solutions for this unreliability is to store the excess solar energy after meeting the load and use it if needed later on. This produces the requirement of thermal energy storage.

To ensure maximum retention of solar energy, the adoption of PCM as thermal energy storage (TES) systems has shown considerable potential [21]. In the past two decades, researchers have been exploring the various aspects of utilizing PCMs to increase energy efficiency [22]. When integrated with solar-based systems, PCMs boosted the thermal efficiencies of those systems from 12% up to 87% [23]. Babar et al. conducted a study where the addition of paraffin inside a flat plate solar collector of a dryer cut down the drying time of green chili by 35–38% [24]. Pankaew et al. experimented with a solar greenhouse dryer under two conditions, with PCM and without it, for drying purposes of 500 kg of chilies. In their experiment, paraffin wax was used in a rectangular tube of dimensions 25 mm × 50 mm × 1100 mm. Results demonstrated that when a dryer was operated with PCM, it took 2.5 days to dry chilies, whereas without PCM, it took 3.5 days. It was also found that dryers with PCM had higher energy efficiency ($\cong$ 13.1%) than without PCM ($\cong$ 11.4%) [25]. Srinivasan et al. carried out a study on utilizing heat storage material in various modes of the solar dryer. Improvement was noted briefly for both *latent heat* and *sensible heat* storage materials. The conclusion was that latent heat storage units could absorb 5–14 times more heat per unit volume than conventional sensible heat storage units, such as rock, water, and masonry [26]. Although some discrete studies have already been performed by several researchers regarding the prospect of PCM as a TES material, dedicated study regarding the potential of PCM for the application in solar dryers is still missing. While doing so, we have briefly discussed the potential of PCMs in solar energy based food drying and also recommended some factors in order to optimize the application of PCM in food dryers. We believe that this study will assist the researcher in applying PCM in an appropriate manner, which will inevitably offer low-cost renewable energy based sustainable food-drying techniques.

12.2 Phase Change Materials and Their Properties

PCMs are latent heat storage (LHS) materials that are used for TES systems. These substances alter their physical state by heat absorption or rejection, called latent heat, nearly at constant temperature. TES systems can save energy where thermal energy is wasted as part of an operating cycle, and they can also ensure smooth operation where there is always a disconnect between demand and supply [27]. As an example, for solar dryers, solar power is the main source of energy that is only available during the day, and by integrating TES, operating hours for that type of solar energy based dryer can be extended up to a few hours. An effective heat storage system is required for long-term stability and high heat storage capacity. PCMs are mostly preferred because they can store large amounts of heat per unit volume and maintain nearly constant temperature while changing phases [28].

12.2.1 Basic Mechanism

Usually, for PCM, heat is absorbed or released as sensible heat provided that change in phase does not occur in that process; however, PCM changes its physical state when it absorbs latent heat. Like conventional materials, PCMs initially absorb sensible heat until it reaches melting or freezing temperature. Upon reaching the melting point, it starts to absorb latent heat and continues to absorb heat nearly at constant temperature and melt gradually. The amount of latent heat needed to receive to melt completely depends on the molecular bond and structure of PCM and varies from material to material. After it absorbs the necessary amount of heat, the melting process stops, and again, it continues to heat up sensibly [27, 29, 30]. This process is also reversible, i.e., when the surrounding temperature drops at freezing temperature, they start to solidify, releasing the latent heat, which was absorbed during the melting process. The solidification process also occurs at nearly constant temperature and stops after the latent heat of the material is released completely [7]. This released energy is then utilized for heating and cooling applications, such as for solar dryers, when solar energy is unavailable. During operation, energy stored in PCM is released, and the temperature inside the drying chamber is kept higher than ambient temperature. Thus, the drying period is extended using stored energy.

12.2.2 Classification

PCMs can be classified based on their melting temperature, chemical composition, and mode of phase change. A snapshot of their detailed classification is revealed in Figure 12.1.

12.2.2.1 Based on Melting Temperature

Based on melting temperature, PCMs are classified into categories of low, medium, and high temperature. The melting point of low-temperature PCMs are up to 15°C [31]. These are mostly utilized for air-conditioning and food processing. Paraffin ($C_{14,15}$), formic acid (HCOOH), and potassium hydrogen phosphate hexahydrate ($K_2HPO_4.6H_2O$) are examples of such PCMs. The suitable temperature range for most of the applications falls into medium temperature PCM, having a melting point ranging between 15°C and 90°C [31]. Some PCM from this category include technical grade paraffin (5838, 6035), paraffin (C16-34), and capric acid ($CH_3(CH_2)_8$ COOH) and are used in buildings, solar dryers, electronics cooling, etc. High-temperature PCM has a melting point of over 90°C [31] (e.g. magnesium chloride hexahydrate [$MgCl_2.6H_2O$], methyl fumarate [$CHCO_2(NH_3)_2$]), and is utilized in the aeronautical industry [31].

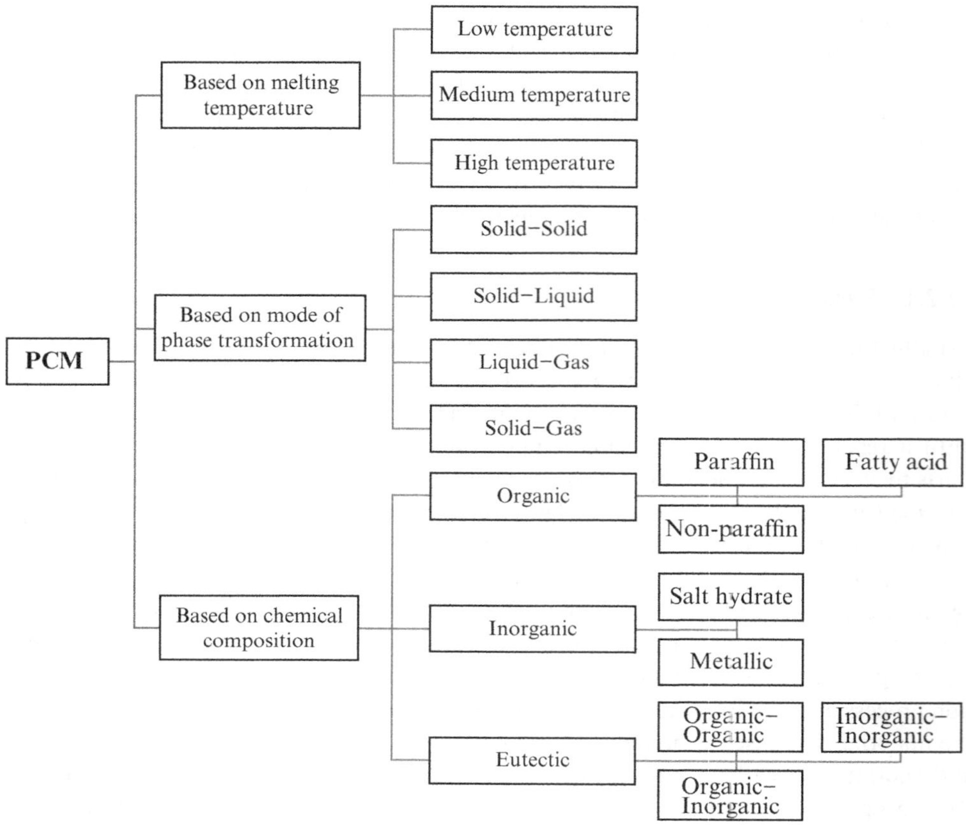

Figure 12.1 PCM classification.

12.2.2.2 Based on the Mode of Phase Change

Another approach to classify PCM is by phase transition modes. PCMs can also be classified as solid–liquid, solid–solid, liquid–gas, or solid–gas, depending on how they shift between phases. However, due to their large phase transition volumes, the final two categories of PCMs, i.e., solid–gas and liquid–gas are not much exploited [32]. The other two PCMs have less volume changes in transition, making them viable for practical use despite their lower *enthalpy* [27]. In solid–solid PCMs, the substance changes its crystal structure, unlike others, while changing their physical state. They have the lowest volume change during the transition, thus offering a small storage area [33]. Lithium sulfate (Li_2SO_4), an organic solution of pentaerythritol ($C[CH_2OH]_4$) and pentaglycine ($C_5H_{12}O_3$) are some examples of solid–solid PCMs [34]. Solid–liquid such as acetic acid (CH_3COOH) PCMs are widely used as they are economically feasible and found in a wide range of fusion heat and freezing temperature [28].

12.2.2.3 Based on Chemical Composition

Apart from the ways of classification as mentioned above, PCMs can be categorized according to their chemical identity. In terms of chemical properties, PCMs might be organic, inorganic, or eutectic. Organic materials include paraffin, fatty acids, and other non-paraffin substances; inorganic materials are salt hydrates and metallics, and eutectics are combinations of organic–organic, inorganic–inorganic, or inorganic–organic compounds [27].

12.2.2.3.1 Organic PCM

Paraffin wax and fatty acids are the most widely used organic PCM. Paraffin has been employed for energy storage because of some significant qualities that include high latent heat and high specific heat, which makes them capable of storing a large amount of heat per unit volume. Moreover, they do not show supercooling properties, and no interference occurs with heat extraction. Their volume changes moderately (10–20%) [35] during phase change with lower vapor pressure [36], which minimizes storage area and containment problems. They are also chemically inert and nontoxic if not digested [37]. They have a melting point of over 37°C [38] and a latent heat ranging between 200 and 220 kJ/kg [27]. However, the major downside of paraffin wax is its limited heat conductivity, and in some cases, they are moderately flammable [39]. Fatty acids are organic compounds characterized by a long aliphatic chain that can be saturated or unsaturated. Their melting point range has been suitable for a few TES systems that ranges between −5°C and 71°C. The heat of fusion for these substances is between 45 and 210 kJ/kg, higher than that of paraffins [40]. Fatty acids have a high nucleation rate; thus they do not show any supercooling and are almost inflammable [41]. However, they have poor thermal conductivity (TC) and are also unstable at high temperatures. The major shortcoming is that these materials cost nearly double that of the commercial-grade paraffin [27].

12.2.2.3.2 Inorganic PCM

Salt hydrates are promising as inorganic PCMs because of their larger latent heat of fusion and TC. They are inexpensive, slightly corrosive, and show little toxicity. However, their major downsides are supercooling and incongruent melting. Incongruent melting decreases the reversibility of the phase transition process and negatively affects their heat storage capability [37]. Supercooling leads to the release of energy lower than the fusion temperature [39]. Another inorganic PCM, metallics, is not being greatly considered due to their heaviness. However, considering per unit volume, they have high latent heat of fusion. So metallics may be considered where the size of the storage area is a great concern rather than weight. Some other properties they exhibit are high TC, low vapor pressure, and low SH [28].

12.2.2.3.3 Eutectic PCM

Eutectic PCMs consist of a mixture of two or more components at their minimum boiling point, which shows congruent melting behavior and contains a mixture of crystallized components in the solid state [42]. They have a high volumetric thermal storage density than organic materials and a low vapor pressure. However, they are costly, have low TC, and are not very stable after repeated melting–freezing cycles [43].

12.2.3 Key Properties

For an effective LHS system, these materials must have certain thermophysical, kinetic, and chemical characteristics. The properties a PCM should possess for being considered as an effective heat storage material are discussed in this section.

12.2.3.1 Thermal Properties

Thermal properties are melting temperature, latent heat of fusion, SH, and TC. When choosing a PCM for a certain operation, the working temperature of the system should be suited to the PCMs' transition temperature. LH and SH should be high to ensure an increased amount of energy stored per unit volume and minimize the size of the storage area. TC should also be high in both phases because this will increase the heat transfer rate and facilitate the charging and discharging process [39]. Paraffin wax is found in a wide range of temperatures (5.5–75.9°C) and have a latent heat of fusion of 200–220 kJ/kg [27]. Therefore, it is one of the most widely used PCM. However, they have a low TC (0.15–0.21 W/mk) [36], so heat transfer enhancement methods (extended surface, *encapsulation*, mixing high conductive material, etc.) are used to negate the effect of low TC. Inorganic materials, in general, have a greater value of TC, such as for calcium chloride hexahydrate ($CaCl_2.6H_2O$), the value is 1.08 W/mK (solid), and 0.56 W/mK (liquid) [44].

12.2.3.2 Physical Properties

Density, vapor pressure, melting behavior (congruent, incongruent, semi-congruent) [28], and volume change during phase transition are the chemical properties. Density of material should be high, whereas change in volume and vapor pressure needs to be low. High density of the material will reduce storage area. Low volume fluctuations on phase transition and minimal vapor pressure at working temperatures will reduce the containment problem. Melting of substance needs to be congruent to ensure phase stability by keeping the same storage capacity throughout each cycle [28, 39]. Salt hydrates are much denser than paraffins, such as for barium hydroxide octahydrate ($Ba(OH)_2.8H_2O$), it is 1937–2070 (kg/m^3) [45], and for paraffin wax (C23–24), it is 769–912 (kg/m^3) [46]. Paraffin has a slightly higher value of volume change during phase change (10–20%) [35] than salt hydrates; however, salt hydrates melt incongruently, which is their major drawback and limits their application in many cases [37].

12.2.3.3 Kinetic and Chemical Properties

Kinetic properties include nucleation rate and crystal growth rate. High nucleation rate will prevent supercooling. Supercooling is undesirable as a couple of degrees of supercooling may hamper proper heat extraction from TES, and 5–10°C of supercooling can completely prevent heat extraction. [39] The material needs to have a high crystallization rate so that it can catch up with the heat transfer rate and change its phase without delay [28]. Paraffins and fatty acids show very little or no supercooling and have a sufficient crystal growth rate [47].

Chemical properties include stability, corrosiveness, toxicity, flammability, and explosiveness. One of the most noteworthy properties of PCM is that it should not degrade after several number of freeze/melt cycles. They need to be nonpoisonous, nonflammable, and non-explosive for safety purposes [28, 39]. It also should be noncorrosive to the construction material [39]. Paraffin (melting point 53°C) shows great stability as its LH and melting point does not change even after the 1500 freeze–melt cycle [48]. Table 12.1 shows some commonly used PCM with their two main thermal properties, i.e.,

Table 12.1 Melting point and latent heat of fusion of commonly used PCMs Adapted from [39].

Material type	PCM	Melting point (°C)	Heat of fusion (kJ/kg)	Group
Organic	6106	42–44	189	II
Technical grade	P116	45–48	210	I
Paraffin	5838	48–50	189	I
	6035	58–60	189	I
	6403	62–64	189	II
	6499	66–68	189	III
Paraffin (no. of carbon atoms)	14	5.5	228	III
	16	10	205	III
	17–21	21–40	200–240	II
	22–25	45–60	230–250	I
	26–28	56–61	236–256	I
	29–31	63–68	240–250	II
	32	69.5	170	III
	33–34	74–75	268	III
Fatty acids	Acetic acid	16.7	184	III
	Capric acid	36	152	III
	Lauric acid	49	178	II
	Eladic acid	47	218	I
	Palmatic acid	55	163	II
	Stearic acid	69.4	199	III
Nonparaffins	Phenol	41	120	III
	Cetyl alcohol	49.3	141	II
	a-Nepthylamine	50	93	III
	9-Heptadecanone	51	213	I
	Thymol	51.5	115	II
	Methyl behenate	52	234	I
Inorganic	$Na_2SO_4.H_2O$	32.4	241	III
Salt hydrates	$Na_2HPO_4.12H_2O$	40	279	II
	$Mg(NO_3).4H_2O$	47	142	II
	$Na_2S_2O_3.5H_2O$	48.5	210	I
	$Ca(NO_3)_2.3H_2O$	51	104	III
	$MnCl_2.4H_2O$	58	151	II
	$CH_3COONa.3H_2O$	58	265	I
	$Fe(NO_3)2.6H_2O$	60.5	126	II

(Continued)

Table 12.1 (Continued)

Material type	PCM	Melting point (°C)	Heat of fusion (kJ/kg)	Group
Metallics	Gallium	30	80.3	III
	Cerrolow eutectic	58	90.9	III
	Bi–Cd–In eutectic	61	25	III
	Cerrobend eutectic	70	32.6	III
Eutectics	NH_2CONH_2(53%) + NH_4NO_3(47%)	46	95	III
	$Mg(NO_3)_3.6H_2O$(61.5%) + NH_4NO_3(38.5%)	52	125.5	II
	$Mg(NO_3)_3.6H_2O$(50%) + $MgCl_2.6H_2O$(50%)	59.1	144	II
	CH_3CONH_2(50%) + $C_{17}H_{35}COOH$(50%)	65	218	II

melting point and latent heat of fusion. The hot air temperature for the food dryer, especially for fruits and vegetables, needs to be between 45°C and 60°C [49]. High amount of latent heat per unit mass is always preferable. Based on only these two properties, they are categorized into three groups for applicability on solar food dryers as (i) group I: most suitable; (ii) group II: suitable; (iii) group III: less suitable. Some of the substances may seem suitable based on melting point and latent heat only, but due to other properties like supercooling, toxicity, and nonstability, mostly, they are not used. Moreover, economic feasibility is also the most concerning parameter for selecting and integrating proper PCM in a food dryer.

12.3 Potential of PCMs in Food Drying

12.3.1 Energy Potential of the Upgrade

While substantial work has been done regarding PCMs as TES devices, very few have studied their direct impact on energy consumption if implemented with a dryer. In an experiment, Mehran et al. noted that a solar dryer could produce the same output with four times less energy if it were powered by natural gas [50]. However, Yahya et al. observed that the solar collectors used would not work in cloudy weather [51]. To circumvent around such weather-related issues, Khadraoui et al. noted that these dryers need to use thermochemical energy, latent, or sensible heat [52]. Even though rock beds can be used to retain heat, considering the operating conditions and suitable characteristics, PCMs are a superior alternative. In their study, Paneliya et al. used paraffin wax as a replacement for an additional burner operated by liquefied petroleum gas equipped with a large-scale greenhouse solar dryer. Experimental studies confirmed the feasibility of this substitution because the result showed that PCMs have substantial energy potential to reduce the cost of the solar dryer and eliminates the mismatch of the energy requirement of the solar dryer in case of bad weather condition. They also recommended paraffin wax as a latent heat energy carrier due to its affordability and chemical stability under 500°C [53]. In an earlier investigation, Devahastin and Pitaksuriyarat mentioned that while drying sweet potatoes, PCM installations could account for an energy savings of 34% and 40% for an incoming air velocity of 2 and 1 m/s, respectively [54]. Although they reflected the energy potential of using PCMs in food drying, the capital cost of integration and the operating cost were not mentioned.

12.3.2 Economy of PCM Integration

Two primary factors determining the technological and economic viability of a solar thermal system are the affordability and performance of solar collectors. Improved energy transformation, lower power intake, reduction in energy dissipation, higher asset utilization, weather conditions, and component costs are all factors that influence economic rationale and play a role in determining the payback period. While retrofitting existing systems with PCM may require an initial investment, it will pay itself off in the form of a highly efficient energy system. In their seminal study, Godarzi et al. designed a PCM storage for a $LiBr–H_2O$ device by utilizing exergo-economic analysis and applying a genetic algorithm. They reported that the payback period increases by six months as PCMs are expensive [55]. On the other hand, PCMs may not become a financially sound investment as long as the market price is not within reach and they are not commercially available. While writing this chapter, prices of paraffin wax ranged from USD1.15 to USD1.25 per kg. Before making a purchase decision, it is up to the user to determine whether they want to have a slightly expensive device with better efficiency or a mainstream, affordable dryer that will eventually cost more, accounting for its operating costs.

12.3.3 Safety Concerns Related to Food and PCMs

One of the crucial aspects of drying is to ensure that the food is not contaminated and does not pose any hazard upon consumption. The US Food and Drug Administration has certified food-grade paraffins for use in food and medication. However, their exorbitant price prevents them from being widely used. Commercial paraffins, on the other hand, are poisonous and should not be consumed. If burned, they release harmful poisons, and prolonged contact with their vapor is dangerous due to compounds, such as toluene, benzene, butanone, and naphthalene, among others. Vegetable oil-based PCMs, however, are much safer and environmentally friendly. Different salt hydrates have different levels of toxicity. Typically, they are nontoxic, but when they come in contact or are consumed, they can cause inflammation of the skin, eye discomfort, and breathing difficulties [37]. Some salt hydrates, such as magnesium nitrate, do not appear to be poisonous. The components that make up eutectics, which might be organic or inorganic, have an impact on their toxicity. They are discovered to have an unpleasant odor, which is also a known problem regarding fatty acids. Nonetheless, since PCMs will be used in encapsulated states, there should be no concern using these as TES in food drying, and further investigation in this area is highly encouraged to explore the effects in case there is a minor leakage.

12.4 Current Status of Utilizing PCMs for Food Drying

Limitation of sunshine hours is a major barrier to the efficient operation of solar-operated food dryers. One of the most suitable methods to eradicate this limitation is to incorporate PCMs as the heat storage media. During the sunshine hours, PCMs absorb heat by changing its state from solid to liquid. After sunset or when the sunshine hours are gone, it releases the stored heat again by changing its state from liquid to solid. The inclusion of PCM results in the better efficient operation of the solar dryer by increasing working hours and ensuring constant drying temperature, which in turn shortens the time of food drying. A comparative overview of the solar dryer with and without PCM is provided in Figure 12.2.

Lately, solar dryers combined with various PCMs have been designed, and their performance is evaluated in various kinds of food-drying purposes. Various experimental studies examined the effectiveness and benefits of using PCM in a solar-assisted food dryer. A summary of the literature reflecting the effect of using PCM combined with various types of solar dryers is presented in Table 12.2.

Although the literature reflects significant improvement of solar dryers using PCM, its optimization is still a big concern. For instance, the lower TC of organic PCM results in a poor heat transfer rate, whereas phase segregation and corrosiveness on metal are the major drawbacks of inorganic PCM. Therefore, for the selection of proper PCM for a specific type of solar dryer, it is necessary to optimize all the deterministic parameters of PCM, such as thermal stability, heat conductivity, economic and environmental aspects, etc., to ensure better output. The following section will give a brief discussion about the optimization parameters that need to be considered during the use of PCM in a solar dryer.

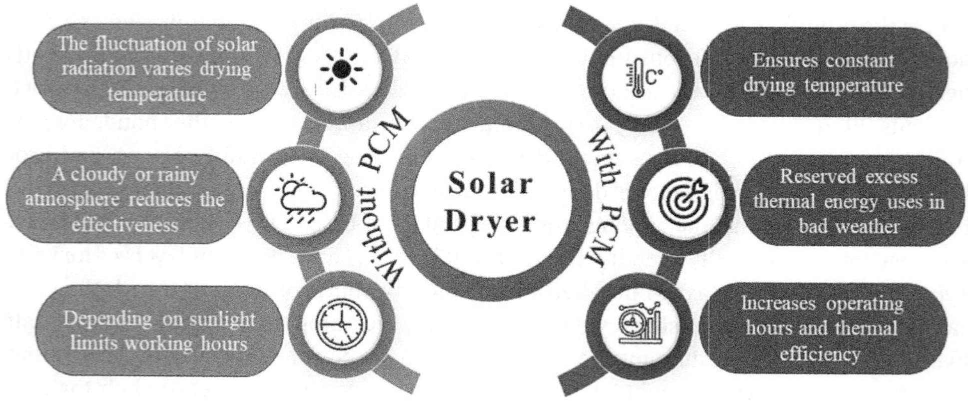

Figure 12.2 Benefits of utilizing PCM in a solar dryer compared with the benefits of not having it.

Table 12.2 Effect of incorporating PCMs in different types of solar energy based food dryers.

Dryer type	PCM type used in the dryer	Drying food	Study area	Summary	References
Indirect type force convective dryer	Paraffin wax having carbon atoms C31–33 Paraffin wax having carbon atoms C23–24	Fish	India	Using paraffin wax(C23-24) as PCM reduces the drying time of fish by about 2 hours	Virbhadra et al. [46]
Mixed-mode solar greenhouse dryer	Paraffin wax	Peppers	Tunisia	The moisture content was reduced by 95% after 30 hours of operation of the dryer with PCM, while it took 55 hours in the absence of PCM. PCM reduced almost 45% drying time	Azaizia et al. [56]
Indirect type force convective dryer	Calcium chloride hexahydrate	Seeded grape	Turkey	Solar dryer with PCM required only 56 hours to reduce the moisture of seeded grape from 75% to 9%, while it took 200 hours in the case of open sun-drying (OSD) method. PCM brought the drying time about one-fourth compared with the OSD method	Çakmak et al. [57]
Indirect type force convective dryer	Paraffin wax	Mushroom	Chile	After incorporating PCM, this dryer took 9 hours and 20 hours for 8 and 12 mm mushroom slices, respectively, to reduce the moisture content from 91% to 5%. The PCM unit increased the dryer's thermal efficiency by 45%	Reyes et al. [58]
Indirect type force convective dryer	Paraffin wax	Basil and yellow oleander	Egypt	In comparison with OSD, solar dryers with PCM took approximately 10 hours less to complete the drying process	Shalaby et al. [59]
Indirect type force convective dryer	Paraffin wax	Kiwi (Hayward variety)	Iran	PCM with the inclusion of nanoparticles Al_2O_3 was used in their study. The addition of nanoparticles improved the PCMs heat conductivity, absorption rate, and storage capacity. As a result, the drying time was reduced by 38.6% compared with using pure paraffin	Bahari et al. [60]

(Continued)

Table 12.2 (Continued)

Dryer type	PCM type used in the dryer	Drying food	Study area	Summary	References
Indirect type force convective dryer	Paraffin wax	Banana and bitter gourd	India	Results demonstrated that the time required for drying with solar dryer with PCM was 10 hours for banana flakes, and in the case of bitter gourd slices, it was 18 hours, which was significantly less time taken than the OSD method	Arun et al. [61]
Indirect type natural convective dryer	Paraffin wax	Mint leaves	India	The operational hours of the solar dryer with PCM were increased about 5–6 hours even after sunset. At that time, drying air temperature was nearly 6°C higher than the ambient temperature	Jain and Tewari [62]
Indirect type natural convective dryer	Paraffin wax	Banana	India	By providing PCM, the working time of this solar dryer was increased up to 5 hours compared with no PCM use	Singh and Mall [63]
Mixed-mode force convective dryer	Paraffin wax	Black turmeric	India	Using solar dryer with PCMs, black turmeric was dried within 18.5 hours, while it took 46.5 hours using the OSD technique. When compared with the OSD, PCMs saved 60.7% time required for drying	Lakshmi et al. [64]
Mixed-mode force convective dryer	Paraffin wax	Bitter gourd	India	Solar dryer with PCM minimized the required drying time of bitter gourd by about eight hours compared with the OSD method. Moreover, it raised the thermal efficiency of the dryer by about 8% than without using PCM	Zachariah et al. [65]
Greenhouse solar dryer	Paraffin wax	Chilies	Thailand	A total amount of 500 kg of chilies were dried in this dryer. It took 2.5 days with a solar dryer having PCM and 3 days when it was done without PCM. In contrast, it took 10 days in the case of the OSD technique. Using PCM reduces drying time by about 12 hours compared with drying time without using PCM	Pankaew et al. [25]

12.5 Recommendation for Optimization of PCM for Use in Solar Dryers

Better functional operation and performance can be achieved from a PCM only when it is running under optimized conditions. Optimization requires the improvement of drawbacks of PCM, such as low TC, poor heat transfer rate, stability during melting/freezing of PCM, and economic sustainability. At the same time, the environmental impact of using PCM is a big issue for the successful implementation of PCM in food dryers. A brief discussion of all these optimization parameters is provided in the following subsection. Moreover, it will provide a clear indication to select the proper PCM for different kinds of solar dryers.

12.5.1 Thermal Conductivity

The TC of all mainstream PCMs, including organic and inorganic, is extremely low. TC defines the heat transfer rate through PCMs, which initiates the phase transformation process. Very low TC delays the process of melting and solidifying, which adversely affects the performance of the solar dryer. Rather improvement of TC will enhance the effectiveness of PCMs and better output of solar dryers. The possible solutions include surface area increment of PCM, such as using fins, improvement of the heat transfer fluid turbulence to increase the contact surface and encapsulation of PCMs in polymer or metal shells. Addition of various higher thermal conductive materials (carbon, graphite, aluminum) that are compatible with PCM has a positive effect on the improvement of TC. Bugaje conducted an experiment about the thermal enhancement of paraffin wax by adding various designs of aluminum conductivity promoters [66]. Results showed that during the thermal cycles, the time for phase change was found to be significantly reduced following such modification [66]. They have mentioned that better performance could be achieved by using internal fins of aluminum rather than using star matrices in the PCM. Moreover, as time progressed, researchers moved on to recent advances, and as a consequence, several authors also investigated the TC performance by adding nanoparticles to PCM [67, 68]. They concluded that carbon-based nanoparticle is the best option for TC improvement. Evidently, there is room for studying the enhancement of TC of PCMs that can be implemented on a commercial scale, and doing so will generate wider acceptance for the materials in drying systems.

12.5.2 Long-Term Stability

The stability of PCM largely depends on time, and chemical, thermal, and physical stability are aspects that can alter over time. The best output can be achieved from the PCM used in the solar dryer until all the properties remain unchanged. Two factors can cause the deterioration of long-term stability. One is the nature of the corrosion between the container and PCM, and the other is the poor performance of stability induced by the materials themselves [69]. Optimization of these factors is essential, whereas instability of PCM results in a shortened service life and increases the additional cost for the replacement of PCM. One of the most effective solutions for increasing corrosion resistance is the proper encapsulation of the PCM. During encapsulation, highly conductive and also economically feasible materials need to be selected. Hawlader et al. did a study on the long-term durability of paraffin that has been adequately encapsulated. After 1000 heat cycles, the capacity of energy storage, as well as the geometrical profile of the encapsulated paraffin, remained unaltered [70]. Moreover, some additives and mixture of several PCMs show better characteristics in case of improving the thermal stability of materials. Kimura et al. experimented with salt ($CaCl_2 \cdot 6H_2O$) PCM by integrating NaCl and concluded that the PCM was stable until 1000 thermal cycles of heating and cooling [71]. Another experimental investigation was done by Mehrali et al. regarding the stability of paraffin by adding graphene oxide and verified that this composite PCM was stable up to 2500 thermal cycles without any kind of chemical reaction [72]. Furthermore, the TC of paraffin was considerably improved. This also ensures the thermal and chemical reliability of paraffin used as PCM. Temperature ranges of melting/freezing (40–70°C) for both organic (paraffins) and inorganic (salt hydrates) PCMs suit more in solar dryers for the purpose of food drying. However, after several thermal cycles, paraffin exhibits superior thermal and chemical stability than salt hydrates. This is due to salt hydrate phase separation and subcooling. So, in the context of stability, it is recommended that paraffins having melting temperature ranges from 40°C to 70°C would be a suitable option for solar-assisted food dryers. Further research about adding additives to paraffins for ensuring long-term stability is required. Prior to industrial application, it is highly recommended that the PCMs' thermophysical stability should be tested utilizing a repeated thermal cycle.

12.5.3 Environmental Impact

According to the International Energy Outlook, the residential and commercial sectors use up to 20% of the total energy supply. From 2012 through 2040, each sector's energy demand is predicted to grow by 1.4% and 1.6% each year, respectively [73]. As a result, greenhouse gas emissions will rise simultaneously. Improvements in the efficiency of heating and cooling devices are required to offset the increase in energy usage. In an experimental study, Llorach-Massana et al. evaluated the environmental impact of employing Rubitherm as PCM on a root zone heating system and reported that it could reduce significant CO_2 emissions [74]. Moreover, PCM has great potential in the reduction of carbon footprint to ensure better environmental conditions as PCM reduces the conventional energy demand. Material compatibility with the environment is another environmental consideration. Biodegradable materials can be decomposed down into natural elements. Theoretically, entirely biodegradable materials should not have any negative environmental impact. Despite that, biodegradable products can have a long-term environmental impact, as it may take years or decades for some biodegradable goods to totally disintegrate. The biodegradability of PCMs has been determined using several accepted testing protocols, including ASTM 6400 and OECD 301. Biodegradability of paraffin products ranges from 21% to 55%, implying that paraffin-based products will take a long time to break down (years or decades) and will continue to have an adverse effect on the environment in the long run. Vegetable-based PCMs, on the other hand, are noncarcinogenic and harmless to plant life and wildlife, without any long-term harmful influence on the environment. Biodegradability testing on biobased PCM products revealed that they decompose at an average of 80% in 28 days, making them environmentally beneficial materials. However, the production cost and proper encapsulation are still a big concern for vegetable-based PCM. Moreover, thermal property and long-term stability of vegetable-based PCM are not good enough compared with paraffins. Therefore, more research regarding the biodegradability of paraffin would be a great prospect for reducing the environmental impact and making more eco-friendly effective solar energy based food dryers.

Industrial Corner

- A study showed that micro-encapsulated paraffin wax with aluminum pipes used in solar dryers saved USD246 annually [65].
- Besides this, if all the other mentioned optimization parameters are considered during PCM implementation in the solar dryer, then there is a huge potential to increase the thermal efficiency and effectiveness of the solar dryer, which will consequently increase the annual savings.
- If implemented commercially, it is highly recommended to consider all the optimization factors that are critically discussed in this chapter.

12.6 Concluding Remarks and Future Perspectives

Due to the usage of fossil fuels that pollute the air and exacerbate global warming through CO_2, CO, and NO_x emissions, drying has a significant impact on the environment. This chapter discussed the prospect of PCM in light of paving the way for energy-efficient dryers and TES systems. Future work into fabricating PCM that has better performance and reliability with reduced cost is highly desired. These improvements will accelerate its adoption in the industry and will help guide future innovations in PCM research. Research is already being performed to increase the efficiency of PCMs by utilizing macro, micro, or nanoscale encapsulation and stabilization [75, 76]. In influencing thermophysical properties, physical methods have been employed. However, the investigation into using chemical methods for this purpose remains unexplored. In addition, PCMs that operate outside the common – 20–100°C range should also be studied. The price, availability, TC, chemical stability, SH capacity, energy storage density, melting temperature, and supercooling and subcooling effects all play a role in the selection criterion for a suitable PCM. We recommend organic paraffin-based PCM for solar drying systems as they are benign, do not corrode easily, and are affordable. Lastly, it is imperative that the science community develop a standardized method for the testing of PCMs to make the process of choosing a material seamless. Gaining ample exposure in the past 20 years, it is reasonable to expect that PCMs will be widely implemented in industries as soon as their drawbacks are addressed, if not completely mitigated.

Glossary

Encapsulation Isolation of PCM from surroundings by covering with a coating material to hold the liquid or solid phase of the material.

Enthalpy It is a thermodynamic property and defines as the total heat content of a system.

Latent heat (LH) The quantity of heat a material absorbs or releases while changing its physical phase at a constant temperature.

Sensible heat (SH) The amount of heat a material absorbs or releases without changing its physical phase.

Thermal conductivity The ability of one material to conduct heat from its one side to the other.

Specific heat The amount of heat needed to increase the temperature of 1 gram of a substance by 1°C.

Phase change material A material in which a large amount of latent heat will be absorbed or released when it undergoes changes in its physical state.

References

1 H. Qu, M. H. Masud, M. Islam, M. I. H. Khan, A. A. Ananno, A. Karim, *Crit. Rev. Food Sci. Nutr.* **2021**, *62*(25), 6872–6886-115.

2 M. U. H. Joardder, M. H. Masud, *Food Preservation in Developing Countries: Challenges and Solutions*, Springer, **2019**, 57–66.

3 M. H. Masud, A. A. Ananno, N. Ahmed, P. Dabnichki, K. N. Salehin, *J. Food Eng.* **2020**, *281*, 110002.

4 M. Hasan Masud, A. Karim, A. A. Ananno, A. Ahmed, *Sustainable Food Drying Techniques in Developing Countries: Prospects and Challenges*, Springer, **2020**, 63–80.

5 M. H. Masud, H. H. Himel, A. M. E. Arefin, A. A. Ananno, M. Rashid, P. Dabnichki, *Environ. Challenges* **2021**, *5*, 100372.

6 M. Hasan Masud, A. Karim, A. A. Ananno, A. Ahmed, *Sustainable Food Drying Techniques in Developing Countries: Prospects and Challenges*, Springer, **2020**, 81–168.

7 M. H. Masud, A. Karim, A. A. Ananno, A. Ahmed, *Sustainable Food Drying Techniques in Developing Countries: Prospects and Challenges*, Springer, 2020.

8 L. S. Kjær, M. Poulsen, K. Sørensen, T. Condra, *Eng. Sci. Technol. An Int. J.* **2018**, *21*, 1047.

9 M. H. Masud, A. Karim, A. A. Ananno, *A. Ahmed, Sustainable Food Drying Techniques in Developing Countries: Prospects and Challenges*, Springer, **2020**, 41–61.

10 K. Chojnacka, K. Mikula, G. Izydorczyk, D. Skrzypczak, A. Witek-Krowiak, K. Moustakas, W. Ludwig, M. Kułażyński, *J. Clean. Prod.* **2021**, *320*, DOI: 10.1016/j.jclepro.2021.128706.

11 A. S. Mujumdar, *Handbook of Industrial Drying, Revised and Expanded*, CRC Press, 1995.

12 A. Menon, V. Stojceska, S. A. Tassou, *Trends Food Sci. Technol.* **2020**, *100*, 67.

13 M. U. H. Joardder, M. H. Masud, S. Nasif, J. A. Plabon, S. H. Chaklader, *AIP Conference Proceedings*, AIP Publishing, **2019**, 40010.

14 L. Ye, H. S. El-Mesery, M. M. Ashfaq, Y. Shi, H. Zicheng, W. G. Alshaer, *Case Stud. Therm. Eng.* **2021**, *26*, 101113.

15 A. Madhlopa, G. Ngwalo, *Sol. Energy* **2007**, *81*, 449.

16 C. Catorze, A. P. Tavares, P. Cardão, A. Castro, M. E. Silva, D. W. Ferreira, S. Lopes, I. Brás, *Energy Reports* **2022**, *8*, 392.

17 M. H. Masud, A. Karim, A. A. Ananno, A. Ahmed, *Sustainable Food Drying Techniques in Developing Countries: Prospects and Challenges*, Springer, **2020**, 169–185.

18 V. Palomba, V. Brancato, A. Frazzica, *Appl. Energy* **2017**, *199*, 347.

19 M. H. Masud, M. Nuruzzaman, R. Ahamed, A. A. Ananno, A. N. M. A. Tomal, *Int. J. Sustain. Energy* **2020**, *39*, 132.

20 M. U. H. Joardder, S. Mandal, M. H. Masud, *Int. J. Ambient Energy* **2020**, *41*, 1664.

21 A. A. Ananno, M. H. Masud, P. Dabnichki, A. Ahmed, *Sol. Energy* **2020**, *196*, 270.

22 S. R. L. da Cunha, J. L. B. de Aguiar, *J. Energy Storage* **2020**, *27*, 101083.

23 B. Gupta, J. Bhalavi, S. Sharma, A. Bisen, *Mater. Today Proc.* **2020**, *46*, 5550.

24 O. A. Babar, V. K. Arora, P. K. Nema, A. Kasara, A. Tarafdar, *Sol. Energy* **2021**, *229*, 102.

25 P. Pankaew, O. Aumporn, S. Janjai, S. Pattarapanitchai, M. Sangsan, B. K. Bala, *Int. J. Green Energy* **2020**, *17*, 632.

26 G. Srinivasan, D. K. Rabha, P. Muthukumar, *Sol. Energy* **2021**, *229*, 22.

27 M. M. Islam, A. K. Pandey, M. Hasanuzzaman, N. A. Rahim, *Energy Convers. Manag.* **2016**, *126*, 177.

28 L. M. Bal, S. Satya, S. N. Naik, *Renew. Sustain. Energy Rev.* **2010**, *14*, 2298.

29 L. F. Cabeza, A. Castell, C. de Barreneche, A. De Gracia, A. I. Fernández, *Renew. Sustain. Energy Rev.* **2011**, *15*, 1675.

30 E. Oró, C. Barreneche, M. M. Farid, L. F. Cabeza, *Renew. Energy* **2013**, *57*, 130.

31 M. Liu, W. Saman, F. Bruno, *Renew. Sustain. Energy Rev.* **2012**, *16*, 2118.

32 A. Abhat, *Sol. Energy* **1983**, *30*, 313.

33 K. K. Pillai, B. J. Brinkworth, *Appl. Energy* **1976**, *2*, 205.

34 H. P. Garg, S. C. Mullick, V. K. Bhargava, *Solar Thermal Energy Storage*, Springer Science & Business Media, **2012**.

35 A. R. Vakhshouri, *Paraffin Overv.* **2020**, 1–23.

36 S. S. Magendran, F. S. A. Khan, N. M. Mubarak, M. Vaka, R. Walvekar, M. Khalid, E. C. Abdullah, S. Nizamuddin, R. R. Karri, *Nano-structures & Nano-objects* **2019**, *20*, 100399.

37 S. S. Chandel, T. Agarwal, *Renew. Sustain. Energy Rev.* **2017**, *67*, 581.

38 M. Freund, R. Csikos, S. Keszthelyi, G. Mozes, 1982.

39 A. Sharma, V. V. Tyagi, C. R. Chen, D. Buddhi, *Renew. Sustain. Energy Rev.* **2009**, *13*, 318.

40 M. Kenisarin, K. Mahkamov, *Renew. Sustain. Energy Rev.* **2007**, *11*, 1913.

41 C. S. Herrick, D. C. Golibersuch, *Qualitative Behavior of a New Latent Heat Storage Device for Solar Heating/Cooling Systems*, General Electric Company Corporate Research And Development, **1977**.

42 N. Khandelwal, M. Sharma, O. Singh, A. K. Shukla, *J. Therm. Sci.* **2020**, *29*, 298.

43 F. S. Javadi, H. S. C. Metselaar, P. Ganesan, *Sol. Energy* **2020**, *206*, 330.

44 E. M. Alawadhi, H. J. Alqallaf, *Energy Convers. Manag.* **2011**, *52*, 2958.

45 G. A. Lane, *Int. J. Ambient Energy* **1980**, *1*, 155.

46 V. M. Swami, A. T. Autee, T. R. Anil, *J. Energy Storage* **2018**, *20*, 310.

47 G. A. Lane, D. N. Glew, E. C. Clark, H. E. Rossow, S. W. Quigley, S. S. Drake, J. S. Best, *Solar Energy Storage Subsystems for the Heating and Cooling of Buildings*, The University Press Of Virginia Charlottesville, **1975**, 43–55.

48 A. Sharma, S. D. Sharma, D. Buddhi, *Energy Convers. Manag.* **2002**, *43*, 1923.

49 V. K. Sharma, A. Colangelo, G. Spagna, *Renew. Energy* **1995**, *6*, 413.

50 S. Mehran, M. Nikian, M. Ghazi, H. Zareiforoush, I. Bagheri, *Sol. Energy* **2019**, *190*, 167.

51 M. Yahya, A. Fudholi, K. Sopian, *Renew. Energy* **2017**, *105*, 22.

52 A. ELkhadraoui, S. Kooli, I. Hamdi, A. Farhat, *Renew. Energy* **2015**, *77*, 1.

53 S. Paneliya, S. Khanna, Utsav, A. P. Singh, Y. K. Patel, A. Vanpariya, N. H. Makani, R. Banerjee, I. Mukhopadhyay, *Renew. Energy* **2021**, *167*, 591.

54 S. Devahastin, S. Pitaksuriyarat, *Appl. Therm. Eng.* **2006**, *26*, 1705.

55 A. A. Godarzi, M. Jalilian, J. Samimi, A. Jokar, M. A. Vesaghi, *Int. J. Refrig.* **2013**, *36*, 88.

56 Z. Azaizia, S. Kooli, I. Hamdi, W. Elkhal, A. A. Guizani, *Renew. Energy* **2020**, *145*, 1972.

57 G. Çakmak, C. Yıldız, *Food Bioprod. Process.* **2011**, *89*, 103.

58 A. Reyes, A. Mahn, F. Cubillos, P. Huenulaf, *Energy Convers. Manag.* **2013**, *70*, 31.

59 S. M. Shalaby, M. A. Bek, *Energy Convers. Manag.* **2014**, *83*, 1.

60 M. Bahari, B. Najafi, A. Babapoor, *J. Energy Storage* **2020**, *28*, 101181.

61 K. R. Arun, G. Kunal, M. Srinivas, C. S. S. Kumar, M. Mohanraj, S. Jayaraj, *Energy* **2020**, *192*, 116697.

62 D. Jain, P. Tewari, *Renew. Energy* **2015**, *80*, 244.

63 D. Singh, P. Mall, *Energy Sources, Part A: Recovery, Utilization, and Environmental Effects*, **2020**, *42*, 1–18.

64 D. V. N. Lakshmi, P. Muthukumar, A. Layek, P. K. Nayak, *Renew. Energy* **2018**, *120*, 23.

65 R. Zachariah, T. Maatallah, A. Modi, *Int. J. Energy Res.* **2021**, *45*, 1879.

66 I. M. Bugaje, *Int. J. Energy Res.* **1997**, *21*, 759.

67 S. Bazri, I. A. Badruddin, M. S. Naghavi, M. Bahiraei, *Renew. Energy* **2018**, *118*, 761.

68 L. Qiu, Y. Ouyang, Y. Feng, X. Zhang, *Renew. Energy* **2019**, *140*, 513.

69 M. M. Farid, A. M. Khudhair, S. A. K. Razack, S. Al-Hallaj, *Energy Convers. Manag.* **2004**, *45*, 1597.

70 M. N. A. Hawlader, M. S. Uddin, H. J. Zhu, *Int. J. Energy Res.* **2002**, *26*, 159.

71 S. Behzadi, M. M. Farid, *Appl. Energy* **2014**, *122*, 11.

72 M. Mehrali, S. T. Latibari, M. Mehrali, H. S. C. Metselaar, M. Silakhori, *Energy Convers. Manag.* **2013**, *67*, 275.

73 A. Sieminski, U. S. Administrator, *Int. Energy Outlook* 2016.

74 P. Llorach-Massana, J. Peña, J. Rieradevall, J. I. Montero, *Renew. Energy* **2017**, *103*, 570.

75 W. Cheng, R. Zhang, K. Xie, N. Liu, J. Wang, *Sol. Energy Mater. Sol. Cells* **2010**, *94*, 1636.

76 S.-G. Jeong, J. Jeon, J.-H. Lee, S. Kim, *Int. J. Heat Mass Transf.* **2013**, *62*, 711.

13

Multi-Functional Properties of Halloysite Nano-Clays in Food Safety and Security

*Satwik Majumder and Saji George**

Department of Food Science and Agricultural Chemistry, Macdonald Campus, McGill University, 21,111 Lakeshore Ste Anne de Bellevue, Quebec H9X 3V9, Canada
** Corresponding author*

Highlights

- Nanotechnology that fulfills constraints of sustainability holds great promise for enhancing food production, storage, and safety.
- Biocompatibility, cost-effectiveness, and multi-functionality make halloysite nanotube (HNT) a suitable candidate for *sustainable nanotechnology* applications in the agri-food sector.
- Suitable morphological and surface chemistry features enable HNT to be incorporated in food-packaging matrices to overcome poor gas-barrier and thermo-mechanical properties.
- The tubular morphology and tunable surface chemistry make HNT valuable for nano-enabled drug and fertilizer delivery in animal and plant agriculture.
- Comprehensive studies on the release kinetics of active substances, issues of HNT agglomeration, and risk assessment of HNTs and HNT-based composite films should be resolved to maximize the application of HNT in the agri-food sector.

13.1 Overview

Ensuring the food security of the growing world population calls for sustainable solutions aimed at enhancing food production and maximizing the safe utilization of produced food [1]. The World Health Organization estimates a need for 60% more food compared with the current requirement to feed the expected 9.5 billion world human population by 2050 [1]. While it may be possible to meet the needed food production, experts warn of the environmental and social implications of current agricultural practices. Similarly, *food spoilage* due to microbial contamination and undesired physicochemical reactions has huge social, economic, and environmental implications across the globe [2]. Recent studies predict a rise in food loss and spoilage to 2.1 billion tons worth $1.5 trillion annually by 2030 without adequate measures [3]. The increasing demand for fresh foods thus requires new technologies to ensure food safety and security but operating within the constraints of sustainability [4]. Food trade and supply chains are becoming increasingly longer and more

Materials Science and Engineering in Food Product Development, First Edition. Edited by Wing-Fu Lai.

complex arising from global trends in urbanization, population migration, increasing appetite for nontraditional food and changing consumer preferences. Consequently, ensuring *food security and safety* demands technological and surveillance intervention at every stage in the food supply chain. At the food production stage, technologies, including nanotechnology, have great promises for warding off infections and improving nutrient delivery and uptake (Industrial Corner I) [5]. For instance, nanotechnology-enabled products are now commercially available to control bacterial, fungal, and viral infections [6]. Similarly, there have been products to improve plant and animal nutrition to enhance agricultural production [7, 8]. After the harvest, until the food reaches the consumer, food comes in contact with different types of containers and food-packaging materials and thus needs protection from microbial and physicochemical issues.

The agri-food industry is a major user of plastic materials – mainly for packaging and transport purposes. Studies in the recent past show the magnitude of environmental issues arising from the use and discard of plastic material in the environment [9]. Therefore, over the past few decades, the sole role of food and beverage packaging for marketing functions has advanced not only to an intelligent/smart role for protection (against spoilage), and detection (of spoilage) but also to ensure the reduction in traditional plastic consumption and encourage higher recycling efficiency [10]. For instance, the Food and Agriculture Organization of the United Nations (FAO) proposed that active packaging is the key technology to prevent microbial or physicochemical spoilage, and improve the quality of fresh foods during storage/transportation as it employs *bioactive components* in the *packaging systems* that interact with the packed food or its surroundings by releasing or absorbing substances [11, 12]. More recently, the popularity of intelligent packaging systems that enable recording, tracking, and logistics relevant to the alterations in the quality of packaged foods to consumers, is on the rise [13]. For instance, active agents, like curcumin and anthocyanin, have been used in intelligent packaging to monitor the freshness of food according to the color changes toward pH fluctuations, whereas essential oils have been used in active packaging as antimicrobial/antioxidant agents [14, 15]. A typical active packaging consists of a film matrix and active agents that contribute to various functional properties [16]. There has been an interest in developing and using biodegradable natural polymers, such as proteins and polysaccharides, as food package films [17]. For instance, chitosan, alginate, starch, cellulose, and zein have been applied as biopolymer materials over the past years as they provide good film-forming and functional properties [18]. However, poor gas-barrier and thermo-mechanical properties are some of the limiting factors that have been restricting their successful widespread commercial application.

Integrating reinforcement agents into the polymer matrix is a promising way of improving thermo-mechanical properties and enhancing the controlled release of active substances. A suitable reinforcement agent should possess a satisfactory combination of intrinsic properties, such as safety, biodegradability, and compatibility, with the matrix [19]. The application of nanomaterials, such as zinc oxide, titanium oxide, carbon nanotube, and nanoclays, as reinforcement agents have substantially improved the performance and functions of both active and intelligent packaging technologies [20]. Among these nanofillers, clays, such as kaolin, montmorillonite, vermiculite, laponite, and saponite, are drawing significant attention in a vast range of industrial applications due to their *biocompatibility* and ability to enhance the physical/thermal/mechanical properties of films [21]. In 2014, the global nanoclay market for food packaging was considered one of the largest segments accounting for USD340 million, and is expected to expand significantly in the years to come, which is due to their superior performance in comparison with other tested nanofillers such as nano-silica, nano-alumina, nano-titanium dioxide, calcium carbonate, etc. [22]. For instance, studies showed that the platy morphology of nanoclays implements better *mechanical properties* in biopolymers than the nano-cellulose with typical spherical or cubical morphology [10]. Moreover, the application of a minimal amount (<10 wt%) of nanoclays in the polymer matrix has been reported to enhance the mechanical, thermal, barrier, and degradation traits significantly [23]. There has been research on the use of clay-based nanomaterials for food-packaging applications. Tang et al. have shown the applicability of starch-montmorillonite clay *nanocomposite* for improving barrier and mechanical properties [24]. Bentonite clay materials have been used in feed to mitigate mycotoxin, whereas zeolite-clay-based nanocomposites are often used for soil and plant conditioning [25]. Thus, while different types of clay particles are used at different stages in the food supply chain, halloysite clay nanotubes (HNTs) distinguish as a material with the possibility of accomplishing all these purposes (Industrial Corner I) [26]. This chapter focuses on a systematic discussion of the multifunctional attributes of HNTs that make them useful for different food and agricultural safety and security applications, limitations of HNTs, and future perspectives to achieve optimal quality of food and reduction in food spoilage.

Industrial Corner I

- There have been major advances in nano-based technologies in food science and research as it aids in the tracking–tracing–monitoring and remediation of toxins, pathogens, and pesticides that can assure the maintenance of food quality.
- HNTs are biocompatible, cost-effective, and easily accessible, making them a suitable candidate for industrial use in the fields of biomedicine, agriculture, and food safety.

13.2 Halloysite Nanotubes (HNT): A Versatile Natural Nanomaterial

HNT is a type of clay mineral with aluminosilicate chemistry that is tubular in structure (Figure 13.1a). Typically, HNTs are 0.2–1.5 μm in length with an inner and an outer diameter of 10–30 and 40–70 nm, respectively (Figure 13.2a) [27]. Mineral forms of HNTs are mined from several countries, including Belgium, China, France, and New Zealand [28]. The morphological features, however, could vary with their source and methods of purification. Studies have shown that the meso/macroscopic superstructure of tubular HNTs results from the clay layers enfolding around themselves forming hollow cylindrical structures that are associated with a mismatch between AlO_6 sheets and oxygen-sharing tetrahedral SiO_4 sheets [29]. Consequently, the exterior surface of HNT comprises siloxane (Si–O–Si) groups, whereas the interior of HNTs contains a gibbsite-like array of aluminol (Al–OH) groups (Figure 13.2a) [30].

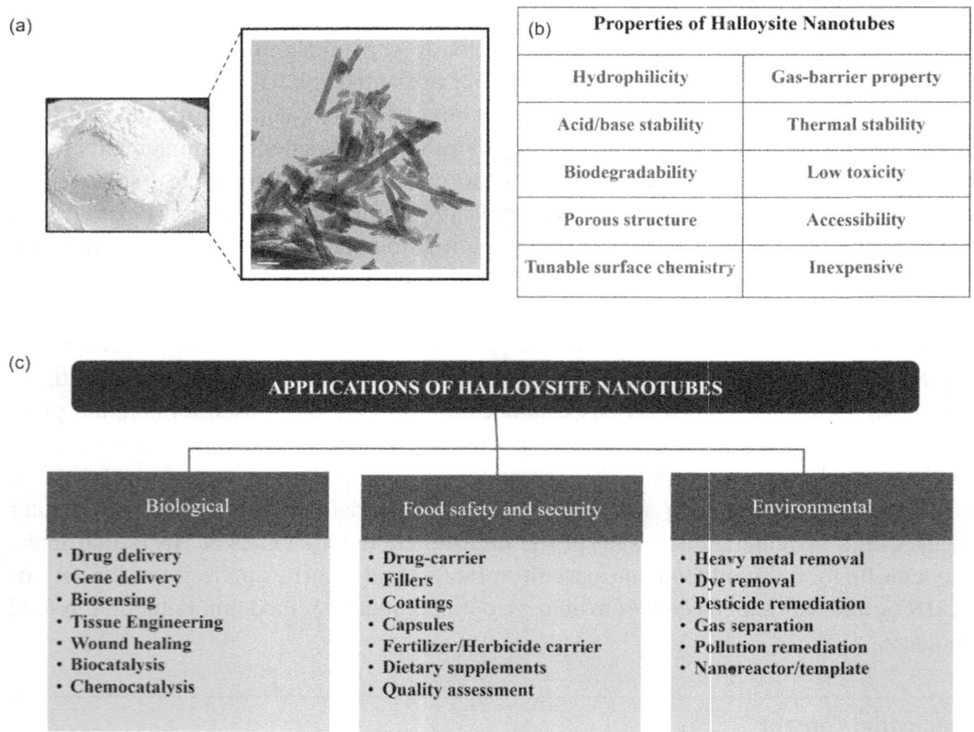

Figure 13.1 The (a) TEM image; (b) multifunctional properties; (c) applications of HNTs in the fields of biology, environment, and food safety and security.

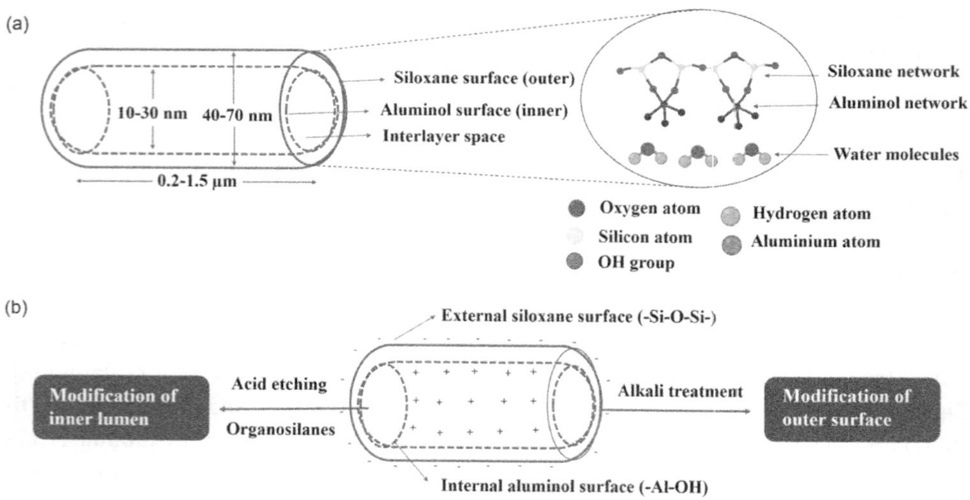

Figure 13.2 (a) The surface chemistry of HNTs; (b) surface modification strategies of inner lumen and the outer surface of HNTs. (Graphic created with Biorender.com.).

HNTs possess many desirable qualities for their use in the agri-food sector. First of all, the tubular shape and nanometric dimension provide an extremely high surface area for a diverse range of applications, such as adsorbent, filler, and encapsulation agents [31]. Different surface chemistries of the lumen and external surface provide opportunities for surface functionalization and selective loading of active agents into the lumen (Figure 13.2b) [32]. The silica chemistry of the surface allows the functionalization of the material with a range of molecular entities, rendering the material hydrophilic, hydrophobic, or preferential, binding to the target surface, etc. (Industrial Corner II) [33]; whereas, the Al_2O_3 interior could be leached by acid etching to widen the lumen volume [34]. Etching by acid treatment and grafting with organosilanes also facilitates enhanced agent loading/adsorption. For instance, Abdullayev et al. used sulfuric acid to increase the average diameter of the HNTs lumen, which resulted in an increase in the loading capacity by 2–3 times [35]. Modification of HNTs with (3-aminopropyl) triethoxysilane (APTES) enhanced the absorption and loading capacity by 25% and improved the release kinetics of the active agents [36]. The most common methods that are used in the loading of active agents in HNTs are by stirring and sonicating to facilitate dispersion and homogeneous mixing of HNTs with the solution of the active compounds or by vacuum pumping in/out to exhaust air and initiate agent loading [19]. Owing to high biocompatibility, cost-effectiveness, easy availability, large specific surface area, and ideal nanotubular structure, HNTs show high potential to develop HNTs-based antimicrobial nanocomposites for biological (wound healing, treatment of microbial infections, etc.) and environmental (wastewater management, surface decontamination, etc.) applications (Figure 13.1b,c) [32, 37]. In addition, favorable biodistribution, stability, controlled cargo release, compatibility with other polymers, and low toxicity are some of the crucial features that make HNTs multi-functional, and desirable nanofiller/carriers for sustainable packaging systems and agricultural applications (Figure 13.1b) [30].

13.2.1 Halloysite as Nanofillers in Packaging Matrix

Natural polymers ensure high availability and good biodegradability; however, they often lack suitable functional properties, which is a necessity in packaging systems. To overcome this challenge, nanocomposites are incorporated as nanofillers in the film matrix as an active ingredient and also to fulfill the reinforcing functions. Owing to the nano-size and large surface area, HNTs can be utilized as nanofillers to occupy the gap in the molecular chains of the film matrix that could contribute to the improvement of the barrier, mechanical, and thermal properties [39]. To withstand stress during product production, handling, and transportation, films with high mechanical properties, i.e. favorable tensile strength and elongation at break are usually considered for food-packaging applications [40]. Reports from previous studies suggest that an increase in HNTs content causes an increase in tensile strength and a decrease in elongation at break, ultimately enhancing the properties of the film matrix [19]. The possible reasons for this phenomenon could be the existence of an intermolecular high number of hydrogen bonds between HNTs and the film matrix that restricted the mobility of the matrix chains, resulting in high strength [41]; electrostatic interaction between the negative charges in the external surfaces of the HNTs and a polycationic matrix [42]; the strain-induced alignment of the HNT layers in the polymer matrix [42], and good interfacial adhesion between the incorporated fillers and the film matrix [40]. However, in some cases, the *agglomeration* of HNTs often results in a decrease in tensile strength and a reduction of effective surface areas that ultimately deteriorate the uniformity and smoothness of films [23]. For instance, Othman et al. reported a reduction in the mechanical properties of a polylactic acid-based film incorporated with ≥5 wt% HNTs due to agglomeration [43]. Therefore, the uniform dispersion of nanofillers in film matrix chains is one of the most critical problems that need to be addressed.

Moisture and gas permeation in the packaging matrix can lead to microbial adhesion and colonization and oxidation of food constituents [12]. HNT-incorporated films for application in food packaging decrease water and gas permeability since the HNTs extend the path of the permeating molecules through the film matrix (Figure 13.3a) [19]. Recently, a bio-nano composite constituting polyvinyl alcohol and starch was prepared that demonstrated good water and gas-barrier properties upon the addition of HNTs to prolong the *shelf life* of avocados and peaches [44]. Ample evidence has been reported by researchers to affirm the nature of HNTs in improving the thermal stability and flame-retardant properties of polymers (Figure 13.3b) [7, 45]. For instance, polymers doped with HNTs (4–5 wt.%) increase the strength and adhesivity of the composite by 40–50% along with improvement in its flame retardancy as HNTs reduce the heat flux of the film and slow down the heat transfer through the film [19, 38]. These characteristics can be explained as a direct consequence of the hydrogen bonds between HNTs and the film matrix at the molecular level [19].

13.2.2 Halloysite as Nano-Carrier in Active Packaging Systems

Functional agents, such as antimicrobials, antioxidants, preservatives, etc. play an important role in food safety and preservation when incorporated into packaging systems. Direct and uncontrolled exposure of active agents in the packaging matrix typically leads to a burst release of the agents, deteriorates the functionality of the materials, and further contributes to toxicity, and migration. The biocompatible nature of HNTs, the hollow tubular interior with a positive charge, and the negatively charged outer surfaces of HNTs facilitate the incorporation (loading in the lumen)/adsorption (onto the external surface) of active

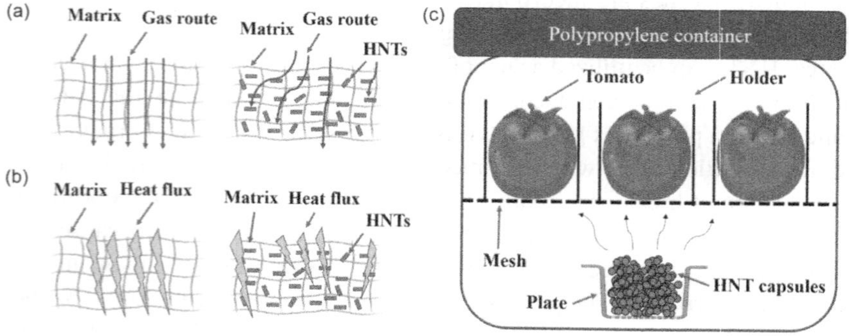

Figure 13.3 Effect of (a) gas and (b) heat flux with or without the application of HNTs in the packaging matrix; (c) illustration of active packaging systems with alkali-treated HNT capsules for preserving cherry tomatoes. (Modified from Lee et al. and Li et al. [15, 19]; graphic created with Biorender.com.).

agents before they are blended into the packaging films that could enhance the biological efficiency and physical performance of the film [23]. The compounds loaded into the positively charged lumen of HNTs acts like a "shield" that assists in the controlled release and stability of the active agents. On the other hand, the negatively charged outer surface owing to the siloxane groups of HNTs can form electrostatic interaction with positively charged molecules [42]. For this reason, HNTs could adsorb positively charged active agents, such as essential oils, and several metal-based nanocomposites, such as silver, copper, zinc oxide, etc. [32]. The tunable surface chemistry of HNTs has attracted significant interest in the agri-food sector, mainly due to their applicability as polymer–clay nanocomposites, such as to facilitate increased adsorption of the active agents. For instance, the functionalization of HNTs with amine groups using organosilanes (such as APTES) has been investigated as an alternative to enhance the dispersion of clay in a polymer matrix, leading to an improvement in the mechanical properties of the resultant matrix [29]. Similarly, the specific surface area and pore volume of HNT could also be increased by acid treatment, as detailed in the preceding section. Selective etching of the inner layer of HNT initiates with the diffusion of hydrogen ions into the lumen of HNTs, followed by interaction between alumina and hydrogen ions and diffusion of the reaction's product out of the lumen [35].

The strategy of loading active agents into HNTs for enhanced stability, controlled-release kinetics, and effectivity has been used in various fields, including medicine, material science, and agriculture [7, 46]. Recently, we have synthesized a nanocomposite constituting silver nanoparticles fabricated on tannic acid-modified HNTs that showed enhanced physicochemical stability, improved bioavailability, and retention time of silver and tannic acid, which resulted in a superior antibacterial efficiency against multi-drug resistant bacteria when compared with pristine silver nanoparticles [47]. A study similar to HNTs loaded with nanosilver embodied into poly-l-lactic acid exhibited a sustained release of silver ions and a long-lasting antibacterial property without compromising the cytocompatibility [48]. These studies suggest the potential application of inorganic nanotubes, such as HNTs as a drug carrier. Preservatives have been widely used in the food industry, but several synthetic compounds have shown harmful effects on the environment and human health, leading to an increase in the demand for preservatives with a natural origin [49]. Essential oils and tannins are typical examples of natural preservatives predominantly possessing antibacterial and antioxidant properties. However, high volatility, low thermal stability, strong flavor, and aroma limit the use of natural preservatives in processed foods or food packaging. Loading of such natural preservatives in HNTs has been reported to get protection from degradation and reduced undesirable flavor and aroma in packaging systems [50]. Further studies are needed in choosing compatible functional agents, and investigating novel controlled-release technologies and methods to achieve improved loading capacity and adsorption [19]. More techniques focused on surface modification, such as layer-by-layer self-assembly, are a perspective direction to achieve nanocomposite with enhanced physicochemical properties [19].

13.2.3 Halloysite as a Colorimetric Indicator in Intelligent Packaging Systems

Nanoclays have promising applications in intelligent packaging systems as well. For instance, montmorillonite nanoclay encapsulating a blueberry extract has been recently introduced as a colorimetric indicator for intelligent packaging by Gutiérrez et al. [51] An anthocyanin-loaded montmorillonite powder demonstrated a distinguishable response to varying pH under acidic and alkaline conditions [51]. The study supported the application of nanoclay-based composites in the polymer matrix as a colorimetric indicator system for fresh meat and fishery products. Similar strategies have been applied to detect milk spoilage where biodegradable starch–clay nanocomposite films integrated with colorimetric indicators, such as methyl orange and bromocresol green, were developed and attached to milk bottles [52]. This technique remarkably lowered the solubility of the film in water and employed controlled-release kinetics of the dyes into the milk [52]. A 4-aminothiophenol-capped HNT/silver nanoparticles were synthesized by Zhang et al. for the determination of nitrite ions in meat products [53]. More research is, however, required in the future to explore the application of HNTs as food-quality indicators of a wide range of foodstuffs.

13.2.4 Halloysite in Food Coatings, Capsules, and Fibrous Films

Apart from the use of HNTs in packaging films, HNTs have potential applications in direct (as coatings) or indirect (as capsules/powder) contact with foodstuffs. The traditional coating technique involves dipping the food by immersing it in a coating solution resulting in a film that is well intact on the food surface [54]. This technique can preserve quality, and increase shelf life, of fresh fruits by reducing the ripening rate and/or inhibiting microbial growth on the surface [19]. For instance, Buendía et al. encapsulated essential oils within cyclodextrins and HNTs and coated them onto cardboard boxes, which prolonged the shelf life of fresh tomatoes without affecting the sensory quality [55]. HNTs in the form of capsules/powder in indirect contact with foodstuff have interesting applications as well. For instance, Jang et al. encapsulated thyme essential oil in HNTs using vacuum-pulling and layer-by-layer surface-coating methods. The thyme burst release was restricted by end-capping [56]. The nanocapsules were further mixed with flexographic ink and integrated on paper for applications in food packaging [56]. The composite showed a controlled release of thyme oil and strong antibacterial activity for about 10 days [56]. Lee et al. reported that the capsule synthesized with NaOH-treated HNTs entrapping thyme oil decreased molds and yeast colonization on the surface of cherry tomatoes and demonstrated a sustained release of thyme oil (Figure 13.3c) [57]. In the future, these techniques need to get tested for a wide range of foodstuffs as a potential material for antimicrobial packaging.

Electrospinning is an effective, cheap, and versatile technique used to produce submicron or nano-scale fibrous films for their fabrication in active food-packaging materials employing antimicrobial, antifungal, or antioxidant properties [58]. For instance, Zou et al. developed chitosan/polycaprolactone electrospun nanofibers with chlorogenic acid-loaded HNTs that showed a prolonged sustained release with enhanced antioxidant and antimicrobial activities [59]. In a similar study, Li et al. fabricated composite fibers constituting polyethylene oxide and chitosan with HNTs and iron oxide nanoparticles. The nanofibrous adsorbents showed wide adaptability and reusability along with superior antibacterial activity against *foodborne pathogens* [60].

13.2.5 Halloysite in Plant and Animal Agriculture

Chemical fertilizers as nutrients for agriculture play an important role in improving crop yields; however, the low uptake rate of the nutrients by the crops has resulted in water pollution and soil compaction [61]. Many metal-based nanoparticles, such as cadmium sulfide (CdS), magnetic nanoparticles (Fe_3O_4), and nano-titanium dioxide (TiO_2), have been introduced over the years as nano-fertilizers that were reported to stabilize the loaded active agent to achieve environmentally responsive and controlled release under a range of stimuli [62]. However, the potential toxicity of such heavy metal elements and the laborious synthesis of such materials limited their application in plants. Therefore, there is an urgent demand for environmentally friendly fertilizers that not only offer good nutrient uptake and increased crop quality but also reduce the toxicity generated by the overapplication of fertilizers to the soil [63]. Bellani et al. revealed that exposure to 1.5 g/L of HNTs has no significant effect on seedling development, germination, or mitotic index: thus, providing enough evidence in support of HNT use for environmental purposes [64]. Owing to their versatile properties, HNTs have been recently used as a delivery vehicle in agricultural applications. For instance, Wang et al. reported a nanocomposite, comprising chitosan assembled to HNTs that increased stability and achieved the controlled release of fertilizers and thus proposed a delivery system for agricultural fertilizers [65]. Similarly, chemical pesticides that are often used in controlling weeds and pests are usually highly toxic and pose a threat to public health and the environment [66]. The issue was addressed by Zeng et al. where botanical herbicide-loaded HNTs were incorporated into poly(vinyl alcohol)/starch composites where the addition of HNTs significantly improved the erosion resistance of the film, delayed the release of herbicide, and prolonged the service life [7].

Over the past decades, consumption of meat-based products has increased significantly and the soaring demand resulted in many efforts to enhance the efficiency of meat production. One of the promising ways to achieve this goal is by increasing feed's nutritional effectiveness using feed additives, such as aluminosilicate minerals [8]. Nanoclays under the aluminosilicate group do not get digested but can bind to toxic metabolites and mycotoxins, and move them through the digestive tract, without causing a detrimental effect on the animal's health [67]. Moreover, they may increase animal growth and vitality, strengthen immune functions, prevent various kinds of diseases, as well as improve the quality of the derived final product [68]. In a recent study, the addition of 1% HNTs to the feed of broiler chickens has caused a reduction in feed consumption and higher utilization of crude protein that led to improved fattening results. Moreover, the consumption and excretion of water by the chickens got reduced, resulting in overall improvement in the chicken's health [8].

In North America, 75–80% of all antibiotics that are regularly applied to livestock are considered to be medically important to humans [69]. Extensive use of such antibiotics in animal agriculture has contributed to the emergence of *antimicrobial resistance* (AMR), which often gets transmitted to humans either through the food supply chain or direct contact with infected animals [69, 70]. Recent studies from our group have shown the prevalence of AMR and virulence in a library of bacteria against a diverse class of antibiotics and metals relevant to veterinary and public health [71, 72]. These emerging challenges of AMR warrant developing sustainable antibacterial strategies that are efficient in remediating *zoonotic infections* in animals. Combination therapy with antimicrobial-adjuvant combinations has been applied for the treatment of infections involving AMR pathogens; however, poor bioavailability, cytotoxicity, stability, release, and overdosing have been matters of concern [72]. Owing to the favorable surface chemistry as discussed earlier, HNTs could also be used as a nano-carrier for multiple active agents with complementary modes of action and could be applied against recalcitrant bacterial infections in livestock.

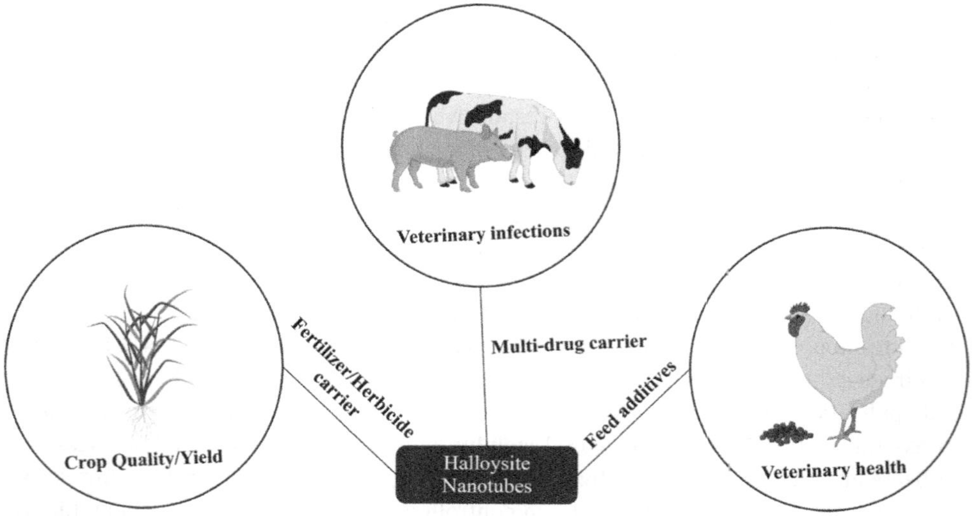

Figure 13.4 The application of HNTs in crop quality/yield, veterinary health, and infections. (Graphic created with Biorender.com.).

However, to achieve that goal, extensive studies on selective drug delivery mechanisms and toxicity need to be explored. Figure 13.4 denotes the application of HNTs in crop quality/yield, veterinary health, and infections.

Industrial Corner II

- The tunable surface chemistry and large specific surface area in the exterior and interior HNT surface provide developers the chance to accommodate multiple active agents.
- HNTs can play a significant role as drug delivery systems, nanofiller, coatings, capsules, and carriers for dietary supplements, fertilizers, and herbicides, thus making them desirable for food and agricultural industries.

13.3 Toxicity and Migration Associated with Halloysite

Nanoclays that are often used in food and beverage packaging films, such as beer bottles or bags for foodstuffs, usually stay in direct contact with them, which could potentially affect public health especially if the clay is organically reformed with quaternary ammonium [73]. Moreover, exposure to nanoparticles in humans could also occur via inhalation of contaminated air, dermal contact, and consuming contaminated food with particles that potentially migrate from the packaging systems [74, 75]. As nanoclays, such as HNTs, are difficult to be segregated into individual airborne nanosheets, the larger particles of micrometer size are often responsible for environmental exposures (Industrial Corner III) [76]. Therefore, assessing the significance and impacts of nanoparticle migration and evaluating the extent of migration is extremely important. Most of the studies that focused on the migration behavior of nanoclays reported the migration from polymer/nanoclay composites to food to be negligible in comparison with other nanocomposites [75, 77]. The possible reason could be the late erosion of the polymer matrix by the food simulants because of the strong and uniform adherence of nanoclays to polymer employing a larger surface contact area than spherical nanocomposites, such as AgNPs [10]. Nasiri et al. reported a reduction in the diffusion rate and migration properties in food packages to foodstuffs after integrating lamella-structured nanoclays in the polymer as a reinforcement agent [78]. Recently, HNTs were tested for toxicity in a nematode model, *Caenorhabditis elegans*, and found to be safe up to 1 mg/mL of concentration [38]. HNTs were concluded as safe when compared with the multi-walled carbon nanotubes for toxicity to human umbilical vein endothelial cells and blood vessels of mice [79]. However, in the future, a more extensive study targeting a broad spectrum of food simulants is necessary for true and productive assessment and regulatory establishments.

13.4 Future Perspectives

The antigen-specific biomarkers are often used to identify food spoilage-causing organisms. Harnessing its utility in the quick and accurate detection of food pathogens, including bacteria, viruses, and mycotoxins through HNT-enabled nanosensors could be an interesting approach in the future [80]. These nanosensors could also be applied to sense the release of gases as a result of food spoilage (Industrial Corner III) [81]. That way, spoilage could be detected in every stage of the food chain, thus reducing food loss and benefiting the food producers, retailers, and consumers [82].

Although HNTs exhibit excellent properties in food-packaging films, comprehensive studies on controlled loading and release kinetics of active substances encapsulated in HNTs are required [19]. One of the promising methods to control the release of active compounds is by coating the two ends of HNTs with responsive polymers [83]. A comprehensive understanding of the dynamics of the release kinetics of active substances will improve the utilization efficacy with minimum doses of preservatives while reducing food spoilage and maximizing the food quality and shelf life [19]. Excessive HNTs can result in a high degree of agglomeration; thus, more research is suggested to focus on reducing the localized accumulation of HNTs and improving the homogeneity of the film [15, 84]. The use of long and thin HNTs, HNTs below 15 wt%, surface modification, fabrication of active agents, and combination with other nanofillers (such as cellulose nanocrystals) could resolve this issue [40, 84].

Packaging from foods contributes to approximately two-thirds of the total packaging waste globally [85]. Thermoplastics like polyolefin face issues related to non-biodegradability as they are inert to the natural environment [10]. Therefore, the degradation of the packaging in the biosphere is highly recommended as it saves time and requires less energy [86]. Photoirradiation or the use of thermal energy could play a significant role in the abiotic degradation of petroleum-based polymers [87]. Few studies have reported the enhancement of UV absorption capability in petroleum and bio-based polymers after the addition of nanoclays, resulting in accelerated biodegradation [88]. For instance, Oliver-Ortega et al.

prepared a composite material based on PLA and reinforced it with bentonite nanoclay (up to 4% w/w), which demonstrated an improved barrier character against water vapor transportation and biodegradable property [89]. However, there is hardly any scientific evidence on the effect of HNTs blended packaging systems concerning biodegradability and thus it opens a new scope for research. One of the strategies to explore could be the encapsulation of hydrolytic enzymes (e.g. phosphatases, esterases, protease) in HNTs, which could play a crucial role in polymer degradation.

HNTs have been listed under generally regarded as safe; however, there are still safety concerns about nanostructured materials and the controversy regarding the global acceptance of nanomaterials. Multiple studies have been conducted to identify the regulatory gaps in the field of nanotechnology, covering different jurisdictions such as the United States, the European Union, and Australia [90]. To date, there is a lack of information on the regulation of migration limit on the use of nanoclays; however, the legislation in Europe (Council Directive 90/128/EEC [1990]) has allowed an overall migration limit of 10 mg/dm [2] surface area to all the substances that are capable of migrating from the packaging material to the food product [10, 90]. Therefore, *risk assessment* on a case-by-case basis and systematically valid and safe metabolic data of HNTs and HNT composite films are urgently needed. Approval from consumers is mandatory for their implementation on a worldwide scale, thus, the government agencies and accredited research institutes should take the initiative and work in co-ordinance and set down common rules and regulations that would develop acceptability among consumers.

Industrial Corner III

- Development of HNT-enabled nanosensors, understanding the drug-release kinetics and biodegradability and solving issues of agglomeration open up new scope for research.
- However, more studies are required, focusing on the toxicity and migration of active agents from HNTs to different types of foodstuffs before they are considered to be developed in industrial setups.

13.5 Conclusive Remarks

Issues on food and agricultural safety and security are growing concerns worldwide. The development of active and intelligent packaging systems has been considered promising to improve food quality and reduce food spoilage. HNT-based nanocomposites can serve as nanofillers, nano-carriers, food coatings, and nanocapsules in such food-packaging systems, to improve physicochemical, antimicrobial activities, and enhance film performances. Meanwhile, HNTs could be also applied in nano-enabled antibacterial therapies to eliminate zoonotic pathogens, and as a delivery system for fertilizers, herbicides, and dietary supplements to improve crop/soil quality, and animal health. However, extensive studies are required to evaluate the toxicity and potential migration associated with HNTs. Comprehensive studies on the release kinetics of active agents and the biodegradability of

HNT-based nanocomposites warrant further exploration as well. The limitations regarding HNT agglomeration and regulatory issues are also required to be addressed. Overall, extensive studies involving microbiology, food science, and material sciences concerning HNTs are required to achieve optimal quality maintenance for the food and agricultural industries in the future.

Glossary

Agglomeration Two or more nanoparticles that have associated into a cluster.
Antimicrobial resistance (AMR) The ability of microorganisms to survive exposure to an antimicrobial agent that was previously an effective treatment.
Bioactive components Biomolecules from food and plants that exhibit the capacity to employ one or more metabolic processes.
Biocompatibility Ability of a nanomaterial to perform with the response of an appropriate active component.
Foodborne pathogens Biological agents that can cause a foodborne illness.
Food security and safety State of having a stable and sufficient or adequate supply of food without any physical, chemical, or microbial hazards.
Food spoilage Undesirable changes of food products in taste, smell, appearance, or texture due to physical, chemical, or microbial interference.
Nanocomposites Nanomaterials with one or more phases at nano-sized dimensions embedded in a ceramic, metal, or polymer material.
Packaging systems Coordinated system of developing products for food transport, storage, protection, logistics, sale, and end-use.
Risk assessment Systematic evaluation of potential risks/hazards in the application of nanomaterials in the agri-food sector.
Shelf life Length of time for which a food product remains fit for consumption.
Sustainable nanotechnology Development of technology within the 1–100 nm scale, with durable economic feasibility, responsible utilization of natural/artificial resources, and minimizing hazards to public/veterinary health and the environment.
Mechanical property Physical attributes that a material exhibits upon the application of forces.
Zoonotic infections Infectious diseases that could transmit between species from animals to humans and vice versa.

References

1 M. van Dijk, T. Morley, M. L. Rau, Y. Saghai, N. Food **2021**, *2*, 494.
2 P. G. Braun, J. P. J. F. S. M. C. W. P. Sutherland, **2006**, 86.
3 Food and Agriculture Organization of the United Nations, **2011**.
4 World Health Organization. Regional Office for South-East Asia, **2015**.
5 B. S. Sekhon, *Nanotechnol. Sci. Appl.* **2014**, *7*, 31.

6 a) A. R. Kirtane, M. Verma, P. Karandikar, J. Furin, R. Langer, G. Traverso, *Nat. Nanotechnol.* **2021**, *16*, 369; b) S. Sharmin, M. M. Rahaman, C. Sarkar, O. Atolani, M. T. Islam, O. S. Adeyemi, *Heliyon* **2021**, *7*, e06456.
7 X. Zeng, B. Zhong, Z. Jia, Q. Zhang, Y. Chen, D. Jia, *Appl. Clay Sci.* **2019**, *171*, 20.
8 M. Nadziakiewicz, M. W. Lis, P. Micek, *Anim. Open Access J.* from MDPI **2021**, *11*.
9 R. Kumar, A. Verma, A. Shome, R. Sinha, S. Sinha, P. K. Jha, R. Kumar, P. Kumar, S. S. Das, P. Sharma, P. V. Vara Prasad, *Sustainability* **2021**, *13*, 9963.
10 N. Bumbudsanpharoke, S. Ko, *J. Nanomater.* **2019**, *2019*, 8927167.
11 A. Arora, G. W. J. J. o. F. s. Padua, **2010**, *75*, R43.
12 C. E. Realini, B. Marcos, *Meat Sci.* **2014**, *98*, 404.
13 a) C. Sharma, R. Dhiman, N. Rokana, H. Panwar, *Front. Microbiol.* **2017**, *8*, 1735; b) S.-K. Lee, M. Sheridan, A. J. C. o. M. Mills, **2005**, *17*, 2744.
14 a) H.-Z. Chen, M. Zhang, B. Bhandari, C.-H. Yang, *Food Hydrocoll.* **2020**, *100*, 105438; b) S. Chen, M. Wu, P. Lu, L. Gao, S. Yan, S. Wang, *Int. J. Biol. Macromol.* **2020**, *149*, 271; c) R. Ribeiro-Santos, M. Andrade, A. Sanches-Silva, *Curr. Opin. Food Sci.* **2017**, *14*, 78.
15 M. H. Lee, S. Y. Kim, H. J. Park, *Food Hydrocoll.* **2018**, *84*, 58.
16 R. Sharma, S. M. Jafari, S. Sharma, *Food Control.* **2020**, *112*, 107086.
17 S. A. Mohamed, M. El-Sakhawy, -M. A.-M. El-Sakhawy, *Carbohydr. Polym.* **2020**, *238*, 116178.
18 a) M. R. Kasaai, *Trends Food Sci. Technol.* **2018**, *79*, 184; P. S. Bakshi, D. b) Selvakumar, K. Kadirvelu, N. Kumar, *Int. J. Biol. Macromol.* **2020**, *150*, 1072.
19 Q. Li, T. Ren, P. Perkins, X. Hu, X. Wang, *Food Control.* **2021**, *124*, 107876.
20 a) G. Lazzara, G. Cavallaro, A. Panchal, R. Fakhrullin, A. Stavitskaya, V. Vinokurov, Y. Lvov, *Curr. Opin. Colloid Interf. Sci.* **2018**, *35*, 42; b) E. P. Rebitski, A. C. Alcântara, M. Darder, R. r. L. Cansian, L. Gómez-Hortigüela, S. B. Pergher, *ACS Omega* **2018**, *3*, 13538; c) P. K. Kharwar, R. K. Verma, *Mater. Today Proc.* **2019**, *18*, 4067; d) Y. Zahedi, B. Fathi-Achachlouei, A. R. Yousefi, *Int. J. Biol. Macromol.* **2018**, *108*, 863.
21 a) J. George, H. Ishida, *Prog. Polym. Sci.* **2018**, *86*, 1; b) H. A. T. Banu, P. Karthikeyan, S. Vigneshwaran, S. Meenakshi, *Int. J. Biol. Macromol.* **2020**, *154*, 188; c) Z. R. Afarani, M. N. Sarvi, M. A. Alavijeh, *J. Taiwan Inst. Chem. Eng.* **2018**, *84*, 19.
22 a) H. C. Voon, R. Bhat, A. M. Easa, M. Liong, A. Karim, *Food Bioprocess Technol.* **2012**, *5*, 1766; b) Y. Zare, H. Garmabi, F. Sharif, *J. Appl. Polym. Sci.* **2011**, *122*, 3188; c) I. Resano-Goizueta, B. K. Ashokan, T. A. Trezza, G. W. Padua, *J. Polym. Environ.* **2018**, *26*, 3817.
23 M. Makaremi, P. Pasbakhsh, G. Cavallaro, G. Lazzara, Y. K. Aw, S. M. Lee, S. Milioto, *ACS Appl. Mater. Interf.* **2017**, *9*, 17476.
24 X. Tang, S. Alavi, T. Herald, *Cereal Chem.* **2008**, *85*.
25 a) A. A. Ghazalah, M. O. Abd-Elsamee, K. E. M. E. Moustafa, M. A. Khattab, A.-E. A. A. Rehan, *Animals* **2021**, *11*, 2129; b) K. Manjaiah, R. Mukhopadhyay, R. Paul, S. Datta, K. Periyamuthu, B. Sarkar, **2019**, *309*.
26 a) G. Cavallaro, G. Lazzara, E. Rozhina, S. Konnova, M. Kryuchkova, N. Khaertdinov, R. Fakhrullin, *RSC Adv.* **2019**, *9*, 40553; b) P. Franciszczak, I. Taraghi, S. Paszkiewicz, M. Burzyński, A. Meljon, E. Piesowicz, *Materials* **2020**, *13*, 4459; c) R. Gao, Q. Meng, J. Li, M. Liu, Y. Zhang, C. Bi, A. Shan, *J. Anim. Sci. Biotechnol.* **2016**, *7*, 14.
27 G. Biddeci, G. Cavallaro, F. Di Blasi, G. Lazzara, M. Massaro, S. Milioto, F. Parisi, S. Riela, G. Spinelli, *Carbohydr. Polym.* **2016**, *152*, 548.

28 A. Stavitskaya, S. Batasheva, V. Vinokurov, G. Fakhrullina, V. Sangarov, Y. Lvov, R. Fakhrullin, *Nanomaterials* **2019**, *9*, 708.

29 P. Yuan, P. D. Southon, Z. Liu, M. E. R. Green, J. M. Hook, S. J. Antill, C. J. Kepert, *J. Phys. Chem. C* **2008**, *112*, 15742.

30 S. Satish, M. Tharmavaram, D. Rawtani, *Nanobiomedicine (Rij)* **2019**, *6*, 1849543519863625.

31 Y. Yang, Y. Chen, F. Leng, L. Huang, Z. Wang, W. Tian, *Appl. Sci.* **2017**, *7*, 1215.

32 S. Saadat, G. Pandey, M. Tharmavaram, V. Braganza, D. Rawtani, *Adv. Colloid Interf. Sci.* **2020**, *275*, 102063.

33 V. Vahedi, P. Pasbakhsh, S.-P. Chai, **2015**, 341.

34 D. Garcia-Garcia, J. M. Ferri, L. Ripoll, M. Hidalgo, J. López-Martínez, R. Balart, *Appl. Surf. Sci.* **2017**, *422*, 616.

35 E. Abdullayev, A. Joshi, W. Wei, Y. Zhao, Y. Lvov, *ACS Nano* **2012**, *6*, 7216.

36 a) H. Li, X. Zhu, J. Xu, W. Peng, S. Zhong, Y. Wang, *RSC Adv.* **2016**, *6*, 54463; b) D. Tan, P. Yuan, F. Annabi-Bergaya, H. Yu, D. Liu, H. Liu, H. He, *Micropor. Mesopor. Mater.* **2013**, *179*, 89.

37 a) Y. Lvov, W. Wang, L. Zhang, R. Fakhrullin, *Adv. Mater.* **2016**, *28*, 1227; b) J. Kurczewska, P. Pecyna, M. Ratajczak, M. Gajęcka, G. Schroeder, Saudi Pharmaceut. J. **2017**, *25*, 911.

38 G. I. Fakhrullina, F. S. Akhatova, Y. M. Lvov, R. F. Fakhrullin, *Environ. Sci. Nano* **2015**, *2*, 54.

39 R. Sothornvit, *Curr. Opin. Food Sci.* **2019**, *25*, 82; S. H. Othman, *Agric. Agric. Sci. Procedia* **2014**, *2*, 296.

40 H. Aloui, K. Khwaldia, M. Hamdi, E. Fortunati, J. M. Kenny, G. G. Buonocore, M. Lavorgna, *ACS Sustain. Chem. Eng.* **2016**, *4*, 794.

41 N. P. Risyon, S. H. Othman, R. K. Basha, R. A. Talib, *Food Packag. Shelf Life* **2020**, *23*, 100450.

42 D. Huang, Z. Zhang, Y. Zheng, Q. Quan, W. Wang, A. Wang, *Food Hydrocoll.* **2020**, *101*, 105471.

43 S. Othman, H. Ling, R. A. Talib, M. N. Naim, N. Risyon, M. Saifullah, *J. Nano Res.* **2019**, *59*, 77.

44 Z. W. Abdullah, Y. Dong, N. Han, S. Liu, *Composit. Part B Eng.* **2019**, *174*, 107033.

45 a) G. L. Fakhrullina, E. Khakimova, F. Akhatova, G. Lazzara, F. Parisi, R. Fakhrullin, *ACS Appl. Mater. Interf.* **2019**, *11*, 23050; b) B. Lecouvet, J. Gutiérrez, M. Sclavons, C. Bailly, *Polym. Degradation Stabil.* **2011**, *96*, 226.

46 a) M. Massaro, G. Cavallaro, C. G. Colletti, G. D'Azzo, S. Guernelli, G. Lazzara, S. Pieraccini, S. Riela, *J. Colloid Interf. Sci.* **2018**, *524*, 156; b) D. Xu, C. Lou, J. Huang, X. Lu, Z. Xin, C. Zhou, *Prog. Organic Coat.* **2019**, *134*, 126.

47 S. Majumder, C. Viau, A. Brar, J. Xia, S. George, *App. Clay Sci.* **2022,** 106569.

48 W. Guo, W. Liu, L. Xu, P. Feng, Y. Zhang, W. Yang, C. Shuai, *J. Mater. Sci. Technol.* 2020, *46*, 237.

49 F. Froiio, A. Mosaddik, M. T. Morshed, D. Paolino, H. Fessi, A. Elaissari, *Ind. Eng. Chem. Res.* **2019**, *58*, 20932.

50 M. Krepker, R. Shemesh, Y. D. Poleg, Y. Kashi, A. Vaxman, E. Segal, *Food Control.* 2017, *76*, 117; G. Gorrasi, *Carbohydr. Polym.* **2015**, *127*, 47.

51 T. J. Gutiérrez, A. G. Ponce, V. A. Alvarez, *Mater. Chem. Phys.* **2017**, *194*, 283.

52 S. Pirsa, I. Karimi Sani, S. Khodayvandi, *Polym. Adv. Technol.* **2018**, *29*, 2750.

53 H. Zhang, H. Lai, G. Li, Y. Hu, *Talanta* **2020**, *220*, 121366.
54 S. M. Nor, P. Ding, *Food Res. Int.* **2020**, *134*, 109208.
55 L. Buendía, M. J. Sánchez, V. Antolinos, M. Ros, L. Navarro, S. Soto, G. B. Martínez, A. López, *Food Control.* **2020**, *107*, 106763.
56 S. H. Jang, S. R. Jang, G. M. Lee, J. H. Ryu, S. I. Park, N. H. Park, *J. Food Sci.* **2017**, *82*, 2113.
57 M. H. Lee, H. S. Seo, H. J. Park, *J. Food Sci.* **2017**, *82*, 922.
58 R. Becerril, C. Nerín, F. Silva, *Molecules* **2020**, *25*, 1134; F. Topuz, T. Uyar, *Food Res. Int.* **2020**, *130*, 108927.
59 Y. Zou, C. Zhang, P. Wang, Y. Zhang, H. Zhang, *Carbohydr. Polym.* **2020**, *247*, 116711.
60 L. Li, F. Wang, Y. Lv, J. Liu, D. Zhang, Z. Shao, *Appl. Clay Sci.* **2018**, *161*, 225.
61 a) P. Wang, E. Lombi, F.-J. Zhao, P. M. Kopittke, *Trends Plant Sci.* **2016**, *21*, 699; b) R. Raliya, V. Saharan, C. Dimkpa, P. Biswas, *J. Agric. Food Chem.* **2017**, *66*, 6487.
62 a) G. Zhang, L. Zhou, D. Cai, Z. Wu, *Carbon* **2018**, *129*, 711; b) H.-Y. Chiu, D. e. Gößl, L. Haddick, H. Engelke, T. Bein, *Chem. Mater.* **2018**, *30*, 644.
63 F. Zulfiqar, M. Navarro, M. Ashraf, N. A. Akram, S. Munné-Bosch, *Plant Sci.* **2019**, *289*, 110270.
64 L. Bellani, L. Giorgetti, S. Riela, G. Lazzara, A. Scialabba, M. Massaro, *Environ. Toxicol. Chem.* **2016**, *35*, 2503.
65 C. Wang, Z. He, Y. Liu, C. Zhou, J. Jiao, P. Li, D. Sun, L. Lin, Z. Yang, *Appl. Clay Sci.* **2020**, *198*, 105802.
66 M. Nuruzzaman, M. M. Rahman, Y. Liu, R. Naidu, *J. Agric. Food Chem.* **2016**, *64*, 1447.
67 a) W. X. Peng, J. L. M. Marchal, A. F. B. van der Poel, *Anim. Feed Sci. Technol.* **2018**, *237*, 129; b) P. Vila-Donat, S. Marín, V. Sanchis, A. J. Ramos, *Food Chem. Toxicol.* **2018**, *114*, 246.
68 a) R. Slamova, M. Trckova, H. Vondruskova, Z. Zraly, I. Pavlik, *Appl. Clay Sci.* **2011**, *51*, 395; b) P. Zhou, Y. Q. Tan, L. Zhang, Y. M. Zhou, F. Gao, G. H. Zhou, *Asian-Australas J. Anim. Sci.* **2014**, *27*, 1311.
69 M. J. Martin, S. E. Thottathil, T. B. Newman, *Am. J. Public Health* **2015**, *105*, 2409.
70 P. Boerlin, R. J. Reid-Smith, *Anim. Health Res. Rev.* **2008**, *9*, 115.
71 S. Majumder, D. Jung, J. Ronholm, S. George, *BMC Microbiol.* **2021**, *21*, 222.
72 A. Brar, S. Majumder, M. Z. Navarro, M.-O., Benoit-Biancamano, J. Ronholm, S. George, *Nanomaterials* **2022**, *12*, 2179.
73 a) G. Janer, E. Fernández-Rosas, E. Mas Del Molino, D. González-Gálvez, G. Vilar, C. López-Iglesias, V. Ermini, S. Vázquez-Campos, *Nanotoxicology* **2014**, *8*, 279; b) S. Lordan, J. E. Kennedy, C. L. Higginbotham, *J. Appl. Toxicol.* **2011**, *31*, 27.
74 Q. Chaudhry, L. Castle, *Trends Food Sci. Technol.* **2011**, *22*, 595.
75 M. Farhoodi, S. M. Mousavi, R. Sotudeh-Gharebagh, Z. Emam-Djomeh, A. Oromiehie, *Packag. Technol. Sci.* **2014**, *27*, 161.
76 M. Ellenbecker, S. Tsai, *J. Clean. Prod.* **2011**, *19*, 483.
77 a) J.-Y. Huang, Y. Chieng, X. Li, W. Zhou, *Food Bioprocess Technol.* **2015**, *8*, 382; b) Y. Echegoyen, S. Rodríguez, C. Nerín, *Food Addit. Contamin. Part A* **2016**, *33*, 530.
78 a) A. Nasiri, N. Gontard, E. Gastaldi, S. Peyron, *Appl. Clay Sci.* **2017**, *146*, 27; b) A. Nasiri, S. Peyron, E. Gastaldi, N. Gontard, *J. Mater. Sci.* **2016**, *51*, 9732.
79 B. Wu, M. Jiang, X. Liu, C. Huang, Z. Gu, Y. Cao, *Nanotoxicology* **2020**, *14*, 1017.
80 Y. Cho, C. Kim, N. Kim, C. Kim, B. Park, *Biochip. J.* **2008**, *2*, 183.

81 S. Das, L. Jagan, R. Isiah, B. Rajesh, S. Backianathan, J. Subhashini, *Indian J. Pharmacol.* **2011**, *43*, 409.

82 M. Sahoo, S. Vishwakarma, C. Panigrahi, J. Kumar, *Food Front.* **2021**, *2*, 3.

83 a) S. H. Jang, S. R. Jang, G. M. Lee, J. H. Ryu, S. I. Park, N. H. Park, *J. Food Sci.* **2017**, *82*, 2113; b) S. Hendessi, E. B. Sevinis, S. Unal, F. C. Cebeci, Y. Z. Menceloglu, H. Unal, *Progr. Organic Coat.* **2016**, *101*, 253.

84 L.-F. Wang, J.-W. Rhim, *Appl. Clay Sci.* **2017**, *150*, 138.

85 K. Marsh, B. Bugusu, *J. Food Sci.* **2007**, *72*, R39.

86 P. K. Annamalai, D. J. Martin, *Mater. Sci. Technol.* **2014**, *30*, 593.

87 Z. Cibulková, A. Vykydalová, A. Chochulová, P. Šimon, P. Alexy, L. Omaníková, *J. Therm. Anal. Calorim.* **2018**, *131*, 1491.

88 H. Qin, Z. Zhang, M. Feng, F. Gong, S. Zhang, M. Yang, *J. Polym. Sci. B Polym. Phys.* **2004**, *42*, 3006.

89 H. Oliver-Ortega, V. Vandemoortele, A. Bala, F. Julian, J. A. Méndez, F. X. Espinach, *Polymers* **2021**, *13*, 2741.

90 Official Journal of the European Union (OJ) **2011**.

14

Electrospinning Technologies for Encapsulation of Probiotics

*Seethu, B.G., Aditya Sukumar P., Devikrishna P., Harshvardhan Kulkarni, Magdaline Eljeeva Emerald, Chandram Grover, and Heartwin A. Pushpadass**

ICAR-National Dairy Research Institute, Southern Regional Station, Bengaluru – 560030
** Corresponding author*

Highlights

- *L. rhamnosus* CRD11 was encapsulated by electrospinning using food-grade pullulan, prebiotics, and lyoprotectants
- Encapsulation of *L. rhamnosus* resulted in the electrospun fibers appearing beaded
- Less than 1 log reduction in the population of *L. rhamnosus* was observed after electrospinning with the use of isomalto-oligosaccharide as prebiotic and trehalose as lyoprotectant, which was comparable to conventional freeze drying
- The fermentation ability of encapsulated *L. rhamnosus* was not affected
- Encapsulated *L. rhamnosus* of high viability in electrospun fibers could be used for the development of probiotic functional foods.

14.1 Introduction

The term 'probiotic' comes from the Greek word '*pro bios*' meaning 'for life'. Probiotics are defined by FAO/WHO as 'live microorganisms that when administered in adequate amounts confer a health benefit on the host' [1]. The role and share of probiotics in human diet is increasing due to their proven health benefits. In 2021, the global market potential for probiotics was estimated as USD61.1 billion. At a compounded annual growth rate of 8.3%, it is forecasted that the market potential could reach USD91.1 billion by 2026 [2]. The increase in demand for probiotics is due to better consumer awareness on the consumption of probiotic foods that confer health benefits past basic nutrition. This immense growth has been catalyzed by the recent COVID-19 pandemic. The unforeseen impact of the pandemic is a shift in the consumption pattern of the global population toward foods promoting immunity from diseases. It is also anticipated that the demand for probiotic foods could be much higher than the estimated numbers as a lot of such foods are consumed locally by people without marketing by an organized sector.

The human gastrointestinal (GI) tract comprises more than 500 species of microorganisms, which could be beneficial or pathogenic. The balance of microorganisms in the GI

Materials Science and Engineering in Food Product Development, First Edition. Edited by Wing-Fu Lai.

tract can be altered by the consumption of probiotics. Regular consumption of probiotics and probiotics-added foods establishes a positive balance in the population of beneficial microbes in the intestinal microbiome. Probiotics are used to mitigate GI and non-GI diseases, such as diarrhea, colon cancer, obesity, diabetes, and inflammation. Health benefits are active digestion capacity, modulation of gut beneficial microflora, antagonistic activity against pathogens, improved colon integrity, immunomodulation, down-regulated allergic responses, anti-carcinogenic properties, reduction of serum cholesterol, mitigation of lactose intolerance, and nutritional enrichment.

Probiotic products contain a single microorganism or a mixture of several species. To achieve the intended health benefits, the probiotic bacteria should remain viable, active, and have a population of at least 10^6–10^9 cfu/g at the time of consumption of the product [3]. The most widely used probiotics are *Lactobacillus* and *Bifidobacterium* species. The important characteristics of probiotic organisms are their ability to withstand the harsh and unfavorable conditions in the GI tract, colonize and reproduce in the gut, adhere to the intestinal epithelium, and stabilize the balance of gut microflora. They also possess non-pathogenic properties and remain viable during their shelf life. The beneficial effects produced by probiotics are strain-specific. The common probiotic microorganisms are *Lactobacillus acidophilus, Lactiplantibacillus plantarum, Limosilactobacillus reuteri, Lacticaseibacillus rhamnosus, Bifidobacterium lactis, Bifidobacterium longum*, etc. [4]. A single strain may exhibit different benefits when used individually and in combination. Yue et al. observed that *L. plantarum* strain YYC-3 showed inhibition against colon cancer cell lines [5]. Similarly, three strains of *L. plantarum* with 1.2×10^9 cfu population were used in combination to evaluate their effect on hypercholesterolemic patients [6]. The cholesterol level in the body reduced by 13.6% after consumption for 12 weeks. Similarly, fermentation of apple juice with *L. plantarum* improved the overall antioxidant capacity of apple juice by modifying the phenolic composition [7].

For a long time, cryopreservation was used as the standard method to preserve probiotics and bacterial starter cultures. But it was not commercially feasible due to sub-zero transportation and storage. It also resulted in bacterial cell damage during freezing and thawing. Therefore, drying techniques, such as spray drying, freeze drying, vacuum drying, and fluidized bed drying superseded its place. Drying of probiotics into powder provides encapsulation, ease of handling, high stability, less space requirement for storage, and extension of viability. The potential health merits of probiotics are often not realized due to significant reduction in their viability during storage and transit in the GI tract. One of the biggest challenges encountered by probiotics is surviving the harsh conditions during GI transit and effectively colonizing the intestine.

In recent years, encapsulation has evolved as a promising approach to offer protection and enhance their ability to resist such unfavorable conditions. Encapsulation involves entrapment of probiotics within another matrix, which is often referred to as wall or shell material. In this process, the probiotics are packaged only by physical entrapment, thereby allowing the microorganisms to be immobilized within the semipermeable and biocompatible materials. An effective encapsulation thus creates a protective layer around probiotic cells, aiding them to sustain the unpropitious conditions of thermal, oxidative, and osmotic stresses, high levels of pH, and relative humidity encountered. Furthermore, probiotics in their native form often show significant loss in their viability during storage, which renders them less beneficial when consumed. It is recognized that encapsulation effectively offers protection to probiotics against detrimental conditions of acids and bile salts, maintains their viability through the GI tract, and helps in delivering them to specific targets at controlled release rates.

14.2 Major Methods for Encapsulation of Probiotics

The familiar encapsulation techniques are spray drying, freeze drying, extrusion, coacervation, etc. The mode of preservation influences the viability of encapsulated bacteria. Selection of a particular technique among them requires analysis of criteria, such as probiotic viability, process conditions encountered during food processing, storage conditions prior to consumption of end product, size and density of encapsulate needed, mechanism of release, and economic constraints. Freeze and spray drying are commercially used for encapsulation and production of dried probiotics (Figure 14.1).

Industrial Corner

- Proper selection of methods for encapsulation of probiotics is a challenge always encountered during functional food production.
- Encapsulation and drying of probiotics can be done by using different methods, though the pros and cons of each of these methods have to be considered.
- The characteristics of each probiotic strain have to be taken into consideration when the encapsulation method is chosen.

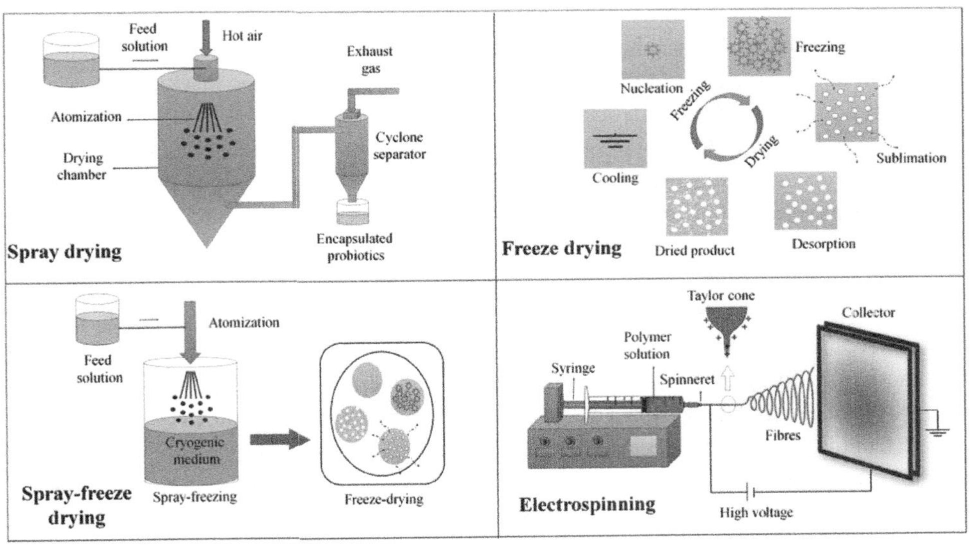

Figure 14.1 Methods of encapsulation of probiotics.

14.2.1 Freeze Drying

Freeze drying (lyophilization) is an invaluable drying and encapsulation technique for probiotics. The main principle behind it is sublimation of water (Figure 14.1). This process consists of three steps namely, low temperature freezing, primary drying, and secondary drying. The shortest step in this process is freezing, which could cause damage to probiotics and denaturation of their proteins. The freezing temperature is thus a critical parameter of this encapsulation technique. It depends on the strain and type of protectants used. Wang et al. evaluated the effect of pre-freezing temperature during freeze drying on the survivability of *L. plantarum* strains [8]. Pre-freezing of *L. plantarum* strains AR113, AR307, and WCFS1 in phosphate buffer saline and sorbitol at −196°C, −40°C, and −20°C provided the highest survival rates. When trehalose was used as a protectant, pre-freezing at −20°C gave the best survivability for AR113, whereas for AR307 and WCFS1, the highest survivability was obtained at −60°C. The difference in survivability of strains during pre-freezing was due to its effect on their cell membrane integrity, membrane permeability, and lactate dehydrogenase activity.

Primary and secondary drying remove frozen and unfrozen water, respectively from the product. Freeze drying results in long-term preservation coupled with convenience in handling and storage. It also produces very high encapsulation efficiency and viability. The survival of microorganisms depends on drying media, growth factors, intrinsic factors, and storage conditions. Both direct and indirect methods of freeze drying can be used. The direct method involves mixing of cryoprotectants in media containing probiotics, followed by freeze drying the mixture to remove water. In the indirect method, probiotics are first immobilized within the polymer matrix and transferred to the freeze dryer for lyophilization. Recently, Rajam and Anandharamakrishnan studied the effect of wall materials on the encapsulation efficiency and viability of spray-freeze-dried *L. plantarum* [9]. The encapsulation efficiency ranged from 87.92% to 94.86%, which was only 3.16% to 6.73% lower than freeze-dried *L. plantarum* due to the effect of atomization pressure and freezing conditions. Combinations of whey protein isolate (WPI), sodium alginate, denatured WPI (DWPI), and fructo-oligosaccharides (FOS) were used as encapsulants. The authors observed that combinations of DWPI-FOS and DWPI-sodium alginate produced microcapsules with porous sponge-like structures during freeze drying, while WPI-sodium alginate-based microcapsules had a smoother outer surface. The authors concluded that spray-freeze-dried microcapsules based on WPI-sodium alginate exhibited superior properties in terms of morphology, flowability, loose bulk density, and viability.

The negative effect of freezing could be overcome to some extent with the addition of cryoprotectants in the drying media. They protect the viability of cells by providing high osmotic pressure. Saarela et al. freeze-dried *Bifidobacteria* using cryoprotectants, such as sucrose, betaine, and reconstituted skim milk [10]. Sucrose provided excellent stability for the bacterium for 5–6 months of storage at refrigerated and frozen temperatures. *Bifidobacteria* also had good survival at 37°C for 2 months even at acidic pH of 3.0 or in the presence of bile acids. Similarly, Siaterlis et al. analyzed the effect of trehalose, sucrose, and sorbitol on the viability of freeze-dried *L. plantarum* and *L. rhamnosus* [11]. Among the three sugars used, sucrose at concentration between 5% and 10% (w/v) offered the best protection during freeze drying of both strains. *L. rhamnosus* had a survival of 60% with 10%

(w/v) sucrose, whereas *L. plantarum* showed a survival of 90% with 5% (w/v) sucrose. In recent years, Reddy et al. evaluated lactose, skim milk, and maltodextrin as cryoprotectants during freeze drying of lactic acid bacteria [12]. The freeze-dried and encapsulated probiotic using maltodextrin exhibited 60–83% retention in survival rate, and it was very effective in protecting the viability and properties of encapsulated bacteria. Similarly, encapsulation of *Lactocaseibacillus paracasei* by spray freeze drying was performed with trehalose and maltodextrin matrix [13]. The encapsulated *L. paracasei* had viability above 60%, and trehalose concentration was the critical factor governing it. Maltodextrin and trehalose prevented changes to the cell membrane during drying by reducing its glass transition temperature (T_g) and enhancing their viability.

Apart from the studies mentioned above, Estilarte et al. reported that freeze drying of *Enterococcus durans* using lactulose maintained viability for 92 days at 4°C with less than 2 log reduction in population [14]. The dried culture also showed resistance to GI conditions and acid and bile treatments, but grew in the presence of preservatives, such as potassium sorbate and sodium chloride. The protective effect of disaccharides and others was due to their ability to lower the T_g of dry membranes by replacement of water between the lipid headgroups and preventing unfolding and aggregation of proteins (of probiotic cell membrane) by hydrogen bonding with their polar groups. However, disaccharides could be detrimental to the survival of probiotics due to their crystallization if stored at high vapor pressure. Recently, Wang et al. studied the effect of macromolecular cryoprotectants and their combinations on bacterial activity during freeze drying [15]. The survival rate of freeze-dried *L. plantarum* AR113 was 19% higher with soy polysaccharides than with trehalose as cryoprotectant. Similarly, a 90.52% survival rate of *L. plantarum* WCFS1 was achieved using the mixture of soy polysaccharides and trehalose, which was respectively 31.48% and 36.47% higher as compared with the use of trehalose and soy polysaccharides alone. It is thus evident that combinations of cryoprotectants might work better during freeze drying of probiotics.

Even divalent and monovalent cations could be used as cryoprotectants. Tan et al. studied the effect of monovalent and divalent cations with sodium alginate on the survival of *L. rhamnosus* GG freeze-dried at −84°C [16]. It was observed that divalent cations, such as Ca^{2+}, Sr^{2+}, Mg^{2+}, and Ba^{2+}, exhibited antagonistic effects and caused viability loss as compared with monovalent cations on *L. rhamnosus* because divalent cations slowed down the freezing rate. In a study by Dimitrellou et al., the effect of the cooling rate prior to freeze drying on the viability of *Lacticaseibacillus casei* was evaluated [17]. The cells harvested were frozen to −45°C at freezing rates of 1, 3, and 5°C/min, and freeze-dried at the same temperature for 24 h. At a freezing rate of 5°C/min, the survival of *L. casei* was 3.3 and 5.3 times higher than at 3 and 1°C/min, respectively. However, the optimal cooling rate should be slow enough to allow sufficient time for water to flow out of the cells, preventing freezing of water within the cells, and thereby avoiding intracellular ice crystal formation.

14.2.2 Spray Drying

Spray drying is an alternative and popular technique for encapsulating and drying probiotics (Figure 14.1). The main challenge associated with spray drying, however, is the production of viable cultures from heat-sensitive probiotic strains [18]. This is because bacterial inactivation could be caused by heat, osmotic, oxidative, and dehydration stresses. The temperature of hot air used in spray drying is critical because low temperature reduces the rate of evaporation and results in aggregates of microspheres, while high temperature damages the bacterial membrane, resulting in reduced viability of encapsulated probiotics. In particular, the survival of bacterial cells encapsulated by spray drying was influenced by the outlet temperature of air than its inlet temperature. Recently, Arepally et al. spray-dried *L. acidophilus* using 20% maltodextrin and 8.5% gum arabic as wall materials [19]. The viability of probiotic ranged from 67.5% to 87.7% and encapsulation efficiency ranged from 65% to 80%. The encapsulation yield decreased with the increase in inlet air temperature, whereas it increased with increasing concentration of gum arabic. The decrease in the survival of bacterial cells with the increase in inlet air temperature was attributed to thermal and dehydration inactivation of cells. Rokka and Rantamäki recommended inlet and outlet air temperature, ranging from 100°C to 170°C and 45°C to 105°C, respectively for the microencapsulation of probiotic cultures [4].

Similar to the research works mentioned above, Broeckx et al. studied the effect of initial cell concentration and growth phase on the viability of spray-dried *L. rhamnosus* GG [20]. The outlet temperature of air was maintained between 50°C and 60°C to obtain dried probiotics with the lowest reduction in viability. It was found that *L. rhamnosus* GG cells in their stationary growth phase were more resistant to spray drying than mid-log grown cells due to the encountered stress. Leylak et al. spray-dried *L. acidophilus* LA-5 at optimum inlet air temperature, feed rate, and whey powder to gum arabic ratio of 140°C, 0.485 L/h, and 0.83:0.17, respectively [21]. The viability of microencapsulated *L. acidophilus* was significantly higher than that of free form at the end of 28 days of storage at 4°C. This was because microencapsulation provided a protective shield to *L. acidophilus* from the low pH environment of yogurt. Thus, it is possible to circumvent the effects of such deteriorative stresses during spray drying of probiotics to a large extent.

The survival of probiotics during spray drying is also influenced by the type of bacteria and encapsulating material used. When heat-adapted cultures of *Lactococcus lactis* subsp. *cremoris*, *L. rhamnosus* GG, and *L. acidophilus* were spray-dried [22], the survival of *L. lactis* subsp. *cremoris* and *L. acidophilus* increased by 21.0% and 13.7%, respectively, whereas the increase shown by *L. rhamnosus* was relatively less at 9.9%. Similarly, WPI, pullulan, trehalose, and sodium glutamate were used while spray drying of *L. plantarum* [23]. The ratio of bacteria to protective agents was 3:1 at the feed flow rate and inlet air temperature of 240 mL/h and 115°C, respectively. The viability and survival rates achieved for *L. plantarum* were 12.22 ± 0.27 log cfu/g and 85.12%, respectively. The viability of spray-dried *L. plantarum* was more than 10^{10} cfu/g after 200 days of storage at −20°C.

Spray freeze drying is a combination of spray and freeze drying that yields a better product than spray drying and faster processing than freeze drying. Yoha et al. used spray drying and spray freeze drying techniques to encapsulate *Lactiplantibacillus plantarum* [24]. FOS, whey protein, and maltodextrin were used as encapsulating agents at different ratios. FOS–whey protein–maltodextrin in the ratio of 2:0.5:0.5 gave higher encapsulation efficiency of 89.21% and 96.16% for spray and spray freeze drying, respectively. Despite the seemingly encouraging potential of encapsulation as presented above, each of the above encapsulation techniques for probiotics has some limitations. Considerable inactivation of bacteria could occur during spray drying by heat, osmotic, oxidative, and desiccation stresses, while cold shock and thawing cause cell damage in freeze drying. The energy and time required for sublimation in freeze drying is much higher, and it may take 2–4 days for complete drying of probiotics. Therefore, it also incurs high capital and operating costs. Furthermore, freeze-dried probiotics contain a combination of dead, viable, and sub-lethally injured cells in variable proportions, depending on the strain's sensitivity to environmental stresses. Freezing also causes severe changes in pH, which need to be counteracted.

14.2.3 Electrospinning

Electrospinning is an alternative technique to conventional spray and freeze drying methods for the encapsulation of probiotics with a high survival rate. The main advantage of electrospinning is that it is done at ambient temperature. It is an electrohydrodynamic process wherein high voltage in the order of kilo volts (kV) is applied to form nano to microscale fibers from the feed solution. The electrospinning setup consists of four main components namely (i) a high voltage source; (ii) a blunt-ended stainless steel capillary; (iii) a syringe pump that controls the flow rate; and (iv) a grounded collector (Figure 14.2). Fibers are formed from the electrically charged jet of polymer solution emanating from a pendant droplet, which gets collected into a grounded collector. The electrospinnability of the feed solution depends on the operating conditions, type of polymer, and solvent used.

For encapsulation of probiotics by electrospinning, they need to be dispersed in a polymer(s) solution and loaded into the spinneret. The bacteria are pulled into the jet of wall material, wherein they get encapsulated due to the sink-like flow of the Taylor cone. That is, when the electrostatic forces exceed surface tension, the droplet at the spinneret tip elongates to form the 'Taylor cone' (Figure 14.2) [25]. A charged liquid jet of the polymer is ejected, which undergoes instability and elongation and segregates into long and thin fibers [26]. The solvent evaporates, and the charged polymer fiber is deposited on the grounded collector. The distance from the tip to the collector decides the strength of the applied electrostatic field and the time taken by the jet (at the spinneret tip) to reach the negatively charged collector. The normal range for electrospinning of probiotics is 100–150

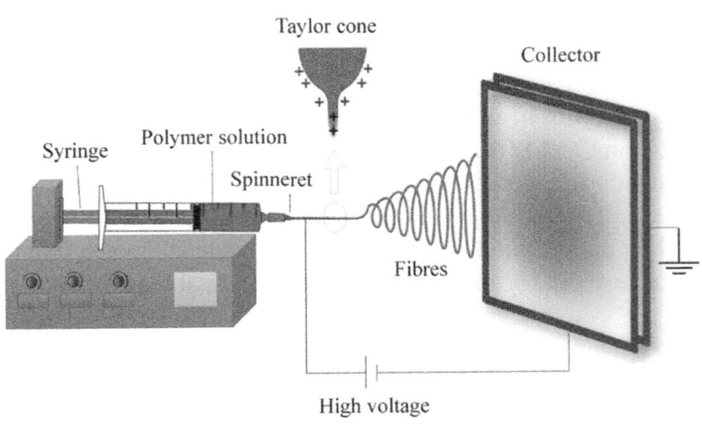

Figure 14.2 Electrospinning setup.

mm [27]. If the distance from the tip to the collector is insufficient, it could lead to inadequately dried, sticky, and overlapped fibers.

A number of factors, such as the molecular structure of the biopolymer, viscosity, surface tension, electrical conductivity, flow rate, and applied voltage affect electrospinning. Altering these parameters makes it possible to achieve fibers with the desired structure and morphology. Hence, it is essential to understand the role of these factors on electrospinning. Selection of biopolymer(s) for electrospinning is difficult as many do not have the electrospinning ability. Similarly, viscosity is one of the critical factors to obtain beadfree nanofibers during electrospinning. As the viscosity of the feed solution decreases, it cannot produce a stable jet and discontinuous fibers are produced. Tiwari and Venkataraman reported that the viscosity of the feed solution directly affected the morphology of electrospun fibers, and added that if the viscosity of the feed solution were higher, unstable jet flow would be observed at the spinneret tip and fibers with uneven diameter could be obtained [28].

There is an optimal range of surface tension, viscosity, and net charge density of the liquid jet for the formation of clean and uniform fibers, and beads are formed when they are out of the normal range. Among all, the most influencing factor for encapsulation of probiotics in biopolymers is the applied voltage. On one hand, a minimum voltage is required for the transformation of the spherical droplet to the conical shape at the end of spinneret. Devaraju et al. reported that applied voltage below 14 kV was insufficient to overcome surface tension during the electrospinning of pullulan, while unstable jets were formed at voltages above 25 kV [29]. The critical voltage (V_c) for electrospinning can be calculated using Equation (14.1), where V_c is the critical voltage required for electrospinning (kV), H is the distance between electrodes (m), L is the length of the spinneret (m), γ is the surface tension of the solution (N/m), and R is the jet radius (μm).

$$V_c^2 = 4\frac{H^2}{L^2}\left(ln\frac{2L}{R} - 1.5\right) \times \left(0.117\pi\gamma R\right) \tag{14.1}$$

Conversely, application of high voltage may be a detrimental factor for the survival of probiotics. For example, the population of *L. plantarum* 423 cells encapsulated in nanofibers decreased from 2.3×10^{10} cfu/mL to 4.7×10^{8} cfu/mL after electrospinning [30]. In contrast, Škrlec et al. studied the survival of *L. plantarum* after electrospinning at 10, 15, and 20 kV applied voltage, and reported that the probiotic showed the highest survival at 15 kV, while the population decreased at both 10 and 20 kV [31]. Similarly, Feng et al. developed a fibrous mat of *L. plantarum* using 2.5% FOS and 6% PVA by electrospinning, and observed that applied voltage had no significant impact on cell viability [32]. The authors added that the addition of prebiotic to feed solution improved the viability and thermal stability of electrospun *L. plantarum*. Thus, optimization of applied voltage is challenging as it should not affect bacterial viability but it should be high enough to produce clean nanofibers from biopolymers.

About a decade ago, López-Rubio et al. fabricated nanofibers of *Bifidobacterium animalis* using polyvinyl alcohol (PVA) at an applied voltage of 11 kV [33]. The encapsulated bacteria remained viable for 40 days at room temperature and 130 days at refrigerated temperatures. Fung et al. used soluble dietary fibers from agro-wastes, okara, oil palm trunk, and oil palm frond for electrospinning of *L. acidophilus* [34]. PVA at 8% concentration was used along with soluble dietary fibers to assist electrospinning. Okara-based feed solution produced finer electrospun fibers due to its higher pH of 5.39 and viscosity of 578 mPa.s. The diameter of electrospun *L. acidophilus*-loaded fibers ranged from 229 to 703 nm. Similarly, fibers of *L. plantarum* ATCC 8014 were developed using polyethylene oxide (PEO) and PEO-lyoprotectant combinations with a high loading rate of 7.6×10^8 cfu/mg [31]. The flow rate and applied voltage of the electrospinning process was 0.4 mL/h and 15 kV, respectively. The most critical parameter for high viability during electrospinning was the initial load of probiotics in the feed solution than applied voltage and relative humidity. However, these probiotics were electrospun using nonedible polymers, and hence, cannot be used in foods.

For food applications, proteins and carbohydrates are to be used as wall materials for the encapsulation of probiotics. The commonly used biopolymers are pullulan, chitosan, pectin, alginate, and proteins, such as zein, gelatin, and WPI. Some of these wall materials decrease the osmotic and dehydration stresses, thereby increasing the survival of bacteria during electrospinning and long-term storage. Similarly, while selecting the biopolymers, the source (plant, animal, and microbial) from which it is obtained is also a concern for food applications, particularly among the vegan population. In an earlier study, Hu et al. electrospun cellulose acetate nanofibrous membranes loaded with *L. plantarum* as the starter culture for the production of fermented milk [35]. The authors reported that the GI resistance and the population of viable cells increased after a three-hour incubation in simulated GI media. The viable cells in fermented milk remained at 11 log cfu/g, which warranted a shorter fermentation time for milk and indicated higher survival of probiotics. Similarly, Yu et al. showed that the viability of lactic acid bacteria could be improved after encapsulation by coaxial electrospinning [36]. In another study, Yilmaz et al. encapsulated *L. paracasei* within electrospun alginate nanofibers of 305 nm mean diameter [37].

Coaxial electrospun nanofibers of *L. plantarum* were developed using 8% (w/v) polylactic acid (PLA) and 2.5% (w/v) FOS [36]. The mean diameter of nanofibers was 676 ± 162 nm with a survival rate above 72%. Similarly, *Lactobacillus* strains were electrospun using biopolymers, such as gum arabic and pullulan [38]. Addition of gum arabic to pullulan solution resulted in increased electrical conductivity, reduced surface tension, and viscosity, which resulted in the formation of smaller diameter nanofibers. Atraki and Azizkhani electrospun probiotic strains of *L. acidophilus*, *L. rhamnosus*, *Bifidobacterium bifidum*, and *B. animalis* using corn starch and sodium alginate [39]. The electrospun cultures of *Lactobacilli* and *Bifidobacteria* showed viability of 90% and 84.1% in simulated gastric, and 71.7% and 77.8% under intestinal conditions, respectively. These results showed that electrospinning could be a good technique to encapsulate probiotics without considerably affecting their viability.

14.2.4 Our Approach to Encapsulate *L. Rhamnosus* CRD11 by Electrospinning

We have encapsulated *L. rhamnosus* CRD11 by electrospinning. In the first study, pullulan of 150 kDa molecular weight (Mw) was used as the biopolymer for encapsulating *L. rhamnosus*. It is a straight-chain glucosic polysaccharide, white in color, dissolves readily in cold or hot water; it is odorless, tasteless, edible, and nontoxic and has wide applications in foods [40]. López-Rubio et al. reported that smooth and round capsules with ultrafine fibers were formed during the electrospinning of *Bifidobacterium* using pullulan as wall material at 12%, 14%, and 16% (w/w) concentrations [41]. Considering its good fiber-forming ability without additives, and being edible, it was chosen as the encapsulant for the electrospinning of *L. rhamnosus*. As there could be reduction in the viability of *L. rhamnosus* by the high voltage applied during electrospinning, pullulan solutions were added with lyoprotectants.

Lyoprotectants are substances that protect bacterial proteins from denaturation during encapsulation and mitigate the loss of biological activity of probiotics during storage. Commonly used lyoprotectants are sucrose, trehalose, lactose, mannitol, inulin, etc. Basholli-Salihu et al. opined that lyoprotectants, such as saccharides, sugar alcohols, and amino acids protect probiotics during electrospinning, and the optimal lyoprotectant depends on the strain encapsulated [42]. Škrlec et al. postulated that amorphous trehalose as lyoprotectant promoted cell survival in nanofibers due to its interactions with cells [31]. Similarly, Hirsch et al. reported that bacterial survival increased above 80% by the use of trehalose, saccharose, and skim milk as lyoprotectants during electrospinning [43]. Therefore, the lyoprotectants used in our study were expected to improve the survival rate of *L. rhamnosus* during encapsulation by electrospinning. The feed solutions were prepared by dissolving pullulan and one of the lyoprotectants namely sucrose, lactose, and maltodextrin at 10% (w/w) in Milli-Q water, and autoclaving them at 121°C for 15 min. The solutions were cooled before adding probiotics at 20% concentration of wall material and electrospinning them. The electrical conductivity, viscosity, and surface tension of feed solutions were measured. Fabrication of probiotics-loaded nanofibers was done in the electrospinning machine (Model: H/VPD40CH, Royal Enterprises, Chennai, India), consisting of a variable high voltage supply (1–30 kV), a syringe pump with spinneret and a stainless steel drum collector of 7.5 cm diameter rotating at 70 rpm. The electrospinning parameters were flow rate of 0.75 mL/h, tip to collector distance of 10 cm, RH of 50% and applied voltages of 18, 21, and 24 kV. The electrospun fibers were collected on an 18-μm-thick aluminum foil wrapped around the collector. Preliminary trials suggested that the feed solutions should have electrical conductivity of at least 0.9 mS/cm and viscosity of 125 mPa.s for the formation of clean and uniform electrospun nanofibers. Similarly, it was observed that feed solutions with a surface tension range of 30–70 mN/m were only easily electrospinnable into clean and uniform fibers. Lower surface tension is preferred for electrospinning as it reduces the applied voltage required and the diameter of electrospun fibers obtained. In our study, the pullulan-based feed solutions had surface tension in the electrospinnable range of 30–71.88 mN/m. Devaraju et al. reported that electrical conductivity of 0.9–2.0 mS/cm and viscosity range of 150–600 mPa.s were ideal for electrospinning of pullulan solutions, while the critical value of electrical conductivity was 1.0 mS/cm [29]. A higher electrical conductivity is desirable for electrospinning as it lowers the applied voltage required.

The electrical conductivity of feed solution depends on the type of polymer, solvent, and the presence of probiotics. As some biopolymers are polyelectrolytic in nature, the ions present in them increase the charge-carrying capacity of feed jets and surface tension under an electric field [44]. But, pure pullulan solution had electrical conductivity of 0.1–0.2 mS/cm, depending on the concentration, which did increase only moderately even after the addition of lyoprotectants. However, when *L. rhamnosus* was added to the feed solution, the electrical conductivity increased to the bare minimum of 0.9 mS/cm required for electrospinning. This is because bacteria possess overall negative charge because of the presence of select proteins that are rich in carboxy and amino groups. The critical voltage calculated for all feed solutions was in the range of 10.11–15.25 kV. Hence, applied voltage in excess of 15.25 kV (18, 21, and 24 kV) was selected for application across the terminals during the electrospinning of *L. rhamnosus*. The morphology of encapsulated *L. rhamnosus* was examined using a field emission gun scanning electron microscope (FESEM) (Model: Gemini Ultra 55, Zeiss, Jena, Germany) with a resolution of 1 nm. The addition of lactose as lyoprotectant did not yield satisfactory fibers of *L. rhamnosus*, while the addition of other lyoprotectants did not have any negative impact on the formation of electrospun fibers loaded with *L. rhamnosus*. The polymer–probiotic solutions were electrospun into microcapsules and nanofibers, with a mean diameter of 86.61 nm. In the electrospun fibers, *L. rhamnosus* had a size range of 2–4 μm with a mean diameter of 2.46 μm and was successfully encapsulated by the wall material. The FESEM images also revealed that *L. rhamnosus* that were randomly distributed in the feed solution started to orient at the Taylor cone during electrospinning, and thus were aligned in the jet that solidified into nanofibrous mats. Salalha et al. and Fung et al. stated that due to sink flow at the Taylor cone, the randomly oriented bacteria in the feed solution became gradually aligned along the streams of fibers [34, 45]. Similar observations were reported by Alehosseini et al., who used electrospraying to encapsulate *L. casei* in transglutaminase cross-linked whey protein concentrate–WPI matrix [27]. Spherical water-resistant capsules of *L. casei* having a mean diameter of 3.09 μm with encapsulation efficiency of nearly 100% were obtained by the authors. The diameter of the fibers ranged between 31 and 170 nm, excluding the encapsulated bacterium. The blank fibers were clean, uniform, and beadfree, while those encapsulating probiotics appeared to be beaded. As the probiotic was encapsulated within electrospun fibers, it led to their local widening. For comparison, when *B. animalis* subsp. *lactis* Bb12 was electrospun using chitosan and PVA as wall material and inulin as prebiotic, Mojaveri et al. obtained a mean fiber diameter of 117.5 ± 70.6 to 217.6 ± 62.7 nm [46]. The fiber diameter was also on par with the value of 135 ± 25 nm reported by Zupančič et al. for *L. rhamnosus*-loaded nanofibers [47].

Industrial Corner

- The use of lyoprotectants is a feasible method to enhance the viability of probiotics during electrospinning, in which the use of high voltage may be involved.
- Pullulan can serve as the encapsulant for electrospinning of *L. rhamnosus* because of its good fiber-forming ability without additives.
- Factors such as polymer type, solvent used, and presence of probiotics affect the electrical conductivity of the feed solution.

14.2.4.1 Determining the Viability and Properties of Encapsulated *L. Rhamnosus*

The encapsulated fibrous mat of *L. rhamnosus* was peeled off from the aluminum foil, and the viability was determined using pour-plate assay. The electrospun fibers were rehydrated to its initial mass in the feed solution with sterile water for 30 min, and the population was enumerated in MRS agar. The initial viable population in the feed solution ranged from 10.08 to 10.65 log cfu/mL, which after electrospinning decreased from 8.16 to 9.14 log cfu/mL. It was evident that 1-2 log reduction in the viability of *L. rhamnosus* occurred after encapsulation by electrospinning. The results were in accordance with those of Fung et al. who stated that the viability of *L. acidophilus* decreased after electrospinning [34]. The reduction in the viability of *L. rhamnosus* was attributed to the sudden change in the osmotic pressure created by the rapid evaporation of water, applied voltage, and shear stress experienced during electrospinning. Applied voltage was the most influencing factor affecting the viability of encapsulated *L. rhamnosus* (Figure 14.3).

The survival rate was found to be between 80.35% and 89.26%. The reduction in the viability of *L. rhamnosus* during electrospinning was, however, not very high but comparable to established techniques like freeze drying. Ma et al. reported that the survival rate of *Lactobacillus* was found to be higher in electrospinning (85.38–97.38%) as compared with freeze drying [38]. The authors suggested that the cell viability was significantly influenced by the encapsulation method. Previous studies also reported that preparation of dry probiotic cultures by conventional spray and freeze drying techniques also had comparable viability loss [48, 49]. For higher viability of encapsulated *L. rhamnosus* by electrospinning, the optimized conditions were applied voltage of 18 kV, pullulan concentration of 14% and maltodextrin as lyoprotectant. These results also established the superiority of the electrospinning method, as well as, the biocompatibility of pullulan to encapsulate *L. rhamnosus*

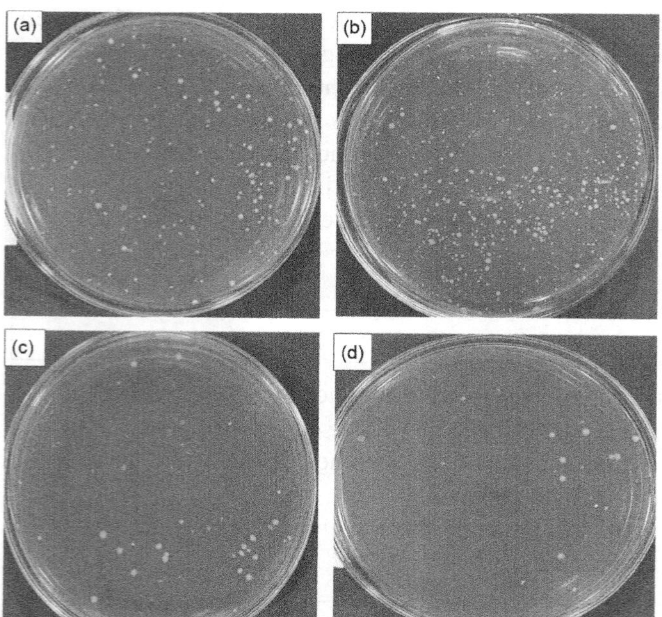

Figure 14.3 Colonies of *L. rhamnosus* (a) before and after electrospinning at voltages of (b) 18; (c) 21; and (d) 24 kV.

in terms of survival rate as compared with other encapsulation techniques discussed above. Such electrospun fibers with higher viability of probiotics can be used for product development. The survival of *L. rhamnosus* in electrospun fibers incorporated with lyoprotectant was found to be higher than those without lyoprotectant (Table 14.1). For example, *L. rhamnosus* encapsulated in electrospun fibers containing 40% isomalto-oligosaccharides (IMO), with and without lyoprotectant, was found to be 10.81 and 10.15 log cfu/mL, respectively. Thus, only a log increase of 0.66 in the population of *L. rhamnosus* in the electrospun fibers was observed after the inclusion of lyoprotectants. Maltodextrin provided better protection for *L. rhamnosus* as compared with lactose and sucrose.

Maltodextrin and other lyoprotectants can stabilize the protein structure because of their stability and high T_g, which could limit molecular interactions by forming glassy structures. The glassy matrix protects the bacterial cells by restricting molecular movement during storage, slowing down detrimental diffusion-controlled processes and bacterial metabolism. The protective effect could also be explained by their hydrogen bonding with biomolecules and their ability to modify the solvation of proteins without affecting the diameter of electrospun fibers. Škrlec et al. reported that lyoprotectants such as sucrose and trehalose had no significant effect on fiber diameter when the lyoprotectant to PEO

Table 14.1 Viability of *L. rhamnosus* after electrospinning.

Applied voltage (kV)	Concentration of pullulan (%)	Type of lyoprotectant	Viability before electrospinning (log cfu/mL)	Viability after electrospinning (log cfu/mL)	Survival rate (%)
18	12	Sucrose	10.11	8.65	87.81
18	12	Maltodextrin	10.14	8.93	88.06
18	12	Lactose	10.48	9.01	85.95
18	14	Sucrose	10.08	8.65	85.88
18	14	Maltodextrin	10.24	9.14	89.26
18	14	Lactose	10.65	8.98	84.44
18	16	Sucrose	10.21	9.09	89.08
18	16	Maltodextrin	10.23	9.09	88.88
18	16	Lactose	10.52	8.82	83.86
21	12	Sucrose	10.11	8.56	87.00
21	12	Maltodextrin	10.14	8.84	88.34
21	12	Lactose	10.48	8.78	83.72
21	14	Sucrose	10.08	8.38	83.14
21	14	Maltodextrin	10.24	8.97	87.57
21	14	Lactose	10.65	8.74	82.06
21	16	Sucrose	10.21	8.55	83.78
21	16	Maltodextrin	10.23	8.75	84.58
21	16	Lactose	10.52	8.74	83.06

ratio was maintained at 1:1 [31]. The purity of encapsulated *L. rhamnosus* was examined based on the production of the catalase enzyme, which is an indicator of pathogenicity. The encapsulated *L. rhamnosus* was found to be catalase negative. There was no formation of effervescence on the application of H_2O_2 on to the microorganisms smeared on a microscopic slide, indicating that contamination of the probiotic did not occur during encapsulation in the nanofibers. The morphological features of *L. rhamnosus* were observed under a light microscope after Gram staining. The purple color of stained bacteria showed that *L. rhamnosus* was a Gram-positive bacterium, rod-shaped and occurred single, in pairs and short chains. The colony characteristics were milky white and round in varying sizes. Jose et al. reported that *L. rhamnosus* appeared as round, opaque, and creamy white colonies on the surface of MRS agar [50].

The interactions between constituents and functional groups in the nanofibers were examined by Fourier transform infrared spectroscopy (FTIR) (Model: Frontier, Perkin Elmer, Singapore). The spectra of pure pullulan, sucrose, lactose, maltodextrin, *L. rhamnosus*, and electrospun fibers were examined. Pure pullulan exhibited peaks in the wave number regions of 3435, 2937, 1630, 1363, 1158, and 1024 cm^{-1} due to O–H stretching, C–H stretching, O–C–O stretching, C–O–H stretching, C–O–C stretching, and C–O stretching, respectively. In the spectrum of *L. rhamnosus*, the peak in the 3100–2800 cm^{-1} region was assigned to the stretching of membrane fatty acids, while the peak observed in the 1800–1500 cm^{-1} region represented amide I and II regions of the peptides in bacterial proteins. Similarly, the stretching vibrations of phosphate bonds led to the formation of peaks in the 1500–1200 cm^{-1} region, while the stretching vibrations of carbohydrates were represented as peaks in the 1200–900 cm^{-1} wave number region.

The spectrum of sucrose indicated major bands in the region of 1000–800 cm^{-1}, relating to the stretching vibrations of carbohydrates, while those in the region of 3800–3500 cm^{-1} were assigned to the OH groups. The FTIR spectra of other lyoprotectants, such as lactose and maltodextrin, also showed similar peaks. The presence of similar spectral peaks in blank and *L. rhamnosus*-loaded nanofibers indicated that there was no presumable degradation in the structure of *L. rhamnosus* during electrospinning. The results of FTIR and FESEM confirmed that *L. rhamnosus* was successfully encapsulated within the fibers of pullulan and lyoprotectant.

To evaluate the fermentation ability after encapsulation, skim milk was fermented into curd using electrospun *L. rhamnosus*, and the acidification characteristics was compared against the curd fermented by unencapsulated bacterium (free form). For a period of 12 h, the acidity, pH, population, and viscosity were determined at regular intervals (Table 14.2). The initial pH of both milks was 6.62, and after 12 hours, it decreased to 4.59 in curd fermented by encapsulated *L. rhamnosus* and to 4.44 in normally produced curd. Correspondingly, acidity increased from 0.18% lactic acid (LA) at the time of inoculation to 0.80% and 0.83% LA in curd fermented by electrospun and normal *L. rhamnosus*. The fermentation kinetics and viscosity data showed that encapsulated *L. rhamnosus* was able to produce curd from skim milk after 8 hours of fermentation itself. However, there was significant difference in the pH ($p < 0.05$) and acidity ($p = 0.042$) of curd prepared from the two cultures. The fermentation kinetics of skim milk were in agreement with those of Minj, who reported that the single strain *L. rhamnosus* was able to ferment milk into curd at a pH of 4.01–4.85 and acidity rate in the range of 0.71–0.82% LA [51].

Table 14.2 Fermentation ability of encapsulated *L. rhamnosus*.

	Electrospun *L. rhamnosus*				Unencapsulated *L. rhamnosus*			
Time (h)	pH	Acidity (% LA)	Set viscosity (mPa.s)	Probiotic count (log cfu/mL)	pH	Acidity (% LA)	Set viscosity (mPa.s)	Probiotic count (log cfu/mL)
0	6.62	0.18	1.95	2.60	6.62	0.189	1.66	2.54
4	5.09	0.50	20.43	5.19	5.0	0.630	17.35	5.69
8	4.92	0.63	2921.33	8.18	4.78	0.648	2484.67	8.60
12	4.59	0.80	4598.33	8.62	4.44	0.828	3264.70	9.30

At the time of inoculation, the viscosity of skim milk added with encapsulated and normal *L. rhamnosus* was 1.95 and 1.66 mPa.s, respectively (Table 14.2). After four hours of fermentation lag, the viscosity increased steeply with the initiation of curd formation in both samples. After 12 hours, the viscosity of curd made from encapsulated *L. rhamnosus* was 4598.33 mPa.s, while that of the control was just 3264.70 mPa.s. The viscosity of curd prepared from encapsulated probiotic was much higher due to the presence of pullulan from the nanofibers, which acted as a stabilizer. The population of *L. rhamnosus* increased from 2.60 to 8.62 log cfu/mL in curd fermented using encapsulated bacterium, while in the normal curd, the count increased from 2.54 to 9.30 log cfu/mL. The population of *L. rhamnosus* in both curds did show noticeable difference ($p = 0.039$) due to the reduction in its viability in the former after electrospinning. The amount of whey separation was relatively less in curd fermented using encapsulated *L. rhamnosus* as compared with that prepared using normal bacterium. Being a polysaccharide, pullulan involved in casein–polysaccharide interactions, which increased the gel-forming properties of casein, enhancing its viscosity and water-holding capacity. Therefore, syneresis was less in curd fermented using encapsulated *L. rhamnosus*. Thus, it could be clearly established that electrospun *L. rhamnosus* was able to ferment skim milk into curd, without sacrificing its physico-chemical properties.

14.2.4.2 Determining the Effect of Prebiotics on the Electrospinning of *L. Rhamnosus*

In our study, prebiotics such as IMO, polydextrose (PD), and FOS were added with pullulan as wall material for encapsulating *L. rhamnosus*. Concentration of pullulan was maintained at 14% (w/w), while prebiotics were added at 20%, 30%, and 40% (w/w) of pullulan. Trehalose was added as the lyoprotectant at 10% of wall material, while *L. rhamnosus* was added at the rate of 20%. Other experimental conditions remained similar to the previous study. The viscosity of prebiotic-added feed solutions ranged between 154.39 and 561.29 mPa.s, which was found to be in the suitable range for electrospinning. The surface tension of feed solutions decreased with the addition of trehalose. For example, the surface tension of 14% pullulan–20% IMO solution was observed as 52.31 mN/m, which decreased to 45.32 mN/m. If the viscosity of feed solutions increased and surface tension reduced simultaneously, it favors the formation of continuous and uniform fibers [52]. Also, the electrical conductivity of feed solution increased with the addition of prebiotics and trehalose. Owing to the increase in electrical conductivity and viscosity, and reduction in surface tension, the critical voltage required (calculated using Equation 14.1) for electrospinning reduced to 10.22–11.4 kV. Therefore, the applied voltage for electrospinning was brought down to the 14–18 kV range with the aim of improving the viability of *L. rhamnosus* in the electrospun fibers.

Regardless of the concentration of feed solutions, it was observed from the FESEM images that FOS did not produce clean and uniform fibers in the applied voltage of 14–18 kV. Mostly dripping (electrospraying) of feed solution was observed at the spinneret tip. Pullulan–FOS feed solutions loaded with *L. rhamnosus* could not produce uniform fibers because the feed solution did not meet the electrohydrodynamics of electrospinning. In contrast, feed solutions containing PD and IMO produced uniform and continuous fibers (Figure 14.4). Between IMO and PD, pullulan added with IMO was found to be better in producing clean, thin, and uniform electrospun fibers. The mean diameter of IMO- and PD-added electrospun fibers was 135.52 and 189.58 nm, respectively. Similarly, the fibers produced at 18 kV were relatively uniform and continuous as compared with those produced at 14 kV.

As evidenced before, encapsulation of *L. rhamnosus* made the electrospun fibers to appear beaded (Figure 14.4). The beads appeared as approximately oblong, and uniformly distributed along the fibers. Feng et al. reported that the addition of probiotics into feed solution, resulted in beaded fibers, leading to local widening [32]. Pullulan-based electrospun fibers incorporated with 40% PD as prebiotic and loaded with *L. rhamnosus* manifested more widening (beads) as compared with pullulan-based electrospun fibers incorporated with IMO as prebiotic. The mean length and width of *L. rhamnosus* in these two types of electrospun fibers were 3.60 and 0.549 μm and 3.63 and 0.508 μm, respectively. From FESEM micrographs, it was observed that *L. rhamnosus* was successfully encapsulated into the prebiotic-added electrospun fibers.

The addition of prebiotic in feed solution also had a positive effect ($p = 0.03$) on the viable population of *L. rhamnosus* after electrospinning. The population of *L. rhamnosus* in

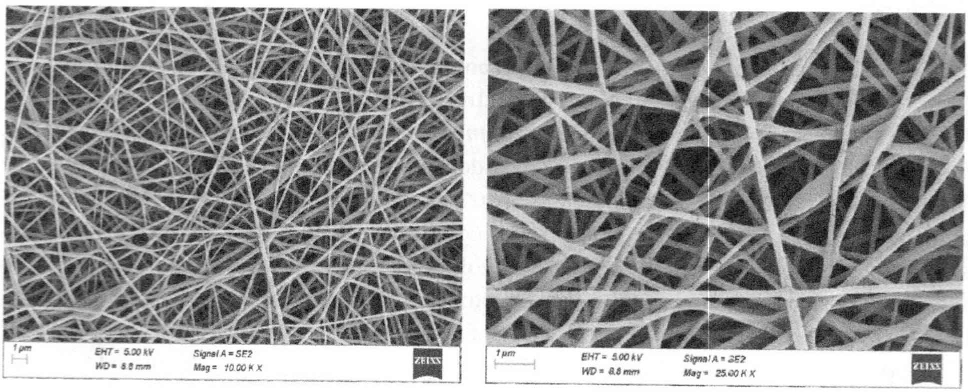

Figure 14.4 FESEM images of pullulan-based electrospun fibers incorporated with 40% IMO and encapsulating *L. rhamnosus*.

electrospun fibers containing FOS, with and without lyoprotectant, was in the range of 6.00–6.30 and 7.35–7.97 log cfu/mL, respectively. In comparison, the mean population of *L. rhamnosus* in electrospun fibers made from pullulan–IMO solutions, with and without lyoprotectant, ranged from 10.42 to 11.15 and 9.60 to 10.15 log cfu/mL, respectively (Table 14.3). It could be stated that *L. rhamnosus* had a lower survival rate in pullulan–FOS capsules than in the electrospun fibers produced from IMO and PD as prebiotic. The viability of *L. rhamnosus* in pullulan–FOS was less because clean and uniform fibers were not formed and only electrospraying occurred, which did not encapsulate the probiotic and exposed it to various stresses. As IMO produced the highest viability, it was the recommended prebiotic for retaining the viability of *L. rhamnosus* after electrospinning, and for the formation of clean, uniform, and beadfree electrospun fibers.

From Table 14.3, it could be observed that trehalose enhanced the viability of *L. rhamnosus* by at least 0.7 log population in the electrospun fibers. Agudelo et al. concluded that the addition of trehalose was more effective in improving the viability of *L. plantarum*, regardless of drying by lyophilization or electrospinning [53]. The survival of electrospun *L. rhamnosus* improved with the addition of trehalose because it was effective in basically carrying the electrical charges in the feed solution, thereby freeing the

Table 14.3 Viability of *L. rhamnosus* in electrospun fibers.

Pullulan concentration (%)	Prebiotic	Prebiotic concentration (%)	Lyoprotectant (Trehalose)	Survived population (log cfu/mL)
14	PD	20	-	9.6021
14	PD	30	-	9.6990
14	PD	40	-	10.0410
14	IMO	20	-	9.8451
14	IMO	30	-	9.6990
14	IMO	40	-	10.1461
14	FOS	20	-	6.0000
14	FOS	30	-	6.3010
14	FOS	40	-	6.3010
14	PD	20	10%	10.4225
14	PD	30	10%	10.4058
14	PD	40	10%	10.8116
14	IMO	20	10%	10.5114
14	IMO	30	10%	10.8997
14	IMO	40	10%	11.1440
14	FOS	20	10%	7.3521
14	FOS	30	10%	7.5396
14	FOS	40	10%	7.9722

IMO: isomalto-oligosaccharides; FOS: fructo-oligosaccharides; PD-polydextrose.

bacteria from the impact of high voltage. Zussman postulated that viability of encapsulated bacteria could be improved by controlling viscous stresses and evaporation rates during electrospinning so that the osmotic pressure level was maintained within the tolerance limit of bacterial cells [54]. Dai et al. also showed that the viability of probiotics could be maintained during electrospinning by using suitable lyoprotectant–prebiotic combinations [55].

The properties of feed solution used also have a major role in the viability of the encapsulated probiotic. Viscosity and surface tension affect the residual water content, the degree of geometrical confinement, the permeability of polymer matrix, and the interactions between bacterial cells and wall material, thereby governing the viability of probiotics in electrospun fibers. In this study, the viscosity and surface tension of feed solutions were modified by the addition of IMO and trehalose, which consequently reduced the applied voltage required for electrospinning of *L. rhamnosus*. Deng et al. reported that the reduction of surface tension of gelatin solutions from 37.86 to 19.76 mN/m by the addition of Tween 80 resulted in the formation of beadfree fibers [56].

The survival rate of *L. rhamnosus* encapsulated in electrospun fibers was found to range between 69.98% and 93.89% (Table 14.4). Feed solutions containing IMO as prebiotic and trehalose produced encapsulated *L. rhamnosus* with the highest survival rate of 88.30–93.62%. In contrast, the survival rate of *L. rhamnosus* in electrospun fibers made from pullulan–FOS solution ranged only from 69.90% to 74.86%. With IMO, the highest survival rate of 93.62% was observed at 40% concentration in the feed solution. Thus, it could be stated that the addition of prebiotics (particularly IMO) and trehalose improved the encapsulation of *L. rhamnosus* owing to the changes in feed solution properties, thereby enhancing their survival in the electrospun fibers. These results were closer to the findings of Alehosseini et al. who estimated the survival rate of *L. casei* as 99.4% after electrospinning [27]. These authors obtained a higher survival rate because they used high-resistant protein capsules from aqueous solution of whey protein containing transglutaminase enzyme.

Table 14.4 Survival rate of *L. rhamnosus* in electrospun fibers.

Prebiotic	Prebiotic concentration (%)	Lyoprotectant	Lyoprotectant concentration (%)	Viability before electrospinning (log cfu/mL)	Viability after electrospinning (log cfu/mL)	Survival rate (%)
PD	20	Trehalose	10	11.9514	10.4225	87.19
PD	30	Trehalose	10	11.7782	10.4058	88.35
PD	40	Trehalose	10	11.8451	10.8116	91.27
IMO	20	Trehalose	10	11.9031	10.5114	88.30
IMO	30	Trehalose	10	11.9542	10.8997	91.17
IMO	40	Trehalose	10	11.9542	11.1440	93.62
FOS	20	Trehalose	10	10.5051	7.3521	69.98
FOS	30	Trehalose	10	10.5798	7.5396	71.26
FOS	40	Trehalose	10	10.5563	7.9722	75.52

IMO: isomalto-oligosaccharides; FOS: fructo-oligosaccharides; PD: polydextrose; cfu/mL stands for colony-forming units per milliliter.

The encapsulated *L. rhamnosus* in electrospun pullulan–IMO fibers were stored at 10°C to evaluate the viability during storage. The population of *L. rhamnosus* in electrospun fibers on the 1st, 8th, and 15th day was enumerated (Table 14.5). The population of *L. rhamnosus* in the electrospun fibers declined by 0.371–0.432 log on the 8th day and by 0.637–0.748 log on the 15th day. However, it could be seen that as the concentration of the prebiotic increased, the population of *L. rhamnosus* in electrospun fibers during storage also increased. Estilarte et al. reported that the viability of freeze-dried *E. durans* decreased by not more than 2.5 log at 4°C after 92 days, while a dramatic loss of viability was observed at 25°C during the same period [14]. Loss of encapsulated probiotic during storage is inevitable, but the conditions need to be optimized so as to keep the viability loss at a minimum.

The fermentation activity in curd prepared using *L. rhamnosus* encapsulated in IMO-added electrospun fibers and of freeze-dried *L. rhamnosus* (control) was compared. The acidity and pH of reconstituted skim milk just inoculated with bacteria (after 0 hour of fermentation) and curd (after eight hours of incubation) were determined. The acidity increased from 0.18% to 0.57% LA, while the pH decreased from 6.56 to 4.80 after eight hours of fermentation. The viable population of *L. rhamnosus* in electrospun fibers increased from 3.18 to 9.27 log cfu/mL in the curd. It was observed that there was no significant difference in acidity, pH, and viable population of curd fermented using encapsulated *L. rhamnosus* and control ($p > 0.05$). The fermentation ability of *L. rhamnosus* encapsulated in electrospun fibers and that of freeze-dried culture was nearly the same. Thus, it could be concluded that the addition of prebiotic, particularly IMO, and trehalose as lyoprotectant improved the survival of *L. rhamnosus* in the electrospun fibers, maintaining their fermentation ability after encapsulation.

Industrial Corner

- The properties of feed solution used plays an important role in determining the viability of the encapsulated probiotic.
- Addition of trehalose can improve the survival of electrospun *L. rhamnosus* by carrying the electrical charges in the feed solution and freeing the bacteria from the impact of high voltage.
- Addition of prebiotics and trehalose as lyoprotectants can maintain the fermentation ability of *L. rhamnosus* after encapsulation into electrospun fibers.

Table 14.5 Viability of encapsulated *L. rhamnosus* during storage in electrospun fibers.

Prebiotic	Prebiotic concentration (%)	Lyoprotectant	Lyoprotectant concentration (%)	0th day population (log cfu/mL)	8th day population (log cfu/mL)	15th day population (log cfu/mL)
IMO	20	Trehalose	10	10.5114	10.0791	9.7634
IMO	30	Trehalose	10	10.8997	10.4861	10.1682
IMO	40	Trehalose	10	11.1440	10.7731	10.5075

IMO: isomalto-oligosaccharides; cfu/mL: colony-forming unit per milliliter of sample.

14.2.5 Comparisons of Electrospinning with Other Encapsulation Methods

Previously, Alehosseini et al. observed the survival of electrosprayed and freeze-dried cultures of *L. casei* as 99.4% and 84.2%, respectively [27]. The higher survival rate of *L. casei* encapsulated by electrospraying was attributed to the lower stresses experienced in comparison with freeze drying. The major cause of viability loss in freeze drying was cell surface damage with extensive cell wall pitting, wrinkling, and disruption of cell membrane by ice crystal formation during freezing. As the salt concentration within and around probiotic cells increased, damage of cell membrane occurred. Furthermore, the bound water that was removed during freeze drying could have resulted in the destabilization of cell wall (loss of structural integrity) of *L. casei*. Similarly, Ma et al. electrospun *Lactobacillus* strains using gum arabic/pullulan (20:80), and showed better survivability of 85.38–97.83% by electrospinning as compared with 80.92–89.84% by freeze drying [38]. Mojaveri et al. also electrospun *B. animalis* in chitosan/PVA/inulin fibers, with reduction in viability from 9.47 log cfu/mL to 8.37–8.44 log cfu/mL (1 log reduction) on electrospinning [46]. In contrast, *B. animalis* capsules prepared by the emulsion/extrusion technique had 0.9–2.8 log reduction in survival [57].

Recently, Arepally et al. spray-dried *L. acidophilus* using maltodextrin and gum arabic at inlet air temperature of 150°C, and achieved survival of 67.5–87.7% [19]. In comparison, electrospun *L. acidophilus* in PVA fibers had a higher survival rate of 78.6–90% for 21 days at refrigeration temperature [34]. Thus, it could be stated that the survival of probiotics is on a par or even higher in the case of electrospinning as compared with other encapsulation techniques. Zussman proposed that the electrical charges applied to feed solution during electrospinning were mainly distributed at the free surface of the liquid jet, and the encapsulating microorganisms were not affected drastically [54]. By making a proper selection of prebiotic, lyoprotectant, and electrospinning conditions, the viability of probiotics in electrospun fibers could be improved. Encapsulation of probiotics by electrospinning also improves their viability during storage and their thermal stability as well. Hu et al. proposed that electrospun nanofibrous membranes proved to be an excellent scaffold for bacteria biofilm with high stability, wherein the biofilms were interlocked to form a cohesive structure [35].

14.3 Conclusions

The potential of electrospinning in encapsulation of probiotics and food applications is considerably less explored. Therefore, encapsulation of *L. rhamnosus* by electrospinning was attempted in this study using food-grade pullulan, prebiotics, and lyoprotectants. From the results, it was evident that *L. rhamnosus* was effectively encapsulated in the electrospun fibers. By a careful selection of prebiotics, lyoprotectant, and electrospinning conditions, it was possible to achieve high viability of *L. rhamnosus*, which was comparable to conventional freeze drying. The fermentation ability of encapsulated probiotic was also not affected, and it was comparable to that of normal culture. Electrospinning proved to be a viable alternative to costlier freeze drying for encapsulation of probiotics without much impact on their survival rate. Further extensive experiments are required to validate the functional attributes of encapsulated *L. rhamnosus* before commercialization of the electrospinning technique.

Glossary

Electrospinning A process that generates fibers from a polymer solution under high electric field voltage.

Probiotic A substance produced by microorganisms that gives health benefit(s) to the host by improving the intestinal microbial balance.

Prebiotic A nonviable food component that gives health benefit(s) to the host via modulation of the microbiota.

Lyoprotectant A substance added to a formulation during freeze drying or water removal so as to protect an active ingredient from degradation.

Trehalose A nonreducing disaccharide composed of two glucose molecules.

Wall material The material used to encapsulate an active ingredient during spray drying.

Spray drying A process in which a solution is atomized into a drying chamber to transform the solution from the liquid state into the dried particulate form.

Freeze drying A process that dries a frozen solution via sublimation.

References

1 FAO/WHO, Probiotics in food. Health and nutritional properties and guidelines forevaluation. FAO food and nutrition paper no. 85. Rome, Italy: World Health Organization and Food and Agriculture Organization of the United Nations, **2006**.
2 Markets and Markets, Probiotics Market by Application (Functional Food & Beverages, Dairy Products, Non-dairy Beverages, Infant Formula, Cereals, Dietary Supplements, Feed, Ingredient (Bacteria, Yeast), Form (Dry, Liquid)), End User, & Region - Global Forecast to 2026, **2021**.
3 S. N. Casarotti, A. L. B. Penna, *Int. Dairy J.* **2015**,*41*, 1–6.
4 S. Rokka, P. Rantamäki, *Eur. Food Res. Technol.* **2010**, *231*, 1.
5 Y. K. Ye, J. Lu, X. Wang, S. Zhang, L. Liu, B. Yang, K. Nassar, X. Xu, X. Pang, J. Lv, *Biomed. Pharmacother.* **2020**,*127*, 110159.
6 C. Fuentes, T. Lajo, J. M. Carrion, J. Cune, *Br. J. Nutr.* **2013**,*109*, 1866.
7 L. J. Teng, Y. Lyu, X. Hu, Y. Zhao, M. Wang, *Molecules* **2019**,*24*, 51.
8 G. Q. Wang, J. Pu, X. Q. Yu, Y. J. Xia, L. Z. Ai, *J. Dairy Sci.* **2020**,*103*, 3066.
9 R. Rajam, C. Anandharamakrishnan, *J. Food Eng.* **2015**,*166*, 95.
10 M. Saarela, I. Virkajärvi, H. L. Alakomi, T. Mattila-Sandholm, A. Vaari, T. Suomalainen, J. Mättö, *J. Appl. Microbiol.* **2005**,*99*, 1330.
11 S. G. Deepika, D. Charalampopoulos, *Lett. Appl. Microbiol.* **2009**,*48*, 295.
12 B. P. K. Reddy, S. P. Awasthi, A. N. Madhu, S. G. Prapulla, *Food Biotechnol* **2009**,*23*, 243.
13 S. O. Ramon, Z. Kaplun, L. Levin-Brener, N. Gurevich, E. Shimoni, *Food Bioprocess Technol.* **2010**,*43*, 193.
14 M. L. Estilarte, E. E. Tymczyszyn, M. D. L. Á. Serradell, P. Carasi, *LWT-Food Sci. Technol.* **2021**,*137*, 110496.
15 G. Wang, L. Luo, C. Dong, X. Zheng, B. Guo, Y. Xia, L. Tao, L. Ai, *J. Dairy Sci.* **2021**,*104*, 2606.

16 L. L. Tan, K. Sampathkumar, J. H. Wong, S. C. J. Loo, *Food Bioprod. Process.* **2020**,*124*, 369.

17 D. P. Kandylis, Y. Kourkoutas, *LWT-Food Sci. Technol.* **2016**,*69*, 468.

18 B. D. Vandenheuvel, I. J. Claes, S. Lebeer, F. Kiekens, *Int. J. Pharm.* **2016**,*505*, 303.

19 R. S. Arepally, T. K. Reddy, Goswami, *Curr. Res. Food Sci.* **2020**, *3*, 235.

20 B. S. Kiekens, K. Jokicevic, E. Byl, T. Henkens, D. Vandenheuvel, S. Lebeer, F. Kiekens, *Drying Technol.* **2020**,*38*, 1474.

21 K. S. Leylak, G. C. Özdemir, Z. B. O. Gurakan, *Int. Dairy J.* **2021**,*112*, 104865.

22 H. N. Fu, H. Ndiaye, M. W. Woo, R. Jeantet, X. D. Chen, *Food Bioprocess Technol.* **2021**,*14*, 120.

23 S. X. Hua, M. Zhang, Y. Wang, Y. Chen, J. Zhang, C. Wang, Y. Wang, *Food Sci. Anim. Resour.* **2020**,*40*, 118.

24 K. S. Yoha, J. A. Moses, C. Anandharamakrishnan, *Drying Technol.* **2020**,*38*, 610.

25 M. A. Alav, C. Karakas, E. Dertli, M. T. Yilmaz, Z. Ceylan, *LWT-Food Sci. Technol.* **2019**,*113*, 108292.

26 B.G. Seethu, H. A. Pushpadass, F. M. E. Emerald, B. S. Nath, N. L. Naik, K. S. Subramanian, *Food Bioprocess Technol.* **2020**,*13*, 341.

27 A. M. Sarabi-Jamab, B. Ghorani, R. Kadkhodaee, *LWT-Food Sci. Technol.* **2019**,*102*, 150.

28 S. K. Tiwari, S. S. Venkatraman, *Mater. Sci. Eng.* **2012**,*32*, 1037–1042.

29 D. Rajanna, H. A. Pushpadass, F. M. E. Emerald, N. V. Padaki, B. S. Nath, *J. Sci. Food Agric.* **2021** *102*, 1684-1698.

30 D. J. Heunis, M. Botes, L. M. T. Dicks, *Probiotics Antimicrob. Proteins* **2010**, *2*, 46–51.

31 K. Škrlec, Š. Zupančič, S. P. Mihevc, P. Kocbek, J. Kristl, A. Berlec, *Eur. J. Pharm. Biopharm.* **2019**, *136*, 108.

32 M. Y. Feng, Z. Y. Zhang, R. J. Linhardt, M. H. Zong, L. Li, H. Wu, *J. Agric. Food Chem.* **2018**,*66*, 10890.

33 L.-R. E. Sanchez, Y. Sanz, J. M. Lagaron, *Biomacromolecules* **2009**,*10*, 2823.

34 Y. Fung, K. H. Yuen, M. T. Liong, *J. Agric. Food Chem.* **2011**,*59*, 8140–8147.

35 M. X. Hu, J. N. Li, Q. Guo, Y. Q. Zhu, H. M. Niu, *J. Agric. Food Chem.* **2019**,*67*, 3198.

36 Y. W. Liu, D. Li, C. Liu, Z. Feng, B. Jiang, *Polymers* **2020**,*12*, 1565.

37 M. T. Yilmaz, O. Taylan, C. Y. Karakas, E. Dertli, *Carbohydr. Polym.* **2020**,*244*, 116447.

38 M. C. Xu, H. Yu, Z. Feng, W. Yu, L. Gu, Z. Liu, L. Chen, Z. Jiang, J. Hou, *Food Hydrocoll.* **2021**,*111*, 106381.

39 A. M. Azizkhani, *Innovative Food Sci. Emerging Technol.* **2021**,*72*, 102750.

40 C. Cheng, A. Demirci, J. M. Catchmark, *Appl. Microbiol. Biotechnol.* **2011**,*92*, 29.

41 L.-R. E. Sanchez, S. Wilkanowicz, Y. Sanz, J. M. Lagaron, *Food Hydrocoll.* **2012**,*28*, 159–167.

42 M. Basholli-Salihu, M. Mueller, S. Salar-Behzadi, F. M. Unger, H. Viernstein, *LWT-Food Sci. Technol.* **2014**,*57*, 276–282.

43 H. E. Pantea, P. Vass, J. Domjan, M. Molnar, A. Suhajda, S. K. Andersen, T. Vigh, G. Verreck, G. J. Marosi, Z. K. Nagy, *Food Bioprod. Process.* **2021**,*128*, 84.

44 Z. K. Kim, D. Fang, S. Ran, B. S. Hsiao, B. Chu, *Polymer* **2002**,*43*, 4403.

45 S. J. Kuhn, Y. Dror, E. Zussman, *Nanotechnology* **2006**,*17*, 4675–4681.

46 J. Mojaveri, S. F. Hosseini, A. Gharsallaoui, *Carbohydr. Polym.* **2020**, 116278.

47 Š. Zupančič, K. Škrlec, P. Kocbek, J. Kristl, A. Berlec, *Pharmaceutics* **2019**,*11*, 483.

48 N. Sosa, E. Gerbino, M. A. Golowczyc, C. Schebor, A. Gómez-Zavaglia, E. E. Tymczyszyn, *Front. Microbiol.* **2016**,*7*, 584.

49 T. K. C. Capela, N. P. S. Hay, *Food Res. Int.* **2006**,*39*, 203–211.

50 M. Jose, C. R. Bunt, M. A. Hussain, *Microorganisms* **2015**,*3*, 198–212.

51 Minj, Doctoral Dissertation, ICAR-NDRI (Deemed University), Karnal **2011**.

52 P. Gupta, C. Elkins, T. E. Long, G. L. Wilkes, *Polymer* **2005**,*46*, 4799–4810.

53 J. Agudelo, A. Cano, C. González-Martínez, A. Chiralt, *J. Funct. Foods* **2017**,*37*, 416–423.

54 Zussman, *Polym. Adv. Technol.* **2011**, *22*, 366.

55 M. Dai, A. Senecal, S. R. Nugen, *Nanotechnology* **2014**,*25*, 225101.

56 L. Deng, X. Kang, Y. Liu, F. Feng, H. Zhang, *Food Chem.* **2017**,*231*, 70–77.

57 M. Kumherová, K. Veselá, K. Jokešová, I. Klojdová, Š. Horáčková, *Czech, J. Food Sci.* **2020**,*38*, 57.

15

Three-Dimensional Printing in Food Manufacturing and Mechanics

Stefania Chirico Scheele[1], *Martin Binks*[2], *and Paul F. Egan*[1,*]

[1] *Department of Mechanical Engineering, Texas Tech University, Box 41021, Lubbock, TX 79409*
[2] *Department of Nutritional Sciences, Texas Tech University, Box 41270, Lubbock, TX 79409*
[*] *Corresponding author*

Highlights

- 3D food printing is a potentially disruptive food technology that provides food creativity, user control of shape, structure, taste, and delivery of on-demand customized meals that are personalized for a consumer's individual needs.
- 3D food printing enables customized shape fabrication without requiring multi-step mold development, thereby allowing less skilled users to economically produce novel shapes.
- Printed foods include fruits, vegetables, proteins, and desserts that can all be processed in an automated manner to form complex shapes.
- Mixing additives with 3D printed foods to improve printability results in taste, texture, and nutrition alterations that present complex trade-offs to be considered for consumer satisfaction.
- Future research for food printing technology will lead to a greater number of types of foods printed and improved usability for success in the food industry.

15.1 Introduction

Three-dimensional (3D) printing processes are part of an emerging area of additive manufacturing that provides unprecedented freedom regarding design, manufacturing, and innovation [1, 2]. It is often synonymous with additive manufacturing, and relies on the additive deposition of materials to form parts that are widely used in mechanical engineering, biotechnology, biomedical engineering, aeronautics, pharmaceutics, and, more recently, the food industry [3, 4]. The 3D printing process begins with specifying a digital design, often created using computer-aided design (*CAD*) software, that is used to direct hardware to deposit material, layer upon layer, to form a physical recreation of the digital design. In the food industry, 3D printing enables the recreation of complex and intricate shapes with food ingredients that bypass the need for traditional manufacturing techniques, such as molds, that are typically cost-prohibitive for small-batch designs [5, 6]. Further, 3D food printing allows for the customization of ingredients and shapes for each

Materials Science and Engineering in Food Product Development, First Edition. Edited by Wing-Fu Lai.

new food fabrication that provides high levels of personalization for achieving desirable food shape, *texture*, taste, and nutrition [7–9]. In this chapter, recent advances in 3D food printing research are explained in relation to design processes that lead toward practical adoption for the food industry, which provides the foundation for translating current research into successful products [10, 11]. The approach provides an overview of design processes to aid users in fully leveraging 3D food printing technology by tuning food ingredients, manufacturing, and mechanics to achieve printable designs that are desirable to consumers.

3D food printing borrows its fundamental technology from more common metal/polymer additive manufacturing processes, with the most common food printing techniques being: (i) material extrusion; (ii) powder bed fusion; (iii) binder jetting [12–14]. Material extrusion is by far the most widely used and affordable food printing approach that works by depositing soft food in specified patterns via flow through a nozzle. Due to material extrusion's simplicity and utility, it is the focus of this chapter since it is also the most practical for translation to widespread industry use [15]. Powder bed fusion and binder jetting, on the other hand, typically require large vats of powdered food material that necessitate greater initial capital and also limit material selection, with common foods produced being sugar-based. However, extrusion food printing still has challenges in technological implementation, particularly since food processing has distinct needs that make a direct translation of the technology from metal/polymer processing infeasible. For instance, commonly printed thermoplastics possess higher hardness and viscosity properties than food and have much less variation in physicochemical properties that make printing conditions more consistent and defined [16]. In addition to difficulties in achieving consistent physical food properties, food printing has further challenges based on the need to satisfy the complex, and often subjective, needs of consumers that benefit from printed food.

Although 3D food printing can widely provide value to the public, the technology currently excels in customizing foods for the needs of specific demographics, such as tuning food shapes, textures, and nutritional profiles, to produce more appealing foods for young children or the elderly. There are also medical applications for 3D food printing, such as producing soft foods with desirable tastes and textures to provide aesthetically pleasing meals to those with dysphagia [17]. Personalized nutrition is also enabled by 3D food printing since it is possible to mix food ingredients during fabrication (i.e. at print time) for each new food preparation, thereby delivering specified nutrient profiles (e.g. food fortification) that may help address concerns for cardiac diseases, cancer, and obesity [5, 18, 19]. Additionally, 3D food printing enables the delivery of fresh foods that are blended and pureed on-site to provide nutrient-rich whole foods to consumers, which can further benefit from the use of locally sourced ingredients and the potential creation of novel foods with nonconventional ingredients. Achieving success in 3D food printing thereby requires multifaceted considerations related to the food ingredients themselves, the printing parameters to process the food, the food's resulting mechanics, and the consumer's approval of the printed food, which make up the key phases in the food printing design presented in Figure 15.1.

The initial material preparation step requires determining the desired ingredients and material preparation to ensure a steady flow of food through a nozzle and consistent attachment of a new food layer to each deposited food layer [20]. Food ingredients for 3D printing require consideration of mechanical and *rheological* characteristics, such as viscosity, complex modulus, *firmness*, and phase behavior, which have the greatest influence on the resulting *printability* [21, 22]. Printability, in the context of this chapter, refers to the food's capabilities for flowing through the nozzle, adhering to surfaces/previously printed food layers, achieving dimensional *accuracy*, and retaining its shape over time [23]. If a base food material does not initially have suitable printing qualities, it may be modified through the addition of water, starches, proteins, and gums to reach a consistency suitable for printing and holding a shape afterward.

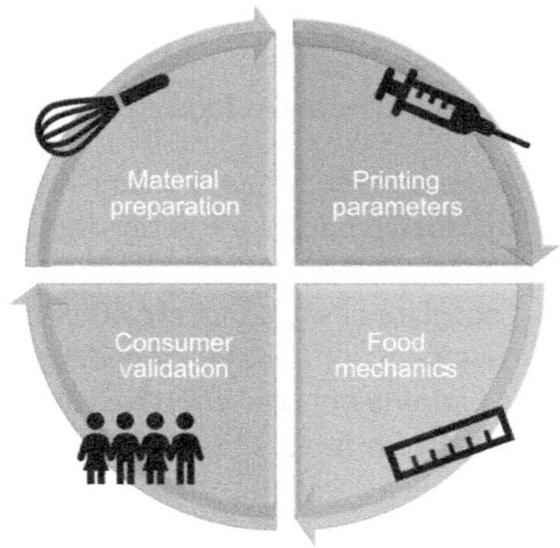

Figure 15.1 3D food printing design process for successful food fabrication with mechanics and ingredients that appeal to consumers.

The printing parameters phase refers to altering the settings and build process of a printer that controls the flow and deposition of food, including printing speed, flow rate, nozzle diameter, nozzle height, and temperature, which are all commonly manipulated to improve the structure and stability of printed foods [24, 25]. After printing, food mechanics investigation is necessary to measure properties that are related to how well the printed food facilitates repeatability, while also indicating textural properties that affect a consumer's eating experience. The consumer validation phase considers how consumers perceive the 3D printed food fabrication *fidelity* and their assessment of its taste experience. Fabrication fidelity is the match of a 3D printed shape to its CAD image [26], and the taste experience is related to the taste, texture, and mouthfeel of food, and is defined as chemicals that generate gustatory sensations [27–29].

The design process covered in this chapter provides the foundations for translating research findings in food printing research to practical applications in the industry, which has led to many new business models in the industry, as described in Industrial Corner 1. The process provides key links to improving food printability by considering decisions across all Figure 15.1 phases and how they relate to one another for delivering a high-quality food product with consumer appeal with regard to texture, taste, and shape [30]. This chapter is structured with the next section providing a practical description of how to create foods with 3D printing followed by four sections covering each of the Figure 15.1 design phases using a case study with mashed potatoes with pea protein additive to illustrate key principles. The chapter concludes with a section summarizing the state of research and future considerations for 3D food printing in the food industry. As a whole, the chapter provides practical considerations for leveraging 3D food printing as a potentially disruptive technology for automating personalized meal fabrication that aims to surpass the conventional one-size-fits-all approach to food preparation.

Industrial Corner 1

How does 3D-printed food contribute to gastronomy and the culinary arts?

- Additive manufacturing processes enable the creation of truly sophisticated, one-of-a-kind food products; the design freedom and customization offered by 3D printing enable intricate and replicable designs difficult to recreate efficiently with any other process.
- Chocolate designs that are 3D printed can have complex geometries that exceed the constraints of traditional models. It is possible to print each design uniquely, which is typically not cost-effective when using molds.
- Recently, a factory in Belgium called Miam has incorporated 3D printing to create ready-to-eat edibles from milk chocolate, dark chocolate, and white chocolate [31].
- Complex entrées are achievable with multi-nozzle extrusion systems, such as automated pizza production by depositing dough, sauce, cheese, and toppings sequentially. Such processes are extendible for extruding cake batters to create elaborate baked goods.

15.2 Print Process

The 3D printing process consists of preparation steps for readying a digital design and food materials for the printer, followed by loading the printer, specifying printing parameters, and printing the food in a desired shape. These steps are illustrated for a Procusini food printing system in Figure 15.2 and are representative of general steps that will vary based on the exact build/brand of the printing system.

An initial preparation step requires obtaining or generating a 3D model that represents the desired shape of the printed food (Figure 15.2a). Designs are typically generated using a three-dimensional CAD system, such as Inventor, Creo, SolidWorks, and AutoCAD,

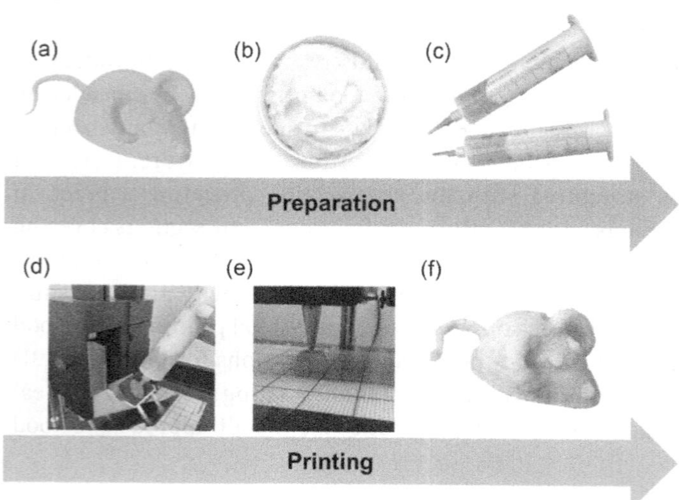

Figure 15.2 Three-dimensional (3D) food preparation process steps for (a) digital design creation; (b) food mixing [*Source:* New Africa/Adobe Stock Photos]; (c) cartridge filling; 3D food printing process demonstrating; (d) cartridge loading; (e) deposition; and (f) final product [*Credit:* The American Society of Mechanical Engineers].

among many other options, including those that are open source and free to users, such as TinkerCAD. The generated model is converted to an *STL-file* that stores information concerning the edges and surface of the design that is then transferred to the printer or slicing software to generate a *G-code* file that directs the printer on how to regulate its temperature, flow speed, and layer-by-layer deposition [32].

There are restrictions on 3D models suitable for printing since overhangs and top-heavy designs can make the shape unstable, and exact design constraints may differ based on the food and printer used. For example, in polymer printing, a printer may produce a part without the need for support material when the overhang angle is at most 45° [33, 34], but research in 3D food printing has demonstrated that a minimum overhang angle of 55° is needed for marzipan material while a minimum overhang angle of 40° is needed for chocolate to produce a 3D print without wall support [26].

Before printing, there is also the need to prepare a food mixture (Figure 15.2b) with suitable mechanical and rheological properties to flow through the printer nozzle, followed by filling the cartridge that holds the material and has one end for the printer extruder to enter and a nozzle on the other (Figure 15.2c). In the case of mashed potatoes, the food may be prepared by hydrating flakes or from fresh potatoes by mashing and blending. The solid pieces of potato require removal by straining; otherwise, they risk clogging the nozzle during printing. Additives, such as corn starch, xanthan gum, or protein powders, are then added to improve consistency before printing.

The printing process is initiated by inserting the cartridge in the printer (Figure 15.2d) followed by specifying printing parameters according to the inserted food and selected design to ensure a proper build. Printing parameters include the initial distance of the nozzle from the build surface, extrusion speed, and movement speed of the print head [35]. For the Procusini printer and many other systems, a dosing step is required to prime the printing process by extruding some food through the nozzle to ensure a steady flow. Priming also provides an initial compression of food in the nozzle and cartridge to reduce air bubbles that may have formed. The printing process then begins with extruder movement that results in food flowing through the nozzle as the print head moves in horizontal directions to lay down a single layer of food, followed by the print head moving up to initiate the next layer (Figure 15.2e). The final printed product should resemble the initial digital design (Figure 15.2f), with high degrees of resemblance being assessed as a higher fidelity print that is visually appealing to the consumer while also retaining its shape over time.

Some 3D printed foods may also benefit from pre- and post-processing techniques, such as heating and cooling. In particular, heating can soften food to improve flow through the nozzle while cooling helps food harden and retain its shape after printing, which is crucial for printing chocolate. Microwaving foods, such as potatoes before printing, has been demonstrated to decrease moisture content to improve printability [36] while microwaving mushroom structures results in increased hardness after printing [37]. Infrared heating has also been used as a post-processing step to improve rigidity and structural stability [38, 39]. Modification to the printer can also improve the process, such as using multi-nozzle extrusion to print multiple foods or the same food multiple times to reduce the overall printing time. In addition to altering the print process, simply altering the food itself by adding thickeners, gums, and proteins can significantly improve the quality of a print and ensure suitability for a variety of printers and processes.

15.3 Material Preparation

Preparing materials for food printing is one of the most important factors impacting print success and ensuring the resulting food will have nutrition, taste, and texture that is desirable to consumers. Suitable food materials should flow through the nozzle and maintain their shape after deposition [21]; however, not all food components are natively printable and proper preparation through the inclusion of additives and agents is not always straightforward due to the need to consider complex food matrices during the design process. Although many foods are easily extruded through simply pureeing, straining to remove solids, and adjusting the water content, many food materials require the addition of additives, such as hydrocolloids and starches, that can alter the material's viscosity [40–42]. Currently, researchers have prepared and successfully printed a variety of foods for fruits and vegetables [43], including mashed potato [44, 45], pumpkin [46], lemon gel [47]; proteins, including eggs [48], shrimp/fish [49, 50], chicken [51]; and desserts, including chocolate [52, 53], marzipan [54], and cookies [55, 56], as demonstrated in Figure 15.3.

Further foods that have been 3D-printed include dairy-based products, such as cheese [59, 60], and unconventional foods, including insects [61, 62] and microalgae [56, 63]. The technology can process unconventional foods to take on more desirable forms, allowing the consumer to take advantage of their beneficial nutritional profiles. Some examples might include insects that are a highly available/affordable source of protein and algae that are rich in cellulose. One study found the addition of chlorella and spirulina in dough led

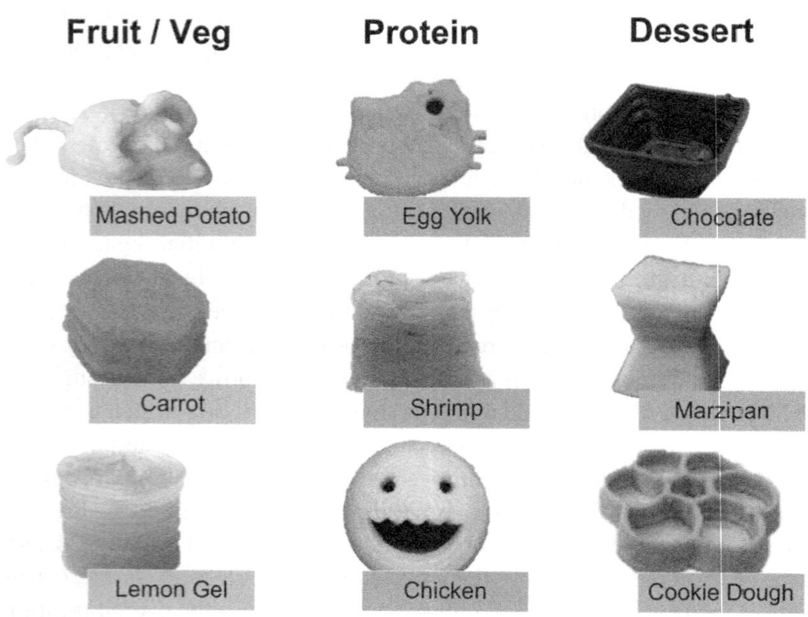

Figure 15.3 Sample fabrications of fruit/vegetable, protein, and dessert-based food materials, including mashed potato [57], carrot [40], lemon gel [47], egg yolk [48], shrimp [50], chicken [58], chocolate [26], marzipan [26], and cookie dough [55]. Images adapted from [40, 47, 48, 58] with permission from Elsevier; Images adapted from [26] with permission from ASME.

to a greater mechanical resistance during the 3D printing process, while also improving the printability in terms of dimensional accuracy and providing the 3D printed structures with more stability and resistance to baking [56]. Mechanical and baking resistance refers to the ability of the printed sample to retain its shape and geometric accuracy while printing, after printing, and during post-processing treatments. Ensuring mixed materials are printed accurately and retain their shape is linked to measurable properties that can inform what additives to include in food mixtures according to the mechanical and rheological properties of food materials [11, 23].

Successful printing requires foods with high enough viscosity for self-supporting and stackable layers, but low enough viscosity to ensure flow through the nozzle. Starch has been extensively used in food applications to improve the processing and final quality of products. Hydrocolloids, such as xanthan gum and gelatin, can improve printability as they have been postulated to inhibit swelling and consequently decrease the viscosity of starch by retarding water accessibility to the starch granules [64, 65]. One study investigated the effect of agar, alginate, glycerol, and lecithin additives on the rheological properties of mashed potatoes [66], with two different concentrations of each additive considered (0.5% and 1%). The rheological properties analyzed were the storage modulus G' (indicator of the elastic behavior), the loss modulus G'' (an indicator of the viscosity behavior), and complex viscosity η^* (related to the global viscoelastic response). The study found that in most cases agar and alginate could moderately affect and stabilize potato puree, while lecithin and glycerol showed a destabilizing effect even at very small concentrations.

Structural components, such as carbohydrates, proteins, and fats, also affect how additives mix with foods and printability, meaning there is not a single solution for improving the printability of all foods [22]. For example, dietary fiber can increase the water-holding capacity, and thus, impact the textural properties, such as the firmness and the rigidity of the food material, and taste experience, such as mouthfeel [67]. On the other hand, understanding fat crystallization is crucial to printing chocolate as cocoa butter is the main ingredient responsible for chocolate's structural stability and shape retention [52]. The triglycerides in cocoa butter can crystallize into six different main forms with increasing melting points that impact the characteristics of the final product and can influence parameters, such as self-supporting layers, gloss, taste, and shelf life [68]. The differences in the material properties between food ingredients and combinations have led to a complex matrix of possible printing parameters, such as nozzle diameter, flow rate, and printing speed. Thus, an intricate and complicated relationship exists between material properties and printing parameters that consequently suggests these aspects of the design process require a simultaneous study and consideration to improve food printability and ensure a successful final fabrication.

15.4 Printing Parameters

Once the desired set of food ingredients is mixed, printing parameters, such as temperature, printing speed, nozzle height, and nozzle diameter, enable further tuning of printing processes to improve success [21]. Temperature is a critical parameter that affects material properties and influences viscosity with an inverse relationship. Higher temperatures are used to decrease the viscosity to allow for improved flowability, and in the case of chocolate, the optimal extrusion temperature for forming the most stable crystalline chocolate melt is about 32°C [52, 68].

Printing speed is a crucial mechanical parameter that directly affects structural stability. There are two aspects to printing speed: the first is the speed of material flow through the nozzle, known as extrusion speed, and the second is the speed of the print head, known as nozzle movement speed. A strong coupling exists between extrusion and movement speed. If the extrusion speed is faster than the nozzle movement speed, slurry accumulation will occur producing larger diameter depositions and wavy lines. If the extrusion speed is slower than the nozzle movement speed, the low feed pressure and low flow rate will lead to rupture lines. The nozzle movement speed and the material extrusion speed require a suitable matching range for successful prints [38, 47]. For example, one study observed that printing rice starch at lower nozzle movement speeds (800–1500 mm/min) resulted in the best printability for ensuring a properly printed shape with high retention [69].

Nozzle height refers to the distance between the bottom of the nozzle and the work surface in the printing process. Studies have shown that the ideal printing process, without shrinkage or expansion, has the extruded material diameter equal to, or slightly smaller, than the nozzle diameter to ensure that the material adheres to the previous layer and avoids inaccuracies caused by delayed deposition [32, 47].

Nozzle diameter mainly affects the printing precision and the surface roughness of the sample. For example, studies have determined the ideal consistency for cream cheese was obtained when using a 0.84 mm nozzle diameter [70], for fish surimi gels when using a 2 mm nozzle [49], and for chocolate, the ideal nozzle diameter was determined as 1.25 mm [53]. The smaller the nozzle diameter, the higher the degree of refinement of the printed sample. Figure 15.4 demonstrates lattice structure shapes that were printed using only mashed potatoes with two nozzle sizes, 0.8 and 1.5 mm, with imaging used to investigate the microstructure and build accuracy [71, 72]. The microscopic details are highlighted in the figure, where for the 0.8 mm nozzle, walls are homogeneous as the nozzle passes multiple times with small spacing, while for the 1.5 mm nozzle, the nozzle passes two times and introduces gaps.

Selecting a proper nozzle size has multiple trade-offs that require consideration for improving printability and efficiency. Smaller nozzle diameters have the benefit of improving the printing resolution and the surface finish. However, a decrease in the nozzle diameter also increases the printing time and the feed pressure, which may result in the discontinuous deposition of the printable food material [49]. Therefore, when selecting the nozzle diameter, considering the viscosity of the food material and the maximum pressure that the equipment can sustain is necessary since the material is not extrudable under some circumstances, potentially creating pressures that damage the printer and introduce unnecessary wear [73, 74]. The longer printing time required for smaller nozzles requires balancing with cost factors and demands for the rate of food delivery necessary for consumers. Once the food is successfully printed based on selected food materials and processing parameters, the next major considerations in the design process are mechanics/texture post-printing that affect shape retention over time and how consumers experience food.

Figure 15.4 Imaging for 3D printed mashed potato lattices using a 0.8 and 1.5 mm nozzle tip size; 1 mm scale bar.

15.5 Food Mechanics

Mechanical and rheological properties of foods require consideration throughout the entire design process, with texture being a key concern for consumers as noted in Industrial Corner 2. Mechanical properties are those that a material exhibits upon the application of forces, while rheological properties are those that govern the material's deformation or flow behavior. Generally, for food materials, some common mechanical properties used for assessment are firmness, adhesiveness, and tensile strength, while common rheological properties include viscosity, storage modulus, and loss modulus. Thus, knowledge of the material properties is critical to obtaining high-quality 3D printed food products. The measured properties of printed food are influenced by their ingredient composition, how they are processed, and the design's shape. For instance, infill density, which refers to the density of material layered as a lattice inside the print, affects the hardness of a printed food since higher density results in a greater force required to deform the print. Research has demonstrated that the hardness and gumminess of brown rice were strongly related to infill density, followed by nozzle size [75]. Additionally, the hardness and fracturability of air-fried potatoes significantly increase as the infill level increases due to a greater amount of material extruded to fill the inner part of the samples [76]. Since infill density increases the amount of food deposited to form a design, it also affects the total nutrients and calories of the printed food, thereby necessitating its consideration for both successful printing and ensuring suitable nutritional content of a food product.

Mechanical and rheological food properties are measured using a texture analyzer or a rheometer, which generally uses a small sample of the food material that is representative of the expected properties of the food after 3D printing. Rheometers have been broadly used to characterize material properties for 3D printing, but in recent years, the texture analyzer has become increasingly used due to its ease of use, time-effectiveness, and affordability for measuring a variety of mechanical properties [77]. Generally, a rheometer is used to measure the properties of food under shear deformation while a texture analyzer uses compression/penetration testing, although there are modifications to each piece of equipment that further expands on these core capabilities. Both rheometric and texture analysis measurements provide an objective means of describing food that correlates strongly with the sensory judgments of consumers [78].

There is often a high correlation in values among food properties; for instance, in the case of firmness and viscosity, a firmer food will generally have a higher viscosity. Texture analyzers are particularly useful since they can relate a diverse number of material properties through compression or penetration for quantifying firmness, relaxation, swelling, resilience, springiness, and cohesiveness, among others [79, 80]. One study showed that springiness, which indicates the elasticity of the material, and the cohesiveness, which indicates the adhesive forces between internal granules, of alginate increased greatly when pea protein was added compared with a 100% alginate sample [81]. Recently, texture analysis has been applied to 3D printed foods [82], which is demonstrated in Figure 15.5 for a case study with mashed potato and pea protein additive.

The analysis demonstrates how pea protein addition affects the firmness and work of extrusion of mashed potato for a pure mixture and when two different concentrations of pea protein are added. The mechanical response was measured using standard procedures for food compression testing [81]. The mashed potato mixture was prepared using 50 grams of "Great Value Instant Mashed Potatoes" and 230 g of water. Later, 100 g of mashed potatoes were mixed with 15 g or 30 g of "Now Sports Pea Protein Powder Unflavored." Firmness and work of extrusion were measured by the compression test with 100 g of material using a texture analyzer TA.XTPlus (Figure 15.5a). The maximum peak force measured was described as firmness and the area under the force–time curve was described as a work of extrusion. The force needed to compress the material per unit time is shown in Figure 15.5b, which demonstrates the necessary force for displacement increased as pea protein concentration increased. Similarly, the firmness and the work of extrusion increased as pea protein concentration increased (Figure 15.5c and 15.5d). Both the firmness and the work of extrusion resulted in a linear trend with pea protein concentration, with $R^2 = 0.95$. Similar trends have been observed in food engineering research, where the hardness of potato snacks increased linearly as sucrose content increased [83]. Overall, these results provide a set of measurements that enable the prediction of food texture and links to consumer appeal based on food material properties and food combinations. Additionally, firmer foods also generally indicate a higher degree of shape retention for 3D printed foods, which is additionally important for consumers' satisfaction with the printed food form [50].

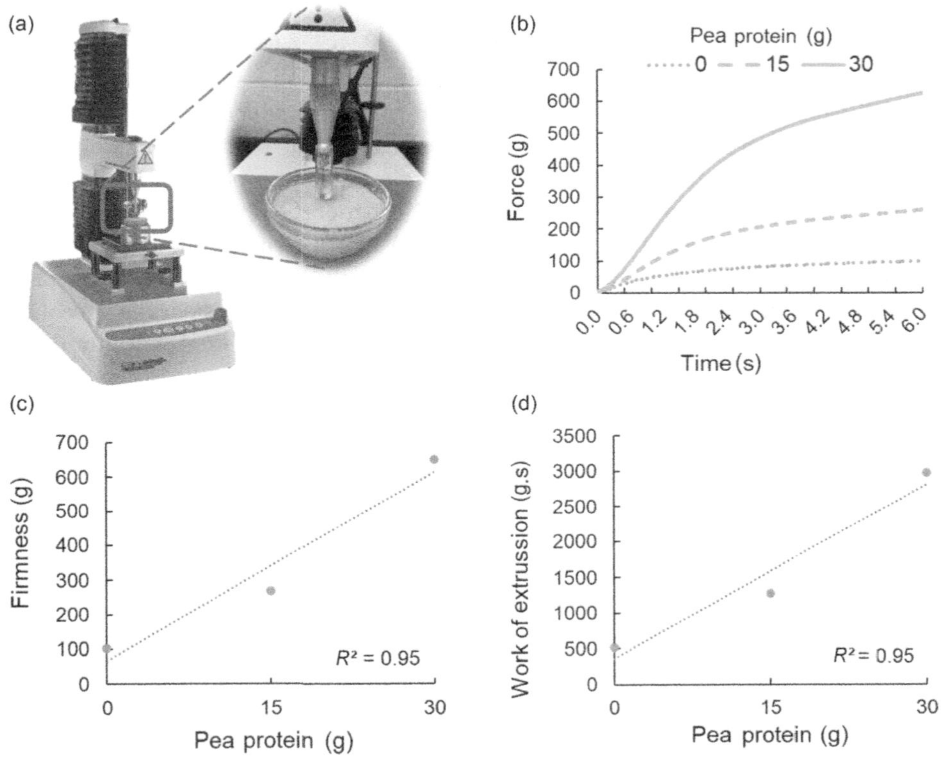

Figure 15.5 (a) Instrument used for the texture analysis [*Source:* Texture Technologies Corp]; (b) force vs. time graph; (c) firmness; (d) work of extrusion for mashed potato with the addition of 0, 15, and 30 grams pea protein powder. Values are the average of five replicates.

Industrial Corner 2

Why is texture analysis important in food product development?

- Texture is a crucial parameter in product development that affects processing and handling, shelf life, and consumer acceptance of products.
- Major manufacturers regularly apply texture analysis techniques in new product development and as part of quality control in all stages of manufacturing to analyze raw materials, semi-finished goods, packaging, and finished products.
- Texture analysis is a cost-effective method to determine the effects of raw material quality or the adjustment of formulation on end product acceptability that helps quantify the overall physical properties of "gold standard" products for future comparison.
- Manufacturers should invest in physical product testing (texture analysis instrumentation) during the earlier stages of development and production to ensure efficiency, optimize recipes, and most importantly, ensure desirable textural quality for consumer satisfaction.

15.6 Consumer Validation

Consumer preferences and attitudes regarding the use of 3D food printing technology are essential because they inform the market launch of new products to better address consumer desires and expectations [84], which is highlighted in Industrial Corner 3. Although there is a large amount of research on printing food products with different ingredients, there is little research on consumers' experiences and perceptions of 3D printed foods [85–88]. Primary areas of investigation that inform consumer appeal for printed foods include the fabricated fidelity achieved for the food design and consumer taste experience. Figure 15.6 demonstrates the results of a study to consider the appeal of foods suitable for 3D printing. The study used mashed potato with different concentrations of pea protein. Initially, a set of prints were created to ensure printability and consistent fidelity with shape retention (Figure 15.6a) [57]. The "MP" printed shape represents a sample that is pure mashed potato; "15P" represents a sample made out of 100 g mashed potato, 15 g pea protein, and 45 g water, and "30P" represents a sample made out of 100 g mashed potato, 30 g pea protein powder, and 90 g water.

Next, approximately 10 g samples of the foods were provided for participant ($N = 30$) consumption. Participants were asked to rate two questions on a Likert scale from 1 to 9: (1) "How much do you like the taste of the sample?" and (2) "How much do you like the mouthfeel of the sample?" Results aggregated in Figure 15.6b demonstrate that taste and mouthfeel decreased as pea protein and water increased. When comparing the taste and mouthfeel of the 15P food mixture to the MP food mixture as a control, each one is significantly lower according to an unpaired two-tailed t-test (both $p \leq 0.001$). When comparing the 30P food mixture to the MP food mixture for taste and mouthfeel, they are additionally significantly lower (both $p \leq 0.001$). The results demonstrate that although the food is printable and the pea addition offers more protein compared with MP foods, the satisfaction of the user decreases. Additionally, altering pea concentration has been demonstrated to alter printability that could constrain the possible library of printed shapes available to

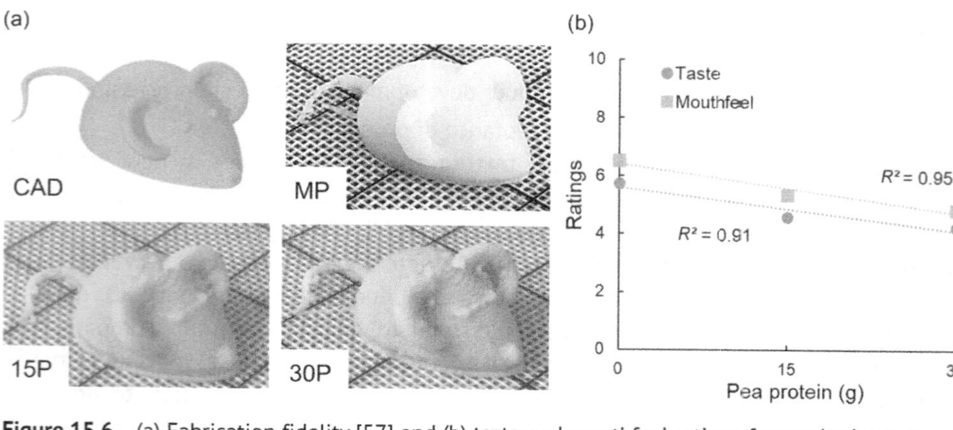

Figure 15.6 (a) Fabrication fidelity [57] and (b) taste and mouthfeel ratings for mashed potato with 0, 15, and 30 g addition of pea protein powder at a water to protein ratio of 3:1. Images adapted from [57] with permission from ASME.

consumers [57], which is an important consideration since shape complexity and fidelity play a role in consumer appeal [54]. Further studies are needed to explore the product/consumer interface with all types of decisions throughout the food design process, as material preparation, printing processing, and food mechanics play a role in consumer satisfaction together with the end product. In summary, achieving the highest consumer appeal requires tuning of food ingredients and mechanics to ensure printability and the taste experience are optimized. Furthermore, this approach may lend itself well to creating individualized end products that are personalized to each consumer's preferences.

Industrial Corner 3

What might future applications of 3D food printing have for personalized nutrition?

- Consumers are increasingly interested in maintaining control over ingredients and experimenting with novel food combinations, both for health and gustatory preference.
- 3D food printing technology may give rise to local business opportunities to create personalized foods at the point of sale.
- As 3D food printer technology evolves to become more mainstream and affordable, home applications for personalization may become possible.
- Studies show that involving children and adults in healthy food preparation increases acceptability and willingness to try novel foods. The technology may have applications in research and consumer environments to increase nutritional diversity in the diet.
- Provides an expedient way to nutritionally tailor (e.g. fortification) foods in manufacturing applications.

15.7 Concluding Remarks

3D food printing offers a wide range of new possibilities within the food industry, from the realization of complex food designs to the automated preparation of personalized meals. As an emerging food processing technique, there is an immense design space enabled by 3D food printing that allows the creation of novel food products that exceed conventional processing techniques. However, 3D printing is currently limited primarily to soft foods and further research is necessary to fully leverage the technology's capabilities. Research is needed to discover the most advantageous ingredient combinations and food creations to satisfy consumers, which can be supported through taste and texture studies. In addition, dissemination of information and increasing knowledge on the part of consumers is needed to raise awareness on 3D food printing. Developing a relevant and rational scientific evidence base could improve understanding and acceptance of the technology both with consumers and industry partners.

Although several food printing technologies are available, further research and tuning of the technology are needed to ensure cost-effective delivery and ease of use. Food printing technology is still in the growth stage and early adopters are working towards successful business models using the core values of 3D printing to demonstrate value in market niches, such as personalized desserts. In terms of future growth, this technology has the potential to become an everyday cooking technology that improves people's lifestyles through easily prepared and automated fresh foods that are customized for every person in a household. Food printing could provide unprecedented dietary advantages that ensure meals meet the precise nutritional needs and preferences of individuals. Partnerships with industry are needed to fully realize the potential of this emerging technology for widespread application. The future of 3D food printing includes domestic cooking, catering services, and industrial manufacturing processes, to name a few. As 3D food printing continues to mature, the technology is poised to deliver high-quality, personalized nutrition, while encouraging the creation of new shapes, flavors, and textures to create an entirely new dining experience.

Glossary

3D Printing Technology that automates deposition of material to form three-dimensional objects.

CAD Computer-aided design, which is a tool used to generate digital design representations with parameterized dimensions.

Texture The feel, appearance, or consistency of a surface or substance.

Rheology The branch of physics that deals with the deformation and flow of matter, especially the non-Newtonian flow of liquids and the plastic flow of solids.

Firmness The maximum force required to compress food.

Printability The ability of a material to make a 3D object and support its structure once printed.

Accuracy The degree of how close a measured value is to the true value.

Fidelity The degree of exactness with which an object is copied or reproduced.

STL-file A file with raw, unstructured triangulated surfaces defined by the unit normal and vertices of the triangles using a three-dimensional Cartesian coordinate system.

G-code Computer code that instructs a 3D printer where to move, how fast to move, and what path to follow.

References

1 J. I. Lipton, M. Cutler, F. Nigl, D. Cohen, H. Lipson, *Trends Food Sci. Technol.* **2015**, *43*, 114.

2 A. M. Arefin, N. R. Khatri, N. Kulkarni, P. F. Egan, *Polymers* **2021**, *13*, 1499.

3 V. Jayasena, S. M. Nasar-Abbas, *J. Texture Stud.* **2012**, *43*, 153.

4 B. Sandeep, T. T. M. Kannan, J. Chandradass, M. Ganesan, A. John Rajan, *Mater. Today Proc.* **2021**, *45*, 6941.

5 J. I. Lipton, *Curr. Opin Biotechnol.* **2017**, *44*, 198.

6 D. Kokkinis, M. Schaffner, A. R. Studart, *Nat. Commun.* **2015**, *6*, 8643.

7 A. Le-Bail, B. C. Maniglia, P. Le-Bail, *Curr. Opin. Food Sci.* **2020**, *35*, 54.

8 J. Sun, Z. Peng, W. Zhou, J. Y. H. Fuh, G. S. Hong, A. Chiu, *Procedia Manuf.* **2015**, *1*, 308.

9 C. Guo, M. Zhang, B. Bhandari, *Compr. Rev. Food Sci. Food Saf.* **2019**, *18*, 1052.

10 P. Egan, J. Cagan, C. Schunn, F. Chiu, J. Moore, P. LeDuc, *J. Mech. Des.* **2016**, *138*.

11 R. Mahmoud, Q. Nguyen, G. Christopher, P. F. Egan, 3D Printed food design and fabrication approach for manufacturability, rheology, and nutrition trade-offs. Presented at *International Design Engineering Technical Conferences and Computers and Information in Engineering Conference,* Virtual, Online, **2021**.

12 S. Vock, B. Klöden, A. Kirchner, T. Weißgärber, B. Kieback, *Prog. Addit. Manuf.* **2019**, *4*, 383.

13 M. Ziaee, N. B. Crane, *Addit. Manuf.* **2019**, *28*.

14 G. D. Goh, Y. L. Yap, H. K. J. Tan, S. L. Sing, G. L. Goh, W. Y. Yeong, *Crit. Rev. Solid State Mater. Sci.* **2019**, *45*, 113.

15 J. Sun, Z. Peng, L. Yan, J. Fuh, G. S. Hong, *Int. J. Bioprint.* **2015**, *1*.

16 J. Y. Zhang, J. K. Pandya, D. J. McClements, J. Lu, A. J. Kinchla, *Crit. Rev. Food Sci. Nutr.* **2021**, *1*.

17 C. Tan, W. Y. Toh, G. Wong, L. Lin, *Int. J. Bioprint.* **2018**, *4*.

18 C. Severini, A. Derossi, *J. Clin. Gastroenterol.* **2016**, *50*, S175.

19 P. Policastro, C. Harris, G. Chapman, *Appetite* **2019**, *139*, 42.

20 B. Perez, H. Nykvist, A. F. Brogger, M. B. Larsen, M. F. Falkeborg, *Food Chem.* **2019**, *287*, 249.

21 C. Feng, M. Zhang, B. Bhandari, *Crit. Rev. Food Sci. Nutr.* **2019**, *59*, 3074.

22 H. Jiang, L. Zheng, Y. Zou, Z. Tong, S. Han, S. Wang, *Crit. Rev. Food Sci. Nutr.* **2019**, *59*, 2335.

23 F. C. Godoi, S. Prakash, B. R. Bhandari, *J. Food Eng.* **2016**, *179*, 44.

24 I. Dankar, A. Haddarah, F. E. L. Omar, F. Sepulcre, M. Pujolà, *Trends Food Sci. Technol.* **2018**, *75*, 231.

25 A. Gholamipour-Shirazi, M. A. Kamlow, T. N. I, T. Mills, *Foods* **2020**, *9*.

26 S. Chirico Scheele, M. Binks, P. F. Egan, Design and manufacturing of 3D printed foods with user validation, Presented at *International Design Engineering Technical Conferences and Computers and Information in Engineering Conference*, Virtual, Online, **2020**.

27 C. Velasco, A. T. Woods, L. E. Marks, A. D. Cheok, C. Spence, *PeerJ* **2016**, *4*, e1644.

28 C. Spence, M. K. Ngo, *Flavour* **2012**, *1*.

29 J. Prescott, *Curr. Opin. Food Sci.* **2015**, *3*, 47.

30 S. Jayaprakash, J. Paasi, K. Pennanen, I. Flores Ituarte, M. Lille, J. Partanen, N. Sozer, *Foods* **2020**, *9*.

31 L. M. Factory, 3D printing, https://www.miamfactory.com/chocolate-3d-printing/?lang=en.

32 F. Yang, M. Zhang, B. Bhandari, *Crit. Rev. Food Sci. Nutr.* **2017**, *57*, 3145.

33 R. Brockotter, Key design considerations for 3D printing, https://www.3dhubs.com/knowledge-base/key-design-considerations-3d-printing.

34 A. Garaigordobil, R. Ansola, E. Veguería, I. Fernandez, *Comput.-Aided Des.* **2019**, *109*, 33.

35 A. Derossi, M. Paolillo, R. Caporizzi, C. Severini, *J. Food Eng.* **2020**, *275*.

36 I. Dankar, A. Haddarah, F. Sepulcre, M. Pujola, *Foods* **2019**, *9*.

37 K. Keerthana, T. Anukiruthika, J. A. Moses, C. Anandharamakrishnan, *J. Food Eng.* **2020**, *287*.

38 E. Hertafeld, C. Zhang, Z. Jin, A. Jakub, K. Russell, Y. Lakehal, K. Andreyeva, S. N. Bangalore, J. Mezquita, J. Blutinger, H. Lipson, *3D Print. Addit. Manuf.* **2018**, *6*.

39 M. Lille, A. Nurmela, E. Nordlund, S. Metsä-Kortelainen, N. Sozer, *J. Food Eng.* **2018**, *220*, 20.

40 A. Pant, A. Y. Lee, R. Karyappa, C. P. Lee, J. An, M. Hashimoto, U. X. Tan, G. Wong, C. K. Chua, Y. Zhang, *Food Hydrocoll.* **2021**, *114*.

41 A. Dick, B. Bhandari, X. Dong, S. Prakash, *Food Hydrocoll.* **2020**, *107*.

16

Techniques for Characterization of Food-Packaging Materials

Shubham Sharma[1,2,3$], *Kalpani Y. Perera*[1,2$], *Dileswar Pradhan*[1,2$], *Brendan Duffy*[3], *Amit K. Jaiswal*[1,2], *and Swarna Jaiswal*[1,2,*]

[1] *School of Food Science and Environmental Health, Technological University Dublin – City Campus, Central Quad, Grangegorman, Dublin D07 ADY7, Ireland*
[2] *Environmental Sustainability and Health Institute (ESHI), Technological University Dublin – City Campus, Grangegorman, Dublin D07 H6K8, Ireland*
[3] *Centre for Research in Engineering and Surface Technology (CREST), FOCAS Institute, Technological University Dublin – City Campus, Kevin Street, Dublin D08 CKP1, Ireland*
* *Corresponding author*

Highlights

- The optimal selection of packaging material is a critical aspect in ensuring quality and food safety.
- Characterization techniques are required to better understand the structure of the packaging material.
- Electron microscopy is a technique for high-resolution structural and chemical characterization.
- UV–Vis spectroscopy, ColorQuest spectrophotometer, dynamic light scattering (DLS), and zeta potential measures the optical properties of the packaging material.
- FTIR determines the chemical structure of the packaging material.
- The thermal stability of a food-packaging material is determined using thermogravimetric analysis (TGA) and differential scanning calorimetry (DSC).
- The standard testing protocols of biodegradability testing establish the intrinsic biodegradability of plastics in soil under controlled conditions.

16.1 Introduction

Food packaging is a complex system that includes the food product, packaging material, visual and structural designs, as well as transportation and distribution. For various food products, a variety of packaging materials are introduced to provide required functions within the packaging system [1]. The optimal selection of packaging material is a critical aspect of ensuring quality and food safety. Common properties of a food-packaging system

$ Authors contributed equally

Materials Science and Engineering in Food Product Development, First Edition. Edited by Wing-Fu Lai.

are morphological, mechanical, rheological, optical, chemical, thermal, physical, and gas barrier properties. Characterization techniques are required to better understand the structure of the packaging material [2]. Different characterization techniques, such as electron microscopy (EM), spectroscopy, TGA, water contact angle, etc., are being used to study the packaging material. These characterization techniques help understand the polymer–polymer interaction, distribution of the active agent in the polymer matrix, and the effect of the system parameters on the final characteristics of the packaging material.

Food has always been packaged in a variety of materials, including plastics, wood, papers, glass, biopolymers, earthenware, plant leaves, and metals [1]. To provide the optimal packaging solution for different food products, a blend of two or more packaging materials is generally required. Polymers have become increasingly popular as food-packaging materials in recent decades as a result of their benefits over other traditional materials [2]. Identification of polymers can be determined by a balance of physical and chemical properties, as well as their ease of processing. Polymers reinforced with micrometer fillers are now commonly utilized to increase strength and rigidity, improve solubility, or fire protection, or simply cut costs. The use of these micro-sized fillers, on the other hand, has several disadvantages, such as brittleness and transparency [3]. Currently, studies on biodegradable materials are mostly focused on synthetic and natural polymers. Various synthetic polymers are commonly used as biodegradable materials in everyday life, including polycaprolactone (PCL), polylactic acid (PLA), polybutyl succinic acid-butyl adipate, polyvinyl alcohol, and polyadipate butylene terephthalate (PBAT). A variety of natural polymers, such as proteins (soy protein, whey protein), polysaccharides (starch, cellulose, chitosan), and lipids (beeswax, lauric acid) are also employed as raw ingredients in the composition of biodegradable materials [4]. To study the adaptation of food-packaging material, various characterization techniques were used for observing the change in many obtained polymeric composites such as homogeneous and heterogeneous polymer systems. In this chapter, various characterization techniques and their importance for the food-packaging material are reviewed and the use of these techniques in recent research is discussed. Industrial aspects of the use of these techniques in the food-packaging industry are also discussed.

Industrial Corner 1

- A significant trend in the packaging sector is the search for innovative technologies capable of replacing products with lower environmental impacts.
- The food industry is constantly looking for novel and smart biodegradable packaging materials that can improve the shelf life of food products and retain quality.
- Suitable packaging material for the food industry should have high thermal stability, outstanding mechanical property, good optical property, excellent physical characteristics, and desirable chemical properties.
- Thus, characterization techniques and instruments are vital for developing a packaging material for the food industry. In this regard, instruments, such as SEM, TEM, TGA, DSC, FTIR, etc., have been extensively utilized by both the research community and industry during the development of packaging materials.

16.2 Characterization of Food-Packaging Material

16.2.1 Microstructure of Food-Packaging Material

Microstructures are very small and cannot be seen with the naked eye. Optical microscopes have the resolution and range to observe micron-structured material. Several techniques, including scanning electron microscopy (SEM), transmission electron microscopy (TEM), and atomic force microscopy (AFM), were developed to study and characterize the structural properties of microstructured materials.

16.2.1.1 Scanning Electron Microscope

The scanning electron microscope is a powerful electron microscope that is used to image surfaces. The SEM scans the surface of a relatively thick sample with a high-energy electron beam. The resolution of a SEM is 1–10 nm. When an incident electron beam collides with the surface of a sample, secondary electrons, backscattered electrons, diffracted electrons, and particularly X-rays are formed, which are highly dependent on the surface morphology, topography, and chemical composition. Secondary electron imaging produces a very high-resolution image that can catch features as small as five nanometers. Secondary electron imaging creates an image and detects the elemental structure of a sample by using particular X-rays and backscattered electrons from the sample. SEM's high-resolution abilities make it suitable for detecting nanoscale characteristics of materials that are important to their properties and applications. Numerous authors applied SEM for the characterization of the developed film (Table 16.1). Figure 16.1 shows the microstructure of a poly(lactide) (PLA)–poly(butylene adipate-co-terephthalate) (PBAT) composite film incorporated with ferulic acid at different concentrations. A low concentration of ferulic acid in the packaging film has shown a smooth structure (Figure 16.1 a–c) whereas a high concentration has an agglomerated structure (Figure 16.1d) [5].

Recently, Roy and Rhim [6] had developed a gelatin and curcumin film for food industry application. After casting nanoparticles on the microscope slide and putting them in the specimen stub, they utilized a very thin gold coating to avoid charging under the electron beam. A smooth surface was observed and curcumin was found to be uniformly distributed in the gelatin matrix, demonstrating high compatibility and intermolecular interaction. A thermosensitive antibacterial biopolymer film composed of pectin matrix and modified halloysite nanotubes containing essential peppermint oil was demonstrated by Biddeci et al. [7].

16.2.1.2 Transmission Electron Microscopy

TEM is a technique for high-resolution structural and chemical characterization [8]. An electron beam passes through an ultrathin sample and interacts with it as it passes through. By collecting information about the particle's size, crystallinity, shape, and interaction, TEM enabled direct imaging of the structure of the atom of the solid and the surface. The TEM produces an extremely high-quality image with a resolution of roughly 0.1 nm. The electron beam traveling through a nanocrystal with a very small diameter (less than 3 nm) is also a significant approach for chemical characterization [9].

Table 16.1 Characterization techniques used in various research.

	Techniques	Material	Active agent	Results	References
Spectro-photometer	UV–Vis	Pectin and agar	Zinc sulfide nano petals (ZnSNP)	• Significant reduction in UV light transmittance with the addition of ZnSNP • The addition of ZnSNP did not significantly reduce the film's transparency	[18]
		Poly(lactic acid)	Curcumin	• The exponential decrease in light transmittance at 280 nm with increasing concentration of curcumin • Significant decrease in light transmittance at 660 nm with the addition of curcumin concentrations above 0.5 wt%	[6]
	FTIR	Poly(lactide) / poly(butylene adipate-co-terephthalate)	Ferulic acid	• The C = C aromatic ring was observed around 1515 cm^{-1}, which was not seen in the control blend film • Ferulic acid was defined by the characteristic peak of C–H stretching at 2951–3016 cm^{-1}	[5]
		Chitosan film	Gallic acid	• The peaks of GA at 3500 and 3285 cm^{-1} correspond to modes of OH groups • The peak at 1324 cm^{-1} was due to in-plane OH bending, those at 1268 and 1216 cm^{-1} to Ph–O stretching and C–O stretching	[19]
	Zeta potential	Oxalic acid	Lignocellulose	• The highest surface charge was achieved at 1 h modification, with zeta potential of −51.8 mV and carboxyl group content of 0.43 mmol/g fibrils	[20]
		Moringa oil	Chitosan	• High zeta potential and high encapsulation efficiency could be considered as a stale system	[21]
	DLS	Silver nanoparticle	Chitosan	• A thin dipole layer of the solvent is formed on the surface of AgNPs due to the solvent adhered to the surface of AgNPs • DLS analysis gives the diameter of the core molecule together with the solvent layer	[22]
	Color Quest	Poly(vinyl alcohol) and corn starch	Pineapple peel extract (PPE)	• A minor decrease in L value on increasing PPE concentration • Increase in greenness and yellowness with an increase in PPE concentration	[23]
		Gelatin	*Lepidium sativum* seed extract	• Decrease in film's lightness value and a* value • Increase in film's yellowness value	[24]

Microscopy	SEM	Pullulan and carboxylated cellulose nanocrystal	Tea polyphenol (TP)	• Films become more compact with the addition of 3% TP • Agglomeration of TP particles and presence of micro-voids with the addition of 5% TP	[17]
		Chitosan	*Plectranthus amboinicus* essential oil	• No phase formation and film was found to be continuous • No cracks or bubbles on the film's surface	[25]
	AFM	Gelatin	Grapefruit seed extract and TiO_2	• Increase in surface roughness with the increase in TiO_2 concentration • The film containing 3 wt% and 5 wt% had the highest roughness	[26]
Permeation properties	Water contact angle (WCA)	Pectin and agar	Zinc sulfide nano petals (ZnSNP)	• WCA increased with increasing concentration of ZnSNP • A film containing 2 wt% ZnSNP had maximum WCA (63.9 ± 2.8)	[18]
		Sodium alginate and carboxymethyl cellulose	Shallot onion waste extracts (SOWE)	• Control film had the lowest WCA (37.29°) • WCA increased from 37.29° to 63.26° with the addition of SOWEs	[27]
	water vapor permeability (WVP)	Chitosan	Pomegranate peel powder (PPP)	• Pure chitosan film had lowest WVP ($3.72 \pm 0.04 \times 10^{-11}$ g m^{-1} s^{-1} Pa^{-1}) • WVP increased remarkably with the addition of more than 3% of PPP	[28]
		Gelatin	*Lepidium sativum* seed (LSE) extract	• Control gelatin films had the highest WVP value (4.72×10^{-11} g m^{-1} s^{-1} Pa^{-1}) • Decrease in WVP with the addition of LSE extract	[24]
Thermal properties	TGA	Pullulan and carboxylated cellulose nanocrystal	Tea polyphenol	• The first step of thermal degradation 80–150°C • Maximum decomposition temperature at around 230–400°C	[17]
		Gelatin	Grapefruit seed extract and TiO_2	• Initial weight loss was at around 80–120°C, second weight loss at around 200–300°C, and third step weight loss at around 320°C • The addition of TiO_2 increased the final residues significantly to 20–26.3%	[26]
	DSC	Poly(lactic acid)	Curcumin	• A slight change in glass transition temperature, cold crystallization temperature, and melting temperature with the addition of curcumin • Crystallization degree increased from 25.4% to 29.9% with the addition of 1.5 wt% of curcumin	[29]
		Sodium alginate and carboxymethyl cellulose	Shallot onion waste extracts (SOWE)	• Increase in glass transition temperature, melting temperature, and enthalpy change with the addition of SOWEs • No occurrence of exothermic peak	[27]

(Continued)

Table 2.1 (Continued)

	Techniques	Material	Active agent	Results	References
Mechanical properties	Tensile strength (TS)/ elongation at break (EB)/ electron microscopy (EM)	Chitosan	Silica nanoparticles and gallic acid	• Chitosan film had the lowest TS (101.30 MPa) and highest elongation at break (EB) (4.78%) • Significant improvement in TS and decrease in EB with the addition of active substances	[19]
		Poly(lactic acid)	Curcumin	• Slight increase in TS with the addition of curcumin up to 0.5 wt% • Decrease in EB with the addition of curcumin up to 1 wt% • Increase in EM with the addition of curcumin up to 1 wt%	[29]
Oxygen barrier	Permeability meter	Sodium alginate and carboxymethyl cellulose	Shallot onion waste extracts	• Decrease in oxygen permeability from 8.50×10^{-6} g/day · Pa · m to 1.31×10^{-6} g/day · Pa · m in a film containing onion peel extract • Decrease in oxygen permeability from 8.89×10^{-6} g/day · Pa · m to 2.90×10^{-6} g/day · Pa · m in a film containing onion stalk extract	[27]
Thickness	Digital micrometer	Chitosan	*Plectranthus amboinicus* essential oil	• Thickness in the range of 30 to 50 μm	[25]
		Poly(vinyl alcohol) and corn starch	Pineapple peel extract	• Increase in film thickness from 0.055 mm to 0.062 mm with an increase in PPE concentration from 0% to 20%	[23]

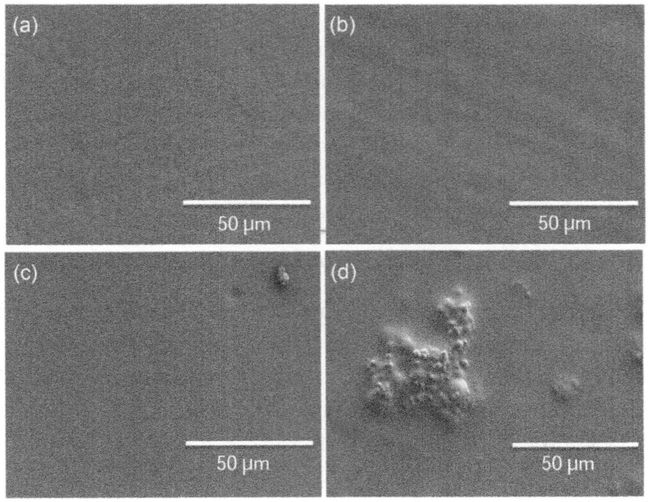

Figure 16.1 Scanning electron microscopy micrographs of (a) PLA/PBAT; (b) PLA/ PBAT-FA1%; (c) PLA/ PBAT-FA5%; (d) PLA/ PBAT-FA10%. [5] S. Sharma et al (2020), ELSEIVER.

The two condensers allow an electron beam created by the emitter to pass through, making the beam thin, tiny, and coherent. All high-angle electrons are eliminated by the condenser aperture just before the electron beam interacts with the sample. The interaction of the electron beam results in the detection of scattered and unscattered electrons [10] by the detector. When performing TEM studies on nanoparticles including materials, it is important to consider that they can be susceptible to the TEM instrument's highly energetic electron beam. Biswal and Misra [11] biosynthesized and characterized silver nanoparticles for evaluating their application in food packaging. The study showed the adhesion of phytochemicals on the nanoparticle surface and secondary metabolites by TEM images. The nanoparticles are determined to be between 2 and 30 nm in size. Sobhan et al. [12] had developed an activated carbon-based nanocomposite film as smart food packaging. Under TEM, the observed black dotted activated carbon particles densely clustered cellulose nanofiber edges in large parts of cellulose nanofiber with longer edges.

16.2.1.3 Atomic Force Microscopy

A scanning probe microscope, the AFM, is used to measure local parameters. Friction, magnetism, and height are the local attributes. Scanning probe microscopy (SPM) measures the local characteristics of a small area of a sample. It is a form of SPM with a demonstrated resolution of fractions of a nanometer, which is more than 1000 times greater than the optical diffraction limit [13]. Force measurement, imaging, and modification are three advantages of AFM. AFM is used to measure the forces between the sample and the probe as a function of their mutual separation in force measurement. This can be used to conduct force spectroscopy to determine the sample's mechanical properties, such as its Young's modulus, which is a measurement of stiffness. The probe reaction to the force applied by the sample is utilized to generate the topography (three-dimensional shape) of the sample surface at a very high resolution for imaging. This might be accomplished by raster scanning the sample positions in relation to the probe's height and tip, which would result in continual probe sample contacts.

By exerting a force between the tip and the sample during manipulation, the properties of the sample can be modified or controlled [14, 15]. Atomic manipulation, scanning probe lithography, and local cell stimulation are some instances of manipulation. Other qualities, such as an image, can be measured with AFM, and with the same great resolution. Mechanical traits like stiffness and adhesion strength, as well as electrical properties like conductivity and surface potential, are examples of such properties.

16.2.2 Optical Properties of Food-Packaging Material

The appearance of the food product and consumer acceptability are influenced by packaging properties, such as color, transparency, gloss, and UV barrier properties. Different techniques such as ultraviolet–visible spectroscopy, ColorQuest XE spectrophotometry, DLS spectroscopy, and zeta potential spectroscopy are used to measure the optical properties of the packaging material.

16.2.2.1 Ultraviolet-Visible Spectroscopy

Absorption or reflectance spectroscopy in the UV–visible spectral band is referred to as ultraviolet–visible or UV–Vis spectroscopy [16]. Absorption spectroscopy is a technique for determining how much light is absorbed as a function of wavelength or frequency. In the visible range, absorption influences color identification. Transitions occur in atoms and molecules, which are visible in the electromagnetic spectrum. The molecules that contain n-electrons and electrons can absorb energy in the form of ultraviolet or visible light to excite these electrons into higher orbitals. The capacity to absorb the wavelength of light is directly proportional to the electron excitation propensity.

UV–Vis spectroscopy is commonly employed for the quantitative analysis of highly conjugated chemical compounds, biological macromolecules, and transition metal ions (Table 16.1) [5, 6, 17–29]. Barrier attributes that enable the transfer of low molecular weight substances at a higher or lower level through penetration, absorption, and migration processes are the properties that polymeric materials should have to be used in food-packaging applications [16]. However, some low molecular weight components may migrate from plastic container walls, causing product damage or deterioration; thus, low molecular weight chemicals should not be transferred via polymers in food-packaging applications. When electrons excite from one state to another after absorbing light, transition metal solutions change color.

Food-packaging film transparency and UV–Vis light transmission are important elements that impact consumers' packaged food choices. Chen and Chi evaluated light transmission of the bionanocomposite film of pullulan/carboxylated cellulose nanocrystal/TP using UV–Vis-spectroscopy. They observed that as the TP concentration was increased, the UV–Vis light transparency of the bionanocomposite films dropped, indicating that the film's light barrier capacity was improved significantly [17]. Sharma et al. had incorporated clove oil in the PLA–PBAT composite film and observed 80% enhancement in UV barrier property [30].

Industrial Corner 2

- During manufacturing, packing, distribution, storage, display, and consumption, food, and beverages are exposed to natural and artificial light.
- Light absorbers or light blocking agents are added into or coated onto food-packaging materials to prevent photooxidation of the food, which is one of the main causes of food degradation, making active packaging with a light barrier one of the most promising alternatives to traditional packaging.

16.2.2.2 ColorQuest XE Spectrophotometer

The color of a product's packaging can distinguish it from other products' visual qualities. It is more appealing to draw the attention of the customers, leading them to associate with the product. Therefore, packaging and color quality are becoming increasingly important in the manufacturing process. Various instruments characterize the color of the packaging material, such as ColorQuest XE, a handheld colorimeter, and so on. It is an optical detection device that replicates the eye's color vision mechanism. The degree of difference in packaging color can be described mathematically using the CIE color space approach, in which the symmetry of the color space and the coordinate system used to identify points within that space differ. The human eye has three color receptors: red, green, and blue, according to CIE principles, and all colors are combinations of these. The value can be used to precisely calculate the color of the packaging, fine-tune the manufacturing process, and monitor package color quality [31].

In the food sector, the HunterLab L*, a*, b* and the modified CIE system known as CIELAB color scales were opponent-type systems. The L*, a*, and b* CIELAB coordinates were read directly. It was referred to as the CIELAB uniform space, and it was used to measure two color variables, a* and b*, and a perceptual index of brightness, L*. The parameter a* is positive for reddish colors and negative for greenish colors, whereas the parameter b* is positive for yellowish colors and negative for bluish colors. The value of the distance vector between the initial color values and the actual color coordinates can be used to calculate color changes. The total color difference is the name given to this notion [31]. The total color difference between the stored and control samples reveals the magnitude of the color difference [32]. The color difference between the standard plate and the total color difference (ΔE) is determined as

$$\Delta E = \sqrt{(\Delta L)^2 + (\Delta a)^2 + (\Delta b)^2}$$

According to Dash et al., as the concentration of TiO_2 nanoparticles in composite films increases, the L* and b* values increase significantly, whereas the a* value decreases significantly [33]. Sharma et al. had observed that in the clove oil incorporated poly(lactide) / poly(butylene adipate-co-terephthalate) blend film's greenness and yellowness values (PLA/PBAT-clove oil films) are significantly different from one another, with the highest *a* and *b* values, indicating less green and more yellow color [30].

16.2.2.3 Dynamic Light Scattering Spectroscopy

The technique of DLS can be used to examine the size distribution profile of tiny particles in any solution. The magnitude of the intensity or the function of photon autocorrelation can be used to readily determine and analyze the fluctuations. In terms of the time-domain analysis, the autocorrelation function begins to decay at zero delay time, and faster dynamics owing to smaller particles result in faster decorrelation of the dispersed intensity trace [34]. DLS involves directing a monochromatic laser beam through a biomolecule solution and analyzing fluctuations in scattered light intensity. The experiment is noninvasive, takes only 12 μL of material, and can offer information regarding the size and homogeneity of biomolecules in a couple of minutes.

As a function of various preparations, solvent conditions, temperature, or time, sample features, such as aggregation, folding, or conformation, can be examined. The instrument may be used at temperatures ranging from 4°C to 60°C. A monochromatic light source, usually a laser, is fired into a sample through a polarizer. The scattered light is gathered by a photomultiplier after passing through a second polarizer, and the image is projected onto a screen [35].

Light strikes all the molecules in the solution, and all of the molecules diffract light in all directions. All the molecules' diffracted light can either interact constructively (bright regions) or destructively (dark regions) [15]. An autocorrelator analyzes the intensity of light at each point over time and the resulting set of speckle patterns is analyzed.

16.2.2.4 Zeta Potential Spectroscopy

The "electrokinetic potential" is referred for the zeta potential. When nanoparticles or colloidal particles are suspended, they have a surface charge. The interaction between the charged particle and the applied field causes the particles to move when an electric field is applied. The motion's direction and velocity are determined by particle charge, suspending medium, and electric field strength. The Doppler shift in the dispersed light is then used to determine particle velocity. The particle velocity is proportional to the zeta potential, which is the electrical potential of the particle at the shear plane.

Electrophoresis is the study of particle mobility in the presence of an electric field. "The sample particles are suspended in a solvent with a known refractive index n, velocity η, and dielectric constant Ɛ." Laser light of wavelength λ [15, 35] is used to irradiate the material. A strong electric field (E) is applied, which enables the particles to move. The scattered light at angle θ is detected because of the moving particles, and the particle velocity V is calculated from the frequency shift. The ratio of velocity to electric field strength, V/E, is then easily calculated. The most prevalent approach, known as the Smulochowski model, is used to calculate zeta potential.

$$U = \frac{(\lambda.V.d)}{2.E.n.\sin\left(\frac{\theta}{2}\right)}$$

In recent times, the utilization of various nanoparticles in the development of nanocomposite materials for food-packaging applications has been extensively reported [35–39]. During nanocomposite development, generally, high stability of the nanoparticles in the film-forming solution is required to prevent particle aggregation. In general, a colloidal suspension with a zeta potential of more than 30 mV or less than − 30 mV is considered to be stable, where the nanoparticles have enough surface charges to repel one another while preventing agglomeration [40], for instance, [21] the stability of moringa oil-loaded chitosan nanoparticles (MO@CNPs) using zeta potential value. The maximum zeta potential value of 45.1 ± 4.2 mV was observed in the MO@CNPs containing 5 mg/mL of chitosan; however, MO@CNPs containing 3 and 4 mg/mL of chitosan were also stable as they had a zeta potential of 32.5 ± 1.8 mV and 38.4 ± 3.6 mV, respectively. Finally, based on the encapsulation efficiency and zeta potential, MO@CNPs containing 3 mg/mL of chitosan were embedded in gelatin nanofibers for the development of an active packaging system for the preservation of cheese (Table 16.1) [20, 21].

16.2.3 Chemical Properties of Food-Packaging Material

The chemical structure of packaging material can be determined by attenuated total reflectance Fourier-transform infrared spectroscopy (ATR-FTIR). This spectroscopic process, which includes infrared light, enters the prism at an angle that exceeds the critical angle for internal reflection. At the surface, this creates an evanescent wave (reflecting). The evanescent wave's change is measured, resulting in a spectrum that is then subjected to a Fourier transform. The critical component of this technique is its infrared radiation, which is an analytical method for obtaining spectra from various states of matter. Passing infrared radiation straight through the material has been utilized to analyze solids, liquids, and gases. The sample is either liquid or solid, and the thickness of the sample determines the strength of the spectral characteristics. This sample thickness is often limited to a few tens of microns [41]. Furthermore, faster sampling, improved sample-to-sample reproducibility, and reduced spectral variance have all revolutionized IR solid and liquid sampling.

FTIR technique has been used extensively by researchers to assess the chemical characteristics of materials intended for food-packaging applications (Table 16.1) [5, 19]. For instance, Shankar et al. [42] used the FTIR technique to study the chemical characteristics of silver nanocomposite films with polylactic acid and lignin. At a resolution of 4 cm^{-1}, the spectra were captured as 32 scans per sample. Further, FTIR spectra were used by Sharma et al. to characterize ferulic acid incorporated PLA–PBAT films. An ATR–FTIR spectrophotometer was used to collect ferulic acid in the range of 4000–500 cm^{-1}. The presence of ferulic acid was found by the characteristic peak of C–H stretching at 2951–3016 cm^{-1} [5].

16.2.4 Thickness of Food-Packaging Material

The thickness of the film is usually measured using a digital micrometer with an accuracy of ± 1 μm [43]. The film thickness was measured at 10 different locations on the film and reported as a standard deviation of the measurements [44]. The thickness of the film is linked with the mechanical properties, and it is essential to determine the tensile strength (TS) and elongation of the film [43]. Further, the barrier properties such as the transmission rate and permeance alter with the change of film thickness [45]. The thickness of the film is determined by both the composition of the film and the processing parameters (Table 16.1) [23, 25].

16.2.5 Oxygen Barrier Properties of Food-Packaging Material

The barrier properties are crucial properties that are extremely important in food packaging to avoid value loss, reduction of quality, and increase in the shelf life of the food product. The reduced oxygen barrier properties may lead to the gases, smell, flavor, moisture, and aroma exchange, which will gradually lead to the reduction of moisture and weight loss. Structural features, polarity, hydrogen bonding, molecular weight, the amount of crosslinking, the size of volume holes, the processing method, the thickness, the synthesis procedures, and the level of crystallinity aspects influence the gas permeability of packaging materials. Recently, many effects have taken place to increase the oxygen permeability of packaging materials by blending different polymers, crosslinking, and addition of nanoparticles [46].

The oxygen barrier properties are usually measured using three parameters: transmission rate, permeance, and permeability. Under testing settings, the oxygen transmission rate (OTR) is defined as the volume/weight of oxygen moving through a film per unit of surface area and time [45]. The units of OTR are $cm^3/m^2 \cdot day$ and the equation is:

$$\text{OTR} = (\text{Volume passed through}) / (\text{Area} \times \text{Time})$$

The transmission rate is divided by the partial pressure differential of the permeant across films to calculate permeance. Permeance multiplied by thickness equals oxygen permeability (OP) [45]. The unit for OP is $cm^3 \cdot \mu m/m^2 \cdot day \cdot atm$ and the equation is:

$$\text{OP} = (\text{OTR} \times \text{thickness}) / (\text{Oxygen partial pressure difference})$$

The OTR is measured using a permeability meter according to the ISO15105-1 standards [47] or using an automated OP tester according to the ASTM D3985 standard [48], or Oxygen Permeation Analyzer 8001 following ASTM D 3985–95 and ASTM F 1927–98 [49]. OP of a material is measured by the permeation tester at 25°C and 55% RH [50], or manometric method (ISO/DIS 15 105 1) on permeability testing appliance [51].

16.2.6 Mechanical Properties of Food-Packaging Material

Food-packaging materials must maintain their integrity during storage, distribution, processing, and handling; thus mechanical properties such as TS, EB, and elastic modulus are crucial in packaging. The food package has to have the ability to withstand mechanical stress at all stages [52]. The TS of packaging material is the maximum stretch the film can withstand before breaking. The composition of the packaging material and the environmental factors influence the TS. The addition of different fillers and plasticizers can improve the TS of the materials. EB is the flexibility of the packaging material. EM is the stiffness of the film, and it is denoted by the Young modulus. The mechanical properties of the packaging film are dependent upon the nature of the polymer, type of plasticizer, crosslinking effect, aging effect, crystallinity, and macrostructure [43, 52, 53].

The TS of the films can be calculated using the following equation and the units are MPa [43]:

$$\text{TS} = \frac{\text{Force of film at break}}{\text{Film thickness} \times \text{Width of sample}}$$

EB of the film can be calculated using the following equation and the unit of EB is the percentage (%) [43].

$$\text{EB} = \frac{\text{Film elongenation at break} - \text{Original grasping length of film}}{\text{Original grasping length of film}} \times 100$$

The EM (GPa) measures the resistance of the film from being elastically deformed. The stress–strain curve in the region of elastic deformation defines the EM that corresponds to the stress divided by the strain of the film sample. The TS and the EB of packaging material are determined using the Instron universal testing machine by the universal standard ASTM D882-10, 2000 [53] or electronic tensile testing machine (Tinius Olsen Testing Machine) using the standard method ASTM D882/12 [54] (Table 16.1) [19, 29].

16.2.7 Thermal Properties of Food-Packaging Material

Evaluating the thermal characteristics is important to understand the limitations of a food-packaging material. The packaging film possessing high thermal stability can help in maintaining the integrity of the film and its functional characteristics, ensuring excellent protection and preservation of the packaged food. The thermal stability of a food-packaging material is generally determined using TGA and DSC [55].

16.2.7.1 Thermogravimetric Analysis

TGA measures a material's mass change as a function of time or temperature when the sample is exposed to a temperature-controlled program. Information regarding a material's thermal degradation temperature is acquired in this analysis [56]. TGA can help in examining physical processes like the second-order phase transitions, including sublimation, vaporization, adsorption, absorption, and desorption. Furthermore, TGA can provide information regarding chemical processes, including solid–gas reactions (such as reduction or oxidation), desolation, chemisorption, and decomposition [57]. The gain or loss in mass of a material because of decomposition, oxidation, or loss of volatiles can be measured using TGA [52, 57].

TGA has been used extensively by researchers to investigate the thermal characteristics of materials intended for food-packaging applications. In a study [5] that used TGA for thermal analysis of a ferulic acid incorporated PLA/PBAT film, the first stage of weight loss was observed at 60–80°C. They exhibited the second stage of weight loss for PLA/PBAT film at 280–290°C, but the PLA/PBAT film containing FA of 1, 5, and 10 wt% exhibited weight losses at 290–300°C, 330–340°C, and 370–380°C, respectively (Figure 16.2). These findings confirmed that the addition of FA increased the thermal stability of the

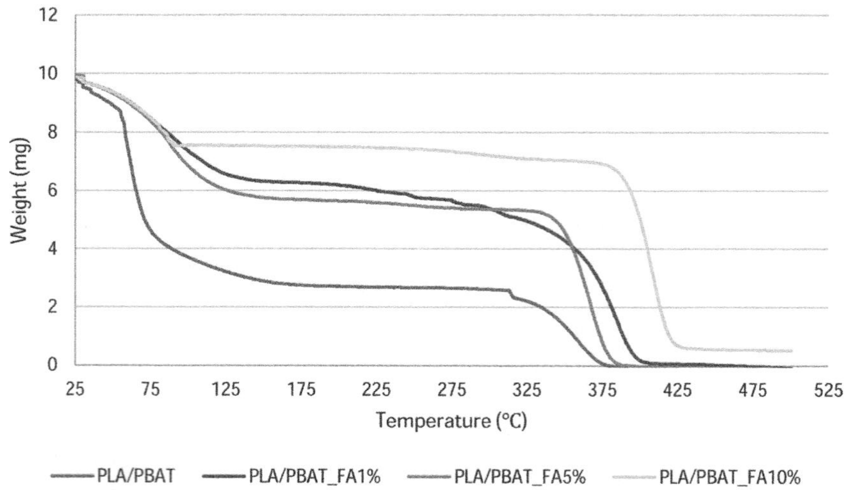

Figure 16.2 TGA of PLA/PBAT incorporated with ferulic acid [5] / Elsevier / CC BY 4.0.

PLA–PBAT composite film. Similarly, using TGA [58], the initial thermal degradation temperature (T_{onset}) and highest degradation temperature (T_{max}) of the pure starch film observed were 307°C and 335°C, respectively. The CNC–starch film, on the other hand, showed lower T_{onset} and T_{max} compared with the pure starch film, and both T_{onset} and T_{max} decreased as the CNC concentration in the film increased (Table 16.1) [17, 26].

16.2.7.2 Differential Scanning Calorimetry

The DSC instrument is used for determining the difference in the amount of heat needed to raise the temperature of a sample and reference as a function of temperature and time. All through the analysis, the reference and sample are maintained at equal temperatures. Moreover, a DSC analysis program is typically designed in such a manner that the temperature of the sample and reference holders rises linearly as a function of time [57]. Further, a DSC measurement can determine the energy changes that occur during continuous cooling and heating. This allows for the determination of transition temperatures such as the glass transition temperature (T_g), melting temperature (T_m), and crystallization temperature (T_c) [52]. In a study [29], a DSC analysis of PLA and PLA/curcumin composite films was carried out and it was observed that the addition of curcumin increased the $T_{(g)}$ from 60°C (PLA film) to 64.9°C (film containing curcumin of 0.25 wt%). Also, a slight drop in T_c and a slight increase in T_m were noticed after curcumin addition. In addition, the crystallinity degree increased from 25.4% (PLA film) to 29.9% (film containing curcumin of 1.5 wt%).

16.2.8 Permeation Properties of Food-Packaging Material

16.2.8.1 Water Contact Angle

Research has led to the development of a number of sophisticated technologies for analyzing the surface properties of food packaging. Surface hydrophobicity is a feature of a composite polymer that impacts the packaging's function in an aqueous environment and provides information about the food packaging's transit [59]. The hydrophobicity of an active substance affects the chemical absorption and toxicity [15, 59]; therefore, determining it aids in analyzing the nanoparticle's interaction and risks. However, many papers on wetting still provide only a single value of a static contact angle, resulting in a lack of accuracy and reliability when describing how the liquid acts on the surface. It is critical to perform research using common and reliable standards. Also, the contact angle must be measured as soon as possible to avoid any inaccuracy.

Surface hydrophobicity is measured using a variety of approaches, including biphasic partitioning, contact angle measurement, hydrophobic interaction chromatography, and probe absorption. Recent techniques, such as X-ray photon correlation spectroscopy, have been used to measure the hydrophobicity of a surface and to identify specific chemical groups on a nanoparticle's surface [60]. Food emulsions have a complicated composition. Proteins, small-molecule surfactants (emulsifiers), and, in some conditions, polysaccharides (hydrocolloids), all help to stabilize their droplets to variable levels (Table 16.1) [18, 27].

16.2.8.2 Water Vapor Permeability

The packaging film's water vapor permeability (WVP) is generally determined gravimetrically using the ASTM Standard Test Method [61]. The film specimen is subjected to a relative humidity gradient over different film sides to measure the WVP [56]. Determining the WVP of a food-packaging film is critical since it is directly associated with the food product's shelf life [62]. To enhance the shelf life, a packaging film must avoid moisture interaction during film handling, transportation, and application [63]. Furthermore, a lower WVP value of the packaging film is essential to prevent moisture loss within food as well as migration of water from the external environment [55]. However, a packaging film's permeability depends on several parameters, such as polymerization, molecular weight, crosslinking degree, density, polarity, chemical structure, crystallinity, and the presence of other plasticizers. In addition, the barrier to water vapor is primarily associated with the hydrophobicity and hydrophilicity of the film composition [62, 63].

In an earlier study, Duan et al. [64] reported a WVP value of 13.33×10^{-11} $g{\cdot}m{\cdot}Pa^{-1}{\cdot}m^{-2}{\cdot}s^{-1}$ in the k-carrageenan (KC)/konjac glucomannan (KGM) composite film. However, when 5 wt% and 7 wt% of TiO_2 nanoparticles were added to the film, the WVP of the KC/KGM based film was reduced to 9.18×10^{-11} $g{\cdot}m{\cdot}Pa^{-1}{\cdot}m^{-2}{\cdot}s^{-1}$ and 9.02×10^{-11} $g{\cdot}m{\cdot}Pa^{-1}{\cdot}m^{-2}{\cdot}s^{-1}$, respectively. Similarly, Kan et al. [65] reported a WVP value of 0.80×10^{-10} $g\ m^{-1}\ s^{-1}\ Pa^{-1}$ for the chitosan–gelatin blend film. Chen and Chi used the solution casting approach to create active films based on pullulan and carboxylated cellulose nanocrystal (C–CNC) combined with TP [17]. When TP was added up to 5%, the WVP values of the PC–TP bionanocomposite increased from 1.91 to 1.96 $g\ m^{-1}\ s^{-1}\ Pa^{-1}$, indicating that an excessive amount of TP would be aggregated in the PC matrix. This aggregation decreased the effective amount of TP and fragmented the bionanocomposite films' continuous internal structure, resulting in a loss in compactness and so allowing water vapor to enter more easily [17]. These observations show that WVP is an important factor affecting the performance of a packaging film, and should be characterized during the development of food-packaging films (Table 16.1) [24, 28].

16.2.9 Biodegradability of Food-Packaging Material

The need for biodegradable and renewable packaging materials has increased significantly in recent years. This increase in demand is linked to rising environmental concerns over the widespread use of synthetic and nonbiodegradable polymeric packaging, particularly polyethylene. Standards for determining the degradability of plastics under aerobically composted conditions (ISO 17088, ASTM D6400) and the soil environment (ISO 17556, ASTM D5988) are based on macroscopic disintegration and organic carbon conversion to CO_2 but do not necessitate the measurement of micro- and nano-residues accumulated in soils [66].

The standard testing protocols are intended to establish the intrinsic biodegradability of plastics in soil under ideal controlled settings that may or may not be indicative of any specific environmental conditions, but they ensure reproducibility. The conventional test techniques' normalized parameters differ in a number of ways. Several parameters of the testing techniques are affected by these differences: (a) soil medium (natural soils from specific locations, laboratory mixtures of soils, mixture of natural soil, and mature compost); (b) test sample (intact, cut into pieces, or pulverized); (c) soil pH (natural or adjusted); (d) C/N ratio (natural or adjusted to 10:1–20:1 for added organic C in the test sample or at least 40:1 for sample organic C to soil N) [67].

Soil is regarded as a biodegradation environment because it is the core of extensive biological activity and a valuable resource for food and fiber production, as well as the medium in which soil-biodegradable plastics are disposed of. The rate of disintegration of bio-based substances in the soil varies greatly, based not only on molecules of the substance, but also on soil properties and variables such as temperature, water, and availability of oxygen, most of which vary considerably and impact microbial activity. To assess the activity of the soil, a controlled biodegradable substance (e.g. starch) must also be tested. If the control sample's biodegradation rate is less than 70% after 6 months, the test is inconclusive and therefore should be repeated with fresh soil [67].

16.3 Conclusion and Prospects

The characteristics of food-packaging materials are the most critical parameters that influence their effectiveness in extending the shelf life and retaining the quality of food products during handling, transportation, and storage. Therefore, the utilization of numerous characterization techniques during packaging material development is necessary for obtaining a material with desirable characteristics. Different characterization techniques and instruments have been utilized to assess various physical, chemical, mechanical, optical, and thermal characteristics of packaging material. Structural characterization of the packaging material is critical and is determined by several techniques, such as SEM, TEM, and AFM. Optical properties such as UV barrier property and transparency can be determined using a UV–Vis spectrophotometer. A ColorQuest XE spectrometer is used to determine the color of the packaging material. The mechanical properties of the food packaging, such as TS, EB, and EM, could also be determined using the ASTM D882 standard method. The thermal stability of a food-packaging material is generally determined using TGA and DSC. Various standard protocols could be used to study the biodegradability of the packaging material. All these techniques are crucial for the study of new food-packaging materials.

Glossary

Characterization The broad and general process by which a material's structure and properties are probed and measured.
Nanoparticle A small particle that ranges between 1 and 100 nanometers in size.
Electron microscopy (EM) Technique for obtaining high-resolution images of biological and nonbiological specimens.
Spectrophotometer Instrument that measures the number of photons (the intensity of light) absorbed after it passes through the sample solution.
Tensile strength Stress that is measured as force per unit area.
Permeation Movement of a gas or vapor through a semipermeable barrier, such as the wall of a film, the whole package, a medical device, or even a flexible electronic display.
Biodegradation Degradation of the materials into environmentally acceptable products, such as water, carbon dioxide, and biomass, by the action of naturally available microorganisms under normal environmental conditions.

References

1 Y. T. Kim, B. Min, K. W. Kim, *Innovations in Food Packaging*, ed: Elsevier, **2014**, pp. 13–35.
2 M. Farhoodi, *Food Eng. Rev.* **2016**, *8*, 35–51.
3 P. Kumar, K. Sandeep, S. Alavi, V. Truong, *Polym. Packag. Appl.* **2014**, *307*.
4 H. Cheng, L. Chen, D. J. McClements, T. Yang, Z. Zhang, F. Ren et al., *Trends Food Sci. Technol.* **2021**, *114*, 70–82.
5 S. Sharma, A. K. Jaiswal, B. Duffy, S. Jaiswal, *Food Packag. Shelf Life* **2020**, *24*, 100491.
6 S. Roy, J. W. Rhim, *Colloid. Surf. B Biointerf.* **2020**, *188*, 110761, 2020.
7 G. Biddeci, G. Cavallaro, F. Di Blasi, G. Lazzara, M. Massaro, S. Milioto, *Carb. Pol.* **2016**, *152*, 548–557.
8 J. Mock, M. Barbic, D. Smith, D. Schultz, S. Schultz, *J. Chem. Phy.* **2002**, *116*, 6755–6759.
9 M. Joshi, A. Bhattacharyya, S. W. Ali, Characterization techniques for nanotechnology applications in textiles, **2008**.
10 S. Chaturvedi, P. N. Dave, *Nanoengineering in the Beverage Industry*, ed: Elsevier, **2020**, pp. 137–162.
11 A. K. Biswal, P. K. Misra, *Mater. Chem. Phys.* **2020**, *250*, 123014.
12 A. Sobhan, K. Muthukumarappan, L. Wei, T. Van Den Top, R. Zhou, *Mater. Today Commun.* **2020**, *23*, 101124.
13 J. E. Sader, R. Borgani, C. T. Gibson, D. B. Haviland, M. J. Higgins, J. I. Kilpatric, *Rev. Sci. Instrum.* **2016**, *87*, 093711.
14 K. Mollaeian, Y. Liu, S. Bi, Y. Wang, J. Ren, M. Lu, *Int. J. Mol. Sci.* **2018**, *19*, 3461.

15 S. Sharma, S. Jaiswal, B. Duffy, A. K. Jaiswal, *Bioengineering* **2019**, *6*, 26.
16 H. M. Fahmy, R. E. S. Eldin, E. S. A. Serea, N. M. Gomaa, G. M. AboElmagd, S. A. Salem, *RSC Adv.* **2020**, *10*, 20467–20484.
17 F. Chen, C. Chi, *Int. J. Biol. Macromol.* **2021**, *186*, 405–413.
18 S. Roy, J. W. Rhim, *Colloids Surf. B* **2021**, *207*, 111999.
19 W. Dong, J. Su, Y. Chen, D. Xu, L. Cheng, L. Mao, *Food Chem.* **2022**, *373*, 131414.
20 W. Wang, F. Gu, Z. Deng, Y. Zhu, J. Zhu, T. Guo, *Carbohydr. Polym.* **2021**, *255*, 117431.
21 L. Lin, Y. Gu, H. Cui, *F. Pack Shelf Life* **2019**, *19*, 86–93.
22 V. K. Pandey, S. N. Upadhyay, K. Niranjan, P. K. Mishra, *Int. J. Biol. Macromol.* **2020**, *157*, 212–219.
23 P. Kumar, R. Tanwar, V. Gupta, A. Upadhyay, A. Kumar, K. K. Gaikwad, *Int. J. Biol. Macromol.* **2021**, *187*, 223–231.
24 A. Salem, M. Jridi, O. Abdelhedi, N. Fakhfakh, M. Nasri, F. Debeaufort, *Heliyon* **2021**, *7*, e08099.
25 N. V. Priya, U. Vinitha, M. M. Sundaram, *Biocatal. Agric. Biotechnol.* **2021**, *34*, 102021.
26 Z. Riahi, R. Priyadarshi, J.-W. Rhim, R. Bagheri, *Food Hydrocoll.* **2021**, *112*, 106314.
27 P. Thivya, Y. Bhosale, S. Anandakumar, V. Hema, V. Sinija, *Int. J. Biol. Macromol.* **2021**, *188*, 790–799.
28 J. Zeng, X. Ren, S. Zhu, Y. Gao, *Int. J. Biol. Macromol.* **2021**, *192*, 1160–1168.
29 S. Roy, J.-W. Rhim, *Int. J. Biol. Macromol.* **2020**, *162*, 1780–1789.
30 S. Sharma, S. Barkauskaite, B. Duffy, A. K. Jaiswal, S. Jaiswal, *Foods* **2020**, *9*, 1117.
31 P. B. Pathare, U. L. Opara, F. A. J. Al-Said, *Food Bioprocess Technol.* **2013**, *6*, 36–60.
32 S. U. Kadam, S. Pankaj, B. K. Tiwari, P. Cullen, C. P. O'Donnell, *F. Pack. Shelf Life* **2015**, *6*, 68–74.
33 K. K. Dash, N. A. Ali, D. Das, D. Mohanta, *Int. J. Biol. Macromol.* **2019**, *139*, 449–458.
34 F. Araújo, R. Kelmann, B. Araújo, R. Finatto, H. Teixeira, L. Koester, *Eur. J. Pharm. Sci.* **2011**, *42*, 238–245.
35 S. Bhattacharjee, *J. Controll. Release* **2016**, *235*, 337–351.
36 M. A. Sani, M. Tavassoli, S. A. Salim, M. Azizi-lalabadi, D. J. McClements, *Food Hydrocoll.* **2022**, *124*, 107324.
37 M. Alizadeh-Sani, E. Moghaddas Kia, Z. Ghasempour, A. Ehsani, *J. Polym. Environ.* **2021**, *29*, 588–598.
38 P. Ezati, Z. Riahi, J. W. Rhim, *ACS Sustain. Chem. Eng.* **2021**, *9*, 9300–9307.
39 S. Sherafatkhah Azari, A. Alizadeh, L. Roufegarinejad, N. Asefi, H. Hamishehkar, *J. Polym. Environ.* **2021**, *29*, 1143–1152.
40 V. Gupta, D. Ramakanth, C. Verma, P. K. Maji, K. K. Gaikwad, *Biomass Convers. Biorefin.* **2021**, 1–12.
41 P. Elmer, FT-IR spectroscopy attenuated total reflectance (ATR), Technical Note, Retrieved April, **2005**, 15, 2013.
42 S. Shankar, J.-W. Rhim, K. Won, *Int. J. Biol. Macromol.* **2018**, *107*, 1724–1731.
43 R. Arham, M. Mulyati, M. Metusalach, S. Salengke, *Int. Food Res. J.* **2016**, *23*, 1669, 2016.
44 K. Li, J. Zhu, G. Guan, H. Wu, *Int. J. Biol. Macromol.* **2019**, *122*, 485–492.
45 J. Wang, D. J. Gardner, N. M. Stark, D. W. Bousfield, M. Tajvidi, Z. Cai, *ACS Sustain. Chem. Eng.* **2018**, *6*, 49–70.

46 M. Zabihzadeh Khajavi, A. Ebrahimi, M. Yousefi, S. Ahmadi, M. Farhoodi, A. Mirza Alizadeh, *Food Eng. Rev.* **2020**, *12*, 346–363.

47 A. Pietrosanto, P. Scarfato, L. Di Maio, M. R. Nobile, L. Incarnato, *Polymers* **2020**, *12*, 804.

48 H.-L. Nguyen, T. H. Tran, L. T. Hao, H. Jeon, J. M. Koo, G. Shin, *Carbohydr. Polym.* 2021, *271*, 118421.

49 J. M. Kim, M. H. Lee, J. A. Ko, D. H. Kang, H. Bae, H. J. Park, *J. Food Sci.* **2018**, *83*, 349–357.

50 G. Singh, S. Singh, B. Kumar, K. K. Gaikwad, *J. F. Measur Character.* **2021**, *15*, 585–593.

51 A. Reščček, Z. Katančić, L. Kratofil Krehula, M. Ščetar, Z. Hrnjak-Murgić, K. Galić, *Adv. Polym. Tech.* **2018**, *37*, 837–842.

52 S. Jafarzadeh, S. M. Jafari, *Crit. Rev. Food Sci. Nutr.* **2021**, *61*, 2640–2658.

53 N. Bumbudsanpharoke, P. Wongphan, K. Promhuad, P. Leelaphiwat, N. Harnkarnsujarit, *Food Control* **2022**, *132*, 108541.

54 S. Yadav, G. Mehrotra, P. Dutta, *Food Chem.* **2021**, *334*, 127605.

55 J. Huang, Z. Hu, L. Hu, G. Li, Q. Yao, Y. Hu, *Trends Food Sci. Technol.* **2021**, *118*, 167–178.

56 K. S. Mikkonen, M. Tenkanen, *Trends Food Sci. Technol.* **2012**, *28*, 90–102.

57 F. Garavand, M. Rouhi, S. H. Razavi, I. Cacciotti, R. Mohammadi, *Int. J. Biol. Macromol.* **2017**, *104*, 687–707.

58 C. C. de Souza Coelho, R. B. S. Silva, C. W. P. Carvalho, A. L. Rossi, J. A. Teixeira, O. Freitas-Silva, *Int. J. Biol. Macromol.* **2020**, *159*, 1048–1061.

59 Y. Xiao, M. R. Wiesner, *J. Hazard. Mater.* **2012**, *215*, 146–151.

60 S. Bhatia, *N. Poly. Drug Del. Sys.* **2016**, 33–93.

61 A. S. F. Testing and Materials, *Standard Test Methods for Water Vapor Transmission of Materials*, ASTM International, **2013**.

62 W. Zhang, Y. Zhang, J. Cao, W. Jiang, *Int. J. Biol. Macromol.* **2021**, *166*, 288–296.

63 S. S. Ahankari, A. R. Subhedar, S. S. Bhadauria, A. Dufresne, *Carbohydr. Polym.* **2021**, *255*, 117479.

64 N. Duan, Q. Li, X. Meng, Z. Wang, S. Wu, *Food Chem.* **2021**, *364*, 130441.

65 J. Kan, J. Liu, H. Yong, Y. Liu, Y. Qin, J. Liu, *Int. J. Biol. Macromol.* **2019**, *140*, 384–392.

66 H. Y. Sintim, M. Flury, *Environ. Sci. Technol.* **2017**, *51*,3, 1068–1069, ACS Publications.

67 D. Briassoulis, F. Degli Innocenti, *Soil Degradable Bioplastics for a Sustainable Modern Agriculture*, Springer, **2017**, pp. 139–168.

17

Development and Use of Edible Materials for Food Protection and Packaging

Lin Lin[1,], Mohamed Abdel-Shafi Abdel-Samie[2], Sherif M. Abed[2], and Haiying Cui[1]*

[1] *School of Food and Biological Engineering, Jiangsu University, Zhenjiang, China*
[2] *Department of Food and Dairy Sciences and Technology, Faculty of Environmental Agricultural Sciences, Arish University, El-Arish, Egypt*
[*] *Corresponding author*

Highlights

- Edible packaging materials are used to protect food against microbial contamination and oxidation through edible antimicrobial and antioxidant materials.
- Antimicrobial sources, including natural extracts, essential oils, peptides, organic acids, enzymes, bacteriocins, metals, and many other sources.
- Packaging materials include polymers, edible plastics, and proteins.
- Antimicrobial and antioxidant agents are incorporated into biopolymers of natural resources to form totally natural packaging systems that can be edible.
- Antimicrobial and antioxidant agents are twisted together to form packaging systems.
- Electrospinning and liposomal formation are among the novel methods that convert natural antimicrobial and antioxidant bioactive compounds into packaging systems that could validate the application of these bioactive compounds in food systems even if they are not suitable or cannot survive food conditions.

17.1 Introduction

Foods spoilage and *quality deterioration* are mainly caused by microorganisms and oxidation and some other factors. Food processing, including thermal processing, drying, freezing, fermentation, and salting, can prolong the shelf life of food, but it does not eliminate nor inhibit microorganisms, and treatments that can kill microorganisms reduce the biological value of food [1, 2]. Recognizing the hazards of microbial contamination and growth in foods urged food producers to extensively apply synthetic *antimicrobial agents* (AMA) to protect foods in addition to the extensive application of antibiotics in healthcare, which developed antibiotics resistance in microorganisms to an extent that make it inactive in achieving the purpose they are used for, causing

Materials Science and Engineering in Food Product Development, First Edition. Edited by Wing-Fu Lai.

profound a serious hazard to health and to food as well. This created the need for finding new sources of AMA, especially from natural resources [3, 4]. In addition to microbial contamination, lipid oxidation decreases quality attributes that can be inhibited by *antioxidant agents* (AOA) by playing the role of free radical scavengers, metal chelators, ultraviolet absorbers, oxygen scavengers, and singlet oxygen extinguishers [5]. Aromatic and medicinal plants that have been used in herbal folkloric medicine for decades are among the many interesting sources of AMA and AOA activities. The addition of aromatic and medicinal plants known for their antimicrobial and antioxidant activities not only protect foods against microbial contamination and oxidation; they also add some other functionalities due to their composition and bioactivity. These added functionalities that turn food into a functional product, which is nowadays a new demand from consumers because of their awareness of the nutritional value and functionalities of functional foods [6]. Medicinal and aromatic plants play their role as AMA and AOA through some *bioactive compounds* (BACs) or bio-constituents that suffer physical and chemical instabilities, are not water-soluble in some cases, and are heat- and light-sensitive, limiting their application in food production. In order to overcome the instability and validate the application of BACs in food production and packaging, encapsulation must be applied through different techniques [7].

On one hand, food packaging traditionally played a passive role as it was designed not to interact with foods and to be as inert as possible [8]. On the other hand, "novel packaging" technologies serve very important functions, such as preventing quality deterioration, improving distribution, sale, and consumption efficiency, making it easier to use [9]. Novel packaging systems are designed to protect food from external deterioration factors, such as temperature, light, and humidity changes, as well as environmental factors, such as odors, microorganisms, impacts, dust, vibrations, and pressure forces [10]. In this chapter, we make an in-depth analysis of the edible packaging systems and food-contact packaging-safe materials that involve AMA and AOA that are used to protect food systems against spoilage and quality deterioration (oxidation). We present edible packaging materials based on three categories of information: bioactive materials (AMA and AOA), carriers of active agents, and/or barriers that form insulators of foods and techniques used in the application of food-packaging systems. We will also discuss the release of BACs and techniques to control them to assure the presence of their activity in foods with the purpose of protection whenever needed.

Industrial Corner

- Food must be protected against microorganisms and oxidation because they both alter food quality and safety.
- Microorganisms can let food become toxic when food contains the microorganisms or their products while oxidation of lipids can reduce quality.
- Inhibition of microorganisms and oxidation in food systems must not be done through synthetic or chemical materials and antibiotics because they cause critical problems to food and to human health.

17.2 Antimicrobial and Antioxidant Active Agents Used in the Field of Food Packaging

AMA are defined as agents that could inhibit the contamination or growth of microbes in food systems. Commercially, synthetic and chemical AMA and AOA were the most used and only applied to food production systems to protect foods against contamination or growth of microorganisms [11]. Consumers' expectation for healthy foods urged food producers and scientists to formulate food preservatives of natural resources instead of synthetic/chemical AMA. The natural AMA into food systems has been widely applied after enhancing their stability and food unsuitability through encapsulation [12]. In addition to consumers' desires in natural food preservatives, *resistant microorganism* phenomena also created the need to find AMA of different sources, specially of a natural origin, to include essential oils (EOs), peptides, organic acids, natural extracts, metals, and other AMA and AOA sources.

17.2.1 EOs Used as AMA and AOA in Food Packaging

EOs are able to extend a food item's shelf life with no probability to let microorganisms form resistance because they do not follow a single inhibition mechanism. EOs nowadays are trendy specially to consumers because of the growing awareness of their benefits with no side effects. EOs were introduced to the public as the power of nature. EOs are biodegradable, very disposed to oxidation by light, oxygen, and high temperature; some of them have a strong aroma and bitter taste and are probably not soluble in water, and to overcome those characteristics, encapsulation was proposed [13]. Encapsulation could also increase the antimicrobial activity in some cases, such as nanoencapsulation, which transports the EO to the cell membrane. The knowledge about these facts and inhibition mechanisms of microorganisms is a cornerstone in the application of EOs and increase its viability in food systems [14, 15].

Thymus capitatus and *Origanum vulgare* were nano-encapsulated in polymeric nanocapsules and simultaneously showed better antimicrobial activity against foodborne pathogens with stability [16]. The instability and water insolubility of guava leaves EO was treated by its encapsulation in hydroxypropylβ-cyclodextrin. Engineered guava leaves EO improved stability to sunlight and its solubility increased and inhibited the growth of *S. aureus* [17]. Thyme EO was used in the liposomal system to improve stability and release behavior with an increased antimicrobial [18, 19]. Clove EOs were encapsulated in different systems to overcome their strong aroma and to increase the antimicrobial activity against *S. aureus, E. coli, L. monocytogenes*, and *S. Typhimurium* in meat, meat products, and scrambled eggs [20–22]. Apricot kernel EOs (rich in oleic acid and N-methyl-2-pyrrolidone, which is a strong antioxidant and antimicrobial) was formed as a coating film with chitosan. Films showed both antimicrobial activity against fungal growth and antioxidant activity and it was good in packaging bread slices [23]. Thyme EO was incorporated in a liposomal system containing chitosan-based emulsion and it was active in extending the shelf life of kerish cheese [24]. Oregano EO and cinnamon EO were mixed with chitosan to form a film with antimicrobial activity and could preserve duck slices under a modified atmosphere [25]. Marjoram EO with mung bean protein isolate was used against

S. aureus and *E. coli* as a film on minced beef [26]. Lemon EO incorporated in a chitosan/zein complex film was used as a mushroom packaging system against microorganisms and oxidation for 12 days [27]. All reported EOs in packaging systems proved to have beneficial activities, including antimicrobial and antioxidant properties, and were reported to have potential applications in food systems.

17.2.2 Peptides and Organic Acids Used as AMA and AOA in Food Packaging

Antimicrobial peptides have been incorporated into many delivery systems to protect food against pathogenic and spoiling microorganisms [28]. Nisin and lactoferrin were incorporated in sodium diacetate, sorbic acid, and potassium sorbate and Saran resin, and the film was antimicrobial against the growth of *L. monocytogenes* [29]. Nisin Z was embedded in corn starch polyvinyl alcohol (PVA) to form a film that can inhibit *L. monocytogenes* [30]. H. Nisin, lysozyme, and EDTA chelating agent were incorporated in a starch-based edible packaging system with improved antimicrobial activity [31]. *Edible antimicrobial packaging systems* (EAMPSs) were formed using ε-polylysine hydrochloride with starch or gelatin and their characteristics were developed without the need for extrusion at high temperatures [32]. EAMPSs formed of the bioactive peptides generated by the lacto-fermentation of palm kernel cake was incorporated in different polysaccharides as antifungal agents on mango without changing its sensory acceptability [33].

Apart from peptides, organic acids were commonly used for decades as antimicrobial preservatives. The antimicrobial activity of organic acids was studied briefly and concluded that it may have a wide range of applications in food systems. Parabens were incorporated into low-density polyethylene (LDPE) films to preserve cheese and toasted bread [34]. Sorbic acid and benzoic acid, individually or combined, were active in the inhibition of *Saccharomyces cerevisiae*, *Pichia anomala*, *Issatchenkia occidentalis*, and *Candida diddensiae*, isolated from table olives [34]. Tannic acid enhanced the antimicrobial activity of chitosan, gelatin, and methylcellulose films in the preservation of cherry tomatoes and grapes against *E. coli* and *S. aureus* up to 14 days [35]. Citric acid was added to a xylan and hydroxyethylcellulose and polyethylene glycol film to inhibit *S. aureus* [36].

17.2.3 Natural Extracts and Other Agents Used as AMA and AOA in Food Packaging

All parts of *Pistacia terebinthus* (seeds, leaves, and stem) are high in phenolic compounds, and they were extracted and incorporated in chitosan-based films to impute antioxidant and antimicrobial activity [37]. Fe III hollow metal-organic frameworks were used as carriers of capsaicin in the gelatin/chitosan matrix to preserve fresh apple by inhibiting *E. coli* [38]. The quercetin–starch bio-composite film with chitosan–gelatin inhibited *E.coli* and *B. subtilis*, and it was also an antioxidant [39]. The curcumin–gelatin composite film with an evenly distributed curcumin prevented the growth of *E. coli* and *L. monocytogenes*, in addition to being an antioxidant [40]. Functional properties of pure chitosan were improved by the addition of curcumin-loaded mesoporous silica nanoparticles to inhibit *S. aureus* and *E. coli* [41]. Pomegranate peel powder and pomegranate seed juice were individually mixed with fish gelatin to form EAMPSs against *S. aureus*, *L. monocytogenes*, and *E. coli*, and the film was an antioxidant [42, 43]. Wild garlic extract was incorporated into polylactic acid (PLA) films to be applied as EAMPS against *E. coli* and other pathogens [44]. Olive pomace was added to a chitosan-based film to add antioxidant activity to preserve nuts against oxidation for 31 days [45]. An antioxidant film was formulated using electrospun zein fibrous and PLA with the addition of carvacrol to preserve wheat bread [46].

Since the turn of the past several decades, nanotechnology had transformed metals to an edible grade that could inhibit the growth of some microorganisms and prolong a food's shelf life as a result. Application of nano metals can reduce the application of food preservatives and the higher rate of reactions to hinder microbial growth [47]. Polypropylene (PP) embedded with silver nanoparticles (AgNPs) and LDPE were used as a food-packaging system to inhibit *E. coli* and *S. aureus* [48]. Calcium crosslinked alginate films were spiked with sulfur nanoparticles to form an antimicrobial packaging system of frozen meat and meat products against *E. coli* and *L. monocytogenes* [49]. AGNPs reinforced in tragacanth/hydroxypropyl methylcellulose/beeswax (novel trinary bio-composite) were antimicrobial against a wide range of pathogens in the contact surface zone [50]. AGNPs were used as AMAs in films in direct contact with chicken sausages against lactic acid bacteria for 30 days. Home-cooking processes showed the ability to eliminate AgNPs easily on washing or cooking [51].

Industrial Corner

- Natural antimicrobial and AOA from medicinal and aromatic plants or any other natural resources suffer food and environmental conditions and that could be overcome through the application of encapsulation in carrier materials that validate their application in food systems of a different nature.
- Natural antimicrobial and antioxidant materials include EOs, peptides, organic acids, natural extracts, metals, and their micro or nano derivatives.
- The carriers that could protect these active agents include starches, chitosan, cellulose, bioplastics, PLA, some proteins, and many other materials.
- Placing the active agents in carriers can modify their characteristics and could also modify their release kinetics by prolonging the release time or delaying the release to a specific time or specific conditions.

17.3 Carriers Applied in Food-Packaging Applications

Over the years, different agents have been used as carriers of AMA and AOA used in food packaging. Many of these carriers are natural polymers, partly because of their high biocompatibility and high safety profiles in food applications. In this section, major agents used as carriers in food packaging are presented.

17.3.1 Starch and Chitosan

Thermoplastic corn starch was added to chitosan oligomers (diluted in citric acid) to develop a perishable food products packaging system. Films were formed as sachets to a pack of strawberries, ricotta, and flavored breads to protect against molds and yeast for 7 days under controlled conditions. Sachet films containing thermoplastic corn starch were better than spraying antimicrobial materials on the food products [52]. Starch, cationic starch, and sodium alginate formed films showing antimicrobial activity against a wide range of microorganisms [53]. Starch was used with polycaprolactone and pomegranate rind as a hybrid EAMP system. Starch decreased the cost of formation of the films and also improved its rigidity and antimicrobial activity and provided releasing channels for the polyphenols [54]. Starch–gelatin films with *N*-α-lauroyl-l-arginine ethyl ester monohydrochloride formed films to preserve chicken breast against oxidation and microbiological contamination at 4°C. The oxidation of starch during the formation of the film further increased the antioxidant activity through the Maillard reaction during products formation [55]. Thermoplastic starch formed a new hydrogen bond with the hybrid of carvacrol EO and montmorillonite that improved the film's stability. Starch allowed the diffusion of the EO through the film that increased the antimicrobial activity against *E. coli* [56]. Cationic starch with PLA, polybutylene adipate without the addition of AMA and was used as EAMP and with the incorporation of cinnamaldehyde or tea polyphenols to preserve soy protein-based meat at 4°C for 10 days against *E. coli* and *S. aureus* [57].

While starch has been extensively used as one of the polysaccharides, the most common aminopolysaccharide polymer occurring in nature is actually chitin, which helps in building the structure of the exoskeletons of crustaceans, insects, and the cell walls of fungi. Chitosan (the most abundant chitin derivative) can be derived from chitin by enzymatic or chemical deacetylation. The main source of chitin and chitosan are shrimp and crab shells that are considered as food-processing byproducts. Chitin and chitosan are characterized as biocompatible, biodegradable, nontoxic, and able to form films that could carry antimicrobial active agents within their structure, enabling the potential application as EAMPSs in addition to being AMAs by themselves. Furthermore, nanotechnology increased the antimicrobial activity and the applicability of chitin and chitosan as EAMPSs [58]. Chitosan was reported as AMA against the growth and contamination of *E. coli*, *L. monocytogenes*, *S. typhimurium*, and *C. jejuni*. EAMPSs that consist of chitosan have been reported by many studies to be useful through spraying, dipping, coating, or wrapping. It has been reported that applications of chitosan-based films or coatings or other treatments have resulted in the extension of the shelf life of fresh produce, meat products, bread, and dairy products [59].

Chitosan was blended with poly(ε-caprolactone) as a carrier of grapefruit seed extract to form EAMPS of salmon to protect it against *E. coli* and mold for seven days [60]. Chitosan was used to carry zinc oxide and AgNPs to form EAMPS against *S. aureus* and *E. coli* [61]. Chitosan, corn starch, and nano-clay bio-composite was used to incorporate potassium sorbate as AMA to inhibit fungal growth, including *A. niger* on packed bread [62]. *E. coli*, *S. aureus*, *C. albicans*, and *A. niger* were inhibited by a nano-hybrid film formed of chitosan, mesoporous silica nanoparticles with a silver nanoshell [63]. Chitosan–gelatin film, with the incorporation of lemongrass EO and zinc oxide, was considered as EAMPS against the growth of *E. coli*, *B. subtilis*, and *S. aureus* on fresh raspberries for eight days at refrigerator temperatures compared with four days without the packaging [64]. Chitosan films containing *Plectranthus amboinicus* EO were active against oxidation and microbial contamination. The addition of the EO improved the tensile strength, opacity, and water vapor barrier of chitosan films and increased their antimicrobial activity against pathogenic microorganisms [65]. Chitosan, of different molecular weights, was used to carry the poly(β-pinene) film to protect food against oxidation. Chitosan added its antimicrobial activity to the film, and the film thus formed had both properties (antioxidant and antimicrobial) that extended the shelf life [66]. Red cabbage and clove bud oil were loaded on chitosan/carrageenan capsules as an AMA and spoilage indicator. Films protected fish against spoilage [67].

17.3.2 Cellulose and Edible Bioplastics

Cellulose, the structural component in plants, has potential use as a biodegradable filler in composite manufacturing, coating, and self-standing thin films. Nanocellulose was used severally in preparing food-packaging systems that could be edible. Nanocellulose is of a great quality as a film-forming component to give good mechanical and barrier properties [68]. Nanocellulose was spiked with AgNPs to improve the physical, mechanical, and thermal properties of polyvinyl alcohol nanocomposite films with a strong antimicrobial activity against *E. coli* and *S. aureus* [69]. Hydroxymethyl cellulose, citric acid, and polyethylene glycol together formed a carrying system to retain β-cyclodextrin/sodium benzoate on their surfaces as an AMA against *S. aureus* [36]. Cationic hydroxymethyl cellulose was applied in the production of polyelectrolyte-structured food-packaging material without the addition of AMA together with hydroxymethyl cellulose and sodium alginate. The formed film had good thermal and antimicrobial properties [70]. Nanocrystalline cellulose has a high specific surface area that makes it ready to receive N-halamine on its surfaces to form a strong antimicrobial film against *E. coli* and *S. aureus* [71].

Apart from cellulose, bioplastics have been extensively used. Traditional plastic is a petroleum-based raw material that was produced in billion tons in the latest decades and most of it imparts a serious hazard to our environment. It was reported that millions of tons of petroleum-based plastics are trashed into our oceans. All these hazards have pushed researchers to find other sources of packaging that can help avoid the disadvantages associated with petroleum-based plastics. PVA, LDPE, PE, polyvinyl chloride, and PP are prepared through a green route to substitute petroleum-based plastics. PVA films containing polyhexamethylene guanidine (PHMG) were reported to be transparent and non-leaching of antimicrobial agent due to the hydrogen bond. The films were active against *E. coli* and *S. aureus* [72]. The PVA/pluronic/zinc nanocomposite film was prepared through a green route and was used against a wide range of microorganisms [73]. LDPE with silver-loaded zeolite was active against *E. coli* and *S. aureus* [74]. PE with an antimicrobial thin film, with halloysite nanotubes, nanoparticles, and carvacrol, was prepared. About 85% of pathogens were inhibited by the film formed on chicken meat surfaces, which extended their storage life [75].

17.3.3 Other Materials Used as Carriers in Food Packaging

PLA is a biodegradable material prepared by polymerizing lactic acid. PLA is safe for food-contact applications; its production is fast and cost-effective. Films prepared using PLA as a carrier are characterized by being compatible and they are comparable with petroleum-based polymers [76]. PLA was employed with pure modified potato starch through a casting method to form a crosslinked film. Films had more thermal stability and better mechanical properties and they are biodegradable [77]. Films formed of PLA and poly(ε-caprolactone or PLC) loaded with thymol, carvacrol, and their mixture, using supercritical CO_2 at 40°C and 10 MPa for five hours, were formed. PLA/PLC/BACs films showed good film properties with both antioxidant activity [78]. EO (black cumin) was added into the PLA matrices to preserve food against *S. aureus*, *Vibrio parahaemolyticus*, and *E. coli* [79]. Shelf life of meats, fruits, and vegetables were extended through their packaging using PLA films with the addition of thymol, kesum, and curry. The water vapor barrier was improved by using the AMA against *E. coli*, *S. aureus*, and *A. brasiliensis*, and the film prolonged the shelf life of chicken meat up to 15 days [80].

Besides PLA, natural biomacromolecules, such as proteins, play an important role in serving as carriers. Animal, plant, and microbiological proteins are able to form films with abundance, biodegradability, and high nutritional value. But its quality as a film-forming agent is poor and needs some aiding factors that could improve its barrier activity, such as plasticizers, crosslinkers, nanotechnology, and some composite films [81]. Algal proteins were applied to pack foods with the addition of lysozyme (AM protein) against foodborne pathogens [82]. Whey protein isolate was used with cell-free supernatant of *Lactobacillus rhamnosus* against many bacterial strains [83]. Soy protein isolate with nisin, sodium lactate, and ethylenediamine tetraacetic acid (EDTA) formed an antimicrobial packaging film against the growth of *E.coli*, *Salmonella*, and *B. cereus*, which extended the shelf life of meat up to six days [84]. EO of *Zataria multiflora* Boiss. was also added as AMA to inhibit *L. monocytogenes* and *E. coli* in a protein film formed of zein [85]. Cotton seed protein hydrolysate was incorporated in alginate films with good mechanical properties, and antimicrobial activity against *Colletotrichum gloeosporioides*, *Rhizopus oligosporus*, and *S. aureus* was noticed. The formed film was also an antioxidant with a high DPPH, FRAP, and ABTS scavenging activity that makes it more important in fatty foods [86]. Soy protein isolate with gelatin proteins were used with layered silicates as nanofillers to increase strength, improve resistance, and increase barrier properties [87].

In addition to starch, chitosan, and cellulose as the most abundant carriers used in the formation of food-packaging films, many other biopolymers were used with AMA and AOA to form EAMPSs. Basil seed gum–based film was used with *Zataria multiflora* EO, which extended the shelf life by retarding the release of volatile constituents [88]. Polyhydroxybutyrate-*co*-valerate was designed to incorporate triclosan as AMA against *E. coli* and *S. aureus* [89]. Edible agar with zinc films were casted as a food-packaging film against the growth of *S. aureus* [90]. Sodium alginate was used with oregano EOs and mesoporous silica nanoparticles as post-harvest films. Antioxidant activity of the formed films showed high DPPH scavenging activity of 95%, and the film was also antimicrobial against *C. lunata*. [91]. Composite films made of sodium alginate, glycerol, and gallnut extract at different weight ratios were prepared. The formed film showed high antioxidant activity and was also antimicrobial against *S. aureus*, *E. coli*, *Aspergillus niger*, and *Penicillium digitatum* [92]. A wide range of carriers of film-forming materials were used to produce antimicrobial and antioxidant packaging systems, which are also edible because they are formulated using all natural edible resources. Some of them had the ability to insulate food against water and moisture changes and others needed some aiding materials to prevent moisture changes in food. Some of these carriers had good properties, such as the mechanical properties of a film, and some others had some other uses, such as plasticizers and many other materials that could enhance the physical and mechanical and barrier properties.

17.4 Methods of Fabrication or the Enhancement Activity of Edible Packaging Films

As mentioned before, EAMPSs consist of two main components, AMA or AOA and carriers, twisted together through many methods and strategies, which can also enhance the antimicrobial and antioxidant activities of films. Films can be elaborated through many mechanisms, including simple coacervation (a process of hydrocolloid precipitation from an aqueous solution by solvent evaporation through the addition of a nonelectrolyte solute, or by pH modification of the solution), complex coacervation (a process that mixes two hydrocolloids that have opposite electric charges that form a complex polymer through their interactions), melting and solidification (processes performed for solid materials that could be melted and then, AMA is added to them and let to solidify again), and gelation or thermal coagulation (a process involving denaturation, gelation, and precipitation of macromolecule through heating).

17.4.1 Coacervation

Coacervation is a successful method to produce polymer droplets in a suspension through a chemical reaction in which two liquid phases form one concentrated colloidal (coacervate) (Figure 17.1). Coacervation is the separation of a single polyelectrolyte from a solution and deposition of the agglomerated colloidal particles on an immiscible active core that results in the formation of a simple coacervate or a complex coacervate, respectively [93]. Two different charged biopolymers in a liquid–liquid separation phenomenon represents the complex coacervation process [94]. Blends of whey protein and pectin EAMPS were developed by complex coacervation. Pectin electrostatically interacted with whey protein isolate at pH 3.0 [95]. The preparation of low methoxy pectin/sodium caseinate complexes was possible through complex coacervation [96]. Complex coacervation was between pea protein isolate and high methoxy citrus pectin [97]. Coacervation in all reported studies was pH-responsive and the balance between components was critical.

Phase change materials (PCMs) were applied in edible food-packaging films when spiked using antimicrobial and antioxidant active materials. Accuracy and the control of the process of simulation is very critical in getting good films. Charging and discharging of the thermal energy storage of the PCM decide its state (liquid or solid). In the real processing and application area, phase changes are altered by many factors all of which should be optimized [100]. Nisin was incorporated into a film of poly (butylene adipate-*co*-terephthalate) through a melting and solidification process. The addition of nisin did not alter the melting temperature and crystallization. Formed films showed antimicrobial activity against *L. monocytogenes*. Melting process showed a great potential in the field of the formation of antimicrobial films that could enhance food safety [101].

Casting liquid homogeneous solutions containing AMA and AOA onto fresh cut vegetables and fruits is a potential application as it is easy to solidify when all food surfaces are covered [102]. Mango kernel starch-based heat sealable pouches were developed using the casting technique in the presence of glycerol and sorbitol to preserve red chili powder for six months [103]. PVA based active food-packaging material reinforced with AgNPs was prepared by solvent casting [104]. Okara dietary fibers, pectin, sodium

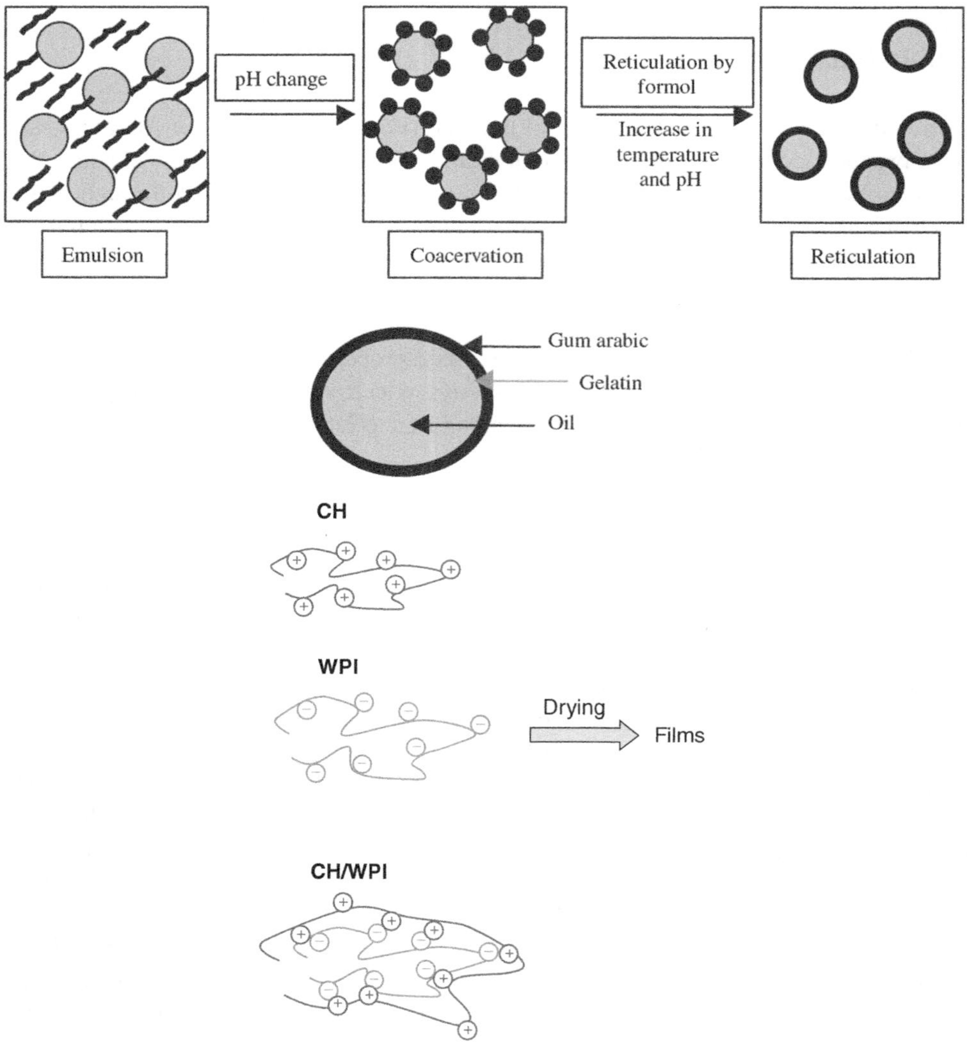

Figure 17.1 Principle of the complex coacervation method. *Source*: Madene et al. [98]/John Wiley & Sons and adapted from Tavares et al. Adapted from [99].

carboxymethyl cellulose, and thyme EO were cast as edible films to inhibit *E. coli* and *S. aureus* against oxidation [105]. With the addition of pectin to form a matrix, following the internal breakdown of mango pulp, films were formed by using the solution casting method [106].

17.4.2 Electrospinning

Electrospinning or sometimes referred to as the electrospraying method uses high-voltage electricity current to evaporate solvents used to prepare the film by encapsulating BACs

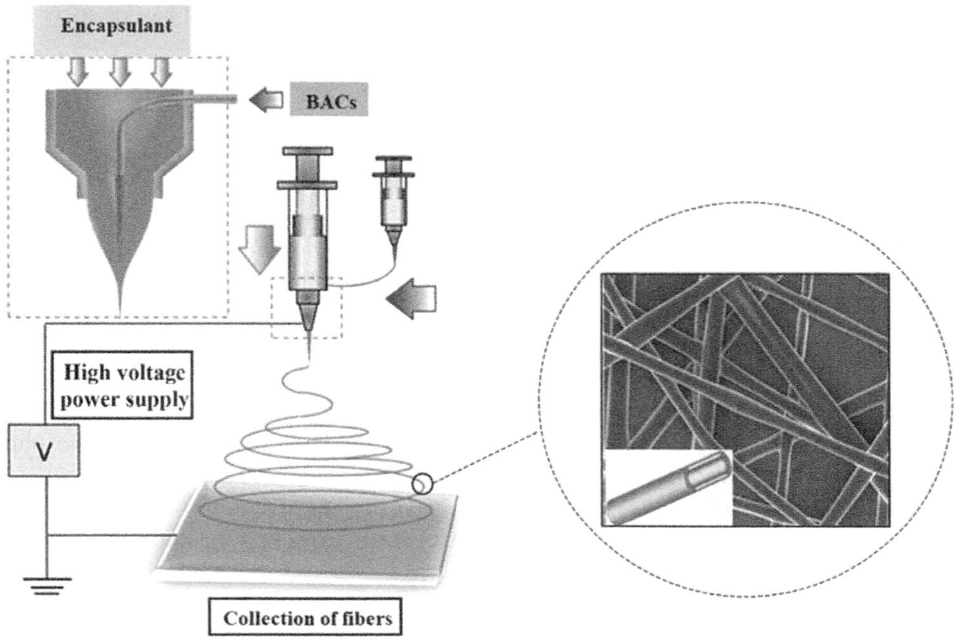

Figure 17.2 Electrospinning technique diagram.

(Figure 17.2). Electrospinning produces fibers with an incorporated activity using any solvent because all of it is likely to evaporate during the electrospinning process, but only food-approved solvents were used to prepare fibers that were meant to be used as EAMPSs [18, 107]. Electrospraying was used in the encapsulation of BACs (AMA and AOA) to protect them and increase their bioavailability [108, 109]. Electrospinning is simple and flexible with a low to cost-effective impact and does not involve any severe conditions. Electrospun-prepared fibers have the advantages of having a micro- to nano-sized diameter, a high surface-to-volume ratio, suitable porosity, tunable fiber diameters, and tailored morphology that makes them of a high encapsulation efficiency with regard to BACs, stability, bioavailability, and can engineer and control the release time and behavior [110].

Electrospinning was used to prepare TiO_2 nanoparticles and nanofibers with uniform and optimal stability to pack fruits and to remove the emitted ethylene that extended the fruit's shelf life [111]. Gelatin–glycerin–ε-poly-lysine electrospun nanofibers were of a good tensile strength, uniform diameter, and strong thermal stability, and it was antimicrobial against *L. monocytogenes* for 10 days at 4°C with no alteration of sensory characteristics or color [112]. Gelatin electrospun nanofibers containing thyme EO/β-cyclodextrin ε-polylysine were used to preserve poultry surfaces against the propagation of *C. jejuni* [113]. Electrospinning was used to prepare chitosan encapsulated in PVA to effect antimicrobial activity against *E. coli* and *S. aureus* [114]. Electrospinning was used to engineer zein nanofibers with thyme EO, citric acid, and nisin to protect food against *E. coli* and *L. innocua* [115]. These studies reported that electrospun antimicrobial and antioxidant fibers have potential application in food systems with naturally derived biopolymers and AMAs and AOAs.

17.4.3 Liposomes

Formation of liposomes is based on the phospholipid's amphipathic properties to bind both polar solvents and nonpolar materials (Figure 17.3). Unilamellar or multilamellar vesicles are formed through dissolving phospholipids in polar liquids as they go through a micellization process that results in liposomes. Liposomes consist of two layers: an inner layer binds the nonpolar material, while the outer layer dissolves in water. This validates the solubility of the nonpolar materials in water, which lets liposomes be considered a very important delivery system. Liposomes are safe, nontoxic, biologically compatible, and do not have immunogenicity effects that make it a perfect delivery system [116]. Lecithin from egg yolk or soybeans and cholesterol were used as phospholipids from natural resources in the formation of liposomes [117]. Silica nanoparticles were used to encapsulate eugenol to treat its volatility and instability in food systems. Formed eugenol silica liposomes were high in its antioxidant activity and antimicrobial activity [118].

Industrial Corner

- There are many methods that could be used to encapsulate the bioactive compounds that can be applied on industrial scales.
- Simple and complex coacervation is based on the nature of bioactive compounds and the carrier itself. Melting and solidification is based on materials that can be melted; then, bioactive compounds are added and mixed, and let to solidify.
- Gelation or thermal coagulation is based on the nature of liquid carriers that could go through denaturing or gelation or precipitation processes with bioactive compounds within them. All these strategies are easy to be applied on industrial scales with no complex processes needed.

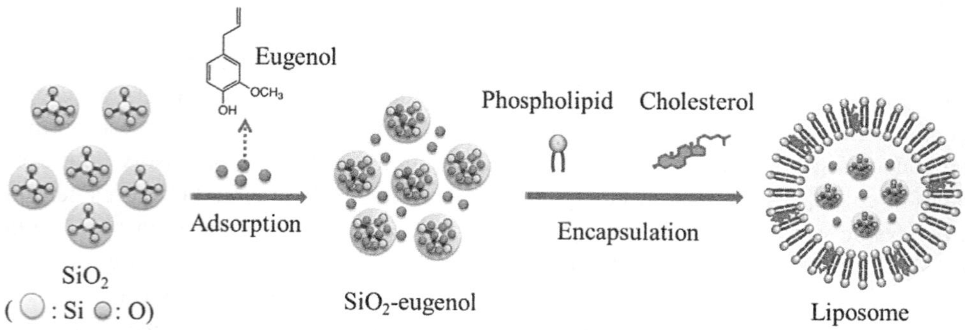

Figure 17.3 Liposome formation: example (liposomes-encapsulated SiO_2-eugenol).

- Novel methods could also be applied to form edible food-packaging systems with very good and active characterizations but with comparably expensive equipment needed.
- Electrospinning is one of these novel methods that uses high-voltage electricity to form nanofibers that could cover food surfaces using bioactive compounds.

17.5 Controlled Release of the BACs from Encapsulation Materials

All encapsulation systems (used to validate the application of BACs in food systems) can be engineered to control the release behavior by prolonging the presence of BACs in the food system in the needed time through optimizing the ratio between carriers and BACs and the other structure-aiding materials. Controlled release is defined as the change of the time or location of the release of an encapsulated substance using materials with specific barriers [119]. Control of the release time or location is done in two ways: delayed release or sustained release (Figure 17.4). Delayed release is defined as delaying the release of the BAC from a finite (lag time) to a favored release time or specific conditions. Sustained release, on the other hand, is defined as maintaining constant concentration of the BAC through the storage time. The release of the BACs might be self-released as the shell is decomposed through time or needs a trigger to aid the release that may be one or a combination of more than one.

The release of tea tree oil from a rice bran solid carrier and PE/acrylic acid films was designed with sustained antimicrobial activity [120]. A pH-responsive polyacrylic acid (PAA)/lysozyme (LYS) complex incorporated within a hydrophilic whey protein isolate film matrix was engineered to control the release of lysozymes with long-lasting antimicrobial activity [121]. A PLA with a thyme EO film was designed to control its release in the food system with a thermal-dependency behavior [122]. Electrospun basil oil in a zein and alginate was designed to postpone the release of basil oil (under gastric conditions) [123]. A sustained release of BACs of the cellulose nano-crystals through a film-casting technique performed a sustained release due to the presence of cellulose nano-crystals [124].

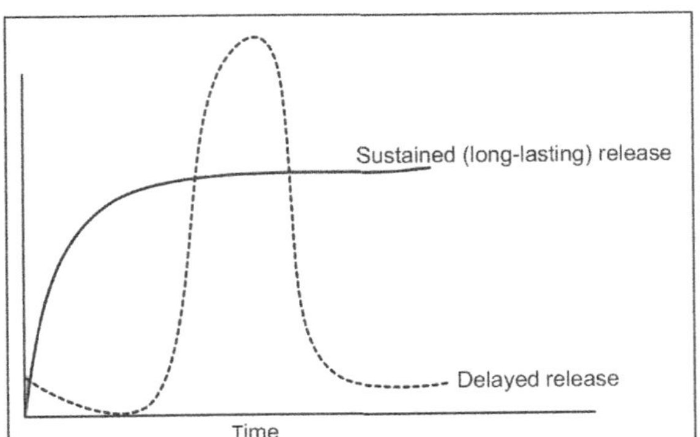

Figure 17.4 Controlled release systems. Source: Lakkis [119] / with permission of John Wiley & Sons.

17.6 Conclusion

AMA and AOA have been applied in food systems for decades but have encountered many challenges, including unsuitable environmental and food conditions in addition to their nature as they are probably volatile and, in some cases, have a strong pungent aroma and flavor. Encapsulation technologies were applied to form a delivery system of the BACs to the food systems. AMA and AOA and packaging materials are combined using a variety of techniques. In some cases, active compounds are applied to the food surface at the same time as packaging; other times, active compounds are applied at a later time, and, in some cases, active compounds are released with the help of a trigger to preserve food under specific conditions. All active compounds, carrier packaging materials, and formation and packaging techniques have been particularly reviewed in this chapter. Novel technologies and carrier materials might be used to apply AMA and AOA on the surfaces of food materials as packaging systems against microbial spoilage and food quality deterioration.

Glossary

Edible antimicrobial packaging systems (EAMPS) Food-packaging systems that use antimicrobial active agents specially from natural resources, naked or encapsulated, which is suitable to pack foods to protect against microbiological contamination or growth.

Food spoilage A process or any change that alters the safety of foods or even makes it less desirable or unacceptable to consumers. Food spoilage is a heterogeneous action that happens in food involving the biochemical activity of microorganisms.

Food quality deterioration A process or a change that alters the quality of food that leaves the food less desirable to consumers but to an extent that is less than spoilage. The food is still safe to be consumed, but its quality characterization is less and its nutritional and economical values have diminished.

Resistant microorganisms Microorganisms that cannot be inhibited using antimicrobial agents (AMA) of common sources, including synthetic or chemical AMAs. If a microorganism is resistant to multiple AMAs, it is defined as multi-drug resistant.

Antimicrobial agent An agent that could kill or inhibit the contamination and/or the growth of microorganisms. For example, the antimicrobial agent that could kill bacteria is named as bactericidal and the antimicrobial agent that can inhibit the growth of bacteria is bacteriostatic. Bacteria is inhibited through antibiotics; fungus is inhibited through antifungal agents; an antiviral can inhibit virus, and so on.

Antioxidant agent A substrate that could relegate or prevent oxidation reactions in food or in the body. Antioxidant materials can remove electrons or hydrogen from media. During oxidation of biological compounds, free radicals are produced and because they are reactive, they can start a chain of reactions that could lead to the damage or death of cells. That is why AOA are needed to prevent oxidation processes or eliminate free radicals.

Bioactive compounds A compound that has action in the food systems or in the body that can protect foods or add some functionality to it and can also promote health.

Encapsulation A process that is used to validate the use of sensitive bioactive compounds or those that have a strong aroma or flavor such as EOs. Bioactive compounds are covered by carrier materials that form an envelope or capsule on the bioactive compounds and could protect it and cover its unwanted flavor or aroma. Capsules will then release the bioactive compounds automatically or could be through a trigger such as pH or other conditions.

Carriers Materials that can form capsules that could contain the bioactive compounds within it through many properties that enable it to protect their activities against environmental conditions, such as food conditions, light, thermal processing, and many other conditions.

Essential oils Concentrated hydrophobic volatile compounds that are produced by plants to protect themselves against insects and other enemies and for many other purposes. They are extracted by distillation, expression, solvent extraction, or many other methodologies. EOs were used in folkloric medicine for decades, and nowadays, it is used to inhibit many microorganisms involved in food spoilage because it has the ability to penetrate cells because of its hydrophobic nature.

Electrospinning The technique uses high-voltage electricity to evaporate solvents from a solution containing antimicrobial-encapsulated bioactive compounds to be converted to a nanofiber. Electrospun nanofibers are very successful as antimicrobial packaging systems. Electrospinning is referred to as electrospraying.

Liposomes Liposome formation is based on the phospholipid's amphipathic properties to bind both polar solvents and nonpolar materials. Unilamellar or multilamellar vesicles are formed through dissolving phospholipids in polar liquids as it goes through a micellization process that forms liposomes.

References

1 X. Sun, J. Wang, M. Dong, H. Zhang, L. Li, L. Wang, *Trends Food Sci. Technol.* **2022**, *119*, 122.

2 F. Chemat, N. Rombaut, A. Meullemiestre, M. Turk, S. Perino, A. S. Fabiano-Tixier, M. Abert-Vian, *Innov. Food Sci. Emerg. Technol.* **2017**, *41*, 357.

3 V. Ahmad, M. S. Khan, Q. M. S. Jamal, M. A. Alzohairy, M. A. Al Karaawi, M. U. Siddiqui, *Int. J. Antimicrob. Agents* **2017**, *49*, 1.

4 A. MacGowan, E. Macnaughton, *Medicine (Baltimore)* **2017**, *45*, 622.

5 S. Smaoui, H. Ben Hlima, L. Tavares, O. Ben Braïek, K. Ennouri, S. Abdelkafi, L. Mellouli, A. Mousavi Khaneghah, *Prog. Org. Coatings* **2022**, *166*, 106780.

6 D. R. Dias, D. A. Botrel, R. V. D. B. Fernandes, S. V. Borges, *Curr. Opin. Food Sci.* **2017**, *13*, 31.

7 M. Vinceković, M. Viskić, S. Jurić, J. Giacometti, D. Bursać Kovačević, P. Putnik, F. Donsì, F. J. Barba, A. Režek Jambrak, *Trends Food Sci. Technol.* **2017**, *69*, 1.

8 D. Dainelli, N. Gontard, D. Spyropoulos, E. Zondervan-van den Beuken, P. Tobback, *Trends Food Sci. Technol.* **2008**, *19*, doi:10.1016/j.tifs.2008.09.011.

9 J. H. Han, in *Innov. Food Packag.*, Second edn. (Ed.: J. H. Han), Academic Press, San Diego, **2014**, pp. 3–12.

10 M. Carocho, P. Morales, I. C. F. R. Ferreira, *Trends Food Sci. Technol.* **2015**, *45*, 284.

11 J. H. Han, *Nov. Food Packag. Tech.* **2003**, *8*, 50.

12 R. Delshadi, A. Bahrami, E. Assadpour, L. Williams, S. M. Jafari, *Food Control* **2021**, *128*, 108180.

13 J. Fernández-López, M. Viuda-Martos, *Foods* **2018**, *7*, doi:10.3390/foods7040056.

14 V. Marcillo-Parra, D. S. Tupuna-Yerovi, Z. González, J. Ruales, *Trends Food Sci. Technol.* **2021**, *116*, 11.

15 M. Jalali-Jivan, F. Garavand, S. M. Jafari, *Adv. Colloid Interf. Sci.* **2020**, *283*, 102227.

16 G. Granata, S. Stracquadanio, M. Leonardi, E. Napoli, G. M. L. Consoli, V. Cafiso, S. Stefani, C. Geraci, *Food Chem* **2018**, *269*, 286.

17 J. Rakmai, B. Cheirsilp, J. C. Mejuto, J. Simal-Gándara, A. Torrado-Agrasar, *Ind. Crops Prod.* **2018**, *111*, 219.

18 L. Lin, M. A.-S. Abdel-Samie, H. Cui, *Encycl. Food Secur. Sustain.* **2018**, *1*, Elsevier.

19 H. Cui, W. Li, L. Lin, *RSC Adv.* **2016**, *6*.

20 M. Radünz, M. L. M. da Trindade, T. M. Camargo, A. L. Radünz, C. D. Borges, E. A. Gandra, E. Helbig, *Food Chem.* **2019**, *276*, 180.

21 M. Mulla, J. Ahmed, H. Al-Attar, E. Castro-Aguirre, Y. A. Arfat, R. Auras, *Food Control* **2017**, *73*, 663.

22 H. Cui, C. Ma, L. Lin, *Food Control* **2016**, *66*, 8.

23 R. Priyadarshi, S. B. Kumar, F. Deeba, A. Kulshreshtha, Y. S. Negi, *Food Hydrocoll.* **2018**, *85*, 158.

24 M. Al-Moghazy, H. S. El-sayed, H. H. Salama, A. A. Nada, *Food Biosci.* **2021**, *43*, 101230.

25 X. Chen, W. Chen, X. Lu, Y. Mao, X. Luo, G. Liu, L. Zhu, Y. Zhang, *Food Res. Int.* **2021**, *147*, 110491.

26 N. Haghighatpanah, M. Omar-Aziz, M. Gharaghani, F. Khodaiyan, S. S. Hosseini, J. F. Kennedy, *Int. J. Biol. Macromol.* **2022**, *201*, 318.

27 X. Wang, Y. Sun, Z. Liu, X. Huang, F. Yi, F. Hou, F. Zhang, *Int. J. Biol. Macromol.* **2021**, *192*, 635.

28 J. C. P. Santos, R. C. S. Sousa, C. G. Otoni, A. R. F. Moraes, V. G. L. Souza, E. A. A. Medeiros, P. J. P. Espitia, A. C. S. Pires, J. S. R. Coimbra, N. F. F. Soares, *Innov. Food Sci. Emerg. Technol.* **2018**, *48*, 179.

29 P. Limjaroen, E. Ryser, H. Lockhart, B. Harte, *J. Plast. Film Sheeting* **2003**, *19*, 95.

30 T. V. de Oliveira, P. A. V. de Freitas, C. C. Pola, J. O. R. da Silva, L. D. A. Diaz, S. O. Ferreira, N. de Ff Soares, *Food Packag. Shelf Life* **2020**, *24*, 100459.

31 S. Bhatia, A. Bharti, *J. Food Sci. Technol.* **2015**, *52*, 3504.

32 Y. Cheng, S. Gao, W. Wang, H. Hou, L.-T. Lim, *Carbohydr. Polym.* **2022**, *278*, 118990.

33 F. H. Ranjith, B. Adhikari, B. J. Muhialdin, N. L. Yusof, N. K. Mohammed, S. H. Ariffin, A. S. Meor Hussin, *Food Control* **2022**, *135*, 108789.

34 J. Dobiáš, K. Chudackova, M. Voldrich, M. Marek, *Food Addit. Contam.* **2000**, *17*, 1047.

35 A. Kamari, E. Phillip, *Int. J. Biol. Macromol.* **2018**, *120*, 1119.

36 Y.-C. Yang, X.-W. Mei, Y.-J. Hu, L.-Y. Su, J. Bian, M.-F. Li, F. Peng, R.-C. Sun, *Int. J. Biol. Macromol.* **2019**, *134*, 122.

37 M. Kaya, S. Khadem, Y. S. Cakmak, M. Mujtaba, S. Ilk, L. Akyuz, A. M. Salaberria, J. Labidi, A. H. Abdulqadir, E. Deligöz, *RSC Adv.* **2018**, *8*, 3941.

38 J. Zhao, F. Wei, W. Xu, X. Han, *Appl. Surf. Sci.* **2020**, *510*, 145418.

39 S. Yadav, G. K. Mehrotra, P. Bhartiya, A. Singh, P. K. Dutta, *Carbohydr. Polym.* **2020**, *227*, 115348.

40 S. Roy, J.-W. Rhim, *Colloids Surf. B Biointerf.* **2020**, *188*, 110761.

41 C. Wu, Y. Zhu, T. Wu, L. Wang, Y. Yuan, J. Chen, Y. Hu, J. Pang, *Food Chem.* **2019**, *288*, 139.

42 Z. A. N. Hanani, F. C. Yee, M. A. R. Nor-Khaizura, *Food Hydrocoll.* **2019**, *89*, 253.

43 A. Valdés, E. Garcia-Serna, A. Martínez-Abad, F. Vilaplana, A. Jimenez, M. C. Garrigós, *Molecules* **2020**, *25*, 166.

44 T. Radusin, A. Tomšik, L. Šarić, I. Ristić, M. Giacinti Baschetti, M. Minelli, A. Novaković, *Polym. Compos.* **2019**, *40*, 893.
45 T. de Moraes Crizel, A. de Oliveira Rios, V. D. Alves, N. Bandarra, M. Moldão-Martins, S. Hickmann Flôres, *Food Hydrocoll.* **2018**, *74*, 139.
46 A. Altan, Z. Aytac, T. Uyar, *Food Hydrocoll.* **2018**, *81*, 48.
47 M. Hoseinnejad, S. M. Jafari, I. Katouzian, *Crit. Rev. Microbiol.* **2018**, *44*, 161.
48 Y. Jo, C. V. Garcia, S. Ko, W. Lee, G. H. Shin, J. C. Choi, S.-J. Park, J. T. Kim, *Food Biosci.* **2018**, *23*, 83.
49 R. Priyadarshi, H.-J. Kim, J.-W. Rhim, *Food Hydrocoll.* **2021**, *110*, 106155.
50 A. Bahrami, R. Rezaei Mokarram, M. Sowti Khiabani, B. Ghanbarzadeh, R. Salehi, *Int. J. Biol. Macromol.* **2019**, *129*, 1103.
51 N. G. Marchiore, I. J. Manso, K. C. Kaufmann, G. F. Lemes, A. P. de Oliveira Pizolli, A. A. Droval, L. Bracht, O. H. Gonçalves, F. V. Leimann, *LWT - Food Sci. Technol.* **2017**, *76*, 203.
52 L. A. Castillo, S. Farenzena, E. Pintos, M. S. Rodríguez, M. A. Villar, M. A. García, O. V. López, *Food Packag. Shelf Life* **2017**, *14*, 128.
53 F. Şen, I. Uzunsoy, E. Basturk, M. Kahraman, *Carbohydr. Polym.* **2017**, *170.*
54 S. Khalid, L. Yu, M. Feng, L. Meng, Y. Bai, A. Ali, H. Liu, L. Chen, *Food Packag. Shelf Life* **2018**, *18*, 71.
55 O. Moreno, L. Atarés, A. Chiralt, M. C. Cruz-Romero, J. Kerry, *LWT* **2018**, *97*, 483.
56 A. G. de Souza, N. M. A. Dos Santos, R. F. da Silva Torin, D. D. S. Rosa, *Int. J. Biol. Macromol.* **2020**, *164*, 1737.
57 L. Wang, J. Xu, M. Zhang, H. Zheng, L. Li, *Food Chem.* **2022**, *380*, 132022.
58 M. R. Kasaai, *Carbohydr. Polym.* **2022**, *280*, 119027.
59 S. Kumar, A. Mukherjee, J. Dutta, *Trends Food Sci. Technol.* **2020**, *97*, 196.
60 K. Wang, P. N. Lim, S. Y. Tong, E. S. Thian, *Food Packag. Shelf Life* **2019**, *22*, 100396.
61 J. Kaur, K. Sood, N. Bhardwaj, S. K. Arya, M. Khatri, *Mater. Today Proc.* **2020**, *28*, 1904.
62 P. Jha, *Int. J. Biol. Macromol.* **2020**, *160*, 571.
63 S. F. Abaza, N. S. Elbialy, N. Mohamed, *Int. J. Biol. Macromol.* **2021**, *189*, 792.
64 J. Jovanović, J. Ćirković, A. Radojković, D. Mutavdžić, G. Tanasijević, K. Joksimović, G. Bakić, G. Branković, Z. Branković, *Prog. Org. Coatings* **2021**, *158*, 106349.
65 N. Vishnu Priya, U. G. Vinitha, M. Meenakshi Sundaram, *Biocatal. Agric. Biotechnol.* **2021**, *34*, 102021.
66 P. R. Rodrigues, L. M. Junior, W. F. C. de Souza, H. H. Sato, R. M. V. Alves, R. P. Vieira, *Int. J. Biol. Macromol.* **2021**, *193*, 425.
67 K. J. Park, J.-S. Lee, H. J. Jo, E. S. Kim, H. G. Lee, *Int. J. Biol. Macromol.* **2022**, *196*, 163.
68 A. K. Bharimalla, S. P. Deshmukh, N. Vigneshwaran, P. G. Patil, V. Prasad, *Polym. Plast. Technol. Eng.* **2017**, *56*, 805.
69 M. S. Sarwar, M. B. K. Niazi, Z. Jahan, T. Ahmad, A. Hussain, *Carbohydr. Polym.* **2018**, *184*, 453.
70 F. Şen, M. V. Kahraman, *Polym. Adv. Technol.* **2018**, *29*, 1895.
71 Y. Zhang, Y. Liu, R. Li, X. Ren, T. Huang, *J. Appl. Polym. Sci.* **2019**, *136*, 47101.
72 J. Chen, D. Wei, W. Gong, A. Zheng, Y. Guan, *ACS Appl. Mater. Interfaces* **2018**, *10*, 37535.
73 K. M. Amin, A. M. Partila, H. A. Abd El-Rehim, N. M. Deghiedy, *Part. Part. Syst. Charact.* **2020**, *37*, 2000006.
74 L. O. de Araújo, K. Anaya, S. B. C. Pergher, *Coatings* **2019**, *9*, 786.
75 B. A. Tas, E. Sehit, C. E. Tas, S. Unal, F. C. Cebeci, Y. Z. Menceloglu, H. Unal, *Food Packag. Shelf Life* **2019**, *20*, 100300.

76 N. S. A. Malek, M. Faizuwan, Z. Khusaimi, N. N. Bonnia, M. Rusop, N. A. Asli, *J. Phys. Conf. Ser.* **2021**, 12037, IOP Publishing.
77 N. Gürler, S. Paşa, H. Temel, *J. Taiwan Inst. Chem. Eng.* **2021**, *123*, 261.
78 I. Lukic, J. Vulic, J. Ivanovic, *Food Packag. Shelf Life* **2020**, *26*, 100578.
79 A. Misaghi, A. Khanjari, F. Talebi, *Int. J. Infect. Dis.* **2020**, *101*, 308.
80 N. Mohamad, M. M. Mazlan, I. S. M. A. Tawakkal, R. A. Talib, L. K. Kian, H. Fouad, M. Jawaid, *Int. J. Biol. Macromol.* **2020**, *163*, 1451.
81 S. J. Calva-Estrada, M. Jiménez-Fernández, E. Lugo-Cervantes, *Food Eng. Rev.* **2019**, *11*, 78.
82 S. Benelhadj, N. Fejji, P. Degraeve, H. Attia, D. Ghorbel, A. Gharsallaoui, *Algal. Res.* **2016**, *15*, 43.
83 S. C. Beristain-Bauza, E. Mani-López, E. Palou, A. López-Malo, *Food Control* **2016**, *62*, 44.
84 Y. Liu, H. Zhang, Y. Chi, Y. Wu, W. Cao, T. Li, L. Xu, *Emirates J. Food Agric.* **2017**, *589*.
85 M. Kashiri, J. P. Cerisuelo, I. Domínguez, G. López-Carballo, V. Muriel-Gallet, R. Gavara, P. Hernández-Muñoz, *Food Hydrocoll.* **2017**, *70*, 260.
86 J. G. de Oliveira Filho, J. M. Rodrigues, A. C. F. Valadares, A. B. de Almeida, T. M. de Lima, K. P. Takeuchi, C. C. F. Alves, H. A. de Figueiredo Sousa, E. R. da Silva, F. H. Dyszy, *Food Hydrocoll.* **2019**, *92*, 267.
87 M. Zubair, A. Ullah, *Crit. Rev. Food Sci. Nutr.* **2020**, *60*, 406.
88 H. H. Gahruie, E. Ziaee, M. H. Eskandari, S. M. H. Hosseini, *Carbohydr. Polym.* **2017**, *166*, 93.
89 P. K. Sabharwal, S. Chattopadhyay, H. Singh, *J. Appl. Polym. Sci.* **2018**, *135*, 46862.
90 I. Malagurski, S. Levic, A. Nesic, M. Mitric, V. Pavlovic, S. Dimitrijevic-Brankovic, *Carbohydr. Polym.* **2017**, *175*, 55.
91 W. Lu, M. Chen, M. Cheng, X. Yan, R. Zhang, R. Kong, J. Wang, X. Wang, *Food Packag. Shelf Life* **2021**, *29*, 100691.
92 H. Aloui, A. R. Deshmukh, C. Khomlaem, B. S. Kim, *Food Hydrocoll.* **2021**, *113*, 106508.
93 C. G. de Kruif, F. Weinbreck, R. de Vries, *Curr. Opin. Colloid. Interf. Sci.* **2004**, *9*, 340.
94 N. Eghbal, R. Choudhary, *LWT* **2018**, *90*, 254.
95 K. S. Silva, T. M. R. Fonseca, L. R. Amado, M. A. Mauro, *Food Packag. Shelf Life* **2018**, *16*, 122.
96 J. Wang, E. Dumas, A. Gharsallaoui, *Food Hydrocoll.* **2019**, *88*, 163.
97 P. K. S. Pillai, B. E. Morales-Contreras, L. Wicker, M. T. Nickerson, *Food Chem.* **2020**, *305*, 125433.
98 A. Madene, M. Jacquot, J. Scher, S. Desobry, *Int. J. Food Sci. Technol.* **2006**, *41*, 1.
99 L. Tavares, H. K. S. Souza, M. P. Gonçalves, C. M. R. Rocha, *Food Hydrocoll.* **2021**, *113*, 106471.
100 Z. Andrássy, Z. Szantho, *J. Therm. Anal. Calorim.* **2019**, *138*, 3915.
101 G. Zehetmeyer, S. M. M. Meira, J. M. Scheibel, R. V. B. de Oliveira, A. Brandelli, R. M. D. Soares, *J. Appl. Polym. Sci.* **2016**, *133*.
102 S. Santonicola, V. García Ibarra, R. Sendón, R. Mercogliano, A. Rodríguez-bernaldo de Quirós, *Coatings* **2017**, *7*, 177.
103 A. Nawab, F. Alam, M. A. Haq, M. S. Haider, Z. Lutfi, S. Kamaluddin, A. Hasnain, *Int. J. Biol. Macromol.* **2018**, *114*, 626.
104 S. Mathew, J. Mathew, E. K. Radhakrishnan, *J. Polym. Res.* **2019**, *26*, 223.

105 D. Lin, Y. Zheng, X. Wang, Y. Huang, L. Ni, X. Chen, Z. Wu, C. Huang, Q. Yi, J. Li, W. Qin, Q. Zhang, H. Chen, D. Wu, *Int. J. Biol. Macromol.* **2020**, *165*, 1241.

106 F. C. A. Oldoni, M. P. Bernardo, J. G. Oliveira Filho, A. C. de Aguiar, F. K. V. Moreira, L. H. C. Mattoso, L. A. Colnago, M. D. Ferreira, *LWT* **2021**, *145*, 111339.

107 Y. P. Neo, S. Ray, C. O. Perera, in *Role of Materials Science in Food Bioengineering* (Eds: A. M. Grumezescu, A. M. Holban), Academic Press, **2018**, pp. 109–146.

108 M. R. I. Shishir, L. Xie, C. Sun, X. Zheng, W. Chen, *Trends Food Sci. Technol.* **2018**, *78*, 34.

109 A. M. Pisoschi, A. Pop, C. Cimpeanu, V. Turcuş, G. Predoi, F. Iordache, *Eur. J. Med. Chem.* **2018**, *157*, 1326.

110 P. Wen, M. H. Zong, R. J. Linhardt, K. Feng, H. Wu, *Trends Food Sci. Technol.* **2017**, *70*, 56.

111 Z. Zhu, Y. Zhang, Y. Shang, Y. Wen, *Food Bioprocess Technol.* **2019**, *12*, 281.

112 L. Lin, Y. Gu, H. Cui, *Food Packag. Shelf Life* **2018**, *18*, 21.

113 L. Lin, H. Cui, Y. Zhu, *Leb. + [i.e. und] Technol.* **2018**, *97*, 711–718.

114 Y. Liu, S. Wang, W. Lan, *Int. J. Biol. Macromol.* **2018**, *107*, 848.

115 Z. Aytac, R. Huang, N. Vaze, T. Xu, B. D. Eitzer, W. Krol, L. A. MacQueen, H. Chang, D. W. Bousfield, M. B. Chan-Park, *ACS Sustain. Chem. Eng.* **2020**, *8*, 15354.

116 H. Y. Cui, J. Wu, L. Lin, *J. Dairy Sci.* **2016**, *99*, 6097.

117 H. Cui, W. Li, L. Lin, *J. Food Saf.* **2017**, *37*, e12302.

118 H. Cui, L. Yuan, W. Li, L. Lin, *Food Packag. Shelf Life* **2017**, *11*, 49.

119 J. M. Lakkis, *Encapsulation and Controlled Release Technologies in Food Systems*, 2007.

120 A. E.-A.-A. El-Wakil, H. Moustafa, A. M. Youssef, *J. Thermoplast. Compos. Mater* **2020**, *35*, 938–956.

121 B. B. P. Ozer, M. Uz, P. Oymaci, S. A. Altinkaya, *Food Hydrocoll.* **2016**, *61*, 877.

122 I. S. M. A. Tawakkal, M. J. Cran, S. W. Bigger, *LWT-Food Sci. Technol.* **2016**, *66*, 629.

123 S. Dede, O. Sadak, M. Didin, S. Gunasekaran, *J. Food Eng.* **2022**, *319*, 110914.

124 J. T. Orasugh, N. R. Saha, G. Sarkar, D. Rana, R. Mishra, D. Mondal, S. K. Ghosh, D. Chattopadhyay, *Carbohydr. Polym.* **2018**, *188*, 168.

18

Packaging Design as Part of a Holistic Food Quality Assurance Process

Agnieszka S. Cholewa-Wójcik and Agnieszka K. Kawecka*

Cracow University of Economics, Institute of Quality Science and Product Managements, Department of Products Packaging, 27 Rakowicka Street, Cracow, Poland
** Corresponding author*

Highlights

- Packaging is an inherent component of an integrated product, which is offered on the market in line with its prevailing trends.
- An integrated product-designing process must take into account care for the quality of both a product and its packaging.
- Product quality assurance is not possible if the designing process is not based on a holistic view taking into account the diversified needs and requirements of consumers.

18.1 Introduction

For the development and shaping of foodstuffs that require packaging, it is necessary to consider and adapt to changing market trends that considerably impact the consumer goods market, and consequently, products offered thereon. This is manifested through a changed structure and consumption, as well as a greater need for products along with their packaging, which, to a large extent, will be adapted to suit the requirements of individual supply chain entities, and specifically, to meet the needs and expectations of consumers. Given that approximately 95% of foodstuffs, in order to be offered on the market, must come with packaging, the integrated product concept – for which the final product is treated as an inherent combination of two products: the product itself and its indispensable immediate packaging, while maintaining the equivalence between them – is assuming special importance. In consideration of the foregoing, the issues related to integrated product designing are very much in the foreground and give a response to the growing heterogeneity of consumers' needs and requirements for foodstuffs. Hence the chapter aims to exhibit the essence of integrated product designing and its role in the foodstuff quality assurance process.

Materials Science and Engineering in Food Product Development, First Edition. Edited by Wing-Fu Lai.

18.2 Essence of Quality-Oriented Product Designing and Its Role in Quality Assurance

Quality assurance constitutes one of the four components of quality management [1]. Quality assurance is a concept that is most often associated with the application of a specific product-manufacturing technology ensuring that products meet the needs and requirements of a customer [2, 3]. However, it is impossible to manufacture a product possessing specific, predefined features, if no properly planned and completed designing process is incorporated into a quality assurance process [4]. In broad terms, designing is understood as a substantive and creative type of human activity, which represents conceptual and pragmatic preparation for cognitive functions [5, 6]. According to the definition proposed by Milton and Rodgers [5], "designing is an activity that introduces changes to a given environment or context." Eyal [6], on the other hand, understands designing as the efficient and effective development of ideas in the course of a process leading to new products. The essence of designing is also portrayed as creating products and predicting the conditions under which the product manufactured will be used.

Designing may also be seen as a cognitive activity aimed at formulating such a theory that will be applied to resolve practical problems relating to the satisfaction of needs [7]. At a designing stage, while taking the economical aspect into consideration, important decisions are taken, the effects of which affect technosphere (technical designing), ecosphere (eco-designing), and the level at which consumers' needs are satisfied. Hence designing must be perceived as a comprehensive process that covers social, technical, ergonomic, and economic matters, as well as issues related to environmental protection and so on [8–10]. In view of the foregoing, designing necessitates the integration of knowledge from various fields associated with the subject of designing. Jackson termed such an approach to the designing process "specialization without isolation" [11]. Based on the aforesaid deliberations, one may assert that designing is a term that includes many different meanings and carries obvious quality-oriented connotations. Therefore, it is closely linked to product quality assurance, and in particular, quality assurance relating to foodstuffs [12].

Industrial Corner

- Designing covers all activities aimed at developing ideas in the course of a process leading to new products.
- Reaching the objectives of quality-oriented foodstuff designing is to ensure the proper quality of final products.
- Quality-oriented foodstuff designing should take into account both consumers' entire gamut of expectations identified in marketing research and newly created customers' needs.

18.3 Quality-Oriented Product-Designing Process

In a quality-oriented product-designing process, all decisions made before any action is taken are analyzed in order to assess whether it is possible to deliver and maintain required, declared product quality, determined based on customers' needs and requirements [13]. The designing of products that are based on social needs and requirements is a process divided into stages, meaning that it continues gradually (stage by stage), moving through consecutive project phases, and finally, leading to an assessment during which results are analyzed and decisions concerning further steps in the process are taken. Designing process flow schemata, developed, for instance, by various researchers [14, 15], provided a basis for a general product designing and development model, which is shown in Figure 18.1. The product-designing process flow demonstrated in the aforesaid model comprises four phases [16]. These phases include specific process stages during which a product concept is developed. At each stage of the product-designing process, attention must be devoted not only to social, technical, and economical, but also ecological aspects. Such a perspective ensures the optimization of decisions and a comprehensive approach to the iterative designing process [17].

The problem identification and preparation phase includes the design objective planning and analysis stage. The needs of all supply chain entities, which have been identified and must be satisfied by the final outcome of designing, provide a starting point for formulating a design objective [18]. This phase also encompasses the design objective analysis stage, at which, based on preliminary designing assumptions and a prepared checklist for design assumptions, design requirements defined as restrictions imposed on design solutions are set out and optimized. The determination of design requirements may be aided by the

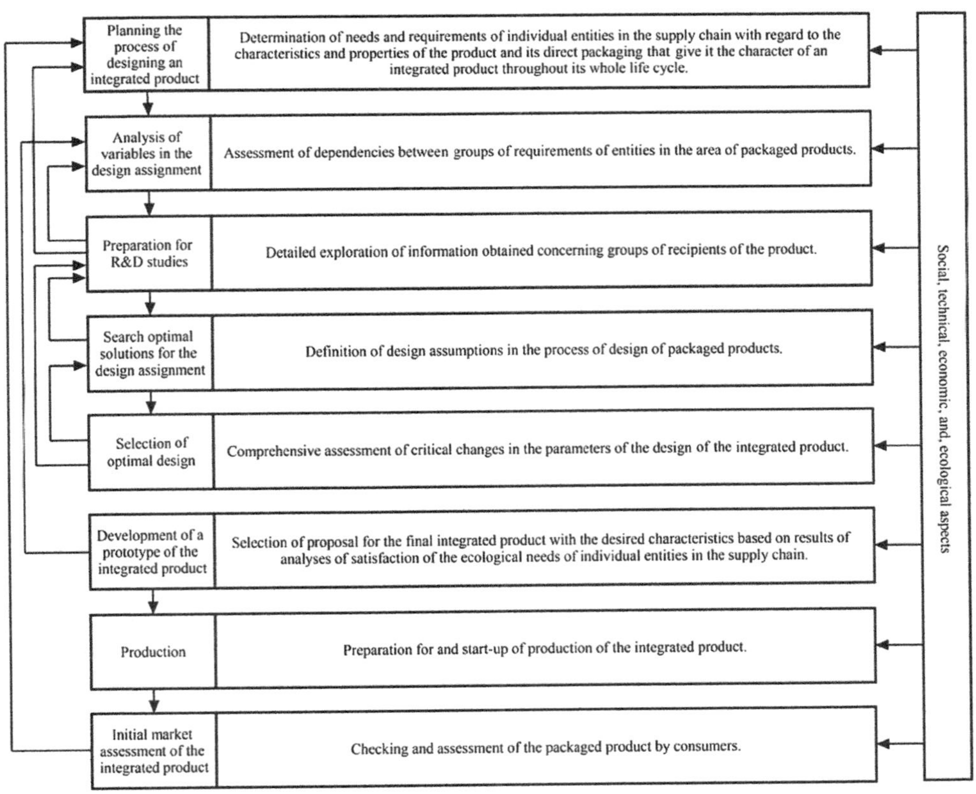

Figure 18.1 Model of the product-designing process. *Source:* own elaboration based on [15].

following specific methods: space of events and interactions, a goal tree, as well as an examination of a user and the process of use [19]. With a view to optimizing design requirements, it is crucial to define a so-called goal function and determine the values of decision variables so that the maximum value could be reached for that function. Design requirements, which, in this case, are decision variables, have been referred to as optimum requirements, selected based on the analyses of an integrated product, the target market on which the product is supposed to be put, a consumer profile, potential product-manufacturing methods, and the manner of distribution. Precise requirements set out as design assumptions that incorporate requirements characterizing a product serve as a basis for seeking solutions to achieve design objectives, being a stage of another phase called concept and solution development [20].

Looking for solutions to achieve a design objective aims to address a problem framed in the design objective, which, until this stage, may be seen as a problem having multiple variants [21]. A broad variety of solution variants available for a design objective necessitates their initial selection. Data presented in the literature on the subject, for instance, by such authors as Grunet [22], demonstrates that when seeking solutions to achieve a design objective, three approaches are available. Under the first approach, existing solutions are used and adapted to a given design objective. This type of designing, which is connected with improving/streamlining a product and/or its packaging, is termed "an action that entails the selection of design information in response to a need identified." The second approach aims to refine existing solutions or adopt them in order to complete partial objectives. Such designing, which is an example of introducing incremental innovations, is referred to as the application of various known designing methods with a view to creating products that satisfy the identified needs [23]. And finally, the third solution-focused approach, namely, the development of new solutions, leading to breakthrough innovations, includes actions aimed at developing such products that allow for discovering needs that have been unidentified so far [24].

In the stage during which solutions to attain a design objective are sought, the following methods are applied most often: a trial-and-error method, questioning, brainstorming, a morphological method, a solution tree, a systemic method, or PechaKucha [25]. Apart from the foregoing methods, the following are also used: synectics, looking for the ideal solution and, e.g., a new combination method. The methods mentioned above allow for the analysis and verification of information obtained from research and make it possible to take decisions framed as conceptual assumptions based on which a design is selected and optimized. Product designs proposed are assessed using such criteria as legal, technical, aesthetic, ergonomic, economical, and marketing aspects [26]. This stage serves to select a solution, which, from the perspective of a criterion or a system of criteria adopted, is the best and meets all requirements laid down in a design objective. During the analysis of design solutions, either preferential or generative methods are applied. The generation and selection of a design solution is thus a focal point of a designing process, in which the cognitive process turns into the creative process, entering the phase where a solution is verified and adopted. Subsequently, in the next phase, design solutions are verified and a product prototype is made. Prototyping enables the development, assessment, and optimization of designs. A prototype of a product along with related documentation takes into consideration its recipe and incorporates its characteristic features and attributes. An efficient prototype of packaging created, e.g. using such software as ArtiosCAD, DVS4CAD, or Eng View, and combining the components and features of packaging, should provide supply chain entities with performance-related, hedonistic, and semiotic benefits [27].

A prototype developed and verified under laboratory conditions is then assessed using techniques and methods that include the examination of risk related to the satisfaction of supply chain participants' needs [28]. The result of that assessment provides a basis for starting the mass production of a product. At this stage, which is the last phase of the designing process, a product is commercialized. According to Kim et al. [29], the commercialization process for a new product should be accompanied by the observation of consumers' behavior and their attitude to the features of a new product [30, 31]. Strict requirements for designing create a need for using and incorporating integrated computer systems into the process. The systems not only provide a uniform common platform for all phases of product and packaging designing and development, but also enable the verification of results obtained at each process stage. The interface between science and techniques of computer-aided designing process contributes to its greater effectiveness, which translates into the degree to which supply chain entities' needs are satisfied [32].

Industrial Corner

- Foodstuff designing is a step-by-step process, after which results need to be verified.
- In quality-oriented designing, attention must be given not only to social and economic issues, but also to technical and ecological aspects.
- The optimization of design-related decisions is possible when aspects concerning many different areas have been considered.

18.4 Integrated Product Designing as the New Approach to Packaged Product Designing Process

The development and shaping of new foodstuffs that must have – given their features and properties – packaging in order to be offered on the market require that an approach focused on developing integrated products be incorporated into the designing process. It is assumed that the essence of an integrated product is that immediate packaging constitutes an inherent component of a packaged product [33]. Hence it is necessary for the concept of integrated product designing to abandon the traditional approach to products and their packaging, under which they are both developed separately, and instead, to shift to such an understanding of designing, where a product and its packaging are inherent. Integrating two elements that are independent of each other, i.e. a product and packaging, is the reason in the majority of cases, only a packaged product may be treated as a final product. Indeed, packaging that is in direct contact with a foodstuff, e.g. a drink, gives the latter the character of a product that may be offered on the market. Accordingly, when dealing with the designing of products that require immediate packaging, an attempt must be made to manufacture an integrated product. Due to the fact that interactions between these inherent components must be taken into account, the coincidence of individual stages and basic designing process activities included in these stages and relating both to a product itself and its packaging is of paramount importance. Figure 18.2 shows a diagram illustrating the impact of factors on integrated product components.

Criteria elements that impact the quality of an integrated product being designed, which comprises a product and its packaging, include, without limitation: the quality of raw materials, the method and conditions of a technological production process, packing techniques and systems, type of packaging (with attention given to its structure and a cryptoclimate inside of it), packaging material properties, and conditions under which it is stored and used [34, 35]. Issues relating to product quality have received considerable attention in the literature on the subject. For example, food quality defined as the sum of properties of a foodstuff, determining whether that product is capable of satisfying the needs of a contemporary consumer, was thoroughly discussed in the literature on the subject, by different authors [36–40]. Food quality is a positive combination of nutritive value and sensory attractiveness, for which the risk associated with microbiological safety and the assessment of a given foodstuff are considered [41–43].

Food safety, including microbiological safety, the nutritive value and sensory attractiveness of food, as well as the convenience of use, are those component parts of food quality, which are most often pointed out by other authors [44]. Consumers' perception of food quality, existing definitions, and various aspects and component parts of the term in question were extensively reviewed by Rijswijk and Frewer [45]. Despite differences in specific phrasing, all definitions are consistent with each other, as regards the following

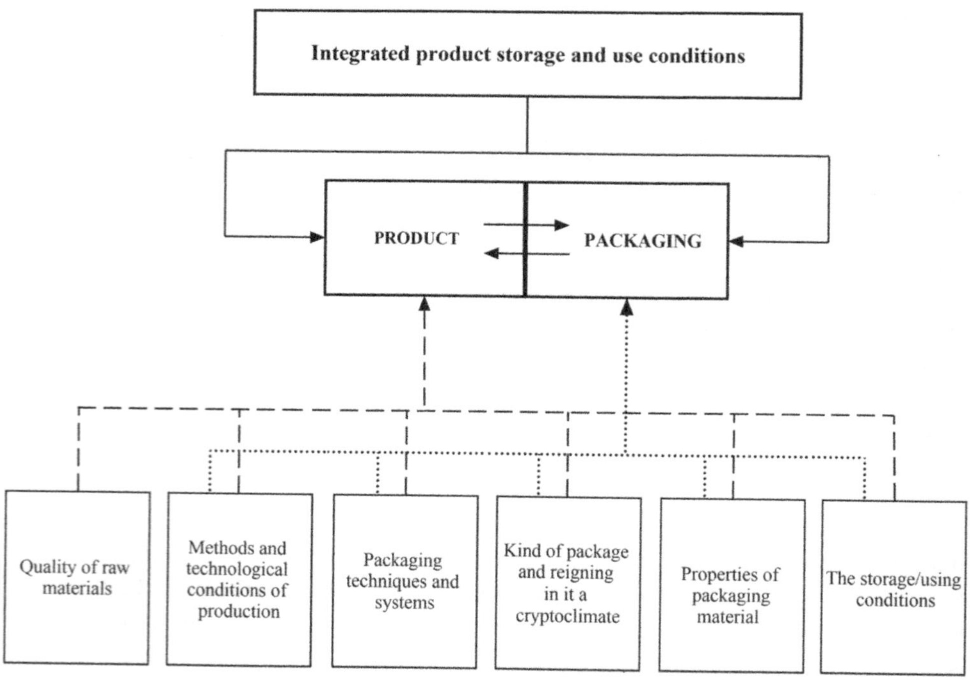

Figure 18.2 Interrelations of the impact of factors on integrated product components. *Source:* own elaboration based on [33].

points: first, food quality is a comprehensive concept that embraces such aspects as safety, nutritive value, and sensory features, and second, sensory features of food are seen in relation to consumers' perception as a very important, if not even the most important, part of the general quality of food [46]. Apart from a product of certain quality, an element that is indispensable in order for an integrated product to exist is packaging and its quality. Packaging quality must be understood as the degree to which a group of packaging properties meets requirements imposed on packaging regarding the safety, storage, carriage, delivery, and presentation of a product on a route from the manufacturer's site to the user's or consumer's door. The type of packing material used substantially affects the quality of finished packaging, and consequently, of packaged products [47]. Most important requirements imposed on these materials include physical and chemical properties of a packaged product, the structure of packaging, fitness for processing in specific packing systems, aesthetic features and possibility of placing printed information on it, time and conditions of storage, method of using packaging, warehousing, and transport conditions, as well as the price and availability of materials [48]. Given all the factors mentioned above, considerable significance must be attached to the properties of a packaged product. The understanding of the features and properties of available packaging materials and packaged products makes it possible to predict interrelations between them under various storage and use conditions [49]. For the above reasons, when designing packaging for specific use, manufacturers have to pay attention to the multifaceted nature and complexity of the problem of selecting proper packaging materials that best match the type and features of packaged products [50]. The scope of requirements laid down for materials intended for food packaging is clearly determined and reflected in legal acts [51].

Apart from the type of packaging material used, packaging quality is also conditional on such factors as the proper production process, adequate storage, transport, and distribution conditions in companies [52]. A comprehensive analysis of packaging quality revealed that it must be considered twofold. On the one hand, packaging quality means that packaging has the necessary technical features and properties characterizing its use value (for example, barrier capacity of packaging material, mechanical strength of packaging, air tightness of packaging). On the other hand, packaging quality must be analyzed from the angle of social demands of people using packaging and packaged products [53].

The satisfaction of technical requirements is linked to the use value of packaging, understood as the degree to which a group of packaging properties meets requirements imposed on packaging regarding its use [54]. Where the utility of packaging is to be seen as a selection criterion, out of all properties that characterize use value, the following deserve special attention: technical features associated with product properties (performance) and related to availability (e.g. reliability). In the case of packaging, huge importance is attached to its technical features. In fact, packaging must have stable performance-related parameters of a certain value, as securing the quality of packaged products depends, first and foremost, on them. On the other hand, as regards features related to packaging availability, use value can be assessed based on the analysis of packaging reliability understood as the ability to satisfy, in a given time and under specific conditions, requirements imposed on packaging. Whereas, the analysis of packaging quality, which is focused on social demands and expectations, must take into consideration prevailing market consumption trends reflecting the needs of today's consumers. Given the foregoing, packaging should be selected in such a manner that ensures its beneficial impact on a packaged product along the entire logistic chain [55]. Moreover, attention must also be given to the role and significance of the impact of external factors on the changing quality of a packaged product already offered on the market. As regards major exogenous factors leading to changes in the quality of packaged products, the following must be mentioned: air temperature and humidity, as well as the influence of sunlight and chemical composition of the surrounding atmosphere. These factors have a bearing on the physico-chemical, biochemical, and microbiological processes taking place in a packaged product. Furthermore, they are treated as catalysts for the aging process occurring both in a product and its packaging [56].

After reconsideration of data contained in the literature on the subject, it must be stated that criteria elements affecting the quality of an integrated product being designed include: the quality of a foodstuff and its packaging. The aforesaid component parts of an integrated product must be considered comprehensively along with criteria elements used to assess their selection for a given product that has specific features and properties. Hence in order to create an integrated product of proper quality, none of the principal components and factors affecting it may be omitted. An integrated product-designing process must therefore take into account care for the quality of both the product and its packaging. This is because packaging is indeed an inherent component of an integrated product, which is a commodity traded and offered on the market in line with prevailing market trends. Figure 18.3 shows the proposal of a theoretical model of integrated product designing, which allows for the diversification of market offer by choosing features and components of both the product and its packaging.

The theoretical model of integrated product designing put forward above provides an opportunity for the diversification of a market offer by selecting the features and components of both a product and its packaging. It further allows for a synthetic description of the integrated product-designing process and indicates the need to regard packaging as an inherent component of a product and its indispensable immediate packaging while emphasizing the fact that equivalence between them must be maintained. Furthermore, an analysis of designing stages makes it possible to determine which stages of designing operations are common and which are divergent. Despite the fact that an integrated product is a whole, going through individual designing stages that are included in the problem identification and preparation phase and the designing process concept and solution development phase requires an individual approach, not only for a product itself, but also for its packaging. Indeed, an in-depth analysis of the features and attributes of food allows us to verify opportunities and limitations that must be considered when selecting and using such types of packaging that will fulfill their functions at each and every stage of a supply chain. Requirements that are characteristic of both a product and its packaging may also serve as guidelines to be followed while developing an integrated product prototype. As regards the incorporation of a product and its packaging into an integrated product, the production stage that depends on the type of a packing system used for the product-manufacturing process must be seen as a focal point. Where a product-manufacturing process uses a packing system in which packaging is considered part of the packing process, this is the stage at which packaging, once it has been integrated with a product, becomes an integrated product. However, if a different packing system has been applied, under which packaging is made, in whole or in part, outside the packing process, then two separate

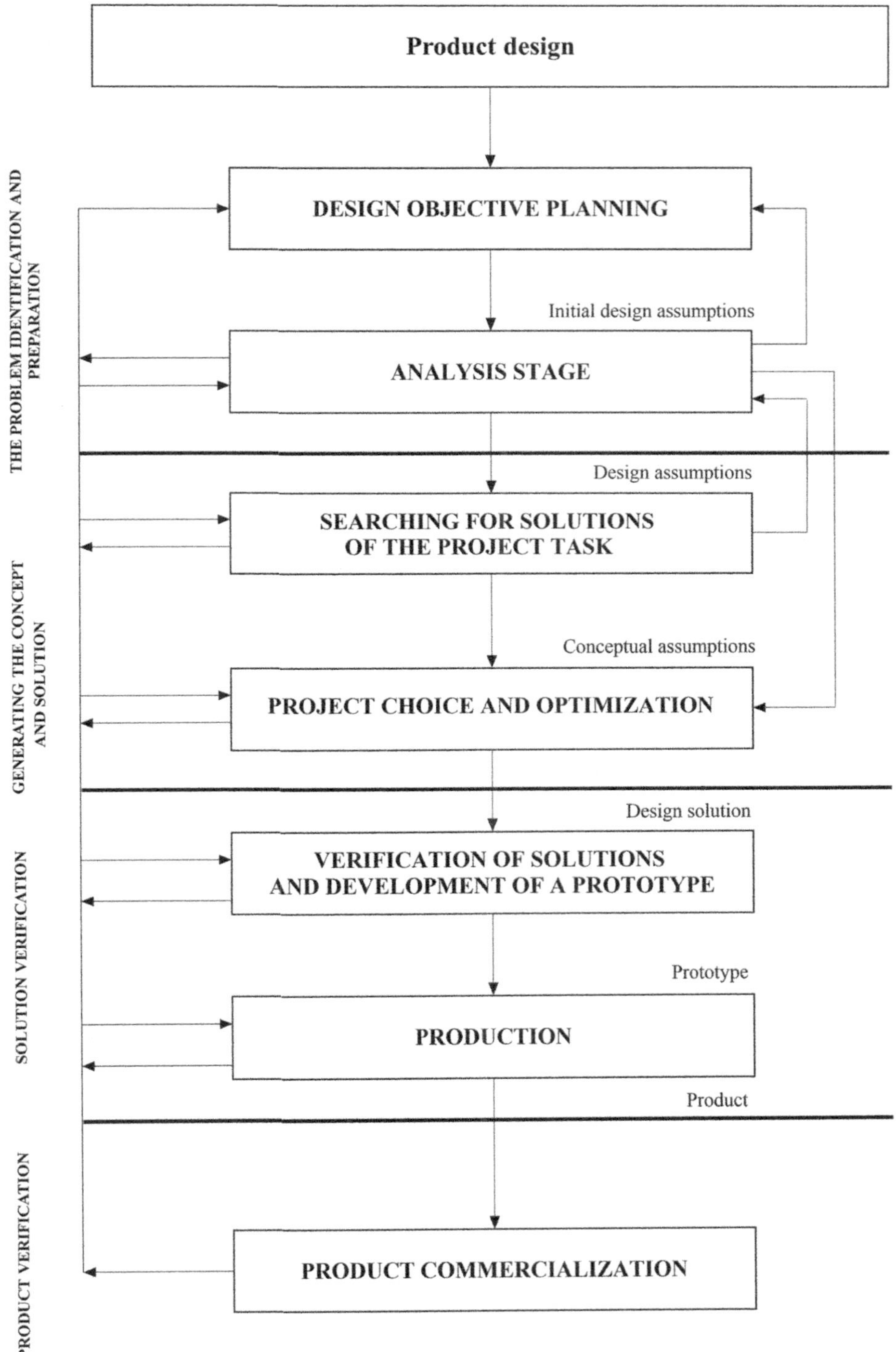

Figure 18.3 Theoretical model of integrated product designing allowing for a diversification of the market offer by choosing features and components of both product and its packaging. *Source:* own elaboration based on [57].

items – namely, a product and packaging, which are independent of each other, are integrated only at the last stage of the phase where a solution is verified and adopted. An integrated product made in the manner described above passes through the subsequent common stages of designing operations carried out as part of the product verification phase [57]. It has been ensured that the model developed embraces social, technical, economic, and ecological aspects at every stage of designing. In fact, such a perspective guarantees the optimization of decisions and a comprehensive approach to the iterative integrated product-designing process. The authors, proposing their own integrated product-designing model, highlighted that it must be perceived as a course of action that contributes to the diversification of a market offer by creating the possibility to select such features of a product and packaging that are desired by supply chain entities.

Industrial Corner

- The development and shaping of new foodstuffs should be based on an approach under which a product and its indispensability are treated as a whole.
- The quality of an integrated product is the outcome of foodstuff quality and packaging quality.
- The theoretical model of integrated product designing put forward above provides an opportunity to diversify a market offer by selecting such features of a product and its packaging, which are desired by customers.

18.5 Methods to Aid Shaping of Quality of Products Being Designed

Quality-oriented methods are employed in order to identify factors that have the greatest impact on product quality with regard to:

- a product design that should guarantee resistance to various exposures affecting a product
- achieving, throughout a production process, such a level of quality that was determined and complies with a design
- quality control while manufacturing and using a product.

Special importance in a designing process is given to quality-oriented methods, including design quality management (DQM), quality function deployment (QFD), and design of experiment (DoE) [58, 59]. The DQM method is applied to enhance the management of designing assumptions and allows for an objective evaluation of the quality of designing solutions. On the one hand, this method makes it possible to record and structure all designing assumptions and control them in order to verify whether the assumptions are not contrary to the objective of the task to be done. On the other hand, by using the method in question, it is further possible to determine customer-specific requirements, identify critical project elements, and control designing guidelines that have been implemented, checking whether expectations are met. A special benefit from using the DQM method is the fact that divergent assumptions can be eliminated and the scope of works, which is appropriate for a given design stage may be determined [60]. The QFD method is about designing a product, with special attention given to features and properties that are needed to deliver its quality. It is used to specify and systematize customers' requirements and expectations, which are then translated into technical requirements defining the features and properties of a product being designed [61–63]. The assumptions of the so-called QFD road map, put forward by Hofmeister [64], can be implemented for products and their packaging. According to this concept, a designing process of a product that needs packaging must consider consumers' requirements and needs relating to a product itself on the one hand and expectations regarding its packaging on the other.

Apart from the QFD method, another quality-oriented designing method, to which particular significance must be attached, is the DoE. This is a set of methods used for designing a product and/or a process by developing a mathematical model that describes how a system being examined functions when product or process parameters are changed. The purpose of these methods is to obtain as much valuable and reliable information on a product being analyzed as possible based on a limited number of experiments. For products that need packaging, such information may provide data, e.g. on the impact of the type of packaging used or a percentage share of a cryptoclimate gas mixture composition on the durability of the basic component of a packaged product. Based on such information, parameters (of a product and its packaging) and a production process are optimized by defining the value of input parameters characterizing the features and properties of a product and its packaging, which will ensure the maximum product quality, reduce the variability of its production process, or guarantee that it will not be sensitive to changes to uncontrolled parameters [65]. The DoE is one of the most effective and, at the same time, most advanced methods applied to examine the impact of diverse input variables in a process on the result of the process being analyzed. The method in question may be employed to extend our knowledge of newly designed products and their packaging [66, 67].

18.6 Concluding Remarks and Future Perspectives

Product quality assurance is one of the major areas that require the attention of manufacturing companies and constitutes a key factor behind their competitive advantage on the market. In this context, performing actions that aim to ensure the desired quality of products by taking a quality-oriented approach to product designing seems to be a proper solution. That view stems from the fact that designing is a very important stage of quality assurance, as errors made at this level will lead to long-term consequences, which at further stages might not be eliminated so easily. The concept of integrated product designing taking into consideration the needs and requirements of consumers requiring a quality-oriented approach, which is characterized by a close relation to the set of actions and activities aimed at satisfying consumers both in terms of products and their packaging. This is because all consumers' needs and requirements, which are closely correlated with market trends, form – in addition to modern technologies and companies' pursuit of competitive advantage – a sound foundation for product design, while providing an area for seeking innovative solutions. The integrated product-designing model developed makes it possible to formulate original and effective theoretical concepts on the one hand and put forward specific practical solutions on the other. Future research into the designing of integrated products must focus on the identification of imperfections in foodstuffs and their packaging dedicated to individual consumer groups with a view to implementing actions designed to streamline the quality assurance process.

Acknowledgments

The publication was financed from the subsidy granted to the Cracow University of Economics. The research was conducted in the frame of Potential Program No. 68/ZJO/2020/POT, grant title: "Managing the safety and quality of food and its packaging in the system aspect."

Glossary

An integrated product It is a product along with an immediate (unit) packaging.

Designing It is to create, fashion, execute, or construct according to plan.

Quality assurance A set of planned and systematic actions that are necessary to ensure that quality requirements for an end product are satisfied throughout its creation process.

Quality function deployment method (QFD) It is a structured method that uses the seven management and planning tools to identify and prioritize customers' expectations quickly and effectively.

Quality-oriented methods These are understood as a general course of action to be followed in order to achieve a goal set for the quality of a product being created.

New product development It covers the complete process of bringing a new product to market, renewing an existing product, or introducing a product in a new market.

Packaging It shall mean all products made of any materials of any nature to be used for the containment, protection, handling, delivery, and presentation of goods, from raw materials to processed goods, from the producer to the user or the consumer. "Non-returnable" items used for the same purposes shall also be considered to constitute packaging.

Packaging design It is the process that allows your packaging to fulfill a wide range of functions.

Product It is an article or substance that is manufactured or refined for sale.

Research and Development (R&D) It invents the finished product to meet the customer's expectations.

References

1 N. R. Tague, *The Quality Toolbox*, 2nd edn, ASQ Quality Press, **2013**.

2 P. Karipidis, K. Athanassiadis, S. Aggelopoulos, E. Giompliakis, *Food Control*. **2009**, https://doi.org/10.1016/j.foodcont.2008.02.008.

3 J. M. Juran, J. A. De Feo, *Juran's Quality Handbook*, Sixth edn, McGraw Hill, **2010**.
4 X. Niu, S. Qin, H. Zhang, M. Wang, R. Wong, *Adv. Mech. Eng.* **2022**, https://doi.org/10.1007/s00170-021-07945-z.
5 A. Milton, P. Rodgers, *Product Design*, Laurence King Publishing, **2011**.
6 A. Eyal, *Hooked: How to Build Habit-Forming Products*, NirAndFar, **2014**.
7 D. Montgomery, *Design and Analysis of Experiments*, 8th edn, Wiley & Sons, New York, **2012**.
8 A. Fuad-Luke, *The Eco-design Handbook*, Thames Hudson, **2002**.
9 J. Liedtka, R. Salzman, D. Azer, *Design Thinking for the Greater Good: Innovation in the Social Sector*, Columbia Univers Press, **2017**.
10 S. Tornincasa, *Technical Drawing for Product Design*, Springer, **2020**.
11 M. Jackson, *Critical System Thinking and the Management Complexity*, John Wiley & Sons, **2019**.
12 R. Dumpala, M. J. Bhavsar, C. Patil, *Int. Jour. Of Trend in Scient. Res. And Dev. (IJTSRD).* **2020**, 4, 32985.
13 K. Mertins, R. Jochem, *Qual. Orient. Des. Bus. Proc.* **1999**, https://doi.org/10.1007/978-1-4615-5071-6_3.
14 L. Bix, J. Fuente, R. P. Sundar, H. Lockhart, Packaging design and development, in *The Wiley Encyclopedia of Packaging Technology* (Ed: K. L. Yam), Wiley and Sons, **2009**, Ch. 30.
15 B. Bergman, B. Klefsjö, *Quality from Customer Needs to Customer Satisfaction*, Gazelle Book Service, Germany, **2010**.
16 F. Tiago, *Product Design Process*, Cute Ediciones Srl., **2019**.
17 S. Selikoff, *The Complete Book of Product Design, Development, Manufacturing, and Sales*, Product Development Academy, **2021**.
18 L. Di Lucchio, Presented at 2nd *International Conference on Designing Food and Designing for Food*, London, Nov **2012**.
19 P. Rodgers, *Research Methods for Product Design*, Laurence King Publishing, **2013**.
20 S. Stummerer, M. Hablesreiter, *Food Design XL*, Springer, Wien **2010**.
21 K. Grunert, W. Verbeke, J. Kügler, F. Saeed, J. Scholderer, *Meat Science.* **2011**, 89, 21605939.
22 K. Grunert, *Encyclopedia of Agriculture Food Systems* **2014**, https://doi.org/10.1016/B978-0-444-52512-3.00062-0.
23 R. Hernández, R. Cooper, B. Tether, E. Murphy, *J. The Journal of Design, Economics, and Innovation.* **2018**, https://doi.org/10.1016/j.sheji.2018.06.001.
24 T. Toshiharu, N. Yukari, *International Journal of Design Creativity and Innovation.* **2017**, https://doi.org/10.1080/21650349.2017.1313132.
25 R. Traill, *Thinking by Molecule, Synapse, or Both? — From Piaget's Schema, to the Selecting/Editing of ncRNA*, Ondwelle Press, Melbourne, **2008**.
26 J. Fuente, S. Gustafson, C. Twomey, L. Bix, *Packaging Technology Science.* **2014**, https://doi.org/10.1002/pts.2087.
27 P. H. Bloch, *Journal of Product Innovation Management.* **2011**, https://doi.org/10.1111/j.1540-5885.2011.00805.x.
28 C. Elverum, C. Welo, S. Tronwoll, *Procedia CIRP.* **2016**, https://doi.org/10.1016/j.procir.2016.05.010.
29 M. Kim, H. Park, Y.-W. Sawng, S.-Y. Park, *Sustainability.* **2019**, https://doi.org/10.3390/su11226267.
30 M. Halagarda, *British Food Journal.* **2017**, http://doi.org/10.1108/BFJ-07-2016-0338.

31 D. C. Wynn, P. J. Clarkson, Research in Engineering Design. **2018**, https://doi.org/10.1007/s00163-017-0262-7.

32 A. Rudawska, N. Čubaňova, K. Pomarańska, D. Stanĉeková, A. Gola, *Advances in Science and Technology Research Journal*. **2016**, http://doi.org/10.12913/22998624/62513.

33 A. Cholewa-Wójcik, *Packaging and Its Role in the Integrated Product Design Having Considered the Aspect of Consumers' Needs and Requirements*, Polish Food Technologists Publishing House, Cracow, **2018**.

34 K. Stylidis, C. Wickman, R. Söderberg, *Journal of Engineering Design*. **2019**, http://doi.org/10.1080/09544828.2019.1669769.

35 A. Cholewa-Wójcik, A. Kawecka, Presented at *Conference: Business Logistics in Modern Management*, Osijek **2014**.

36 G. L. Robertson, *Food Packaging and Shelf Life. A Practical Guide*, CRC Press, Ney York, **2009**.

37 G. L. Robertson, *Food Packaging: Principles & Practice*, CRC Press, New York, **2013**.

38 K. Grunert, H. C. M. van Trijp, Consumer-Oriented New Product Development, in *Encyclopedia of Agriculture and Food Systems*, Vol. 2 (Eds: K. Neal, V. Alfen), Academic Press, San Diego, CA, **2014**, Ch. 10.

39 D. Man, *Shelf Life*, second edn, Wiley & Sons, London, **2015**.

40 D. Baiardi, R. Puglisi, S. Scabrosetti, *Food Quality and Preference*. **2016**, https://doi.org/10.1016/j.foodqual.2015.11.011.

41 R. Shewfelt, A. Ota-Ramirez, A. Clarke, *Introducing Food Science*, CRC Press, 2016.

42 M. Joardder, C. Kumar, M. A. Karim, *Critical Reviews in Food Science and Nutrition*. **2017**, https://doi.org/10.1080/10408398.2014.971354.

43 D. Petrescu, I. Vermeir, R. Petrescu-Mag, *International Journal of Environmental Research and Public Health*. **2020**, https://doi.org/10.3390/ijerph17010169.

44 A. Sameen, A. Sahar, F. Tariq, U. M. Khan, T. Tariq, B. Ishfaq, Food Microbial Hazards, Safety, and Quality Control: A Strategic Approach, in *Food Safety Practices in the Restaurant Industry* (Eds: S. Nurhayati Khairatun, A. Zakiah Abu Bakar, N. Azira Abdul Mutalib, U. Fatimah Ungku Zainal Abidin), IGI Global, Pakistan, **2022**, Ch. 4.

45 W. Rijswijk, L. J. Frewer, *British Food Journal*. **2018**, https://doi.org/10.1108/00070700810906642.

46 T. Sadilek, *European Research Studies Journal*. **2019**, https://doi.org/10.35808/ersj/1407.

47 J. Kaspar, D. Baehre, M. Wielhaber, *Procedia CIRP*. **2016**, https://doi.org/10.1016/j.procir.2016.04.181.

48 H. Jayan, J. A. Moses, C. Anandharamakrishnan, Testing methods for packaging materials, in *Bio-based Materials for Food Packaging* (Ed: S. Ahmed), Springer Nature, Singapore, **2018**.

49 K. Young Teck, M. Byungjin, K. Kyung Won, General Characteristics of Packaging Materials for Food System, in *Innovations in Food Packaging*, 2nd edn. (Ed: J. H. Han), Academic Press, **2014**, Ch. 2.

50 D. Heldman, D. B. Lund, C. Sabliov (Eds), *Handbook of Food Engineering*, 3rd edn, Routledge, **2019**.

51 Regulation (EC) No 1935/2004 of the European Parliament and of the Council of 27 October 2004 on materials and articles intended to come into contact with food and repealing Directives 80/590/EEC and 89/109/EEC.

52 S. Mc Curdy, J. Peutz, G. Wittman, *Storing Food for Safety and Quality*, University of Idaho Press, Idaho, **2009**.

53 A. Cholewa-Wójcik, A. Kawecka, C. Ingrao, V. Siracusa, *Journal of Entrepreneurship, Management and Innovation.* **2019**, https://doi.org/10.7341/20191519.

54 K. Marsh, B. Bogusu, *Journal of Food Science.* **2007**, https://doi.org/10.1111/j.1750-3841.2007.00301.x.

55 A. Cholewa-Wójcik, A. Kawecka, M. Wojnarowska, Presented at *International Scientific Conference on Collaborative Innovation Development*, Bialystok, **June 2019**.

56 S. B. Keller, B. C. Keller, *The Definitive Guide to Warehousing: Managing the Storage and Handling of Materials and Products in the Supply Chain*, 1 edition, Pearson FT Press, New Jersey, **2013**.

57 A. Cholewa-Wójcik, A. Kawecka, in *Sustainable Products in the Circular Economy Impact on Business and Society* (Eds: M. Wojnarowska, M. Ćwiklicki, C. Ingrao), Routlegde, London, New York, **2022**.

58 A. Bolar, S. Tesfamariam, R. Sadiq, *International Journal of Sustainable Built Environment.* **2017**, https://doi.org/10.1016/j.ijsbe.2017.02.002.

59 M. Eldermann, A. Siirde, J. Gusca, *Energy procedia.* **2017**, https://doi.org/10.1016/j.egypro.2017.09.060.

60 C. Shu, *Differential Quadrature and Its Application in Engineering*, Springer, Berlin, Germany, **2000**.

61 M. Cauchick, *International Journal of Quality & Reliability Management.* **2005**, https://doi.org/10.1108/02656710510573002.

62 A. Pinto, C. Paiva, Food Science and Technology (Campinas). **2019**, https://doi.org/10.1590/S0101-20612010000500007.

63 N. Muda, N. Roji, Journal of Industrial Engineering. **2015**, https://doi.org/10.1155/2015/789362.

64 K. R. Hofmeister, Quality function deployment. Market success through customer-driven products, in *Food Product Development from Concept to the Marketplace* (Eds: E. Graf, I. Saguy), Van Nostrand Reinhold, New York, **1999**.

65 G. Taguchi, S. Chowdhury, Y. Wu, *Taguchi's Quality Engineering Handbook*, John Wiley & Sons Press, London, **2005**.

66 Y. Moubachir, D. Bouami, *Procedia CIRP.* **2015**, https://doi.org/10.1016/j.procir.2014.07.172.

67 K. Murata, *Procedia Manufacturing.* **2017**, https://doi.org/10.1016/j.promfg.2017.07.232.

19

Determinants of the Quality and Safety of Food Packaging

Agnieszka K. Kawecka and Agnieszka S. Cholewa-Wójcik*

Cracow University of Economics, Institute of Quality Science and Product Managements, Department of Products Packaging, 27 Rakowicka Street, Cracow, Poland
** Corresponding author*

Highlights

- Packaged food products should be considered as an integrated product with their packaging. Due to the interactions and protective functions of the package, there is a need for a comprehensive view on that matter.
- Food-contact materials, including packaging, are subject to legal regulations in many countries, and although these regulations differ from each other, their primary goal is to ensure an appropriate level of protection of the health and life of consumers.
- Technical aspects of the safety of packaging, including their chemical, sensory, or physical characteristics, are just as important as the systemic aspects that allow for identification, guaranteeing the reliability of the information provided and carrying out activities aimed at ensuring food-contact materials safety.

19.1 Introduction

Ensuring food safety is one of the basic problem areas of both scientific and practical considerations; consumers also pay attention to this issue and expect safe food products. Providing consumers with healthy, tasty, nutritious, and safe food is a priority for food companies operating around the world. This problem is addressed in legal acts by governments, scientific considerations, tasks of market surveillance institutions, nongovernmental organizations, education, and many others.

Packaging is one of the elements that make possible to provide consumers with food products of the right level of quality. Packaging is often treated as a barrier, considered mainly in the context of protecting the product against external factors, contamination, and loss of food properties. However, improper packaging can endanger packaged goods. On the one hand, it may not be a sufficient tool to protect the product; on the other hand, it may itself be a hazard or a carrier of a threat to food safety. Therefore, it has become important to develop a set of requirements, activities, and supervision measures ensuring the appropriate level of safety of

Materials Science and Engineering in Food Product Development, First Edition. Edited by Wing-Fu Lai.

packaging intended for contact with food. This chapter aims to indicate the most important determinants of the safety of food packaging, conditions, and actions that must be met in order for the packaging on the market to be an element supporting ensuring food safety.

19.2 Literature Review Concerning Food-Packaging Safety

Packaging can be considered as a stand-alone product present in the market. However, it seems more common and deeply justified to approach as to which packaging is considered together with the packaged product as an integrated product [1]. In particular, such an approach is justified while considering food packaging; in packaging, many works are devoted to considering (i) the impact of packaging on food and its interaction [2, 3]; (ii) research during storage of food products in polymer packaging [4]; (iii) overall and specific migration of polymer components into food [5]; (iv) plastic packaging materials and their chemical reactions [6]; (v) performance of different materials as a packaging for food products and pondered bio-based polymer materials as food packaging [7]. Different effects of packaging on the quality of the packaged product are considered: one of the most serious influences addressed by a large number of studies is the migration of packaging material components to packaged food [8], considered mostly testing methods used for migration research [9]; in those considerations also, chemical contamination was raised as well as in publication [10]; considered also migration in beverage packaging [11].

However, the effect of the interaction of the product with its packaging may also be changed in its sensory characteristics [12]; changes in taste and smell were investigated [13, 14]. The problem of flavor scalping caused by food packaging was also investigated [15]. Some authors in the field of packaging also draw attention to the fact that packaging has become a guarantee of the integrity of packaged products [16–18]. The full spectrum of interactions between food, packaging, and environment is described in Figure 19.1.

The protective function of the packaging consists primarily in protecting against quantitative losses, maintaining food quality, nutritional value, and sensory characteristics [19–23]. The overall properties of packaging materials and packaging significantly extend the shelf life of food products [24–26]. It is impossible to indicate all publications in the field of packaging safety; the scope presented in the literature is very diverse; the issues raised in the considerations are treated with varying degrees of detail.

Industrial Corner

- Packaging is seen as one of the basic means of extending the shelf life of packaged food products. Properties of packaging materials and packages have the ability to protect its quantity and quality.
- Food and packaging interact, so the features of packaged food may change. Awareness of changes in food products should affect the selection of appropriate packaging.
- The proper selection of packaging for the packed product is determined by many variables, and the conditions in which it will be used.
- It should not be forgotten that there are interactions between the product and the packaged product during storage. Negative effects on the safety of packaged food should be supervised and limited.

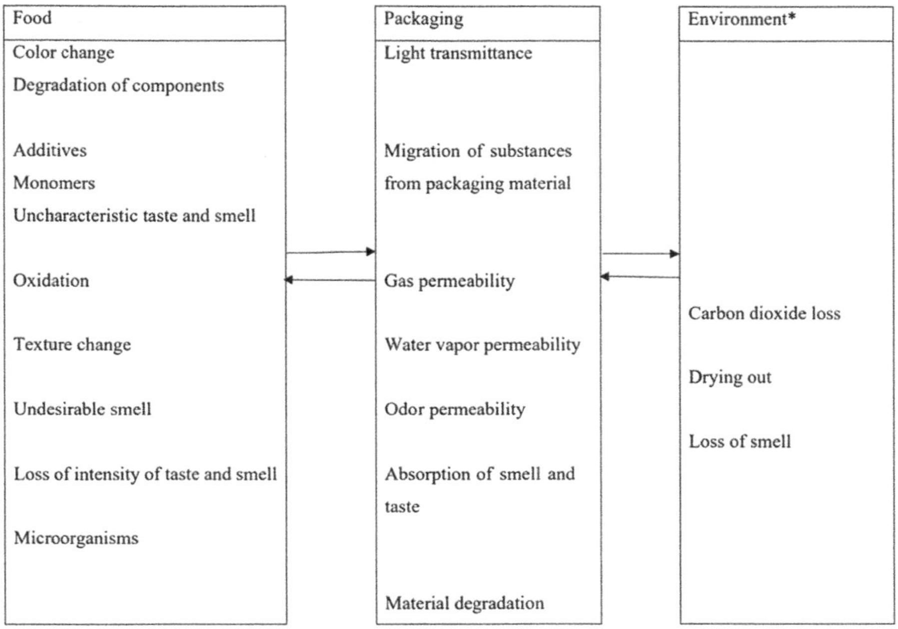

* The impact of the environment on the packaging has been deliberately disregarded. This is not directly under consideration.

Figure 19.1 Interactions of packaging material and food.

19.3 Packaging Safety Hazards

The hazard is the potential to cause harm, which may be biological, chemical, or physical [27]. Among hazards to the safety of food packaging, two groups of them should be mentioned. The first is due to the presence of factors on or in the packaging and on or in the packed product that may cause damage. The factors may be biological, physical, or chemical in nature, derived from:

- surroundings
- the outer surface of the packaging
- the inside of the packaging material
- the inner surface of the packaging
- printing inks (in the case of printed materials)
- adhesive layers (in the case of laminated packaging materials).

The packaging can be a source of biological contamination – pathogenic, saprophytic microorganisms, or molds. Microorganisms from food packaging can be transferred through contact of the surface of contaminated packaging material with food. Research on the presence of microorganisms on the surface of packaging has been carried out by many researchers [28–30]. The adhesion and detachment of bacteria and spores on food-contact surfaces have been investigated [31]. It is known that bacteria die on the surface of packaging due to the influence of the environment, but the rate of their dieback varies and depends on many factors [32]. Among the threats of a biological nature, the presence of pests can also be indicated [33, 34]. The main physical hazard is the presence of foreign bodies in packaged food. In food, foreign bodies can come from the raw material or the packaging material. Packaging may be a source of glass, plastic, metal, and wood fragments, the presence of which in the product may result from breaking or tearing off pieces of the packaging [35].

Chemical hazards indicating that the packaging may be a source of contamination are an extremely important problem. Various chemical compounds may migrate from the packaging material to the packed food, including those remaining in the material after the production process, heavy metals, or other additives [36]. The most frequently indicated hazards in the field of chemical hazards, researched and studied in the literature, is the migration of packaging material components to food [37, 38]. The phenomenon of migration is particularly important in the case of polymer materials, but it also occurs in other materials, regardless of the origin of the raw materials. The ability to migrate is demonstrated mainly by low molecular weight substances [39]. Also environmental and product factors influence the migration of packaging material components, such as contact time, temperature, water activity, pH, and food composition [40].

The second group of threats may be the improper condition of the packaging, causing negative health effects. This condition may result from the lack of a guarantee with regard to the integrity of the packaging, the lack of authenticity of the product, or defects of the packaging critical for safety and consumer requirements [41, 42]. Hazard classification for the purposes of ensuring the safety and certification of enterprises in the food-packaging industry has been proposed in the BRC Global Standard for Packaging and Packaging Materials. It makes it possible to make a decision on the hazard category that the packaging poses for a packed product through a decision tree (Figure 19.2). By answering the questions, it is possible to determine whether the packaging requires special hygienic conditions during production, storage, and transport. In addition, thanks to this classification, it is possible to adapt production procedures to the more stringent conditions for packaging intended for direct contact with very sensitive products, such as foodstuffs [27].

Industrial Corners

- Identification of hazards related to packaging is individual and should be carried out in each enterprise, taking into account the conditions of its activity.
- The determination of the risk level should take into account all the necessary aspects of the company's operations, the type of packaging, and the type of packed products.
- Hazard analysis should take into account different groups of hazards, such as biological, chemical, and physical threats coming from different packaging elements.
- Hazard analysis is a very individual procedure and each company should carry it out taking into account its own raw materials, auxiliary materials, processes, and operating conditions.

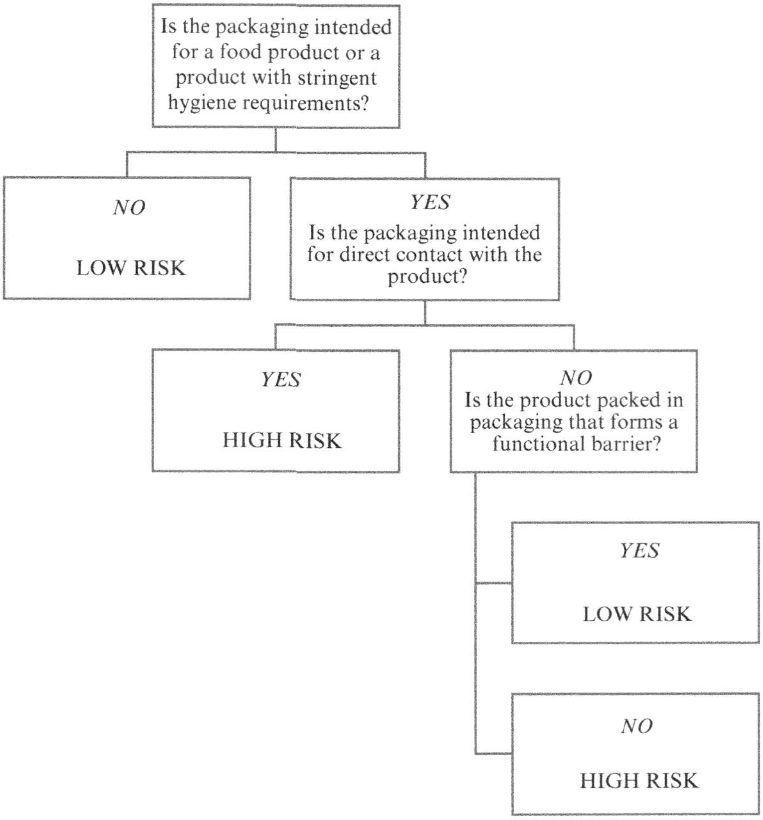

Figure 19.2 Decision tree supporting the definition of the packaging category [43].

19.4 Legal Requirements for the Safety of Food Packaging

Detailed legal requirements have been developed for food packaging, the fulfillment of which should guarantee the safe use of these packaging. The European Union, the United States, and Canada have adopted appropriate legal regulations in the field of production and marketing of materials intended for contact with food. In the United States, in accordance with the Food Drug and Cosmetic Act (FD&C Act), each substance intended for the production of packaging materials for food must be on the list of approved substances. Substances generally recognized as safe can be safely used as components of materials that contact food under conditions of good manufacturing practice. In the United States, a document such as the declaration of compliance is not compulsory for food-contact materials but customers may ask for a letter of guarantee certifying that material is acceptable for the intended food-contact use [44].

In Canada, the regulatory framework for packaging materials is established by the Canadian Food and Drugs Act (F&D Act) and the Food and Drug Regulations. Section 4 of the F&D Act prohibits the sale of food that [45]:

- contains any poisonous or harmful substance
- is unfit for human consumption
- consists in whole or in part of any filthy, putrid, disgusting, rotten, decomposed, or diseased animal or vegetable substance
- is adulterated
- was manufactured, prepared, preserved, packaged, or stored under unsanitary conditions.

The safety of all materials used for packaging foods is controlled under Division 23 of the F&D Act and Regulations, Section B.23.001, which prohibits the sale of foods in packages that may impart any substance to the contents that may be harmful to the consumer of the food. While it is the responsibility of the food seller (manufacturer, distributor) to ensure the safety of packaging material and compliance with those requirements, in Canada, there is no positive list of substances or additives that are permitted for use in manufacturing food-contact materials. Food-contact notification or testing requirements are also nonmandatory. For food-contact materials on the Canadian market, there is a need to assure customers that they are compliant with relevant food-contact regulations. On a voluntary basis, a pre-market assessment of manufacturers and suppliers can submit for evaluation products, to get a letter of no objection as confirmation of conformity. Moreover, there is a need for packaging consideration in the case of the safety evaluation process with regard to infant formula, foods for special dietary use, and novel food.

In China, food-contact materials are regulated as food-related products. The China Food Safety Law prohibits the importation, use, or sales of food-related products, including food-packaging materials, not compliant with national Food Safety Standards. GB 4806.1–2016 National Food Safety Standard: General Safety Requirements on Food Contact Materials and Articles establishes general safety requirements that apply to all kinds of food-contact materials. A positive list of accepted additives exists, and it can be used to produce food-contact materials, their scope of use, and restrictions, by standard GB 9685–2016. China has also adopted standards for 10 material groups and standard testing and evaluation methods [46, 47].

In the European Union, the basic legal requirements for food-contact packaging are included in Regulation (EC) No. 852/2004 of the European Parliament and of the Council of April 29, 2004 on the hygiene of foodstuffs. Fundamental requirements are [48]:

- The material used for the production of unit packages and collective packages must not be a source of contamination.
- Materials for the production of unit packaging must be stored in such a way that it is not exposed to the risk of contamination.
- Work related to unit and collective packaging must be carried out in such a way as to prevent contamination of the products. The integrity of the packaging structure and its cleanliness must be ensured.
- The reusable material used for the production of the unit and collective packaging for foodstuffs must be easy to clean and, if necessary, disinfect.

In addition, Regulation (EC) No. 1935/2004 of the European Parliament and of the Council lays down requirements [49]:

- only substances on the appropriate lists are used in the production and processing of materials and articles intended to come into contact with food;
- community authorities granting authorizations for the use of a new substance – applications for authorization in Poland should be submitted to the Chief Sanitary Inspector, who addresses them to the European Food Safety Authority (EFSA);
- labeling of materials and products;
- declaration of compliance of materials and products with the applicable legal regulations;
- traceability;
- supervision and official control measures.

Regulation (EC) No. 1935/2004 also introduces the obligation to apply the principles of good manufacturing practice (GMP) in the production of food-contact materials, including packaging. Due to the high importance of the application of these principles, the Commission Regulation (EC) No. 2023/2006 of December 22, 2006 on good manufacturing practice for materials and articles intended to come into contact with food was adopted; this act defines GMP. In addition, further regulations concerning specific material groups are introduced [50].

It depends, therefore, that ensuring the safety of food packaging relates primarily to:

- appropriate chemical composition of packaging materials, free from or maintaining the content of potentially hazardous compounds at a sufficiently low level,
- production of food-packaging materials in accordance with GMP to ensure the right level of quality,
- maintaining sanitary conditions when packing and storing packed goods,
- allowing for unambiguous identification of the packaging and its intended use as a material for contact with food.

The multitude of legal requirements in various countries means that consumers all over the world are not protected to the same level, and their health and life may be at risk. Such differences are also a serious trade barrier, as producers have to adapt their activities to the regionally applicable regulations. However, it does not seem feasible to create a universal set of rules for the safety of materials intended for contact with food at the international level.

Industrial Corner

- Food-packaging company entering the international market should familiarize itself with the legal regulations in force in the importer's country and adapt its activities to the applicable legal regulations prevailing on the selected market.
- Regulations in different countries have different levels of detail; different types of documents are required, and rules regarding basic properties, such as chemical composition, may differ.
- The universal rule is the fact that food packaging should not pose any dangers. However, legal remedies used by governments may vary in this regard.

19.5 The Process of Ensuring Security – the Supply Chain

Packaging safety threats can occur at different stages of the packaging supply chain. The physical flow in the packaging supply chain takes place from suppliers of raw materials and packaging components to manufacturers of packaging materials and packaging or without their intermediation to packaging users who form packaging or fill ready-made packaging in the packaging process. Packaged products are delivered to consumers, who, after the products have been used up, hand over the packaging waste to companies in the waste management sector. The participants of the packaging supply chain are influenced by the environment, which is constituted by legislative bodies, state administration institutions, national and international organizations, associations of producers, distributors, and consumers [51]. The structure of the packaging supply chain is created by enterprises related to the entire life cycle of the packaging [52], which is presented in Figure 19.3.

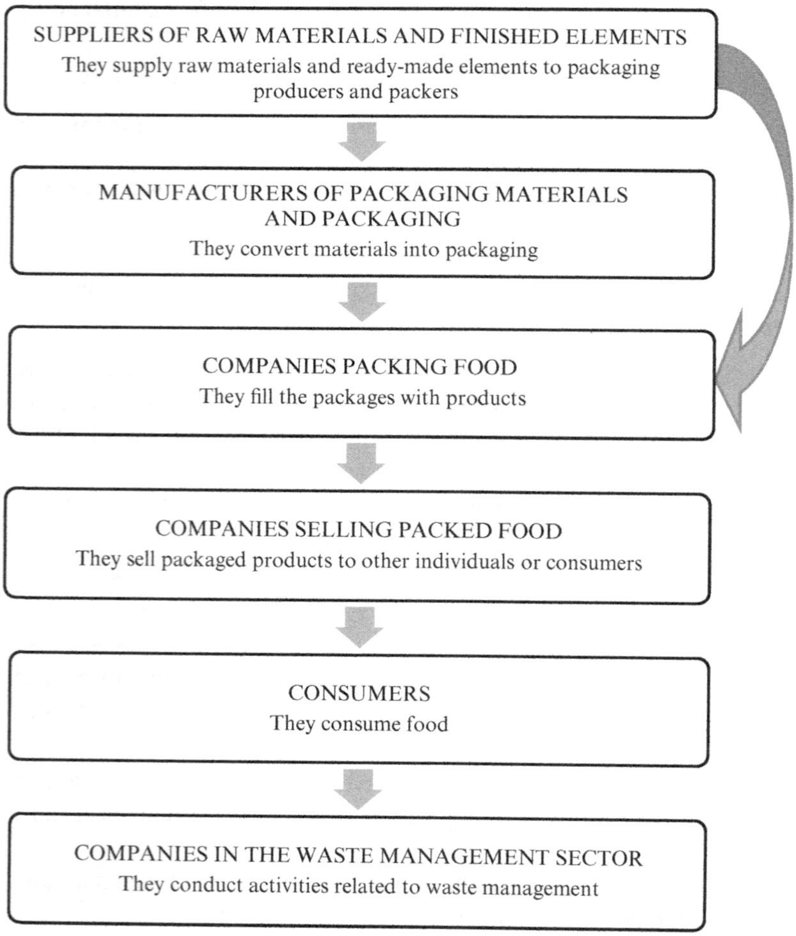

Figure 19.3 Structure of the packaging supply chain.

According to the ISO 22000: 2018 standard, the food chain is defined as: "the sequence of stages and processes taking place in the production, processing, distribution, storage and handling of food and its ingredients, from primary production to consumption." Organizations belonging to the food chain form "a range from producers of feed and primary products, through producers of foodstuffs, transport and storage operators and their subcontractors, to the retail sale of food products and services (including organizations indirectly related to the sector, such as producers of machinery and equipment, packaging materials, cleaning agents, ingredients and additives)." Therefore, the food-packaging supply chain should be considered as part of the food chain [53].

The packaging supply chain can be analyzed in terms of logistics. This is because there is a physical flow, the accompanying flow of information, a sequence of processes, relations between entities, entry and exit in the system, or in the system, and the environment influencing it [43]. Different stages in the packaging supply chain have a different impact on the safety of the packaging materials and packages transferred within. In the initial stage of raw material suppliers and packaging components, entrepreneurs are responsible for the supply of raw materials with appropriate chemical properties, as well as the delivery of packaging elements of appropriate design and quality, free from defects. Attention should be paid to the role of suppliers of raw materials for the production of packaging, who are responsible for the chemical properties of these raw materials (e.g. by proper conduct of the process of obtaining the material), and suppliers of elements, who are responsible for the safe use of these elements.

In the next stage, producers of packaging materials and packaging are responsible for the proper selection of substances for the production of packaging materials and packaging, e.g. the use of lists of substances approved for contact with food. Packaging manufacturers must ensure the appropriate quality of the packaging so that, for example, during use, the packaging does not have sharp edges. In addition, it is the responsibility of the manufacturers of packaging materials to ensure appropriate labeling to facilitate the "traceability" of packaging materials and packaging. Producers are also obliged to protect the produced packaging materials and packaging against contamination, in particular, to ensure proper hygienic conditions to prevent contamination of packaging materials with pathogenic microorganisms or accelerating food spoilage processes.

The proper selection of packaging materials and packaging, which is crucial for the safety of packaged products, rests with the companies that pack the products. Their tasks concern the selection of packaging materials, taking into account the features of the packed products and possible interactions between the product and the packaging, as well as the influence of the packaging material on the organoleptic properties of the packed products. Manufacturers of packaged products must in their assessments, in addition to the properties of the packaged product, also take into account the contact time of the packaging with the product, the conditions under which this contact occurs and the way of using the packaging during product use (e.g. cooking food in microwave ovens in packaging). Producers of packaged products are responsible for the amount, content, and scope of information placed on the packaging, some of them have a very significant impact on the safe use of packaged products. Some of this information is necessary for the safe use of the products. Warnings, often expressed in the form of pictograms, graphic signs or combinations of graphic signs, and verbal information, constitute a very important element of the safe use of products. All information should be provided in a legible, visible, and understandable form. Products with special requirements may be equipped by manufacturers with additional elements, ensuring safety. Such elements include, among others, closures protecting against the unauthorized opening, e.g. by children, or guarantees of the first opening, which significantly affects the safety and guarantees safe use of the product for the consumer. In addition, manufacturers, in order to protect their own economic interests, their brand, and, above all, to protect consumers, should also guarantee the authenticity of products through appropriate elements that will protect the packaging against counterfeit products.

Retailers should ensure proper storage, transport, and display conditions of the packed products, as indicated by the manufacturer, and follow the information on the packaging. It is particularly important that consumers also adhere to any information on the packaging. Proper education and awareness of risks among consumers is essential so that the products they use do not endanger their health. By analyzing the packaging supply chain, it is possible to notice the importance and responsibility of each of the links in order to provide the end user with safely packaged food. The obligations of entrepreneurs forming the supply chain may also vary depending on local and applicable legal requirements. It should also be noted that in the packaging supply chain, some entities implement and use safety or quality assurance systems. In some countries, these systems are mandatory, such as the HACCP system in food production companies in both the United States and European Union countries. Some companies implement voluntary systems and standards, including ISO 22000, BRC Food, IFS Food, and many others. Important activities of safety and quality management systems were shown in the studies of Luning et al. [54] and Kawecka et al. [55]. The implementation of such systems significantly improves the quality and safety of the activities carried out, and, as a result, increases the level of safety of the final product. Each stage has assigned responsibilities, the proper implementation of which contributes to the achievement of a complex goal.

Industrial Corners

- The process of ensuring the safety of food packaging takes place in all stakeholders of the supply chains; each of them has a different scope and different activities in this area.
- Packaging material producers and packaging users are primarily responsible for ensuring the proper selection of raw materials, the proper conduct of the production process, and the selection of appropriate packaging for packed products.
- The purpose of implementing systems and standards such as GMP or BRC is to ensure the basic safety of the produced packaging materials and packaging.

19.6 Packaging Safety Features and Attributes of Food Packaging

All packaging available on the market should be safe, that is "under normal or other reasonably foreseeable conditions of its use, taking into account the time of use of the packaging, as well as depending on the type of packaging and the type of the packed product, it does not pose any risk for the consumer or poses a negligible risk, compatible with its normal use and taking into account the high level of requirements for the protection of human health and life" [41]. This means that the following areas should be considered when assessing the safety of packaging:

- features of the packaging;
- interaction with the packaged product;
- labeling, warnings, and instructions for the use of the packaging and the packed product, and all information about the packed product (also regarding its authenticity);
- categories of product users exposed to possible danger (children, elderly, people with sight disabilities).

Among the important properties of packaging, first of all, one should mention their chemical, physical, or mechanical properties. They must comply with the legal regulations adopted in a given country/region (such as the content of potentially dangerous ingredients) and meet the requirements for the protection of packaged goods. The analysis of the legal requirements appearing in various countries, the review of hazards indicated in the literature on the subject, and the analysis of the role and responsibility of various entities in the packaging supply chain led to a comprehensive presentation of the characteristics of safe packaging for food products. These features are called the safety attributes of food packaging. Food-packaging safety attributes relate to three areas: packaging and packaging materials' technical attributes, secondary packaging, and documentation technical attributes, and system attributes, which are presented in Figure 19.4.

To the group of packaging and packaging materials' technical attributes belong:

- appropriate chemical composition,
- compliance with the total and specific migration limits,
- low level of surface contamination with microorganisms,
- no foreign bodies,

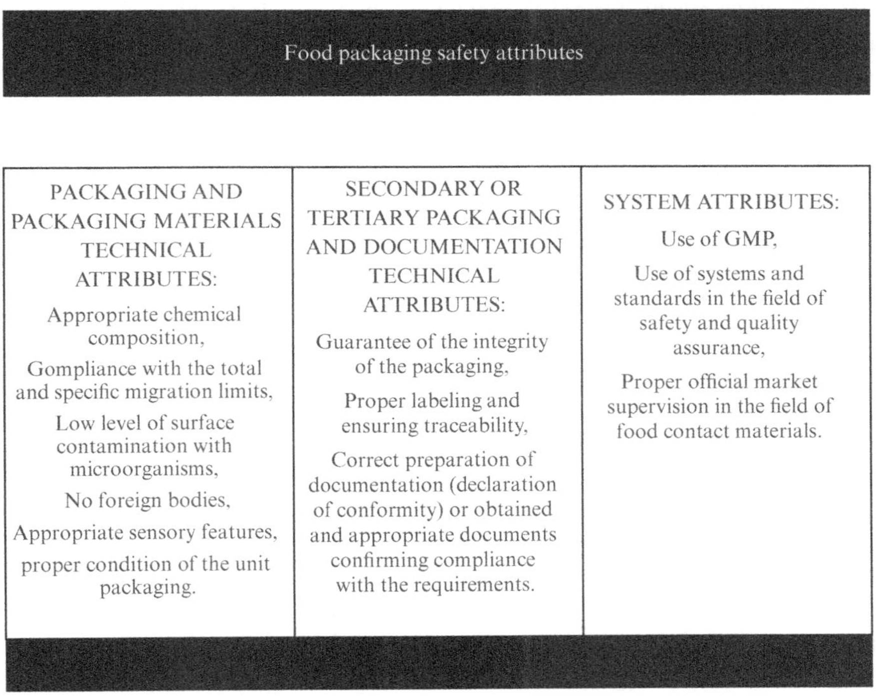

Figure 19.4 Food-packaging safety attributes. *Source*: own elaboration on the basis of [43].

- appropriate sensory features,
- proper condition of the unit packaging.

Those features and properties are basic for safety; they are mostly chemical, physical, and biological attributes. However, these attributes are not sufficient to fully take care of and ensure safety; also, it is important to be sure of authenticity, identity, and traceability, hence identifying secondary or tertiary packaging and documentation technical attributes, such as

- guarantee of the integrity of the packaging
- proper labeling and ensuring traceability
- correct preparation of documentation (declaration of conformity) obtained and appropriate documents confirming compliance with the requirements.

For comprehensive safety assurance, the system attributes should also be filled in; they are: the use of GMP, use of systems and standards in the field of safety and quality assurance, and proper official market supervision in the field of food-contact materials.

19.7 Concluding Remarks

Based on the presented considerations and the attributes of packaging safety, it can be assumed that the determinants of packaging safety are of a technical and organizational nature. In addition, the impact of legal and regulatory requirements in the field of requirements for the receipt, production, circulation, and use of packaging materials and packaging is also clear. Compliance with legal requirements and proper supervision over their compliance, as well as raising awareness among entrepreneurs from the packaging and food industries about their impact and responsibility for the safety of an integrated product, is a challenge for countries, associations, and producer organizations, as well as the world of science. Identification of hazards emerging on the international market, for example, in the Rapid Alert System by Food and Feed, shows that still in terms of the safety of packaging, efforts should be made to reduce the risk of dangerous situations for consumers. In-depth knowledge of the nature of packaging safety determinants, both in the technical sense with the use of advanced chemical, physical, or biological methods, as well as organizational and management methods, is still a challenge for business and science.

Acknowledgments

The publication was financed from the subsidy granted to the Cracow University of Economics.

Glossary

Food chain The sequence of stages and processes taking place in the production, processing, distribution, storage, and handling of food and its ingredients, from primary production to consumption.

Food packaging In its finished state, it is intended to be brought into contact with food or can reasonably be expected to be brought into contact with food or to transfer their constituents to food under normal or foreseeable conditions of use.

Food safety Assurance that food will not cause adverse health effects to the consumer when it is prepared and/or eaten according to its intended use.

Good manufacturing practice (GMP) Those aspects of quality assurance that ensure that materials and articles are consistently produced and controlled to ensure conformity with the rules applicable to them and with the quality standards appropriate to their intended use by not endangering human health or causing an unacceptable change in the composition of the food or causing a deterioration in the organoleptic characteristics thereof.

Hazard Potential to cause harm, which may be biological, chemical, or physical.

Packaging All products made of any materials of any nature to be used for the containment, protection, handling, and delivery and presentation of goods, from raw materials to processed goods, from the producer to the user or the consumer. 'Non-returnable' items used for the same purposes shall also be considered to constitute packaging.

Primary packaging or sales packaging Packaging conceived so as to constitute a sales unit to the final user or consumer at the point of purchase.

Risk A function of the probability of an adverse health effect and the severity of that effect, consequential to a hazard.

Secondary packaging or grouped packaging Packaging conceived so as to constitute at the point of purchase a grouping of a certain number of sales units whether the latter is sold as such to the final user or consumer or whether it serves only as a means to replenish the shelves at the point of sale; it can be removed from the product without affecting its characteristics.

Tertiary packaging or transport packaging Packaging conceived so as to facilitate handling and transport of a number of sales units or grouped packagings in order to prevent physical handling and transport damage. Transport packaging does not include road, rail, ship, and air containers.

Traceability The ability to trace and follow a material or article through all stages of manufacture, processing, and distribution.

References

1 A. Cholewa-Wójcik, A. Kawecka, in *Sustainable Products in the Circular Economy Impact on Business and Society* (Eds: M. Wojnarowska, M. Ćwiklicki, C. Ingrao), Routlegde, London, New York, **2022**.

2 J. H. Hotchkiss, *An Overview of Food and Food Packaging Interactions*, American Chemical Society, Washington, USA, **1988**.

3 J. H. Hotchkiss, *Food Addit. Contam.* **2009**, https://doi.org/10.1080/02652039709374572.

4 K. Barnes, C. Sincair, *Chemical Migration and Food Contact Materials*, Woodhead Publishing Limited, Cambridge, United Kingdom, **2007**.

5 O. G. Pringer, A. L. Baner, *Plastic Packaging Materials for Food*, Wiley, Weinheim, Germany, **2008**.

6 D. Lee, K. Yam, I. L. Piegiovanni, *Food Packaging Science and Technology*, CRC Press, Boca Raton, New York, USA, **2008**.

7 S. K. Vimala Bharathi, J. A. Moses, C. Anandharamakrishnan, in *Bio-based Materials for Food Packaging. Green and Sustainable Advanced Packaging Materials* (Ed: S. Ahmed), Springer, Singapore, **2018**. https://doi.org/10.1007/978-981-13-1909-9_2.

8 R. Frantz, in *Food Packaging Materials. Testing & Quality Assurance* (Eds: P. Singh, A. A. Wani, H.-C. Langowski), CRC Press, Taylor & Francis Group, Boca Raton, New York, USA, **2017**.

9 C. V. Garcia, G. H. Shin, J. T. Kim, *Trends Food Sci. Technol.* **2018**, https://doi.org/10.1016/j.tifs.2018.09.021.

10 R. M. S. Cruz, B. P. M. Rico, M. C. Vieira, in *Food Quality and Shelf Life* (Ed: C. M. Galanakis), Academic Press, London, United Kingdom, **2019**.

11 P. Schmid, F. Welle, *Beverages.* **2020**, https://doi.org/10.3390/beverages6020037.

12 S. E. Duncan, J. B. Webster, *Adv. Food Nutr. Res.* **2009**, https://doi.org/10.1016/S1043-4526(08)00602-5.

13 L. Piergiovanni, L. Amoroso, in *Food Chemistry, Function and Analysis* (Ed: M. Suman), Royal Society of Chemistry, Cambridge, United Kingdon, **2019.**

14 K. A. Thompson-Witrick, E. R. Pitts, J. L. Nemenyi, D. Budner, *Beverages.* **2020**, https://doi.org/10.3390/beverages7020036.

15 T. Nielsen, M. Jägerstad, *Trends Food Sci. Technol.* **1994**, https://doi.org/10.1016/0924-2244(94)90212-7.

16 R. Jansen, G. Schelhove, *Packag.* **1999**, https://doi.org/10.1002/(SICI)1099-1522(199911/12)12:6<255::AID-PTS480>3.0.CO;2-M.

17 D. A. Carus, C. Grant, R. Wattie, M. S. Pridham, *Packag.* **2006**, https://doi.org/10.1002/pts.721.

18 S. Popek, J. Swida, *Eng. Sci. Technol.* **2015**, Nr 1(16), 61–71. https://doi.org/10.15611/nit.2015.1.06.

19 F. A. Paine, H. Y. Paine, *A Handbook of Food Packaging*, Springer, Boston, MA, **1992**.

20 H. Brown, J. Williams, in *Food Packaging Technology* (Eds: R. Coles, D. McDowell, M. J. Karwin), CRC Press, London, United Kingdom, **2005**.

21 H. Brown, J. Williams, M. Kirwan, in *Food and Beverage Packaging Technology*, 2nd edn., Wiley-Blackwell, Hoboken, **2011**.

22 A. Emblem, H. Emblem, *Packaging Technology. Fundamentals, Materials and Processes*, Woodhead Publishing Limited, Cambridge, United Kingdom, **2012**.

23 G. L. Robertson, *Food Packaging. Principles and Practice*, 3rd edn., CRC Press, Boca Raton, USA, **2013**.

24 M. L. Entrup, *Advanced Planning in Fresh Food Industries: Integrating Shelf Life into Production Planning*, Heidelberg, **2005**.

25 K. L. Yam, D. S. Lee, in *Emerging Food Packaging Technologies: Principles and Practice* (Ed: K. L. Yam, D. S. Lee), Woodhead Publishing Limited, Cambridge, United Kingdom, **2012**.

26 R. K. Singh, N. Singh, in *Innovations in Food Packaging* (Ed: J. H. Han), Elsevier Ltd, San Diego, USA, **2014**.

27 *Global Standard Packaging Materials* (Issue 6), British Retail Consortium, London, United Kingdom, **2019**.

28 C. R. Tindale, F. B. Whitfield, S. D. Levingston, T. H. Ly Nguyen, *J. Sci. Food Agric.* **1989**, https://doi.org/10.1002/jsfa.2740490406.

29 G. Bureau, *Food Packaging Technology*, Vol. 1 (Eds: G. Bureau, J.-L. Multon), Wiley, New York, **1996**.

30 O. M. Zacheus, E. Klivanainen, T. K. Nissinen, M. J. Lehtola, P. J. Martikainen, *Water Res.* **2000**, https://doi.org/10.1016/S0043-1354(99)00113-X.

31 C. K. Bower, J. McGuire, M. A. Daeschel, *Trends Food Sci. Technol.* **1996**, https://doi.org/10.1016/0924-2244(96)81255-6.

32 F. Pérez-Rodríguez, A. Valero, E. Carrasco, R. M. García, G. Zurera, *Trends Food Sci. Technol.* **2007**, https://doi.org/10.1016/j.tifs.2007.08.003.

33 X. Hou, P. Fields, W. Taylor, *J. Stored Prod. Res.* **2004**, https://doi.org/10.1016/S0022-474X(02)00063-2.

34 J. Riudavets, I. Salas, M. J. Pons, *J. Stored Prod. Res.* **2007**, https://doi.org/10.1016/j.jspr.2007.03.006.

35 R. A. Marsch, R. E. Angold, in *Detecting Foreign Bodies in Food* (Ed: M. Edwards), Woodhead Publishing Limited, Cambridge, United Kingdom, **2004**.

36 S. Arvanitoyannis, L. Bosnea, *Crit. Rev. Food Sci. Nutr.* **2010**, https://doi.org/10.1080/10408690490424621.

37 R. Golden, *Food packaging safety*, Redtech Report, March/April **2007**.

38 R. Franz, A. Strömer, in *Plastic Packaging Materials for Food* (Eds: O. G. Pringer, A. L. Baner), Wiley, Weinheim, Germany, **2008**.

39 J. H. Petersen, in *Food Safety: Contaminations and Toxins* (Ed: J. P. F. D'Mello), CAB International, Oxon, **2003**.

40 K. Cooksey, in *Intelligent and Active Packaging for Fruit and Vegetables* (Ed: C. L. Wilson), CRC Press, Boca Raton, London, New York, **2008**.

41 M. Lisińska-Kuśnierz, *Społeczne aspekty w opakowalnictwie (Social aspects in packaging)*, Cracow University of Economics Publishing House, Cracow, Poland, **2010**.

42 J. Plimmer, in *Trends in Packaging of Food, Beverages and Other Fast-Moving Consumer Goods (FMCG). Markets, Materials and Technologies* (Ed: N. Farmer), Woodhead Publishing, Cambridge, United Kingdom, **2013**.

43 A. Kawecka, *Czynniki determinujące bezpieczeństwo opakowań przeznaczonych do kontaktu z żywnością (Factors determining the safety of packaging intended for contact with food)*, Cracow University of Economics Publishing House, Kraków, Poland **2014**.

44 Food Drug and Cosmetic Act (FD&C Act), United States of America.

45 Food and Drug Regulations (C.R.C., C. 870), Canada.

46 GB 4806.1-2016 *National Food Safety Standard: general Safety Requirements on Food Contact Materials and Articles, National Standard of the People's Republic of China.*

47 GB 9685-2016 *National Food Safety Standard: standard for the Use of Additives in Food Contact Materials and Articles*, National Standard of the People's Republic of China

48 Regulation (EC) No 852/2004 of the European Parliament and of the Council of 29 April 2004 *on the hygiene of foodstuffs*, OJ L 139, 30.4.2004, pp. 1–54

49 Regulation (EC) No 1935/2004 of the European Parliament and of the Council of 27 October 2004 *on materials and articles intended to come into contact with food and repealing Directives 80/590/EEC and 89/109/EEC*, OJ L 338, 13.11.2004, pp. 4–17

50 Commission Regulation (EC) No 2023/2006 of 22 December 2006 *on good manufacturing practice for materials and articles intended to come into contact with food*, OJ L 384, 29.12.2006, pp. 75–78.

51 M. Lisińska-Kuśnierz, A. Kawecka, *Handel wewnetrzny* **2012**, 1, 60–68.
52 L. Bix, J. De La Fuente, R. P. Sundar, H. Lockhart, in *The Wiley Encyclopedia of Packaging Technology* (Ed: K. Yam), 3rd edn., John Wiley & Sons, Hoboken, NJ, **2009**.
53 ISO 22000:2018 *Food safety management system – requirements for any organization in the food chain.*
54 P. A. Luning, W. J. Marcelis, J. Rovira, M. Van der Spiegel, M. Uyttendaele, L. Jacxsens, *Trends Food Sci. Technol.* **2009**, https://doi.org/10.1016/j.tifs.2009.03.003.
55 A. Kawecka, A. Cholewa-Wójcik, T. Sikora, *Eastern Eur. J. Enterp. Technol.* **2021**, https://doi.org/10.15587/1729-4061.2021.239858.

Index

a

acceleration 49
accessories 64
acetylation 8, 163, 230, 236, 239
acoustic 43, 49, 229
acryl 82
active ingredient 5, 7, 13, 30, 32, 93, 94, 95, 96, 104, 117, 134, 135, 139, 145, 149, 161, 164, 165, 166, 183, 265, 299
active material 6, 7, 149, 155, 349
active packaging 125, 262, 266, 329, 331
additives 15, 16, 19, 82, 144, 164, 190, 216, 229, 234, 235, 238, 256, 269, 288, 289, 303, 308, 309, 380, 382, 385
adequate 60, 172, 220, 261, 273, 279, 367
adhesion 82, 109, 112, 197, 228, 230, 266, 327, 328, 379
adhesive forces 312
adhesive layers 379
adjoining 64, 65
adsorption 8, 44, 57, 63, 64, 66, 67, 69, 86, 103, 105, 106, 110, 111, 163, 265, 266, 267, 333
aeronautical industry 245
aeronautics 303
aesthetic 6, 78, 364, 367
affinity 57
affordability 130, 134, 251
aforementioned 5, 186, 232
agar 81, 197, 290, 292, 309, 324, 348
agglomeration 124, 126, 144, 261, 266, 271, 272, 273, 331
aggregation 5, 31, 50, 61, 63, 83, 103, 110, 173, 201, 283, 330, 331, 335
agricultural 15, 36, 83, 217, 261, 262, 263, 269, 270, 272, 273
agro-wastes 286
aiding factors 348
albedo 9
albumin 66, 112, 141
aldehydes 216
alkaline hydrolysis 10
alternative 18, 43, 63, 79, 130, 134, 152, 156, 164, 192, 198, 201, 211, 220, 232, 234, 250, 267, 283, 285, 298
aluminol 263
aluminosilicate 263, 269
aluminum chloride 5
aluminum pipes 257
amalgamation 6
amino 8, 10, 78, 140, 201, 230, 288, 289
Amorphous 29
amphiphilic 5, 57, 60, 61, 66
amplitude 45, 48, 53, 61, 63, 64
amylopectin 81, 228, 233, 234
amylose 8, 228, 230, 233, 234, 236, 239
anhydride 81, 103, 229, 230
anionic 9, 20, 81, 166, 185, 219, 233, 240
annealing 229, 230, 236, 238
antagonistic activity 280
anthocyanin 79, 120, 121, 131, 174, 188, 196, 262, 268
antibacterial therapies 272
antibiotics 269, 341, 342, 354
anticancer 114, 120, 188, 189, 201
anti-carcinogenic properties 280
anticoagulant 32, 173
antidiabetic 32
antigens 128, 214
anti-inflammatory 32, 99, 114, 187, 220
antimicrobial activity 14, 147, 191, 343, 344, 346, 347, 352, 353
antimicrobial agents 119, 124, 130, 134, 191, 198
antioxidant activities 129, 191, 342, 349
antioxidant agent 129, 187, 262, 341
antioxidant molecule 155

Materials Science and Engineering in Food Product Development, First Edition. Edited by Wing-Fu Lai.

anti-reflectivity 3
aperture 327
approved substances 381
aquatic 219
aqueous environment 123, 334
aqueous fluids 223, 224
aqueous media 75, 96, 113
aqueous solution 17, 44, 79, 81, 87, 123, 175, 187, 190, 239, 296, 349
arabic gum 172, 175
aroma 33, 74, 267, 332, 343, 354, 355
aromatic 187, 324, 342, 345
Arrhenius 34
artificial sweeteners 7
assembling 35, 215
assembly 35, 59, 78, 105, 149, 150, 153, 188, 190, 193, 212, 267
assurance 362, 371, 372, 388
astaxanthin 78, 86, 99, 103, 134, 175, 190, 220
atom 82, 323
atomic force microscopy (AFM) 323
atomization pressure 144, 282
attitudes 314
authenticity 380, 386, 387, 388
autocorrelation 330
automated fresh foods 316
auxiliary material 380
availability 1, 2, 16, 28, 48, 95, 107, 131, 132, 190, 220, 244, 257, 265, 336, 367
axisymmetric 62

b

backbone 81, 89, 216, 224
backscattered electrons 323
bacterial infections 269
bacterial metabolism 291
bacterial protein 288, 292
bacterial starter cultures 280
barrier capacity 140, 328, 367
beadfree nanofibers 286
beads 12, 82, 176, 180, 182, 218, 220, 221, 286, 294
beeswax 322, 345
beneficial microbes 280
bentonite nanoclay 272
benzoic acid 187, 344
bespoke 82
beverage packaging 262, 270, 378
bibliography 58
bicone 65, 66
bilayer 5, 83, 118, 122, 135
bile salts 103, 281
bimodal 51, 52
binder jetting 304
bioactive components 7, 32, 74, 83, 88, 125, 126, 127, 129, 146, 183, 262
bioactive peptides 68, 184, 344
bio-based polymer 271, 378
biochemical 34, 110, 113, 131, 186, 354, 368
biocidal activity 163, 200
biocompatible 4, 14, 17, 78, 83, 106, 129, 140, 145, 163, 188, 189, 211, 263, 266, 346
biodegradability testing 321
biodegradable goods 256
Biodegradable materials 256
biological activity 34, 48, 68, 189, 190, 288, 336
biological agents 32, 212, 220
biological macromolecules 10, 328
biological molecules 211, 212, 219
biological nature 379
biomass 244, 337
biomedical applications 198, 215
biomedicine 163, 216, 263
biomolecule solution 330
biomolecules 48, 212, 220, 291, 330
bionanocomposite film 15, 328, 335
biopeptides 10
biopolymeric material 161
biphasic partitioning 335
blend 139, 140, 141, 142, 322, 324, 329
bloodstream 10, 105, 106
blunt-ended stainless steel capillary 285
brainstorming 364
bread slices 343
broccoli 147
bulk levels 28
buoyant 49
burst 85, 94, 105, 148, 154, 155, 266, 268
butter phase 36

c

cake batters 306
calcium chloride 5, 15, 165, 248
cancer 32, 86, 173, 174, 185, 186, 188, 189, 190, 220, 280, 304
canthaxanthin 220
capillary 43, 62
capsule 6, 19, 74, 79, 139, 144, 145, 146, 149, 153, 154, 155, 183, 184, 185, 268, 354
carbohydrate chain 2, 19
carbon atoms 113, 249, 253
carbon-based nanocomposite film 327
carbonless 7
carboxy 197, 289
carboxylated cellulose nanocrystal (C–CNC) 335
carboxymethyl cellulose 52, 140, 185, 350
cardiac diseases 304

cardiovascular 32, 132, 183, 185, 190
carotene 11, 86, 101, 103, 119, 120, 130, 131, 134, 151, 190, 220
carotenoid 99, 151, 190, 220
carrageenan 9, 30, 82, 165, 172, 213, 221, 237, 240, 335, 347
carry antimicrobial 346
cartridge 306, 307
casein 17, 30, 66, 79, 93, 99, 100, 106, 146, 183, 190, 293
Casein micelles 98, 99
caseinate 9, 11, 221, 222, 349
casting method 347, 350
catalase enzyme 292
catalysis 35
catalyst 230
catalysts 186, 368
catalyzing 2
categories 2, 18, 31, 34, 36, 96, 139, 146, 150, 214, 230, 245, 246, 342, 387
catering services 316
cationic 8, 219, 230, 239, 240, 346
cationization 230, 231, 236, 239
cavitation 43, 229, 236
cell membrane integrity 282
cell wall pitting 298
cellulose nanofiber 327
chamber 74, 144, 164
characterization techniques 322
chelation 9
chemical characterization 323
chemical composition 2, 27, 32, 245, 368, 383, 387
chemical compounds 223, 328, 380
chemical contamination 378
chemical degradation 183, 220
chemical method 230, 257
chemical modifications 231
chemisorption 333
chewiness 227, 228, 235
chicken 14, 104, 146, 269, 308, 345, 346, 347, 348
chickpea starch 233, 241
chili 100, 244, 349
chitin 82, 162, 163, 197, 201, 346
chitosan complex 152
chitosan nanofibers 196
chitosan–gelatin blend film 335
chloride 197, 229, 230, 245, 253, 346, 347
chlorogenic 12, 189, 268
cholesterol 5, 95, 96, 98, 100, 103, 131, 132, 352
chromatography 33
ciprofloxacin 79
circulation 37, 88, 105, 108, 110, 111, 113, 388
classification 31, 150, 211, 245, 246, 380
clay 15, 263, 267, 268, 270, 346
cleanliness 382
cleavage 229
close relation 371
clove essential oil (CEO) 191
Coacervation 4, 11, 12, 19, 142, 143, 145, 146, 148, 164, 201, 349
coalescence 44, 51, 53, 58, 62, 105
coarser 42
coating layer 176
coaxial electrospinning 287
Coaxial electrospun nanofibers 287
cocoa butter 36, 309
cognitive activity 362
cognitive process 364
cohesiveness 227, 228, 234, 235, 312
coincidence 365
cold paste viscosity 232
colloid 42, 43, 44, 53, 164
colloidal suspension 113, 331
colloidosomes 48
colon integrity 280
colorimetric 196, 267, 268
ColorQuest spectrophotometer 321
combination 62, 67, 77, 85, 86, 87, 124, 140, 184, 191, 200, 219, 231, 233, 262, 271, 280, 284, 285, 353, 361, 364, 366
commodity traded 368
comparative 58, 252
complementary 57, 67, 269
complex coacervation 5, 95, 142, 146, 149, 153, 164, 166, 185, 188, 349, 350, 352
complexity 28, 125, 223, 315, 367
compliance 48, 106, 381, 382, 383, 387, 388
component 6, 13, 51, 53, 57, 59, 61, 74, 77, 86, 89, 99, 105, 117, 118, 124, 127, 128, 130, 140, 165, 201, 214, 218, 220, 223, 234, 235, 236, 251, 273, 331, 347, 366, 368, 371
Composite films 348
compression 35, 61, 62, 65, 307, 312
computer-aided design (CAD) 303
computer-aided design (CAD) software 303
concealing 74
conceptual assumptions 364
concurrent 81
condensing 232
conductive materials 255
confocal 31, 49
conformation 59, 330
congruent 247, 248
conical shape 286
consistency 142, 186, 232, 233, 307, 310, 316
constitution 33

consumer appeal 312, 314, 315
consumer satisfaction 303, 313
consumers' perception 367
consumption pattern 279
contamination factors 243
continuous phase 41, 42, 44, 45, 49, 50, 51, 52, 53, 142, 165
contrary 68, 370
convective dryer 253, 254
conventional spray 285, 290
coordinate system 316, 329
co-precipitation 8
core material 3, 6, 73, 139, 142, 143, 144, 146, 147, 148, 149, 152, 153, 154, 155, 167, 184
corn starch 8, 229, 237, 238, 239, 240, 242, 287, 307, 324, 326, 344, 346
corrode 257
corrosion 255
corruption 14
cost-effective 8, 35, 134, 145, 223, 232, 261, 263, 265, 306, 313, 315, 347, 351
cost-prohibitive 303
cosurfactant 83
covalent bonding 181
covalent crosslinkers 214
coverage 66
crayfish 66
creaming 44, 49, 50, 122
creep test 66
crop quality 269, 270
cross-linked polymer chains 211
cross-linked structure 212
crosslinking agents 5, 18, 168, 175, 176, 177, 179, 181, 182, 216, 230, 231
crude protein 269
cryopreservation 280
cryoprotectants 282, 283
cryptoclimate 371
crystal growth rate 248
crystallinity degree 334
crystallization 29, 42, 125, 140, 156, 248, 283, 309, 325, 334, 349
culinary 9, 306
curd 292, 293, 297
curvature 50
curve 46, 47, 53, 145, 333
customization 36, 303, 306
cyclodextrin 83, 156, 193, 199, 222, 343, 347, 351
cytocompatibility 267
cytotoxicity 86, 189, 222, 269

d

dairy-based products 308
dampness 232
deacetylation 32, 163, 167, 171, 174, 175, 177, 179, 181, 182, 201, 346
deamidation 78
decimal-reduction 34
declaration 381, 383, 387, 388
decomposition 10, 325, 333
decorations 36
defoaming 77
deformation 42, 46, 48, 53, 60, 61, 62, 63, 66, 69, 311, 316, 333
dehydration 14, 17, 145, 215, 284, 287
demineralization 163
demographics 304
denaturation thermal temperature 2
denaturing 352
density 13, 33, 49, 51, 62, 144, 153, 185, 186, 247, 248, 257, 281, 282, 286, 311, 335, 344
deproteinization 163
derivatization 232
desiccation 284
desiccation stresses 284
design objective 363, 364
design of experiment (DoE) 370
design quality management (DQM) 370
desirability 73
destabilization 41, 42, 44, 49, 50, 51, 52, 298
destabilization mechanisms 41, 42, 49, 50, 52
detachment 379
deteriorates 266
deterioration 13, 20, 74, 130, 131, 133, 192, 255, 328, 342, 389
detrimental 15, 50, 269, 281, 283, 286, 291
deviatoric 66
dextran 83, 106, 111, 141, 213, 222
diameter 3, 5, 19, 44, 46, 50, 63, 143, 155, 286, 287, 288, 289, 291, 294, 309, 310, 323, 324, 351
dictate 78
dieback varies 379
dielectric 241, 330
dietary supplements 11, 78, 129, 200, 270
differential scanning calorimetry (DSC) 321
diffraction 327
digestive tract 186, 269
diglycerides 42, 69
dihydrazide 216
dilatational 58, 60, 61, 62, 63, 66, 67, 69
dilute 78, 164, 216
dimensional accuracy 309
disaccharides 283
discontinuous fibers 286
discontinuous phase 165
disgusting 382
disinfect 382
disintegration 16, 124, 126, 218, 336

displacement 46, 47, 48, 125, 312
disruption 43, 51, 122, 298
disruptive food technology 303
dissemination 315
dissolved 87, 103, 127, 154, 164, 169, 171, 175, 197
distance vector 329
distinct needs 304
disturbance 62, 63
divalent cations 283
divergent 368, 370
diversification 368, 369, 370
domestic cooking 316
down-regulated allergic responses 280
doxorubicin 81
draining 233
drawback 15, 62, 87, 248
drug delivery mechanisms 270
drying chamber 3, 164, 245, 299
drying media 282
dual modification 231, 239
DWR 65, 66
dysphagia 304

e

early adopters 315
economic viability 251
ecosphere 362
Edible antimicrobial packaging systems (EAMPS) 354
effervescence 292
eicosanol 66
ejection 4
elaboration 363, 366, 369, 387
elastic modulus 53, 69
Elasticity 29
electrical conductivity 49, 286, 287, 288, 289, 294
electrohydrodynamic 74, 75, 285
electrokinetic 330
electromagnetic spectrum 328
Electron microscopy 321, 337
electron microscopy (EM) 322
electronic tensile testing machine 333
electrophoresis 18, 49
electro-sensitive 214
electrospinnability 285
electrospun fibers 4, 279, 287, 288, 289, 290, 291, 292, 294, 295, 296, 297, 298
electrospun zein fibrous 345
electrostatic forces 285
electrostatic interaction 5, 18, 31, 140, 141, 151, 165, 218, 219, 265, 267
elemental structure 323
eliminate 17, 243, 272, 341, 345, 354
elongation 265, 285, 326, 331
embraces 367, 370
emerging area 95, 303
emerging technology 165, 316
emitter 327
emulsification 8, 35, 42, 43, 44, 118, 122, 125, 141, 145, 146, 149, 164, 168, 171, 176, 184, 187
emulsifier 42, 46, 48, 57, 58, 59, 69, 81, 101, 103, 105, 164, 165
emulsifying 51, 63, 78, 79, 140, 183, 239
emulsion matrix 30
encapsulated bacteria 281, 286
encapsulated probiotics 133, 148, 177, 284
encapsulation application 18, 140, 211
encapsulation efficiency 3, 5, 8, 9, 74, 75, 78, 86, 120, 126, 131, 132, 140, 144, 151, 152, 153, 184, 189, 282, 284, 289, 324, 351
encapsulation technique 13, 74, 132, 145, 156, 281, 282, 291
energy demand 244, 256
energy dissipation 251
energy potential 251
energy transformation 251
enormous 58, 63, 113
entrapment 29, 30, 81, 83, 120, 130, 180, 280
entrepreneurs 385, 386, 388
environmental aspects 252
environmental impact 255, 256, 257, 322
environmental implications 261
environmental sensitivities 212
enzymatic modification 234, 236
enzyme immobilization 198
epichlorohydrin 230
epigallocatechin 78, 155, 188, 190
epoxypropyltrimethylammonium 230
equation 34, 62, 171, 332, 333
equilibrium 48, 61, 62, 63, 69, 220
erosion mechanism 155
erroneous 47
ester 346
esterification 217, 229
etching 265, 267
ethylene vinyl alcohol copolymer (EVOH) 191
eugenol 125, 352
eutectic 246, 250
evanescent wave 331
evaporation 15, 16, 125, 149, 183, 217, 243, 284, 290, 296
exactness 316
exergo-economic analysis 251
expansion 61, 62, 65, 69, 140, 154, 232, 310
expedient 315
experimental 45, 46, 47, 53, 63, 139, 153, 187, 252, 256, 293

extensive 1, 50, 51, 186, 197, 199, 270, 271, 272, 298, 336, 341
exterior surface 263
external environment 8, 42, 127, 128, 146, 335
extracted 2, 3, 8, 15, 62, 89, 103, 113, 140, 151, 223, 344, 355
extrusion 8, 29, 36, 74, 76, 81, 118, 140, 142, 144, 155, 192, 241, 281, 298, 304, 306, 307, 310, 312, 344

f

fabric materials 33
fabrication 15, 122, 200, 268, 271, 303, 304, 309, 317
fabulous 2, 3
factorization 47
fast-acting hydrogels 214
fattening results 269
fava bean 63, 66
fenugreek 233, 234
fermentation kinetics 292
fermented milk 287
fertilizer delivery 261
fertilizers 16, 186, 268, 269, 270, 272
ferulic acid 134, 323, 331, 333, 334
fibroblast 222
fibroin 79
fibrous films 268
fibrous mat 3, 197, 286, 290
fidelity 307, 314, 315, 318
film matrix 262, 265, 266, 353
film specimen 335
film-forming 9, 14, 155, 230, 262, 331, 347, 348
filthy 382
firmness 234, 305, 309, 311, 312
first-order 153
fish oil 7, 8, 12, 119, 122, 146, 167, 168, 183, 184, 194, 222
flakes 254, 307
flame retardancy 266
flavonoids 187, 188
flavor scalping 378
Flavourzyme 10
flaxseed 82, 146, 233, 240
flexibility 3, 15, 19, 77, 122, 227, 332
flexographic ink 268
flocculation 43, 46, 50, 51, 122, 124, 126, 164, 220
floccules 50, 53
flow rate 43, 167, 168, 172, 181, 192, 284, 285, 286, 287, 288, 309, 310
flow speed 307
fluctuations 248, 262, 330
fluid turbulence 255
fluid-bed coating 142, 144
fluidized 94, 95, 140, 144, 181, 280
fluidized bed coating 95
foam 57, 78, 79, 141
food additive 13, 32, 96, 126, 131, 134, 166, 188, 200, 221, 231
Food and Drug Administration 8, 95, 113, 150, 187, 251
food application 4, 8, 35, 78, 88, 95, 124, 142, 161, 163, 185, 193, 220, 223, 233, 234, 287, 298, 309, 345
food carrier 198
food chain 271, 385, 392
food coatings 272
food component encapsulation 77, 117, 119, 130, 134, 212, 217, 220
food dryers 243, 244, 250, 252, 253, 255, 256, 257
food emulsions 37, 41, 42, 44, 49, 52, 58, 60
food fabrication 31, 304
food layer 305
food manufacturing 18
food matrices 30, 73, 106, 308
food mechanics 315
food mixture 307, 309, 314
food packaging 1, 2, 3, 13, 14, 15, 16, 17, 18, 197, 198, 199, 263, 266, 267, 268, 327, 334, 335, 336, 342, 345, 367, 378, 379, 381, 383, 386, 387
food packaging materials 15, 16
food pathogens 14, 271
food preservation 13, 14, 19, 125, 243
food processing 6, 8, 60, 74, 184, 198, 245, 281, 304, 315
food production 27, 33, 86, 192, 235, 261, 262, 281, 342, 343, 386
food quality 2, 13, 35, 134, 263, 271, 272, 342, 354, 366, 367, 378
food quality deterioration 354
food safety 2, 3, 14, 36, 261, 263, 264, 266, 321, 349, 377, 378
food security 261, 262
food seller 382
food spoilage 17, 20, 261, 263, 271, 272, 355, 385
food stability 27, 29
food supply chain 28, 88, 262, 263, 269
food systems 10, 17, 64, 80, 341, 342, 343, 344, 345, 352, 353, 354
food-approved solvents 351
foodborne illness 273
foodborne pathogens 2, 15, 348
food-contact 342, 377, 379, 381, 382, 383, 388
food-contact materials 377, 381, 382, 383, 388

food-contact materials safety 377
food-contact surfaces 379
food-grade 15, 42, 52, 57, 103, 106, 134, 179, 183, 197, 199, 218, 251, 279, 298
Food-Packaging 321, 322, 323, 327, 328, 329, 330, 331, 332, 333, 334, 335, 336, 338, 345, 347, 378
food-packaging films 271, 348, 349
food-packaging industry 192, 322, 380
food-printing technology 192
food-processing 122, 139, 146, 243, 346
foodstuffs 186, 268, 270, 271, 272, 361, 362, 365, 370, 372, 380, 382, 385, 391
force spectroscopy 327
force–time curve 312
form films 346, 348
fossil fuels 244, 257
Fourier transform infrared spectroscopy (FTIR) 292
fraction 47, 58, 94, 95, 103, 104, 107, 112, 113, 163
fracturability 311
framework 233
free-flowing 146
freeze drying 8, 146, 150, 153, 156, 279, 280, 281, 282, 283, 284, 285, 290, 298, 299
freezing temperature 245, 246, 282
fresh foods 261, 262, 304
fresh produce 346
friction 53
fucoxanthin 190
fulfillment 381
functional foods 32, 68, 86, 95, 113, 163, 184, 185, 190, 342
functional product 68, 342
functionalities 29, 217, 232, 342
fundamental 13, 34, 35, 74, 145, 304
fungal growth 343, 346
fungi 82, 162, 189, 192, 201, 346
future perspectives 263

g

Garcinia kola (GK) 3
gas emissions 256
gastrointestinal 32, 73, 80, 83, 95, 97, 98, 99, 100, 101, 102, 103, 108, 112, 114, 126, 147, 174, 175, 176, 184, 185, 186, 190
gastronomy 306
gelatin complexes 153
Gelatin electrospun nanofibers 351
gelatin matrix 155, 323
Gelatin–glycerin–ε-poly-lysine 351
gelatinization 8, 30, 228, 229, 232, 233, 235, 236, 238, 239, 240, 242
gelation 5, 6, 8, 9, 11, 82, 94, 95, 118, 126, 140, 156, 164, 165, 171, 174, 175, 183, 187, 188, 189, 190, 217, 237, 240, 349, 352
gelling 9, 44, 63, 77, 122, 140, 212, 232, 241
gels 2, 18, 29, 30, 79, 81, 141, 214, 216, 229, 233, 234, 237, 310
generative 364
genetic algorithm 251
geometric accuracy 309
geometrical confinement 296
geometrical profile 256
geometries 306
geometry 64, 65, 66, 155
globular 30, 61, 78
globular proteins 30, 61, 78
globulin 78
gloss 309, 328
glucopyranose 83
glucosamine 20, 162, 163
glucose 81, 82, 110, 201, 214, 229, 230, 299
glutaralde 216
glycerol 14, 167, 309, 348, 349
glycosidic linkage 2, 9, 19, 20, 81, 229
gold standard 112, 313
good manufacturing practice (GMP) 383
gradient 50, 127, 154
gradients 61
granular 229, 230, 236, 237, 240
granules 144, 218, 219, 227, 228, 229, 230, 231, 233, 235, 236, 237, 312
grapefruit seed 346
graphene 14, 15, 16, 256
graphic signs 385
gravity 33, 50, 51, 62, 83
greasy 232
greenhouse 244, 251, 253, 256
grinding 35
grounded collector 285
growth stage 315
gum arabic 9, 10, 133, 141, 142, 146, 149, 151, 153, 232, 240, 241, 284, 287, 298
gum arabic (GA) 232
gumminess 228, 234, 311
gustatory preference 315
gut microflora 280

h

Halloysite Nano-Clays 261, 262, 264, 266, 268, 270, 272, 274, 276
halloysite nanotubes 323, 347
hardness 227, 228, 233, 234, 235, 304, 307, 311, 312
hardware 303
harsh conditions 184, 280

Hazard classification 380
hazardous compounds 383
health-aware 73
healthy foods 343
heat-sensitive probiotic strains 283
heavy metals 380
hedonistic 365
helices 82, 142
heptadiene 88, 223
herbal folkloric medicine 342
herbicide 269
hesperetin 188
hesperidin 79, 188
heterogeneity 361
heterogeneous 51, 139, 322, 354
heteropolysaccharides 2
hexahydrate 245, 248, 253
hexametaphosphate 10, 11
hexose 82
hierarchical 35
high capital 285
high nutritional value. 348
high-energy electron beam 323
highly conjugated 328
high-resolution structural 321, 323
high-voltage 4, 350, 353, 355
holistic 361
hollow 8, 113, 190, 263, 266, 344
homogeneous 94, 124, 139, 265, 310, 322, 349
homogeneous solutions 349
homogenization 29, 42, 43, 122, 125, 130, 149, 175
hordenine 78
horizontally 47
household 316
human body 2, 16, 19, 68, 79, 134, 149, 150, 152, 186, 200
human gastrointestinal (GI) 279
human health 15, 31, 37, 150, 151, 163, 166, 187, 267, 342, 386, 389
humidity gradient 335
Hunteria umbellata (HU) 3
hyaluronic acid 5, 165, 213
hybrid 65, 79, 83, 190, 212, 243, 346
hydration 5, 11, 215, 228
hydraulic 36
hydrocolloid 4, 5, 164, 233, 241, 349
hydrocolloidal 141
hydrocolloids 4, 13, 19, 33, 142, 197, 232, 233, 235, 240, 308, 335, 349
hydrogenated 103, 228
hydrolysis 5, 104, 181, 183, 197, 201, 216, 219, 227, 230, 231
hydrolytic enzymes 272
hydrolyzing 155, 234
hydrophilic 10, 17, 18, 51, 57, 69, 74, 75, 81, 104, 114, 118, 122, 126, 135, 141, 165, 185, 197, 201, 215, 216, 220, 265, 353
hydrophobic interaction chromatography 335
hydroxide 169, 174, 230, 231, 248
hydroxy 88, 223, 344
hygienic conditions 380, 385
hysteresis 47

i

identified needs 364
imaginary 61
immiscibility 87
immiscible 5, 37, 41, 42, 49, 53, 57, 58, 69, 121, 135, 349
immune functions 269
immunogenicity 79, 352
immunological assays 150
immunomodulatory 32, 96
impediment 233, 244
impermeable 87
improvement 2, 16, 78, 131, 132, 133, 186, 252, 255, 265, 266, 267, 269, 326
incident electron beam collides 323
incorporation 84, 111, 125, 178, 184, 212, 241, 266, 346, 368
incubation 103, 120, 132, 194, 287, 297
India 173, 253, 254, 288
indispensability 370
induces 62, 236
industrial 1, 2, 7, 17, 19, 28, 33, 34, 50, 51, 52, 58, 60, 73, 74, 83, 86, 122, 130, 143, 145, 150, 163, 192, 199, 201, 217, 221, 228, 232, 237, 256, 262, 263, 272, 316, 352
industrial setups 272
infant formula 96, 382
infill density 311
inflammation 102, 106, 114, 183, 186, 251, 280
inflammatory 106, 114, 173
infrared ligh 331
ingredient 13, 17, 29, 32, 37, 79, 94, 95, 103, 127, 134, 139, 147, 150, 154, 156, 183, 192, 197, 199, 200, 311, 315
inherent component 361, 365, 368
injectable 81
inlet air temperature 168, 284, 298
Innovating 34
innovative solutions 371
innovative technologies capable 322
inoculation 292, 293
inorganic 15, 105, 163, 198, 214, 215, 223, 246, 247, 251, 252, 255, 256, 267
insolubilization 78

integrated product-designing process 361, 368
integration 16, 124, 251, 362
integrity 17, 53, 60, 144, 183, 216, 298, 332, 333, 378, 380, 382, 387, 388
intelligent packaging systems 262, 267, 272
intended food-contact use 381
intensity 43, 130, 229, 236, 330, 337
interaction 5, 33, 37, 57, 141, 142, 191, 212, 218, 228, 235, 267, 322, 323, 327, 330, 335, 378, 387
interdisciplinary 28, 31
interface 42, 43, 48, 53, 57, 58, 59, 61, 62, 63, 64, 66, 68, 69, 131, 315, 365
interior 75, 82, 123, 218, 263, 266, 270
intermediation 384
intermolecular interaction 17, 62
interrelations 367
intestinal epithelium 108, 280
intestinal microbial balance 299
intestine 88, 93, 96, 102, 103, 104, 105, 106, 108, 109, 110, 112, 113, 132, 148, 200, 220, 280
intramolecular bonds 230
intricate shapes 303
intrinsic biodegradability 336
inulin 63, 171, 175, 182, 288, 289, 298
Ionic gelation 5, 11, 12, 152, 156, 173, 174, 175, 176, 177, 179, 188, 193
Ionotropic 8, 165, 174, 175
IR thermography 36, 37
irradiate 330
IRRAS 59
irreversibly 53, 66
isolation 29, 163, 186, 362
isomalto-oligosaccharides (IMO) 291
isoprenoids 9
isothermally 45
iterative 363, 370

k

kinetic 11, 31, 34, 42, 51, 52, 58, 60, 110, 122, 126, 247
kinetics 66, 155, 261, 265, 267, 268, 271, 272, 292, 345
kofter 234, 235, 242

l

L. acidophilus 284, 287, 290, 298
L. rhamnosus 133, 279, 282, 283, 284, 287, 288, 289, 290, 291, 292, 293, 294, 295, 296, 297, 298
laboratory conditions 365
lactate dehydrogenase activity 282
lactic acid 197, 267, 283, 287, 292, 324, 326, 345, 347
lactic acid (LA) 292
lactulose 186, 283
lactylates 42
latent heat storage (LHS) 244
lateral mobility 59
latter 45, 57, 58, 59, 99, 189, 212, 233, 365, 389
lattice structure shapes 310
lauric acid 322
layer-by-layer deposition 118, 192, 307
leaching 155, 230, 236, 347
lecithin 42, 51, 96, 103, 131, 132, 169, 171, 309
legal acts 367, 377
legal regulations 377, 381, 383, 387
legal remedies 383
legal requirements 381, 382, 383, 386, 387, 388
legible 386
legume 63
lemon 119, 171, 308
light-sensitive 131
lignin 331
Likert scale 314
limitation 65, 199, 252, 366
linear trend 312
linoleic 9
Lipid emulsions 29
lipid formulations 150
lipid oxidation 87, 129, 152, 191, 193, 194, 342
lipophilic 5, 51, 57, 78, 87, 88, 122, 123, 129, 130, 135, 156, 165, 187
liposomal system 343
Liposome 5, 19, 122, 135, 352, 355
liposomes 5, 6, 9, 10, 83, 84, 85, 95, 96, 99, 103, 110, 111, 117, 118, 127, 130, 132, 133, 352, 355
liquid–liquid separation 349
livestock 269
LMWEs 57, 59, 61, 62, 63, 69
loading capacity 85, 94, 105, 110, 130, 143, 152, 265, 267
locally sourced ingredients 304
logistic 368
logistic chain 368
long-term consequences 371
long-term stability 60, 66, 67, 120, 245, 255, 256, 257
low molecular weight substances 328, 380
low vapor pressure 247
low-methoxylated 81
low-viscosity 43, 45
lumen 110, 264, 265, 266, 267
lutein 79, 99, 175, 190, 220
lycopene 151, 190, 220

lyophilization 79, 174, 188, 282, 295
Lyoprotectants 288
lysozyme 66, 344, 348, 353

m
machinery 385
macroemulsion formulations 5, 6
macromolecular 35, 283
macromolecules 1, 3, 35, 201, 212, 232
macroscopic 18, 42, 59, 68, 263, 336
macroscopically 45
magnetic rod 66
magnetism 327
magnitude 65, 66, 262, 329, 330
mainstream 251, 255, 315
maintenance 1, 3, 149, 171, 263, 273
maize 8, 78, 81, 104, 231, 239, 240
major agents 345
maltodextrin 8, 9, 12, 167, 168, 183, 184, 241, 283, 284, 288, 290, 292, 298
mandatory 272, 386
mangosteen 81
manipulate 30, 228
Manipulation 28, 232, 233
mannitol 288
manufacturing practice 381, 383, 389, 391
manufacturing processes 304, 306, 316
margarine 41, 60
market launch 314
market niches 316
market surveillance institutions 377
marzipan 307, 308
mashed potatoes 192, 307, 309, 310
material adheres 310
material extrusion 304
material science 267, 273
matrices 30, 79, 83, 124, 139, 240, 255, 261, 348
matrix chain 265, 266
maximum value 364
mayonnaise 44, 57, 241
mechanical properties 15, 16, 29, 140, 197, 198, 199, 262, 263, 265, 266, 267, 311, 312, 327, 331, 332, 336, 348, 387
mechanical stress 48, 94, 332
mechanism 9, 44, 49, 50, 51, 63, 94, 103, 104, 112, 127, 128, 132, 144, 146, 148, 149, 150, 154, 155, 281, 329, 343
medicinal plants 342
melting point 14, 16, 125, 135, 154, 155, 245, 247, 248, 250, 309
melting process 154, 245
membrane permeability 282
membranes 10, 123, 150, 283, 287, 298
mesoporous 344, 346
metals 214, 223, 269, 322, 341, 343, 345
methoxyphenyl 88, 223
methyl cellulose 196
methylcellulose 197, 344, 345
micelle 79, 87, 99
micro monomers 28
microalgae 308
microbial growth 15, 191, 268, 345
microbial interference 273
microbial spoilage 354
microbiological 346, 348, 354, 366, 368
microbiological safety 366
microcapsule 3, 106, 148, 149, 153
microencapsulated probiotic 175
microencapsulation 7, 9, 13, 32, 93, 94, 95, 96, 106, 127, 134, 140, 144, 145, 146, 147, 148, 149, 161, 172, 183, 184, 186, 284
microjets 43
micro-nanofiber 75
Micronutrients 37, 113
microparticles 18, 80, 105, 144, 145, 155, 172, 190, 193
microscale fibers 285
microscopic details 310
microscopy 31, 49, 59, 323, 326, 327
microstructure 28, 35, 41, 42, 45, 48, 60, 214, 237, 310, 323
microwires 66
migrate 270, 328, 380
migration limits 387
mineral 134, 186, 263
miniemulsion 5, 6
missiles 30
mixtures 48, 61, 63, 85, 140, 141, 230, 232, 234, 240, 241, 242, 336
modified chitosan 172
modulation 280, 299
moduli 42, 43, 46, 53, 61, 64, 65, 66, 67
moisture 7, 13, 80, 81, 96, 131, 139, 140, 145, 148, 154, 155, 164, 217, 221, 229, 236, 237, 238, 253, 307, 332, 335, 348
molasses 234, 235, 242
molds and yeast 268, 345
molecular bond 245
molecular mass 155
monochromatic 330
monoglycerides 48, 119
monomers 1, 219
monosaccharide 2, 89, 224
monovalent cations 283
montmorillonite 15, 262, 263, 267, 268, 346
morphological 19, 238, 239, 241, 261, 263, 292, 322, 364
morphologically 3

morphology 126, 145, 148, 149, 187, 239, 261, 263, 282, 286, 289, 323, 351
mouthfeel 36, 192, 309, 314
mulberry 79
multifaceted nature 367
multilamellar vesicles 352
multilayer 48, 80, 93
multiple monosaccharides 2, 19
mushroom packaging system 344
mustards 41

n
naked eye 323
nano metals 345
nano-based technologies 263
nano-carrier 269, 272
nanocellulose 221
nanoclay 14, 263, 267, 268, 271
nanocomposite 263, 267, 268, 269, 330, 331, 347
nanocomposites 188, 263, 265, 267, 271, 272, 273
nanocrystal 323, 328
nanoemulsion 5, 6, 121, 122
nanoencapsulation 7, 9, 132, 139, 140, 142, 149, 150, 151, 152, 153, 188, 191, 199, 343
nanofabrication 223
nanofillers 265, 266, 271, 272, 348
nanogel 104, 111, 169
nanomaterial 273
nanometer 149, 327
nanometer scale 149
nanometers 74, 77, 323, 337
nanometric 3, 41, 265
nanoparticle 5, 9, 124, 125, 174, 188, 255, 270, 324, 327, 335
nanoprecipitation 150, 187
nanoscale molecule 156
nanoscopic 59
nanosensors 271, 272
nanosilver 267
nano-sized dimensions 273
nanostructured 1, 2, 3, 18, 85, 118, 129, 272
Nanotechnology 2, 21, 34, 160, 206, 261, 300, 301
naringin 79, 188
native starch 227, 228, 229, 230, 231, 232, 234, 236
natural edible resources 348
natural origin 163, 191, 267, 343
natural polymers 4, 19, 74, 201, 262, 322, 345
natural preservatives 267
natural resource 141, 341, 342, 343, 345, 352, 354
negative health effects 380
niques 58, 59, 139, 145, 153, 236, 267, 298, 303, 304, 366
Nisin 193, 344, 349
NLC 84, 85, 86, 118, 119, 125, 126, 127, 129, 130
NMR 49
nonbiodegradable 336
noncarcinogenic 257
noncommunicable 32
nonconventional ingredients 304
noncorrosive 248
nonelectrolyte solute 349
nonencapsulated 152, 176, 221
non-hygroscopic 77
nonionic 51, 103, 110, 221
non-Newtonian flow 316
non-paraffin substances 247
nonpoisonous 248
nonpolar materials 352, 355
non-solvent 5, 142
nontoxicity 1, 2, 18, 198, 201
nonviable food component 299
novel food 122, 133, 304, 315, 382
novel shapes 303
nozzle 3, 36, 43, 144, 192, 197, 304, 305, 306, 307, 308, 309, 310, 311
nucleation rate 247, 248
nutraceutical 18, 85, 166, 167, 185, 200, 218, 220
nutrient-rich whole foods 304
nutrients 13, 17, 31, 35, 41, 68, 74, 75, 81, 88, 89, 113, 127, 128, 184, 201, 218, 219, 220, 223, 268, 311
nutritional diversity 315
nutritional enrichment 280
nutritive value 366, 367

o
obligations 140, 155, 386
odorless 81, 288
official market supervision 388
oil in water emulsion 30
oil palm frond 287
oil palm trunk 286
oil-soluble vitamins 30
oily 41, 48, 58, 125
okra gum 233, 234
oleic 14, 49, 103, 119, 129, 130, 190, 343
oligo 44
oligosaccharide 2, 19, 112, 190, 279
omega fatty acids 183
on-demand customized meals 303
operating condition 250, 285, 380

operating costs 251, 285
opponent-type system 329
optical detection device 329
optical properties 3, 31, 148, 321
optimal extrusion temperature 310
optimization 2, 13, 52, 252, 255, 257, 286, 365, 370
optimize 236, 244, 252, 313
optimum 59, 179, 180, 284, 364
organic acid 14, 341, 343, 344, 345
organic solution 246
organoleptic 42, 44, 52, 57, 74, 184, 385, 389
organoleptic properties 42, 44, 52, 57, 74, 385
orientation 78, 153
oscillation 61, 63, 66
oscillatory 45, 61, 64, 69
osmotic 31, 154, 176, 280, 282, 284, 287, 290, 296
osteoarthritis 222
outer surface 113, 264, 266, 267, 379
outright 32
ovalbumin 66, 79
ovalbumin-derived 79
overhang angle 307
overhangs 307
oxidative 14, 32, 37, 68, 99, 100, 117, 131, 133, 167, 184, 188, 189, 221, 222, 239, 280, 284
oxide 14, 15, 16, 193, 198, 231, 256, 262, 267, 268, 346
oxygen transmission rate (OTR) 332

p

package color quality 329
packaged goods 377, 387
packaged product 365, 367, 368, 371, 378, 384, 385, 387
packaging category 381
packaging film 16, 41, 191, 267, 268, 323, 328, 332, 333, 335, 348
packaging system 125, 191, 265, 266, 267, 270, 272, 321, 341, 342, 344, 345, 347, 353, 354, 355
paraffin 244, 245, 246, 247, 248, 250, 251, 253, 255, 256, 257
paraffin wax 244, 247, 248, 250, 251, 253, 255, 257
parameter 46, 50, 53, 61, 130, 148, 250, 282, 287, 309, 310, 313, 329
paramount 52, 185, 366
pascal 62
pasting 227, 228, 229, 230, 231, 233, 234, 235, 236, 237, 238, 239, 240, 241, 242
patatin 78
patents 199, 200
pathogenicity 292
pathogens 169, 222, 263, 268, 269, 273, 280, 345, 347
Paulownia tomentosa (PTEO) 191
payback 251
payload state 6
pea addition 314
pea protein additive 312
peak viscosity 227, 229
pectin 10, 13, 81, 105, 106, 148, 165, 172, 175, 183, 197, 213, 221, 222, 239, 240, 287, 323, 349, 350
pendant droplet 285
penetration 96, 97, 99, 110, 126, 187, 312, 328
pentaerythritol 246
pentaglycine 246
PEO-lyoprotectant combinations 287
peptides 103, 106, 111, 112, 171, 184, 185, 186, 292, 341, 343, 344, 345
performance-related 365, 367
performance-related parameters 367
perimeter 64
perishable food products 345
permanent 18, 216
permeability 15, 16, 18, 103, 104, 105, 108, 110, 111, 122, 123, 126, 127, 141, 266, 296, 326, 332, 335
permeable 10, 82, 141, 229
permeance 331, 332
personalized meals 315
personalized nutrition 113, 315, 316
pesticides 263, 269
pests 269, 379
petroleum-based polymers 271
pharmaceutical 3, 5, 9, 34, 74, 75, 83, 86, 88, 96, 112, 118, 124, 148, 189, 190, 216, 223
Phase angle 69
phase inversion 51, 122, 125
phase transformation 255
phase transition modes 246
phenolic compounds 34, 119, 134, 187, 188, 189, 344
phenomenon 49, 51, 265, 349, 380
phosphate bonds 292
phosphate buffer saline 282
phospholipid 5, 9, 19, 83, 119, 352, 355
phosphoryl 230
photochemical 9, 11, 171
photomultiplier 330
photons 337
photopolymerized 219
photosynthetic 223
photothermal 81
pH-responsive polyacrylic acid 353

pH-sensitive 189, 214
phycocyanin 9
physical and chemical 7, 15, 28, 37, 74, 78, 83, 154, 214, 223, 231, 232, 234, 239, 322, 342, 367
physical flow 384, 385
physical methods 229, 230, 257
physical state 148, 244, 245, 246, 258
physico-chemical 293
physico-chemical 368
physicochemical properties 1, 2, 13, 32, 58, 150, 211, 236, 237, 238, 239, 240, 267, 304
physico-chemical properties 293
physicochemical reactions 261
physicochemical spoilage 262
Phytochemicals 32, 37
pictograms 385
piezoelectric 36, 62
plastic container walls 328
plastic flow 316
plastic material 262
plasticizer level 153
plasticizers 16, 29, 77, 332, 335, 348
plateau 45
platinum 63
poisonous 251, 382
polar solvent 352, 355
polarity 4, 48, 332, 335
polarizer 330
poly (ethylene oxide) nanofibers 193
polybutylene adipate 346
polycaprolactone 268, 346
polycationic 165, 166, 183, 200, 265
polycationic matrix 265
polydextrose (PD) 293
polydispersity 75
polyelectrolyte 4, 8, 102, 154, 155, 165, 166, 176, 177, 347, 349
polyelectrolyte layers 154
polyethylene 101, 132, 200, 268, 287, 336, 344
polyethylene oxide (PEO) 287
polygonal 47
polymer droplets 75, 349
polymer materials 380
polymer matrix 154, 201, 262, 263, 266, 267, 268, 271, 282, 296, 322
polymer printing 307
polymer solution 4, 75, 144, 197, 216, 285, 299
polymerization 18, 94, 95, 216
polyphenol 11, 88, 93, 99, 100, 102, 105, 109, 188, 193, 223
polyphenolic 187, 198
polysaccharide 9, 14, 20, 35, 44, 81, 82, 99, 100, 106, 163, 165, 166, 185, 197, 200, 201, 222, 241, 287, 293
polyunsaturated fatty acids 221, 222
polyvinyl 4, 266, 286, 322, 344, 347
pomegranate rind 346
porosity 4, 74, 95, 144, 153, 155, 215, 351
porous starch 8
potassium sorbate 240, 241, 283, 344, 346
potential application 122, 212, 267, 268, 344, 346, 349, 352
powder bed fusion 304
powder form 149, 150, 155
powdered food material 304
power intake 251
pragmatic 362
prebiotic-added feed solutions 293
prebiotics 112, 119, 130, 166, 175, 178, 185, 186, 279, 293, 294, 296, 297, 298
precipitation 4, 118, 128, 142, 181, 190, 349, 352
precise 17, 48, 73, 112, 316
precursors 219
prediction 33, 66, 312
preferences 27, 33, 44, 140, 155, 262, 314, 315, 316
preferential 265, 364
pre-market assessment 382
preparatory 78
preservation 18, 19, 73, 89, 151, 152, 164, 191, 198, 199, 223, 266, 281, 282, 331, 333, 344
preservatives 200, 221, 266, 267, 271, 283, 343, 344
preserve soy 346
pressure-sensitive 7, 214
pressurization 229, 236
prevalence 67, 107
prevalent 8, 10, 75, 192, 330
printability 197, 303, 305, 307, 309, 310, 314, 315
printable 192, 304, 308, 310, 314
printable designs 304
printed thermoplastics 304
printer extruder 307
printing parameters 192, 304, 306, 307, 309
probiotic functional foods 279
probiotic organisms 280
probiotic strain 281, 287
probiotics-added foods 280
procedure 46, 47, 50, 53, 139, 141, 145, 153, 164, 187, 380
processability 200, 217
product-manufacturing 362, 364, 368
prolamine 78, 104

prolonging 20, 221, 243, 345, 353
prominently 45
property 31, 69, 144, 190, 212, 215, 220, 257, 258, 267, 272, 273, 322, 328, 336
protective layer 17, 280
prototype 364, 365, 368
pseudo-equilibrium 64
pseudoplastic 47
public health 113, 269, 270
pullulan–IMO solution 295
pungent 354
putrid 382

q
quality assurance 361, 362, 371, 372, 386, 388, 389
quality deterioration 341, 342, 354
quality function deployment (QFD) 370
quality-oriented connotation 362
quality-oriented designing 371
quaternary ammonium 270
quercetin 80, 83, 86, 101, 104, 130, 134, 188, 199, 344
questioning 364

r
radiation 7, 89, 113, 216, 243, 331
raw materials 28, 35, 48, 75, 126, 199, 217, 313, 366, 380, 384, 385, 386, 389
ready-made elements 384
ready-made packaging 384
ready-to-eat edibles 306
reagent 51, 181
rearrangement 61
recoverable 61, 69
redox balance 150
refinement 35, 310
refractive 330
refrigeration temperature 298
regulatory framework 381
regulatory issues 273
reinforcement agent 262, 271
relaxation 45, 46, 61, 66, 312
releasing channels 346
renovating 34
reorientation 63
repeating units 1, 2, 19, 81, 163
repulsive 58, 218
residual water content 296
resistant starch 105, 179, 238
resolution 289, 310, 323, 327, 328, 331, 337
respiratory 34
resveratrol 78, 79, 86, 99, 101, 102, 104, 105, 106, 119, 187, 189
retail sale 385
retaining 20, 73, 295, 305, 307, 336
retention 12, 93, 99, 111, 129, 131, 187, 218, 219, 224, 244, 267, 283, 309, 310
retention time 267
retrofitting existing systems 251
retrogradation 229, 232, 236, 239, 240, 242
reversibility 247
reversible 81, 142, 214, 216, 245
rheological properties 42, 59, 66, 185, 238, 240, 241, 242, 307, 309, 311
rheology 35, 37, 42, 44, 45, 49, 52, 57, 58, 59, 60, 61, 62, 63, 64, 66, 67, 68, 69, 71, 164, 240, 242, 317
rheometer 65, 66, 232, 312
rigidity 307, 309, 322, 346
ripening process 10
rosmarinic 189
rotten 382
rupture lines 310

s
saccharose 288
safe food products 377
safety concerns 272
safety profiles 345
sales packaging 389
salmon 346
salt hydrate phase separation 256
salt hydrates 247, 248, 251, 256
scattered light 330
scattering 31, 37, 49, 232, 321
schematic 75
scratch 28
SDS 50, 51
sea buckthorn oil 9
seaweeds 213
secondary drying 282
secondary electrons 323
secondary packaging, 387
sedimentation 50, 52, 124, 126, 164
segregation 252
selection criterion 257
semblance 1
semiconductors 223
semi-finished goods 313
semiotic 365
semipermeable barrier 337
semipermeable membrane 154
semisolid 30, 33, 37, 215
sensory characteristics 242, 378
sensory properties 33, 35, 178, 191
separates 58, 154

separation 4, 19, 30, 49, 50, 52, 83, 87, 140, 141, 142, 155, 327, 349
sequence 35, 223, 385, 388
sequential 82, 242
sericin 79
serum cholesterol 280
setback viscosity 229
shape retention 310, 312, 314
shear deformation 312
shear-mixing 43
shell material 3, 7, 94, 106, 127, 199, 280
Sigma-Aldrich 173, 174, 175
signaling 10
silica nanoparticles 344, 346, 348
siloxane 263, 267
silver nanoparticles 267, 268, 327, 345
simultaneously 28, 62, 65, 103, 215, 256, 343
sinusoidal 62, 63, 69
slicing software 307
SLN 84, 85, 86, 118, 124, 126, 127, 130, 131, 132
slurry accumulation 310
small-batch designs 303
social demands 367
sodium alginate 9, 15, 141, 172, 175, 180, 183, 190, 282, 283, 287, 346, 347, 348
sodium chloride 229, 283
sodium glutamate 284
sodium trimetaphosphate 230, 231, 239
soft food 304, 315
soil medium 336
solar collectors 250, 251
solar dryer 244, 245, 250, 251, 252, 253, 254, 255, 256, 257
solar energy 244, 245, 253, 257
solar power 244, 245
solidification 36, 349, 352
solidification process 349
solidifying 36, 255
soluble dietary fibers 286, 287
solute-material 153
solution 3, 4, 19, 53, 63, 65, 73, 74, 78, 110, 142,
solution tree 364
solvent evaporation 197, 349
solvent type 4, 19
sonication 43, 169, 174, 175, 238
sophisticated 7, 95, 200, 216, 306, 334
sorbic acid 191, 344
sorbitan 42
sorbitol 15, 282, 349
soy polysaccharides 98, 99, 283
space differ 329
specimen stub 323
speckle 330
spectra 45, 46, 292, 331
spectrophotometer 336
spectroscopic 59, 331
spectroscopy 49, 59, 141, 156, 321, 322, 328, 331, 335, 338
spectrum 45, 46, 47, 73, 156, 183, 271, 292, 331, 378
spherical 41, 57, 62, 78, 263, 271, 286
spherical droplet 41, 286
spinneret 4, 197, 285, 286, 288, 294
spinneret tip 285, 286, 294
spirulina 12, 308
spoilage 18, 129, 192, 261, 262, 271, 273, 341, 342, 347, 354
spores 379
SPR 59
spray drying 8, 9, 95, 140, 142, 145, 150, 153, 156, 164, 185, 187, 202, 280, 281, 283, 284
Spray-drying 3, 9, 172
spray-freeze-dried 282
springiness 227, 228, 235, 312
stabilizer 15, 57, 132, 227, 232, 293
starch granules 30, 81, 228, 229, 233, 234, 235, 236, 309
starch modification 8, 229, 230, 235
starches 33, 81, 227, 230, 231, 232, 234, 237, 238, 239, 240, 241, 242, 305, 308, 345
static contact angle 335
steadiness 232
stearoyl 42
steric barrier 51
sterile water 290
sterilization 229
stickiness 42
stiffness 327, 328, 332
stimuli 34, 69, 145, 213, 214, 220, 223, 269
stored energy 245
strain-specific 280
strategy 35, 77, 177, 187, 198, 267
stress removal 48
stretching vibrations 292
structure-aiding materials 353
subcategories 60
subcontractors 385
subcooling 256, 257
sub-lethally injured cells 285
sublimation 282, 285, 299, 333
submicron 74, 77, 93, 94, 95, 96, 100, 103, 104, 106, 112, 143, 197
sub-millimetric 62
submissiveness 8
suboptimal 32
subphase 64, 65
subsection 255

substantial 82, 133, 192, 250, 251
substrate 217, 354
succinic 81, 103, 229, 322
sucrose 42, 141, 147, 181, 229, 240, 242, 282, 283, 288, 291, 292, 312
sugar 2, 29, 31, 36, 162, 197, 201, 235, 236, 242, 288, 304
suitable nutritional content 312
sulfur nanoparticles 345
sunlight 343, 368
sunset 252, 254
sunshine hours 252
superabsorbent 217
supercapacitors 217
supercooling 247, 248, 250, 257
supercritical 95, 150
supercritical fluids 150
superimposing 47
supervision measures 377
supply chain 261, 361, 363, 365, 368, 370, 384, 385, 386, 387
supply chain entities 361, 363, 365, 370
supramolecular 35, 241
surface area 35, 65, 66, 255, 265, 266, 267, 270, 272, 332, 347
surface chemistry 261, 264, 267, 269, 270
surface-to-volume 4, 74, 351
surfactant 42, 50, 51, 57, 63, 83, 87, 103, 110, 113, 122, 125, 132, 135, 164, 165, 171, 176, 187, 221
surroundings 216, 258, 262, 379
survivability 97, 98, 132, 148, 282, 298
survival rate 96, 99, 147, 177, 186, 282, 283, 284, 285, 287, 288, 290, 291, 295, 296, 298
susceptible 73, 96, 184, 186, 243, 327
suspended 74, 142, 144, 330
suspension 36, 99, 100, 104, 144, 145, 155, 232, 349
sustainability 15, 52, 128, 255, 261
swelling 18, 105, 139, 154, 155, 220, 228, 229, 230, 234, 235, 236, 237, 309, 312
symmetry 329
syneresis 232, 236, 293
synthetic description 368
syringe pump 197, 285, 288
systemic method 364

t

tannic acid 12, 112, 267
tapioca starch (TS) 231
tarragon 152
Taylor cone 285, 289
tea polyphenols 8, 185, 346
technical features 367
technological implementation 304
technosphere 362
temperature-controlled 333
temperature-sensitive 74, 75, 83, 214
tensile strength 141, 265, 266, 311, 331, 346, 351
tensile strength (TS) 331
tensiometer 62
tertiary packaging 388
testing protocols 256, 321, 336
textural properties 27, 227, 228, 229, 234, 235, 236, 237, 239, 241, 309
texture analyzer 312
theoretical model 368, 370
thermal characteristics 333, 336
thermal cycles 255, 256
thermal energy storage (TES) systems 244
thermal processing 341, 355
thermal properties 239, 248, 265, 347
thermo-mechanical properties 261, 262
thermophysical stability 256
Thermoplastic corn starch 345
thermoreversible 16
thickening 15, 44, 53, 63, 140, 230, 231
threat 269, 377
three-dimensional hydrogels 1
tissue inflammation 150
top-heavy designs 307
topography 59, 323, 327
tortuosity 81
toxins 263
traceability 383, 385, 387, 388
trade barrier 383
transglucosidase 234
transglutaminase enzyme 296
transient 64
transit 111, 280, 335
transition metal ions 328
transmission 15, 17, 81, 323, 328, 332
transmission electron microscopy (TEM) 323
transmittance 16, 324
transparency 94, 322, 324, 328
transparent 16, 81, 85, 347
transportation 7, 262, 265, 321, 335, 336
trehalose 173, 175, 279, 282, 283, 284, 288, 291, 293, 294, 295, 296, 297
trial-and-error method 364
triethoxysilane (APTES) 265
trigger 95, 100, 101, 102, 103, 105, 107, 128, 353, 354, 355
triglycerides 125, 309
trimetaphosphate 220, 230
tripolyphosphate 5, 165, 170, 172, 174, 188, 189, 230

tripolyphosphate (TPP) 188
Tuna oil 10
turbidity 49, 99
turmeric 129, 130, 220, 254

u

ultrafine 4, 288
ultrasonic 8, 44, 49, 167, 175, 184, 238
ultrasonication 42, 174
ultrasound 43, 44, 49, 68, 100, 110, 122, 228, 229, 230, 235, 236, 237, 238, 243
ultrasound treatment time 229
ultrathin sample 323
ultraviolet 328, 342
unambiguous 112, 383
undesired 49, 63, 69, 147, 149, 201, 261
universal set 383
unprecedented 303, 316
unprecedented freedom 303
unsanitary 382
untangled 67
utility 35, 188, 271, 304, 367
UV–Vis spectroscopy 328

v

vapor 15, 16, 141, 247, 248, 251, 272, 283, 335, 337, 346, 348
vaporization 333
variants 168, 364
vegan population 287
vegetal 63
velocity 36, 65, 251, 330
verbal information 385
vesicle 83
veterinary health 270, 273
viability 60, 80, 125, 132, 133, 176, 186, 222, 279, 280, 281, 282, 283, 284, 286, 287, 288, 289, 290, 291, 293, 294, 295, 296, 297, 298, 343
viable cells 287
vice versa 5, 41, 51, 146, 218, 273
vigorous 78
viscoelasticity 41, 45, 61, 62, 63, 216
vitamin 8, 81, 86, 93, 99, 107, 110, 113, 119, 132, 134, 155, 166, 172, 183, 190
volatile oil 187
volatiles 15, 333
voltage 75, 285, 286, 287, 288, 289, 290, 294, 296, 297, 299
volume fraction 43, 46, 47, 49

w

warehousing 367
water contact angle 322
water vapor permeability (WVP) 335
wavelength 3, 89, 328, 330
wavy lines 310
wet spinning 165, 197, 201
wheat bread 345
whey protein isolate 104, 111, 183, 185, 282, 353
whey separation 293
wrinkling 298

x

xanthan gum 2, 44, 52, 178, 233, 234, 240, 241, 242, 307, 309

y

Young modulus 332

z

zeaxanthin 152, 220
zein 9, 16, 78, 104, 185, 190, 262, 287, 344, 348, 351, 353
zeta potential 126, 129, 189, 321, 324, 328, 330, 331
zoonotic infections 269
zoonotic pathogens 272

Heidegger Thinking

Heidegger Thinking

Being and Time Explained

SIMON CRITCHLEY

Edited by Ben Olson

nyrb **New York Review Books** New York

This is a New York Review Book

published by The New York Review of Books

207 East 32nd Street, New York, NY 10016

www.nyrb.com

Library of Congress Cataloging-in-Publication Data
Names: Critchley, Simon, 1960– author | Olson, Ben editor
Title: Heidegger thinking : Being and time explained / Simon Critchley ; edited by Ben Olson.
Description: New York : New York Review Books, 2026.
Identifiers: LCCN 2026013848 (print) | LCCN 2026013849 (ebook) | ISBN 9798896231080 paperback | ISBN 9798896231097 ebook
Subjects: LCSH: Heidegger, Martin, 1889–1976. Sein und Zeit
Classification: LCC B3279.H48 S4616 2026 (print) | LCC B3279.H48 (ebook)
LC record available at https://lccn.loc.gov/2026013848
LC ebook record available at https://lccn.loc.gov/2026013849

ISBN 979-8-89623-108-0
Available as an electronic book; ISBN 979-8-89623-109-7

The authorized representative in the EU for product safety and compliance is eucomply OÜ, Pärnu mnt 139b-14, 11317 Tallinn, Estonia, hello@eucompliancepartner.com, +33 757690241.

Printed in the United States of America on acid-free paper.
10 9 8 7 6 5 4 3 2 1

Contents

Preamble 1
Abbreviations 7
Concordance Table 8

1. Heidegger's Project in *Being and Time* 11

Making sense of Heidegger – Being is time, and time is finite – We are a thrown-open clearing – Hamlet's question – Reawakening the taste for perplexity – Mineness – Authenticity – Average everydayness – Phenomenology – Dasein is being-in-the-world – Ecstasy – Categories and existentials – Not seeing the world with philosophers' spectacles – Facticity – Leaving the cabinet of consciousness – Having a fascination with the world, and not just staring at it.

2. World 31

What is the world, and how does the world show up? – Avoiding all ideas of subjectivity – Against epistemology – There is no problem of perception – We are the worldhood of the world –

Our inside is outside – Praxis, not theory – Why science and the concept of nature do not describe the primary phenomenon of the world – Handiness and thrusting aside our interpretative tendencies – The pragmatics of stuff – The primacy of concern and environment – When the hammer breaks – Signs and Heidegger's funny examples – Situation and the meaningfulness of the world – Wallace Stevens on the idea of order.

3. Space 53

What is wrong with Descartes's picture of the world – Extension renders the world invisible – Phenomenology and the destruction of the modern world picture – Why is this book not called *Being and Space*? – Space as room and roomliness – The unavoidable character of spatial metaphors – Nearness, place, and neighborhood – Art and space – The sacred, temple, and *templum* – The near, clearing away, and bringing things close – Churches, graves, radios, telephones – In Dasein, there is a collapsing of distance – Heidegger's theory of social media – Against the metrical measurement of space – Freeing place and why architects love Heidegger – Fernando Pessoa and lessons in unlearning.

4. Being-with-Others 75

We are defined by a "who," not a "what" – Dasein is the with – Everyone is the other, and no one is himself – I am nobody (Dickinson) – Networked being – They think me (Rimbaud) – Displacing the Cartesian starting point for philosophy, the "I" – Moments in the precipitation of the self in *Being and Time* (anxiety, death, conscience, resoluteness, the people) – How others show up in the workshop of the world – Being alone – Caring for, leaping in, and leaping ahead of the other – Considerateness and forbearance – Do we really understand others? (Levinas) – An encoded social psychology of modernity (Sloterdijk) – The dictatorship of the They – Every secret loses its force – How to become authentic.

5. Thrown Projection 101

The difficulty of stating the obvious – Not splitting the phenomenon – We are ourselves the clearing – The movement of disclosure and closing over as learning the rhythm of existence – The music of *Being and Time* – The four key concept clusters – State-of-mind, always already finding oneself there – Mood, Heidegger on the passions – Into this world we're thrown – Dasein exists factically – Projection, understanding, potentiality for being, freedom – Discourse is the condition of possibility for the articulation of the clearing that is us – *Logos* – We are the being that talks – Falling as horizontal, not vertical, not Satanic – Falling at the world – Authenticity as a modified way in which everydayness is seized upon – Fear – Giant bears and my mother – Interpretation as the layout of understanding – Meaning is not deep, but a question of surfaces – Being is not the rabbit in the hat – Learning to see the hat without wanting the rabbit.

6. Language and Interpretation 125

Interpretation as taking something as something prior to propositions – The hermeneutic circle – Virtuous and not vicious – Judgment is not the primary locus of truth – Cribbing Aristotle – Comparison with Wittgenstein's *Tractatus* – Important nonsense and silence – Taking things hermeneutically – The distinction between discourse and language – Communication – Language is not inside the head, it is ecstatic – Dasein has to have clean ears – The voice of the friend – Hearing, hearkening, silence, and reticence – Keeping quiet – Proposal for a reticence bar called Helsinki – The human being shows itself as a being that talks – Nonhuman language – Excursus on the octopus.

7. Inauthentic Life 147

Back to everyday life – This is *not* a philosophy of culture – Idle talking, gossiping, scribbling, texting, updating – Curiosity, the delight in sight – Concupiscence, eye-lust – The horror of liveliness (Hedda Gabler) – Flirting and the phenomenology of the sexy – Ambiguity, the speed of life – Everything means something else – Being against one another – Falling is not bad – Temptation (groundless floating) – Tranquilization (cultural exoticism, Western Buddhism) – Alienation (self-deception) – Entanglement (self-dissection, the triumph of the therapeutic) – Turbulence (groundless floating) – The structure of thrownness – Not a night view of Dasein, but daytime.

8. Anxiety and Care 165

Bringing the whole thing together – Anxiety is the way in which the whole of Dasein is disclosed – Defining the whole as care – The twelve steps of the care structure – Articulated thrown projection – We are not simple – We are a hinge – We are a unified but multifaceted stretch (future, past, and present) – Temporality as the meaning of care – A stunning personal revelation about my anxiety – Why I do philosophy –

Approaching Heidegger aquatically – Opening oneself to the tender indifference of the world – Anxiety as the mood of the novel (*Robinson Crusoe*) – Falling out of the homely – Party-being – Time and tide – Being alone with others – Excursus on Terrence Malick – Anxiety, calm, and courage – Anxiety reveals the Nothing – Darkness can help – The Christian background to anxiety (Kierkegaard) – Saint Paul and anxious waiting – Heidegger tells us a story and gives us a happy ending – Marrying Aristotle and Christianity within atheism.

9. Reality 199

Four incorrect theses on reality – Empirical idealism (Berkeley) – Transcendental realism (from Plato to Meillassoux) – Empirical realism (Samuel Johnson kicks a stone) – Transcendental idealism (Kant) – Why the question of reality is the wrong question – Why Kant is wrong (about the subject, representation, and the priority of the present-at-hand) – Heidegger's existential deepening of Kant – The scandal of philosophy is expecting that reality requires proof – Being-in-the-world is prior to both proof and faith – Epistemology is a symptom of Dasein's falling – Both realism and idealism are consistent with being-in-the-world, and both are wrong – The wrinkle in the fabric of Heidegger's argument: the Real – The hard core to reality: physical nature, matter, the earth – A possible dualism of reality and the Real: Doesn't this split the phenomenon?

10. Truth 219

Heidegger's reefer moment – Ek-sistence – The history of truth – Deconstructing the traditional concept of truth – Getting beneath the theory of truth as correspondence – Being-uncovering – *Alētheia* as unconcealment – Heraclitus on sleeping like sheep – Preserving the force of the most elemental words – We are in the truth – How to live in a totalitarian lie (Havel) – Dasein is in the truth and the untruth, the two paths (Parmenides) – Truth is relative to Dasein – No Dasein, no truth – Newton's laws – There are no eternal truths – Skepticism is unintelligible because a skeptic has never existed – Heidegger's refutation of suicide – Finishing division 1 and why we need division 2 of *Being and Time*.

11. Death 239

Full-metal Heidegger in lockdown – Why you need to become an existential phenomenologist in a pandemic – Authentic being a whole, the themes of division 2 – Unfairness to Kierkegaard – Death as one's ownmost, nonrelational, and not to be outstripped – Virility and impotence in relation to death – How others die (Tolstoy) – The certainty and indeterminacy of death – I am my not-yet – Authenticity and inauthenticity spin around in Heidegger's analysis – The difference between expecting and anticipating death – Anticipation as the possibility of the possible – Violent language – Shattering oneself against death as a wrenching away from the They – I am possibility – Enter Nietzsche – Enter freedom (towards death) – Having death in your mouth – Fruit – Corpse – Mourning is secondary – Dying for the other (Levinas's discretion) – Sacrifice, holocaust – Undermining Heidegger on death – The relational character of finitude – How I see being-towards-death – The ontology of weakness.

12. Conscience 274

Exoteric and esoteric (Leo Strauss) – The ten-minute exoteric version of conscience – The much longer esoteric version: watching the ice floe of *Being and Time* break up – The null basis-being of a nullity (the shape of the argument) – The peculiar character of the call of conscience – It says nothing (Cordelia in *King Lear*) – It is silent – Me, myself, and I – The call comes from me and yet beyond me – Stranger voice (Nietzsche) – The nothing of the world – The self divided between two nothings – The three *R*'s: repetition, repetition, repetition (Mark E. Smith) – Guilty! – Heidegger's genealogy of morals – Mortgage, death-pledge, money – Dasein is a nullity that does not have power over itself – Double impotence – Lagging behind one's possibilities – An antiheroic Heidegger (Beckett) – Dasein is a double nullity – The experience of freedom – Guilt is the pre-moral source for any morality (Kafka) – Double zero – Quasi-transcendental philosophy (Derrida) – Weakness is the brand (Maria Bamford) – Wanting to have a conscience – Action! – Resoluteness – Stillness (Stars of the Lid) – Situation as the seizing hold of space – Resoluteness as both taking action and the passivity of resistance.

13. Anticipatory Resoluteness 310

What do death and the concrete situation of action have in common? – A puzzling footnote on sin – Holding-for-true as the certainty of conscience – Resoluteness is repetition – How death has power over Dasein's illusions – Sober anxiety – The presence of Freud(e) in *Being and Time* – Joy – Discourse on the method – How we must do violence to ourselves in order to become authentic – Virtuous circularity – How do we conceive the unity of the self? – Heidegger's relation to Kant's "I think" – How Kant is both right and wrong – Kant's response to Hume on the self – *Cogito* without an *ergo sum* – Self as activity (Fichte) – Heidegger on the constancy of the self – Dasein is itself in the silent resoluteness of action in a situation – The constancy of the self consists in repetition – Performance – Descartes's dilemma with the *cogito* and Heidegger's attempt to think the self without ground – We are our acts, nothing more.

14. Ecstatic Temporality 333

Unwavering discipline – Authentic Dasein is the unity of the three *ecstases* of time – The three-step rhythm of temporality – The future as coming towards itself – What comes out of the future is my having-been, the past – The past is null, but not annulled – The curtains lift on action – The future opens up the past and releases itself into the present – The moment of vision, the glance of the eye – *Kairos* as the fulfillment of time (without Christ) – How can time be said without metaphysics? – Verbalizing nouns, speaking in tautology – Temporality temporalizes itself – Dasein is the ecstatic one (Taminiaux's story) – The meaning of finitude – End-ly time – Towards oneself! Existing as the nullity that I am – Sketch of a critique of Heidegger on time: the primacy of the past over the future.

15. The Time of the Everyday World.................. 350

The repetition of average everydayness from the standpoint of authentic temporality – Awaiting – Rapture – More unfairness to Kierkegaard – Making present – Forgetting – Repetition and memory – Bewildered fear, a burning house – Beckett – Resolute Dasein has no fear – Anxiety in the face of the nothing of the world – A taxonomy of moods – Hope comes in to tickle your feet – Equanimity – Being dragged along in thrownness – The authentic and inauthentic *ecstases* do not come in succession – The future is not later than the past – The genesis of the theoretical view of the world – An existential conception of science – Lifeworld practices and science – Phenomenology as pre-science – Deliberation is not absent from Heidegger (Ernst Tugendhat) – Different ways of philosophizing with a hammer – The importance of place – How nature is seen by modern science – Mathematical projection, data, modeling – Dasein is its world existingly – How the transcendence of being-in-the-world gets covered over in subject-object dualism – Horizonal schemas defined (with Kant) – The time of spatiality – Taking space in, literally – Leeway, play-space – Back to the everyday – The relation between authenticity and the everyday – There is no breakout from the cave – The enigma of everydayness – Philosophy and the enigmatic a priori.

16. Historicity and Politics 386

Death-obsessed male philosophers and the question of birth (Arendt) – The power of beginning – Stretchedness, persistence, movedness – Historizing as happening – History is rare – World history (a Greek temple) – The past lies ahead of us – Natural history – The ring – The loyalty of existence – Inauthentic history and the craving for the new – Recurrence and the love of fate – How to approach the science of history on the basis of existence – Monumental, antiquarian, and critical (Nietzsche on history) – Heidegger's debt to Dilthey – Philosophy as a manifestation of life – Fatigue – Why paragraph 74 of *Being and Time* is so important – Heidegger's reply to Löwith – "See you later, alligator": excursus on Philippe Lacoue-Labarthe – Heidegger's revolutionary conception of history – Fate as individual, destiny as collective – My struggle – The People (*Volk*) – "The inner truth and greatness of National Socialism" – Young Habermas – Who are the Germans? – Heidegger's thinking of the One (henology) – Against authenticity – Defusing the link between philosophy and politics – Radicalizing facticity – Tragic and comic Heidegger – Powerlessness – Heritage as the counter-concept to tradition – History as repetition – A weird thought: we repeat from out of the future – History is not the past – History comes to be from out of the future (Sandinista).

17. Existential Time and Vulgar Time 426

Three conceptions of time: primordial time, world-time, and vulgar time – The genesis of the tyranny of the present – Time-reckoning – Datability – Duration or the span of world-time – Inauthentic Dasein has no time – Authentic Dasein is its time and always has time – Take your time – Time first shows itself in the sky – Brightness, night – Tide – Heidegger is not a thinker of the sea – "Primitive" Dasein, the peasant's clock, shadow, and the sundial – "Advanced" Dasein and electricity – Footnote on the theory of relativity – Time and *physis* – The genesis of linear now-time – Time in Aristotle's *Physics* – Counted nows, the flow, the stream of time – The error of eternity – Primordial time is finite, reversible, ecstatic, and episodic – How time becomes leveled off – Time and the soul – Spirit falling into time (Hegel)

– Why Hegel is wrong – The long, final footnote to *Being and Time* (Derrida) – The endurance of Aristotle's conception of time in Bergson – Time-consciousness in Husserl.

18. Coda: Opening and Presence 458

The relation between construction and destruction in *Being and Time* – How the incompletion of Heidegger's book has been exaggerated – The fetishization of failure and the provisional – Heidegger's marginalia on his own book – What happens when we destroy the tradition – Presencing out of the source – The end of philosophy and the task for thinking – Being as opening, time as presence – The last words of *Being and Time*; or, WTF, Martin? – The risk of self-paralysis in false philosophical modesty – Against towardsism – Having the courage of your convictions – The massive amount that we have learned from *Being and Time* – Polemic: why this is the right path to be taken in philosophy – Finitude restated – Naming the experience of the appropriative event – There is no turning in Heidegger's work, just a deepening – The importance of building castles in the air (Gadamer) – All that there is, is air – How *Being and Time* changes one's life and approach to philosophy – Leaving the climate of Heidegger's thinking (Levinas).

Thanks 483

Bibliography 485

Index 499

Preamble

I THINK THAT READING Heidegger's *Being and Time* can change one's life. What I am going to try to do in this book is explain how that might happen.

Heidegger Thinking is a long-form deep dive into the most important philosophical book of the last one hundred years. The year 2027 marks the centenary of the publication of *Being and Time*. *Heidegger Thinking* is divided into eighteen chapters. Each chapter presents one of the key concepts in Heidegger's philosophy. Taken together, this book lays out the entirety of Heidegger's project in *Being and Time*.

To be clear, this book is not intended for specialists, experts, or those few souls lucky enough to have day jobs teaching philosophy. It is aimed at ordinary, intelligent people who are curious, serious, and interested. People who simply don't have the time to sit down and read the hundreds of densely written pages of the book. People who rightly intuit that there is something mightily important going on in *Being and Time* but who are intimidated by its strange jargon and get lost in the baroque intricacy and ornamentation of its architecture. It is my hope that this book will show how Heidegger's thinking might be applied to one's life in ways that are illuminating, elevating, and beneficial.

Heidegger Thinking is not intended for everyone. I am not seeking to make philosophy simple or to offer patronizing banalities about life. There is plenty of that available in the gaudily decorated bazaar of pop philosophy. The eighteen chapters are not TED Talks. In many ways, they are the opposite. They are slow, clear, extended, and intimate explorations of Heidegger's ideas in *Being and Time*. Not the highlights. Not the facile summaries. But *all* the ideas. It is my conviction that genuine philosophy can be explained simply and clearly. But it takes the time that it takes. And that can't be rushed. To be a philosopher is to take one's time.

The aim here is to explain in lay terms the entirety of *Being and Time* in ways that can be understood and—most importantly—*used*. I am asking readers to slow down, take their time, open their eyes and ears, and think deeply. What is said in the eighteen chapters will hopefully be crystal clear and helpful. But it will not be easy. I am not interested in easy. Nor are you. Let's try something else for once.

•

We will get started very soon. But allow me a word on the particular and peculiar circumstances that led to *Heidegger Thinking*. I first taught *Being and Time* to graduate students at the University of Essex in 1989. This was during my first year of full-time teaching at the same institution where I had received a PhD in 1988. The experience was terrifying, as I didn't know the book that well. I was in my late twenties, and the students were just a little younger than I was. The philosophical standards at Essex were very high at the time, and I was far from sure that I could reach them. Somehow I muddled through.

That was thirty-seven years ago. I have taught *Being and Time* in various ways, shapes, and forms every three to five years or so since. But the prospect of teaching the book always gives me a deep sense of trepidation, like preparing for a long and arduous mountain ascent. Because I had been teaching the book for so long, my interpretation had shifted significantly—indeed dramatically—over the course of time. My notes were a mess. I had strata of texts from a number of different periods lying in piles and covered in spidery, handwritten emendations. I kept revising and refining my interpretation of the book. The notes began to look like sketches of an ancient archaeological site or the ravings of a madman.

I knew that I would never write a book on *Being and Time* in the usual way. It was only the adrenaline and anxiety of giving weekly lectures on the book that forced me to get the notes together into a coherent whole. I would do this week by week, give the lecture and then rapidly move on to the next chapter.

Some years passed. In 2019 I had the idea of recording the lectures for some kind of posterity. I approached my friend and producer, Zenon Marko, and we began to record episodes in his studio in Tribeca in Lower Manhattan in January 2020, usually after I had given the lecture at The New School. I was pleased enough with the results, and Zenon and I recorded the first ten episodes without much drama.

Then the COVID pandemic happened. Teaching was suspended and I was unsure what to do. Early one evening in late March 2020, a kind student of mine left a really nice microphone and preamp on my doorstep in Brooklyn. I slowly figured out how to use GarageBand and off I went. I would sit at my desk, surrounded by notes, with the back door open, with the sound of sirens, birdsong, and applause every evening for the medical workers saving lives and helping ease the distress of those whose lives could not be saved. I would wind myself up and then do each episode in one long take, which Zenon would then carefully edit.

Something interesting happened next. The Onassis Foundation, whose cultural work in Greece and around the world focused on live arts, found itself unable to use any of its performance spaces during the pandemic. Discussions quickly turned towards the creation of an Onassis channel for the dissemination of film and audio material. They wanted to release my lectures as a podcast, initially on YouTube and then on the other streaming platforms. I called it, slightly tongue in cheek, *Apply-Degger*, a title I stole from my good friend Christian Madsbjerg. With much hard work from my Greek friends, we released the eighteen episodes of *Apply-Degger* between early May and late June 2020. I even recorded a house track called "Eat Your Funky Dasein" to accompany the podcast. No wonder some people question my seriousness.

The reaction was sudden and surprising. People really seemed to like *Apply-Degger*. It somehow caught the mood of that truly eventful, disturbing, and strange summer of 2020, which has come to define the years that have followed in ways that we are only beginning to understand. As the months and years passed, I kept getting emails and messages from total strangers saying nice things about the podcast. But there was one persistent and overwhelming complaint: listeners wanted a text or a book so that they could follow what was being said and refer back and forth without losing track of the line of thought. They wanted a written version that they could follow. They told me that they were fascinated by Heidegger's ideas, but they needed the shape of a book, complete with full textual citations, footnotes, and a scholarly apparatus so that they could continue reading and thinking on their own.

So, in October 2023, I began to transcribe the podcast in earnest. This proved to be a Herculean labor. I transcribed a few chapters on my own, but the task was beyond me and I needed help. I found it in the person of Ben Olson, who had been a student in the 2020 class, returning to philosophical study after more than fifteen years away working. It was at this point that everything began to click into place very fast. Ben and I worked seamlessly and tirelessly together for many months, months that eventually turned into years. He produced brilliantly accurate transcriptions, and then we would both edit the text, making improvements, clarifying obscurities, and adding references.

The book you have in your hands, now retitled *Heidegger Thinking*, is based on the podcast, but it is much more than a straight transcript. It is a complete rewrite. It is much better, clearer, and more coherent than the original, and we have added a full scholarly apparatus so that readers can continue to explore on their own. Ben and I had the good fortune to test-drive versions of this manuscript with graduate and undergraduate students in classes at The New School throughout 2024 and make innumerable improvements. We taught it one last time in fall 2025. And now it is in your hands.

This is not simply the book of the pod but a freestanding work that found its form through the medium of audio, which—it turns out—is a rather interesting and liberating way to write. I would like to thank Ben for his extraordinary work on *Heidegger Thinking* and Edwin Frank of New York Review Books for taking the bold and frankly surprising step of publishing it.

The table of contents is rather detailed. It provides a series of subheadings and tags that should be used as signposts or memory aids for reading. Hopefully this will allow readers to move back and forth with ease. There is also, thanks to Hugh Verrier, a concordance that allows readers to map the chapters of this book onto the relevant paragraphs of *Being and Time.*

Unlike any other work of modern philosophy that I know—with the possible exception of Hegel's *Phenomenology of Spirit*—*Being and Time* has a dynamic power, a unified energy field of moving concepts that keeps unfolding and deepening. Learning to read *Being and Time* is about learning the rhythm of that movement, which is nothing less than the meaning of our existence. When we learn to move with that rhythm, everything changes in terms of how we think of the world, ourselves, and others.

So, let's begin.

Abbreviations

for Commonly Cited Works by Heidegger

EM/IM = *Einführung in die Metaphysik* / *Introduction to Metaphysics*

GM/FCM = *Die Grundbegriffe der Metaphysik* / *Fundamental Concepts of Metaphysics*

BH/LH = "Brief über den 'Humanismus'" / "Letter on Humanism"

SZ/BT = *Sein und Zeit* / *Being and Time*

UK/OWA = "Der Ursprung der Kunstwerkes" / "Origin of the Work of Art"

WMK/WMC = "Was ist Metaphysik?" / "What Is Metaphysics?"

ZS/TB = "Zeit und Sein" / "Time and Being"

Concordance Table

Being and Time *is organized into two divisions, each with six chapters. The text is also divided into 83 paragraphs, each several pages long. The chapters of* Heidegger Thinking *follow the order of* Being and Time *indicated here.*

Chapters of ***HEIDEGGER THINKING***	Paragraphs	Divisions and Chapters of ***BEING AND TIME***
		DIVISION 1: PREPARATORY FUNDAMENTAL ANALYSIS OF DASEIN
1. Heidegger's Project in *Being and Time*	9–11	1. Exposition of the Task of a Preparatory Analysis of Dasein
	12–13	2. Being-in-the-World in General as the Basic State of Dasein
2. World	14–18	3. The Worldhood of the World
3. Space	19–24	
4. Being-with-Others	25–27	4. Being-in-the-World as Being-with and Being-One's-Self: The "They"
5. Thrown Projection	28–31	5. Being-in as Such
6. Language and Interpretation	32–34	
7. Inauthentic Life	35–38	
8. Anxiety and Care	39–42	6. Care as the Being of Dasein
9. Reality	43	
10. Truth	44	

11. Death	45	DIVISION 2: DASEIN AND TEMPORALITY
	46–53	1. Dasein's Possibility of Being-a-Whole, and Being-Towards-Death
12. Conscience	54–60	2. Dasein's Attestation of an Authentic Potentiality-for-Being, and Resoluteness
13. Anticipatory Resoluteness	61–64	3. Dasein's Authentic Potentiality-for-Being-a-Whole, and Temporality as the Ontological Meaning of Care
14. Ecstatic Temporality	65–66	
15. The Time of the Everyday World	67–71	4. Temporality and Everydayness
16. Historicity and Politics	72–77	5. Temporality and Historicality
17. Existential Time and Vulgar Time	78–82	6. Temporality and Within-Time-ness as the Source of the Ordinary Conception of Time
18. Coda: Opening and Presence	83	
	8	Design of the Treatise [Introduction]

1. Heidegger's Project in *Being and Time*

Making sense of Heidegger — Being is time, and time is finite — We are a thrown-open clearing — Hamlet's question — Reawakening the taste for perplexity — Mineness — Authenticity — Average everydayness — Phenomenology — Dasein is being-in-the-world — Ecstasy — Categories and existentials — Not seeing the world with philosophers' spectacles — Facticity — Leaving the cabinet of consciousness — Having a fascination with the world, and not just staring at it.

MARTIN HEIDEGGER, WHO LIVED from 1889 to 1976, was the most important, the most influential, and the most infamous philosopher in the Continental tradition in the twentieth century. *Being and Time*, which was first published in 1927, was his magnum opus.[1] There's no way of understanding what took place in philosophy, particularly in what we call "Continental philosophy," after Heidegger without coming to terms with *Being and Time*. Whether Heidegger's influence is felt positively (in thinkers like Jean-Paul Sartre, Maurice Merleau-Ponty, Hannah Arendt, Hans-Georg Gadamer, Michel Foucault, and Jacques Derrida) or negatively (in figures like Rudolf Carnap, Theodor Adorno, Jürgen Habermas,

1. Martin Heidegger, *Sein und Zeit* (Max Niemeyer, 1979); *Being and Time*, trans. John Macquarrie and Edward Robinson (Blackwell; Harper & Row, 1962); hereafter SZ/BT with corresponding page numbers.

Gilles Deleuze, and others), Heidegger casts a very long shadow over philosophy right down to the present day. Furthermore, unlike many Anglo-American philosophers, Heidegger has exerted a huge influence outside philosophy, in areas as diverse as architecture (as felt in the work of Peter Eisenman), contemporary art, social theory, political theory, psychotherapy and psychiatry (beginning with his later-career collaborator Medard Boss), theology, and, indeed, the world of business.

The purpose of these chapters is to try to explain Heidegger's project in *Being and Time*, paragraph by paragraph. Along the way I'll explain the huge significance of Heidegger's ideas for philosophy and for our life outside philosophy.

Being and Time is a work of considerable length (437 pages in the German original) and of legendary difficulty. The difficulty is caused by the fact that Heidegger sets himself the task of what he calls a "destruction" (*Destruktion*) of the history of philosophy, a de-structuring of the philosophical tradition (SZ 23 / BT 44). We'll see some of the implications of this destruction in future chapters. But the initial consequence of this is that Heidegger refuses to avail himself of the standard terminology of philosophy, particularly modern philosophy after René Descartes, with its talk of epistemology, subjectivity, consciousness, mind, representation, objectivity, and the rest. Heidegger has the audacity to go back to the drawing board and invent a new philosophical vocabulary. If you try to translate his terminology—such as "care," "reticence," or "falling"—back into the familiar language of traditional philosophical concerns, you're already on the wrong path, and the way forward will only become steeper and more tangled.

For example, Heidegger thinks that all conceptions of the human being as a subject, as a person, as consciousness, or, indeed, as a mind-brain unity are hostages to a tradition of thinking whose presuppositions have not been thought through radically enough. Heidegger is not just a thinker, but a radical thinker, in the sense that he tries to dig down to the roots of our lived experience of the world rather than accept the authority of tradition. That's what often makes his work difficult to understand, particularly in translation. Heidegger won't talk about the subject, consciousness, mind, or body. His name for the human being is what he calls *Dasein*. And that term can be variously translated but is usually rendered as "being-there." Consciousness, mind, body: these terms just put a new limit between Dasein and the world, whereas for Heidegger we are already in our world.

The basic and very simple idea is that the human being is first and foremost not an isolated being, not an individual subject or self, cut off from a realm of objects that it wishes to know about. Rather, we are beings who are always already in the world, outside and alongside a world from which, for the most part, we do not distinguish ourselves. We'll come back to this point repeatedly. For Heidegger, it's hugely important to understand that we're always already outside in the world. Dasein, as Heidegger says, is "being-in-the-world" (SZ 53 / BT 78). And what goes for Dasein, being-there, also goes for many of Heidegger's other concepts. Sometimes this makes *Being and Time* a very tough read, which is not helped by the fact that Heidegger, more than any other modern philosopher, exploits the linguistic possibilities of his native language—in his case, German. Although John Macquarrie and Edward Robinson, with their 1962 Blackwell English translation of *Sein und Zeit*, produced, in my view, one of the classics of modern philosophical translation, reading *Being and Time* can sometimes feel like wading through a conceptual mud of baroque and unfamiliar concepts. That's what we're going to try and work through and work against. We're seeking to bring out Heidegger's basic thoughts in a way we can make sense of and use in our day-to-day lives. That's the purpose of this book.

Heidegger's text is superficially difficult, even unpleasant to read in translation. That said, the basic idea of *Being and Time* is extremely simple. The book is called *Being and Time*. The thesis of the book is *Being is time!* That's it. What it means for a human being to "be" is to exist temporally in the stretch between birth and death. Being is time, and time is finite. *Finite*: it comes to an end with our death. Therefore, if we want to understand what it means to be a human being, and what Heidegger will call an "authentic" human being, then it is essential that we constantly project our lives onto the horizon of our death—what Heidegger calls "being-towards-death." Crudely stated, for thinkers like Saint Paul, Saint Augustine, Martin Luther, and Søren Kierkegaard (thinkers who were very much on Heidegger's mind in this period and who influenced *Being and Time*, despite his tendency to downplay and erase his influences), it is through the relation to God that the self finds itself. The way this works for someone like Augustine is through asking the question "Who am I? Who am I, O God?"[2] The question of the self is opened in relation to the experience of divine perfection, which usually entails finding myself imperfect and sinful. Now, Heidegger raises the question of the self and what the self is, but without reference

2. Augustine, *Confessions*, 2nd ed., trans. F. J. Sheed, ed. Michael P. Foley (Hackett Publishing, 2006), book 9, chapter 1, section 1.

to God. For Heidegger, the self can only become what it truly is through a confrontation with death, by *making a meaning out of our finitude*. If our being is finite, then what it means to be human consists in grasping this finitude in, as Heidegger will say, quoting Friedrich Nietzsche, "becom[ing] what you are" (SZ 145 / BT 186).[3]

On the subject of meaning, I will mention here that, following Thomas Sheehan, I think the concept of being can become something of a red herring, as it sounds too metaphysical, too lofty, and too otherworldly. It distracts us from the real concern of *Being and Time*, which is actually *meaning*.[4] It should be remembered that the question that frames Heidegger's work is that of the *meaning* of being. His contention is that we do not know what being *means*. So, the celebrated question of being cannot be separated from issues of meaning, of the significance of human being-in-the-world.

Ultimately, I want to show how this insight into our finitude is deepened in Heidegger's thought by its relationship to his concepts of 1) conscience, which we'll get to in later chapters, and then, towards the end, the idea of 2) ecstatic temporality. But the thing to note here at the beginning is that Heidegger's thought is very simple, and it is perhaps difficult to grasp because of its simplicity. His thought is that we are a *thrown-open clearing of finitude*. Existence is finite. It is without ground, without metaphysics, without any basis. There is nothing behind the scenes and no one apart from us pulling the strings. We're simply out there with the world and

3. Friedrich Nietzsche, *Ecce Homo: Nietzsche's Autobiography*, trans. Anthony M. Ludovici, vol. 17 of *The Complete Works of Friedrich Nietzsche*, ed. Oscar Levy (Macmillan, 1911), 49.

4. Thomas Sheehan, *Making Sense of Heidegger: A Paradigm Shift* (Rowman & Littlefield, 2015), xi–xii.

with others. We're a purely finite stretch of time. Whatever it means to be us, to be human, has to be understood entirely without ground. This simple thought is also a vertiginous, dizzying thought that will take some time to land and settle and become clear.

Let's press on a little bit and look at the project of *Being and Time*. Heidegger makes clear from the opening page of the book—an untitled page that begins with a quotation from Plato's late dialogue *The Sophist*—that what's at stake is the question of the meaning of being (SZ 1 / BT 19).[5] This is the question that Aristotle raised in an untitled manuscript written two and a half thousand years ago, which became known at a later date as *The Metaphysics*. Aristotle didn't use that term; that was just how the book was cataloged in a library in Alexandria much later on, namely as the "meta"-*Physics*, or that which comes "after" *Physics*. For Aristotle in these lectures, there is a science or a form of knowledge that investigates what he calls "being as such," without regard to any specific realms of being such as the being of living things, what we now call biology, or the being of the natural world, physics.[6] Metaphysics is the area of inquiry that Aristotle himself calls πρώτη φιλοσοφία (first philosophy), which comes before anything else. It is the most abstract, universal, and undefinable area of philosophy. But it is also the most fundamental.

Now, with admirable arrogance (he was still relatively young when he was writing the book), it is this same question that Hei-

5. Plato, *Sophist*, trans. Nicholas P. White, in *Plato: Complete Works*, ed. John M. Cooper (Hackett Publishing, 1997), 244A–B.

6. Aristotle, *Metaphysics*, trans. C. D. C. Reeve (Hackett Publishing, 2016), 1028b2–7.

degger tasks himself with inquiring into in *Being and Time*. He begins with a series of rhetorical questions: Do we have an answer to the question of the meaning of being? "Not at all," he answers. Do we even experience any perplexity about this question? "Not at all," Heidegger repeats. Therefore, the first and most important task of *Being and Time* is to recover our perplexity (in Greek ἀπορία, or *aporia*) for this question of questions, this question of being, which we might know, for example, as Hamlet's question: "To be, or not to be."[7] The task in *Being and Time* couldn't be bigger: the question of the meaning of being. A keyword in the early paragraphs of the book is *access*. Given that the matter Heidegger is trying to raise is the question of the meaning of being, the question that follows from this is, What is our access (*Zugang*) to the question of the meaning of being?

Although Heidegger will try to access the question of being in different ways in his later work (he'll do that through brilliant, extended readings of major philosophical texts, through socio-historical ruminations, and by experimenting with different forms of writing, like dialogues and even poetry), in *Being and Time* access to the question of the meaning of being is given through that being who can raise such a question. That is, the question of the meaning of being is a question for the being whose being is *in* question. Whose being is in question? Well, that's where *we* come in. For Heidegger, what defines the human being is this capacity to be perplexed by the deepest and most enigmatic of questions. "Why is there something rather than nothing?" "Why are there beings at all,

7. William Shakespeare, *Hamlet*, ed. Burton Raffel (Yale University Press, 2003), 3.1.55–6; hereafter HAM followed by line numbers.

and what does that mean?" The task of *Being and Time* is reawakening in us a taste for perplexity, a taste for questioning. Questioning as a movement into perplexity, into aporetics. And questioning, as Heidegger will say much later on in his writings, is the piety of thinking.[8]

If we turn to the first line of part 1 of *Being and Time*, paragraph 9, where the project of analyzing Dasein, *us*, really begins, we read the following: "We are ourselves the entities to be analysed" (SZ 41 / BT 67). This is the key to the first concept that is introduced in *Being and Time*, the concept of *mineness* (in German, *Jemeinigkeit*). This is how the book begins: If I am the being for whom being is a question—to be or not to be—then the question of being is *mine* to be. It is mine to be, one way or another. In what, then, does the being of being human consist, insofar as that is my being? Heidegger's answer is *existence*. For Heidegger, it is not consciousness, *Bewusstsein*, that our being consists in, but our existence as a being that we *have*. Therefore, the question of the meaning of being, this big question that he frames the book with, is to be accessed by way of what Heidegger calls an "existential analytic," or an analysis of our existence (SZ 41 / BT 65).

But what sort of thing is human existence? What can we say about it? Well, obviously, human existence is defined by *time*. We are creatures with a past who move through a present and who have available to us a series of possibilities, or what Heidegger calls "ways to be"

8. Martin Heidegger, "Die Frage nach der Technik" (1949), in *Vorträge und Aufsätze*, ed. Friedrich-Wilhelm von Herrmann, vol. 7 of *Gesamtausgabe*, ed. Friedrich-Wilhelm von Herrmann (Klostermann, 2000), 36; "The Question Concerning Technology," in *The Question Concerning Technology and Other Essays*, trans. William Lovitt (Garland Publishing, 1977), 35.

(*Seinsweise*)—namely, ways to be in the *future* (SZ 11, 20 / BT 31, 41). To say that the being of being human is existence is to say that the being of being human is defined by time, by the past, by the present, and by ways to be (i.e., by the future). Heidegger's point here is wonderfully simple, almost deceptively so, although the language that he uses to describe it is rather rebarbative. The human being is not definable by a *what*, like a table or a chair, but rather by a *who*, and by a who that is shaped by existence in time. Questions about what it means to be human are not what-questions but who-questions. What it means to be human is to exist with a certain past—a personal, social, and cultural history—and an open series of possibilities that I can seize hold of or not.

My existence is temporal. It is shaped by my personal and social history, what Heidegger is going to call, later on in *Being and Time*, "thrownness" (*Geworfenheit*; SZ 135 / BT 174). I'm thrown into the world, and in my thrownness into that world, a series of possibilities opens up for me, which I can grab hold of or not. And this brings me to an important point. If the being of being human is defined by mineness, by the fact that it is mine, then my being is not a matter of indifference to me. A table or a chair cannot recite Hamlet's soliloquy or undergo the experience of self-questioning and self-doubt that such words express. You can try this for yourselves at home, if you like. Ask whichever object is around you—a table, a chair, a bed—whether it experiences Hamlet's question, "To be, or not to be." Ask them, and you'll hear nothing in reply. But *we* can, *we* can ask this question. Dasein is always mine to be, one way or another. We can choose ourselves and win ourselves, or not and thus lose ourselves (SZ 42–43 / BT 68). And the "we" that is able to make this choice is distinct, for Heidegger, not only from inanimate objects, but from nonhuman creatures. It can be asked (and we shall pursue later in these chapters) whether Heidegger's analysis really precludes such beings, and whether this restriction is warranted or fair.

If what it means to be human is to exist, to exist in time shaped by past, present, and future, then that being is mine to be—mine to choose or to lose. This idea of choosing ourselves or losing ourselves is the kernel of Heidegger's idea of "authenticity," which is one of the most hotly debated and influential concepts that came out of Heidegger's *Being and Time*. "Authentic" was *Merriam-Webster*'s word of the year in 2023. We're currently living in a plague of authenticity, of demands to be authentic, and that paper trail leads back to *Being and Time*. But we shouldn't think of authenticity so much as a "good," or as something that is morally superior but, more accurately, as what is proper to the human being, what is its own. For Heidegger, there are two dominant modes of being human: authenticity and inauthenticity. Furthermore, he will say (we're still in the first paragraphs of *Being and Time*) that we have to make a choice between these two modes (SZ 42–43 / BT 68). The choice is whether to be oneself or not to be oneself, to be author of oneself, and self-authorizing, or not. Authenticity means coming into your own. It is often confused with accuracy, but that is not what Heidegger means by it. Heidegger insists, as he will do throughout *Being and Time*, that inauthenticity does not signify a lower or lesser being (SZ 43 / BT 68). Alternatively, authenticity does not mean a higher state of being. Rather, Heidegger insists this distinction between authenticity and inauthenticity is not a moral distinction, or what philosophers would call a "normative distinction." Let's just say that some philosophers have had doubts about that move in Heidegger's thinking. Notably, a philosopher like Theodor Adorno, who detects that there is a "jargon of authenticity" in Heidegger's work.[9] But let's call that a moot point and move on.

Now, we begin with this idea of mineness: existence is mine to be. Regardless of these twin modes of authenticity and inauthenticity, of choosing myself or losing myself, Heidegger insists early in *Being and Time* that the human being must first be presented in its "undifferentiated" character (SZ 43 / BT 69). We must first present, have access to, the human being in an undifferentiated way, prior to any choice of being authentic or not. In words that soon become like a mantra throughout the text, Heidegger seeks to describe the human being as it is presented, as he says, "proximally and for the most part" (SZ 21 / BT 43). The German here is *zunächst und zumeist*, which can perhaps be better translated as "most closely and mostly." Heidegger seeks to describe the human being as it is presented most closely and mostly. Then the question is, Well, how are we presented most closely and mostly? The next move that Heidegger makes is devastating, although he makes it in a very straightforward way. He claims that the way in which the human being is presented most closely and mostly is in terms of the obviousness of everydayness, in terms of everyday existence. And the character of this everydayness is *averageness*.

We should just take a moment to think about the radicality of this move. Heidegger in *Being and Time* is doing philosophy. He insists that philosophy begins and, in a sense, *ends* with average everydayness.

9. Theodor W. Adorno, *The Jargon of Authenticity*, trans. Knut Tarnowski and Frederic Will (Northwestern University Press, 1973).

There's a lovely phrase from paragraph 9 where he says average everydayness is that out of which all existence arises and into which all existence goes: "This undifferentiated character of Dasein's everydayness is *not nothing*, but a positive phenomenal characteristic of this entity. Out of this kind of Being—and back into it again—is all existing, such as it is" (SZ 43 / BT 69). We begin with average everydayness, and we end up with average everydayness. What's subtly devastating about that move is that philosophy might traditionally begin from average everydayness—this is what, say, Plato describes in the middle of *The Republic* as "the cave," the netherworld of average, everyday existence, where we see only shadows and illusions. But the task of philosophy, as illustrated in Plato's allegory, is to leave the cave and have access to the truth of the forms—and then to return later on to the cave, to be killed.[10] Heidegger wants to stay in the cave and to see the cave differently. For him, we begin thinking and we end thinking within the framework of average everydayness.

Philosophy, in Heidegger's radical reinterpretation of the tradition, is not some otherworldly speculation. As to whether the external world exists, or whether human beings are really human beings—whether they're really robots or AI programs or whatever—these are the pseudo-questions that philosophers for the most part amuse themselves with. Rather, philosophy begins, for Heidegger, with the description of human beings in their average, everyday existence. The description of average, everyday existence is what Heidegger calls "phenomenology" (SZ 27–28 / BT 50). There is a

10. Plato, *The* Republic *of Plato*, trans. Allan Bloom (Basic Books, 1968), 514A–517A.

very technical way of understanding phenomenology, which I won't bore you with. A very simple way of understanding phenomenology is that it is the description of existence, the description of average, everyday existence as it shows itself and the way in which that which shows itself can be transformed into language. As a description of existence, phenomenology works well in literature, and maybe even better in relation to music and movies. Phenomenology is the letting-be-seen of that which shows itself, and language is what lets us see that showing (SZ 44 / BT 70).

We begin from average, everyday existence, and on that basis, Heidegger seeks to derive certain general structures, which he calls "existentials." Heidegger's program in *Being and Time* begins with the proposition that he will repeat over and over again: Dasein is being-in-the-world (SZ 53 / BT 78). The task that he sets himself is to deduce the common a priori structures of our being-in-the-world that always already characterize our existence. The key insight to hang on to here is that Dasein *is* being-in-the-world. That is, there is no distinction between us and the world. We and the world are a unity. And what defines the human being, what defines Dasein, is the *transcendence* that it has with regard to the world. We are for Heidegger first and foremost transcendent beings, meaning we are *out there*. Our relationship to what is given, to what shows itself, is *ecstatic*, meaning we are outside ourselves in unity with, and mostly indistinguishable from, the world. Life takes place in this dimension of ecstasy, openedness, or out-thereness. We'll come back to ecstasy in later chapters.

We should note here the difficulty of the task that Heidegger has set himself. What he's trying to describe, what he wants philosophy to describe, is the way in which human existence shows up most closely and mostly, in the way that is most obvious. But this is fiendishly difficult to describe. Nothing is closer to me than myself in my average, everyday, indifferent existence. So how do I describe this? How do I describe this closeness of myself to myself? Heidegger was fond of quoting from Saint Augustine, particularly from the *Confessions*, where Augustine writes, "Assuredly I labor here and I labor within myself. I become to myself a land of trouble and inordinate sweat."[11] What Heidegger is trying to do—and when you're working through these pages of *Being and Time*, you can feel that you're often sweating—is to bring out, and our labor here is to bring out, the character of the obviousness of average, everyday existence and to stay with that obviousness.

When you're reading *Being and Time*, and I'm talking here about paragraphs 9–13, the first two chapters of *Being and Time* proper, it can be a pretty unpleasant and bruising experience. I have tried to give you an overview of the shape of Heidegger's thought. But there are a couple of other things I need to mention. Heidegger makes a distinction between *categories* and *existentials*. Categories in philosophy are metaconcepts. The idea begins with Aristotle: they are the metaconcepts that are used to divide up our experience of things in the world. For Heidegger, categories are ways in which we can understand things that are not us: as Heidegger will say, entities that are not Dasein.

The distinction he's going to make in terms of categories is a distinction between the ready-to-hand and the present-at-hand.

11. Augustine, *Confessions*, book 10, chapter 16, section 25. See SZ 43–44.

I just want to say a word about that distinction—I'll make it clearer in the next chapter, but let's flag it now. The ready-to-hand is really "the handy," the way we use things, where things are available to us in routine life. The present-at-hand is the way in which those things show up when we look at them theoretically, when we look at them with philosophers' spectacles. Heidegger's view is that the fundamental category by which we have an experience of the world is handiness, the ready-to-hand, and on that basis is built the present-at-hand, the theoretical way of viewing things. And he thinks that the error of philosophy is to confuse this relationship between the handy and the theoretical, between the practical and the intellectual.

If categories describe things that are not Dasein, then the structure of Dasein is explained in terms of concepts that Heidegger calls existentials. These existential structures—which are appropriate, Heidegger thinks, for entities like us, entities described with a who and not a what, who have personal pronouns and are not like tables and chairs—are prior to everything that we can find in anthropology, psychology, biology, or now in an area like neuroscience (SZ 42 / BT 68). Therefore, to get a sense of our embeddedness in the world and to be able just to describe that, we have to leave to one side all traditional ideas of the subject, of consciousness, spirit, person, man, or life. All these terms have to be put aside. Introducing them gives us a metaphysical basis by which to understand what it means to be human. Heidegger does not want to understand our humanness in metaphysical terms but in terms of our lived, existential experience. Going back to a point I made earlier, Heidegger thinks that the vocabulary that we've inherited from philosophy gets in the way, and the task of philosophy is to *destroy* that terminology by finding a new language. That new language is the one that he invents in *Being and Time*.

The next chapter of *Being and Time*—chapter 2, paragraph 12—is titled "Being-in-the-World in General as the Basic State of Dasein" (SZ 52 / BT 78). Heidegger did not have a gift for titles, but the good thing about that title is that it is absolutely accurate. What he's trying to argue for in this chapter is that being-in-the-world is a basic state or fundamental constitution of Dasein. The essential thing—as he will say repeatedly—is that being-in-the-world is understood as a whole, as a unitary phenomenon. However, our being-in-the-world can be broken up formally and analytically. It can be broken up in terms of a set of philosophical distinctions, even though it is lived as a whole. And that's what Heidegger is going to do in the next chapters of *Being and Time*. Chapter 3 is about the "in-the-world" of being-in-the-world, where he describes the structure of the world and a concept that he calls "worldhood." Then Heidegger says, in chapter 4, that we need to talk about the *who* of Dasein. If Dasein is defined by having a "who" character rather than a "what" character, then *who is* Dasein in the manner of average everydayness? Heidegger defines that "who" character in relation to Dasein's being-with-others. Chapter 5 is where things really get going in *Being and Time*, where Heidegger asks the question, What does it mean to be "in" at all? What does the "in" of being-in-the-world really mean? That's going to take us into a discussion of the central thought in *Being and Time*, which is the idea of *thrown projection*, Dasein as "thrown into the world" and having the capacity to project out of that thrownness. The other way Heidegger expresses this thought is to say that "Dasein exists factically" (SZ 179 / BT 223).

Allow me another little terminological distinction, a bit of housekeeping before we finish. Heidegger makes a distinction in chapter 2 between a factual understanding that human beings can have of themselves and the factuality of objective things in the world (tables, chairs, and so on). The fact that I'm sitting on a chair to speak to you, at a table that I've got my papers on, these are *facts*. But *we* are also facts, my existence is a fact. There is a factual character that we possess in our everyday lives, and this is what Heidegger calls "facticity." We exist factically; namely, we didn't just come into existence this morning. We are thrown into existence. We arrive here in the world with all sorts of baggage: linguistic baggage, cultural baggage, personal baggage, all of that clutter. Our lives take place in a net and network of facticity that we need to be able to make sense of.

Heidegger insists in *Being and Time*, and indeed earlier in his work, that what we need in philosophy is a *hermeneutics of facticity*. This sounds rather exotic and complicated. But a hermeneutics of facticity is just the way of interpreting our everyday being-in-the-world. Our everyday life is first and foremost experienced as dispersal, as being split up, spread out amongst things in the world. For Heidegger, we exist as a temporal movement of dispersal in an interrelated field of things that makes sense to us. Our being is networked. The way we are *in* the world, Heidegger says (and he will make this point more clearly in the next chapter), is characterized by what he calls "concern" (*Besorgen*). But such concern must not be understood as worry or fretting. Concern is being occupied with or being busy with something. The way in which I'm factically dispersed in the world is in the way in which I am occupied by things: I'm looking at my phone, I'm looking at the street, I'm looking where to go, I'm looking for the signs that will take me to the subway, or whatever it might be. I am concerned, preoccupied, by the world. I know my existence is out there in a dispersed way.

The key thing about that insight, which again sounds rather banal, is that first and foremost, or most closely and mostly, the human being is not independent of its world. The human being is dispersed out there in the world, as Heidegger will say, "always already." And that's what has to be described phenomenologically.

Let me tell you a story by way of conclusion. I knew a student when I was an undergraduate at the University of Essex in England, in the early 1980s. Her name was Fiona Hetherington. She was a graduate student, and she seemed very clever to me, and I was terribly impressed. She was working on a PhD dissertation on Heidegger, and I asked her a question one day as we were sitting outside in the courtyard near the campus shop: "So what is it about Heidegger that is really powerful and important to you?" And she said it could be summarized in the following quotation from *Being and Time*, which she quoted from memory, I recall: "When Dasein directs itself towards something and grasps it, it does not somehow first get out of an inner sphere in which it has been proximally encapsulated, but its primary kind of being is such that it is always 'outside' alongside entities which it encounters and which belong to a world already discovered" (SZ 62 / BT 89). Why is that important? Because our fundamental orientation, the primary directedness of our being-in-the-world, is to be out there, always already amongst and alongside things. We do not first discover ourselves by turning inwards, by asking ourselves the questions, How can I know for sure whether the table that I'm leaning on is really existent or not? How can I really know that the person I meet in the street is a human being and not a robot? These are secondary questions. Our first relationship to the world is one of being out there alongside it. The word Heidegger uses to describe that is "fascination" (SZ 61, 76 / BT 88, 107), a concept that is insufficiently emphasized in writing on Heidegger. The word in German is *Benommenheit,* which I like to think of as being *benumbed* by the world, or fascinated in the way in which we are taken by the world. We move with the world, we flow with the world's flow. That's where we first find ourselves.

This benumbed fascination could be thought of in terms of an intoxication with existence. Children experience this when they are lost in prereflective play. We could also imagine being benumbed as being sleepy, when you've just woken up and the world seems fascinating, and you're dazedly going through the motions, in what Hubert Dreyfus called "background coping."[12] Once again, this seems like an innocent move, but it's really important. Heidegger is saying that if we think of ourselves, as modern philosophers have tended to, as being locked up in a "cabinet of consciousness," looking out at a world that is foreign and alien, then we are fundamentally misdescribing our existential being-in-the-world (SZ 62 / BT 89). The thing for Heidegger is simply not to do what modern philosophers normally do but to break ourselves out of this imagined cabinet of consciousness and restore the unitary phenomenon of being-in-the-world. He says, in the same passage: "Proximally, this being-already-alongside is not just a fixed staring at something that is purely present-at-hand. Being-in-the-world, as concern, is *fascinated by* the world with which it is concerned. If knowing is to be

12. Hubert L. Dreyfus, *Being-in-the-World: A Commentary on Heidegger's* Being and Time, *Division I* (MIT Press, 1991), 3–8.

possible as a way of determining the nature of the present-at-hand by observing it, then there must first be a *deficiency* in our having-to-do with the world concernfully" (SZ 61 / BT 88). The key inference to draw from that passage is that the way in which philosophy normally proceeds—as, for example, Descartes does in his *Meditations*—is by a fixed staring at things, an impassive, cold perception, which is usually how philosophers use their beloved examples, medium-sized dry goods like tables and chairs. On the basis of this fixed, impassive staring at things, we can apparently deduce the nature of things and of beings like us, which is allegedly defined by consciousness. Heidegger thinks that the theoreticist approach of modern philosophy fundamentally misses the point. That is why the philosophical tradition has to be destroyed. For Heidegger, the way in which the world first and foremost shows up is not through a perception of it but through a concern for it, absorption in it, fascination with it.

The implication of this line of thought is that knowledge of the world in the way in which it's normally understood by that area of philosophy called epistemology, which is concerned with how a subject can know a world of objects, is not false but rather *secondary*. Knowing arises only from a deficiency in our lived concern. Epistemology is secondary to our lived experience of being-in-the-world. It is the latter that has to be described first. That's what we are going to try to do in this reading of *Being and Time*. The first step on that adventure is to try to get clear on the phenomenon of "world" and to ask the questions, What is the world, and how does the world show up? This is the topic of our next discussion.

2. World

What is the world, and how does the world show up? — Avoiding all ideas of subjectivity — Against epistemology — There is no problem of perception — We are the worldhood of the world — Our inside is outside — Praxis, not theory — Why science and the concept of nature do not describe the primary phenomenon of the world — Handiness and thrusting aside our interpretative tendencies — The pragmatics of stuff — The primacy of concern and environment — When the hammer breaks — Signs and Heidegger's funny examples — Situation and the meaningfulness of the world — Wallace Stevens on the idea of order.

DASEIN—HEIDEGGER'S name for us, for humans—is being-in-the-world. In chapter 3 of division 1, Heidegger is going to analyze the phrase "in-the-world" and what "world" means. Now, if Dasein is being-in-the-world, then this entails that the world is part of the fundamental constitution of Dasein. While that sounds like an easy thing to say, it's a more difficult thing to grasp. The thought here is that *we are world*, which is to say that I am not standing here (or sitting here, as the case might be) as a free-floating subject, over against a world of objects that oppose me and that are independent of me. Rather, Dasein and the world form a unity.

As Heidegger says in paragraph 10 of *Being and Time*, any idea of the subject appeals to an idea of ground, a metaphysical ground, what he calls *hypokeimenon* (ὑποκείμενον; SZ 46 / BT 72). Heidegger says we therefore need to avoid all conceptions of the subject, subjectivity, mind, consciousness, spirit, soul, person—that whole language. The standard way this is argued for in philosophy is that we think of ourselves as subjects and that what oppose subjects are objects; and those objects stand over against me. Interestingly, the German word for object is *Gegenstand*. The object is that which *stands* against me: *Gegen-Me*, we might say. The idea of the world as broken down into a subject-object split is the view that Heidegger is trying to oppose throughout *Being and Time*. This is what we can call, following Charles Taylor, the "epistemological construal" of the world.[1]

Knowing, the epistemic relationship to the world, Heidegger says in paragraph 13, is a *founded* mode of being-in-the-world (SZ 61 / BT 88). That is, it is a *deficient* mode of relating to the world, and out of this deficient mode, the perception of the world as present-at-hand is born. This is a fascinating thought. Once we begin to relate to the world as a world of knowledge, or potential knowledge, then we apprehend the world as a realm of objects that we perceive. And perceptions, according to the traditional philosophical parlance, can be captured in propositions. Those propositions are things we can assert the truth or falsity of based on evidence, and science can be understood as the totality of true propositions.

Heidegger sets out to undermine that whole picture—objects, subjects, knowledge, perception, representation, propositions. We have to forget about the problem of perception—there is no such problem—and forget about the primacy of epistemology. Rather,

1. Charles Taylor, "Overcoming Epistemology," in *Philosophical Arguments* (Harvard University Press, 1995), 1–19.

for Heidegger, I *am* my world. The world is part of my being, part of the structure of my existence. *I am part world.* The main claim that Heidegger is going to make in chapter 3 is that the worldhood of the world is an *existential.* What he means is that the worldhood of the world—namely *us*, Dasein—is an a priori structure, a constitutive feature of the being of Dasein. A priori in the sense of both ontological priority, in terms of what is philosophically primary according to Heidegger, and a temporal priority, in terms of what comes first in the order of experience. Heidegger makes a link between the a priori and the prior, the previous, or what Heidegger will call, over and over again, the "always already."

The claim he's going to make in this chapter—which might sound like an abstract claim, but I'll break it down in the simplest possible terms—is that we are always already involved in the situation of the world, and I *am* that situation. Therefore, I'm not a thinking thing over against a realm of extended things. That picture of the world is what you can find in Descartes, where Descartes divides up the world into two types of things: thinking things, like us, and extended things, like cups and tables and objects in the world (beings unable to recite Hamlet's soliloquy). That Cartesian view of the world is the picture that Heidegger is going to go on to destroy, or *deconstruct*, in part B of chapter 3 (paragraphs 19–21). We'll come to that next time. The point to retain here is that the human being is not a thing, whether thinking or extended: the human being is not a present-at-hand item, some sort of substance. The world's thingly quality is not disclosed theoretically in terms of what Heidegger will call an "ontology of the present-at-hand" (SZ 95–100, 129, 165 / BT 128–34, 167, 209). What he means by that is a way of conceiving things theoretically as they show up to a perceiving subject. Rather, the thingly quality of the world is disclosed *practically*, in being fascinated or benumbed, being alongside it, being inside its outside. To put this somewhat gnomically, a little bit enigmatically (and I'll try to make good on this), the thought here is that we are *inside the world's outside*, or we're outside on the world's inside, thus breaking with any opposition between the inside and the outside. (To reference the allegory of the cave in Plato's *Republic*, there is no cave—or, *it's all cave*.) This is what the great comedian Eddie Izzard would call a "stroky beard" moment in Heidegger. But it's less mysterious than it sounds.

Does this mean that Heidegger makes the world subjective? He raises this issue later in *Being and Time*. To make the claim that the world is subjective for Heidegger is to make the mistake of understanding Dasein, the human being, as a subject divorced from the reality of the world. That's the Cartesian picture that he's pushing against: for Heidegger, the subject-object distinction is an abstraction that arises at a secondary level. It is, as he puts it, a founded mode of being-in-the-world. Dasein is a priori, always already, worldly. Something like a world has always already announced itself to Dasein; and that verb, to announce itself (the German is *sich melden*), is a verb that's used very often in these pages. The question for Heidegger, as I said last time, is a question of *access*, access to the world. But on the other side of that, "How does the world announce itself?" is very much the way Heidegger puts this thought. The key thought at the core of what he is doing in these pages is that Dasein is *openness*, Dasein is a thrown-open clearedness. Dasein is always already given over to the world in which it finds itself. Our inside, our "self" (whatever that means), is always already outside in the world that has announced itself.

An early translation of "Dasein" in French was by a philosopher called Jean Beaufret. He was very important for the reception of Heidegger in France. (He was the teacher of my teacher in France, Dominique Janicaud.) Heidegger's "Letter on Humanism" was a letter written to Beaufret. And Beaufret's first French translation of "Dasein" was "*l'ouverture*," "the opening."[2] And that's a very good thing to keep in mind. We, Dasein, are the opening. The radicality of Heidegger's first step in chapter 3, in the analysis of world, is that he claims that we cannot get at the *phenomenon* of the world—that which shows itself—by adducing or adding together all the things that there are. We could add together human things, animals, artifacts, and natural things by putting together all those elements and then tracing them back to their ultimate foundation in nature and then providing a scientific determination of nature itself. This is a temptation in many ways, as it is the dominant view of the world in what some people call "the West." (I'm not thrilled with that term, but there we are.) If you adopt that as a philosophical view, then you are a naturalist. Naturalism, which is the most common view of the

2. Martin Heidegger, "Brief über den 'Humanismus,'" in *Wegmarken*, ed. Friedrich-Wilhelm von Herrmann, vol. 9 of *Gesamtausgabe*, ed. Friedrich-Wilhelm von Herrmann (Klostermann, 1976), 313–64; "Letter on Humanism," trans. Frank A. Capuzzi, in *Basic Writings*, rev. ed., ed. David Farrell Krell (HarperCollins, 1993), 213–65; hereafter cited as BH/LH with corresponding page numbers. Beaufret's French interpretation of Heidegger was first published in a series of articles for the journal *Confluences*. See Jean Beaufret, "A propos de l'existentialisme," *Confluences* 2–6 (1945).

world amongst philosophers, claims that being, or that which is, is explained by natural science or by the methods of natural science.

Heidegger's view is not that naturalism is wrong, but that it doesn't give us a satisfactory definition of the way the world shows up. For Heidegger, we can't get at the phenomenon of the world by engaging in a naturalistic, or scientific, determination of a person-independent, causally interacting nexus of natural objects. The scientific conception of the world conceives of the world as present-at-hand—as objectively present to the intellect, to theoretical perception—and disregards the existential context of our lived life in the world. That's what Heidegger is really concerned with: how things show themselves first and foremost in our lived existential context.

Now, the major claim in this analysis of the world, a claim made in paragraphs 14–18, is that the world announces itself, or we become aware of it proximally and for the most part, most closely and mostly, as a world that is *handy and useful*. This is the world of common, average, everyday experience, the world that we're already in a relationship with. My proximal encounter with this table I happen to be leaning my arms on is not as an object made of a certain definable substance existing in a geometrically ordered space-time continuum, as for Descartes or Kant. Rather, this is the table that's just "right here" or "over there," which is useful, handy for arranging my papers and books when I'm giving a talk. In trying to stay close to that which shows itself, we have to "[thrust] aside our interpretative tendencies," which cover over the encounter with the world (SZ 67 / BT 96). To read Heidegger, we have to mistrust and push aside our usual interpretative tendencies. The problem with getting clear on who we are, what a world is, what others are in that world—what keeps blocking it—is our standard habits, opinions, and received views, the views that are handed down by tradition and not sufficiently questioned or examined. They keep getting in the way. And what Heidegger is trying to do in *Being and Time* is to give us another vocabulary for approaching these issues.

How does the world show itself? Heidegger's thought in chapter 3 is that the world shows itself as useful, as full of handy things, what he calls the ready-to-hand (*Zuhandenheit*). This is what Heidegger means by the term "equipment" (a seemingly weird word that always reminds me of camping equipment). The German term is *das Zeug*, and that can be thought of in a much more simpleminded way. Think of *Zeug* as *stuff*. Heidegger's thought is that the world is full of stuff: of *pragmata* (πράγματα) in Greek, of things (SZ 67–68 / BT 96). And those things show themselves first and foremost as handy. If it makes things easier for you, then please go ahead and think that Heidegger is advancing a *pragmatic* ontology. He is certainly a quasi-pragmatist, and in many ways the thinker who is closest to Heidegger in orientation would be someone like William James in his late unfinished philosophical program for a radical empiricism that abandoned both the concept of consciousness and a static conception of the objective world.[3] For James, self and world form a fluent relational and mobile ocean of experience. James was a much better prose stylist than Heidegger was.

Stuff, equipment, is the name for the things encountered in our everyday concern, what Heidegger calls *Besorgen*. Heidegger's

3. William James, *Essays in Radical Empiricism*, ed. H. G. Callaway (Lexington Books, 2022).

thought is that the world is a realm of usefully handy stuff, and all of that stuff has a structure. That structure is what he calls the "in order to"; namely, I use things "in order to" do something. I use a hammer in order to bang nails into a piece of wood. (Admittedly, I'm not very good at examples!) I use things in order to produce a certain outcome. Now, another important point here is that equipment, stuff—which is why "stuff" is a really good translation—is never singular. It is always, as Heidegger says, "a totality of equipment" (*Zeugganzheit*; SZ 68–69 / BT 97–98). And we can think of a totality of equipment in much more simpleminded terms as *a whole of stuff*, a mess of things. And the whole of stuff has, Heidegger says, a referential structure. What does he mean? This is an important point for him, but a very simple point; namely, that stuff always refers to other stuff. It's never alone. It's never one piece of stuff. Stuff always comes together with other stuff. If I use the medium-sized dry goods that philosophers like to use as examples: I have my glasses, I have my glasses case, I have a pen, I have a book, I have a table, I have a chair, I have a room I'm sitting in, there's a heating system, there is lighting, there are floors, there are windows, it's in a building, the building is on a street, the street is in a city, and so on and so forth. Stuff is always a whole of stuff. And that stuff refers to other stuff, and that *whole of stuff* makes up the context in which I find myself.

Heidegger's view is that my primary access to the phenomenon of the world is not through scientific inquiry or through a philosophy of nature but rather through a phenomenological, descriptive elucidation of that which shows itself. And what shows itself is the average, everyday world that is the source of my concern and my fascination. To translate this into another idiom, what Heidegger is performing in his analysis of the world is an inversion of the theory-practice distinction. Heidegger wouldn't be happy with this; he would have these terms in quotation marks with multiple qualifications. But let's just use that distinction because it might be helpful. My primary access to the world is not theoretical, it is not the bare perception, sensation, or the spectatorial relation of the theatergoer in the theater of the world. The word *theory* derives from the word *theoros* (θεωρός) in Greek, which was a term for a spectator in a theater, someone who was looking at a play. Heidegger's point is that our primary access to the world is not that of spectators in a theater looking at a world that is somehow value-neutral and objectively present, which we discern theoretically. Rather, my primary encounter with the world is practical, a relation to a world of things that are useful to me and that are imbued with significance, imbued with meaning. The world *makes sense*. It hangs together. I'll come back to this idea.

For Heidegger, our theoretical view of things is founded upon a practical insight, and this insight is what he calls *Umsicht*, which is translated as "circumspection," literally "around-seeing" or "seeing-around." We are practically involved in the world. As I said last time, the being of Dasein is existence, and the a priori structures of Dasein are existentials. Heidegger has two categories for things which are *not* Dasein. And those two categories, under which we can apprehend things, are the present-at-hand (objectively present things as potential objects of scientific inquiry or theory) and the handy, the ready-to-hand (those very same things cited as useful, as imbued with practical significance). The basic claim that he's making is that those things that are ready-to-hand are not only what is closest to us, proximally and for the most part, but also what is meaningful for us. They're also ontologically *prior*, prior in terms of the order of that which is. *Praxis*, that lovely old Greek word, is prior to and has priority over theory. Although—as Heidegger was very acutely aware—this proposition or this thought of praxis being prior to theory is something that is made as a theoretical proposition in a philosophy book. Ay, there's the rub, as Hamlet would say (HAM 3.1.65).

But the thought is that who I am as a human being, as Dasein, is inextricably bound up and bound together with the complex network of social practices that make up my world, a world that is part and parcel of who I am (this is where Heidegger gets close to Hegel on communal ethical life, or *Sittlichkeit*). The world is the surrounding world, the world that I care about. This is what Heidegger means when he talks about the world as *Umwelt*, as environment, literally as the "around-world," the world that surrounds us, that environs us. There's no way I can cut myself off from my world: I *am* my world. The fabric of Dasein's *openness* is one piece.

What we're going to see now, after I've given you the shape of the discussion of the world in *Being and Time*, is Heidegger breaking down this idea of Dasein as being-in-the-world into a series of analytically distinguishable elements. Analytically or formally distinct: we can make those demarcations philosophically, but they are lived experientially as a whole, as a unity. In this chapter, we are looking at the concept of world, and after a discussion of space, we'll be looking at the relationship to others. The world is not just a world of things, it's also a world where there are other Daseins. How do we account for that? How do we describe *social being*? This is the topic of chapter 4.

What I have said so far provides an overview of the claim that Heidegger's making in this chapter on world. Now I want to give a survey of the particular claims that he's making in specific paragraphs

(15–18). Let's go through the four paragraphs where Heidegger really lays out his discussion of world. The first one is paragraph 15. Heidegger has really ugly titles for these chapters, but they are actually entirely accurate. Paragraph 15 is called "The Being of the Entities Encountered in the Environment." That's pretty, isn't it? What is the being of those things that we encounter in the environment? Well, their being is ready-to-hand: handy, useful, practically available. And the things that we encounter in the environment are what I've called "stuff," or *Zeug*. And this stuff always has the structure of an "in order to." *Zeug* is never singular, it always exists with reference to other stuff, and it makes up a *whole* of stuff, a totality of references or assignments, which we view with circumspection, Heidegger says.

We then begin to encounter a fascinating theoretical problem, namely that the ready-to-hand *as* ready-to-hand, the handy *as* handy, is not grasped theoretically: the peculiarity of the ready-to-hand is that it must withdraw in order to be grasped as such. It is only when the ready-to-hand recedes, the handy breaks down, the useful becomes useless, that it grabs our attention. The question is, then, how does the world as ready-to-hand announce itself? *It announces itself as present-at-hand*, in the breakdown of the ready-to-hand. When the hammer breaks, we begin to see it as such; that is, we begin to see it as a hammer. But when we're seeing it as a hammer—namely, as a hammer that no longer functions, that we can't just forget about and use—then we're seeing it under the category of the present-at-hand. And that's going to be developed in paragraph 16.

Two more thoughts on paragraph 15. First, every ready-to-hand thing for Heidegger has what he calls a "towards-which," or a "what-for," *das Wozu*. I'm wearing a watch (actually I'm not wearing a watch, but let's imagine I am), and I'm using it *in order to* tell the time. And the watch has a what-for structure; namely, it's *for me* to tell the time. Dasein is that towards which, or for what, things are. We'll come back to that thought: the world is a world *for us*, for Dasein.

Second, before we leave this paragraph behind, we should notice that there are little moments here where Heidegger begins to anticipate a concept that we can find in his later work, which is the theme of the *earth*. Ready-to-hand things are not just parts of the world. They're also made out of material stuff and are composed of earth, of matter (SZ 70 / BT 100). The relation between world and earth is something that Heidegger will really begin to analyze in a lecture called "The Origin of the Work of Art," from the mid-1930s, where he'll say something like this: An artwork is always world-constituting, an artwork "sets up a world," but it sets up a world out of the earth from which it is composed. And in setting up a world, the earth juts through, showing itself and withdrawing, which is how Heidegger understands the ancient Greek concept of *physis* (φύσις), as hiding and revealing.[4] Nature likes to hide. But we'll come back to that thought later (see ch. 17, "Existential Time and Vulgar Time").

Let's move on to paragraph 16. The thesis here (again, thinking of the really ugly titles Heidegger uses for these paragraphs: "How the Worldly Character of the Environment Announces Itself in Entities Within-the-World") is that the ready-to-hand announces itself in the breakdown of the totality of assignments or references

4. Martin Heidegger, "Der Ursprung der Kunstwerkes" (1935–36), in *Holzwege*, ed. Friedrich-Wilhelm von Herrmann, vol. 5 of *Gesamtausgabe*, ed. Friedrich-Wilhelm von Herrmann (Klostermann, 1977), 1–74, esp. 31–36; "The Origin of the Work of Art," in *Poetry, Language, Thought*, trans. Albert Hofstadter (Harper & Row, 1971), 15–86, esp. 44–47; hereafter UK/OWA followed by page numbers.

in the ready-to-hand. This occurs when the hammer breaks. How does this happen? Heidegger introduces three terms in this paragraph: he says that the worldly character of things is revealed when the ready-to-hand becomes "conspicuous"—that is, when it becomes unusable—when it becomes "obtrusive," and when it becomes "obstinate." Conspicuous, obtrusive, and obstinate (SZ 73–74 / BT 102–4). For example, when things are working perfectly well—say, when I'm tapping away at my laptop in the local coffee shop (not that I ever do that, but you get the idea)—and things are going swimmingly, and then suddenly the screen goes black and I've got no idea what's happening. My life has evaporated, and so on and so forth. At that point, the computer becomes conspicuous, obtrusive, and obstinate: it shows up as something that no longer works in the way that it did.

The question that Heidegger then raises is: How does this clarify the phenomenon of the world? The thought here is that, with the breakdown of stuff, the world of references into which, say, that hammer or computer fitted, becomes conspicuous, and the world announces itself. The point that Heidegger is making is that the world shows itself up, is disclosed, as ready-to-hand in the breakdown of the ready-to-hand, when the ready-to-hand becomes present-at-hand. And then he raises the following question: If when I'm using my computer I'm practically engaged with it, and then the damn thing breaks and it becomes conspicuous, obtrusive, and obstinate, and I begin to see that computer as something present-at-hand that needs to be fixed, is there a way in which the ready-to-hand can show up as ready-to-hand *without* becoming present-at-hand? Is there a way in which I can apprehend the handiness of the world without falling into the present-at-hand? And the answer to that question is going to be through the concept of signs in paragraph 17.

Signs, Heidegger says, are equipment that disclose the world of the ready-to-hand in a way that is available for circumspective concern, the way in which I orientate myself in the ready-to-hand, practical world. A sign, Heidegger says, is something which *indicates* something, and in its indicating, it refers *to* something. There's a relationship between the sign and that to which the sign refers. This paragraph is full of wonderfully anachronistic examples, and in many ways one of the most charming features of *Being and Time* is its use of outmoded examples, a little like watching a silent movie from the 1920s. For instance, Heidegger says, "An example of a sign would be the south wind for a farmer. When the south wind is blowing, then rain is on the way" (SZ 80 / BT 111). I've got no idea whether that's right or not. I'm not a farmer living in southwestern Germany, so let's just take it as it is. Another charming example he gives of a sign is a knot in a handkerchief. I neither have a handkerchief nor tie knots in them, but let's imagine that we all have handkerchiefs with knots, and that the knot is a sign that we must remember to call our mothers or buy some exotic cheese from the shop or whatever it might be. In that case, the knot in the handkerchief is a reminder, which shows up as something that I see. It's a sign, but it's still ready-to-hand. And then Heidegger gives another example. It's not meant to be funny, but I think it appears a little bit comical. He says, and imagine I'm speaking in a slightly ridiculous German accent, "As an example of a sign, we have chosen one which we shall use again in a later analysis, though in another regard. Motorcars are sometimes fitted up with an adjustable red arrow" (SZ 78 / BT 108). What is Heidegger talking about? "Adjustable red arrow"—I can just about remember this when I was a kid. Really old cars wouldn't have indicator lights, they'd have little arrows that would pop out the side of the car. It would zoom up and indicate whether the car was going to turn left or right. Just imagine that slightly antique world.

> Motorcars are sometimes fitted up with an adjustable red arrow, whose position indicates the direction the vehicle will take—at an intersection, for instance. The position of the arrow is controlled by the driver. This sign is an item of equipment which is ready-to-hand for the driver in his concern with driving, but not for him alone: those who are not traveling with him—and they in particular—also make use of it, either by giving way on the proper side or by stopping. (SZ 78 / BT 108–9)

It's a rather silly example: the sign, the indicator, signals that the driver is indicating whether they are going to turn left or right. And that sign is also something that indicates to fellow drivers traveling on the road and looking at the car: it lights up for them (and for pedestrians too). The point is, Heidegger continues, that "this sign is ready-to-hand within-the-world in the whole equipment-context of vehicles and traffic regulations" (SZ 78 / BT 109). "World of vehicles and traffic regulations"—I bet you never thought you would be reading a philosophical discussion on such gripping topics. But actually, these signs are vitally important—turning signals, crossing lights, traffic lights—and we are masters or experts at reading them.

The point of signs is that the sign is something ready-to-hand, something practical, which indicates the whole of stuff, the "referential totality" (*Verweisungsganzheit*) of the ready-to-hand (SZ

69–70 / BT 99). In this way, for Heidegger, the being of the ready-to-hand is revealed as constituted by a totality of references, a whole network of references. The car signal refers to the car. Obviously, the motion of the car. The car refers to the road. The road refers to other vehicles on the road. The road refers to the transportation network of the town. The town refers to the landscape, the environment in which you find that town, and so on and so forth. The world as handy is constituted by a network of references of one thing to another.

Now, where does this leave us? We are now in paragraph 18, which is where Heidegger tries to pull things together. It's called "Involvement and Significance." What is meant by those two concepts? As I said previously, the Macquarrie and Robinson translation of *Being and Time* is a classic, one of the masterpieces of English philosophical translation, in my humble opinion. When Macquarrie and Robinson decide to use a certain word, there's a long footnote that explains why they made that decision and also other possible meanings of the term in German. So, there is this term, "*Bewandtnis*," which is translated as "involvement." There are a number of moments in *Being and Time* where the choice of the translation actually says nothing. What is "involvement"? In a sense, who knows? But what it means in the context of the argument is that each thing, each bit of stuff, as handy, ready-to-hand, and available, is always already related to other things in a totality of relations or a whole of stuff. A total context or situation is disclosed in our intentional, concernful relation to things.[5] Think about involvement as the *situation*, as

5. This is explained in a translators' note by Macquarrie and Robinson. See BT 115n2.

the total context that we find ourselves in, as the *world* that we happen to be in, which is constituted by things: glasses, glasses case, pens, book, table, chair, room, windows, the whole thing. We're always already *immersed in* this network of relations. As I mentioned last time, Dasein is a networked being.

The world of the ready-to-hand—the handy, practical world—is a structure of involvements, Heidegger says, where everything is defined by what he calls a *Wozu?*, a "what for?" But this structure of involvements has its end in us, in Dasein. Dasein is the being *for the sake of which* (*das Worumwillen*) these ready-to-hand things are destined. The world is for the sake of Dasein. Otherwise said, we are the a priori condition of possibility for the structure of involvements that we call the world. What characterizes us is that we have an understanding of the world: *Weltverstehen*. And the world, as a world *for us*, is a world that relies on us, a world that we *make*.

This takes us to the second concept in this paragraph—namely, "significance." Dasein's familiarity with the totality of involvements in the world is one where the world is *significant*. In other words, the world makes *sense*. Significance makes up the structure of the world. As Merleau-Ponty used to put it in another connection, criticizing Sartre's thought, it's not that we're condemned to freedom, "we're condemned to meaning."[6] The world just hangs together, makes sense. That's just the fact of the world. There's no particular

6. Maurice Merleau-Ponty, *Phenomenology of Perception*, trans. Colin Smith (Routledge, 1962), xxii. See also Jean-Paul Sartre, "Existentialism," trans. Bernard Frechtman, in *Existentialism and Human Emotions* (Philosophical Library, 1957), 22–23; and Jean-Paul Sartre, *Being and Nothingness: A Phenomenological Essay on Ontology*, trans. Hazel E. Barnes (Washington Square Press, 1956), 186.

moral value attached to significance. Meaning is not deep. It's just that things make sense, unavoidably.

Now, I want to conclude this discussion by reading you a poem. And this poem, in a different, lyrical register, elucidates the thought that Heidegger is trying to advance in *Being and Time*. The poem is by someone who, for me, is the greatest philosophical poet in the English language: Wallace Stevens. It's a poem called "The Idea of Order at Key West," which was written in 1934 and first published the following year. Let's read it, and then I'll make a couple of remarks about the poem. This is a particularly apropos thing to do in relation to Heidegger, because *Being and Time* is a work of philosophy in a phenomenological style, in a classical style, what philosophers call a transcendental approach, where we are seeking the a priori conditions of possibility of experience. But as Heidegger proceeded with his path of thinking he became increasingly dissatisfied with standard philosophical approaches to matters and became very interested in the way in which poetry could disclose and illuminate the world and the human being's life in the world. Let's read the poem.

THE IDEA OF ORDER AT KEY WEST[7]

She sang beyond the genius of the sea.
The water never formed to mind or voice,
Like a body wholly body, fluttering
Its empty sleeves; and yet its mimic motion

7. Wallace Stevens, "The Idea of Order at Key West," in *The Collected Poems of Wallace Stevens* (Knopf, 1954), 128–30.

Made constant cry, caused constantly a cry,
That was not ours although we understood,
Inhuman, of the veritable ocean.

The sea was not a mask. No more was she.
The song and water were not medleyed sound
Even if what she sang was what she heard,
Since what she sang was uttered word by word.
It may be that in all her phrases stirred
The grinding water and the gasping wind;
But it was she and not the sea we heard.

For she was the maker of the song she sang.
The ever-hooded, tragic-gestured sea
Was merely a place by which she walked to sing.
Whose spirit is this? we said, because we knew
It was the spirit that we sought and knew
That we should ask this often as she sang.

If it was only the dark voice of the sea
That rose, or even colored by many waves;
If it was only the outer voice of sky
And cloud, of the sunken coral water-walled,
However clear, it would have been deep air,
The heaving speech of air, a summer sound
Repeated in a summer without end
And sound alone. But it was more than that,
More even than her voice, and ours, among
The meaningless plungings of water and the wind,

Theatrical distances, bronze shadows heaped
On high horizons, mountainous atmospheres
Of sky and sea.

It was her voice that made
The sky acutest at its vanishing.
She measured to the hour its solitude.
She was the single artificer of the world
In which she sang. And when she sang, the sea,
Whatever self it had, became the self
That was her song, for she was the maker. Then we,
As we beheld her striding there alone,
Knew that there never was a world for her
Except the one she sang and, singing, made.

Ramon Fernandez, tell me, if you know,
Why, when the singing ended and we turned
Toward the town, tell why the glassy lights,
The lights in the fishing boats at anchor there,
As the night descended, tilting in the air,
Mastered the night and portioned out the sea,
Fixing emblazoned zones and fiery poles,
Arranging, deepening, enchanting night.

Oh! Blessed rage for order, pale Ramon,
The maker's rage to order words of the sea,
Words of the fragrant portals, dimly-starred,
And of ourselves and of our origins,
In ghostlier demarcations, keener sounds.

I don't want to reduce this beautiful poem to a philosophical example, which is usually how philosophers deploy artworks (which is a hideous practice). But there is philosophical thinking going on here. The poem describes a situation where there is someone, "she," who's singing, and she's singing, walking alongside the sea. She and the sea: those are the two initial characters in the poem. And "She sang beyond the genius of the sea" is where we begin. It is not at all clear who this "she" might be: a person, a woman, a spirit, an angel, a siren. We have no physical description of her. She's just described with a pronoun. She is observed by two other characters: let's say the poet, or at least the narrative voice in the poem (which we could flat-footedly identify with Wallace Stevens), and another character called Ramon Fernandez. And the action of the poem is that Ramon and the poet are observing this figure, "she," walking by the sea, singing. The location is Key West, in Florida, in the 1930s. (It must have been nicer then. I went there once for a day about fifteen years ago and thought it was a bit tacky.)

The conclusion the poetic voice comes to is that "she was the single artificer of the world / In which she sang. And when she sang, the sea, / Whatever self it had, became the self / That was her song, for she was the maker." "There never was a world for her / Except the one she sang and, singing, made." The world is a world that is constituted through the song. In this case, the song of the woman, the spirit. And whatever meaning the world has is dependent upon her and dependent upon her song. The raw idea here is that the world is the world. The sea is the sea. They exist as a brute material reality, what Stevens calls "the meaningless plungings of water and the wind." Brute matter is meaningless. But that stuff, that natural stuff, becomes composed into a world of meaning through the activity of song, through the activity of "her." "There never was a world for her / Except the one she sang and, singing, made." The world has *meaning* for us and is dependent *upon us* for meaning. A meaningful world, a world of value (to speak in the traditional terms that Heidegger opposed), is a world that is Dasein-dependent. If we did not exist, if Dasein did not exist, then the world would not be there. The world, as a meaningful world, would just not show up without us. We are the worldhood of the world.

What happens in the poem is that Ramon Fernandez and the poetic voice turn away from the woman singing and the sea by which she is singing, and they turn back towards the town. Then a question is asked: "Tell why the glassy lights / The lights in the fishing boats at anchor there / As the night descended, tilting in the air / Mastered the night and portioned out the sea." Why is there an order? Why, when we turn away from the woman singing by the sea, and we turn back towards the town and we look, why does it appear that the town is ordered, is organized, hangs together in a certain way, is composed in a certain way? Well, the answer is, that's just what we do. We are ordering beings. We give an idea of order to experience. We are being-in-the-world with involvement and significance, and we are the beings who always bring meaning *to* the world. The next step is going to pick up and develop that thought in relation to the theme of space. Ask yourselves, Why is this book not called *Being and Space*? Why time and not space? We will pick that up next time.

3. Space

What is wrong with Descartes's picture of the world — Extension renders the world invisible — Phenomenology and the destruction of the modern world picture — Why is this book not called Being and Space? *— Space as room and roomliness — The unavoidable character of spatial metaphors — Nearness, place, and neighborhood — Art and space — The sacred, temple, and* templum *— The near, clearing away, and bringing things close — Churches, graves, radios, telephones — In Dasein, there is a collapsing of distance — Heidegger's theory of social media — Against the metrical measurement of space — Freeing place and why architects love Heidegger — Fernando Pessoa and lessons in unlearning.*

What is the relationship between being and space? The pages that we've been looking at—namely, chapter 3 of *Being and Time*, division 1—are in three parts. In the previous chapter I went through the first part of that, paragraphs 14–18. Now I want to focus on the last part, paragraphs 22–24, which is on space. But first let me say a few words about the second part, paragraphs 19–21. It's a discussion of René Descartes that is meant to provide a negative confirmation of Heidegger's fundamental ontology of the world. The point is to show how a certain picture of the world is formed that we can find in Descartes and in the philosophers that come after him—notably, in Kant. This is a picture of the world that Heidegger believes is impoverished or deficient and that needs to be deconstructed or destroyed in order for the Heideggerian view of the world to get a grip.

Very crudely, we could ask ourselves, How does Descartes think about the world? Well, he thinks about the world in relation to one concept: *extension*. For Descartes, the world is a totality of things that are extended in space. As you will know if you have even a cursory sense of what he is up to, Descartes distinguishes between two kinds of "things." There are, on the one hand, *extended* things (things in the world, including natural things and things that are made, artifacts). On the other hand, there are things like us, and we are *thinking* things.[1] The question that Heidegger is interrogating is what happens with this idea of the world as extension, and the point he makes is that Descartes's "ontology of the world," as a totality of things extended in space, does not seek the *phenomenon* of the world at all. It passes over, leaps over, the world as it shows itself and interprets the world in terms of the ontology of the present-at-hand. Descartes sees the world with philosophers' spectacles, as a pure, intellectual beholding, what Heidegger calls *noein* (νοεῖν), or "intellection" (SZ 96 / BT 129). The claim he makes is that the being of things or entities is determined by Descartes as "constant presence-at-hand." Things are objectively observable by the intellect, available to knowledge based upon intuition or perception. As you

1. René Descartes, *Principles of Philosophy*, in vol. 1 of *The Philosophical Writings of Descartes*, trans. John Cottingham, Robert Stoothoff, and Dugald Murdoch (Cambridge University Press, 1985), part 1, principle 53 (p. 210): "A substance may indeed be known through any attribute at all; but each substance has one principal property which constitutes its nature and essence, and to which all its other properties are referred."

might know, Descartes is suspicious of the evidence of sensation and perception. He moves back from empirical sensation towards a more abstract, more intellectualist idea of extension, which can be determined mathematically. Mathematics, Descartes thinks, is the only form of thinking through which secure knowledge can be gained (SZ 95 / BT 128).

Heidegger's thought is that the being of the world for Descartes is apprehended as present-at-hand, and if you begin from the idea of seeing the world theoretically and intellectually, then you *miss* the phenomenon of the world, the practical, living realm in which we dwell. It's that Cartesian picture of the world that Heidegger thinks has to be destroyed, radically deconstructed. Descartes does not see the world as it shows up practically in a handy, average every-dayness. If one adopts a Cartesian ontology of the world, then the livedness of the world will be overlooked, *invisible*.

Furthermore, this passing over of the phenomenon of the world in Descartes is not accidental. It's not a simple mistake that can be rectified. Rather, there is what we might call a tendency towards *falling into the tradition* that happens in Descartes. Heidegger will show how Descartes's idea of extension, of world as being composed of extended things, draws from a medieval understanding of substance, and that goes all the way back to the Greek idea we find in Aristotle of substance as *ousia*, οὐσία, the essence that defines the nature of things. The point that Heidegger is making is that Descartes just takes over a determination of being from the tradition without examining its meaning, and that taking over is part of a tendency towards falling. For Heidegger, this falling into tradition defines the modern "world picture," or what philosophers like to call "modernity." People love to talk about modernity. I don't like to enthuse about it so much, as—with Bruno Latour—I'm not really sure if we have ever been modern.[2]

Yet the idea developed in Heidegger's later work is that there's a modern world picture, a Cartesian world picture, the world as composed of objective reality that can be best understood scientifically, and the paradigm for science in the modern worldview is mathematics. What we see in the Cartesian picture of the world is the emergence of the mathematization of nature. As with the epistemological construal of the world, it's not that Heidegger thinks that this conception is *wrong*, it's just working with a set of assumptions that need to be carefully examined as they are taken over from a tradition that has not been thought through. Descartes's ontology of the world as extension takes the mathematization of the world to be primary rather than secondary, foundational rather than founded. We have to deconstruct the Cartesian picture of the world in order to see that which shows itself as a practical, handy world that we live in and find meaningful but that has, in its phenomenal unity, become invisible, obscured by our handed-down interpretations of it.

Now, this raises a question, which I'll just mention here, though I'll come back to it later, especially in the coda to this book. What Heidegger is doing in *Being and Time* is phenomenology. Phenomenology, understood simply, is a description of existence and the attempt to deduce certain structures from the description of existence, what he calls the "existentials." And he characterizes those

2. See Bruno Latour, *We Have Never Been Modern*, trans. Catherine Porter (Harvard University Press, 1993).

existentials as a priori. That's *constructive* phenomenology. But phenomenology has to be linked, for Heidegger, to a destruction of the tradition. And indeed, in the introduction to *Being and Time*, he says that the question of the meaning of being, as he understands it, will only become "concrete" once we have proceeded through a complete destruction of the ontological tradition (SZ 22–23 / BT 44). This ontological tradition is the one that, in the modern period, finds its shape in the work of Descartes's mathematization of nature.

This has two consequences. First, what Heidegger is doing in *Being and Time* is a *formal* analysis, a formal phenomenology of the world. In order for the latter to become concrete, to become actual in our experience, we require a deconstruction of the world picture that dominates our thinking and that we find paradigmatically in Descartes. Second, what Heidegger is doing is close to ideology critique. Heidegger is trying to dismantle the ideology of the modern world picture that he thinks finds its real shape and force in the work of Descartes. And that Cartesian picture is then refined but inherited and reproduced in the work of Immanuel Kant, who is very much on Heidegger's mind as the entire philosophical context of the Germanophone (and indeed Francophone) world was dominated by forms of neo-Kantianism. Heidegger will trace Kant's thinking back to Descartes. The version of Kant that Heidegger traces to Descartes is the Kant of the neo-Kantian revival of interest in Kant as a philosopher of science. The neo-Kantian context is evident from the discussion in paragraph 21, where the present-at-hand conception of the world of entities as things of nature is supposedly rounded out by adding value-predicates like "useful," "beautiful," or "ugly" to material objects and claiming that this completes the picture. For Heidegger, it gets the whole thing back to front: it is not that "value" is added to material nature but rather that we first and foremost inhabit a practical world of "value," and then the scientific conception of nature is founded upon it. If we insist on speaking of the world as nature somehow *clothed* with value, then we simply pass over the phenomenal character of the world and render it invisible. The scientific conception gets it backwards.

Let's leave those large, intriguing questions to simmer in the background while we focus on the question of space in part C of the chapter. The first move that we should make is to note that the words for "space" and "spatiality" in German are *Raum* and *Räumlichkeit*, which are linked (as is evident from the way they sound) to the ideas of "room" and "roomliness." In many ways, Heidegger's point in these very interesting paragraphs of *Being and Time* is to understand the idea of space in terms of room, and spatiality in terms of roomliness—namely, that space is experienced by me right now as the room that I'm in or that we might be in together. And the roomliness of this room may consist in the fact that it is a nice temperature, it feels pleasant, the room is well lit, and so on. I first become aware of space as being inside a room. The point to make in relation to this thought is that the *insideness* of Dasein is not the insideness of consciousness or the mind, which is somehow in the head, but rather, it is a sense of being in a room, spatially spread out in the meaningfulness of a place that we inhabit with its referential totality of stuff and people. I am not "in here," enclosed within my skull and skin, but "out there" with you, open to the shared space of the room we inhabit.

Now, there's a vast and hugely important question about the relationship between time and space in Heidegger's work and about the connection of both these concepts to being. In the early work of Heidegger, it's fair to say that there is a subordination of space to time. Indeed, this is suggested by the title of the book that we're reading: *Being and Time*. The book is not called *Being and Space*. Being is time, and time is finite. The first thing to note is that the words we use to describe Dasein, terms like *openness*, *ekstasis*, *transcendence*, *disclosure*, *clearing*, *outsideness* —these are all spatial terms, spatial metaphors. If Dasein is a transcendent being, if being is to be in the world, if it's always already *out there* in the world, then the way I'm understanding the temporal being of Dasein is spatially.

There's a bigger question here about whether time can be understood temporally or whether time always has to be understood in terms of spatial metaphors. At the very beginning of *Being and Time*, on its untitled first page that we discussed previously, Heidegger says that time is the *horizon* for the understanding of the meaning of being (SZ 1 / BT 19). Again, *horizon* is a spatial term, let's say the line that separates the sea from the sky if I am looking at the ocean. Time is more important than space in Heidegger's early work. But in Heidegger's later work—and this comes out in his brilliantly difficult 1962 lecture "Time and Being," where he explicitly tries to complete the line of thought begun in *Being and Time*—there is a move to what he calls "Time-Space" (*Zeit-Raum*), which has a unitary structure.[3] But let's not get ahead of ourselves. I hope to speak and write about the later Heidegger in a future work.

3. Martin Heidegger, "Zeit und Sein" (1962), in *Zur Sache des Denkens*, ed. Friedrich-Wilhelm von Herrmann, vol. 14 of *Gesamtausgabe*, ed. Friedrich-Wilhelm von Herrmann (Klostermann, 2007), 18–19; "Time and Being," in *On Time and Being*, trans. Joan Stambaugh (Harper & Row, 1972), 14; hereafter ZS/TB followed by page numbers.

In paragraphs 22–24, the question of space is analyzed in three moments. Let me just mention the key concepts, and then I'll try and demonstrate them. First, in paragraph 22, the question is, What is the spatiality of inner-worldly, ready-to-hand things? Again, this is ugly, baroque Heidegger talk. What is the space of ready-to-hand, handy stuff? He identifies three terms: the first is "nearness," or "the near"; the second is "place"; and the third is "region." The German word for "region" is *Gegend*, which can indeed be translated as region, but it means something closer to "neighborhood" or "vicinity." Focus on these terms: the near, place, and region or neighborhood.

In paragraph 23, he talks about the spatiality of being-in-the-world, the spatiality of us, of Dasein. What do we do with space? He introduces two concepts here, which I'll explain. The first concept is that our relation to space is one of what he calls "de-severance" or "de-distancing" (SZ 104–5 / BT 138). Again, he's working with a German term that just means distance, *Entfernung*, and he's breaking that term up in a particular way and giving it a new sense. The prefix is a negation, *ent-*, and *fernung* means, one could say, "far-ing," or distancing. The idea of *Ent-fernung* is an idea of *de-faring*. What does that mean? It means that our relationship to space is one where we bring things into the near, we bring things close. Furthermore, we bring things close with *direction*—the second concept in this paragraph—or with a towards-which. Note the linkage here between bringing things close and direction. I'll say some more about that.

The third paragraph that discusses space is paragraph 24, and what Heidegger is concerned with is the genesis of the three-dimensional, objective understanding of space—namely, the concept of space that we associate with modern science, with the Newtonian picture of space. How does that three-dimensional, objective understanding of space get formed? His claim is that the objective idea of space arises as a *de-worlding*, as an increasing impoverishment in our lived existential understanding of space, which is what gives rise to the modern worldview.

In Heidegger's later work there is a preoccupation with space, as I mentioned. There's a late text called "Art and Space," from 1969. In this very short essay, Heidegger discusses space as a "clearing-away." Instead of the conception of space that we find in early modern science in Galileo Galilei and then in Isaac Newton, space as a homogenous expanse of cosmic space, we should think of space as a clearing-away. In this text, written forty-six years after *Being and Time*, Heidegger writes:

> Clearing-away [*räumen*] is uttered therein. This means: to clear out, to free from wilderness. Clearing-away brings forth the free, the openness for man's settling and dwelling. When thought in its own special character, clearing-away is the release of places toward which the fate of dwelling man turns in the preserve of the home or in the brokenness of homelessness or in complete indifference to the two. Clearing-away is release of the places at which a god appears, the places from which the gods have disappeared, the places at which the appearance of the godly tarries along. In each case, clearing-away brings forth locality preparing for dwelling. Secular spaces are always the privation of often very remote sacred places.[4]

4. Martin Heidegger, "Die Kunst und der Raum" (1969), in *Aus der Erfahrung des Denkens*, ed. Hermann Heidegger, vol. 13 of *Gesamtausgabe*, ed. Friedrich-Wilhelm von Herrmann (Klostermann, 1983), 203–10; "Art and Space," in *The Heidegger Reader*, trans. Jerome Veith, ed. Günter Figal (Indiana University Press, 2009), 305–9. [trans. mod.]

Thought of *existentially*, space is a clearing-away, the thrown-open clearing that we are. This shouldn't be confused with Marie Kondo or anything daft like that. It's not about cleaning things up, throwing out your books and other trash and living in a minimal Japanese or mid-century Scandinavian way. It's an idea of clearing-away as being based in locality and dwelling, and connecting those terms to the idea of the sacred, which is always linked for Heidegger to the idea of the *sacrum*, a sacred space, and as the sacred space of the area in and around a temple, a consecrated space. (In my view, the sacred is best experienced as cluttered, as crowded with holy objects, like relics, icons, and tombs.) The sacred, and appeals to the sacral, are always appeals to a place, a locality, a privileged site: the temple. A temple is a very good example of somewhere time and space are unified, where the temple *temples* and the sacral is a holding together of the space and the intergenerational time of a people. (The academy, too, at its best, allows this intergenerational communication.) But that is later Heidegger, much later.

Let's come back to *Being and Time*, beginning with paragraph 22. Heidegger is talking about the spatiality of the ready-to-hand in relation to what he calls "closeness," the near. The near is linked to place (*Platz*), which is the "where" of equipment, of stuff (SZ 102 / BT 136). Equipment always has its place. It's always somewhere: the hammer is in the hammer drawer, or wherever. The near is a "where," a place. Ready-to-hand equipment always has a place. This idea of closeness and place is linked to region or neighborhood. There is a *whereabouts* of stuff. In our average, everyday experience, space is "the above," "the below," "the behind." Don't mistake this for an abstract, three-dimensional directionality. This "above," "below," "behind," and so on, is how things show up right here (SZ 103 / BT 136–37)—that is, in the local space around us. When we're asked where the bathroom is, we don't give precise coordinates, we say "down there on the left" or whatever. Space is round about us—that's the key thing for Heidegger's conception of space. It is round about us in the same way that the environment, or *Umwelt*, is around us.

Heidegger offers the example of the sun, which is characterized by certain places. The sun is not objectively present-at-hand. Rather, the sun is experienced in terms of its place in relationship to our environment. Sunrise. Midday. Sunset. Midnight (SZ 103 / BT 137). This is why the geocentric view of the universe, with the earth at the center, is intuitively right, phenomenologically true.[5] It doesn't feel to us, proximally and for the most part, that the earth is rotating around the sun in some obscure corner of a vast galaxy of many millions of other galaxies in this ever-expanding cosmos. It feels like we're at the center of things, and things, natural things, show up in relation to where we are. It feels like the sun rises and sets. And it's that everyday, intuitive understanding of space that Heidegger wants to focus on here. He also talks about the way space is organized in dwellings and houses: he'll talk about how the house has its sunny side and its shady side, the way it's divided up into rooms, certain rooms that are sunny and others where the sun doesn't shine at a certain point or at all. We organize our dwelling lives spatially around rooms. He talks in these pages about churches and graves—

5. A related argument can be found in a posthumously published work by Husserl. See Edmund Husserl, "Foundational Investigations of the Phenomenological Origin of the Spatiality of Nature," trans. Fred Kersten, in Edmund Husserl, *Shorter Works*, ed. Peter McCormick and Frederick A. Elliston (University of Notre Dame Press, 1981), 222–23.

and certainly in western Europe, churches and graves point towards the East, they point towards Jerusalem. They're organized in accordance with a spatio-theological orientation.

The point Heidegger is making in paragraph 22 is that we do not perceive or apprehend space abstractly. We inhabit places, we inhabit our whereabouts, and the things that we have around us also inhabit those whereabouts. Things show up as present-at-hand only once they're no longer in their place. For example, if my phone or wallet is missing, I say to myself, "Where is my damn phone?" or "Where is my bloody wallet?" In that moment of the phone or wallet being missing from its place, there is a categorial switch from the ready-to-hand to the present-at-hand. But first and foremost, most closely and mostly, I experience the world as a world that surrounds me. I am that world. I do not exist apart from it.

This takes us to paragraph 23, where the "condition of possibility" for the spatiality of the ready-to-hand is *us*: Dasein. And this Kantian language of transcendental argumentation, in the phrase "condition of possibility," is scattered across the pages of *Being and Time*. In the same way that Dasein is the worldhood of the world, spatially *we* are the condition of possibility for this understanding of roomliness, spatiality. How is that idea of space experienced? The spatiality of being-in-the-world is, as I already said, de-faring, or de-severance, or de-distancing. And this de-distancing is a relation to space that makes farness disappear. Our relation to space is one where we *bring it close*. There's an extraordinary three-word sentence in this paragraph, which makes no sense in English. In German it is "*Ent-fernung entdeckt Entferntheit*," which can be literally rendered as "de-faring dis-covers distance." It's translated as "de-severance discovers remoteness" (SZ 105 / BT 139). Which says pretty much nothing. But the point he's making is really simple: it's that the way in which we discover space is by reducing distance, and Dasein is characterized, Heidegger says, by a tendency towards the near. Space shows up in terms of that towards which I direct myself. And here, continuing on with the anachronistic metaphors in *Being and Time*, we have some wonderfully antiquated technological metaphors. We get a discussion of radio: "All the ways in which we speed things up, as we are more or less compelled to do today, push us on towards the conquest of remoteness. With the 'radio,' for example, Dasein has so expanded its everyday environment that it has accomplished a de-severance of the 'world'" (SZ 105 / BT 140). Let's take the example of radio. When I turn on the radio—and I listen to the radio a lot, I'm a real radiohead—that radio could have its source in New York; in Washington, DC; in London, Paris, Dakar, or Vladivostok; or wherever. Streaming, on Spotify or YouTube, doesn't have the same relation to a source. There's a greater collapsing of distance because there is no explicit source. That closeness is closer; it never shuts off. But when I turn on the radio, I pull it into closeness. I'm in a relationship of complete intimacy, proximity, to the radio program. And this is how it is with the media, for Heidegger. He didn't have the good fortune—or bad fortune, depending on which way you see it—to live in a time, certainly when he wrote *Being and Time*, where there were televisions around. But television in German is *Fernseher*, a "far-seer." This makes Heidegger's point. The television allows us to see far. But in seeing far, the far is drawn in close. The *media are a collapsing of distance*. And obviously we could develop that thought in a way that would be pretty alarming to Heidegger in relation to social media. The intimacy of radio is one thing, the company of television and the broadcast media is another. But think about the intimacy of our relationship to social media, to our Instagram feed or whatever mindless bollocks we numb ourselves with or trigger ourselves into empty rage with. That's a different and more pernicious form of intimacy. The closeness of our digital devices creates dangerous effects, a pathological collapsing of distance.

There's a great example where he talks about spectacles and telephone receivers and streets: "When, for instance, a man wears a pair of spectacles which are so close to him distantially that they are 'sitting on his nose,' they are environmentally more remote from him than the picture on the opposite wall" (SZ 107 / BT 141). The spectacles through which I'm looking at my notes and your faces are further away from me than what I see with them. If one of you raises a question, then my lived space stretches out in your direction. Space is malleable, it shifts between the close and the far. The telephone rings and I'm pulled into proximity with someone who is far away. And if I'm on the street when the phone rings—the street that "slides itself... along... the soles of my feet," as Heidegger says—I'm pulled from the street I'm walking along into a phone conversation (SZ 107 / BT 141–42). Dasein brings things close, reduces distance.

Our fundamental relationship to space is one of *estimation*. We do not measure things objectively or scientifically. We measure things in terms of estimation and approximation. If Dasein's relation to space is one of bringing things close, then that closeness is measured in terms of how we guess that closeness. And this could bring us into a wider discussion of measurement, about which, for brevity's sake, I'll just say the following. This isn't something I believe, but let me just say it as a fun illustration. There's an argument for measurements in England, which is where I'm from, although I take no pride in the fact. It's an argument about measurement. "How long is an inch? An inch is an Englishman's thumb. A foot is an Englishman's foot. A yard is an Englishman's stride." The point is that measurement is estimation—guesswork or guesstimation—based upon the dimensions, in this case, of an Englishman. And this is then opposed to an objective understanding of measurement, which you can find, say, on the European continent, which for some English people (not me, of course) is a long way away and to be treated with much suspicion. And over there, in Paris, on the facade of the Ministry of Justice, they have the *meter rule*, placed there as the objective, absolute measurement of space. The metrical system exists in Cartesian space. Heidegger's point would be that actually we begin measuring by estimation and that the idea of an absolute measure or an objective measure is secondary to that, and that's the key thing that needs to be remembered. Our primary measurement of space is estimation. The lake is a "stone's throw" from here, prior to any exact spatial measuring. The farm is a "pipe smoke away" (SZ 105–6 / BT 140).

The second element of this paragraph, linked to the idea of bringing close, is the idea of "direction." It's not just that we bring things close, we bring them close by directing ourselves *towards* them. If the phone rings, I then direct myself towards the phone and bring the phone in close and speak to the person on the other end. There is a question of spatial orientation here. One philosopher Heidegger is thinking about in these pages, as elsewhere in *Being and Time*, is Kant: the problem of orientation in thinking, and how we orientate, say, between left and right. Kant thought that left and right had to be understood in terms of the relationship to the body and thereby to some conception of the subject.[6] Heidegger's point is that the problem of orientation has to be understood in terms of the directedness of being-in-the-world. We have to engage in a *phenomenology of direction* and bringing things close in order to understand the problem of orientation. Space is aquatic or viscous; we wade through it as we direct our attention to it.

This brings us to paragraph 24, which is where Heidegger argues for the priority of existential spatiality, roomliness, the roundaboutness of space and the derived, or founded, nature of scientific spatiality. And Heidegger often, as *Being and Time* proceeds, comes up with really elegant summaries of his argument, and they're well worth reading. For example, he says:

> As Being-in-the-world, Dasein has already discovered a "world" at any time. This discovery, which is founded upon the worldhood of the world, is one which we have characterized as freeing entities for a totality of involvements. Freeing something and letting it be involved, is accomplished by way of referring or assigning oneself circumspectively, and this in turn is based upon one's previously understanding significance. (SZ 111 / BT 145)

What's new in that quotation is the language of "freeing." Our relationship to the world, our relationship to space, is one where we clear away and we free things up and free ourselves up. The question now becomes, How does this three-dimensionality of pure space, how does the Newtonian or Galilean or physicist's view of space,

6. See Immanuel Kant, *Concerning the Ultimate Ground of the Differentiation of Directions in Space (1768)*, in *Theoretical Philosophy, 1755–1770*, trans. and ed. David Walford and Ralf Meerbote (Cambridge University Press, 1992), 361–72.

arise out of existential spatiality? It arises, as will not be surprising by now, through a relation of derivation: the cognitive understanding of space, the scientific understanding of space, the idea of space as what Kant calls the form of "outer-sense" for the subject, are derived from this existential understanding of space. It is the latter that comes first and that we have to focus on. Interestingly, in these moments in *Being and Time*, there are references to building, surveying, and calculation. This takes us back to what I was saying about the temple and the experience of the sacred. Architects love Heidegger because he dignifies their practice with a philosophical jargon by linking building and dwelling with thinking. He makes architects feel smart, which is fine. Like Christopher Nolan movies, Heidegger has a way of making stupid people feel smart. But I didn't say that.

If we are to understand what a building is, what it is to construct a place, then this has to be approached in relation to our being-in-the-world and not in relation to some abstract idea of spatiality, because that abstract idea of spatiality will not yield dwelling. It will actually accentuate what Heidegger sees as the homelessness of modern experience. We could imagine in this relationship to space the possibility of a Heideggerian architecture that would be based upon our lived experience of the environment, of the space as it is around us. And it's not that we have to reject the objective, scientific understanding of space. But we have to understand the genesis of that conception of space out of being-in-the-world.

I'd like to finish this discussion as I did the last one, with a poem that I think makes Heidegger's point. Actually, there are two poems, and they're by the Portuguese poet Fernando Pessoa. Pessoa is the least known and perhaps the greatest of the literary modernists—his work was left in a trunk in Lisbon, and only fifty to seventy percent of it has been published to this day. He wrote in what he called "heteronyms," completely distinct personages with separate biographies and separate lives and different names. In fact, one of the characters in this universe was Fernando Pessoa himself. There are others, many others, like Álvaro de Campos and Ricardo Reis, in various languages. (Pessoa wrote very well in English and French, as well as in his native Portuguese.) But the master heteronym was someone called Alberto Caeiro, a poet-shepherd living on a hill. He writes poetry of enormous simplicity and wry humor. And I think this particular poem by Alberto Caeiro makes Heidegger's point:

> What we see of things are the things.
> Why would we see one thing when another thing is there?
> Why would seeing and hearing be to delude ourselves
> When seeing and hearing are seeing and hearing?
>
> What matters is to know how to see,
> To know how to see without thinking,
> To know how to see when seeing
> And not think when seeing
> Nor see when thinking.
>
> But this . . .
> requires deep study,
> Lessons in unlearning,
> And a retreat into the freedom of that convent
> Where the stars—say poets—are the eternal nuns
> And the flowers the contrite believers of just one day,

But where after all the stars are just stars
And the flowers just flowers,
Which is why we call them stars and flowers.[7]

The thought here is that we have to learn to see things as *things* and to see them in the space in which they appear and not delude ourselves into believing that we're seeing them through philosophers' spectacles or through the lens of some theory or other. We have to learn to see without thinking and to not think when we should be seeing. And that sounds simple, deceptively simple, but, Pessoa goes on (or Alberto Caeiro goes on), this "requires deep study, / Lessons in unlearning."

In many ways, what Heidegger is trying to do in *Being and Time* is to give us lessons in unlearning the ways in which modern philosophy and science have explained the world, unlearning the Cartesian world picture that has got a strong grip on us. Unlearning that and relearning how to see the world as it shows itself, how we truly see things. This is the riddle, or enigma, of Heidegger. On the one hand, what he's saying is incredibly simple. He's trying to describe the absolutely obvious. But the obvious, the truly simple, has been blocked from our view by the world picture that we've come to inhabit in the modern era, and that's what has to be destroyed in order to allow us to see things *as* things.

As I'm in the mood, let me read you another poem by Caeiro, which makes the same point in a different, and maybe even better, way:

7. Fernando Pessoa, "What we see of things are the things," in *Fernando Pessoa & Co.: Selected Poems*, trans. and ed. Richard Zenith (Grove Press, 1998), 57.

To not think of anything is metaphysics enough.

What do I think of the world?
Who knows what I think of it!
If I weren't well then I'd think about it.

What's my idea about matter?
What's my opinion about causes and effects?
What are my thoughts on God and the soul
And the creation of the world?
I don't know. To think about such things would be to shut my eyes
And not think. It would be to close the curtains
Of my window (which, however, has no curtains).

The mystery of things? What mystery?
The only mystery is that some people think about mystery.
If you're in the sun and close your eyes,
You begin not to know what the sun is,
And you think about various warm things.
But open your eyes and you see the sun,
And you can no longer think about anything,
Because the light of the sun is truer than the thoughts
Of all philosophers and all poets.
The light of the sun doesn't know what it does,
And so it cannot err and is common and good.

Metaphysics? What metaphysics do those trees have?
Only that of being green and lush and of having branches
Which bear fruit in their season, and we think nothing of it.

We hardly even notice them.
But what better metaphysics than theirs,
Which consists in not knowing why they live
And in not knowing that they don't know?[8]

People talk about the mystery of things. The mystery of the cosmos or the mystery of consciousness. But what mystery? If we delude ourselves with mystery, we will find ourselves asking paralyzing, empty questions: "What is matter?," "What is a cause?," "What is an effect?," "Who is God?," "Is there a God?," "Where is the soul?," "Is there a soul?," "Was the world created, or has it always existed?" Blah, blah, blah, blah, blah. To think about these things would be to shut one's eyes to the world. It would be to draw the curtains of existence and to live in an abstract, fake world. As the poet says here, the important thing is not to think about mystery. If you're in the sun, you close your eyes, then you can begin to think about it. But the point is not to close your eyes. The light of the sun is truer than the thoughts of all philosophers and all poets.

In many ways, what Alberto Caeiro is saying is similar to what Heidegger is arguing for in his discussion of space and the wider arguments of *Being and Time*. We have to go back to seeing things *as* things, seeing the absolutely obvious *as* obvious, seeing the everyday *as* everyday. We have to unlearn and relearn how to see what's under our noses and to stop asking pointless questions that we confusedly think are profound. As the poet says, "What do I think of the world? / Who knows what I think of it! / If I weren't well then

8. Fernando Pessoa, "To not think of anything is metaphysics enough," in *Fernando Pessoa & Co.*, 49.

I'd think about it." The condition that makes certain kinds of philosophical questions possible is *sickness*. It's a disease, as Ludwig Wittgenstein says in the *Philosophical Investigations*.[9] And we have to learn to *get well*, to show the fly the way out of the fly bottle, to go back to everyday life insofar as we can. That simple lesson is what Heidegger's trying to encourage us to do in his work. But—and this is going to take us to the theme of the next chapter—the world, the existential spatiality of the world, the roomliness of our living world, is not just a world of things, a world of hammers and radios and so on. It's also a world of *persons*. The next theme we're going to look at is the question of being-with-others. What is the relation between the understanding of the world and the understanding of others? We will pick that up next time.

9. Ludwig Wittgenstein, *Philosophical Investigations*, 2nd ed., trans G. E. M. Anscombe (Blackwell, 1958). See especially sections 254, 255, and 593.

4. Being-with-Others

We are defined by a "who," not a "what" — Dasein is the with — Everyone is the other, and no one is himself — I am nobody (Dickinson) — Networked being — They think me (Rimbaud) — Displacing the Cartesian starting point for philosophy, the "I" — Moments in the precipitation of the self in Being and Time *(anxiety, death, conscience, resoluteness, the people) — How others show up in the workshop of the world — Being alone — Caring for, leaping in, and leaping ahead of the other — Considerateness and forbearance — Do we really understand others? (Levinas) — An encoded social psychology of modernity (Sloterdijk) — The dictatorship of the They — Every secret loses its force — How to become authentic.*

WHO IS DASEIN? Who are we? The first thing to say is that when I introduced this character—Dasein, the human being—we saw that Dasein was a being defined by a "who" and not by a "what"; namely, we're not like tables and chairs, which can be defined in terms of an essence, a whatness or substance. Rather, we're existential beings defined by a "who," who can ask ourselves the question, Who am I? So, who *is* Dasein? That's the question that Heidegger addresses in chapter 4, division 1, of *Being and Time.*

Dasein is being-in-the-world, as we now know well. In chapter 3, we saw what "in-the-world" meant for Dasein. The world is not, most closely and mostly, proximally and for the most part, *zunächst und zumeist*, experienced as something objectively and constantly present-at-hand, or *vorhanden*. The world is not an objective realm over against a thinking subject, a world of extended things over against a thinking thing. The theoreticist picture of the relation between self and world that gives rise to the epistemological construal of the world is not primary. It is misleading, secondary, and deficient. The entire Cartesian picture of the world that feeds into Kant and modern philosophy is wholly inadequate. Epistemology is not primary or first philosophy. My BA adviser and PhD supervisor, Robert Bernasconi, while he was teaching at University of Essex, argued with the department to get rid of the epistemology requirement, on the basis that Heidegger had demonstrated that this field was unnecessary. He lost that argument.

Rather, the world is that referential totality, as Heidegger calls it, that *whole of stuff* that is connected together in our *practical* dealings. The different things in the world are intrinsically related, each to the other, and they make a whole that hangs together. The referential totality of the handy, of the ready-to-hand, with which Dasein is involved, has significance for Dasein. The world is a meaningful whole. Dasein is its world, and the world is for the sake of Dasein.

Think back to that poem from Wallace Stevens I cited: "There never was a world for her / Except the one she sang and, singing, made." "She was the single artificer of the world / In which she sang." Dasein is *with* its world. It is fascinated or benumbed—this word *benommen* in German, which can almost mean being slightly *drunk* or *lost* in the world, *intoxicated* by the world. We are fascinated by the average, everyday world, lost in it thoughtlessly. We are both thrown in and sucked into it. We're lost to the world (abandoned to it) and lost in it (mesmerized by it).

Now, within this world into and in which we are lost, we discover not only things but also other Daseins. Heidegger says, "Dasein 'is' essentially for the sake of Others" (SZ 123 / BT 160). He then makes another compelling remark: "[Dasein] *is* in the way of Being-with" (ibid.). Our being, our who-ness, is *with*. The world *is* for the sake of Dasein ("there never was a world for her / Except the one she sang and, singing, made"), and Dasein *is* for the sake of others. But most importantly, Dasein is *with* others. Dasein indeed *is* the with. Being-with (*Mitsein*) is a simple thought to say, a difficult thought to live and think. Dasein is *Mitsein*; we are fundamentally *with* others. Try and keep that uppermost in your mind: *We are the with*.

This brings us to the question of this chapter: Who is Dasein? Or more specifically—because for Heidegger the issue is always the question of access—what is our *access* to Dasein, to ourselves and to others like us? Well, our access to being-with is average, everyday life. Then the question is, Who is Dasein in this mode of average, everyday existence? Towards the end of the chapter, Heidegger says in an extraordinary passage, "Everyone is the other, and no one is himself. The '*they*' [*das Man* or "the anyone"], which supplies the answer to the question of the '*who*' of everyday Dasein, is the '*nobody*' to whom every Dasein has already surrendered itself in Being-among-one-another" (SZ 128 / BT 165–66). The question of who Dasein is in average everydayness is going to be answered by saying that the "who" of everyday Dasein is "they," is "them." I'm going to lay that claim out a bit more thoroughly later. But the thought is that Dasein is *nobody* in the sense of nobody in particular. I am just one of them, anyone. "Everyone is the other, and no one is himself."

As I like poems, I thought I would read you another little poem, by Emily Dickinson. It's called "Life."

I'm nobody! Who are you?
Are you nobody, too?
Then there's a pair of us!—don't tell!
They'd banish us, you know.

How dreary to be somebody!
How public, like a frog
To tell your name the livelong day
To an admiring bog![1]

I don't know where the frog came from, talking to a bog. But the idea of "I'm nobody" is the thought that Heidegger begins with. Life in the livelong day is dreary. Gray on gray, as Hegel famously described philosophy.[2] This is the meaning of publicness. Heidegger took a dim view of publicness. In its everyday mode, the mode in which it encounters the world first and foremost, proximally and for the most part, as a handy, signifying, referential totality, Dasein is not itself, it is the Other. It is They. It is nobody in particular. And we're going to try to get to the bottom of this peculiar state of affairs in this chapter.

Dasein is with others. That's the thought to hang on to. In the same way that we are with the world or we are with the stuff in the world in the way that I bring close through de-severance and directionality, we are also with others. Others are *lit up* in the world. And another way of putting this, which would be a word that Heidegger

1. Emily Dickinson, "Life," in *Poems, Second Series*, ed. Mabel Loomis Todd and T. W. Higginson (Roberts Brothers, 1891), 21.

2. G. W. F. Hegel, *Philosophy of Right*, trans. T. M. Knox (Oxford University Press, 1967), 13.

doesn't use because it's more Roman than Greek—the Romans taught us nothing, for Heidegger—is to say that Dasein's being is a *social* being. Heidegger prefers talk of the *polis* to that of the *socius*, in a way that Arendt inherits and turns into the questionable distinction between the political and the social.[3] But what Heidegger's doing in *Being and Time* is a *social ontology*. Heidegger doesn't like the word "social" because it's a term taken over from the tradition as self-evident without sufficient examination or destruction. The concept of the social needs to be loosened up or de-structured. Let's try another expression we've alluded to before: we could say that the human being, Dasein, is a *networked being*. We are beings who exist in a network of relations to things, a relationship to persons, and we *are* the practices of those networks. Our identity is shared *with* others, not apart from them. The network is a field of forces in which our existence takes place.

This phrase, "Everyone is the other, and no one is himself," also puts us in mind of another famous expression, from Rimbaud. Not the boxer from Philadelphia but the French poet who ended up as a shady, colonial, commercial trader in Africa and who wrote the *Illuminations*, poems that light up the dark sky like a meteor shower. If you haven't read *Illuminations*, then you haven't really lived.[4] It leaves you feeling cleansed and a little worried about yourself. It has a force of violent wrenching, both satisfying and troubling. In a letter from 1871, Rimbaud says, "*Je est un autre*" (I is another). He continues that quote by saying, "I owe myself to society. It's false to

3. See, e.g., Hannah Arendt, *The Origins of Totalitarianism* (Harcourt, 1968).

4. Start with "Mystical" and "Democracy" in Ashbery's translation. See Arthur Rimbaud, *Illuminations*, trans. John Ashbery (W. W. Norton, 2011).

say, 'I think,' one should rather say, 'they think me'"—"*On me pense*," as he says in French.[5] We can think about this in relation to another category, taken from the sociologist Émile Durkheim, who writes of *anomie*—or normlessness, related to the feeling of "anonymity," in the sense of lacking an identifying limit on desire—as the condition of social life.[6] Heidegger is trying to think about that way of being. In its average everydayness, Dasein is anonymous. You are they, and they are a crowd. We could indeed link the They to the social emergence of the crowd, and to the indifferent brutality and also strange effervescence of crowd-being.[7] There is a uniquely social media style of crowd-being, where we follow stories, people, ideas with the pressing, threatening urgency of the crowd, only to too often forget the whole thing a week later.

The radicality of this opening move in chapter 4 is that Heidegger does not begin his analysis of the *who* of Dasein by first positing an individual human being as given and then go on to construct that human being's relation to others and the surrounding world. No, for Heidegger, Dasein *begins* as being-with. Therefore, the standard way in which philosophers will talk about the relationship to

5. Arthur Rimbaud, *Complete Works, Selected Letters: A Bilingual Edition*, trans. Wallace Fowlie, rev. Seth Whidden (University of Chicago Press, 2005), 371–75.

6. See Émile Durkheim, *Suicide: A Study in Sociology*, trans. John A. Spaulding and George Simpson, ed. George Simpson (Routledge, 2002), esp. 201–39; and Durkheim, *The Division of Labour in Society*, trans. W. D. Halls (Macmillan, 1984), esp. 291–309.

7. See Elias Canetti on being taken by the crowd, losing yourself in the social contagion of the crowd, with its latent brutality. Elias Canetti, *Crowds and Power*, trans. Carol Stewart (Viking Press, 1962).

others—"I am a thinking subject. How is my subjectivity related to other subjectivities?"—this overstated and overrated problem of *intersubjectivity* in philosophy, even in Hegel, is entirely avoided by Heidegger. I *is* with. And this first move does nothing less than problematize the Cartesian starting point for philosophy, as Christopher Fynsk noted.[8]

The Cartesian starting point in Descartes's *Meditations* and its procedure of radical doubt is that there is an I that thinks, and this ego is indubitable. If I doubt, I exist. Therefore, I am. That's where Descartes begins from. And from that "I am, I exist," we raise the question of the relationship between that self and the world of objects standing over against the subject. For Heidegger, the self of everydayness is not a Cartesian subject defined in terms of the constant present-at-handness of itself as a thinking thing over against a world of extended substance. Rather, we are beings who are constitutively together with others, who share their space with others. Existential spatiality is full of things and Daseins. And if we think about that for a bit, we can ask, If I am *with*, then where do *I* end? Where does the other really begin? Is it so obvious? We're used to thinking of ourselves as individual subjects, as distinct identities, moving in a world of other individual atoms, who are also full of their little identities. Heidegger's thought is rather that we're always already, constitutively, *with each other*. As The Pixies said in 1988, "Where is my mind? / Way out in the water. See it swimming." Not that Black Francis read Heidegger. Thus something like extended mind theory also becomes unnecessary in a Heideggerian account of human

8. Christopher Fynsk, *Heidegger: Thought and Historicity*, exp. ed. (Cornell University Press, 1986). See, e.g., p. 29.

existence. We're always already in our world, *with* our world—we're *part world*—as well as fundamentally *with* one another. For Heidegger, the supposed boundaries between myself and my surrounding (social, physical) environment are not blurred but preempted.

To give a sense of the journey that we're going to take in *Being and Time*, I'd like to look way down the road at the shape of the self that will come more clearly into view as we progress. Afterwards, we'll go into the text of chapter 4, division 1, and pull out the key concepts. Heidegger begins his analysis of the question "Who is Dasein?" by putting into question the Cartesian starting point for philosophy, as I said. Then the question is, Does the self disappear in Heidegger's analysis? Is there no self? That is not the case, as we'll see. The self emerges in *Being and Time*, but it emerges much later. Heidegger will say, at the end of chapter 4, that authentic selfhood, authentic being-a-self, is a modification of the inauthentic they-self (SZ 131 / BT 168). What does that mean? It means that the "who" of Dasein in average everydayness is essentially held together, experienced together, with others. Yet I can modify that inauthenticity and become an authentic self. But what is fundamental, what is primary and a priori for Heidegger is being-with. The question of how the self takes shape in *Being and Time* is the story that we're going to tell in this book, especially as it develops out of the discussion of death (see ch. 11, "Death"). The self emerges in a series of moments that I want to mention now, and then we can analyze them at our leisure as we read further.

The first way in which the self emerges in *Being and Time* is through the experience of *anxiety*, or *Angst*. In anxiety, I pull myself out of the average, everyday world, and I'm revealed to myself as an individuated, singular, isolated being. I like to think about this with maritime metaphors, mostly because I like maritime metaphors, and I am very happy to live in a port. (I could get Melvillean here, but I won't just yet.) To be a self in the world, for Heidegger, is to be with *them*. It's like being immersed in the sea, an anonymous social sea. And then what happens in anxiety, it's as if the tide goes out on the seashore and reveals a self that's there like a rock on the beach. The world slips away, and the self is precipitated from out of the salty brine of the They. Anxiety, this disturbance of familiarity, of at-homeness, is the first moment of the emergence of the self.

The second moment where the self begins to take shape is in being-towards-death. The relationship to death is a relationship to *my* death and to a death that is coming to *me* and for *me*. What Heidegger's got in mind here is that the experience of mortality, of being-towards-death, *individuates* Dasein. The death that appears in my existence is my death. And that death shapes me. Death picks me out and singularizes me.

The third moment, after anxiety and being-towards-death—this is cheerful stuff, isn't it?—is conscience. The call of conscience summons me and renders me powerless, a little like that moment in Franz Kafka's *The Trial* when Josef K. is in a cathedral, and he hears his name called out and accused of guilt.[9] That's what conscience feels like. It plucks you out in the way Josef K. is picked out as an individual self. "*Guilty!*" it calls.

Anxiety, being-towards-death, conscience. When I have experienced those three moments, I become what Heidegger calls "resolute,"

9. Franz Kafka, *The Trial*, trans. Mike Mitchell (Oxford University Press, 2009), 149–52.

Entschlossen in German. I become resolute Dasein, a decided openness in the face of death, and I'm free to assume my fate as an individual.

And then there's a fourth moment that we will get to in a decisive paragraph towards the end of *Being and Time* in paragraph 74, which is discussed at length (see ch. 16, "Historicity and Politics"). This is where the connection can be made between Heidegger's existential analytic of individual Dasein, on the one hand, and the community, the collective, on the other. This is perhaps the most political moment in *Being and Time*, a very *questionable* political moment. The fourth way in which I can shape myself is that I can become a self, an authentic self, as an individual with others within what Heidegger calls "the destiny of the community" (SZ 384 / BT 436). In one passage he refers to this community of destiny as a people, a *Volk*.[10]

If I can pass through these moments of the emergence of the self, then I can achieve a form of *constancy*. This is a theme that comes back in the later parts of *Being and Time*, which is why just reading division 1 and treating division 2 as an existentialist supplement to the book, like Hubert Dreyfus does, really makes no sense. This emergence is the constancy of the self, a self-constancy that is not an Archimedean point in the way it was for Descartes but an authentic constancy that is achieved existentially through the experience of anxiety, being-towards-death, conscience, and being-with-others.

10. Jean-Luc Nancy and Philippe Lacoue-Labarthe track this moment in *Being and Time* to Heidegger's Nazism. See Jean-Luc Nancy, *Being Singular Plural*, trans. Robert Richardson and Anne O'Byrne (Stanford University Press, 2000); and Philippe Lacoue-Labarthe, *Typography: Mimesis, Philosophy, Politics*, ed. Christopher Fynsk (Stanford University Press, 1989). See also Simon Critchley, "With Being-With? Notes on Jean-Luc Nancy's Rewriting of *Being and Time*," *Phänomenologische Forschungen* 3, no. 2 (1998): 198–210.

The political moment in *Being and Time* is the affirmation of community in the form of a people.

The self doesn't disappear in Heidegger's analysis. He just doesn't begin with it. The self forms an inner limit within *Being and Time*, which is continually crossed or crisscrossed between the dissolution of the self, the critique of the standard philosophy of the subject into being-with and the they-self, and the attempt to restore, to build a new conception of the self. We might call it a nonmetaphysical idea of the self. The self returns, but it returns as something radically finite, as defined by time, by history. The self is nothing substantial; rather, it is a temporal, historical movement, a temporalized self that is thrown into a past and that can throw off that past in a movement of free projection. The self is utterly groundless, and the meaningful context of being-in-the-world constantly risks slipping away into the meaninglessness of mortality. We live in a meaningful world, then we die and meaning stops. And there will be a series of moments in *Being and Time* where the self dissipates. But we're getting ahead of ourselves.

I'd like to take you through a few moments in chapter 4, on being-with-others, and lay out some of the key concepts that Heidegger introduces here. In paragraph 25, he begins by problematizing the Cartesian starting point for philosophy. He claims that to conceive of the self as something present-at-hand, as something that we become theoretically aware of, is to see it as a substance or a subject. This is the mistake of Descartes, in the first place, but it's also the mistake of Kant, and it's also the mistake of Heidegger's teacher, Edmund Husserl, with his bloviating talk of the transcendental ego.

We cannot begin to ask the question of who Dasein is by assuming that the I is self-evident, that the I is, as Heidegger will put it, "ontically obvious" (SZ 115 / BT 150). The self is not ontically obvious. But neither is it ontologically obvious. It's important to emphasize the unfamiliarity of Heidegger's language, which he is assaulting us with. We have to ask ourselves the question of who Dasein is in a more radical and thoroughgoing way, and we have to do this by putting aside the seeming self-evidence of the tradition and to begin to interpret Dasein *existentially*. Let's put aside that tradition and try to think with Heidegger in his terms and not constantly translate what he is doing into the traditional language of his predecessors or successors. The substance of Dasein is existence: that's where we begin.

So much for assessing and clearing the ground. In paragraph 26, Heidegger writes of the relationship between ready-to-hand things that we find in the world and other Daseins, other beings like us. Note how he expresses it:

> In our "description" of that environment which is closest to us—the work-world of the craftsman, for example [and for Heidegger this is a self-evident example, because for him the world *is* a workshop and Dasein has a keen work ethic]—the outcome was that along with the equipment to be found when one is at work, those Others for whom the "work" is destined are "encountered too." (SZ 117 / BT 153)

We're in the workshop of the world for Heidegger, and in that workshop there aren't just tools, hammers, and nails lying around. There are also other workers, and I encounter them in my encounters with things. In his workshop language we also hear him saying, "Don't ponder your neuroses and grievances, just do something! Bang the hammer, get to work!" He goes on:

> When, for example, we walk along the edge of a field [a very bucolic Heideggerian example] but "outside it," the field shows itself as belonging to such-and-such a person, and decently kept up by him [these are not slacker farmers]; the book we have used was bought at So-and-so's shop and given by such-and-such a person, and so forth [note that Heidegger is thinking of a *used* bookstore, someplace like The Strand in Greenwich Village]. The boat anchored at the shore is assigned in its Being-in-itself to an acquaintance who undertakes voyages with it; but even if it is a "boat which is strange to us," it still is indicative of Others. (SZ 117–18 / BT 153–54)

In those examples, walking by the edge of a field even when there's no farmer plowing (or whatever it is that farmers do), others, like the farmer, still *show up*. In the book that we bought at the store, we apprehend that the bookstore is run by others. The boat, anchored at the shore, is owned by someone. In the boat perhaps we see someone, a boatman, who owns that boat.

Others show up in our relationship to things, and in freeing things up in the world we're also freeing up a relationship to others. This is where Heidegger begins his analysis of being-with, *Mitsein*. As I said before, *we are with*, we are fundamentally being-with-others. And we're with other Daseins even when no Dasein is there. There's a strange remark where Heidegger talks about "being alone." He says, "Being-alone is a deficient mode of Being-with" (SZ 120 / BT 157). We can be alone, I can be in solitude, but the others that constitute that world are still there, in the background. He goes on to say I can be alone in the presence of others and with others in their absence (SZ 120–21 / BT 157).

There's a moment in Terrence Malick's film *The Thin Red Line* that illustrates this point. (We'll come back to this movie in chapter 8, "Anxiety and Care.") Malick was a graduate student working on Heidegger before he went to film school in Los Angeles. He got his BA at Harvard and BPhil at Oxford, where he was a Rhodes Scholar. The character Witt, played by Jim Caviezel, asks Sergeant Welsh, played by Sean Penn, "You ever get lonely?," and Welsh replies, "Only around people." Being alone is only possible because Dasein is essentially being-with.

There are then a couple of little terminological moves I want to mention because they're important, as Heidegger loved to build a linguistic architecture in his work. The being of Dasein is *care*: we will get to that complex and articulated concept later on in these chapters, and it will become crystal clear. The German word for care is *Sorge*. The relation to things in the world is concern, *Besorgen*. You hear the homophony between *Sorge* and *Besorgen*. And our relationship to persons is described by Heidegger as *Fürsorge*, which is translated as "solicitude," and which can be thought of as "caring for." This idea of solicitude is what describes our relation to others. We care for the other, and we care for the other's welfare, because we are essentially the kind of beings who are *with others*. This idea of "caring for" has had lots of influence in all sorts of ways of thinking about care. For example, in relationship to nursing philosophy and those areas of health care that are critical of the domination of a mechanistic, modern worldview that can arguably be found among doctors or surgeons who see the body as a Cartesian machine with engineering problems that can be broken down into a series of parts, like a used car in need of repair. Such an approach is appropriate at times. If you have a torn meniscus, you need knee surgery. Solicitude is seeing the other human being in terms of caring for that other. And this practice of care as it arises in nursing philosophy is something that Heidegger can be said to have informed and influenced, particularly in the work of the nursing theorist Patricia Benner, who worked closely with Dreyfus.[11]

There are two extreme modes of solicitude for Heidegger. There is, on the one hand, what he calls leaping in and dominating the other—*einspringen*, or "springing in" (SZ 122 / BT 158). The most extreme version of this would be a form of sadism where I control the other, I care for the other so much that I am dominating them. We could also think about various kinds of dependency, where we put the other in a situation of reliance on us for their needs, even a form of servitude, such as in a family. And the other extreme mode of solicitude is what Heidegger calls "to leap ahead of the Other," *vorausspringen* in German. It's a rather nice term, leaping ahead of the other, not in order to take care away from them but in order to give it back to them *authentically*. Leaping ahead of the other enables the other to become transparent to themselves and to become free (SZ 122 / BT 158–59). Such an approach is appropriate for caring for someone with addiction issues. Leaping in for them, forcing them to go to a rehab program, is ineffective and not advised.

These are the two extreme modes of solicitude. On the one hand, I can have a relationship with the other that dominates them and

11. Patricia Benner's *From Novice to Expert* adapts the model of skill acquisition developed by Hubert Dreyfus and his brother, Stuart Dreyfus. See Patricia Benner, *From Novice to Expert: Excellence and Power in Clinical Nursing Practice* (Prentice Hall Health, 2001).

takes their freedom away from them. On the other hand, I can have a relationship with the other that frees them up, and that freeing up is the mode recommended by Heidegger. Perhaps to say he "recommends" it is already going too far. But it is clear that Heidegger favors leaping-ahead-of as an authentic form of welfare work for the other, even though leaping-in-for is obviously not without its necessity in life. The more "dominative" forms of care must be entered into freely in order to have a leaping-ahead-of effect for the recipient. Heidegger goes on to say that "everyday being with one another maintains itself between these two extremes of" domination and liberation (SZ 122 / BT 159). Most of the time we're between these two modes of being-for-the-other.

As always (again, within Heidegger's particular linguistic terminological architecture), every mode of care, in this case solicitude, *Fürsorge*, has a mode of *sight*. It is a way of looking at things. We saw already that the way we relate to the world is through concern, which has a sight that Heidegger calls "circumspection," or *Umsicht*. The kind of sight that belongs to solicitude is what Heidegger calls "considerateness and forbearance." The German here is "*Rücksicht und Nachsicht*"—notice the almost singsongy, rhythmic vocabulary. The thought is that considerateness and forbearance can flip over into their deficient modes of inconsiderateness and intolerance. Very quietly Heidegger is making a certain moral case for a form of solicitude, caring for the other, that would enable the other to become free, and where I would have a considerate and tolerant relationship to the other.

There is another point I'd like to make here in a thumbnail way, and we'll come back to it in later chapters. Heidegger says, "Understanding of Being already implies the understanding of Others" (SZ 123 / BT 161). For Heidegger, it is assumed that our relationship to the other, insofar as we are with the other, is a relationship of *understanding*. Remember that what gets the whole mechanism of *Being and Time* up and running, which he talks about in the introduction, is that I have a vague and average understanding of being. This vague understanding can be made *explicit*, can be made conceptual and philosophical, in the existential analytic of Dasein. But that understanding of being implies an understanding of others. Heidegger's presumption, when he's talking about the relationship to others, is that the relationship to others is a relationship of understanding.

Maybe this sounds fine. Maybe it even sounds cool. The question that it raises is the following: Is one's relationship to others always a relation of understanding, a relation of comprehension? Or are there perhaps moments when we experience the other as separate, as someone we don't understand, as someone who is distant from us, someone whom we don't know, or with whom we've undergone a breakdown in understanding, and perhaps whom we have to respect for that reason? Perhaps we thought we understood each other, but then we catch a glimpse of something, like a glimmer in the eye, that tells us we don't, can't, and never did understand each other. And maybe that further, uncontainable otherness is something we want to preserve in our relationships.

Now, I put the thought in this way because I've got somebody in mind, a philosopher who is heavily influenced by Heidegger but becomes deeply critical of his work after his commitment to National Socialism in 1933. I am thinking of Emmanuel Levinas, a French Lithuanian philosopher, whom I've done a lot of work on over the years. And he wrote an essay in 1951, which is called "Is

Ontology Fundamental?" The simple answer to that question is no, it's not fundamental. He has a critique of Heidegger, which we don't need to go into now—it will come out bit by bit as we move through these chapters. The point that Levinas wants to make is that our relationship to the other is not simply one of understanding, where I am *the same* as them. There is something *surprising*, there is something *challenging* about the other, which cannot be reduced to my framework of understanding.[12] Perhaps the otherness or alterity of the other cannot be framed by relations of understanding. Levinas describes the incomprehensibility of the relation to the other as religious and ethical. At the core of Levinas's thinking is a commitment to the holiness, the sanctity of the other.

Let's just take note of that thought, log it in the memory, and put it to one side for now. We're going to move on to the really fascinating part of chapter 4, paragraph 27, the final pages of the chapter, called "Everyday Being-One's-Self and the 'They.'" Let's try and get clear on this seemingly weird idea of "the They." Who is Dasein in average everydayness? Heidegger's answer is that in average everydayness I am "they" (SZ 126 / BT 164). The term that he uses in German is *das Man*. This is a third-person impersonal pronoun, which sounds more forced in English—the way we might say "one says" or "one does" and not "he" or "she." The problem with "one" is that it sounds too posh, too royal, something King Charles might say. Maybe this whole pronominal issue is changing with the contemporary use of

12. Emmanuel Levinas, "Is Ontology Fundamental?," trans. Peter Atterton, Graham Noctor, and Simon Critchley, in *Basic Philosophical Writings*, ed. Adriaan T. Peperzak, Simon Critchley, and Robert Bernasconi (Indiana University Press, 1996), 1–10.

they and *them*. The point is that "the They," or "the one," has *neutrality*.[13] It can be *anyone*, which is how the philosopher John Haugeland translates *das Man*.[14]

This is going to be an important line of thought for Heidegger. But what is he up to in doing this? What's on his mind? Well, he's giving an account of our dealings with others in average, everyday life. One way of thinking about this (and this is not Heidegger's intention) is drawn from Peter Sloterdijk, the German philosopher with a Dutch name who writes a great book, early in his career, called *Critique of Cynical Reason*, very well known in France, Germany, and the Netherlands but not so much in the US.[15] The book has a discussion of the They which is witty, perspicuous, and sociologically interesting. Sloterdijk claims that the They is an encoded critique of the social psychology of modernity. What he means by that is that what Heidegger is describing in this discussion of the They is the social experience of Germany between the First and Second

13. Interestingly, shortly after the publication of *Sein und Zeit*, Heidegger will use similar language to describe Dasein itself: "The peculiar *neutrality* of the term 'Dasein'... indicates that [it] is neither of the two sexes. But here sexlessness is not the indifference of an empty void... but the primordial positivity and potency of the essence." The neutrality of Dasein allows it to be interpreted in advance of its "concrete factual humanity." See Martin Heidegger, *Metaphysische Anfangsgründe der Logik im Ausgang von Leibniz*, ed. Klaus Held, vol. 26 of *Gesamtausgabe*, ed. Friedrich-Wilhelm von Herrmann (Klostermann, 1978), 171–12; *The Metaphysical Foundations of Logic*, trans. Michael Heim (Indiana University Press, 1984), 136–37. Meanwhile, *das Man* is neuter in the way that it "dispose[s] of" and "take[s] away," rather than anticipates, "Dasein's everyday possibilities of Being" (SZ 126 / BT 164).

14. John Haugeland, *Dasein Disclosed: John Haugeland's Heidegger*, ed. Joseph Rouse (Harvard University Press, 2013), 5.

15. Peter Sloterdijk, *Critique of Cynical Reason*, trans. Michael Eldred (University of Minnesota Press, 1987), 195–210.

World Wars: the experience of Weimar Germany, the experiment with democracy after the catastrophe of defeat in the First World War, which led through numerous crises to the rise of National Socialism in the early 1930s. And that's the context of Sloterdijk's book: the drift and decay of democratic life. It's tempting, when reading these pages of *Being and Time*, to see it as a mirror of the democratic anomie in Weimar Germany, to which Heidegger was consistently and implacably opposed.

Yet this is not the way Heidegger thinks about the "they." For him, the They is an *existential*, and it describes the condition for any possible social formation. The question that is raised by the analysis of the They is whether this is necessarily a negative evaluation of "society," in quotation marks, a critique of everyday life. Heidegger insists that it *isn't*. Regardless of what Sloterdijk or other philosophers might think, Heidegger insists that talking about the they-self is not negative. And perhaps one should take him at his word. We assume that Heidegger hates the They. But what if we divest ourselves of that prejudice, and consider inauthentic, average, everyday life as a *positive* mode of human being-in-the-world? That's the way I would choose to think about it. Inauthenticity isn't *bad*. Heidegger insists he's not making a moral evaluation here. Inauthentic life is life with others, where we are related to others and dependent upon them. And that might not be altogether a terrible thing. Sloterdijk says the analysis of inauthenticity is a critique of Heidegger's political context, although Heidegger insists that this is not the case. His insistence becomes much harder to believe in later chapters of *Being and Time*, but it's worth a serious, independent consideration: the meaningful context of life with others, the un-mineness of *Mitsein*, in early Heidegger. Is inauthenticity meant only to be overcome?

Let's look more closely at what Heidegger says in the text at the end of chapter 4. He says that in our relationship to others, there's a constant concern as to how one *differs* from others, which is *troubling*, he says. One is *distant* from others. In being with one another, there is what he calls "distantiality," a "taking a distance" (SZ 127 / BT 164). The thought would seem to be that I am *them*, but in a way I keep *my distance* from them. I think about this in relation to using the subway—say, when I'm on my way to The New School. I am "them" using the 4/5 train to get from Brooklyn to Union Square, but I keep my distance from "them," I don't collapse myself into "them."

That's the first thought, and it's this idea of distance he begins with when he begins to give some examples of life with the they-self, examples that have become quite famous. And you can see why the subway is on my mind. "In utilizing public means of transport and in making use of information services such as the newspaper, every Other is like the next" (SZ 126 / BT 164). In using the bus or the subway, I am like the next person. In reading the newspaper, or reading the same newspaper, our differences dissolve. In the contemporary world this dissolution of difference would be everyone scrolling on their phones. Heidegger goes on to say:

> This Being-with-one-another dissolves one's own Dasein completely into the kind of Being of "the Others," in such a way, indeed, that the Others, as distinguishable and explicit, vanish more and more. In this inconspicuousness and unascertainability, the real dictatorship

> of the "they" is unfolded. [Strong word, *dictatorship*.] We take pleasure and enjoy ourselves as *they* [*Man*] take pleasure; we read, see, and judge about literature and art as *they* see and judge; likewise we shrink back from the "great mass" as *they* shrink back; we find "shocking" what *they* find shocking. (SZ 126–27 / BT 164)

This is the condition of averageness. Every Dasein is like the next in the "dictatorship of the 'they.'" Now, in the 1920s, the question of dictatorship was very much in the air. The famous conservative German jurist Carl Schmitt, who became a Nazi fellow traveler, published an influential book on dictatorship in 1921.[16] Democracy has an inbuilt flaw. And let's face it, no philosopher is really philosophically in favor of democracy until John Dewey. Once you let the They take charge, a danger will emerge, the figure of the tyrant. The Russian Revolution of 1917 had taken place under the sign of the "dictatorship of the proletariat" (and Schmitt was a big fan of Vladimir Lenin). It's hard not to interpret this reference to the "dictatorship of the 'they'" negatively. Yet, Heidegger insists, this *isn't* a negative evaluation. In using public transport, in reading the newspaper, in finding shocking what *they* find shocking, in enjoying what *they* enjoy, we all become one, anyone, an anonymous mass, but we remain retrievable from it. Heidegger says:

> Overnight, everything that is primordial gets glossed over as something that has long been well known. Everything gained by a struggle becomes just something to be manipulated. Every secret loses its

16. Carl Schmitt, *Dictatorship: From the Origin of the Modern Concept of Sovereignty to the Proletarian Class Struggle*, trans. Michael Hoelzl and Graham Ward (Polity, 2014).

force. This care of averageness reveals in turn an essential tendency of Dasein which we call the "levelling down" of all possibilities of Being. (SZ 127 / BT 165)

I find that phrase, "every secret loses force," particularly intriguing. Everything gained by struggle (*Kampf*—a very heavy word in these years), gained with *difficulty*, becomes leveled down in our life with *them*. And speaking of *Kampf*, this leveling down of every detail of life is evident in Karl Ove Knausgård's monumental, indeed interminable, six-volume work of autofiction, *My Struggle*. I don't personally care about the endlessly minute goings-on of Norwegian men. But that's my struggle.

What characterizes the they-self is distantiality, averageness, and leveling down. And this is what makes up "publicness" or "public life" (*Öffentlichkeit*). Heidegger, unlike Hannah Arendt, is not a fan of publicness. He finds it shallow.[17] The They, or the public, is always right and always already knows everything. Everything is instantly familiar and accessible. Heidegger is thinking about newspapers or public transport, but we could think about this in relation to social media updates and ceaseless, exhausting news flashes. Social media makes everything accessible, familiar, and utterly banal. Every secret loses its force. Everything that is important gets turned into something we always already know about, and the whole thing spins around, flares up, and is then forgotten.

On this view, publicness is leveling down. Think of a movie you

17. The fetish of public life is a bit nauseating. The idea of the "public philosopher" makes me internally cringe.

think you might like or might believe to be important but that you've not seen. Before you see that movie, you read the Wikipedia entry on it, you scan the reviews, look at the Rotten Tomatoes rating of the movie, and you know it already—spatially, we pull it in close; everything loses its remoteness. We form an opinion without needing to see the movie. It is as if in our social lives, we're terrified by the prospect of something that would surprise us, that would alarm us, that would challenge us, that would require *struggle*. We level down, we make things average. This is why Sloterdijk says that what Heidegger is describing in our life with "them," the they-self, is *democratic life*. Democracy is a dictatorship where everything is decided by public opinion, and where nobody actually makes a decision, nobody exercises a sovereign act, as Schmitt would say.[18] There is no answerability, there's no responsibility, in our life with "them."

We're coming to the end, so hopefully by now it's becoming clear why Heidegger says "everyone is the other, and no one is himself." The they-self—which is the answer to the question of who Dasein is—is nobody, the anonymous anyone. That's because, in our inauthentic life with others, we level down, we make things average, we make things familiar, accessible, banal. And this leads us into the dictatorship of the "they." Again, it's hard to think that what Heidegger is describing is not to be understood negatively, but he insists on this point.

Let's stay with that last thought and move on to the final claim in this chapter. The philosophical claim, the fundamental ontolog-

18. Carl Schmitt, *Political Theology: Four Chapters on the Concept of Sovereignty*, trans. George Schwab (University of Chicago Press, 1985), esp. 5–15.

ical claim that ends chapter 4, is that the "they," *das Man*, is an *existential*. It is a primordial phenomenon that belongs to Dasein's positive constitution (SZ 129 / BT 167). The extent of the dominion of the They might change in the course of history, says Heidegger; the They might change historically. But the structure of being-with as being-with-them, the they-self, is a primordial phenomenon. How does one get from this primordial they-self to authenticity? How does one move from the dictatorship of the They to an authentic Dasein?

To answer that, I want to go back to a term that we've already seen in an earlier chapter, where Heidegger talks about Dasein as being *dispersed* amongst the world, strewn out there, scattered in its being-in-the-world. That dispersal is also a social dispersal. My being is dispersed amongst those other Daseins with whom I share the subway, read the newspaper, and follow updates on whichever social media platform I'm browsing. To get from that *dispersal*, that lostness amongst them, to authenticity requires a *modification* in our understanding. And that modification is going to happen through a series of operations, a series of steps, which we're going to analyze in future chapters but which I've already mentioned: anxiety, being-towards-death, conscience, resoluteness, and a community of decision.

For Heidegger inauthentic, average, everyday being-with-others, the "they-self," is what is existentially primordial. And authenticity, he says, does not rest upon an exceptional condition of the subject. A condition that has been detached from the They. He says it's an "*existentiell* [ontic, or if you like, empirical] modification of the 'they'" (SZ 130 / BT 168). The way we become authentic for Heidegger is by *modifying* the way in which we understand our they-self and by taking it into our own possession. Authenticity for Heidegger means something like coming into one's own, *freely*. This is what Heidegger is imagining at the end of this chapter.

Let me say a word about where we're going next, which is division 1, chapter 5, of *Being and Time*. We've already covered a lot of ground. But chapter 5 covers a lot more. It is rather vaguely called "Being-in as Such," but it lays out the core concepts with which Heidegger is going to establish the *rhythm* of the existential analytic. What we've done so far in this book is to lay out and clarify a series of concepts that is perhaps initially bewildering: the relationship to the world (or concern) and the relationship to others (or solicitude)—concepts that appear static, formal, and fixed. Heidegger's thought in this book is that being is *time*, time is *movement*, and we *are* movement, we *are* kinetic. So, what is the *rhythm* of that movement? How do we learn the *music* of the movement of our being? How do the concepts in *Being and Time* begin to dance? Our first step towards answering that will be the introduction of an interrelated, powerfully mobile cluster of concepts—state-of-mind, understanding, discourse, interpretation, thrownness, and projection—which we will get to in the next chapter.

5. Thrown Projection

The difficulty of stating the obvious — Not splitting the phenomenon — We are ourselves the clearing — The movement of disclosure and closing over as learning the rhythm of existence — The music of Being and Time *— The four key concept clusters — State-of-mind, always already finding oneself there — Mood, Heidegger on the passions — Into this world we're thrown — Dasein exists factically — Projection, understanding, potentiality for being, freedom — Discourse is the condition of possibility for the articulation of the clearing that is us —* Logos *— We are the being that talks — Falling as horizontal, not vertical, not Satanic — Falling at the world — Authenticity as a modified way in which everydayness is seized upon — Fear — Giant bears and my mother — Interpretation as the layout of understanding — Meaning is not deep, but a question of surfaces — Being is not the rabbit in the hat — Learning to see the hat without wanting the rabbit.*

NEXT WE'RE GOING to be looking at chapter 5, division 1 of *Being and Time*. What we've seen so far, in chapter 3, was an analysis of being-in-the-world. Chapter 4 was an analysis of being-with-others. And what we have learned in those chapters is vital. But chapter 5 gives us a conceptual infrastructure or undergirding to everything that we've looked at in the previous chapters and that will determine the entirety of *Being and Time*. Chapter 5 is a long chapter with a lot of moving parts. We're going to try and get the basic structure of it clear in this chapter and continue with chapter 5 of *Being and Time* in our discussion in chapter 6, "Language and Interpretation."

Then we will finish chapter 5 with an analysis of inauthentic life. All together, we will devote three discussions to this important chapter of Heidegger's book.

Once we've got a grasp on the concepts laid out in chapter 5, then we're already prepared—or "always already," as Heideggerians love to say—for the definition of the being of Dasein as care. This is the whole thrust of division 1 of *Being and Time*: the being of Dasein is care, and care is a threefold, articulated structure. In order to understand that thought, we have to grasp a whole run of argument that begins in chapter 5, paragraph 28, through to chapter 6, paragraph 42, and to see that argument as a single arc. After we've surveyed that arc (see ch. 8, "Anxiety and Care"), we'll look at the function of anxiety in revealing the nature of care. And then, in two separate discussions, we will look at paragraphs 43 and 44, which are really two mini-essays at the end of division 1, on the questions of reality and truth, respectively. That's the menu.

We're moving now into the core of what Heidegger is up to in this book. And I want to make that as clear as I possibly can. Everything—even in Heidegger and even in *Being and Time*—can be said clearly. Anyone who says otherwise is bullshitting you. Let's make a start with the beginning of chapter 5, in paragraph 28. The question here—and again, it sounds stupidly obvious when it's put in the way Heidegger describes it himself—is, What is the "Being of the 'there'" for Dasein? (SZ 133 / BT 171). Dasein is being-there. But what is the *being* of the "there"? What does it really mean to be *there* anyhow? That's the obviousness of the question that Heidegger is asking. We know a number of things already. We know that the "there" of Dasein's place is the "where" of being-in-the-world. Dasein is in the world, and that world is a world that absorbs Dasein, benumbs Dasein, a world with which Dasein is fascinated. We also know, from paragraph 12, that the "in" of Dasein's being-in-the-world is not the "in" of water in a glass or a piece of cheese in a fridge or a hamster in its cage. It's not the in-ness of a present-at-hand thing inside another thing (SZ 53–54 / BT 79). The way we are in the world is not like that. The being-in of being-in-the-world is a dwelling, residence, familiarity, rather than the in-ness of present-at-hand objects. It's the in-ness of involvement: situation, context.

Some of the argument has been prepared, but what you feel at the beginning of this chapter is both the difficulty of what Heidegger wants to say and also its sheer obviousness. For example, he says, "Being-in is . . . an essential kind of being of [Dasein] itself" (SZ 132 / BT 170). But then he says it would not be correct to talk about the being of Dasein as a subject in the world of objects. We know already that epistemology is the wrong approach. It is a founded and not found*ing* mode of being-in-the-world—secondary and not primary. What is essential to our being will be a different kind of being-in.

He suggests that it might be more accurate to describe Dasein as the "being of the 'between,'" in quotation marks (ibid.). But then he adds, well, that's not right either, because to say that the "there" of Dasein is between selves, or between humanlike things and world-like things, is to assume that there is a *between* in the first place—namely, that there is a difference or initial separability between these two sets of entities—what philosophers call "mind" and "world." And Heidegger doesn't want to begin from that premise. He wants to begin from the idea of the *unity* of Dasein as being-in-the-world. On the same page, he makes a very interesting remark: "What is decisive for ontology is to prevent the splitting of the phenomenon" (ibid.). Heidegger wants to think about Dasein as being-in-the-world as an undivided unity and to hang on to that unity without splitting it into parts.

And then, as he's clearing his throat, as philosophers do, and clearing the ground in this chapter, you can see him wrestling with this thought. Dasein is there-being, its being is *there*. And this "there" has always been *disclosed* (SZ 132 / BT 171). He just drops this term into the beginning of the chapter, "disclosure," which is going to become a really important concept in Heidegger's vocabulary, because it's going to define the nature of truth. The thought is that we are *there* and this is disclosed to us always already.

Let me say a few more words on this topic. Heidegger talks about the "natural light" in man—the *lumen naturale*, or "natural light of reason," as it's normally discussed in, say, Descartes. He's going to take that idea of "natural light" and shift its sense:

> When we talk in [a] . . . figurative way of the [natural light] in man, we have in mind nothing other than the existential-ontological structure of this entity that it *is* in such a way as to be its "there." . . . To say that it is "illuminated" [*erleuchtet*, he says in German] means that *as* Being-in-the-world it is cleared [*gelichtet*] in itself, not through any other entity, but in such a way that it *is* itself the clearing [*Lichtung*]. (SZ 133 / BT 171)

Listen to that: *Dasein itself is the clearing*. Dasein is disclosed, it is lit up, it is cleared, and it is itself the clearing: "Only for an entity which is existentially cleared in this way does that which is present-at-hand become accessible in the light or hidden in the dark. By its very nature, Dasein brings its 'there' along with it. If it lacks its 'there,' it is not factically the entity which is essentially Dasein; indeed, it is not this entity at all. *Dasein is its disclosedness*" (ibid.). Dasein is its disclosedness, and this disclosedness is described with a spatial metaphorics of light. Dasein is lighting, is clearing. Furthermore, Dasein is not cleared or disclosed through any other entity. We're not cleared or revealed, or we don't come into the truth, through a relationship to God, or to nature, or to reason, or to capital, or to power, or any metaphysical conception of being. Whether being is a unity, as it was for Plato and the Neoplatonic thinkers, or being is a form of multiplicity, as it is in figures like Gilles Deleuze and Alain Badiou—all of that is just metaphysical talk. We are the sort of beings who are cleared through ourselves. We are the clearing. That is the thought that we have to begin with.

The question in this chapter, awkward as it sounds, is, What is the "in" of Dasein's being-in? What is the "there" of Dasein's being-there? Well, that "in" and that "there" are our disclosedness in the clearing, a clearing that we ourselves are and that isn't given through any external agency. The structure of this chapter, like many of the chapters in *Being and Time*, is twofold. In part A, Heidegger advances the three key existential concepts. Those concepts are state-of-mind, understanding, and language or discourse. Those are the three existentials, or a priori structures. Then, in part B of the chapter, if Dasein is disclosed then it is also closed over. Dasein is covered up insofar as it exists in average, everyday inauthenticity. Part B of the chapter, which we'll get to in chapter 7, "Inauthentic Life," is described under the concept of falling (*das Verfallen*). So state-of-mind, understanding, and discourse are mirrored in three "fallen" or inauthentic concepts: idle talk, curiosity, and ambiguity.

Chapter 5 is a movement between parts A and B, between existential concepts and factical concepts. Between things being disclosed and closed over. Dasein is disclosure; it is itself the lighting, the clearing. But it's a clearing that can be covered up in average, everyday life. What Heidegger is trying to describe in this chapter is the movement, the *kinesis* (*Bewegung*), the rhythm of disclosure and covering up, or unconcealment and concealment, that describes the being of Dasein. And this movement also defines the concept of truth for Heidegger. Truth is a movement, a rhythmical oscillation between disclosure, or unconcealment, and covering over, concealment, the movement between the ancient Greek concepts of *alētheia* (ἀλήθεια) and *lēthē* (λήθη).

I'm emphasizing this because, as I said at the end of the last chapter, learning to read Heidegger and think with Heidegger is about learning the rhythm of his thinking, which is the rhythm of our becoming. The rhythm of our existence is a motion between disclosure and closure, existence and facticity, or—as we'll see in a moment—projection and thrownness. The point of reading Heidegger is to learn that rhythm of thinking and to think Heidegger musically. There's a music to *Being and Time*, a percussive dance of disclosure and closure. The book can be thought of in terms of our experience of music. The problem that he faces is that music—an experience of rhythm, of motion, cadence, and counterpoint—becomes frozen in the static language of philosophy. How does frozen philosophy get loosened up, become rhythmical? How do we make philosophy funky? How do we make philosophy play and dance in this space between revealing and concealing? It's a hard thing to do, as philosophers are usually lousy dancers.

Having cleared the ground, let me now run through the key claim in these pages, which is the idea of thrown projection. I'd like to try and lay out the concept clusters that run throughout chapter 5. Let's back up and move towards an overview of the terrain, as we're going to cover a lot of ground.

Heidegger seeks to reawaken perplexity about the question of being, this huge issue, which has been the defining issue of metaphysics for three thousand years. In *Being and Time*, he pursues this question of being through an analysis of the human being, what he calls Dasein. This by now is self-evident to us. The being of Dasein is the average, everyday existence of our life in the world. Really, the question in chapter 5 is, How might we give some more content, more motion and emotion, to this rather formal idea of existence? The central claim of this chapter, which is only deepened in the remainder of *Being and Time*, is that Dasein is thrown projection. Once you've got this structure of thrown projection clear in your mind, then all that Heidegger is going to do in the remainder of the book is to burrow down further into that structure. This thought doesn't change, it just becomes more profound and far-reaching. In German it sounds nicer: "*Dasein ist geworfener Entwurf*" (Dasein is a thrown throwing off; SZ 148 / BT 188): thrownness and projection are contrasting aspects of one movement. Let me try and unravel this thought.

The first concept cluster presented in chapter 5 contains three items: state-of-mind, mood, and thrownness. Now, what does state-of-mind mean in English? It doesn't really suggest anything, other than *état d'âme* in French, a state of soul or disposition of the sentiments. It's a rather questionable rendering of the German term *Befindlichkeit*. You can say in German, "*Wie befinden Sie sich?*" (How are you?), which really means "How do you find yourself?" The idea of *Befindlichkeit* is really a question of how exactly one finds oneself. An early attempt to render this concept in English was given by the great philosopher Father William Richardson, who died some years ago, in 2016. I knew him well. He was a lovely man, a Jesuit priest and a psychoanalyst. In his huge book on Heidegger from 1961—the first book on Heidegger—Richardson translates *Befindlichkeit* as "already-having-found-itself-there-ness."[1] Which is absolutely accurate, but incredibly ugly. But Heidegger's thought here is that the human being is always already found or disclosed somewhere, it is always disclosed in the there of its being-in-the-world, the *Da* (there) of Da-sein. How did you find yourself this morning? "I just rolled back the sheets, and there I was," quoting the words of the great Vivian Stanshall from The Bonzo Dog Doo-Dah Band, from a long, long time ago.

The idea is that we're always already found, retrieved, located somewhere. And how are we found? This introduces the second concept. We're always found in a *mood*. The word in German is *Stimmung*, which can mean mood but is closer to an idea of *attunement*. It can be thought of as the tuning of a musical instrument, like a guitar. The guitar has to be tuned to itself, every string has to have the right relationship to the others. If it is a solo instrument, it must be in tune in a way that works for a particular performance.

1. William J. Richardson, *Heidegger: Through Phenomenology to Thought*, 4th ed. (Fordham University Press, 2003), p. 64.

If it is in a group, it must be tuned in relation to the other instruments as well. In Heidegger's phenomenology of existence, we basically face that same difficulty in our day-to-day lives and pursuits on a constant basis. We constantly have to readjust to one another's moods, making sure to balance our attunements to ourselves, to our world, and to our many others.

A mood is a way in which we can be attuned to the world, in which we constantly encounter others. I am always found in a mood. And this is meant in the strong Aristotelian sense of *pathos*, of affection, of passion, the passions of the soul, the passions of the human being. We are always found in a passion, an affection that befalls us. But for Heidegger, the passions are not some psychological coloring for an essentially rational agent. They are not colorful clothing on the skeleton of the mind. One way of thinking about passions is to think of ourselves as essentially rational beings organized around the center of a mind or an intellect, and then gathered around that intellect are the passions that we fall into at different times. Heidegger's not going to take that view. For him, everything is a passion, everything is a mood, and indeed *reason itself is a form of passion*, a certain kind of mood (SZ 135–36 / BT 175). A reflective mood, perhaps. These moods are the fundamental ways in which we are attuned to the world and attuned to others. I am excited, I'm bored, I'm distracted, I'm attentive, I'm sleepy, I'm wired, whatever it might be. One is always attuned to the world most closely and mostly through moods, *Stimmungen*, and this idea is also musicologically linked to tuning and pitch, being tuned to, or in the pitch of, our world. In this "state-of-mind" we first encounter ourselves, and we do so always already in a mood, an attunement. It's here that a person's personality is to be found.

One of the compelling aspects of Heidegger's work is his attempt to provide a phenomenology of moods, of the affects and emotions that make up our everyday life in the world. Rather than denigrate moods or subordinate the passions to reason, Heidegger gives us a rich description of the life of moods. What do those moods disclose? For Heidegger, those affections reveal the human being as thrown into the "there" of my being-in-the-world. This is the third concept in the cluster. After state-of-mind, or always-already-having-found-oneself-thereness, and mood, or attunement, we get thrownness (*Geworfenheit*). Moods disclose me as thrown. As Jim Morrison intoned many decades ago, "Into this world, we're thrown."

Thrownness is the simple awareness that we always find ourselves somewhere, delivered over to the world with which we are fascinated. A world that we share with others. I know this is a really obvious thought, but we didn't just decide to come into being, did we? We didn't snap our fingers and suddenly wake up into consciousness. No, we were always already there and we found ourselves already existing. Even in our earliest memories, we were already here. There's no way to get behind that thrownness (SZ 209, 284 / BT 253, 330). When we wake up in the morning, which is always a nice surprise to have (it beats being dead), we're factically there in a certain mood, a mood of sleepiness or wild wakefulness or mild irritation. We haven't quite caught up with ourselves. Thrownness is also linked to the dimension of the past. We'll come back to that. Like the line in Paul Thomas Anderson's film *Magnolia*, "We may be through with the past, but the past isn't through with us." Moods are linked to the dimension of the past. In our moods, we pull along our past.

At the end of chapter 5, Heidegger writes that for "*as long as*

[Dasein] is . . . [it] remains in the throw" (SZ 179 / BT 223). We are always caught up in our everyday life in the world, in the throw, or in the throes of various moods, whether that's fear, boredom, excitement, or, as we'll see later, anxiety, the most revelatory mood, the shadowy queen amongst moods—the fundamental mood (see ch. 8, "Anxiety and Care"). But Heidegger insists we're not just thrown, we're always *in the throw*. We can't throw ourselves off. We're stuck with ourselves. We're riveted to ourselves.[2] We are facts to ourselves. But he insists we're not just thrown into the world. Dasein can also throw off that thrownness in a movement that he calls "projection."

This introduces the next concept cluster in chapter 5. We begin with projection (*Entwurf*). Think about projection as an idea of throwing off or throwing forth rather than as someone saying, "Well, I like to think of myself as a person that has projects," or something daft like that. It is also distinguished from the psychoanalytic term *projection* as displacing my intolerable impulses onto others (SZ 124 / BT 162). It's less an interpersonal projecting (which would, wrongly for Heidegger, presuppose a problematic of intersubjectivity) than a temporal anticipation that precedes my comportment by projecting an understanding in the form of my being-there. Projection is an activity of throwing off that is always a throwing forth from thrownness. This comes up in an important discussion of the concept of understanding in paragraph 31 of *Being and Time*.

What does "understanding" mean? By "understanding," Heidegger means understanding how to do something, like being able to

2. See Emmanuel Levinas, *Totality and Infinity: An Essay on Exteriority*, trans. Alphonso Lingis (Duquesne University Press, 1969), 37.

operate something. The German is *Etwas können*, being capable of something, like hammering or farming or using technology in a classroom (or not, in my case). It is strongly linked to a term that Heidegger uses in his discussion of authentic Dasein in division 2 of *Being and Time*, which is characterized by the ability or potentiality to be. Understanding how to do something should be linked to this larger idea of ability, which is going to be described in deeper ways later in the book, as potentiality for being, what he calls *Sein-können*. We are the sort of beings who *can*, who have capability, who have potentiality as their being.

The thought here is that the human being is not just a being defined by being thrown into the world. It can also throw off that thrownness in a movement of projection. As we'll see later on, this is going to be a *free* projection, where Dasein acts by seizing hold of its possibilities in a concrete situation. The movement of projection, of potentiality, of ability, is the very experience of what Heidegger will call "freedom." This is important: freedom for Heidegger is not an idea, much less an ideal. It is not an abstraction; it is an experience. It's the experience of the human being enacting its potential through acting in the world. And to act in this way for Heidegger is to be authentic, to exhibit potentiality, capability.

Human existence for Heidegger is defined by thrown projection. Thrownness is linked to our inauthentic, average, everyday life in the world, and projection is the ability to act freely and to become authentic. If all you take away from this discussion is the idea that Dasein is thrown projection, that we find ourselves thrown into a world, and we can throw forth from that thrownness—if you get the rhythm of that thought, which is simply the rhythm of life, that's all you need.

Yet there is another vital concept introduced in chapter 5, a third existential to go along with state-of-mind and understanding. This is what Heidegger calls "discourse" or "language," or—and I think this is clearer—"talk." The word here in German is *Rede*, which was a perfectly good English word until at least the seventeenth century. For example, you can find Ophelia in Shakespeare's *Hamlet*, when she speaks of the hypocrite who "recks not his own rede," where "rede" means discourse or language (HAM 1.3.51). Indeed, Shakespearean English is closer to the Germanic—or Germano-Scandinavian—roots of English than the soft, fluffy stuff that we speak now. Languages often lose their connection to their etymological thrownness, thus we get dog Latin, dog English, or what the French call *Globish*. But that's another story. For Heidegger, talk, or the capacity for expression or communication, is the way in which our life in the world becomes intelligible. It's the way in which the structure of our existence as thrown projection becomes articulable. Heidegger makes two claims in these paragraphs, which stand in a contradictory relation. He will describe these three existential concepts (state-of-mind, understanding, and discourse) as, in his language, "equiprimordial." They are equal, one to the other. But, if you think through the logic of what he's saying, then it is talk or discourse that is the condition of possibility for the articulation of Dasein as thrown projection. If there were no language, there would be no us. We are linguistic beings. Heidegger is part of what is rather superficially called the "linguistic turn" in philosophy, along with the later Wittgenstein and the linguistic philosophy of the 1950s and '60s. If language is equal to state-of-mind and understanding, then it is first amongst equals. And it is more basic than propositions.

The idea here is pretty straightforward. Aristotle famously described the human being as "*zoon logon echon*," ζῷον λόγον ἔχον, which is translated as the rational animal, the *animal rationale*.[3] The Aristotelian Greek in that formulation means an animal with the capacity for λόγος, for *logos* or speech (SZ 165 / BT 208). Heidegger disagrees with the idea of the human being as a rational animal and insists that what Aristotle is saying is that we're the sort of beings that have *logos*. Aristotle's words should be translated as "The human being shows itself as the being that talks" (ibid.). Now, buried in that seemingly straightforward expression is Heidegger's definition of phenomenon. A phenomenon is that which shows itself (SZ 28 / BT 51). And his definition of *logos* is that which lets us see. Language lets us see that which shows itself. This defines very precisely Heidegger's concept of phenomenology. Language, or *logos*, is the human ability to disclose the world, the capacity to reveal a world that is shared and meaningful. To go back to the beginning of this chapter, if what defines us is disclosedness, that we are the clearing of our existence, then what makes that clearing possible is language. "Language is the house of being," as Heidegger says much later (LH 217 / BH 313). (But what's going on beneath the floorboards of the house of being? Are there rats, gnawing and scratching, as in Henrik Ibsen's household of the Allmers in *Little Eyolf*?[4])

We need to think carefully about what kind of language we're

3. Aristotle, *Politics* 1253a8–10, *Nicomachean Ethics* 1097b33–1098a5.

4. Henrik Ibsen, *Little Eyolf*, trans. Henry L. Mencken (John W. Luce, 1909).

dealing with here. Heidegger is going to have some words to say about this vital issue in the middle paragraphs of chapter 5, paragraphs 32 and 33, which we'll get to at the end of this discussion and deal with in more detail next time. There can be authentic language, full language, and there can be empty language, idle talk. There are different ways of experiencing language. And Heidegger thinks that our fundamental activity as linguistic beings is not propositional. It is not something that takes place in philosophical propositions of which we can assert the truth or falsity. Full language, authentic talk, is closer to the primal feeling or meaning of elemental words. You could relate, for instance, to a cat propositionally, or you could relate more primally: "Cat!" or just "Gasp!"

To summarize: Dasein is thrown projection, and thrown projection is articulated through discourse, through talk. The three key concepts in chapter 5 are state-of-mind (meaning mood, thrownness, or facticity), understanding (meaning being able to do something, projection), and language (talk, or discourse). But there is a fourth concept introduced in chapter 5, which I want to say a word about here. And that fourth concept is *falling* (*das Verfallen*). And this concept of falling can also be thought of as an ensnarement, falling into a trap. Its mechanisms are described towards the end of chapter 5.

Having described the three existentials that constitute the disclosure of Dasein, Heidegger goes on to describe those characteristics of our everyday being-in-the-world, given that we live with the They most closely and mostly. We'll get into this in more detail in chapter 7, "Inauthentic Life," but Heidegger describes three modes of falling: idle talk, curiosity, and ambiguity. Now, what does Heidegger mean by "falling"? He insists that this is not a negative evaluation, in the same way as he has insisted that inauthenticity is not a moralism, not a negative value judgment. So, too, falling does not mean that we have fallen from a state of grace into a state of sin. This theological and intensely Christian way of looking at falling is the framework Heidegger is pushing against, and it is more than a moot point whether he succeeds—Heidegger both inhabits the language of Protestant theology and tries to disavow it. Rather, falling describes our usual way of being-in-the-world.

Falling is not a descent into the world, like Satan's plunge from heaven to hell after the war in heaven, into the infernal mud and pits described in John Milton's *Paradise Lost*.[5] It is a different kind of falling. The way I think about this is in the following terms: The usual way we think about falling is vertical. We fall down from one state to another, from grace to sin. Heidegger's idea of falling is that we fall *towards* the world. We fall *at* the world. We fall to it, or through it, alongside it, and we're constantly falling towards the world, falling "forward," or horizontally, because we're absorbed by the world and we're fascinated—benumbed, taken—by it. Now we're right at the end of chapter 5. I'm going to go back to the beginning of the chapter soon, but I want to give you a taste of where all this goes at the end of these many paragraphs.

To quote paragraph 38, in the final pages of the chapter, falling "does not express any negative evaluation, but is used to signify that Dasein is proximally and for the most part *alongside* the 'world' of its concern" (SZ 175–76 / BT 219–20). Falling is not negative. Rather, it represents a positive possibility. In falling, our potentiality for being is still an issue, even if that's in the mode of inauthenticity.

5. John Milton, *Paradise Lost* (Routledge, 1905).

Dasein can fall because it understands and projects. Authenticity, Heidegger insists, is not some state that floats above falling everydayness. Authenticity is simply a modified way in which everydayness is seized upon and thrown off from, like a jumping-off point. Authenticity is a modification of inauthenticity; it's not morally better, it's just different. Or, as John Haugeland understands it, authenticity is self-ownership.[6] And the chapter finishes with this compelling thought. Heidegger writes, returning to the theme of modification:

> In falling, nothing other than our potentiality-for-Being-in-the-world is the issue, even if in the mode of inauthenticity. Dasein *can* fall only *because* Being-in-the-world understandingly with a state-of-mind is an issue for it. On the other hand, *authentic* existence is not something which floats above falling everydayness; existentially, it is only a modified way in which such everydayness is seized upon. (SZ 179 / BT 224 [trans. mod.])

All that authenticity means for Heidegger is a modification of our inauthentic existence, a shift in perspective. A shift that, of course, does not last forever. We can't help but fall back into, or gravitate towards, everydayness. Authenticity is a modification done in mid-air. I can be having the most authentic day of my life, with hot yoga, mindfulness classes, and a long, solitary walk with stops for "journaling" (when did that become a verb?). And then at a certain point,

6. John Haugeland, "Heidegger on Being a Person," in *Dasein Disclosed*: "Insofar as self-understanding critically takes over its roles, it is said to be self-owned (*eigentlich*; Macquarrie and Robinson: authentic).... Self-ownedness, is, roughly, 'getting your act together,'" 14–15.

I'll start binge-watching a Netflix true crime series, working my way through a six-pack, and then fall asleep with my clothes on, only to awake the next morning, finding myself hungover and in a rotten mood. Authenticity keeps sliding back into fallen everydayness.

That is the broad arc of the story in chapter 5 that I want to get across here. But there are a couple more themes I'd like to discuss: *fear* and *interpretation*.

In paragraph 30 of *Being and Time*, there's this little discussion of fear that is often ignored. But it's very revealing. Heidegger talks about fear in relation to Aristotle's *Rhetoric*. It's a very important text for Heidegger because in the *Rhetoric* Aristotle discusses emotions. Why, we might ask, does Aristotle discuss emotions in the *Rhetoric*? Because rhetoric is that activity of speech, of language, that is meant to be persuasive, that is meant to arouse emotions in the hearer. Rhetoric and emotion are tightly linked.[7] We might find ourselves literally moved by a speaker, roused to commit certain acts. Heidegger describes the *Rhetoric* of Aristotle—it's a lovely phrase—as "the first systematic hermeneutic of the everydayness of Being with one another" (SZ 138 / BT 178). And then he goes into a discussion of fear. In many ways, what Heidegger's doing here is that he's got his copy of Aristotle open and he's just making notes. For Aristotle, fear is fear of something happening, fear of something threatening, something possibly destructive. And it is opposed to confidence.[8] Heidegger just borrows this, and echoes this thought:

7. Aristotle, *Rhetoric* 1356a2–21.

8. Ibid., 1382a22–23.

fear discloses us as Dasein in our being-there, as fearfully concerned with ourselves. Fear is a state-of-mind that is concerned with something threatening me in the world.

If you want an example of this, and you know that I am bad at examples, which is why I just choose absurd ones, imagine that there is a giant bear in the room now. A huge grizzly bear appears at the window, eight feet tall, clawing away, its mouth dripping with saliva. He's hungry all right, and I am food. I would feel threatened by that bear, right? He might come in and kill me and eat me. I'd be right to be scared of that bear. That would be a good response. The fear is always fear *of* something, something threatening, in this case a huge bear. Why is that important? Why does that matter? Well, when we get a little bit further on in *Being and Time*, in paragraph 40, Heidegger will make a distinction between fear and anxiety. Fear is fear of some particular thing, like a giant bear. Fear is determinate. Take away the object of fear, and you remove the fear. Anxiety is anxiety in the face of nothing in particular. Which is why Heidegger will say, provocatively and gnomically, in a lecture from 1929 called "What Is Metaphysics?," "Anxiety reveals the nothing."[9] It's revelatory. While fear is always determinate, anxiety is anxiety in the face of *nothing*. Anxiety is reticence for the open moment, readiness. To reduce it to a disorder is to ignore that anxiety is, in its primordial ontological movement, not a disorder but a clarification. And in its clarificatory, revelatory readiness, anxiety has the character of a conversion experience.

Now, there's a lot more that we could say about fear, and it has a

9. Martin Heidegger, "Was ist Metaphysik?" (1929), in *Wegmarken*, 103–22, esp. 112; "What Is Metaphysics?," in *Basic Writings*, 89–110, esp. 101; hereafter WMK/WMC, followed by page numbers.

very fleeting place in Heidegger's analysis. But it's also a way in which we can think about Heidegger's relation to others, to other Daseins. A common way of thinking about Heidegger is that he's ultimately a rather solipsistic thinker who turns his back on the average, everyday character of existence where authenticity is the solitary rising up of the self in the face of the anonymous world of the They. That's a very distorted picture. This discussion of fear is also a place in which fear can be about others. And that can be fear *for* others. I can be afraid for myself, but I can also be fearful that the other might do something reckless with regard to what threatens him or her. There can be a "co-state-of-mind" with the other, he says, a *Mitbefindlichkeit*, which takes place through fear (SZ 142 / BT 181).

So this is one place in *Being and Time* where others light up. There is a relational dimension to what Heidegger's describing in the term "fear." It is compatible or comparable with his talk of solicitude, which I mentioned last time, especially leaping ahead of the other, *vorausspringen*. I think about this in relation to my mother. Before she died, I used to fear for my mother. I feared for her a lot. I had a co-state-of-mind with her, fearing for her. I feared that things would get even worse for her (she suffered from depression and other maladies of the mind). Of course, in fearing for my mother, I'm also fearing for myself. I didn't want to lose my mother. But the point is that fear lights up a relationship with another person or persons. This is a dimension of Heidegger that isn't usually brought to light.

The last thing I want to talk about is interpretation, and I'll return to this theme and clarify it further next time. But let me just say a couple of words here in closing. What I'm thinking about now is paragraph 32, which is called "Understanding and Interpretation." What Heidegger is trying to bring out here is something terribly simple but very difficult to get a grip on. We have an understanding of the world. The world is that bunch of stuff, that referential totality of involvements that are significant to us. The world hangs together and is meaningful. We use things. We have a concern for things. We take things as things. When I'm in the world, moving around, I take the cross light as a cross light, I take the coffee cup as a coffee cup, the door as a door, the table that I'm resting my notes on as a table, and so on. The things that environ us and that make up the stuff of the world are things that we understand. And the understanding is articulated when we take something as something.

This taking of something as something is what Heidegger calls "interpretation." I think that becomes clearer if you think about interpretation in relation to the German term that Heidegger uses, *Auslegung*, which has the sense of "laying out" in English. Interpretation for Heidegger is the laying out of our understanding of being-in-the-world. He says at the beginning of paragraph 32, "In interpretation, understanding does not become something different. It becomes itself" (SZ 148 / BT 188). In interpretation, the things that we take as things become entities that we understand. This is a little bit like the discussion of signs we looked at previously. If there's an exit sign over the door, I'm aware of that sign as a ready-to-hand item of equipment, a piece of stuff in my world, which I can see as a sign, but I'm not aware of that in a present-at-hand way that I look at quizzically. I'm aware of that as a sign that functions in a world that allows me to orient myself in the world. I take something as something, and I lay that out to myself. I project that understanding ahead of time, and that's interpretation.

Now, incidentally, what Heidegger is doing in these pages of paragraph 32 is inventing a whole new field of philosophy that is called *hermeneutics*. The tradition of hermeneutics, which Heidegger adapted from Wilhelm Dilthey, was developed decisively by Heidegger's student Hans-Georg Gadamer in his hugely important book *Truth and Method*.[10] But the core concepts are all in *Being and Time*. Interpretation—and this is what Gadamer picks up—is grounded in a threefold structure of *fore-having*, *fore-sight*, and *fore-conception*. That is, we have the thing in advance of the interpretation that lays out that having. We are always already in the world, understanding it. Interpretation just lays that out, but we already have a grasp on things.

This leads Heidegger—and I find this move really suggestive—to a discussion of meaning. He says, in his terminology, meaning is that "'*upon-which*'... *something becomes intelligible as something*" (SZ 151 / BT 193). But that sounds a bit too fancy. The thought that he's trying to develop is that meaning is not some property that is independent of us. Meaning is not something we can perhaps find or lose. Rather, meaning is an *existential* of Dasein. Meaning is something that we do. We make meaning. Or, we *are* meaning. The world is significant to us. Meaning is that which gets articulated in an act of interpretive understanding. Dasein is meaningful. Only Dasein can be meaningful. If there were no Dasein, there would be no meaning. And where Dasein ceases to be, meaning ceases to be. On a Heideggerian view, death is the nonmeaningful limit to life's meaning, the edge of our meaning.

10. Hans-Georg Gadamer, *Truth and Method*, 2nd rev. ed., trans. Joel Weinsheimer and Donald G. Marshall (Bloomsbury, 2004).

Let's think a little bit more about this idea of meaning in closing, because he makes a very interesting remark in paragraph 32:

> If we are inquiring about the meaning of being, our investigation does not then become a "deep" one, nor does it puzzle out what stands behind Being. It asks about being itself in so far as Being enters into the intelligibility of Dasein. The meaning of Being can never be contrasted with entities, or with Being as the "ground" which gives entities support. (SZ 152 / BT 193–94)

What he's saying here is that meaning is not deep. It just sounds deep. When we're talking about the meaning of being, this sounds really profound. But to construe it as "deep" is to misconstrue meaning metaphysically. It sounds as if we're looking for something that lies behind our life in the world. Some ground or foundation, some God, or something, anything, that can aetiologically explain our experience.

No, Heidegger says: being is superficial. It is phenomenal. It is something that shows up on the surface of things. This takes us back to the poem I quoted two chapters back, when Fernando Pessoa talks about seeing things, and what we see of things *are* those things. Why would we see something and then think about something else? Or why would we think that what we see is not real? That would require, Pessoa says, a sickness of the mind, a form of illness. Philosophy is like an illness a lot of the time. Heidegger's point is that we have to be concerned with that which is, with being, only insofar as it is intelligible to us. And it lies on the surface of things. Another way of putting this thought—maybe you will find this helpful—is that many readers of Heidegger see being as a rabbit in a hat that you can pull out, as with a magic trick. Ladies and gentlemen, boys and girls, and everybody else: *There is no rabbit*. The point is *learning to see the hat without wanting the rabbit.* If you can do that, then you are learning to think in a Heideggerian way.

6. Language and Interpretation

Interpretation as taking something as something prior to propositions — The hermeneutic circle — Virtuous and not vicious — Judgment is not the primary locus of truth — Cribbing Aristotle — Comparison with Wittgenstein's Tractatus *— Important nonsense and silence — Taking things hermeneutically — The distinction between discourse and language — Communication — Language is not inside the head, it is ecstatic — Dasein has to have clean ears — The voice of the friend — Hearing, hearkening, silence, and reticence — Keeping quiet — Proposal for a reticence bar called Helsinki — The human being shows itself as a being that talks — Nonhuman language — Excursus on the octopus.*

LET'S PICK UP where we left off last time. We went through the basic concepts of chapter 5, division 1, of *Being and Time*. Those four basic concepts are state-of-mind, understanding, discourse, and falling. State-of-mind is linked to thrownness. Understanding is linked to projection. Discourse is linked with language or talk. And falling is characterized by idle talk, curiosity, and ambiguity. The key concept is the idea of Dasein as thrown projection, and we ended with little discussions of fear and interpretation.

I'd like to explain interpretation more crisply, more succinctly. After that, I want to talk about Heidegger's understanding of language in *Being and Time*, which is the condition of possibility for the disclosure of Dasein. And then we will turn to our discussion of inauthenticity, looking at part B of chapter 5, where Heidegger talks about idle talk, ambiguity, and curiosity.

Let's start with interpretation. Dasein always begins with an understanding of the world. The world is the "referential totality," as Heidegger calls it, that whole network or field of stuff that hangs together and that forms the meaningful context in which we find ourselves. The world is a totality of involvements that are significant to us. In this world, we use things. We have a concern for things. And we *see* things. Every relationship to things is a mode of sight. Heidegger's point is that our fundamental mode of sight is not one of pure perceptive beholding or quizzical looking in the mode of the present-at-hand. Rather, we see things in the mode of circumspection, practical around-seeing, *Umsicht*. We take things as things, we understand them—we see them—in the context of this network or field of meaning. We take the table as a table, the door as a door, the hamster as a hamster. The things that surround us and that make up the stuff of the world are things we already understand. And that understanding is articulated when we take something *as* something. When I use the table as a table—the table I'm leaning on now without thinking about it—I'm taking it as a table. This taking-as is interpretation, *Auslegung*, the laying out of understanding. Interpretation is the act by which the understanding becomes itself, projects itself or lays itself out, like a light spilled over the world, illuminating it as my world.

Interpretation is the taking of something as something, and it is also the taking apart of something and seeing it for what it is. The crucial point in paragraph 32 is that this taking of things, interpretation, is pre-predicative, it's pre-propositional. I see the hammer as a hammer, a hammer that I use. I do not see it in the mode of the present-at-hand. I do not see it as a theoretical object that I can take apart or doubt. Cartesian, radical doubt happens when you see things as present-at-hand. But this taking-as can be compared to our earlier discussion of signs (see ch. 2, "World"), where signs are those things that we take as things—like an exit sign. We see the exit sign as something ready-to-hand in relation to the context of a particular world, the world of this room, and as a way of escape from that room in the case of some emergency.

Picking up the point I ended with last time, on hermeneutics, we are moving here in a "hermeneutical circle." Interpretation already understands, and what it lays out is already understood. This is different from a scientific inquiry, wherein we discover new things and then work out how to deal with the discovery. But in everyday engagement with the world, we are in a hermeneutical circle where we lay out what is already understood. A standard scientific procedure proceeds by a process or logic of discovery—induction. We cannot presuppose that which it is our task to provide grounds for. On the contrary, we discover it as a new fact. However, with our being-in-the-world, we're always stuck in a circle. And it's a question of entering this circle in the right way and not trying to get out of it. The hermeneutical circle, the fact that interpretation already understands, is a virtuous circle, not a vicious circle (SZ 152–53 / BT 194–95).

Now, if we've understood Heidegger's thought about interpretation, taking-as, laying out, then it has a devastating philosophical consequence. And here we move on to paragraph 33 of *Being and Time*, which is on the idea of assertions or propositions. It'll take a few minutes to get this straight in our minds. The consequence of the argument that Heidegger is developing is that truth is not something we can assert uniquely of propositions. As Heidegger puts it, the proposition or the assertion or judgment (really three terms for the same thing) is not the primary and authentic locus of truth, where truth is understood as the truth of that which is—namely, our being-in-the-world (SZ 156–57 / BT 199). The truth of being-in-the-world is laid out in interpretation, which is prior to assertion, prior to a proposition, in the same way as the ready-to-hand is prior to the present-at-hand (SZ 157 / BT 199).

Heidegger, being such a good scholar of the history of philosophy, recognizes the radicality of what he's doing here. Although it's hard to detect as one leafs through its pages, *Being and Time* was ostensibly a book on Aristotle. And in paragraph 33, we can see the direct influence of one of Aristotle's more tedious, logical works, which is called *Peri Hermeneias*, or *De Interpretatione*, or *On Interpretation*. In this little text, which is just a few pages, Aristotle is concerned with the definitions of word, sentence, and proposition. It's dry stuff, but worth a look if you are interested. But this is why Heidegger, at this point in *Being and Time*, has moved into a discussion of assertions. It's more or less cribbed from Aristotle. The key issue, which Heidegger will deepen in the next paragraph, on discourse, is the following: For ancient ontology—by which Heidegger in this period simply means Plato and Aristotle (this changes radically in his later work when he turns to the pre-Socratics)—*logos*, language, is the means for obtaining access to, and defining the being of, that which is. So far, so good. The assumption of Aristotle's logic, which is definitive for subsequent philosophy, as Heidegger points out, is that propositions are the highest form of *logos*, the highest form of language. Propositions—of the form of "this is X" or "Y is true" or "this is a hammer" or "this is a brown bear"—are the logical form by which we gain access to that which is.[1] For Aristotle, the proposition picks out the structure of reality, whereas Heidegger disagrees fundamentally with this way of building the world.

To get a sense of the radicality of the move that Heidegger is making here, the claim that being-in-the-world is prior to propositions, it might be helpful to think about another philosophical example, that of Wittgenstein's *Tractatus Logico-Philosophicus*, published in 1921, just six years prior to *Being and Time*. The *Tractatus* is an apparently complicated book, but actually it consists of just seven propositions. I convinced my friend Ray Monk to write a book called *How to Read Wittgenstein*, which offers a reading that resists this tendency to wallow in the complexity of the *Tractatus*.[2] If you ignore the fascinatingly beautiful and brilliant tangle of sub-propositions, there are just seven thoughts.

The first proposition is the following: The world is all that is the case.[3] This seems fair enough. The second proposition: What is the case is facts or the existence of states of affairs (TLP 5). Again, who would disagree? The world is all that is the case, and what is the case are facts. The third proposition in the *Tractatus*: A logical picture of facts is a thought (TLP 12). And the fourth proposition is as follows, and this is the key one for our purposes: A thought is a

1. Aristotle, *De Interpretatione* 17a19–20.

2. Ray Monk, *How to Read Wittgenstein* (Granta, 2005).

3. Ludwig Wittgenstein, *Tractatus Logico-Philosophicus*, trans. D. F. Pears and B. F. McGuinness (Routledge, 2014), 5; hereafter TLP with corresponding page numbers.

proposition with a sense (TLP 22). The world, which is all that is the case, can be captured in propositions. A proposition is a thought with a sense. The fifth and sixth propositions of the *Tractatus* establish roughly and readily the truth functions of what Wittgenstein calls "elementary propositions," or protocol sentences, the general form of propositions that are reducible either to being empirically verifiable—certain propositions are true by virtue of being empirically demonstrable (e.g., "There is a glass of water on the table" is a proposition I can test by virtue of there being or not being a glass of water on the table)—or by virtue of their logical form (according to Wittgenstein's understanding of logic), which is either reducible to a tautology, "A is A," or a contradiction, "A is not A" (TLP 43ff.). That's it. And then, the seventh proposition of the *Tractatus*: "Whereof one cannot speak, thereof one must be silent" (TLP 189).

The point of Wittgenstein's argument here is that propositions give us a picture of the facts and access to the totality of the world, and the logical form of the proposition exhausts all that can be meaningfully thought. The rest is nonsense. Philosophy is divided on how we understand that nonsense. But this is very important nonsense for Wittgenstein. This nonsense is the realm of the ethical, the aesthetic, and the religious.

There's a letter that Wittgenstein wrote to Ludwig von Ficker (which is a very rude name if you know German) shortly after he finished the *Tractatus* in November 1919, where he says that the key discoveries of the *Tractatus* are what is *not* said in the book.[4] The book exhausts all that can be said meaningfully that is either reduc-

4. "Letters to Ludwig von Ficker," in *Wittgenstein: Sources and Perspectives*, ed. C. Grant Luckhardt (Cornell University Press, 1979), 82–98.

ible to logical form or empirically verifiable. But for Wittgenstein, what's truly valuable lies elsewhere, and that's something about which we have to be silent. There is no meaningful metalanguage about ethics, aesthetics, and religion. (For early Wittgenstein, moral philosophy is twaddle.) The rest is silence, in Hamlet's words (HAM 5.2.344).

Now, the later Wittgenstein shifts his position and his thinking away from what he called "the crystalline purity of logic"[5] and claims to have been captivated by a picture of language that sees propositions as offering a mirror to nature (as in Richard Rorty's critique of philosophy[6]). In his ever-unfinished later work, it can be argued that Wittgenstein moved towards a position on language that indeed some people have thought is quite close to Heidegger, when Wittgenstein talks about forms of life, language games, and the rest. But for the earlier Wittgenstein, access to that which is is given through propositions. Heidegger is trying to uncover an area of language that is prior to propositions. This is a really radical thought in relation to the history of philosophy.

Let's turn more closely to paragraph 33. Heidegger says, "In concernful circumspection"—namely, the kind of "looking at things" that we have in our relationship to the world—"there are no such assertions 'at first'" (SZ 157 / BT 200). Proximally and for the most part, most closely and mostly, propositional language is not to be found. He goes on:

5. Ludwig Wittgenstein, *Philosophical Investigations*, p. 46e (section 107).

6. See Richard Rorty, *Philosophy and the Mirror of Nature* (Princeton University Press, 1979).

> But such circumspection has of course its specific ways of interpreting, and these, as compared with the "theoretical judgment" just mentioned, may take some such form as "The hammer is too heavy," or rather just "Too heavy!," "Hand me the other hammer!" Interpretation is carried out primordially, not in a theoretical statement, but in an action of circumspective concern. (ibid.)

Like two plumbers in a basement. The key thing is that interpretation is not carried out in a theoretical statement (a proposition) but in an action, the act of being circumspectively concerned (having contextual reference) with something: "too heavy." Heidegger says that interpretation is carried out in "laying aside the unsuitable tool, or exchanging it, 'without wasting words'" (ibid.). Note that phrasing: "without wasting words." We get the idea that we've seen a number of times in Heidegger that the world is a workshop, where hammers are passed from one person to another, and they might be accompanied by the word "hammer," or "Pass me the other hammer." It's a bit like when a surgeon says "Scalpel" or when a movie director says "Cut." These aren't propositions; these are words that are actions. And the fewer words, the better. The phrase "without wasting words" also anticipates his upcoming discussion of reticence: "From the fact that words are absent, it may not be concluded that interpretation is absent" (SZ 157 / BT 200). Heidegger seems protective of workshop talk with its few words. It cannot be assumed that in the absence of words a lack of understanding follows.

On the basis of this claim, Heidegger makes a distinction between two ways of taking-things-as. He says there is an existential and hermeneutic—interpretive—"as," which is what he wants to argue is primary, and what he calls an *apophantical* "as," which is what takes place in an assertion or a proposition. "Apophantical" here means a declarative statement whose truth-value can be assessed either logically ("All penguins are birds") or empirically ("There is a penguin on the table"). In the latter case, that would be a falsehood, as there is no penguin on the table (at least I don't see one). Heidegger writes, "Thus assertion cannot disown its ontological origin from an interpretation which understands. The primordial 'as' of an interpretation which understands circumspectively we call the 'existential-hermeneutical "as"' in distinction from the 'apophantical "as"' of the assertion" (SZ 158 / BT 201). That may be a slightly bewildering quotation. But the thought here is very simple. It is that we take things as things existentially through understanding and interpretation. That's our founding relationship to things. Propositions, theoretical statements, and assertions are founded upon interpretations. Hermeneutics precedes logic and epistemology. Like he did with the bluntness of workshop speak, Heidegger's de-structuring of language here, his reversal of the relationship between propositional thought and the pre-propositional taking-as, seems to also hint at a defense of silence—the absence of words—against the supposed primacy of assertions. The language of the ready-to-hand world is in its straightforwardness, its lack of too many words, prior to the propositional taking-as that reads silence only as a lack. And he describes this apophantical "as" (or the presupposing of its primacy) as attempting to "disown its origins." A similar theme to *verleugnen*, to deny or disown, appears much later in the book, in connection with the theme of repetition—namely, in the form of *Widerruf*: revocation or disavowal (SZ 386 / BT 438). But disavowal is a more hermeneutical disowning, in a word, a repudiation or a reinterpretation. Or even a revision. The insistence in metaphysics that taking-things-to-be is essentially propositional is perhaps being cast by Heidegger as a renunciation of its circumspective origins, but a renunciation that doesn't know itself to be such, an unchosen or passive (perhaps passive-aggressive) denial of its foundedness in a mode of speech it doesn't acknowledge. We'll see better how this de-structive dimension of Heidegger's project in *Being and Time* plays out when we look at his analytic of historicity, which offers an authentic (owned, chosen) version of disavowal (see ch. 16, "Historicity and Politics").

At the end of paragraph 33, Heidegger remarks that the basis for the understanding of language, of *logos*, in Aristotle and ancient ontology is not primordial enough, because Aristotle understands language as something present-at-hand, something theoretical, something that is articulated in the form of a proposition. The Greeks thought propositions were foundational.

Heidegger thinks that our actual understanding, our lived use of discourse, is deeper than that given by ancient ontology and all the philosophy that follows in its wake. This is what he's going to try to lay out in paragraph 34, which is called "Dasein und Rede" (Being There and Discourse), full stop, "Die Sprache" (Language). In the text it's printed just like that: "Dasein und Rede. Die Sprache," with one period separating the two phrases. What is that period doing there? It appears that Heidegger is making a distinction between discourse (*Rede*) and language (*Sprache*) or speech. And one question we have to ask ourselves in this paragraph is what that distinction consists in. We will come back to that question, but Heidegger doesn't really make it clear.

Paragraph 34 is an important moment in *Being and Time*. He writes, "*The existential-ontological fundament of language is discourse*" (SZ

160–61 / BT 203; trans. mod.). Heidegger admits that his previous analyses have been making constant use of discourse, but he's "kept it suppressed" in the "thematic analysis" (SZ 161 / BT 203). Discourse is constitutive for the disclosure of the "there" of being-in-the-world. Discourse is the way in which the "there" of our essential disclosedness becomes intelligible or meaningful. You can feel its importance: we are being-there, we are Dasein, and the way in which our "there" is articulated (disclosed) is through discourse.

Now, discourse, *Rede*, renders the Greek term *logos*. If we go back to the introduction to *Being and Time*, paragraph 7, then *logos* is what lets us see (SZ 32 / BT 56). Discourse is the letting-be-seen, the intelligibility of that which shows itself. What shows itself is a phenomenon, and how we're allowed to see a phenomenon is through discourse in average, everyday life. We might say that discourse is the "-ology" in "phenomenology." This raises an interesting question. As I mentioned in the last chapter, "Discourse is the Articulation of intelligibility," of meaning as such. Discourse "underlies both interpretation and assertion" (SZ 161 / BT 203–4). In order for interpretation to interpret, to lay things out, there already has to be discourse. Being-in-the-world is always discursively articulated. The intelligibility of the world is given through words.

The way in which discourse gets expressed is in language. Language for Heidegger is the totality of words. And here we have the suggestion of a distinction between discourse and language—an existential grammar or structure of language, which is quite tricky to work out, as there are not many clues in the text. If language is words, what I'm speaking right now, then what is discourse, what is *Rede*? Is it the capacity for language that precedes words? Is it a sort of grammar, a basic grammar, a transcendental grammar? It's not clear. But thinking of it as grammar makes sense to me. It's an interesting issue, which has some consequences, one of which I'll get to very soon. However we understand these terms, discourse and language, Heidegger's thought here is that discourse is existential language: "*Die Rede ist existenzial Sprache*" (SZ 161 / BT 204). Discourse is language prior to the apprehension of present-at-hand word units, like sentences. Discourse is the structuring or ordering of the intelligibility of being-in-the-world to which our concernful being with one another belongs. In other words, discourse—in language that Heidegger won't use because he thinks it's too obvious, too bound to the tradition that must be destroyed—is *intersubjective*. Discourse is the articulation of our being-with-others.

Moving on a tiny bit, there's a connection between discourse, understood as being-with-others, and communication. And Heidegger makes a play of the relationship in German between *Mitsein* (being-with-others), and *Mitteilung* (communication, literally "sharing-with"), where communication is grasped as understanding and being with one another, sharing a state-of-mind with others, and sharing a common world. And this also raises a very interesting issue, which I want to emphasize here, which is that *language is not inside the head*. We used to appreciate this better in the heyday of structuralism. To speak is to be outside in symbolic, social structures. It's not just what's running through our minds. It's not dependent on, or explained by, its relationship to the brain or anything like that. That's not the way Heidegger wants to present this argument. Word structure is *world* structure: "In talking, Dasein *ex*presses itself not because it has, in the first instance, been encapsulated as something 'internal' over against something outside, but because as Being-in-the-world it is already 'outside' when it understands" (SZ 162 / BT 205). Language is ecstatic. Language is always already outside with others, in the shared expression of discourse. Insofar as one understands, and we always have understanding of the world for Heidegger, one is already outside, and so is everybody else. There is no such thing as a private language, as the later Wittgenstein would argue.

In an aside, Heidegger says this is also the way we can think about speaking in relation to poetical discourse. Poetry—which becomes such an important theme in his later work but is barely mentioned in *Being and Time*—is a way of laying out our understanding of the world and our understanding of being-with-others (SZ 162 / BT 205). Poetry is the disclosure of the world and the co-disclosure of the world as a world that we share with others. And it is prepropositional. And this is why I emphasized those links to poets like Wallace Stevens and Fernando Pessoa, because they really give us a sense of this world-structuring activity of language. Poetry is not necessarily propositional. It is world-disclosure. This is also true, perhaps even more so, of music.

After Heidegger has made this point about discourse, which is his main point in paragraph 34, we get a beautiful little phenomenology of hearing, of hearkening, of silence, and of reticence. Hearing is essential. Dasein has to have ears and, if you like, clean ears that listen to the other and to the world. The problem with us, with humans, because we constantly fall into—and become hypnotized by—tradition, into customary ways of thinking and speaking, is that our ears are blocked with habits and dubious assumptions. Heidegger is asking us to unblock our ears and clean them out. In many ways, *Being and Time* is a way of getting that wax—that fallen wax —out of our ears and cleaning those channels so we hear more crisply and cleanly. In this way, we can be open to the world and to others. He says, "Listening to . . . is Dasein's existential way of Being-open as Being-with for Others. Indeed, hearing constitutes the primary and authentic way in which Dasein is open for its ownmost potentiality-for-Being—as in hearing the voice of the friend whom every Dasein carries with it. Dasein hears because it understands" (SZ 163 / BT 206). This is a strange phrase, "as in hearing the voice of the friend whom every Dasein carries with it." We might ask, Who is this friend that I carry around with me? (Is this your special friend, Martin? A six-foot rabbit?[7]) We'll come back to that later, when we look at Heidegger's discussion of conscience. It might well turn out that Dasein is its own best friend, which is a slightly sad state of affairs. The point that Heidegger is making here is that in listening to and hearing, we are open and we're with others. We have to clean our ears.

The second aspect of this little phenomenology in relation to discourse is in relation to what he calls hearkening. And he says that hearkening is more primordial than what is called "hearing" in psychology. In a fascinating passage, he says:

> It is on the basis of this potentiality for hearing, which is existentially primary, that anything like hearkening becomes possible. Hearken-

7. Dasein's best friend, for Heidegger, is itself, but it is hard to say this with a straight face and without veering into banalities of childhood make-believe or pop mythology such as the giant rabbits of Henry Koster's *Harvey* (1950) and Richard Kelly's *Donnie Darko* (2001), who keep their hosts aloof from their social surroundings.

> ing is phenomenally still more primordial than what is defined in the... perception of sounds.... What we "first" hear is never noises or complexes of sounds, but the creaking waggon, the motor-cycle. (SZ 163 / BT 207)

Listen, hearken, to these examples. We don't hear a noise, we don't hear a sound-complex. We hear a creaking wagon, and we hear a motorcycle. "We hear the column on the march, the north wind, the woodpecker tapping, the fire crackling" (ibid.). These are strange examples. Where is the column on the march? And where is it on the march to? Where did it come from? And why a woodpecker? "The fire crackling": What's being burned? These are slightly suspicious metaphors (reminiscent of creation narratives), given Heidegger's later political commitments. But let's just pass over that for the nonce. The point that he's making is that when we think about these examples—the column on the march, the fire crackling, the woodpecker tapping—Dasein hearkens to these insofar as it dwells alongside what it understands. It dwells alongside things in the world. Hearkening is similar to spatiality, the discussion of space we saw in chapter 3, where space is what brings things close, this collapsing of distance and remoteness. In space, Dasein has a tendency towards the near, Heidegger says. Dasein brings things close. "De-distancing" is the rather ugly term he uses, but in hearing another, in hearkening to another, we are already with him or her. Then he talks about silence. And this is revealing: "Speaking at length about something does not offer the slightest guarantee that thereby understanding is advanced. On the contrary, talking extensively about something, covers it up and brings what is understood to a sham clarity—the unintelligibility of the trivial" (SZ 164 / BT

208). The unintelligibility of the trivial. I like that. It sounds like my life. For Heidegger, authentic discourse is keeping silent, using words in a limited way. And what he says here about wordiness—that speaking at length doesn't guarantee understanding—echoes his earlier sentence: absence of words is not absence of understanding. He says, "To be able to keep silent, Dasein must have something to say. . . . It must have at its disposal an authentic and rich disclosedness of itself. In that case one's reticence makes something manifest, and does away with 'idle talk'" (SZ 165 / BT 208). We'll introduce idle talk—fast talk—in the next chapter. But notice here the reference to *reticence*. Hearing, hearkening, silence. Heidegger prefers few words rather than many words, and for those few words to come out of a deep, rich understanding. If we wish to become authentic for Heidegger, we need to cultivate reticence.

Now, one of my personal fantasies has been a wish to start a reticence bar. Imagine it: a bar where no one speaks except when they have something to say, where people keep silent, and then when they have something to say it's because they genuinely understand it, and they can say it in a few words. Otherwise, they'll keep quiet. Not out of inhibition or fear—the fear of looking like an idiot—but because of a concern with the truth. I think this would be fantastic. (There used to be a bar like this in the East Village inside a Japanese restaurant above the much-missed and long-gone St. Mark's Bookshop, called Angel's Share. It had rules such as you couldn't speak while standing, and they only served parties of four or fewer.[8]) I spent many happy evenings in Finland over the years, particularly

8. Happily, Angel's Share reopened in mid-2023.

in Helsinki. I remember being with some Finnish friends of mine after a long night discussing philosophy, music, and football, drinking beer and such. The people I was with had known one another their whole lives. (Helsinki has a small philosophical world.) And at a certain point in the evening, one or two a.m., things became quieter and quieter until, eventually, these really, really good friends fell into absolute silence. There was no anxiety. No awkwardness. It's just that nothing was said. And I thought: Well, this is good. This wouldn't happen in New York, with everybody chattering away about themselves and their endless problems, interrupting you just to talk more about themselves. Anyway, I want to open a reticence bar.[9] For those of you who have not been to Helsinki, please take a look at any movies by Aki Kaurismäki. You might begin with his more recent 2023 masterpiece, *Fallen Leaves*, and then work back to *Man Without a Past* from 2002. Lots of reticence and dry, dark humor.

Now, towards the end of paragraph 34, Heidegger cites Aristotle again, and it's the famous phrase I cited previously, that the human being is "*zoon logon echon*," the words that are usually translated into Latin as *animal rationale*, "rational animal." What Heidegger is suggesting here in relation to language is the following: the tradition thinks about the human being as an animal with the use of reason. Heidegger's suggestion is that we understand *logos*, discourse, as the way in which the human being shows itself in an everyday, average

9. I discovered that a bar exists on East 7th Street and 2nd Ave., Burp Castle, where patrons are meant to keep their voices down. Not quite a reticence bar, but close. I have now been there several times and recommend it strongly.

manner. Heidegger translates these words of Aristotle with the German phrase: "*Der Mensch zeigt sich als Seiendes das redet*" (The human being shows itself as a being that speaks or discourses; SZ 165 / BT 208). In other words, the human being is phenomenological. The human being shows itself, discloses itself, as the talker—but preferably in the mode of reticence, not talking too much, and listening. Talk, discourse, or language is the manner in which we are allowed to see the constitution of our being-in-the-world. We are linguistic beings. It is language that picks us out. And language is what allows us to see that which shows itself—namely, our being-in-the-world. And furthermore, this language is pre-propositional. Yet here I want to go back one more time to this distinction between discourse and language. We can experience language as word things, word units, pre-propositionally or propositionally. ("Hammer!" versus "This is the hammer that I need for the job.") This is the distinction between the hermeneutical "as" and the apophantical "as." But discourse is something deeper than that. Discourse is the way in which the world is articulated as such. It is the deep grammar of existence, the rich and complex undergrowth of the world that makes it intelligible.

Let me end by putting a wrinkle in the fabric of Heidegger's argument. If what Heidegger means by language is not the human capacity for vocal utterance alone but rather the ability to disclose a world (in part through silence) and for a world to have significance or intelligibility for the being that discovers it, then by virtue of what does one have to restrict that idea of discourse to *human* discourse? Might the same not be true for nonhuman creatures—I'm thinking here of higher primates and of dolphins and whales. If discourse is a form of grammar that allows for an understanding of the world, then could we not say the same of chimpanzees? Is that not true of dolphins? I don't see how we can escape that conclusion. It seems obvious to me that in the case of higher primates and more developed creatures, they certainly have a social existence. They have a shared world that is intelligible to them, that makes sense. And furthermore, there are forms of communication amongst those animals.

The usual way of reading Heidegger is as a humanist. He would appear to be arguing that Dasein is us, the human being, and we human beings have this distinctive relationship to the world. But you could push this claim a little bit further and say, actually, any species, any group of creatures that has a world that hangs together in a certain way, has discourse and is Dasein too. There's a moment in a 1929 lecture course called *The Fundamental Concepts of Metaphysics* where Heidegger makes a threefold distinction between human beings, which have a world; animals, which are poor in world; and rocks and stones, which have no world at all.[10] Such would be Heidegger's classical chain of being. But on the basis of Heidegger's analysis of discourse, I think we can question this humanism. We can take the idea of discourse and apply it to easy

10. Martin Heidegger, *Die Grundbegriffe der Metaphysik: Welt—Endlichkeit—Einsamkeit*, ed. Friedrich-Wilhelm von Herrmann, vols. 29/30 of *Gesamtausgabe*, ed. Friedrich-Wilhelm von Herrmann (Klostermann, 1992), 263; *The Fundamental Concepts of Metaphysics: World, Finitude, Solitude*, trans. William McNeill and Nicholas Walker (Indiana University Press, 1995), 177; hereafter GM/FCM followed by page numbers.

cases—dolphins and whales, for instance. But why not take it further? Animals have worlds in all sorts of intriguing and compelling ways. Why restrict the concept of the world to the "higher" species? Don't beavers have a world? What about bees, termites, or ants? We could go on. If discourse is more capacious than vocal language use, then I see no reason why it should be restricted to humans.

Permit me a final thought here. Some of you might remember the years from 2016 onwards as those of the octopus boom, or the cephalopod craze. I read all sorts of books about octopuses, squid, cuttlefish, and nautiluses: the four members of the cephalopod group. One outstanding book was *Other Minds: The Octopus, the Sea, and the Deep Origins of Consciousness*, by Peter Godfrey-Smith, an Australian philosopher of science and keen scuba diver.[11] This was one of Richard Bernstein's favorite books. And Dick was hard to please. Cephalopods are fascinating creatures. If we consider the octopus or the squid and think about the world structure of the squid, then just on the basis of empirical research, I would venture to say it becomes very hard to say that an octopus is poor in world. I think we'd have to say that an octopus has a world and that that world makes sense. Octopuses are indeed the worldhood of their world, I would venture. They've got huge brains. Their sensory apparatus is not restricted to eyes stuck in their heads and nasal cavities and ears but is totally dispersed amongst their bodies. And this is a body unrestricted by an interior skeleton like human beings (our great

11. Peter Godfrey-Smith, *Other Minds: The Octopus, the Sea, and the Deep Origins of Consciousness* (Farrar, Straus and Giroux, 2016).

weakness that restricts our malleability). They are creatures that can move through the tiniest holes and gaps. They have large, complex bodies that are rather like our tongues. (Imagine your tongue being your entire body, with its dazzling sensory wherewithal.) They live brief, intense lives of two or three years—strange, given how complex they are; we imagine with that kind of intelligence there would be longevity—and after mating and reproducing they expire like disused plastic bags and usually get eaten, often by other octopuses. These creatures have a world. I think it would be crazy to deny them that.

What's also fascinating about octopuses—I know I'm going off on a tangent here, so forgive me—is that they are not particularly social creatures. This makes them more Heideggerian, anxiety-fueled beings. They disguise themselves brilliantly and spend a lot of time hiding. We tend to think of animal intelligence as social (*Mitsein*) intelligence defined by teleological utility (e.g., the chimp uses the wooden stick in order to break open the bee nest). But with octopuses there's a sheer too-much-ness, a squandering of excessive activity, such as amazing and often apparently pointless light shows they perform with their bodies, often simply for their own delight. We seem to like the idea of animals having a social organization—when we think about a troop of chimpanzees, for instance. We love watching documentaries (such as 2023's *Chimp Empire*) about the hierarchical relationships that chimpanzees have amongst themselves, the alpha male in competition with other males, the female chimps protecting their young, intense affection for insiders, intense aggression towards outsiders, and so on. Chimpanzees are pretty smart, but octopuses are a lot smarter. They have worlds, too, but they experience those worlds largely in solitude, in a strange, Heideggerian authenticity, perhaps even with a sense of mortality. Anyway, I would be delighted to learn more about octopus-Dasein. But that's another topic for a separate occasion.[12]

12. After floating this line of thought publicly, I was made aware of an extraordinary study of Octopus-Dasein and much else besides called *Vampyroteuthis Infernalis*. See Vilém Flusser and Louis Bec, *Vampyroteuthis Infernalis: A Treatise, with a Report by the Institut Scientifique de Recherche Paranaturaliste*, trans. Valentine A. Pakis (University of Minnesota Press, 2012). This book is part fable, part science, and wholly phenomenological.

7. Inauthentic Life

Back to everyday life — This is not *a philosophy of culture — Idle talking, gossiping, scribbling, texting, updating — Curiosity, the delight in sight — Concupiscence, eye-lust — The horror of liveliness (Hedda Gabler) — Flirting and the phenomenology of the sexy — Ambiguity, the speed of life — Everything means something else — Being against one another — Falling is not bad — Temptation (groundless floating) — Tranquilization (cultural exoticism, Western Buddhism) — Alienation (self-deception) — Entanglement (self-dissection, the triumph of the therapeutic) — Turbulence (groundless floating) — The structure of thrownness — Not a night view of Dasein, but daytime.*

THINGS GOT A BIT HEAVY last time. It is not just the hammer that's heavy, but Heidegger's prose is heavy, and *Being and Time* is a weighty tome. But if you get beneath the surface unpleasantness of reading Heidegger in English translation, then the ideas are powerful, remarkably intuitive, and, I think, life-altering. That said, I'm determined to have a little bit more fun in this discussion.

I want to talk about inauthentic life, which I think many of us are experts at in our day-to-day existence, which Heidegger deals with in part B of chapter 5. We're still moving within this huge chapter on being-in as such. We've gone through the key existential concepts—state-of-mind, understanding, discourse. And now we're going to find a set of mirror concepts to the existentials that take place in relation to Dasein's *falling*. Heidegger begins by confessing that in all this fascinating discussion of interpretation, assertion, discourse, language, and so on, maybe he's lost sight of Dasein's everydayness. He's lost sight of the humdrum averageness with which he started his analysis (SZ 166–67 / BT 210). Let's go back to everyday life. Let's go back to publicness. Let's go back to the They. How does everything that Heidegger has said relate to our average, everyday life in the world?

We have three existential concepts: state-of-mind, understanding, and discourse; and then the three counter-concepts of idle talk, ambiguity, and curiosity. Heidegger insists once again that he's not engaged in any moralizing critique of Dasein, of human life, and that these paragraphs do not aspire to a philosophy of culture. He insists on this so much—"the lady doth protest too much, methinks." Heidegger protests so much that we might begin to suspect that perhaps this *is* a philosophy of culture, and maybe it's also a critique of the culture of Heidegger's day, and maybe of our own. That word, *culture*: in the German philosophy of this period, *Kultur* was a very strong word that was, at the time that Heidegger was writing *Being and Time*, associated with the most prominent German philosopher of this period, Ernst Cassirer, who did fascinating work on the philosophy of symbolic forms in the 1920s. The story of the debate between Cassirer and Heidegger in Davos in 1929 encapsulates the tragedy of twentieth-century Germanophone philosophy and the division between Analytic and Continental traditions. But that, as they say, is another story. If you are interested in that story (as you should be), then read Michael Friedman's *A Parting of the Ways:*

Carnap, Cassirer, and Heidegger.[1] I wrote my MPhil thesis on Carnap and Heidegger.[2]

Let's stick tight to what Heidegger's saying, beginning with "Idle Talk," paragraph 35. I think this is fairly straightforward. If discourse, *Rede*, is the disclosure of Dasein's intelligibility of our being-in-the-world, then idle talk, *Gerede*, is the closing off of this disclosure. Heidegger says:

> And because this discoursing [idle talk] has lost its primary relationship-of-Being towards the entity talked about, or else has never achieved such a relationship, it does not communicate in such a way as to let this entity be appropriated in a primordial manner, but communicates rather by following the route of *gossiping* and *passing the word along*. (SZ 168 / BT 212)

He continues:

> Idle talk is constituted by just such gossiping and passing the word along—a process by which its initial lack of grounds to stand on becomes aggravated to complete groundlessness. And indeed this idle talk is not confined to vocal gossip, but even spreads to what we write, where it takes the form of "scribbling" ["*das 'Geschreibe'*"].

1. Michael Friedman, *A Parting of the Ways: Carnap, Cassirer, and Heidegger* (Open Court, 2000).

2. Part of this thesis found its way into chapter 6 of my *Very Short Introduction to Continental Philosophy*. See Simon Critchley, *Continental Philosophy: A Very Short Introduction* (Oxford University Press, 2001), 90–110.

> In this latter case the gossip is not based so much upon hearsay. It feeds upon superficial reading. (SZ 168–69 / BT 212)

Heidegger goes on: "The average understanding, moreover, will not want [to make] any... distinction, and does not need it, because, of course, it understands everything" (SZ 169 / BT 212).

Idle talk is language employed in completely groundless ways. We can think of countless examples of this from conversations in bars, reality TV shows, and endless podcasts. In writing, it becomes scribbling. It becomes tabloid journalism. It becomes what we used to call, all those distant years ago, "blogging." Or it becomes social media updates, tireless updating, sharing photos, and vapid virtue signaling. That's what Heidegger means by idle talk, and it is dominated by average understanding, where averageness understands everything already. Everything is always already understood in the domain of idle talk. There's nothing new. No alarms and no surprises. Everything is known already. In part, this lack of surprises comes from the fact that such media tend to repeat, to parrot sentiments, to aggressively agree with each other, or to just "retweet" someone else's post. But even in the supposedly original posts, the very idea of originality is blocked from the construction of such thought. Idle talk assumes a reassuring already-known-ness about itself. It posits itself as simply the best summary of what it takes to be common sense. It is interesting to watch people engaged in idle chatter politely discovering that they actually have opposite points of view, retroactively expanding the boundaries of their shared "common sense" more and more in order to make room for the broadest possible agreement in order to avoid the blunt impact of opposing perspectives. There can be no room for discovery or disagreement in idle talk.

Think about this in relation to news-flash culture, that our phones are flooded with endless updates about the latest results of whatever event has happened in the world: a war, an election, a natural disaster, some meme about piglets, or whatever. Everything is always already familiar. Everything is already understood; we just talk into it. And that's the condition of public discourse, of publicness. And as Heidegger will say, in publicness we float unattached, in an anchorless tide of empty words (SZ 170 / BT 214).

Which brings us to the next concept: curiosity. This is a fascinating little paragraph, but it's easy to ignore. Paragraph 36 begins with a discussion of the natural light, the *lumen naturale*, which Heidegger discussed at the beginning of chapter 5, when he talks about disclosure, the clearing. The natural light is not the "natural light of reason," as it was for Descartes. It is rather the disclosed light of the clearing that we ourselves are. We covered this previously (see ch. 5, "Thrown Projection"). At the beginning of the paragraph on curiosity, he cites the opening words of Aristotle's *Metaphysics*, when he says, "All men by nature desire to know. An indication of this is the delight we take in the senses and above all, the sense of sight. We prefer seeing to anything else" (SZ 171 / BT 215).[3] Interesting phrase, "the delight we take in the senses." We take delight in sight. And this is a form of curiosity, a philosophical curiosity that human beings have, perhaps even a metaphysical need.

But then things become a little inflected and inflamed. The volume control goes up a tiny bit because Heidegger begins to quote Saint Augustine—you know, the ex-Manichean nightclubber turned

3. Aristotle, *Metaphysics* 980a21–4.

bishop. In the *Confessions*, Augustine talks about *concupiscence*. *Concupiscence* is not a word that's used very much these days. *Concupiscentia* is the meaning of curiosity. And in Augustine that concupiscence, that curiosity, is linked to lust and in particular to sexual desire. Scopic desire, voyeurism, things like that. And if you open a copy of *Confessions*—and I've got my copy in my hands as I speak to you—in book 10, there's a chapter called "Empty Curiosity and Frivolous Interests." This gives you an idea of the tone here. And Augustine says, "Seeing belongs properly to the eyes."[4] This hearkens back to that quote from Aristotle. But the problem of seeing for Augustine is the problem of sense experience in relation to concupiscence of the eyes, the way in which curiosity takes us in, sweeps us away. We become tempted with our eyes. We experience what Augustine calls "lust of the eyes." "*Augenlust*" is the way Heidegger translates it (SZ 171 / BT 215–16). "Eye-lust"—I think we know what this means in our day-to-day lives. Just think of a visit to your local cake shop. Curiosity has that lustful dimension.

Let me quote another tasty morsel from paragraph 36, when Heidegger says, "When curiosity has become free . . . it concerns itself with seeing . . . but *just* in order to see. It seeks novelty only in order to leap from it anew to another novelty" (SZ 172 / BT 216). What he's saying here is that curiosity as this concupiscence, where we're tempted with the eyes, seeks novelty. Imagine you see a delicious cake at the cake shop, then you see another cake, even more appealing than the last. The desiring eye drifts randomly from thing to thing. Curiosity wants newness, and it wants new things all the time; it wants to go from novelty to novelty, to be constantly outside itself,

4. Augustine, *Confessions*, book 10, chapter 35, section 54.

distracted. Curiosity, or *Neugier*, is a greed or lust (*Gier*) for the new (*Neu*). Heidegger says that curiosity "is a specific way of *not tarrying* alongside what is closest.... In not tarrying, curiosity is concerned with the constant possibility of *distraction*. Curiosity has nothing to do with observing entities and marveling at them" (SZ 172 / BT 216). He quotes the Greek here, *thaumazein* (θαυμάζειν), wonder. This is not what is usually meant by curiosity. Curiosity is rather a constant distraction by things. It has nothing to do with wonder, amazement at the existence of the cosmos. Rather, curiosity is the experience of "*never dwelling anywhere*"—it's a loss of dwelling, a homelessness[5]—of being "everywhere and nowhere" (SZ 173 / BT 217). Curiosity is the experience of "constantly uprooting" ourselves because we're distracted by some new thing that comes along, scrolling and swiping and gawking. Heidegger ends this paragraph by saying, "Curiosity, for which nothing is closed off, and idle talk, for which there is nothing that is not understood, provide themselves... with the guarantee of a 'life' which, supposedly, is genuinely 'lively'" (SZ 173 / BT 217). I think about this supposedly genuinely lively life in at least two ways. The first is through Hedda Gabler, Henrik Ibsen's most demonic character. Now, I love Hedda, although I would not want to be married to her. All Hedda wants is liveliness. She craves liveliness. Unfortunately, she finds herself with a husband who's not lively and who bores her to death and who's not going to deliver the kind of life that she thinks she wants. And so she engages herself with other desires, other curiosities, other men who cross her path. She craves liveliness until she slips away into a side room, begins banging

5. A couple of years after the publication of *Being and Time*, Heidegger will describe philosophy as a *homesickness* (*Heimweh*), a "restless" seeking of worldliness and homeliness, a seeking of being at-home (GM 7–13 / FCM 5–9).

away on the piano, and takes out her father's gun and shoots herself through the head.[6] She wants liveliness, and her pursuit of this tortures her semi-failing academic husband. (Welcome to Norway! In my experience, women in Norway always know where the gun is.)

Now, I'm not saying don't be curious on that basis, but maybe liveliness is something we should be a little bit suspicious of. The second thing I think about in relation to curiosity is flirtation. Curiosity is very much the life of flirting, the play of looking, being looked at, the seduction of looking, the whole theater of scopic desire. If we wanted to find a philosophy of flirtation or even a phenomenology of the sexy in Heidegger, we could look at this paragraph. Heidegger is going to do no such thing, but it doesn't mean somebody else might not do it. The craving for liveliness and the wish for flirtation are two symptoms of curiosity.

The third concept in this mini-trilogy is ambiguity. When everything is always already understood, and everyone talks about it all the time, it "becomes impossible to decide what is disclosed in a genuine understanding, and what is not" (SZ 173 / BT 217). Everything becomes ambiguous. Things look as if they're understood, but they're not. Everyone is acquainted with everything, but they understand nothing. Heidegger says, "But when Dasein goes in for something in the reticence of carrying it through or even of genuinely breaking down on it, its time is a different time and, as seen by the public, an essentially slower time than that of idle talk, which 'lives at a faster rate'" (SZ 174 / BT 218). Reticence, which we encoun-

6. Henrik Ibsen, *Hedda Gabler*, trans. Edmund Gosse and William Archer (Lerner Publishing Group, 2015), 178–79.

tered in the discussion of language and interpretation in chapter 6, and which Heidegger seems to prefer, is slow. It's a slowing down of time. Idle talk is fast. If we go back to the reticence bar in Helsinki, or imagine its construction in Brooklyn or somewhere else, then this would be a slow bar, whereas idle talk is speedy, like a boring cocaine monologue. I find that Heidegger's discussions of these fallen concepts are pretty interesting. He goes on in the same paragraph to say:

> Dasein's understanding in the "they" is constantly *going wrong* in its projects, as regards the genuine possibilities of Being. Dasein is always ambiguously "there"—that is to say, in that public disclosedness of Being-with-one-another where the loudest idle talk and the most ingenious curiosity keep "things moving," where, in an everyday manner, everything (and at bottom nothing) is happening.
>
> This ambiguity is always tossing to curiosity that which it seeks; and it gives idle talk the semblance of having everything decided in it. (SZ 174 / BT 218–19)

There is also this idea of "being in on it," this shared knowingness, and being on the scent of "it": "Being 'in on it' with someone [*das Mit-dabei-sein*: being-included-with] when one is on the scent, and so long as one is on it, precludes one's allegiance when what has been surmised gets carried out" (SZ 174 / BT 218). Like idle talk, ambiguity offers a superficial sense of being together, of belonging. But it's always predicated on an allegiance that actively resists the level of scrutiny that's entailed in belonging to a singularly, genuinely decided-upon destiny, a people. As may be the case with social media acquaintances one has never met offline, the common interest is often clung to in times of crisis, including a "crisis of faith" when one questions one's own investment in the "online [pick a topic] community." This could be because we aren't sure who we'd be otherwise and are afraid of losing our sense of identity. Maybe we protect ourselves from losing this allegiance when we, in Heidegger's words, "ensure that what is genuinely and newly created is out of date as soon as it emerges before the public" (ibid.). It's as if, in speaking (or writing) in this already-outdated way, cutting-edge yet constantly passé, conveying this image of the cynical expert, we protect ourselves from any interest that goes deeper than mere curiosity. Just as it's already known, it's already old. It's like we jump out in front of the new, as into moving traffic, and in mock cynicism deride those who tell us to watch out. In ambiguity, we avoid being questioned by pretending we already know, pretending the new is nothing new, that nothing can touch our sense of blithe comfort. Or maybe we protect ourselves from actual novelty by lounging in the comfort of this superficial, nonthreatening and nonthreatened belonging. Unfortunately, as Heidegger points out, it's only once this interest wanes that genuine newness can become possible (ibid.). Heidegger doesn't seem to like idle talk, ambiguity, and chatter, this "falling" language. But it's easy to sympathize with the urge for it. It's an obvious pleasure, the sense of belonging, of fitting in, that is implied in these kinds of interactions. But it's also foolhardy to act untouchable around powerful, influential, potentially dangerous, or simply life-altering ideas.

These three concepts—idle talk (chatter, twitter), curiosity (constantly being engaged by whatever it is that passes in front of one's eyes, scrolling, swiping), and ambiguity (the fact that everything seems to mean this and something else)—feed off each other; they run together, intersect and inflame each other. We could think about this in relation to where we are socially and politically (he said, in a desperate stab at relevance). Everything seems to mean something else. Such is ambiguity. They say climate change is real. They also say that it's a conspiracy created by China. They say pandemics are real threats for which we need the protection of vaccines. They also say that these are "Plandemics" rigged by big pharma and corporate power that will end with "The Great Reset," where human beings end up living in pods and eating insects. I could go on.

They say this; they say that. What we see not only in the comfortable countries of the West but also across the whole globe is a politics of ambiguity that is being used for an arguably malicious intent in order to divide and control opinion. Ambiguity is also a constant watching of one another, listening to gossip and overhearing, where being-with-one-another is really being-against-one-another, secretly plotting against each other, Heidegger says (SZ 175 / BT 219). There's a superficial sociality to ambiguity where secrets are exchanged, people listen in, and people seem to be for one another—like being at a party or something—but then, just beneath the surface, there is a hostility and paranoia towards one another. Everything is understood, everything means something else, and everything and everyone flips around into something else. There is no constancy in the average, everyday world. Everything floats unmoored and drifting. Such is our life with them.

We're now in the final pages of chapter 5, you'll be pleased to hear. Heidegger is going to pull everything together powerfully in paragraph 38, "Falling and Thrownness." The three everyday ways of inauthentic being-there—idle talk, curiosity, ambiguity—make up the falling of Dasein. And this falling is the being of everydayness. Heidegger once again insists that this is not negative: "This term [falling] does not express any negative evaluation" (SZ 175 / BT 220). Falling is being lost in the publicness of the they-self. It is an absorption in the world. Again, "falling" is not a vertical descent from a state of grace into a state of sin, like Satan, but falling *to* the world. Heidegger insists, and rightly, that this inauthentic way of being—falling *at* the world (or falling *at* the cakes)—does not mean *not* being-in-the-world. Rather, inauthentic being is a distinctive way of being-in-the-world that is completely fascinated with the world. Falling is a positive possibility.

Heidegger then picks out five ways of characterizing this movement of falling. Again, this is barely discussed in most accounts of Heidegger, but it is a compelling moment in *Being and Time*. For all the research on Heidegger's account of authenticity in *Being and Time*, there is a lack of work done on falling and inauthentic life. A major aspect of my reading of Heidegger, which I try to show in this book, is that the relationship to the inauthentic is not as straightforward as is usually thought in the Heidegger literature. The five characteristics of falling are that it is *tempting*, *tranquilizing*, *alienating*, *entangling*, and *turbulent*.

Let's take them in turn. Idle talk discloses to Dasein a relation to its world where we are pulled in by it and tempted by falling. "Being-in-the-world is in itself *tempting*" (SZ 177 / BT 221). It is temptation. We float groundlessly in average everydayness, and we're drawn towards falling. There's safety and security in such temptation. He goes on: insofar as life with the They allows for a "full and genuine life," and we live in the "self-certainty and the decidedness of the they," we find what Heidegger calls tranquility (SZ 177 / BT 222). This is intriguing, tranquility. We're tempted by being-in-the-world, we're pulled into falling, and that pulling has a tranquilizing effect.

Heidegger says "a full and genuine 'life,' brings Dasein a *tranquillity*.... Falling Being-in-the-world, which tempts itself, is at the same time *tranquillizing*" (SZ 177 / BT 222). It's as if we're hypnotized. I'm tempted to think about this in pharmacological terms, in terms of the rise of tranquilizers and SSRIs that numb us to the world as a way of getting through it. We are drawn in by the world, tempted, and anesthetized. Pulled in and lulled, satiated. In tranquility, we're also insulated from the exposure to the possibility of failure (*das Scheitern*), or "genuinely breaking down on" the subject matter. This phrase "breaking down on" is a nice construction by the translators, as it echoes a phrase that will come up in division 2, when Dasein "shatters itself against" death in its embrace of finitude, openness to failure.

The third element in this picture of falling is alienation. Heidegger talks about the way in which, for Dasein in average everydayness, "the opinion may now arise that understanding the most alien cultures and 'synthesizing' them with one's own may lead to Dasein's becoming for the first time thoroughly and genuinely enlightened about itself" (SZ 178 / BT 222). This is provocative. What he's suggesting is that an interest in foreign cultures is a way of inauthentically becoming enlightened about ourselves, the way in which, say, in developed Western cultures, there's a fetishization of the non-Western, in particular the allure of the East. This is what

Nietzsche called "European Buddhism."[7] We might say to ourselves: Our Judeo-Christian Western heritage has messed us up with its lies and hypocrisy, and the answer lies elsewhere in foreign or indigenous cultures, if we can synthesize them with our own. Such cultural syncretism leads to mere exoticism and the fascination with fields like comparative mythology and the hugely popular work of figures like Carl Jung and Joseph Campbell. Heidegger says, "When Dasein, tranquilized, and 'understanding' everything, thus compares itself with everything, it drifts along towards an alienation in which its ownmost potentiality-for-Being is hidden from it. Falling Being-in-the-world is not only tempting and tranquilizing; it is at the same time alienating" (ibid.). Tranquilized falling into or at the world is alienating. It alienates us from our ownmost potentiality for being ourselves, Heidegger says. Yet—and this is the fourth element of falling—this alienation leads to entanglement:

> Yet this alienation cannot mean that Dasein gets factically torn away from itself. On the contrary, this alienation drives it into a kind of Being which borders on the most exaggerated "self-dissection."... This alienation *closes off* from Dasein its authenticity and possibility.... The alienation of falling—at once tempting and tranquillizing—leads by its own movement, to Dasein's getting *entangled* in itself. (SZ 178 / BT 222–23)

7. See Friedrich Nietzsche, *The Will to Power*, trans. Walter Kaufmann and R. J. Hollingdale, ed. Walter Kaufmann (Random House, 1967), 16–18. Nihilism at the death of the Judeo-Christian God leads to Western Buddhism ("passive nihilism"), the self-help industry, etc. See also Nietzsche, *On the Genealogy of Morals*, trans. and ed. Walter Kaufmann (Random House, 1967), 19.

The alienation—via "alien cultures"—that Heidegger is talking about does not mean that we get factically torn away from ourselves but rather that we become entirely preoccupied with ourselves in the form of an exaggerated self-dissection, constantly analyzing ourselves. When we "let ourselves go," when we "fall," we insulate from direct challenges, we prevent ourselves from achieving self-understanding since this involves self-mastery through exposure to the threat of failure. But in being so closed off, alienated, we become self-obsessed. Now, although Heidegger doesn't say this, I think this could be understood in relation to a concept you can find in that lovely old curmudgeon Philip Rieff, where he talks about the triumph of the therapeutic.[8] That whatever the truth or falsity of psychoanalytic theory—and we could debate that for a lifetime—Freudianism has led to a culture of therapy, which leads us to become completely entangled with ourselves, preoccupied with ourselves, committed to some idea that there is some story about us, a story of trauma, of abuse, of victimization, however that's understood, with the various hierarchies of trauma. On the subject of psychoanalysis and its widespread appeal, when Freud and Jung first came to the US in 1910 to give talks and to meet William James, they were greeted by enthusiastic followers. Freud said to Jung, "They don't realize we're bringing them the plague."[9] Psychoanalysis becomes a pandemic all its own. For Heidegger, this tempting, tranquilizing, alienating structure of average everydayness leads to an entanglement

8. Philip Rieff, *The Triumph of the Therapeutic: Uses of Faith After Freud* (University of Chicago Press, 1987).

9. This anecdote is told by Jacques Lacan, who claims to have heard it directly from Jung. See Jacques Lacan, *Écrits: The First Complete Edition in English*, trans. Bruce Fink (W. W. Norton, 2006), 336.

in ourselves. And this entanglement can prevent us from appreciating, or really experiencing, the power of something like anxiety, which we'll get deep inside in the next chapter.

Fallen everydayness has the characteristic of groundless floating, for Heidegger, that we are adrift in the world in such a way that we're constantly caught up in its turbulence. The latter is the fifth and final characteristic of the structure of falling. Heidegger says, "Falling is not only existentially determinative for Being-in-the-world. At the same time turbulence makes manifest that the thrownness which can obtrude . . . upon Dasein in its state-of-mind, has the character of throwing and of movement" (SZ 179 / BT 223). We're like corks in a stream, or fallen leaves on a car windscreen, caught up, engulfed by the turbulence of the world that surrounds us, that pulls us this way and that. We might think of ourselves as living concretely, living our best lives ever, being happy, enlightened, living in the present, getting better every day in every way. But for Heidegger, this is just the groundless floating of falling being-in-the-world. Inauthentic falling. The elements that we've identified as temptation, tranquilization, alienation, entanglement, and turbulence make up the structure of thrownness. They are what it means to be in its throes.

The key takeaway from the end of this chapter, which links to what I was saying previously, that Dasein is thrown projection (see ch. 5, "Thrown Projection"), is the phrase that Heidegger introduces here for the first time: "Dasein exists factically" (SZ 179 / BT 223). We are always thrown. We're always in the movement of falling. We exist factically, and for as long as Dasein is, it remains in the throw, in the turbulent movement of thrownness.

Yet, if that's true, then how do we change direction? What about authenticity in relation to what has been said? This picks up on something that we've talked about a couple of times before, from the last lines of chapter 5. Heidegger says, "*Authentic* existence is not something which floats above falling everydayness; existentially, it is only a modified way in which such everydayness is seized upon" (SZ 179 / BT 224). We are thrown-being-in-the-world, always already immersed in an average, everyday existence, defined by idle talk, curiosity, and ambiguity. Yet, because we are the beings for whom our own being is an issue, we can seize hold of that everydayness in a different way. Even in the mode of inauthentic life, we can achieve a modification and seize hold of that in a different way. Such seizing hold is the possibility of authenticity. Heidegger insists in a singular expression that falling is not a "'night view' of Dasein."

It's not the dark, shadowy side of existence. Rather, falling "reveals an essential ontological structure of Dasein" (SZ 179 / BT 224). The human being's movement, its *kinesis*, is falling. This is the structure of everyday life. It is daytime, like daytime TV, aggressively banal. We are always already captivated, fascinated by average everydayness, caught in its movement of ambiguity, curiosity, and chatter. Authenticity is simply a change of movement, a modification of our falling.

The question is, How are we to conceive of that modification? How can we bring about that change in direction in the midst of the turbulence of being thrown? Or rather, given the primacy of moods in Heidegger's account, how is that shift in movement going to be felt? How is that going to take shape, take form, become concrete in Dasein's existence? To answer this pivotal question, we need to see Dasein's existence as a whole. Is there a way to bring the whole structure of being-in-the-world into view? Everything is going to turn on the next big concept we're going to introduce, which is anxiety, *Angst*, that uncanny calm in which Dasein first discloses itself to itself, and comes face-to-face with itself as pure possibility. The phenomenon of anxiety is going to reveal the structure of our being—our articulated structural whole—as care.

We have now gone through the first five chapters of *Being and Time*. Maybe at this point you would like to take a moment and review the ground we have covered so far. With all the concepts that we've analytically delineated, we have everything in place in order to get to a definition of *care*, which is going to define the being of Dasein. That's where we're going next in chapter 6 of *Being and Time*, the final and culminating chapter of division 1.

8. Anxiety and Care

*Bringing the whole thing together — Anxiety is the way in which the whole of Dasein is disclosed — Defining the whole as care — The twelve steps of the care structure — Articulated thrown projection — We are not simple — We are a hinge — We are a unified but multifaceted stretch (future, past, and present) — Temporality as the meaning of care — A stunning personal revelation about my anxiety — Why I do philosophy — Approaching Heidegger aquatically — Opening oneself to the tender indifference of the world — Anxiety as the mood of the novel (*Robinson Crusoe*) — Falling out of the homely — Party-being — Time and tide — Being alone with others — Excursus on Terrence Malick — Anxiety, calm, and courage — Anxiety reveals the Nothing — Darkness can help — The Christian background to anxiety (Kierkegaard) — Saint Paul and anxious waiting — Heidegger tells us a story and gives us a happy ending — Marrying Aristotle and Christianity within atheism.*

WE'RE GOING TO PULL things together in this chapter. All the elements that we've laid out previously are going to come together in a beautiful harmony. It will all make sense. I promise.

Let's dive in. Dasein is being-in-the-world, as I've said repeatedly. And being-in-the-world is primordially and constantly a whole. We must not "split the phenomenon." This phrase that Heidegger uses very early in chapter 5 keeps coming back. We must not divide experience into mind and world, subject and object, and so on. We must keep things together and keep Dasein as being-in-the-world as a whole.

The question in chapter 6, this key chapter on care, is twofold, and it's vital to the whole structure of the work. The first part of the question is, How is the whole of being-in-the-world to be disclosed? And second, How is it to be defined? The first question, that of disclosure, is going to be answered in paragraph 40 through anxiety, *Angst*, which we'll discuss later on (in the second half of this discussion). Anxiety is what Heidegger calls, paradoxically—and he knows it's a contradiction, an antithesis—an *understanding state-of-mind*. Understanding (i.e., projection) and state-of-mind (i.e., thrownness). He uses this double, portmanteau formulation. He says, "Is there . . . an understanding state-of-mind in which Dasein has been disclosed to itself in some distinctive way?" (SZ 182 / BT 226). This understanding state-of-mind is going to be anxiety. The second question is answered in paragraph 41, when Dasein's being is defined as *care*, which says nothing as yet, but we will fill it out. In paragraph 42, which I'll say a word or two about, we get what he calls a pre-ontological confirmation of care in a weird little myth. Heidegger tells us a story, and brings things to an oddly happy ending.

Care Structure (fig. 1)

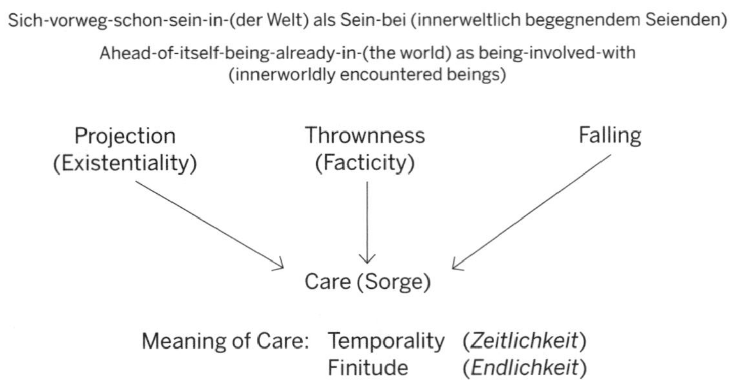

Also in chapter 6—and this is the last chapter in division 1, so we're approaching a climax—the question of being, this question that frames *Being and Time*, slips back into view, and he says something like, What I've done in this analytic of Dasein, of the human being, is what we can call an existentially a priori anthropology. Anthropology: a way of talking about us, human beings. A priori: picking out the formal, transcendental features of the human being insofar as we are defined by existence. The question then is, What does "anthropology" have to do with the question of being as such? The second part of chapter 6, which we'll discuss after this (see ch. 9, "Reality," and ch. 10, "Truth"), has two sections, which are both mini-essays, short and almost independent inquiries, which conclude division 1 of *Being and Time*. One deals with reality, and the other is on the question of truth. In a way, they can be read independently of *Being and Time*, but we will connect them to the whole body of the argument. That's the overall structure of chapter 6.

What I want to do now is to lay out the care structure. The being of Dasein is care, but what is the structure of care? I want to lay this out in twelve steps, a bit like Alcoholics Anonymous.

Step 1: The analysis of state-of-mind—this concept, *Befindlichkeit*, always-already-having-found-oneself-thereness—yielded a determination of thrownness. Thrownness is the fact, or facticity, of being-there, being found somewhere, and disclosed in a certain way in the "there" of our existence. Thrownness describes our way of being delivered over to our place, of being found always already somewhere and disclosed in a specific way. And the way we're disclosed is through mood, through affect, through a form of attunement: state-of-mind, thrownness, mood.

Step 2: In the analysis of understanding in paragraph 31 of *Being and Time*, understanding (*Verstehen* in German) was understood by Heidegger as being able to do something—what Dasein can do—and therefore is linked to possibility, potentiality. Understanding is linked by Heidegger to projection, also understood as throwing forth. Dasein is not just thrown, but by having projects it can throw forth from that thrownness. And Dasein always remains in the throw. We saw this last time: the way in which Dasein remains in the throw and in the throes of idle talk, curiosity, and ambiguity, the everyday time of falling.

Step 3: However, these two existentials—state-of-mind and understanding—are combined in a rhythm, even harmony—or, better, they're interwoven into the structure of Dasein's thrown projection. As I've said, Dasein is thrown projection—the key concept introduced in chapter 5—and the structure of thrown throwing off is described as enigmatic (SZ 148 / BT 188). Thrown projection is an enigmatic structure, and the point, Heidegger says, is to let us see the enigma as an enigma, and not just to try to solve it like a riddle. The third existential is discourse or talk. And this was defined as the articulation of Dasein's enigmatic structure. Discourse is thrown projection. How is that articulated? It's articulated in talk. Discourse, or talk, is the way in which the whole structure of the human being, the ontological structure of our being-in-the-world, is made intelligible. The way in which we are in the world is interpretable, or achieves intelligibility, through discourse.

Step 4: Dasein is articulated, like a hinge. Think of Dasein as what we call in England an articulated lorry, or what in the US is called a semitrailer truck. It's this articulated hinge movement. Dasein is like the hinge of a door. We're articulated around this dual structure of thrownness and projection, which is rendered intelligible through discourse. Which is to say what exactly? It's to say that Dasein is *articulated thrown projection*. I repeat: "articulated thrown projection." Otto Pöggeler translates it this way.[1] This proposition combines the three existentials—state-of-mind, understanding, and discourse—into a meaningful whole. As Heidegger makes clear at the end of paragraph 41, this basic structure of Dasein as articulated through projection is not simple. It is a mobile structure, a complex structure, a structure that tries to render the kinetic quality of human existence. Heidegger writes, "In the end, it will be shown that the idea of Being in general is just as far from being 'simple' as is the Being of Dasein" (SZ 196 / BT 241). The basic existential structure of beings like us is not simple: it is articulated, we are hinged around this mobile structure of articulated thrown projection. We are the spread of that structure. And we are the whole truck, not just the trailer hinged to it. That's step 5.

Step 5, to take it a little bit further: Since care is the being of Dasein, what is the meaning of care? What does it mean? Let's lay this out, piece by piece. First, care means that Dasein projects, that its being is defined by possibility, by potentiality. Dasein is a being who has potential, potentiality for being, and insofar as it has this potentiality, Dasein always *runs ahead of itself*. Our being is being ahead of ourselves. And the term that Heidegger uses here is literally the term "to run ahead," *Vorlaufen* (SZ 191–92 / BT 236). Now to say this, that we are fundamentally beings who run ahead, is also to say or to imply that we are futural, we are orientated towards the future insofar as we understand and we project. Also, and this is a

1. Otto Pöggeler, *Martin Heidegger's Path of Thinking*, trans. Daniel Magurshak and Sigmund Barber (Humanities Press International, 1987), 43.

claim that will become clearer, our being futural is the meaning of freedom. Freedom is futural, the capacity to do something. Freedom is the freedom of being able to disclose our existential structure. Dasein is the being that's capable of the disclosure of its existentiality. This is also to say—by the by—that we're the beings who are capable of what Heidegger calls phenomenology—namely, bringing to sight, letting us see that which shows itself. Remember, the definition of *logos* is "letting see," and the definition of phenomenon is "that which shows itself." Insofar as we have this potentiality, we can *disclose*. We're always ahead of ourselves, letting ourselves see what shows itself. The being of being human is phenomenological. It presses against an enigma, the enigma of who we are.

Step 6: But Dasein only projects on the basis of its thrownness. Dasein can only be ahead of itself insofar as it is already being in a world. And this is the meaning of the portmanteau phrase that Heidegger uses in these pages: that Dasein is being-ahead-of-itself already-in-a-world. Namely, Dasein is thrown, as we've said. Which means that Dasein is concealed. Dasein is therefore unfree in some significant measure. Dasein is a being that's capable of phenomenology and of understanding itself. But that understanding is also capable of being caught up in facticity, futural but continually thrown back, ensnared, caught up in concealment. I am open to the future, but I can conceal that from myself. Freedom is the movement between itself and unfreedom. This point is made often in Heidegger's later work.

Step 7: What are we saying here? The being of Dasein is care. The care structure is the double movement of articulated thrown projection. This is to say that Dasein is *multifaceted*. It faces the future: projection. It faces the past: thrownness. And in relation to the future and past, it turns to the present. Importantly, Dasein is not a point; it does not occupy a temporal *punctum*, a moment, like a *res cogitans*. So much of our thinking reduces us to a point, like a speck of consciousness sitting atop the brain. Rather, Dasein is *stretched* temporally. Dasein is a temporal span, and the link between care and time, care and temporality, is one of the main concerns of division 2 of *Being and Time*, which we will get to. Dasein is this stretch, this tensile expanse of our being-in-the-world, like a bridge. This is care: a span, a suspension bridge where one bank is thrownness and the other bank is projection. A bit like the Brooklyn Bridge, although I'm not sure which side is which (maybe thrownness is Brooklyn and projection is Manhattan). And this is a unitary phenomenon, which imposes unity on Dasein's multifaceted temporal structure: the fact that it's not a simple structure; it's an articulated, hinged, complex structure. This is the sense of what he calls "man's *perfectio*." Man's perfection, *die perfectio des Menschen*, in the sense of completion or fulfillment: that the fulfillment of the human being is accomplished by care. And he says, "In the 'double meaning' of 'care,' what we have in view is a *single* basic state, in its essentially twofold structure of thrown projection" (SZ 199 / BT 243). We're thrown into the past, and we project into the future. Care is the single concept that is meant to capture the twofold articulated structure that we *are*.

Step 8: Dasein is thrown projection. This is also to say that Dasein exists factically. Insofar as both these phrases capture the hinged movement of Dasein—between thrownness and projection on the one hand, and existence and facticity on the other—they are interchangeable. But we've left something important out, and this is the point of step 8. Chapter 5 discusses the three existentials: state-of-mind, understanding, discourse. But in part B of that chapter, he goes on to describe how those characteristics are revealed, how they show up, in everyday being-in-the-world, in inauthentic life. The three concepts that mirror the three existential structures are idle talk, curiosity, and ambiguity, which I discussed last time. And those three concepts make up the structure of falling. And falling is the being of everydayness. Falling is groundless floating. This is the fact that we are beings who always fall to the world, at the world, fall from ourselves into inauthenticity. And that's neither good nor bad. Don't be tempted, Heidegger seems to want to say, by the supposed religious overtones of the terminology. The fall Heidegger is describing is not a vertical fall from grace to sin. It's a horizontal fall, a falling *at*. We fall forward, towards a horizon that we simultaneously push ahead. Such falling-at describes our absorption in the world, our fascination with the world. I take this very seriously: the claim that inauthentic life—succumbing to or falling in with everyday life—is what constitutes us. The diagram below (fig. 2) was presented by Heidegger at a seminar for medical professionals at the University of Zurich psychiatric clinic in 1959 in order to illustrate the human being's capacity to disclose/perceive its world to itself.[2] It also depicts the movement of our falling into or in with the world we apprehend. It's not an upward or downward movement but an *outward* one. We'll come back to that constitutive inauthenticity when we look at anxiety.

Step 9: We reach a full, fulfilled, unified determination of Dasein's structure in terms of the three elements that we've identified: *exis-*

2. Martin Heidegger, *Zollikoner Seminare, Protokolle—Gespräche—Briefe*, ed. Medard Boss (Klostermann, 1987), 3; *Zollikon Seminars: Protocols—Conversations—Letters*, ed. Medard Boss, trans. Franz Mayr and Richard Askay (Northwestern University Press, 2001), 3.

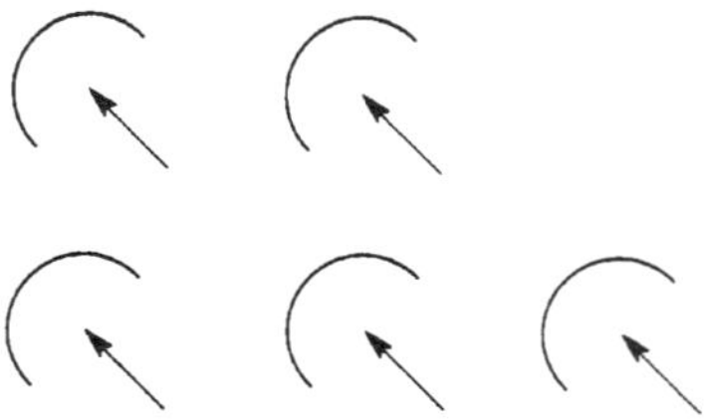

(fig. 2)

tentiality, *facticity*, and *falling*. Existence, projection, potentiality for being, freedom. Facticity, thrownness, unfreedom. And falling, our everyday way of being together with things of concern and with others for whom we feel solicitude. Those are the three elements of care, and it is this unity of those three elements that Heidegger wants to bring together in the concept of care. Dasein is an entity that is ahead of itself. We're fundamentally defined by being able, being potent, having potential, being free, headed towards the future. And we project upon, we project forth from our thrownness. We're always already in a world: existing is always factical. Heidegger goes further in these pages. He says "existentiality is essentially determined by facticity" (SZ 192 / BT 236–37). We're not just thrown projection, we're thrown projection insofar as we are always already absorbed in the world of our concern, absorbed and alongside, out there, within that world, the inauthentic world that we inhabit.

Step 10: Care names the unity of Dasein's complex existential structure. It is an a priori claim. In other words, it is a formal and transcendental claim. It precedes and structures *experience*, a term that Heidegger won't use much in *Being and Time* but that he deploys in an oddly un-deconstructed way in his later work.[3] But is

3. Martin Heidegger, "Aus der Erfahrung des Denkens," in *Aus der Erfahrung des Denkens*, 75–86 (written in 1947; first published in 1954); "The Thinker as Poet," in *Poetry, Language, Thought*, 1–14.

care the last word in the determination of the being of Dasein? The answer is no. But why not? Why doesn't the book end with care? There is something more fundamental still, which we must reveal, something still more primordial that provides an ontological support for the unity of care.

This brings us to step 11. There is a more primordial phenomenon that provides a basis for the unity of the structure of care, the unity of this multifacetedness, and the fact that Dasein faces these three ways at once. We can call this more primordial phenomenon the *meaning* of care, where meaning is understood as that upon which care is based. We've already met this idea of meaning at the end of our discussion in chapter 5, "Thrown Projection," as that upon which our being-in-the-world rests. This "upon which" is *temporality*. It is temporality that makes possible the being of Dasein as care. This claim is laid out in division 2, in a paragraph that is called "Care and Selfhood" (SZ 316ff. / BT 364ff.). I think it is useful to look a little bit down the road to see where we're going in this discussion.

If we look a couple of hundred pages further into *Being and Time*, Heidegger makes the following remark: "Time is primordial as the temporalizing of temporality" (SZ 331 / BT 380). Temporality, he says, is essentially ecstatical. We're going to get to that later on (see ch. 14, "Ecstatic Temporality"). This is the important thing: "Primordial time is finite" (ibid.). The meaning of care is temporality or time. The meaning of time is finitude. The meaning of our being is finitude, and the meaning of being as such is going to be finite as well. If we acknowledge this, if we say the meaning of our being as such is finite, then we're also saying that the being that structures different worlds and different historical periods is also finite. Being is not eternal. Any claims to eternity in Heidegger are a nonstarter.

A determination of being can structure different epochs with distinct discursive schemas. This thought influences many thinkers. This is where Michel Foucault, who was deeply influenced by Heidegger, develops this point into the *épistémè*. The *épistémè* is the fact that there are different historical epochs or discursive regimes that shape the past and the past's presence.[4] We're getting ahead of ourselves, but it's good to keep the big picture in view. Please look carefully at the schema for the care structure (fig. 1) at the beginning of this chapter.

This brings us to step 12. What I've said so far is true in the sense of being textually accurate. It's also true in the sense of a formal truth, an analytic definition of the being of Dasein. But it doesn't really have enough content. And this is where we're going to turn towards the concept of *anxiety*. Anxiety is going to provide the existential filling, or fulfillment, of these formal, ontological claims. The definition of the being of Dasein, what it means to be human, is care. But how is care disclosed to us? In anxiety.

•

Let's talk about anxiety. On a painfully embarrassing personal note, it was anxiety that first really spoke to me in Heidegger, through a text called "What Is Metaphysics?," a famous lecture that Heidegger gave in 1929 (WMK 103–22 / WMC 89–110). I read this essay in a compendium of existentialist texts one summer while living in a grim, extremely run-down council flat in a very unfashionable corner

4. See Michel Foucault, *The Order of Things: An Archaeology of the Human Sciences* (Routledge, 1989).

of East London in 1982. It was anxiety that prompted me to pick up *Being and Time*, and I spent two weeks reading the first division and understanding very little. I was around twenty-two at the time. I went to university late, after a rather misspent youth. University was a very stressful experience, and I had, I guess, what people used to call a nervous breakdown in my first year. It was horrible. I was gripped with terrible anxiety, an anxiety that was linked in particular to a terror of failure, more specifically of failing exams. The thought of sitting for an exam used to make me vomit. The university exam system in England at the time was rather rigorous and punitive, and it induced an awful fear in me. It was meant to. I simply fell apart towards the end of my first year at university. And when I was trying to make sense of that, trying to understand what was happening to me, I discovered Heidegger on anxiety, and it simply spoke to me in a very powerful way. I want you to keep that in mind.

I have no idea how you see me, and I have no capacity to see myself. We're all self-deceived. The gap of not-knowing is often filled with fantasy. The unconscious parts of ourselves, what we don't know of ourselves, may be glimpsed in a roundabout way—for instance, in a psychoanalytic relationship, through transference and countertransference (transferring feelings for others onto each other), which according to Jung is based in love. But you are certainly wrong in whatever you think about me, and I am wrong too. But keep in mind the twenty-two-year-old, anxious, prematurely balding version of me sitting in a fourteenth-floor room of a really ugly, anonymous tower block at the University of Essex in 1982 (a proper shithole of a university, regularly ranked bottom in the UK at the time—unjustly), listening to Echo and the Bunnymen, listening more specifically to "The Cutter":

Conquering myself
Until I see another hurdle approaching
Say we can, say we will
Not just another drop in the ocean.[5]

There you are, young, anxious. You find the world a weird place, and you're feeling strange, uncanny. Heidegger can speak to that condition. And it's important to remember this anxiety. In many ways, in my naiveté, I came into teaching philosophy because I felt anxious and I thought I could help other people with anxiety through teaching philosophy. Was I stupid and naive in that thought? Sometimes I think so. Sometimes I don't. But it's important to recall such powerful feelings. I remember very clearly realizing when I was twenty-two that I simply wasn't cut out for happiness. But I was an expert in anxiety, and it was Heidegger who allowed me to see in that mood something disclosive and revelatory. In a philosophical context, and perhaps in others too, anxiety isn't something to be nullified or tranquilized but *used*. Philosophy's goal is not happiness; anxiety is the motor, the drive of philosophy. Philosophy puts anxiety to work.

Returning to Heidegger, it is anxiety that lets us see being-in-the-world as a whole. It is the basic mood or the fundamental mood. The term in German is quite nice: *Grundstimmung*, the ground attunement. This is what permits us to get a grip on the unitary

5. Music was a sounding board for how people were feeling. In 1982, Echo and the Bunnymen were the best band in Britain. But not in 1983. By then it was Simple Minds or The Cure. Still, Echo and the Bunnymen were from Liverpool, so we were proud.

structure of Dasein's multifaceted, articulated whole, which is care. At one level, which is important for the argument of *Being and Time*, anxiety fulfills a methodological requirement. Heidegger needs a way—"method" just derives from the Greek for "way," "road," or "street," *hodos*—a way of revealing being-in-the-world as a whole. That's done through the concept of anxiety. He'll say something similar about death. Death is also a methodological requirement in *Being and Time*. The argument there is going to be that in order to have a grasp on Dasein as a whole, we have to have a sense of Dasein's *end*. We have to have Dasein's end in view, in our grasp. Conceptually, Dasein's end (its futural end, its end*ing*) is death. Therefore, we formally require a concept of death.

Anxiety and death fulfill methodological functions in the arguments. But—and it's a big "but"—the resonance of anxiety is more than methodological. This basic mood of anxiety is a rare and subtle thing. As Heidegger will say, it is in this mood of anxiety that the self is first precipitated. I use the word "precipitated" deliberately. Like a raindrop from a cloud. This is the aquatic or maritime model of Heidegger that I prefer and that we met previously (see ch. 4, "Being-with-Others"). We're lost in the sea of "them," lost in the sea of our absorption in the world, and the self precipitates in the experience of anxiety. Being with them is being pleasantly lost at sea. Anxiety is the first formation of the self, the free, authentic self, out of the they-self. Not just "out of" but *of* this expanse, not only an emergence but also a taking-shape from the recoiling or shrinking of this expanse. The self is not just covered under the sea of them but is the sudden separating of that sea from itself. A condensation that gathers and then breaks from that aimless floating. As suddenly lost as we may find ourselves in the precipitation of the self, it's in this retreat that we can gather our bearings and make our way. But insofar as we emerge out of the They, we remain "they-selves," modified inauthenticities.

This idea of anxiety as what first shapes the self was the mood that launched a thousand novels, from Sartre's *Nausea* (*La Nausée* [1938]),[6] where the protagonist feels the nausea of existence, through to a book like Albert Camus's *The Outsider*, or *The Stranger* (*L'Étranger* [1942]), which ends with this beautiful line: "I opened myself to the tender indifference of the world."[7] This tender indifference is one way of describing what occurs in anxiety. Or indeed, think of the anxiety that you can find in more recent novels. I think of my good friend Tom McCarthy's novel *Remainder* (2005)—perhaps a self-serving example—which begins with the line "About the accident itself I can say very little. Almost nothing."[8] The mood of the traumatized, obsessive protagonist in *Remainder* is very much the mood of anxiety, someone who holds existence at a distance.

Although this is a very general claim to make, maybe anxiety is the mood of the novel itself. This might sound abstract and absurd, but if we think about the novel, then when does the novel begin, and where does the novel begin? It really begins in Britain in the early eighteenth century, with figures like Daniel Defoe and in books like *A Journal of the Plague Year* and, most famously, *Robinson Crusoe*.[9] If you read *Crusoe*, it is clear that the narrator experiences profound,

6. Jean-Paul Sartre, *Nausea*, trans. Lloyd Alexander (New Directions, 2013).

7. Albert Camus, *The Stranger*, trans. Matthew Ward (Vintage International, 1989), 122.

8. Tom McCarthy, *Remainder* (Vintage, 2005), 3.

9. Daniel Defoe, *A Journal of the Plague Year* (Lerner Publishing, 2022); Defoe, *Robinson Crusoe*, ed. Thomas Keymer (Oxford University Press, 2007).

quasi-psychotic anxiety. He goes crazy on the island in the strangest possible way. The "rise of the novel"—as Ian Watt used to say, in his classic study of the eighteenth-century novel—in many ways perhaps goes together with the idea of anxiety.[10] That's just a speculative thought.

It is in anxiety that freedom is first felt. The anxiety that defines a free individual. And the key thing here is that freedom and anxiety are not first and foremost conceptual or cognitive. They are experiential and affective, moodful and emotional.

Anxiety arrests the movement of falling to the world. We fall to the world, or *at* the world. And that falling is our fundamental movement that is carried by idle talk, curiosity, and ambiguity. Chatting, gawking, flirting. Anxiety arrests that movement of falling and lets us step back and grasp our being-in-the-world as a whole. In anxiety, we are individuated as beings, capable of grasping our being (SZ 191 / BT 235). The existential cost of this understanding, of this grasping of ourselves, is that we fall out of the world. We fall out of the cozy, familiar world where we're concerned with things and have relations of solicitude with others. We move into a world of uncanniness, a world in which we're no longer at home, a world that is *unheimlich*, uncanny, unhomely. Inauthentic selfhood is shared with others, to the point where we are at sea with others in an experience of groundless floating, of ambiguous, curious chatter. And this can be nice. This can be lovely. It's like being at a good party. When you're at a party and you've got over the initial anxiety of party-being: "Nobody

10. Ian Watt, *The Rise of the Novel: Studies in Defoe, Richardson and Fielding* (University of California Press, 1957).

likes me," "I don't know anybody," "Everybody hates me," that full Thom Yorke mode. "I want to leave." "I want to die." Whatever it might be. That Steven Patrick Morrissey mood of going to a place and standing alone and leaving alone and then going home and crying and wanting to die. All that nasty stuff. But maybe you get past that. And maybe you've had a couple of drinks, maybe you're a little bit drunk or a little bit high. And you are caught in the flow of ambiguous, curious chatter. You've got over your initial anxious party-being, now you're in the second version, the second wave of party-being, the acid-house version.[11] You're looking at things, looking at people. You're talking, idly or even wildly. This can be delightful. Groundless floating with the They can be a lot of fun. Everything is ambiguous at a party, and that ambiguity can be pleasant. We're immersed in the sea of others. We're drowned in the world's suffocating banality, its tender indifference.

Anxiety is the experience of that tide of indifference going out; it recedes. What I'm thinking of here is the way in which the words for *time* and *tide* are linked in, say, the Nordic languages, in Dutch, and the Frisian dialect, and also in English, at its roots. The experience of time is an experience of tide, where time is linked to the movement of the tides. And if you've spent time in rather old-fashioned maritime communities, you'll meet people (and I've met people like this in the past, in places like Jersey, in the Channel

11. Party drugs and their corresponding genres of music have changed the way people interact at parties. When amphetamines were the main party drug, when I was a youthful punk, there was a fight at every gig. This changed with the rise of X in the so-called second summer of love, 1988. The concert was a simulacrum of being-together. Now you can remember this party-being in an angsty nostalgia by listening to Frankie Knuckles and acid-house compilations on Spotify.

Islands, between England and France), whose whole day will be shaped by the experience of the tide. And they can smell when the tide changes, when the waters are coming in, and when they are going out. And that's the fundamental marker of time. And we're immersed in it. We're floating in a sea of inauthenticity, idle talk, curiosity, distraction. Time first shows up in the tide, the smell of the sea, and in the sky.

Anxiety is the experience of the tide going out where the water drains away and leaves one stranded on the strand, ashore on the great beach of being. And I'm fond of this image: that the first way in which the self shows up is when the tide goes out. The water of being-in-the-world drains away, and we're revealed as anxious individuals, alone. Heidegger wants us to think of anxiety in a singular way. You can train yourself to use anxiety as a form of clarity. We can think about it as worry or jitteriness or fear. But this is not what Heidegger is trying to get us to think about. For him, anxiety is a sort of calm, an experience of peace. It's a term he uses in "What Is Metaphysics?," a *Ruhe*: calm or peace (WMK 111 / WMC 100). Anxiety is not worry but an experience of detachment, a tranquility that's not tranquilized by the world but awoken from its slumber. Perhaps anxiety is the philosophical mood par excellence. To follow this thought all the way through and really feel its impact in our lives requires first defamiliarizing ourselves with how we normally think about anxiety.

Let me put this a slightly different way. If we look at the first paragraphs of the discussion of anxiety, Heidegger compares anxiety with the movement of falling and fleeing. It's the idea of anxiety as a form of movement that I want to focus on here. In a state-of-mind, in facticity, Dasein discloses the "there" of its being-in-the-world, in its movement of turning away from itself. And we can compare this to the way in which we might wake up in the middle of the night and shift the position of our body. There's a sort of awareness there and a turning away. And we turn, we alter the position of our body, and we try to return to sleep. In a state-of-mind, there is this movement of evasion, where we turn our backs on ourselves because our being is a burden that we don't want to take up. Dasein finds itself fleeing. It finds itself in flight from itself. That's falling. We know that already. Now, in anxiety, Dasein turns away from the turning away of evasion and turns back towards that in the face of which it flees (SZ 184–85 / BT 229). What's it fleeing from? It's fleeing from itself. It's this movement of flight from flight. Anxiety is flight from flight, a movement from falling away to something that pulls away from that falling away. England's erstwhile agony aunt Claire Rayner once said you can try to escape your problems by going on vacation, but you'll meet yourself at the airport.[12] It is

12. Claire Rayner, *Claire Rayner's Lifeguide: A Commonsense Approach to Modern Living* (New English Library, 1980), 685: "To use a career change as a way of running away helps not at all. Because when you get to the new career you find yourself waiting there to do it with you. It's the same as the idea of emigrating as a way of dealing with problems; you take your problems with you to the new country." Travel does not expand one's perspective, it narrows the mind. We tend to become singularly focused on petty concerns while on vacation. Heidegger had a pretty low opinion of tourism, suggesting even that it should be banned. According to the protocols of Heidegger's 1973 *Seminar in Zähringen*, if Dasein is to make itself free to understand itself in its being, it must "surrender [its] determination as *producer*" of a so-called reality limited to consumer goods and constrained within a network of "socio-economic 'imperatives.'" For this to be possible within such constraints, we must renounce things like recreational travel: "'Tourism' would no longer be possible, instead one would restrict oneself and remain at home." See Martin Heidegger, *Seminare*, ed. Curd Ochwadt, vol. 15 of *Gesamtausgabe*, ed. Friedrich-Wilhelm von

here that the self is first revealed. Heidegger even describes anxiety as a *solus ipse*, a self alone, a sort of solipsism (SZ 188 / BT 233). Namely, anxiety individualizes Dasein. Anxiety individualizes us, and it individualizes us in a movement of flight from flight. I hope that makes sense, as it is important: flight from flight. We flee to the world in idle talk and curiosity. In anxiety, we take flight from that flight. Anxiety is therefore a change of movement, a modification of Dasein's *kinesis*, its movement, its motion and emotion.

Falling is nothing more than the privation of a disclosedness that becomes deprived and open in the experience of anxiety. That might sound a little bit gnomic. If falling is nothing more than the privation of a disclosedness—we cover ourselves up in falling—then that disclosedness becomes disclosed, it becomes open, in the experience of anxiety. If, as Heidegger insists, we are ourselves the clearing, then anxiety is the way in which disclosedness is disclosed, the clearing is cleared. Anxiety is the opening of Dasein. Anxiety is the way in which we open to ourselves, in a form of aloneness. But what kind of aloneness is that? Well, we can be alone when we're with others, as we know. Think of David Riesman's classic study *The Lonely*

Herrmann (Klostermann, 1986), 389; *Four Seminars*, trans. Andrew Mitchell and François Raffoul (Indiana University Press, 2003), 74. Sometimes I think Heidegger has a point about tourism. If you're ever at a loose end and in need of a smile, then look at Heidegger's rather ridiculous vacation journals. *Zu Hölderlin—Griechenlandreisen*, ed. Curd Ochwadt, vol. 75 of *Gesamtausgabe*, ed. Friedrich-Wilhelm von Herrmann (Klostermann, 2000), 244; *Sojourns: The Journey to Greece*, trans. John Panteleimon Manoussakis (State University of New York Press, 2005), esp. 54–57, where Heidegger refers to tourism as an "unthoughtful assault" and "an alien power [that] enforces its own commands and regulations over ancient Greece." See also Simon Critchley, *ABC of Impossibility* (Univocal Publishing, 2015), 25.

Crowd, from 1950, and the way in which we can be alone in the crowd of a city.[13] For example, I just came down here on the subway from Greenwich Village to Tribeca, and everyone is there on the subway. It's the they-self, right? I am there with others. But you can be totally alone in those crowds. That's Heidegger's point. Aloneness here does not mean I'm necessarily physically alone. I can be most alone in my being-with-others, and my being-with-others can therefore induce Sartre's "nausea." I sit there on the subway and I'm no longer falling to absorption in the subway. I'm feeling a profound disgust that is gripping me. I think you all know what that's like.

Allow me a couple more remarks on anxiety, as it's a fascinating concept. How do we distinguish it from fear? We got close to this when we looked at fear in the discussion on thrown projection in chapter 5. Let me look at the distinction between fear and anxiety. Fear is that which threatens Dasein. Fear is fear of something, anything. (As I noted, Heidegger takes this from Aristotle's *Rhetoric*.) Let's say a big rabbit with big teeth, or a terrifying badger that appears at the door, threatening me with its badgeriness. Heidegger's point is that my fear is related to some entity in the world. Now, the experience of anxiety reveals that that in the face of which one has fear is Dasein itself. You could say that *I* am that in the face of which I experience fear. I am the rabbit. I am the badger. Sort of. If in fear Dasein shrinks back from an entity in the world, then in anxiety, Dasein shrinks back from this shrinking-back and becomes, we

13. David Riesman, Nathan Glazer, and Reuel Denney, *The Lonely Crowd: A Study of the Changing American Character*, abr. and rev. ed. (Yale University Press, 2020).

could say, courageous. Heidegger doesn't use that word, but I think it makes sense in relation to what Heidegger's thinking about here. What seems at first like an object of fear turns out to be me. And we can face up to that courageously.

Let me give an example here, because it might help. There is a filmmaker you might know called Terrence Malick, whom I mentioned previously (ch. 4, "Being-with-Others"). Malick was a philosopher at Oxford and wanted a career in philosophy as an academic. And he was told by Gilbert Ryle, the most powerful person in philosophy in Oxford and Britain at that time, the mid-1960s, that the topic he wanted to work on—Heidegger, Wittgenstein, and Kierkegaard and their understanding of the world—was not "real philosophy." Ryle, who had reviewed *Being and Time* after its publication,[14] called the project nonsense and essentially told Malick to "grow up." And he became so discouraged by this that he came back to the United States and taught as an adjunct for a year at MIT, replacing the teaching of Hubert Dreyfus, the great Heidegger interpreter, who at that time was in Paris in a reading group on Heidegger (along with my teacher, Dominique Janicaud, and others including Jacques Derrida and Alphonso Lingis. It must have been quite a reading group).

Anyway, Malick is a fully formed young philosopher, a graduate student without a sense of direction who hasn't got permission to write his PhD and is drifting a little. Some of this might sound familiar to some of you. Anyway, some years later, he finds himself

14. Gilbert Ryle, "Review of Martin Heidegger, *Sein und Zeit*," *Mind* 38, no. 151 (1929): 355–70.

at film school at the American Film Institute. This was in 1968. I think it was the first year the institute was receiving fellows. He was around twenty-eight years old at the time. And funnily enough, in that year a fellow student in the intake was David Lynch, who would have been a good deal younger. He came from Philadelphia, from the art school there, and spent many years shooting his first movie, *Eraserhead*, in improvised spaces in the grounds of the film institute. Reading Lynch's autobiographical guide to transcendental meditation, *Catching the Big Fish*, reveals that, perhaps in contrast to Malick, Lynch has very little understanding of what he does, or at least he conceals very well.[15] Another story, a good story to tell about anxiety, but for another time. But Malick, a more mature and fully formed intellect, goes to film school, having realized that philosophy is the wrong medium for his thinking, and releases his first film, *Badlands*, in 1973, and his second film, *Days of Heaven*, in 1978. And then twenty years pass without him making another film. But film became Malick's medium for thinking.

Then, in 1998, Malick released a movie called *The Thin Red Line*, based on a book by James Jones. He filmed about seven hours of footage for the film, just under three hours of which made it into the final cut. Malick's films are all flawed, like Stanley Kubrick's (but don't get me wrong, they are both geniuses). In this film, Malick tells the story of the Battle of Guadalcanal in the South Pacific during the Second World War. It's a fascinating and harrowing story. The central character in the movie, but not in the book, is somebody

15. David Lynch, *Catching the Big Fish: Meditation, Consciousness, and Creativity* (Tarcher, 2006).

called Witt (the name, happily for Malick, evokes Wittgenstein).[16] Witt is played by an angelic-looking Jim Caviezel. He's the hero. And this is what happens: he dies. Or rather, he's killed, towards the end of the film. But in a very particular way. He is with the members of his squadron, a platoon of American soldiers, who are being chased through the jungle by Japanese soldiers. And at a certain point, Witt breaks away from the group and diverts the Japanese soldiers down another path. So he saves his buddies and sacrifices himself. The Japanese soldiers catch up with him and surround him and begin to scream in Japanese. He doesn't understand what they are saying but sort of understands their intent—namely, that they want him to defend himself because they won't kill him if he doesn't defend himself, because that would be dishonorable. He gets the message, half-heartedly raising his gun, and then he's shot dead.

Now, in the context of the film, this brings us back to an earlier moment. When you see Witt being killed in this film, he's not in terror. He's not scared. There's an extraordinary calm at the moment of his demise. And earlier in the film, in flashback, he thinks of his mother's death. He has a memory of his mother's death. He's on the beach at the time in the paradise-like situation of a beach in the South Pacific. And he thinks about his mother dying and thinks about the calm with which she faced death. And he wonders whether he will be able to face death with that same calm. The point of this story is that, here's an example of heroism. And the hero is someone

16. James Jones, in a preface to *Whistle*, writes that Witt was originally based on Prewitt from the previous novel in his World War II trilogy, *From Here to Eternity*. Because of the previous character's death, however, the name had to be changed in *The Thin Red Line*. (Perhaps Prewitt becomes pre-Witt.) James Jones, *Whistle* (Delacorte Press, 1978), xx–xxi.

who shrinks back from shrinking back into the world and has the courage to face themselves and face their imminent disappearance and achieve a calm resoluteness. This is what Heidegger means by anxiety. Anxiety is not fear. It's not fear of something in the world. It is the sense of ourselves. If we can face that sense of ourselves with a certain peace, a certain calm, then that might be a condition for courage.

Anxiety is anxiety before being-in-the-world as such. Anxiety is not anxious in the face of some particular thing in the world but in the face of the world as a whole. The world as something indefinite, the world—and Heidegger begins to dramatize the language here in these pages in ways that will deepen and accelerate as the book unfolds—as something that is "nothing and nowhere" (SZ 188 / BT 233). Nothing and nowhere. The world becomes meaningless. What we've seen so far in *Being and Time* is this insistence on Dasein as being-in-the-world, where the world has the structure of significance. So far, so good. That's all groovy. This is the Heidegger that the pragmatists like. But in anxiety, when I pull back from that world, I suddenly see the world as lacking involvement, lacking significance. The world becomes meaningless. When this happens, the worldhood of the world—and remember, the worldhood of the world is us—becomes obtrusive.

Anxiety—and I still feel this fundamental mood as I'm addressing these words to you, I feel it a lot of the time—is one of the most personal moments in *Being and Time*. I feel Heidegger talking directly to my soul. Anxiety is nowhere. Yet it is so close that it stifles one's breath and becomes oppressive. As the world slips away, we obtrude. The world pulls away and we are revealed. Anxiety is nothing and nowhere. This is why Heidegger will say, in "What Is Metaphysics?," "Anxiety reveals the nothing" (WMK 112 / WMC 101). I listen to those words with my twenty-two-year-old self, my first-year-undergraduate self, after so many youthful misadventures. Those words spoke to me profoundly and very personally. And I still think that's the way in which Heidegger's address in these pages needs to be understood. This is why Heidegger is not Husserl, because Heidegger is intimate and hot where Husserl is analytical and cold. This is where, in anxiety, Heidegger's thinking starts to cook. It's not some cognitive, conceptual enterprise. Philosophy is not some clever intellectual sport. It is the drama of being us. Good and bad. Searing tragedy and comic farce.

Anxiety is not definite, is not linked to some particular threat, as in fear, whether it's the threat of a big rabbit or a big badger. In anxiety, Dasein is thrown back on itself, thrown back onto its potentiality for being, and this is how anxiety individualizes Dasein as a being who is defined by possibility, by freedom, by a sense in which things can be seized hold of, made one's own. But this comes at a cost: individualized Dasein, anxious Dasein, is not at home. It feels uncanny, *unheimlich*, no longer at home in the world. We flee in the face of uncanniness when we fall, when we're back in that party situation I was describing. It's a bit like Franz's crab walking in Sartre's *The Condemned of Altona*, always starting off on the wrong foot, taking a step backwards, momentarily retreating.[17] But in *Angst*, in anxiety, we flee towards that uncanniness. We seek it out. And

17. Jean-Paul Sartre, *The Condemned of Altona: A Play in Five Acts*, trans. Sylvia and George Leeson (Vintage, 1961).

Heidegger says that anxiety is not something that requires darkness and insomnia. It doesn't require these more dramatic moods. It's something that can arise suddenly. He writes, "Anxiety can arise in the most innocuous Situations. Nor does it have any need for darkness, in which it is commonly easier for one to feel uncanny" (SZ 189 / BT 234). We can feel nausea, we can feel anxiety, we can feel this anguish on the subway, wherever we like. Darkness might help, and we might think about that in relation to a disturbing childhood memory: you're in bed at night and it's dark and you can't sleep and there's a hubbub of noise downstairs, your parents or family talking, and maybe the television is on, or something like that. And you want to be there, but you can't be there because you've got to sleep because it's a school night. And the daytime world, the world of familiarity, flips around, and your room is immersed in darkness. The toys and objects in your bedroom might become animated and scary. In the experience of childhood, the intentionality of things in your room undergoes reversal. The toy soldier looks menacing, for instance. Maybe the bold look on a hero's face becomes an expression of anguish, of terror, and you realize your heroes can't protect you; they're just as helpless and vulnerable as you. The whole context can induce acute anxiety. And darkness can help. Anyway, I digress.

Heidegger also insists that true anxiety is rare. It doesn't happen all the time. It is "seldom" (*selten*; WMK 118 / WMC 106). It's not physiologically conditioned. Obviously physiology can and does play a part (palpitations of the heart, shortness of breath, sweaty palms and armpits), but anxiety is ontological. It is the structure of our being, and it has to be faced and affirmed courageously and not just diagnosed and medicated out of existence. Anxiety isn't a disorder, in Heidegger's view, but an order. Anxiety is what human beings do. It is how they order their being-in-the-world. It becomes a question of how we interpret anxiety and use anxiety, use its energy and transform that into action, creative and liberatory action. I think this is desperately important. If you only see anxiety as a disorder, as something diagnosable using the DSM, then the risk is that one would try to medicate it out of existence in order to maintain a numb, falling being-in-the-world. As unpleasant and dysregulating as the experience of anxiety may become for some, to treat it as merely a physiological disturbance that should be eradicated and avoided at all costs is to lose the clarifying, insightful potential of this mood. This is the point Al Pacino's LAPD lieutenant Vincent Hanna makes in *Heat* (1995), written and directed by Michael Mann, when he says, memorably, "I've got to hold on to my angst. I preserve it because I need it. It keeps me sharp."

At the end of the discussion of anxiety, as Heidegger sometimes does in *Being and Time*, he steps back and asks, Where has anxiety been talked about in the tradition? (SZ 190, 492 / BT 234–35). It's a good question, relevant not only to Heidegger's thoughts on anxiety but on tradition as well. Were the Greeks anxious? It doesn't appear so. The Hebraic tradition is certainly full of worry and lament. But there's a footnote where Heidegger talks about the Christian legacy of thinking about anxiety. He mentions Saint Augustine, where anxiety plays a key role in the conversion experience in book 8 of the *Confessions*. He also mentions Luther, and he was reading a lot of Luther in the years prior to *Being and Time* and also speaking about him in public (although there is nothing written, just the published protocol of a seminar on the problem of sin in

Marburg in 1924 with the famous theologian Rudolf Bultmann[18]). He also mentions a thinker from whom he borrows the concept of anxiety but relegates to a couple of innocuous and rather ungrateful footnotes: Søren Kierkegaard (whom Heidegger, so keen on texts in their original languages, always quotes in German, oblivious to the fact that he was thinking in Danish). In many ways, Kierkegaard is a key figure in *Being and Time*. But Heidegger plays down his influence in a way that I think is desperately unfair.

What would Kierkegaard have thought about *Being and Time*? Well, he was already dead, so we couldn't ask him. But I have no doubt that he would have found it as preposterous as the usual German philosophical attempts to synthesize the absolute in abstract, conceptual terms, in the way that Kierkegaard liked to poke fun at Hegel. Kierkegaard chose different tactics. Clever stratagems. He chose to write *indirectly*. He chose to write in the form of *pseudonyms* in a whole number of different voices, because he didn't believe that a direct philosophical discourse, a discourse on that which is, was possible anyway. Indirection and pseudonymity were essential tactics in thinking the truth of Christianity. Kierkegaard would have seen Heidegger as a flat-footed German cobbler of concepts.

In this way, the analysis of anxiety really has its home not so much in philosophy but more in religious traditions. And the religious tradition to which Heidegger was closest was Christianity. Let me

18. Martin Heidegger, "Das Problem der Sünde bei Luther," in Bernd Jaspert, ed., *Sachgemäße Exegese: Die Protokolle aus Rudolf Bultmanns Neutestamentlichen Seminaren 1921–1951* (N. G. Elwert, 1996), 28–33; "The Problem of Sin in Luther (1924)," trans. John van Buren, in *Supplements: From the Earliest Essays to* Being and Time *and Beyond*, ed. John van Buren (State University of New York Press, 2002), 105–10.

add one tiny detail to this picture: another thinker that we could link to this discussion of anxiety is Saint Paul. The true founder of Christianity, whose letters precede the composition of the four Gospels. Paul's letters are the first documents of Christianity. And Paul doesn't even describe himself as Christian. The term didn't exist in his time. He talks about small communities, households really, who are "in Christ." Paul had little interest in the historical Jesus. His obsession is with the *resurrected* Christ. In the Letter to the Romans, Paul asks, "What should the person who has experienced the call of Christ, has experienced the demand of the risen Christ, and who has converted, what should they do? What should their stance be?" And he says, "Remain in the condition in which you were called" (1 Corinthians 7:20).[19] What a strange thing to say after conversion: Remain in the condition in which you were called. But the deep point that Paul is making is that if you are in Christ you should await, remain in the condition of anxiety. There's an anxious waiting that defines Christianity. The anxious waiting for Paul was very real. Christ had been born, had been nailed to a tree in occupied Palestine and then resurrected. But he was coming back, and coming back soon, in *parousia*. What should one do in the meantime? One should wait, anxiously. Sadly, he's not turned up yet, but quite a lot of people are still hoping to see him.

There's a lot more compelling material in these pages of *Being and Time*, which I won't go into, but people like a happy ending. So let me give you one. The happy ending is in paragraph 42. Having laid

19. *The Writings of St. Paul*, ed. Wayne A. Meeks and John T. Fitzgerald (W. W. Norton, 2007), 31.

out this argument about care—the multifaceted structure of care, and anxiety as what discloses care—he does something in paragraph 42 that has always slightly mystified me. He calls it a "confirmation of the existential interpretation of Dasein." A confirmation that is done by way of what he calls a "pre-ontological way of Dasein's interpreting itself" (SZ 196 / BT 241). What on earth does that mean? Well, Heidegger is going to tell us a story. He tells us a tale about care. The definition of the being of Dasein is care. Just to remind you, this means ahead-of-itself, being already in-the-world alongside innerworldly things; that is, Dasein is ahead-of-itself, existential, already in the world, factical, and alongside innerworldly things, falling. This is Heidegger's attempt to redefine the human being. And Heidegger thinks this structure has been demonstrated in a myth of care, the fable of *Cura*, as he calls it. This is a rare "once upon a time" moment in *Being and Time*. There aren't many of them, so enjoy it. Heidegger is quoting somebody called Franz Bücheler, translated by Konrad Burdach in his article "Faust und die Sorge." This is the myth of care:

> Once when "Care" was crossing a river, she saw some clay; she thoughtfully took up a piece and began to shape it. While she was meditating on what she had made, Jupiter came by. "Care" asked him to give it spirit, and this he gladly granted. But when she wanted her name to be bestowed upon it, he forbade this, and demanded that it be given his name instead. While "Care" and Jupiter were disputing, Earth arose and desired that her own name be conferred on the creature, since she had furnished it with part of her body. They asked Saturn to be their arbiter, and he made the following decision, which seemed a just one. (SZ 198 / BT 242)

Quoting Saturn:

> Since you, Jupiter, have given its spirit, you shall receive that spirit at its death; and since you, Earth, have given its body, you shall receive its body. But since "Care" first shaped this creature, she shall possess it as long as it lives. And because there is now a dispute among you as to its name, let it be called "homo," for it is made out of *humus* (earth).[20] (ibid.)

The myth here is that the human being has the stamp of care. The human being is shaped by care out of clay or mud. And this is why we're called "homo," because "*esse factus ex humo*": We are made out of humus (not the good stuff you get in the delicatessen), or earth. Our breath was given to us by Jupiter, or Jove, which basically means "top God." God gives the human being spirit, breath, *pneuma*. We're made of earth and spirit, but we belong to care. And this was decided by Saturn. And Saturn, in mythological terms since at least the early Roman period and probably a lot further back, is linked to time. Temporality is the meaning of care. Our careful sojourn in the world is dominated by time (SZ 199 / BT 243). And that's what Heidegger means by this little story. There's your happy ending: The being of Dasein is care; it's disclosed by anxiety and then confirmed with this weird little myth.

One final thing, in closing. There is another footnote (isn't there always another footnote?). But it makes an interesting point. Hei-

20. Konrad Burdach, "Faust und die Sorge," *Deutsche Vierteljahrsschrift für Literaturwissenschaft und Geistesgeschichte* 1 (1923), 1–60.

degger talks about trying to link together "Augustinian ... Christian anthropology" and "the ontology of Aristotle." He says care can be thought of in these terms (SZ 199n1 / BT 492n7). This reveals something important for understanding *Being and Time*: that book is trying to do at least two things that are in tension; perhaps it is even an impossible balancing act. On the one hand, the book was written as a treatise on Aristotle. That's how it was understood before it was published. This was Heidegger's book for tenure as a professor in Marburg. To put it crudely, Heidegger's understanding of the world, and the theory of interpretation, meaning, and much of what we might call the "pragmatic realism" of *Being and Time* (the parts that philosophers such as Hubert Dreyfus prefer) comes from his reading of Aristotle. But Heidegger's aim is to bring together that Aristotelian realism with the tradition of Christian anthropology that we can find in Paul, Augustine, Luther, and Kierkegaard. And this is why anxiety becomes important, for *Angst* is the basic mood in which the self turns itself around, converts itself in the presence of God. How does the world show up in anxiety? I would argue that for the Christian, the world is fallen, marked by original sin, and essentially *nothing*. That is, nothing that is essential to the soul's redemption.

How can these two traditions be thought of together and even lived together? How can one hold, on the one hand, that the world makes sense like a good, sensible, naturalistically inclined Aristotelian, botanizing and measuring and categorizing and funding research projects on the constitution of cities, while on the other hand believing that meaning lies outside the world in relation to God as revealed through the resurrected Christ and experienced in the life of faith? Heidegger's *Being and Time* is the attempt to bring those two knotty strands of thought together. The ontology of Aristotle and Christian anthropology resound against each other in an awkward counterpoint, perhaps just discord. We might venture to say that Heidegger fuses together the Aristotelian and the Christian in a way that is neither Aristotelian nor Christian but resolutely atheist. At least at this point in Heidegger's itinerary. Let's leave that thought hanging in the air until we get into division 2. In the next discussions, we're going to pick up on the implications of Heidegger's approach for two huge philosophical questions: reality and truth.

9. Reality

Four incorrect theses on reality — Empirical idealism (Berkeley) — Transcendental realism (from Plato to Meillassoux) — Empirical realism (Samuel Johnson kicks a stone) — Transcendental idealism (Kant) — Why the question of reality is the wrong question — Why Kant is wrong (about the subject, representation, and the priority of the present-at-hand) — Heidegger's existential deepening of Kant — The scandal of philosophy is expecting that reality requires proof — Being-in-the-world is prior to both proof and faith — Epistemology is a symptom of Dasein's falling — Both realism and idealism are consistent with being-in-the-world, and both are wrong — The wrinkle in the fabric of Heidegger's argument: the Real — The hard core to reality: physical nature, matter, the earth — A possible dualism of reality and the Real: Doesn't this split the phenomenon?

WITH THE CONCEPTS of anxiety and care, the main argument of division 1 is at an end. These issues, which are central to *Being and Time* as a whole, are about human beings. But doesn't this reduce Heidegger's argument to an anthropology? In the following two discussions, we'll look at the final two paragraphs of division 1, which look at two traditional philosophical concerns from the perspective of the argument he's already laid out. In his treatment of these concerns, he'll try to answer this concern in the negative, but his later criticism of *Being and Time* will be that this analysis was, in the end, indeed too anthropocentric.

What I want to do in our next two discussions is to talk about paragraphs 43 and 44 of *Being and Time*. These are paragraphs on reality and truth. Let's begin with reality. Paragraph 43 is a brief but exceptionally rich paragraph that has significant implications for how we read Heidegger and how we understand his relationship to traditional philosophy and in particular the theses of realism and idealism. To make a start, let me take four theses on the relation of thought to things, or of mind to world. Those four theses are *empirical idealism*, *transcendental realism*, *empirical realism*, and *transcendental idealism*.

The first thesis, empirical idealism, is the claim that empirical objects are simply ideas in the mind. This is the thesis associated with the estimable bishop George Berkeley's *esse est percipi*, "to be is to be perceived."[1] This is what Kant refers to in his refutation of idealism as psychological idealism.[2] Empirical idealism is the claim that knowledge is knowledge of a subject-dependent but unreal world. That all reality is just ideas, and those ideas depend on their being perceived by us, by the subject. That's the first thesis. With a lot of followers of this thesis, this means we're living in a kind of dream, thus subject-dependent, but of a world that's not real.

The second thesis is transcendental realism. Now, this is the dominant position in the history of philosophy, prior to Kant. Prior to Kant's Copernican turn. One finds it in Plato and Descartes. But it persists all over the place with anybody that's claiming some thesis

1. George Berkeley, *A Treatise Concerning the Principles of Human Knowledge*, in *Philosophical Writings*, ed. Desmond Clarke (Cambridge University Press, 2012), 67–150.

2. Immanuel Kant, *Critique of Pure Reason*, trans. Paul Guyer and Allen W. Wood (Cambridge University Press, 1998).

of what we might call external realism, whether they're eliminative materialists or whatever they might be. Now, transcendental realism is the idea that the objects of our cognition are transcendentally real, that they have the constitution that we represent them as having, independently of our mode of cognition, and that things can be known as they are in themselves. Transcendental realism is the claim that knowledge is knowledge of a subject-independent real world (the forms in Plato, extended substance in Descartes). And there are various revivals of this claim of transcendental realism. One fairly recent and at the time fashionable revival was that of Quentin Meillassoux in his book *After Finitude*.[3] That's the second thesis, transcendental realism.[4]

The third thesis: empirical realism. Now, empirical realism is the claim that the world feels real. It feels really real. It feels like a world of solid, palpable objects. And it does not feel as if these things are simply ideas in the mind. This goes back to the great Samuel Johnson's refutation of Berkeley, which he did by simply kicking a stone: "I refute it thus!" By kicking a stone, we show that the stone is real.[5] It feels real. But to explain this thesis of empirical realism, we need to explain the fourth thesis and turn to Kant.

The fourth thesis is transcendental idealism. Now, transcendental

3. Quentin Meillassoux, *After Finitude: An Essay on the Necessity of Contingency*, trans. Ray Brassier (Continuum, 2008).

4. Heavy-lifting realism comes back every twenty or so years. Marx dismissed this in his doctoral thesis. His realism is based in social praxis and thus subject-dependent; see Karl Marx, *The Difference Between the Democritean and Epicurean Philosophy of Nature*, trans. Norman D. Livergood, in Norman D. Livergood, *Activity in Marx's Philosophy* (Martinus Nijhoff, 1967), 61–109.

5. James Boswell, *Life of Samuel Johnson*, ed. Herbert Vaughan Abbott (Scott, Foresman, 1923), 126.

idealism begins from the distinction between objects as things in themselves and objects as appearances. Transcendental idealism is the thesis that the objects of our cognition are appearances. That is, they're empirically real, but they are also transcendentally ideal. That's the key idea. To say that they're transcendentally ideal is to say that things, objects, are not known independently of our subjectivity and its mode of cognition—the constitution that we represent them as having. Rather, our mode of cognition determines this constitution.

Now, this is the meaning of the Copernican turn in Kant. The Copernican turn can be explained very simply with the following thought: Knowledge does not consist in the correspondence of subjects to objects as things in themselves. Rather, knowledge is the correspondence of objects as appearances to a subject that takes them to be as such. Things cannot be known as they are in themselves, Kant argues, because that would require an intellectual intuition. It would require our intellect to see into the nature of things in themselves. And for Kant, that is simply beyond what Scots—and remember that Kant is a fairly common Scottish name, usually spelled with a *C*, from the area around Fife, and Kant was rather proud of that fact—call our *ken*, it's beyond our epistemic ability.

Those are the four theses. We're going to get back to Heidegger in a moment. But I'm setting this up around Kant for reasons that will soon become obvious. The basic advance of Kant's epistemology in the *Critique of Pure Reason*—what's called the "critical turn" in philosophy—is that it does not suppose, as Plato and Descartes both do in quite different ways, that in order for knowledge to be possible there must be a correspondence between thoughts (or mental representations) and things in themselves (whether that means the realm of forms, the metaphysical realities of the soul, God, a material substance, or simply a belief in the radical independence of reality from the mind). This is what the American philosopher Wilfrid Sellars called "the Myth of the Given." There's nothing "given" that our knowledge corresponds to, nothing given in itself. What we see, what appears to us, is what appears to *us*. It's subject-dependent.[6]

After Kant, that which is true is that which is taken to be true—namely, that which appears to a subject or self. Now that which appears *might* indeed refer to a thing in itself. The thing in itself, Kant suggests, is thinkable. It's not knowable. It's thinkable, but we can never be in a position to know this fact independently of how that fact appears to us. In Kant's picture, the realm of sensibility, the realm of our empirical intuitions, is our access to a world that is indeed real for us. But that world is always already shot through with conceptual content. It is articulated through what Kant calls the "categories of the understanding," and is dependent upon the spontaneity of the subject. Later on, Kant will link this, in the middle of the first *Critique*, to the activity of what he calls the "transcendental imagination." But this is why, Kant says in the *Critique of Pure Reason*, the transcendental idealist is therefore an empirical realist. If we accept Kant's proposition that empirical idealism is false, then reality isn't just ideas in the mind; it's really real. Transcendental realism is false because we can't have knowledge of things

6. Wilfrid Sellars, *Empiricism and the Philosophy of Mind* (Harvard University Press, 1997).

as they are in themselves; empirical realism is right, the world feels really real, but the way the world feels really real is dependent upon the activity of the subject. It's a subject-dependent reality. Hence, empirical realism requires transcendental idealism.

What is the relation of Heidegger's thought to these theses? In particular, the thesis of transcendental idealism. To cut to the chase, Heidegger criticizes both realism and idealism for having an inadequate account of being-in-the-world, where the question of the reality of the external world gets raised without any previous clarification of the *phenomenon* of world, the phenomenon of world as that meaningful existential context that is most familiar and closest to us. For Heidegger, in chapter 3 of division 1, something like reality has to be understood in terms of its genesis in our ready-to-hand dealings with things in the world in the way in which they are "experienced." Heidegger won't use that word, but let's use it to be a little bit clumsy. He thinks transcendental idealism is right but Kant is wrong in how he elaborates it.

Reality has to be understood in terms of its genesis in our ready-to-hand dealings with things in the world in the way things are "proximally and for the most part," "most closely and mostly," practically. The ready-to-hand, as we've seen already, gets hardened up into the present-at-hand, and this presence-at-hand is then considered to be the first access to the world. As we've now seen by reading Heidegger, this is the purpose of paragraph 13. Knowing is a "founded mode" of being-in-the-world: "knowing... is founded ontically upon [Being-in-the-world]" (SZ 61 / BT 88). And therefore the question of the reality of the world gets raised without any previous clarification of the phenomenon of the world. Within the epistemological construal of the world, being, Heidegger says, acquires the meaning of reality and that which is gets conceived as a relation between two purportedly real substances: thinking things and extended things, *res cogitans* and *res extensa* in Descartes. Or in Kant, subjects and objects. To quote Heidegger: "The question of whether there is a world at all and whether its Being can be proved, makes no sense if it is raised by *Dasein* as Being-in-the-world; and who else would raise it?" (SZ 202 / BT 246–47). The world, for Heidegger, is disclosed along with the being of Dasein, and the question of the reality of the external world gets raised without any previous clarification of the phenomenon of the world. And for Heidegger, that's the shortcoming in the epistemological construal of the world.

Let's go back to Kant and keep him in mind. Although Kant's position is an advance for Heidegger—this is always the way Heidegger reads Kant,[7] as getting very close to the problem but staring at the "abyss of meaninglessness" (SZ 152 / BT 194) and "shrinking back" (SZ 23 / BT 45), as Heidegger says—it is still problematic, not only because he doesn't have the courage of his convictions but because Kant's argument presupposes two things. The first thing that Kant presupposes is a conception of the subject as what Kant calls the "I think." Now the "I think" has at the very least a family

7. Heidegger begins expanding this reading immediately following the publication of *Sein und Zeit*, first in a 1927–28 lecture course, and then in 1929's *Kant and the Problem of Metaphysics*, also known as the "Kantbuch." See Martin Heidegger, *Kant und das Problem der Metaphysik*, ed. Friedrich-Wilhelm von Herrmann, vol. 3 of *Gesamtausgabe*, ed. Friedrich-Wilhelm von Herrmann (Klostermann, 1991); *Kant and the Problem of Metaphysics*, 5th ed., trans. Richard Taft (Indiana University Press, 1997).

resemblance to Descartes's thinking substance, *res cogitans*, even though for Kant it's a *cogito* without an *ergo sum*. It's an "I think" without an entailment between thinking and being.

This would be a longer form of discussion that would be appropriate in a course on Kant's first *Critique*, but I don't want to do that here. Let's just say that the "I think" in Kant has a logical, rather than an ontological, function. This is what Kant calls "the transcendental unity of apperception." The transcendental unity of apperception—a great topic to raise at a dinner party—is *logically* entailed from the fact that experience has a unity and a coherence.[8] But it does not imply any *ontological* insight into the nature or substantiality of the self or the soul. This is the way Kant replies to David Hume, Scot to Scot. Hume leaves us with distinct bundles of perceptions without any self to tie those perceptions together. This is the way in which Hume decisively transforms the empirical question of personal identity.[9] (I hate the obsession with identity, but I like Hume on this issue, the episodic self. That the self needs to be turned into a Grand Narrative is the work of ideology.[10]) The way Kant will answer Hume is by saying: well, the world, or, as he puts it, the manifold of intuition, hangs together. It makes sense. But in order to explain that, we have to presuppose logically an "I think" that is tying those appearances together.[11] The world would not hang

8. Immanuel Kant, *Critique of Pure Reason*, A106–8, A117.

9. David Hume, *A Treatise of Human Nature*, ed. L. A. Selby-Bigge (Clarendon Press, 1888), 251–62.

10. See Simon Critchley, "Life Doesn't Need a Narrative Arc," *The New York Times*, June 12, 2023.

11. Immanuel Kant, *Critique of Pure Reason*, A350–3.

together unless there was a self making it all hang together, not empirically but transcendentally.

The first problem with Kant's position for Heidegger is that it presupposes an "I think," which although it's not a completely Cartesian position still takes over Descartes's position. The second problem with Kant on Heidegger's reading is that Kant presupposes that the subject's relation to the objective world is mediated through representations, or *Vorstellungen*. This is what Hegel will call "picture thinking,"[12] representations in your mind. The problem with that position in Heidegger's eyes is that if we see the relationship that the self has to the objective world as one that's constituted through representations, then those representations are contained within the ontology of the present-at-hand. That's the problem of representation and what Heidegger wants to get us to question. Kant is working with the ontology of the present-at-hand when he elaborates his transcendental idealism. But the world first shows up as ready-to-hand.

If we place in question those two presuppositions, the "I think" and representation, then it might lead us to abandon the entire epistemological construal of the relation of thought to things and mind to world. The world does not first and foremost, proximally and for the most part, show itself as an object contemplatively and disinterestedly represented by a subject within an ontology of the present-at-hand. Rather, as we've seen by working through *Being and Time* together, the world shows itself as a place in which we are completely

12. G. W. F. Hegel, *Phenomenology of Spirit*, trans. A. V. Miller (Oxford University Press, 1977).

immersed and from which we do not radically distinguish ourselves. And this is Heidegger's thesis of Dasein as "being-in-the-world." We cannot separate Dasein from the world, Dasein *is* world: it is world in a way that is prior to representation and prior to any idea of the "I think," any idea of the subject.

Although Kant's thesis on transcendental idealism is an advance, it is inadequate insofar as it still moves within a traditional epistemological framework—namely, the ontology of the present-at-hand. And that ontology of the present-at-hand is the real target, the persistent target of *Being and Time*. However, this is not to say that Heidegger refuses the Copernican turn, or that he returns to some pre-critical conception of the relation of thought to being or that which is. If Heidegger did that, if he ignored the Copernican turn, then he'd be doing traditional metaphysics, transcendental realism. (And let's face it, there is a lot of dumb external realism and super-hard new materialism out there in the world of pseudo-thinking. Never trust a movement that begins with word "New.") On the contrary, my hunch or hypothesis is that Heidegger's fundamental ontology is what we might think of as an *existential deepening* of the thesis of transcendental idealism. It's a turn away from Kant that moves within the Copernican turn, and hopefully this will become clear.

Let's turn back to the text. I want to signal that there is a wrinkle in Heidegger's argument, and I want to come back to that wrinkle at the end of these remarks on reality. Heidegger refers to Kant's refutation of idealism, from B274 of the B edition of the preface to the *Critique of Pure Reason*.[13] And this little passage in Kant is very

13. Immanuel Kant, *Critique of Pure Reason*, B274–9.

interesting for Heidegger. Kant says that it remains a scandal of philosophy and universal human reason that there is still no proof for the existence of things outside us, for the existence of the external world, and that we have to take this on faith. In what Kant calls his only real supplement to the B edition of the *Critique of Pure Reason* (there's an A edition and a B edition, from 1781 and 1787, respectively), by way of proof, as he calls it, he argues that the mere consciousness of my Dasein—Kant uses that word—is enough to prove the existence of objects in space outside of me.

Now, Heidegger comments on this. He says that by "Dasein" Kant means my consciousness of being present-at-hand in the sense of Descartes's "I think" as a thinking thing—that is, as a *res cogitans* (SZ 204 / BT 247–48). Although Kant seems to offer an ontological proof of the outside world through a self that is temporal—and Heidegger would seem to approve of that—he goes on to add that this is only a semblance. Kant remains mired in the Cartesian ontology of the present-at-hand and its metaphysical dualism between thinking things and extended things, between subjects and objects mediated through representation. Thus, Heidegger writes, "That Kant demands any proof at all for the 'Dasein of Things outside of me' shows already that he takes the subject—the 'in me'—as the starting point for this problematic" (SZ 204 / BT 248). On the contrary, in a classic paragraph, Heidegger writes of "the 'scandal of philosophy.'" Those words are Kant's, from a footnote in the *Critique of Pure Reason* that Heidegger's thinking of.

> The "scandal of philosophy" is not that this proof has yet to be given, but that *such proofs are expected and attempted again and again*. Such expectations, aims, and demands arise from an ontologically

> inadequate way of starting with *something* of such a character that independently *of it* and "outside" *of it* a "world" is to be proved as present-at-hand.... It is not that the proofs are inadequate, but that the kind of Being of the entity which does the proving and makes requests for proofs *has not been made definite enough*. This is why a demonstration that two things which are present-at-hand are necessarily present-at-hand together, can give rise to the illusion that something has been proved, or even can be proved, about Dasein as Being-in-the-world. If Dasein is understood correctly, it defies such proofs, because, in its Being, it already *is* what subsequent proofs deem necessary to demonstrate for it. (SZ 205 / BT 249)

The real scandal of philosophy for Heidegger is to expect that the existence of things outside me is something susceptible to proof. Such an approach begins from an inadequate ontology of Dasein. Dasein "defies such proofs, because, in its Being, it already *is*." Dasein exists on a level deeper than the experience of proof. Proof, the whole problematic of proof, for Heidegger, remains at the level of the ontology of the present-at-hand. "How do I know you're in pain?," and so on. It's simply the wrong level of description. When Heidegger is talking about the a priori existential structures of Dasein as being-in-the-world, this is something earlier than proof. This is a key point in this paragraph and indeed the next one. But does that mean that this is prior to proof? Does that mean that the existence of the external world has to be taken on faith? Heidegger says no. The idea that the subject must have to prove the external world, he continues, means that philosophy begins "with the construct of an isolated subject" (SZ 206 / BT 249–50). For Heidegger, this approach always comes too late because we are always already in a world. Elaborating on this question between faith and proof, he says:

> To *have faith* in the Reality of the "external world," whether rightly or wrongly; to "*prove*" this Reality for it, whether adequately or inadequately; to *presuppose* it, whether explicitly or not—attempts such as these which have not mastered their own basis with full transparency, presuppose a subject which is proximally *worldless* or unsure of its world. (SZ 206 / BT 250)

"Proximally worldless," listen to that. Homeless. Without shelter. All questions of rational proof or indeed faith, and indeed the entire opposition between reason and faith—for Heidegger this comes too late, and all these questions have to be referred to what is earlier, to what is a priori. And what is a priori is our being-in-the-world, whose being, as we saw, is defined as care.

All questions of reason versus faith presuppose a worldless subject. In many respects, the way in which philosophers raise the question of the external world presupposes a self that is worldless. And then we have to think about, well, how do we connect something like a mind with something like a world? For Heidegger, the self—us—is always already worldly, always already being-in-the-world. The division between things for me and in themselves presupposes worldlessness.

How do we explain the persistence of the epistemological construal of the world, the subject-object picture? Why does philosophy keep going down this blind alley of epistemology? Heidegger makes an interesting remark: "Our task is not to prove that an 'external world' is present-at-hand or to show how it is present-at-hand, but to point out why Dasein, as Being-in-the-world, has the tendency to bury the 'external world' in nullity 'epistemologically' before going on to prove it. The reason for this lies in Dasein's falling..." (SZ 206 / BT 250). Nota bene: the reason for this lies in Dasein's falling "and in the way in which the primary understanding of Being has been diverted to Being as presence-at-hand—a diversion which is motivated by falling itself" (ibid.). This is an astonishing claim. Heidegger is saying that epistemology—and it is more than arguable that modern philosophy winds up as epistemology on a certain reading after Kant (introductions to philosophy are usually introductions to epistemology in ways that seek to make philosophy consistent with scientific knowledge [e.g., in W. V. Quine[14]])—or the epistemological construal of the relation of thoughts to things, is a symptom and consequence of Dasein's falling. We saw that falling in the constitution of Dasein, in chapter 5, and again as an element in the care structure. We fall. We fall to the world, *at* the world. We're fascinated with the world. We're absorbed with the world. Heidegger's thought here is that philosophy as epistemology begins with this falling and misinterprets it, transforms it into a conception of the world as present-at-hand. The entirety of philosophy is a misdescription, a misunderstanding of our falling being-in-the-world. Descartes's skepticism towards objects "outside my mind," for example—how do I know that others aren't robots?—is a retreat to the present-at-hand. It misses, overlooks, the *phenomenon* of the world. The entire epistemological construal of the world is a con-

14. See, e.g., W. V. Quine, "Epistemology Naturalized," in *Ontological Relativity and Other Essays* (Columbia University Press, 1969), 69–90.

sequence of Dasein's tendency towards falling. What we have to do is something much more radical, if we're going to think with Heidegger. We have to put that ontology of the present-at-hand to one side, we have to dig into the phenomenon of falling and pose the question in a different way, an existential, ontological, and hermeneutical manner. This is the way in which we've encountered the question in *Being and Time*. Let's face it: *Being and Time* has extremely radical implications for the way we do philosophy at all. Philosophy is a misdescription of being-in-the-world. Prior to reality is care. The question of the reality or unreality of the world is founded on that.

After that, still in paragraph 43, Heidegger then rather delightfully subverts the theses of realism and idealism, showing that he can agree with both realism and idealism *and* that they're both wrong. If Dasein is being-in-the-world, then entities are given along with Dasein (SZ 207–8 / BT 251). And therefore this claim is consistent with realism. Entities are not in the head, as in Bishop Berkeley. They are *really there*. But on the other hand, this thesis—Dasein is being-in-the-world—is also consistent with idealism insofar as Heidegger's claim is that the condition of possibility for any understanding of things lies in Dasein's understanding of being. The worldhood of the world—which is Dasein itself, you will recall—is the condition of possibility for the world as such; "there never was a world for her / Except the one she sang and, singing, made." Hence, reality is subject-dependent or subject "constituted" in quotation marks, big quotation marks. But if that's true, Heidegger concludes—and this is what we might see as an attempt at a joke—if reality is subjectively constituted, then "Aristotle was no less an idealist than Kant" (SZ 208 / BT 251). Huge guffaws. Aristotle was as much of an idealist as Kant; namely, "the soul is, in a certain way, things or entities (*ta onta*)," as Heidegger says, quoting Aristotle very early in the book (SZ 14 / BT 34). Heidegger's thesis of Dasein as being-in-the-world is consistent with both idealism and realism and shows the inadequacy of both theses. It's a brilliant stretch of argument, which I think provokes all sorts of questions about how we think about philosophy. That's the main point I want to make. The idealism/realism debate is a philosophical dead end. Existential phenomenology takes a different direction.

Let's move through the rest of paragraph 43. There's a little bit in part B on reality in Dilthey—who just preceded Heidegger and who is a decisive influence on him—and Max Scheler, who was a contemporary of Heidegger and therefore had to be sort of pushed aside rather unkindly (SZ 210 / BT 253). (There can only be one son in the house. Heidegger had an intense academic narcissism.) But we can put that to one side. It's not terribly important. What is more important is the following. This is part C of paragraph 43, which is called "Reality and Care." This is the wrinkle in the fabric of Heidegger's argument here. The claim that Heidegger is making, a simple claim, is that the question of reality has to be understood with reference to the being of Dasein, which has been defined as care. That's the easy bit. But then he then goes on to say, "*Reality is referred back to the phenomenon of care.* But the fact that Reality is ontologically grounded in the Being of Dasein, does not signify that only when Dasein exists and as long as Dasein exists, can the Real [*das Reales*] be as that which in itself it is" (SZ 211–12 / BT 255).

Listen to that: "Can the Real be as that which in itself it is." The question is, What does he mean by "the Real"? Here we continue with the quotation. "Of course only as long as Dasein is . . . 'is there' Being. When Dasein does not exist, 'independence' 'is' not either, nor 'is' the 'in-itself.' In such a case this sort of thing can be neither understood nor not understood." That's not a particularly helpful sentence, is it? He goes on:

> In such a case even entities within-the-world can neither be discovered nor lie hidden. *In such a case* it cannot be said that entities are, nor can it be said that they are not. But *now*, as long as there is an understanding of Being and therefore an understanding of presence-at-hand, it can indeed be said that *in this case* entities will still continue to be.
>
> As we have noted, Being (not entities) is dependent upon the understanding of Being; that is to say, Reality (not the Real) is dependent upon care. (SZ 212 / BT 255)

There are a number of things here to point out. First, for Heidegger, empirical reality has to be referred back to care. I think this means that reality, as in empirical realism, as taking things as real things, real appearances—is dependent on us, it's Dasein-dependent. Therefore, it makes no sense to speak of the independence of reality from Dasein in the same way as the question of being is dependent on the understanding of being, and hence being is Dasein-dependent. Hopefully you begin to feel the issue here, because he then introduces a distinction between reality and the Real. The Real here is *das Reales* in the text, and its mere mention is enough to give disciples of Jacques Lacan a saucy, wet dream. To say that reality is founded on care does not entail that the Real is what it is only so long as Dasein exists. That is, entities as real have a Dasein-independent existence.

What is one to infer from this run of argument, this vague, slightly ad hoc distinction between reality and the Real? Is it that Heidegger is committed to the claim that the reality of things, in the sense of empirical, lived reality as appearances, is transcendentally conditioned by Dasein? That would seem to be Heidegger's thought here. Dasein as being-in-the-world is an existential deepening of the thesis of transcendental idealism. We indeed live in a world that hangs together, makes sense, but the condition of possibility for that world—the worldhood of that world—is us. But, there is a notion of the Real external to this, some notion of things in themselves beyond their appearances to us, a hard core to reality, what natural scientists would call "matter" or "nature." Interesting, isn't it? It's not clear what Heidegger is up to, and it's not something he comes back to in the book at all. It poses a question that I think moves through Heidegger's work over the decades so that there is this dimension of the Real, this dimension of a Dasein-independent notion of the Real, on the one hand, which is independent of us. Ultimately perhaps there is the dimension of the Real, regardless of whether we exist or not. But, on the other hand, all questions of being are dependent upon us. There can only be a question of being insofar as there can be an understanding of being. The question this little passage raises is, What is the dimension of the Real in Heidegger? What is the status of matter in Heidegger? What is the status of the body, of physical nature, of what he will call, in *The Origin of the Work of Art*, earth (UK 28 / OWA 41), or perhaps even what thinkers like James Lovelock or Bruno Latour call *Gaia*?[15] If Heidegger is deepening Kantian idealism, then perhaps this is another version of the mirror, of the Real as limit. In the 1930s, Heidegger will look for something outside Dasein. In the 1950s, Heidegger will find something like this in what he'll call the fourfold (*das Geviert*), the gathering of earth, sky, mortals, and divinities.

Hopefully you see the force of the problem here. On the one hand, Heidegger solves the problem of realism and idealism by claiming that reality has to be referred back to the phenomenon of care. Care is Dasein's being as being-in-the-world. There is a world that shows itself up and the condition of possibility for that world showing itself up is Dasein, is us. Yet there is a dimension of the Real that exceeds that, a hard core to reality that exceeds our understanding and that is not dependent upon us. This dimension of Heidegger's thought is one that is not explored by him in *Being and Time* and that comes back here and there in his later work and poses, I think, a really interesting question: We are beings for whom being is an issue and being is dependent upon us, but there is a Real out there, a nature, that is not dependent upon us. And what's the relationship between these two issues? Is there not a dualism, ultimately, in Heidegger's work, a dualism between the Real and reality, between being and what lies outside of it? And how might one make sense of that, given that Heidegger persistently, as he says in *Being and*

15. See Bruno Latour, *Facing Gaia: Eight Lectures on the New Climatic Regime*, trans. Catherine Porter (Polity, 2017); see also James Lovelock, *Gaia: A New Look at Life on Earth* (Oxford University Press, 1979).

Time, does not want to "split the phenomenon"? Is this not a little splitting of the phenomenon? Anyway, this is a speculative question and a sidetrack in our exposition of the text.

That will do for the question of reality. Now we're going to move on to the concept of truth.

10. Truth

Heidegger's reefer moment — Ek-sistence — The history of truth — Deconstructing the traditional concept of truth — Getting beneath the theory of truth as correspondence — Being-uncovering — Alētheia *as unconcealment — Heraclitus on sleeping like sheep — Preserving the force of the most elemental words — We are in the truth — How to live in a totalitarian lie (Havel) — Dasein is in the truth and the untruth, the two paths (Parmenides) — Truth is relative to Dasein — No Dasein, no truth — Newton's laws — There are no eternal truths — Skepticism is unintelligible because a skeptic has never existed — Heidegger's refutation of suicide — Finishing division 1 and why we need division 2 of* Being and Time.

WE'RE NOW LOOKING at paragraph 44 of *Being and Time*, which is called "Dasein, Disclosedness, and Truth." This paragraph begins with what I think of as one of the "reefer moments" in *Being and Time*. Heidegger inhales a huge—whatever—and just begins to associate widely and wildly. There's a wonderful passage (SZ 212–13 / BT 256) at the beginning of this paragraph where he talks about Aristotle, Parmenides, and Heraclitus. Half the page is in Greek. And I've got lots and lots to say about it, but I want to focus on the question of being and its relationship to truth insofar as it's related to the analytic of Dasein. Let's assume—and it's a correct assumption—that there's a connection between being and truth that is as old as philosophy, where philosophy is understood as the science or *episteme* of being, where being is to be revealed in its truth. This is the classical definition of metaphysics as the science of being.

Before I get into the argument, I want to read you a passage from another text by Heidegger because it gives you an idea of where we're heading in this discussion. There's a lecture that Heidegger gave in 1930, on the Greek ontology of Plato and Aristotle, called "On the Essence of Truth." It's a difficult lecture, but a really interesting one. It's in his *Basic Writings*. And he says there, at the very end of the lecture, in what appears to be a note added later:

> "Truth" is not a feature of correct propositions which are asserted of an "object" by a human "subject" and then "are valid" somewhere, in what sphere we know not; rather, truth is disclosure of beings through which an openness essentially unfolds [or "essences" (west)]. All human comportment and bearing are exposed in its open region. Therefore, the human being *is* in the manner of ek-sistence [ek-sistence that is ek-static, not "in here" but "out there"].[1]

The takeaway from that passage is that truth is not a feature of propositions, which can be warranted logically or empirically, where, for instance, "This is a black pen" is a "valid truth" because it is empirically true. Truth, Heidegger argues, is not to be thought about in this way. Prior to the propositional conception of truth, there lies an idea of truth as a disclosure of our openness. That's where Heidegger is going to go in paragraph 44.

1. Martin Heidegger, "Vom Wesen der Wahrheit" (1930), in *Wegmarken*, 190; "On the Essence of Truth," trans. John Sallis, in *Basic Writings*, 127.

Let's get stuck in. If the connection between being and truth has been established in the history of philosophy (the history of philosophy is concerned with truth and the relation of truth to that which is, or being), then it's appropriate for an inquiry into truth to have a central place in fundamental ontology. Heidegger has justified himself, as he regularly does in *Being and Time*, like a good, eager, ambitious young academic. This, he says, is a good place to talk about truth.

What is the traditional conception of truth? The traditional conception of truth is *correspondence*: the correspondence between propositions or statements and facts or states of affairs. This is the standard picture of truth. Truth is the correspondence between a proposition and a fact, and science can be understood simply as the totality of true propositions. In the correspondence theory of truth, on the one hand, we have a proposition, "this is a table," and, on the other hand, we have an object, a table. As one of my teachers, Frank Cioffi, used to say, "Truth isn't interesting; if you want truth, open a phone book."[2] Not that we have phone books anymore. But you get the point.

Now, this idea of correspondence, Heidegger says, leans on and draws on an earlier, scholastic definition of truth: in Saint Thomas Aquinas, as the *adequation* of the intellect and the thing. This idea

2. Cioffi, who was slightly eccentric, had a gift for directness when offering his opinions. Once, when I asked him about switching from a seminar on Derrida to one on Foucault, he said the difference was between horseshit and bullshit. He used to walk around campus wearing headphones. When we finally asked him who he was listening to, we discovered it wasn't music at all but recordings of himself delivering his lectures.

of an adequation of thought to thing is a medieval version of correspondence theory. Heidegger says that Thomas gets this from Avicenna, the great tenth-century Persian philosopher, otherwise known as Ibn Sina.[3] (Avicenna lived a saucy life and had an alarming death.) Avicenna in turn takes over this definition from Isaac Israeli's *Book of Definitions*,[4] also from the tenth century, the work of a Jewish scholar written in Arabic in Cairo (SZ 214 / BT 257).[5] This idea of adequation harks back to our friend Aristotle, and his idea of truth as *homoiousios* (ὁμοιούσιος), or "the likeness of the soul or psyche with entities or things" (SZ 214 / BT 257).[6]

This is a typical example of Heideggerian argumentation. The modern correspondence theory of truth (proposition and fact) is a way of taking over the theory of truth as adequation in the scholastics, and it goes back to Aristotle's idea of the *likeness*. Or more accurately, we could think of it as the accordance between psyche or soul and entities and ultimately even back beyond Aristotle. This leads us back to the famous fragment 3 of Parmenides that Heidegger cites at the beginning of paragraph 44 and cited repeatedly in his career: that it is the same thing to think and to be; that being and thinking

3. Ibn Sina, *Ibn Sina's Remarks and Admonitions: Physics and Metaphysics: An Analysis and Annotated Translation*, trans. Shams Inati (Columbia University Press, 2014).

4. Isaac Israeli, *The Book of Definitions*, trans. S. M. Stern, in *Isaac Israeli: A Neoplatonic Philosopher of the Early Tenth Century: His Works Translated with Comments and an Outline of His Philosophy*, trans. Alexander Altmann and Samuel M. Stern (University of Chicago Press, 2009), 3–78.

5. See Thomas Aquinas, *The Disputed Questions on Truth*, trans. Robert W. Mulligan (H. Regnery, 1952), qu. 1, art. 1.

6. See Aristotle, *De Interpretatione*, trans. J. L. Ackrill, in vol. 1 of *Complete Works of Aristotle*, 16a5–9.

are the *same*, where sameness is not the same as a static, fixed identity but a capacious, vibrant, perhaps even changing same (SZ 212–13 / BT 256).[7] Heidegger further developed this discussion of identity, including its distinction from sameness, in the late 1950s.[8]

Having set that up, having established the traditional idea of truth as correspondence, Heidegger then asks, What is the *being* of truth as correspondence? And his inquiry has three steps. The first step is the attempt to establish the ontological foundation of the traditional concept of truth. The second step is to show the primordial phenomenon of truth and the derivative character of the traditional concept. And the third step is to show the being of truth that must be presupposed in philosophical work. Each of these three steps is a different truth: the traditional concept, the primordial concept, and the being that must be presupposed. These three steps are going to make up the three parts of paragraph 44: A, B, and C. These three steps, as we have seen throughout *Being and Time*, are typical of Heidegger. We move back from a traditional, or what Husserl in the mid-1930s will call a "hardened" or *sedimented* idea of truth, and *reactivate* a more primordial ontological foundation.[9] In this reactivation, the derivation of the former from the latter is, in a sense

7. Parmenides of Elea, *Fragments: A Text and Translation with an Introduction*, trans. David Gallop (University of Toronto Press, 1984), 56–57: "...because the same thing is there for thinking and for being."

8. Martin Heidegger, *Identität und Differenz*, ed. Friedrich-Wilhelm von Herrmann, vol. 11 of *Gesamtausgabe*, ed. Friedrich-Wilhelm von Herrmann (Klostermann, 2006); *Identity and Difference*, trans. Joan Stambaugh (Harper & Row, 1969).

9. See Edmund Husserl, *The Crisis of European Sciences and Transcendental Phenomenology: An Introduction to Phenomenological Philosophy*, trans. David Carr (Northwestern University Press, 1970).

that will be disclosed much later in the book, both retrieved and destroyed. Instead of a likeness between proposition and fact, there is a repetition of possibilities revealed in traditions. The movement is from correspondence to disclosure, and in this movement, history points the way from correspondence back to adequation and back still to likeness, accordance, or agreement. In these three steps, Heidegger tracks a genealogy of ideas of truth.

What Heidegger gives us, in his ruminations on philosophical topics throughout this book, are *deconstructions* of traditional concepts (space, time, world, self, reality, etc.). Here he is giving a deconstruction of the traditional concept of truth (namely, the correspondence theory of truth), which shows both the conditions of possibility for that traditional concept, drawing from a deeper idea of the sameness of truth and being, and simultaneously that deconstruction renders that traditional concept *impossible*. Heidegger, as always in his work, is looking into conditions of possibility for concepts that render our understanding of those concepts, our contemporary understanding of those concepts, insofar as we simply take things over from tradition, impossible. Heidegger's technique or method—and Derrida adopts this from him from the beginning to the end of his work—is *quasi-transcendental*: where the standard Kantian conditions of possibility for X are also conditions for X's impossibility. Deconstruction does not destroy or even oppose, it seeks to reactivate what is sedimented in our concepts. In this reactivation there is a restructuring, and thus de-structuring, of the sedimented in traditional conceptions. The genealogical retrieval of these underlying possibilities defamiliarizes the traditions that we presuppose. It doesn't eradicate them, it just rearranges—through a hermeneutical violence—their underlying assumptions, to reveal a deeper historical context. Thus it's not predominantly negative, but it does involve a dismantling of tradition in order to disclose the primordial. This dismantling disclosure is not critique, identifying the ahistorical and a priori conditions of possibility for experience. Critique always implies an act of judgment or decision and requires a prior normative commitment (in Kant's case the primacy of the moral law). The quasi-transcendental method of deconstruction, which Derrida takes over from Heidegger, is historical and proceeds like a genealogy, reactivating those concepts by bringing them to life. (Again, we'll see what this means in a more concrete historical way when Heidegger discusses temporality in division 2.)

Let's back up for a second and look at this traditional concept of truth. Heidegger discusses three theses. He says first, in the traditional notion of truth, the location of truth is in the *assertion*, or, truth is something that can only be asserted *in* propositions and *of* propositions. And we've seen in paragraph 33 a critique of the whole idea of the philosophical priority of assertion that already implies a critique of the propositional idea of truth. That's the first thesis Heidegger's going to discuss. The second thesis is that the essence of truth *precedes* the correspondence or agreement of proposition with object or fact. And it's that *essence* of truth that is going to be linked to an idea of being that Heidegger is going to try to lay forth in these pages. And the third thesis is to go into the *origins* of the correspondence theory of truth and to show that the origins of the correspondence theory of truth lie in Aristotle's idea of *sameness* or *likeness*, "*homoiousios*." With the traditional concept of truth, we are operating at the level of subject-object dualism—subject = proposition, object = fact—and the epistemological construal of the world, which is, as is crystal clear by the end of division 1, based upon an ontology of the present-at-hand that breaks down the unitary structure of the wholeness of Dasein as being-in-the-world and splits the phenomenon. And Heidegger never wants to split the phenomenon.

What I have said so far might well be true to Heidegger's text, but it's a little abstract. Thank God, Heidegger gives us an example here (something he could give us a bit more of in these pages): "Let us suppose that someone with his back turned to the wall makes the true assertion that 'The picture on the wall is hanging askew.' This assertion demonstrates itself when the man who makes it turns around and perceives the picture hanging askew on the wall. What gets demonstrated in this demonstration?" (SZ 217 / BT 260). So, I utter the proposition, "The picture on the wall is hanging askew." In making this proposition, Heidegger says, I'm not concerned with representations or with psychical phenomena corresponding to physical phenomena like objects; rather, I'm concerned with the thing *itself*. When a proposition gets demonstrated, an entity (in this case, the picture on the wall) is *discovered* or *uncovered*. What gets demonstrated in the assertion is the *being-uncovered of the thing*. This is key for what Heidegger is going to do in these passages. This idea of uncoveredness, being-uncovered, is *Entdeckend-sein*. To uncover something is to remove it from being covered up: *ent-deckt*; it's a privative prefix in German. When we demonstrate something, we remove its covering, we pull it into uncoveredness, we pull it into disclosure. Truth for Heidegger is *dis-covered-ness*, it is the discovery of things in speech, the discovery of that which shows itself—in this case, the picture on the wall. Heidegger's claim is that this is a *phe-*

nomenological experience of truth that is more primordial than the correspondence theory of truth.

He tries to make good on this claim in the next couple of pages. When we make assertions about things, those assertions are not lining up with facts or objects in the world. Rather, those things are being *uncovered* in language. Language is the *uncovering* of things. That is the idea of truth that Heidegger wants to hang on to—namely, truth as "Being-uncovering" (SZ 218 / BT 261).

This takes us to part B of paragraph 44, which indeed begins with the line "Being-true means... Being-uncovering." The thought here is that the work of truth is the discovery of entities, taking entities out of their coveredness or their concealedness and showing them as they are in themselves. However, this conception of truth is not arbitrary (and this is his worry at the beginning of part B) because this conception—and this is very important for Heidegger—resonates with the ancient meaning of truth. *Alētheia* (ἀλήθεια), the Greek word for *truth*, is understood by Heidegger as un-concealment, as being-uncovering, taking entities out of their concealedness. And this idea of truth as unconcealment is going to remain constant throughout Heidegger's work and structures much of it. Truth is the movement between unconcealment and concealment. Here Heidegger cites Heraclitus, specifically his longish first fragment when Heraclitus is talking about *logos* (λόγος) (SZ 219 / BT 262). To paraphrase, Heraclitus says, "*Logos* is true evermore. Yet men are as unable to understand it as when they hear it for the first time, as before they've heard it at all. Why? Because they are *asleep*".[10] *Logos*

10. Heraclitus, *The Art and Thought of Heraclitus: An Edition of the Fragments with Translation and Commentary*, trans. and ed. Charles H. Kahn (Cambridge University Press, 1979), fr. 1, p. 29.

is the truth. Truth is *logos*. It is what draws things into unconcealment. Yet human beings refuse to follow *logos* and they are asleep. They live like sheep, inauthentically. And that idea of truth as on the one hand unconcealment and on the other hand being concealed by human activity is going to become hugely important in shaping the entire landscape of Heidegger's work.

Heidegger has not simply shaken off the traditional concept of truth in an arbitrary way. He says he has appropriated its meaning as *alētheia* more primordially. Of course, if truth is *alētheia*, and *alētheia* means unconcealment, then the root of truth, the core of truth, lies in *lēthē*, in forgetfulness, in oblivion, in sleep, in hiddenness. *Alētheia*—which Heidegger is trying to grasp hold of as the primordial experience of truth—is a *bivalent movement* of unconcealment and concealment. In the same way that Heidegger's thought is a source *of* movement and a thought *in* movement, as I've tried to emphasize in these discussions, truth, *alētheia*, is also kinetic, the swaying motion that moves between unconcealment and concealment.

After introducing this concept, which, like I say, is going to be hugely determinative for Heidegger's later work, he makes a really interesting remark. He says, and this might sound surprising:

> In citing such evidence [the testimony of the pre-Socratic and later Greeks] we must avoid uninhibited word-mysticism. Nevertheless, the ultimate business of philosophy is to preserve the *force of the most elemental words* in which Dasein expresses itself, and to keep the common understanding from levelling them off to that unintelligibility which functions in turn as a source of pseudo-problems. (SZ 220 / BT 262)

Ponder that. It's really compelling. We must avoid word-mysticism, and we might wonder whether some Heideggerians do avoid such word-mysticism or end up in a flavorless mixed salad of concepts, or binge-eating in the Being buffet at the Grand Hotel Abyss.[11] But that's by the by. The ultimate business of philosophy is to preserve the force of the most elemental words and to mobilize that force. And if we lose that force, we risk those elemental words like "truth" being *leveled off*, reduced to common understanding: people who are asleep, people who are like sheep. This sheepish somnolence is the basis for pseudo-problems—namely, various supposed timeless problems of philosophy, such as skepticism, as we will see presently.

Heidegger is doing a very important thing here for his work. He's trying to take the traditional idea of truth and connect it to this more primordial experience of truth as unconcealment. Then, as one might expect, he's going to fold this idea of truth as a movement of unconcealment and concealment back into his understanding of the human being, of Dasein, as thrown projection—namely, that the dual structure of truth as unconcealment and concealment mirrors the dual structure, the twofoldness, of thrown projection. This is the reason why Heidegger feels entitled to ask, What is the relationship between truth and Dasein? What's the relationship between

11. In his 1952 polemic against post-Hegelian German thought, Georg Lukács described the systematic philosophy that he saw originating in Arthur Schopenhauer as a "modern luxury hotel on the brink of the abyss" and likened *Being and Time* to Georg Simmel's attempts to "construct a basement underneath historical materialism"; see Georg Lukács, *The Destruction of Reason*, trans. Peter Palmer (Humanities Press, 1981), 243, 501. He later accused Heidegger's great critic Adorno of taking up residence in this "grand hotel abyss"; see Georg Lukács, Preface (1962), *The Theory of the Novel: A Historico-Philosophical Essay on the Forms of Great Epic Literature*, trans. Anna Bostock (MIT Press, 1971), 22.

truth and us? And, he says in italics, to add to the drama of the text, "*Dasein is 'in the truth'*" (SZ 221 / BT 263). What does that mean? As we've seen at the beginning of chapter 5, Dasein is its disclosedness, Dasein is itself the clearing that discloses things to itself, and Dasein has an ahead-of-itself projective relation to its own being-in-the-world. Truth is disclosure.

Dasein understands itself as thrown projection, and in doing so Dasein has understood the *truth* of itself. This is what it means to say that Dasein is in the truth. Incidentally, this idea of living in the truth has been influential on people like the Czech or former Czechoslovak dramatist and writer (and president of Czechoslovakia before it separated into the Czech Republic and Slovakia) Václav Havel, who used to talk about "living in the truth," and living in the truth as a way of hanging on to an experience of personal authenticity in a totalitarian world, in a world that is based on a lie.[12] He picked up this idea from his teacher, the legendary Jan Patočka, who was a student of Husserl and deeply influenced by Heidegger.[13] This idea of living in the truth has been taken up in that political context and indeed in other contexts as well. Let's just keep that in mind, especially in a time where the line separating truth from fakery is harder to draw than ever. To be clear, it's not a question so much of "speaking truth to power" for Heidegger but of being in the truth as an experience of potentiality, as potency and possibility. In that sense, living in the truth *is* power. Living in the truth is assuming for oneself the power that one has but that sleeps within us most of the time.

12. *Václav Havel: Living in Truth*, ed. Jan Vladislav (Faber and Faber, 1989).

13. See James Dodd, *The Heresies of Jan Patočka: Phenomenology, History, and Politics* (Northwestern University Press, 2023), esp. 255–62.

Heidegger takes this idea of truth, this movement of truth as unconcealment and concealment, and then folds it into what we've already learned about Dasein. And as you will recall from chapter 5, we know a number of things about Dasein. The first thing we know about Dasein is its *disclosedness*: it is itself the clearing. The second thing we know about Dasein is that its state of being is one of *thrownness*. Disclosedness is *factically* embedded in a *mood* where things show themselves as thrown. Third, Dasein's state of being has to be understood in relation to *projection*, potentiality for being, the authentic disclosure of what Heidegger is calling now in paragraph 44 "the *truth of existence*." And fourth, we know that to Dasein's state of being belongs *falling*. We fall to the world. We fall to the world through those three characteristics that close it off to us: idle talk, curiosity, and ambiguity.

Where does that leave us? It leaves us in the following situation, which is mightily interesting: Dasein is in the truth, but Dasein is also in untruth. Dasein is in the truth insofar as it projects, insofar as it discloses. And Dasein is in the untruth insofar as it falls, insofar as it covers over. Crucially, Dasein is *both* of these things. Insofar as we are Dasein, we are in the truth, we reveal, we uncover, we unconceal *and* we cover over and conceal, we fall, we err. And this idea of "errance," untruth as errance, as a form of wandering thinking, is something Heidegger is going to pick up in his later meditations on truth in "On the Essence of Truth," the 1930 essay I cited at the beginning.

We have two propositions here: first, Dasein is in the truth, and second, Dasein is in the untruth. And this, of course, echoes the dual structure of Dasein as factical existing or thrown projection. And it also shows the dual, bivalent structure of truth as uncovering and covering up, disclosure and closure, this twofold movement, which is the core of the primordial understanding of truth as *alētheia*. Heidegger goes back again in these pages to the Greeks, to the pre-Socratic Greeks. He cites Parmenides:

> The goddess of Truth who guides Parmenides, puts two pathways before him, one of uncovering, one of hiding; but this signifies nothing else than that Dasein is already both in the truth and in untruth. The way of uncovering is achieved only by κρίνειν λόγῳ—in distinguishing between these understandingly, and making one's decision for the one rather than the other.
>
> The existential-ontological condition for the fact that Being-in-the-world is characterized by "truth" and "untruth," lies in the state of Dasein's Being which we have designated as *thrown projection*. (SZ 222–23 / BT 265)

This is important. There are two paths. Parmenides's poem "On Nature" presents us with the two paths, "of that which is, of being" and "that which is not." The poet-thinker is drawn on a cart led by a horse, and there's a fork in the road and you have to decide which way to go. This idea of κρίνειν λόγῳ, and in particular κρίνειν as crisis, as decision, as incision, as discrimination, as judgment, as cut and crisis are all linked together in these paragraphs. The point is, *we must choose the right path*. And that's for us to decide, or obviously already here in this idea of truth is an idea of *authenticity*. To choose to live in the truth is to choose to live an existence that is, as much as possible, unconcealed.[14]

14. Parmenides of Elea, *Fragments*, pp. 49–53, fr. 1.1–32.

*

Now, there's a lot more we can say about paragraph 44, and it's absolutely fascinating and compelling material, a tour de force by Heidegger. But I want to finish by looking at something he says at the end of the paragraph—this is part C, "The Kind of Being Which Truth Possesses, and the Presupposition of Truth"—because he makes a number of really compelling claims (SZ 226 / BT 269). The first claim he makes—and we are now drawing to the end of division 1, reaching a symphonic climax of sorts—is that truth is Dasein-relative. He'll say in italics at the beginning of this part of the paragraph, "*'There is' truth only insofar as Dasein is and so long as Dasein is*" (ibid.). The truth is relative to Dasein's existence.

The next thing he'll claim is that before there was Dasein, there was no truth. Truth as a truth of existence is only true for as long as Dasein *is*, and Dasein is as being-in-the-world. Nor will there be truth when Dasein is no more. If there is no Dasein, if Dasein is wiped from the face of the planet, then there will no longer be truth. Truth is relative and has this bivalent structure. It is Dasein-dependent and therefore not eternal.

Another thing he says in these pages is that there's no such thing as eternal truths. And in connection with that, he talks about Newton's laws. He says Newton's laws were not true prior to their discovery; rather, they became true by virtue of their discovery. Through the activity of Newton as a natural philosopher, as what we now call a physicist, we're drawn into the discovery of certain propositions about nature. But these truths are "Newton-dependent" rather than eternal verities. Any idea of eternal truth for Heidegger is put to one side, is simply an inadequate understanding of the nature of truth. (There is a possible path to be explored here about developments in theoretical physics in the 1920s, especially the question of the observer dependency of the truths of physics in Werner Heisenberg's quantum mechanics, which is directly contemporary to the developments that led to *Being and Time*. Heisenberg also had some questionable political involvements with National Socialism.)

All of this leads to another startling series of claims at the end of this paragraph, which I'll say a couple of words about. He writes, and you can feel a growing assuredness and intensity to Heidegger's language:

> We *must* presuppose truth. Dasein itself, as in each case my Dasein and this Dasein, *must* be; and in the same way the truth, as Dasein's disclosedness, *must* be. This belongs to Dasein's essential thrownness into the world. *Has Dasein as itself ever decided freely whether it wants to come into "Dasein" or not, and will it ever be able to make such a decision?* "In itself" it is quite incomprehensible why entities are to be *uncovered*, why *truth* and *Dasein* must be. (SZ 228 / BT 271)

Listen to what he is saying here: Insofar as we *are*, insofar as we're factically thrown into a world (and we're always already thrown into a world), then we are tacitly, implicitly, in the truth. We didn't invent ourselves. We didn't call ourselves, summon ourselves, into existence: we found ourselves, in the sense of state-of-mind, always already in existence. We have to *presuppose* truth insofar as we understand truth as the truth of existence. We cannot therefore be *skeptical* about truth.

What Heidegger is doing here is not just refuting skepticism—plenty of philosophers have done that before him—but showing the utter *unintelligibility* of skepticism. He says, "A skeptic can no more be refuted than the Being of truth can be 'proved'" (SZ 229 / BT 271). Truth as the truth of existence is not something that stands in need of proof: it *is*. Truth cannot be refuted because we *are*, we factically exist. Going back to the text, I quote again from paragraph 44:

> And if any skeptic of the kind who denies the truth, factically *is*, he does *not* even *need* to be refuted. In so far as he *is*, and has understood himself in this Being, he has obliterated Dasein in the desperation of suicide; and in doing so, he has also obliterated *truth*. Because Dasein, for its own part, cannot first be subjected to proof, the necessity of truth cannot be proved either. It has no more been demonstrated that there has ever been an "actual" skeptic (though, this is what at bottom has been believed, in the refutations of skepticism, in spite of what these understandings take to do) than it has been demonstrated that there are any "eternal truths." (SZ 229 / BT 271–72)

Philosophy is not meant to refute skepticism. But the problem of skepticism, we could say, is the problem with which philosophy begins with the critique of the Sophists in Plato's dialogues and right the way through the history of philosophy. Earlier, I was talking about Kant's response to Hume's skepticism. Heidegger's radical gesture here is to say a skeptic has never existed. Skepticism does not even need to be refuted. Insofar as we *are*, we are tacitly or implicitly in the truth. Therefore, skepticism about existence would simply be suicide, the self-obliteration of Dasein. It would simply be the obliteration of my being in a suicidal act: that wouldn't prove *anything*, by the way, and Heidegger's certainly not recommending it.

The problem of skepticism and the project of a refutation of skepticism is simply ridiculous as a philosophical proposition. Skepticism is a pseudo-problem. Insofar as I am, I am in the truth. The structure of my being as care is the foundation of truth and reality. Heidegger's procedure in paragraph 44, right at the end of division 1, is to show the *unintelligibility* of skepticism. We are in the truth. We can fall into untruth. We do fall constantly into untruth through the mechanisms of idle talk, curiosity, ambiguity. But we are factically thrown: we are *there*, we are Dasein, we are being-in-the-world. Therefore, skepticism is just a philosopher's pseudo-problem, and we should just grow up and forget about it. That would be Heidegger's thought.

Does this make the problem of skepticism go away? Well, we could save that question for another day, a rainy day perhaps. Let me just say that I am not entirely convinced. What is philosophy without the perpetual return of skepticism? But let's look now at how division 1 of *Being and Time* finishes, how it comes to a finale. He says it is only because Dasein *is*, only because Dasein factically exists, that being can be understood. Truth only is insofar as Dasein is: being and truth are, as he will say, equiprimordial concepts that have the same status, they're linked together in this primal way, they possess the force of elemental words.

But what does it mean to say that that being is? In the last moments of division 1, he says (and he'll repeat these kinds of verbal tics over and over), "The answer to the question of the meaning of Being has yet to be given" (SZ 230 / BT 273). Peculiar. Nearly three hundred pages in, Heidegger insists that we haven't even begun to get close to the question of the meaning of being. We then get some backtracking from Heidegger, what I think of as throat clearing.

Heidegger does a lot of throat clearing in *Being and Time.* I don't much like philosophical throat clearing, and I think that Heidegger risks being dishonest at times. What I mean is that in *Being and Time* or elsewhere in his work, Heidegger will give us hundreds of pages of phenomenological analysis and then say, "But we haven't even begun to think about the question yet." I think he goes too far, protests too much. I will come back to that thought with a more deconstructive intent in chapter 18, a coda on the topic of opening and presence.

But Heidegger does make an interesting remark at the end of division 1 about where we're going to go next in the book. After all, it's important to get a clear sense of the next steps in the Dasein-analytic. Why do we even need division 2? Heidegger gives two reasons. This is the way Heidegger frames it: in the last lines of paragraph 44, he asks the question of whether care has been understood in a truly *primordial* way, or whether there is something deeper to be analyzed. What Heidegger means by this, as we'll see in division 2, chapter 2, is the question of *authenticity*. If we take the question of care and think about that specifically in relation to the question of authenticity, then would that be a more primordial disclosure of who we are? That's the first reason.

The second reason or issue he raises is whether Dasein has really been brought into view *as a whole.* Care is the being of Dasein, which is true, but have we really got Dasein into view as a whole? This is what he will call the problem of *totality* in the introduction to division 2. The problem of totality is what's going to lead him to division 2, chapter 1, on the problem of being-towards-death.

In closing division 1, Heidegger opens two questions: first, the question of authenticity (namely, whether care, as we've shown it, is primordial enough) and, second, the question of totality (namely, whether we've got the whole of Dasein in view). This will lead to being-towards-death. That's what we're going to do when we turn to division 2.

If you've followed me this far in our discussion of *Being and Time*, you should give yourself a big pat on the back. You've got through to the end of division 1 of *Being and Time*, which is way more than most readers or students of Heidegger's thinking manage. And now we're going to move on to division 2, which is called "Dasein and Temporality." We'll pick that up next with a cheery discussion of death.

11. Death

Full-metal Heidegger in lockdown — Why you need to become an existential phenomenologist in a pandemic — Authentic being a whole, the themes of division 2 — Unfairness to Kierkegaard — Death as one's ownmost, nonrelational, and not to be outstripped — Virility and impotence in relation to death — How others die (Tolstoy) — The certainty and indeterminacy of death — I am my not-yet — Authenticity and inauthenticity spin around in Heidegger's analysis — The difference between expecting and anticipating death — Anticipation as the possibility of the possible — Violent language — Shattering oneself against death as a wrenching away from the They — I am possibility — Enter Nietzsche — Enter freedom (towards death) — Having death in your mouth — Fruit — Corpse — Mourning is secondary — Dying for the other (Levinas's discretion) — Sacrifice, holocaust — Undermining Heidegger on death — The relational character of finitude — How I see being-towards-death — The ontology of weakness.

I'M SPEAKING TO YOU from the city of death that New York has become. The first ten episodes of the Apply-Degger *podcast were recorded in the studio of my friend and producer Zenon Marko. But we can't do that anymore because we're in lockdown. And we're all self-isolating.*

I'm going to take you through division 2 of Being and Time, *in a slightly different place, in a different mood. The place is home. In Brooklyn. And the mood is one of fear, anxiety, hypochondria, and the dread of mortality that really defines this pandemic situation into which we've moved. I don't want to keep banging the drum of the contemporary relevance of Heidegger's thought to our time. But I'm going to do it anyway, because if ever there were a moment to become an existential phenomenologist, then it is now. If ever there was a time in which the discussions of fear, anxiety, mortality, and being-towards-death really resounded and made sense, then that time is now.*

New York City
mid-April 2020

•

I want to take you through division 2 of *Being and Time* in the most detailed and faithful way that I can manage. And I want to do it in a more full-metal mode for the remainder of this book. The topic now is being-towards-death. Heidegger has two outstanding problems left, and he's going to try to tackle them in the rest of the book. These problems are those of authenticity and totality. Let's get straight into it. Heidegger claims in paragraph 45 (SZ 231–35 / BT 274–78) that the existential analytic has not been sufficiently primordial because it has been carried out at the level of Dasein's inauthentic average everydayness (SZ 232–33 / BT 275–76). An analysis of Dasein in its authenticity has been absent. The first thing that's missing from the preparatory analysis that's going to be made up for in division 2 is a discussion of authenticity. The second thing that's missing is an analysis of Dasein that truly takes Dasein into view and keeps Dasein in view as a *whole*, in terms of *totality* (SZ 233 / BT 276). In order to see the whole, we need to see the end of Dasein and grasp it. Follow Heidegger's logic here. The issue is a question of the whole or totality. To grasp the whole, we need to see the end. We need to grasp the end of Dasein. That end, Heidegger will say in division 2, chapter 1, is death. And therefore what is required is an analysis of Dasein's relation to its end, its being-towards-death (SZ 234 / BT 277).

These are the two problems that propel the analysis of division 2 forwards: authenticity and totality. We might ask, Why these two themes? We might ask, Why the end of Dasein, and not the beginning of Dasein? Why death and not birth? Later in division 2, Heidegger will touch on the "other end" of Dasein—namely, its "beginning." Being-towards-death is beginning, birth, the rising up into history. We will come to that in detail in chapter 16, "Historicity and Politics."

There's a formal concern driving this chapter, connected to Heidegger's somewhat invented methodological requirements. The problem of wholeness leads to being-towards-death and to the full existential conception of being-towards-death. That's what Heidegger is after: *the full existential conception of being-towards-death* (SZ 255 / BT 299). The problem of authenticity—or authentic totality, authentic being-a-whole—leads to *conscience*. And conscience is ontic or "existentiell" *attestation*, a word that Heidegger uses repeatedly in these pages. This is the attestation, testimony, or evidence of being-towards-death in the form of conscience. The conceptual problems that initiate division 2—authenticity and totality—lead to an analysis of conscience and an analysis of being-towards-death. The relationship between conscience and being-towards-death is a relationship between the ontic (conscience) and the ontological (being-towards-death). We'll come back to that relationship in more detail. Let's just make a note of it now.

With the analysis of being-towards-death and conscience completed, we will, Heidegger assures us, be able to get Dasein into our grasp as an authentic potentiality for being-a-whole. Once he's got these two pieces of authenticity and totality in place, this is going to lead, in the third chapter of division 2, to the concept of *anticipatory resoluteness.* In that discussion, we'll bring being-towards-death and conscience together. Once Heidegger has gone through being-towards-death and conscience, and brought these two matters together, this will lead to a new determination of the being of care seen from the perspective of temporality, which we will deal with in the discussion of ecstatic temporality in chapter 14. Heidegger is going to look at care from the perspective of temporality and then repeat the whole analysis of everydayness that we saw in division 1 (see ch. 15, "The Time of the Everyday World"). And that is what will take place in chapters 3 and 4 of division 2 of *Being and Time*. Then, in chapter 5 of division 2, Heidegger will look at the question of history, by examining the relation of Dasein to historicity. And he will finish the book with an analysis of the relation between the time of Dasein, what he prefers to call the "temporality of Dasein," and vulgar, traditional time, what he calls time-reckoning (see ch. 17, "Existential Time and Vulgar Time").

That's the menu for division 2. Like all menus, it tells you nothing about what you're really going to eat and how it's going to taste. All I will say is that I hope you have a big appetite, because there is a lot of conceptual food to digest.

Before we get started, one of the things I noticed is that there is an entirely unjust little footnote on Kierkegaard sneaked in at the end of paragraph 45, which is worth reading. Heidegger says the following, which catches the eye because it is so unfair. (Philosophers are not usually nice people. And why should they be?)

> In the nineteenth century, Søren Kierkegaard explicitly seized upon the problem of existence as an existentiell problem, and thought it through in a penetrating fashion. But the existential problematic was so alien to him that, as regards his ontology, he remained completely dominated by Hegel and by ancient philosophy as Hegel saw it. Thus, there is more to be learned philosophically from his "edifying" writings than from his theoretical ones—with the exception of his treatise on the concept of anxiety. (SZ 235n / BT 494n6)

Anyway, if you're interested in Kierkegaard, as you should be, you could unpack that, as it's condemning him with faint praise. Heidegger sees Kierkegaard as an existentiell thinker who's still captive to Hegel and Hegel's view of ancient philosophy and philosophically not really up to par. That could easily be flattened as a remark. But these are the ways in which Heidegger repays his debts to the people that he's stolen things from—namely, pretty poorly.

Now we're going to move on to being-towards-death. These pages are extraordinarily exciting. And when I reread them, I always feel strangely enervated, almost overwhelmed by them, which is perhaps a little worrying. But there's something almost malevolent happening in these pages. It's like reading the devil himself, or herself. And as William Blake said, "The devil always gets the best lines."[1] There

1. William Blake, *The Marriage of Heaven and Hell* (University of Miami Press, 1965), 115. Blake argues that Milton was "of the Devil's party without knowing it," because Satan gets the best lines of dialogue in *Paradise Lost*.

is something really powerful in this analysis of being-towards-death. To reiterate, the issue that Heidegger is going to try to raise is how we get Dasein into our grasp as a whole. To get Dasein into our grasp, we need to understand the end of Dasein. The end of Dasein is death. Therefore, we need to have an understanding of death, of being-towards-death, in the Dasein-analytic. The key thing in this chapter—like with much of Heidegger, but maybe particularly here—is to see the general claim he is making and not get bogged down in detail, which is often off-putting and clouded with jargon. My aim is to allow you to look through those details in order to get the big ideas in your mind.

Let's try and pull out the big-picture argument, but let's do it in an unusual way. I want to divide the chapter into two parts: the first part of the chapter, paragraphs 47–49 (SZ 237–49 / BT 281–93), and the second part of the chapter, paragraphs 50–53 (SZ 249–67 / BT 293–311). And I want to look at the second half first. You'll see why I'm doing it this way by the end of this discussion, I promise you. So, let's begin with an overview of Heidegger's conception of being-towards-death based on paragraphs 50–53, the culmination of the chapter. Here Heidegger gives a "*preliminary sketch of the existential-ontological structure of death*" (SZ 249 / BT 293) as "*non-relational*" and "*not to be outstripped*" (SZ 250–51 / BT 294). We'll come back and explain these terms, "non-relational" and "not to be outstripped," but let's just leave them hanging in the air for a second. I want to begin with this key passage:

> Death is the possibility of the absolute impossibility of Dasein. Thus death reveals itself as that *possibility which is one's ownmost, which is*

> *non-relational, and which is not to be outstripped.* As such, death is something distinctively impending. Its existential possibility is based on the fact that Dasein is essentially disclosed to itself, and disclosed, indeed, as ahead-of-itself. This item in the structure of care has its most primordial concretion in Being-towards-death. As a phenomenon, Being-towards-the-end becomes plainer as Being towards that distinctive possibility of Dasein which we have characterized. (SZ 250–51 / BT 294–95)

Now, nonrelationality means that in standing before death, Dasein, us, stands before its ownmost possibility. As such, in standing before our being-towards-death, it has cut off all relations to other human beings, other Daseins. I'm going to question this idea hard at the end of this discussion. More precisely, Dasein has undone those relations. In my view, it is a problem that Dasein's relationship to death is a relationship to its own death, to *my* death. It does not relate primarily to the deaths of others, but to my own death as my ownmost death—that is, my death *for me*. Nonrelationality means that I cannot understand death through the relation to another, but only by relation to my own death. This is a fateful move in my view.

What about the other phrase he uses, "not to be outstripped"? I think that means that death is death. Not to be outstripped means that there's no way of trumping death. Dead is dead is dead, as Gertrude Stein might have said. Death outstrips all possibilities because it is the possibility of the impossibility of Dasein. That is, it is the inability, the limit against which my ability is to be measured. Think about this: Dasein is really understood by Heidegger as, in its authentic mode, potentiality. Dasein is the being that is *able*. Dasein is the being that *can do*. In relation to death, my potentiality, my potency, shatters itself against an impotence, something that I'm not able to do. Death is the limit of the virility of being. But I'm virile towards that impotence. I'm putting this in the language of virility because that's the way in which thinkers like Levinas initially, and also Luce Irigaray, will characterize Dasein, that Dasein's being is being virile.[2] But that virility is always measured against the impotentiality of death. That's the shape of paragraph 50.

In paragraph 51, Heidegger contrasts being-towards-death with everyday death or how death is conceived in the they-self by them. How it's conceived by them is in terms of evasion or fleeing in the face of death (SZ 254 / BT 298). Death is all around us, but we deny it. For *das Man*, for the They, "*man stirbt*" (one dies; SZ 253 / BT 297). Death is a mishap that happens to one, but not to me in particular. Think here of the gravestone of the great artist Marcel Duchamp, which reads, "*D'ailleurs, c'est toujours les autres qui meurent*" (Besides, it's always other people that die). That is to say, with the They, with "the one" or anyone, with average everydayness, we evade death and conceal it in our solicitude and in our consolation. Death happens to us all. It's in the hands of God, or whoever. Namely, the They tranquilizes us in relation to death. They tranquilize us.

There's an interesting moment where we get a footnote reference to Leo Tolstoy's *The Death of Ivan Ilyich*.[3] There's been some discussion of that footnote and the relationship to the death of others in relation to Tolstoy and the way Heidegger cites that book. There's

2. Emmanuel Levinas, *Difficult Freedom: Essays on Judaism*, trans. Seán Hand (Athlone, 1991), 33.

3. Leo Tolstoy, *The Death of Ivan Ilyich and Other Stories*, trans. Rosemary Edmonds (Penguin Books, 1995).

a piece by my old PhD adviser, Robert Bernasconi, on that very topic.[4] When we die, others tell our story. But if you look just before the footnote, Heidegger makes a powerful remark: "The 'they' provides a *constant tranquilization about death*" (SZ 253–54 / BT 298). Think about that tranquilization in terms of the way, say, politicians talk about "prayers and thoughts." Our prayers and thoughts are with X, Y, or Z after another ghastly school shooting—that's tranquilization. We're happy to be shocked by death, just not bothered by it. And he completes that first paragraph by saying, "Indeed the dying of Others is seen often enough as a social inconvenience, if not even a downright tactlessness, against which the public is to be guarded" (SZ 254 / BT 298). The other's death is something inconvenient that gets in the way of our lives, something a little bit awkward and embarrassing that we'd rather not have to deal with. He goes on:

> But temptation, tranquillization, and alienation are distinguishing marks of the kind of Being called "falling." As falling, everyday Being-towards-death is a constant *fleeing in the face of death*. Being-*towards*-the-end has the mode of *evasion in the face of it*—giving new explanations for it, understanding it inauthentically, and concealing it. (SZ 254 / BT 298)

Even in inauthentic average everydayness—which is premised upon the evasion of death, the inconvenience of death, the "denial of death," as Ernest Becker would have said[5]—death is still there, albeit

4. Robert Bernasconi, "Literary Attestation in Philosophy: Heidegger's Footnote on Tolstoy's 'The Death of Ivan Ilyich,'" in *Heidegger in Question: The Art of Existing*, ed. Robert Bernasconi (Humanities Press, 1993).

5. Ernest Becker, *The Denial of Death* (Free Press, 1973).

sleepingly or lethargically. We would simply prefer not to think about death but to forget about it. Heidegger completes this run of thought: "*Even in average everydayness, this ownmost potentiality-for-Being, which is non-relational and not to be outstripped, is constantly an issue for Dasein. This is the case when its concern is merely in the mode of an untroubled indifference towards the uttermost possibility of existence*" (SZ 255 / BT 299). Average everyday existence flees; it evades death.

In paragraph 52, we get the full existential, ontological conception of being-towards-death. Sounds like a death metal album, doesn't it? Maybe some Norwegian death metal or some choice Rammstein. The two criteria that we've already seen, nonrelationality and not-to-be-outstrippedness—death is *my* death, and death is pretty damned important—are supplemented by two further criteria: certainty and indefiniteness. Death is certain, death is indefinite. We know that we're gonna die, but we don't know when. Nobody doubts that they die, do they? Does anybody doubt that they die? I hope not. Maybe Simon Cowell or Elon Musk doubt that they are going to die. But it's certain that we're going to die. Even Barbie gets it. The they-self, what William Blake called the "nobodaddy,"[6] although it can evade and flee in the face of death because it's just inconvenient and tactless, still cannot doubt the certainty of death. We can't avoid death, but we try to avoid thinking about it. Any full conception of being-towards-death will have to admit four criteria. Let's see them spelled out: "The full existential-ontological conception of death may now be defined as follows: *death, as the end of Dasein, is Dasein's*

6. William Blake, "To Old Nobodaddy," in *Collected Poems*, ed. W. B. Yeats (Routledge, 2002), 116.

ownmost possibility—non-relational, certain and as such indefinite, not to be outstripped. Death is, as *Dasein's* end, in the Being of this entity *towards* its end" (SZ 258–59 / BT 303). Those are the four criteria by which Heidegger defines the full conception of being-towards-death. Death is nonrelational, it's not to be outstripped, it is certain, and it is indefinite. I'm going to come back to those four criteria in the second half of this discussion. But you'll already perhaps guess where I'm going to press: on the notion of nonrelationality. What about the death of the other? Does death really show up as *my* death? Or does it show up in relation to another's death? I think that's the real problem in Heidegger. But that's to come when we begin to look at the opening paragraphs of this chapter.

If I am towards-my-end, then I am already in relation to my not-yet. I am living towards my not-yet. If I am towards-my-end, then my being is already a being-towards something not-yet: the not-yet of the future, and the futurity of death in particular. This being my not-yet was already implicit in the notion of being-ahead-of-oneself in the structure of care. The first element in the structure of care is being-ahead-of-oneself, running ahead of oneself. But if this is true, then even inauthentic Dasein has a relation to its running ahead, and even when that's thought in the manner of fleeing. What about an inauthentic understanding of being-towards-death? Heidegger says, and this is a long quotation:

> Meanwhile, it remains questionable whether this problem has been as yet adequately worked out. Being-towards-death is grounded in care. Dasein, as thrown Being-in-the-world, has in every case already been delivered over to its death. In being towards its death, Dasein

> is dying factically and indeed constantly, as long as it has not yet come to its demise. When we say that Dasein is factically dying, we are saying at the same time that in its Being-towards-death Dasein has always decided itself in one way or another. Our everyday falling evasion *in the face of* death is an *inauthentic* Being-*towards*-death. But inauthenticity is based on the possibility of authenticity. (SZ 259 / BT 303)

Did you see what just happened? You might remember that we quoted some lines in the discussion of being-with-others, when we were thinking about the relationship between inauthenticity and authenticity. At the end of chapter 4, Heidegger describes authenticity as a modification of our inauthentic, everyday lives. But here he seems to say the opposite. He says that inauthenticity is based on the possibility of authenticity. These two terms, "authenticity" and "inauthenticity," spin around in Heidegger's analysis. In many ways, what gives *Being and Time* its power, particularly in the second division, is the speed with which it was written.

According to Thomas Sheehan, who knows a lot about this, division 2 was drafted in about six to eight weeks. Heidegger wasn't able to fully control his thoughts and moved quickly. He was moving so fast because he was trying to satisfy the Berlin Ministry of Education's requirements for a full professorship at Marburg to replace Nicolai Hartmann.[7] Luckily for him, Heidegger had Hartmann's support for the chair. Showing so much early talent, he suffered—or

7. Thomas Sheehan, "Heidegger's Early Years: Fragments for a Philosophical Biography," in *Heidegger: The Man and the Thinker*, ed. Thomas Sheehan (Precedent Publishing, 1981), 15. There are drafts of division 1 but not of division 2. There's a virtue to being forced: you make mistakes, but at least they're mistakes you made. You throw things at the wall and see what sticks.

benefited—from Promising Young Man syndrome. (I've seen so many cases in my time in philosophy. They usually spend an age and a career failing to write "the great book.") As lore has it, this is the only time in his life Heidegger didn't shave. There's sometimes a virtue to moving quickly philosophically because certain things are revealed. Perhaps this inversion of the relation between authenticity and inauthenticity is Heidegger's mistake. Or perhaps he was confused about the relation between the two terms and which was a modification of which. In any case, division 2 feels much more loose and less controlled than division 1, which makes it more exciting in my view, like reading the later Nietzsche.

Now we're going to move on to paragraph 53. This paragraph is simply majestic and quite scarily powerful. Heidegger claims that this full conception of being-towards-death is still too formal, and he links it to the possibility of the *authentic* projection of being-towards-death. It's here that Heidegger introduces what will become the key notion of "anticipation." Anticipation, which he distinguishes from awaiting or expecting.[8] The distinction in German is between *das Vorlaufen* (anticipation, running ahead) and *Erwarten* (awaiting or expecting). This is where we get a very interesting specification of possibility, and death as the possibility of impossibility. Throughout *Being and Time*, Heidegger is working with a fully anti-Hegelian distinction between the possible and the actual. For Hegel, the real is the rational, the real is the actual, *wirklich*, and it

8. To worry about the circumstances of my death would be expectation. Heidegger is not really that concerned with death itself. By contrast, Wittgenstein's cancer diagnosis was a relief to him, as his single preoccupation was to die. Three of his siblings killed themselves.

is actuality or *Wirklichkeit* that has to concern us philosophically. For Heidegger, it's exactly the opposite: possibility stands higher than actuality. His claim here is that the expectation of death or the awaiting of death still contains too much of the actual.

Awaiting death, expecting death, expecting it to come to take us away, to sweep us away with a droplet, with a vapor, with a virus or whatever it is, still retains too much of the actual for Heidegger, whereas anticipation is pure possibility. Or anticipation, to speak in a rather ugly way (but I think it's justified at this point), is the *possibilization* of the possible, or the condition of possibility *of* the possible, what makes the possible possible. You can feel the language, the rhetoric, in Heidegger's text, gathering in intensity and momentum:

> But Being towards this possibility, as Being-towards-death, is so to comport ourselves towards *death* that in this Being, and for it, death reveals itself *as a possibility*. Our terminology for such Being towards this possibility is "*anticipation*" *of this possibility*. But in this way of behaving does there not lurk a coming-close to the possible, and when one is close to the possible, does not its actualization emerge? In this kind of coming close, however, one does not tend towards concernfully making available something actual; but as one comes closer understandingly, the possibility of the possible just becomes "greater." *The closest closeness which one may have in Being towards death as a possibility, is as far as possible from anything actual.* The more unveiledly this possibility gets understood, the more purely does the understanding penetrate into it *as the possibility of the impossibility of any existence at all.* Death, as possibility, gives Dasein nothing to be "actualized," nothing which Dasein, as actual, could itself *be*. It is the

> possibility of the impossibility of every way of comporting oneself towards anything, of every way of existing. In the anticipation of this possibility it becomes "greater and greater." (SZ 262 / BT 306–7)

Continuing the quotation, from the beginning of the next paragraph:

> Being-towards-death is the anticipation of a potentiality-for-Being of that entity whose kind of Being is anticipation itself. In the anticipatory revealing of this potentiality-for-Being, Dasein discloses itself to itself as regards its uttermost possibility. But to project itself on its ownmost potentiality-for-Being means to be able to understand itself in the Being of the entity so revealed—namely, to exist. Anticipation turns out to be the possibility of understanding one's ownmost and uttermost potentiality-for-Being—that is to say, the possibility of *authentic existence*. (SZ 262–63 / BT 307)

Anticipation is the possibility of the possible. That is, by projecting towards the possibility of impossibility, Dasein does not actualize any possibility but makes the possibilization of possibility greater and greater. Death increases possibility. *Being and Time* is a long hymn of praise to possibility, which is always higher than actuality. And the highest exemplification of possibility lies in being-towards-death. Dasein—and death—is possibility. Insofar as it authentically projects onto being-towards-death, Dasein comes into its ownmost. That is, what is most one's own is the projection of oneself into the multiplication of possibilities that one is not yet. This is the important thought here. This is what you've got to catch hold of and make use of. Heidegger's thought is that one is that which is not yet. *We are not-yet.* We come into our ownmost in pure possibility, pure anticipation.

Insofar as one is that not-yet, the authentic projection of death as possibility shatters Dasein's inauthentic immersion in the They.

There is a violence to Heidegger's language here, a disconcerting yet oddly compelling violence. He speaks of shattering, of wrenching away: "Death is Dasein's *ownmost* possibility. Being towards this possibility discloses to Dasein its *ownmost* potentiality-for-Being, in which its very Being is the issue. Here it can become manifest to Dasein that in this distinctive possibility of its own self, it has been wrenched away from the 'they'" (SZ 263 / BT 307). Being-towards-death is the countermovement to falling. We are wrenched from tranquilization. Wittgenstein said the First World War saved his life. He liked being shot at. It made him feel alive. If we fall to the world, fall into the world, fall at the world, what happens in being-towards-death is that we're wrenched away, we're torn away. The verb in German is *entreißen*, which has the sense of a tearing, a rending, a ripping of fabric. We're wrenched away from the They and, in that ripping away, we come into our ownmost possibility. In being-towards-death, I arrest the movement of falling. I arrest the impersonal movement of falling to their world, the world of them, and I come into my own, I come into my ownmost, where my own is understood as being open to possibility. I hope this is clear: *I am possibility. I am anticipation.* Insofar as I am not-yet, I am in relationship to the not-yet of my death. And death is what makes possibility possible within me. And that tears me away from average everydayness.

Heidegger then interprets the four criteria for a full conception of being-towards-death from the vantage point of authenticity. Note that being-towards-death as anticipation, as my ownmost possibility, is nonrelational. It lays claim to Dasein alone as an individual, it *singularizes* me in relation to the demand of my finitude, my finite history, which can be retrieved. It picks me out as a singular, unique being. Heidegger writes:

> The ownmost possibility is *non-relational.* Anticipation allows Dasein to understand that that potentiality-for-being in which its ownmost Being is an issue, must be taken over by Dasein alone. Death does not just "belong" to one's own Dasein in an undifferentiated way; death *lays claim* to it as an *individual* Dasein. The non-relational character of death, as understood in anticipation, individualizes Dasein down to itself. (SZ 263 / BT 308)

The individual is claimed by finitude. In this being claimed by death, as we'll see in the next chapter, we can see the effects of the Protestant account of conscience as the work of God in the mind of man. The word "finitude" appears for the first time in the book. It's a keyword: finitude (*Endlichkeit*), endliness. It appears as an adjective in a passage where Heidegger speaks of the link between authentic individuation and the potentiality for the being of others. Consider what has been argued so far: that death is my death, it's my relationship to my death, and that singularizes me. That is the demand of my finitude. So, how do I relate to others on that basis? There is an extraordinary passage that I'd like to quote:

> Anticipation discloses to existence that its uttermost possibility lies in giving itself up, and thus it shatters all one's tenaciousness to whatever existence one has reached. In anticipation, Dasein guards itself against falling back behind itself, or behind the potentiality-for-Being which it has understood. It guards itself against "becoming too old

> for its victories" (Nietzsche). Free for its ownmost possibilities, which are determined by the *end* and so are understood as *finite* [*endliche*], Dasein dispels the danger that it may, by its own finite understanding of existence, fail to recognize that it is getting outstripped by the existence-possibilities of Others, or rather that it may explain these possibilities wrongly and force them back upon its own, so that it may divest itself of its ownmost factical existence. As the non-relational possibility, death individualizes—but only in such a manner that, as the possibility which is not to be outstripped, it makes Dasein, as Being-with, have some understanding of the potentiality-for-Being of Others. (SZ 264 / BT 308–9)

Death singles me out, but I can unleash it as pure possibility. There's a lot to say about this passage, but again, note the verb "shattering," and the very Lutheran, Augustinian violence of the language of wrenching and tearing that Heidegger is employing here. This violent language of separation also echoes Heidegger's remarks about reticence back in chapter 5, division 1. Reticence is a silence towards the They, and death is similarly so: unshared, unexperienceable by those around me, something I go through alone. Reticence characterized discourse, the possibility of disclosing Dasein's intelligibility for others, and death "is based on the fact that Dasein is essentially disclosed to itself" as being-ahead-of-itself (SZ 251 / BT 294). Also note the allusion to Nietzsche's *Zarathustra*: that anticipation guards Dasein against becoming too old for its victories. I interpret this as meaning that Dasein is and must be at the right age. Dasein is and must coincide with itself. I am my age. I am not too young. I'm not too old. (Heidegger is writing a book not for seniors but for heroes. But seniors could revive such heroics.) On this basis—that is, only on the basis of this self-legislating anticipatory aloneness, only as a singular individual that has submitted to the demand of finitude—can I connect with the being of others. I individuate myself on the basis of the demand of my finitude. Alone, I can unleash, wrench myself away from "them." But I could do this alongside others as a people, in a solidarity with those who've affirmed their finitude.

My authentic being-with-others is based upon my individual authenticity. We're going to see that logic reappear later in *Being and Time* when he talks of community and the people. At the level of authenticity, we do not begin with being-with but with individuation. All ideas of authentic community are derived from the way I am singularized by finitude. We could also ponder the significance of the appearance of Nietzsche at this moment in *Being and Time*. He will reappear briefly in the discussion of historicity later on in division 2 (SZ 396 / BT 448), and that's all. But at this point in Heidegger's thinking Nietzsche is linked to an idea of finitude and becoming and seen as an ally in the task of fundamental ontology. It's a superficial reading of Nietzsche. In the course of the lectures on Nietzsche from 1936 to 1940, that alliance will slowly transform into an opposition where Nietzsche came to be seen as the "last metaphysician."[9]

It is here that the theme of anxiety comes back into focus. Being-towards-death is linked back to the experience of anxiety. Anxiety

9. See Martin Heidegger, *Der europäische Nihilismus* (1940), in *Nietzsche*, ed. Brigitte Schillbach, vol. 6.2 of *Gesamtausgabe*, ed. Friedrich-Wilhelm von Herrmann (Klostermann, 1997), 97; *European Nihilism*, in *Nietzsche: Volume 4:* Nihilism, trans. Frank A. Capuzzi, ed. David Farrell Krell, in *Nietzsche: Volumes 3 and 4:* The Will to Power as Knowledge and as Metaphysics *and* Nihilism, ed. David Farrell Krell (Harper & Row, 1982), 58–59.

keeps turning up in these pages like Banquo's ghost in Shakespeare's *Macbeth*, and indeed, later on in the next chapter, Heidegger is going to talk about conscience-anxiety, or *Gewissensangst*, where he'll bring together an anxiety that is being-towards-death with the testimony of conscience. We get a refinement of the thinking of anxiety here: being-towards-death is essential *Angst*. The penultimate line on that page reads, "Being-towards-death is essentially anxiety" (SZ 266 / BT 310). Shattering and wrenching away is a distillation of what happens in anxiety.

Being-towards-death is anxiety. That anxiety, which in chapter 6 was anxiety before the nothing of being-in-the-world, is now anxiety before the nothingness of possibility. Authentic Dasein is a nothing, an anticipation of possibility as possibility. We will sharpen this thought in the next chapter. It is a thought that runs deep through *Being and Time*, so stay with it if you can. Authentic Dasein is nothing actual. It is nothing substantial. Authentic Dasein is a nullity. We didn't inherit actualities but nullities, potentialities. And this nothingness is not a freedom of movement. This nothingness is the movement of freedom. Freedom is wrenching oneself from the They, shattering oneself in relationship to average everydayness and rising up into anticipation, into possibility, into the possibilization of the possible.

Why do I mention freedom? Well, freedom is how the chapter on being-towards-death finishes. Heidegger pulls all the pieces together, in a way that is really powerful, whether you agree with it or not:

> We may now summarize our characterization of authentic Being-towards-death as we have projected it existentially: *anticipation reveals to Dasein its lostness in the they-self, and brings it face to face with the*

> *possibility of being itself, primarily unsupported by concernful solicitude, but of being itself, rather, in an impassioned* freedom towards death*—a freedom which has been released from the Illusions of the "they," and which is factical, certain of itself, and anxious.* (SZ 266 / BT 311)

Freedom has not really been conceptually central before in *Being and Time*. We've alluded to it previously (in chapters 3, 4, and 5, on space, being-with-others, and thrown projection) but not much more than that. And now suddenly it arrives center stage. But notice here what Heidegger does with freedom. Freedom is an *impassioned* liberty towards death. What freedom means is to be released, to be literally emancipated, from the chains and illusions of the They and freed for factical self-certain anxiety. Freedom is not a concept for Heidegger; it's an experience. You can see why I think that this is a little bit like the cave in Plato's *Republic*, that freedom as a freedom from the they-self is a pulling away, a wrenching away from the cave towards something and somewhere else.

But there's also another line of thought here. What Heidegger's drawing on here is a very deep and ancient thought, which I've tried to write about over the years, which is the idea that to philosophize is to learn how to die, an idea we see in Plato's four dialogues on Socrates's trial and execution (the *Euthyphro*, *Apology*, *Crito*, and *Phaedo*). That philosophy is an *ars moriendi*, an art of dying. And there's an essay, a famous essay as you will possibly know, by Michel de Montaigne called "To Philosophize Is to Learn How to Die."[10]

10. Michel de Montaigne, "To Philosophize Is to Learn How to Die," in *The Complete Essays*, trans. M. A. Screech (Penguin Books, 1987), 89–108.

In that essay, Montaigne talks about having formed the habit of having death constantly in his mouth. In the food that he eats and the drink that he imbibes. The thought that Montaigne has, which I find fascinating, is that what characterizes human life is fear. That fear is the fear of death. And that fear of death makes us slaves. We're slaves to the fear of death insofar as we deny death, we flee from it. And Montaigne says he who has learned how to die has unlearned how to be a slave—namely, that freedom consists and can only consist in the acceptance of our mortality, in the acceptance of the framing of our life by necessity, determination by finitude. There is a lot more to say about that. And obviously this line of thought feels particularly apt and maybe particularly disturbing at this point in history, the 2020s, COVID and beyond. Another thing to notice here is that Heidegger's language really loosens up. He's more extreme. He's more unbuttoned. Unbuttoned probably isn't a very good way of putting it. But he's a looser goose, less controlled, more violent, and more direct in division 2. He's letting it rip. He's also improvising.

Fine, we might say. Great stuff, Marty. But why is being-towards-death not enough? Why can't we end the book here? Heidegger concludes the chapter on a very interesting note that turns to the distinction between the existential and the existentiell and the problem of what he calls attestation, or witnessing, or testimony. There's a fascinating passage where he talks about the next step in the argument—namely, that everything that he has said so far is acceptable existentially and ontologically. It's fine at the transcendental ontological level. But it lacks any ontical or existentiell attestation. Testimony. Evidence. Witness. Heidegger's thought in closing this chapter is that being-towards-death is formally right in the way he's described it. But we need to find some evidence, some testimony that allows us to internalize the demand of being-towards-death, to internalize the demand of finitude. This language of demand is italicized and repeated (SZ 266–67 / BT 311).

For those of you who have got an interest in this, I developed a whole theory of ethical experience, what I humbly call "my theory of ethical experience," in a book called *Infinitely Demanding*,[11] where I try to boil down morality to two concepts: the concept of demand and the concept of approval. The first step in any morality consists in binding approvingly to a demand. And the demand that Dasein makes of itself, the demand that Dasein must make of itself, to give ontical evidence to confirm or demonstrate its ontological structure, is its own demand. And that claim is going to be the demand that takes place in the call of conscience. ("Guilty!") And conscience is not God calling to me, is not anybody else calling to me. It's not even the Kantian moral law calling me. It's me, calling to me. And that's what we're going to work out in the next chapter.

That's being-towards-death in Heidegger, paragraphs 50–53. It's a rich brew. The calling of conscience, as the testimony or attestation of mortality, is at the heart of what Heidegger adds to Husserl's thought: *Being and Time* is transcendental phenomenology but combined with a conversion narrative. Luther, whose influence can be felt very strongly in the next chapter, uses the New Testament's absence of church hierarchy to justify the demand of conscience, the work of God in the mind of man. Heidegger replaces God with

11. Simon Critchley, *Infinitely Demanding: Ethics of Commitment, Politics of Resistance* (Verso, 2007), esp. 14–37.

finitude. He says, "Organize yourself around finitude to free yourself from illusions of 'them.'" Lutheranism, perhaps the greatest influence on modern European history (and one could trace this influence through to many forms of modern radical politics),[12] can be tracked through historical "awakening." *Being and Time* is a conversion story about the self. Conversion is never just about you. But the otherness involved here is temporal: the relation between my eventual having-been and the coming of those who are not-yet. And the futural inflection of conversion stories echoes the apocalyptic tone of the book of Revelation: I come quickly, the end time is at hand, the great transformation is about to happen, the New Jerusalem descends bejeweled from the heavens.

•

And yet.

More needs to be said about the earlier paragraphs of Heidegger's account of being-towards-death, and that is what I want to move into now. As we've become accustomed in reading Heidegger, his text is full of exclusions, seemingly baroque distinctions, and sometimes quite tedious and technical stretches of argument, which chime oddly with the heroic, virile, edifying language of the later parts of the chapter on being-towards-death. I want to press a little bit, in the closing parts of this discussion, at the exclusions that Heidegger makes in his analysis of being-towards-death that lead towards the conclusion we have just seen.

Having completed the chapter with the rather muscular mortality

12. The Swiss reformation through Calvin took better root in the US.

of paragraph 53, let's look back at the first part of the chapter and see what has been left out. I'm going to pass over paragraph 48—which is interesting; nothing in the book is *that* uninteresting—where Heidegger gives a formal analysis of the categories of end and totality. And where he even mentions fruit. You might be thinking, "What does Heidegger say about fruit? Why this strange silence about fruit in Heidegger's work?" Well, he talks about fruit in paragraph 48, so you can figure that out for yourself (SZ 243–45 / BT 287–89). He appears silent on love, gender, and vegetables, though.

I want to focus here on paragraph 47. And this brings me to a question that I want to raise. What Heidegger has excluded from the existential ontological analysis of being-towards-death is the possibility that Dasein can gain an understanding of death from the death or dying of others. What happens when another dies? Now, Heidegger sees this quite coolly, quite coldly. He says (and his tone is almost chipper, even upbeat):

> Yet when someone has died, his Being-no-longer-in-the-world (if we understand it in an extreme way) is still a Being, but in the sense of the Being-just-present-at-hand-and-no-more of a corporeal Thing which we encounter. In the dying of the Other we can experience that remarkable phenomenon of Being which may be defined as the change-over of an entity from Dasein's kind of Being (or life) to no-longer-Dasein. The *end* of the entity *qua* Dasein is the *beginning* of the same entity *qua* something present-at-hand. (SZ 238 / BT 281)

In the other's death, they're transformed from Dasein to no-longer-Dasein. And that changeover is a transformation into becoming something present-at-hand. That present-at-hand thing is a corpse. The other, the dead other, becomes a present-at-hand corpse, a mere corporeal thing, "a *lifeless* material Thing" (SZ 238 / BT 282). Heidegger seems to be saying that when you die, you become a present-at-hand corpse. But he admits this approach does not access the "phenomenal content" of the dead one (whatever the phenomenal content of the dead might indeed be). The dead other is experienced proximally as a loss, as a lost loved one. And he moves from the lifeless material corpse as present-at-hand to what he calls the deceased, *der Verstorbene*, who is mourned in funerals and commemorations.

Where is the discussion of grief, mourning, and funerals and commemorations in *Being and Time*? Where does Heidegger think about loss? When we are mourning the dead other, Heidegger says, although we're still with them in a sense, we cannot be said to experience the other's dying in a genuine way. We are "there alongside" or we are "only alongside" (*nur "dabei"*; SZ 238–39 / BT 282) the other in their death. Heidegger acknowledges that there is a place for mourning or grief, but he makes it secondary to "my dying." In other words, the dying of others is not and cannot be a substitute for understanding my death (SZ 239 / BT 283).

Think about that: the dying of others cannot be a substitute for understanding my death. He continues this point: "But above all, the suggestion that the dying of Others is a substitute theme for the ontological analysis of Dasein's totality and the settling of its account, rests on the presupposition which demonstrably fails altogether to recognize Dasein's kind of being" (SZ 239 / BT 283). And so on. That's not going to work for Heidegger. The stranger cannot stand in for my death, and I cannot substitute myself for the other.

I can't hire someone to die for me. In this way, Heidegger introduces the idea of the stranger, as what he calls the "representative" (SZ 239 / BT 283) in the sense of, let's say, a sales representative or a parliamentary representative. The claim he's making, which is incontestable, is that I cannot experience my wholeness, my possibility of being-a-whole, in relation to the other's death.

Now we're going to come to a key passage. Attend carefully, if you would. The passage that I want to draw your attention to was at the heart of a talk that was given in 1987, during what was called the "Heidegger affair" that was all the rage in Paris, when the extent of Heidegger's attachment to National Socialism became clearer and indeed much clearer than it had been viewed previously, set off by Victor Farías's *Heidegger and Nazism*.[13] Jürgen Habermas, who wrote his PhD thesis on F. W. J. Schelling and wrote on the modernism/postmodernism debate, wrote the preface to the German edition of Farías's book. A number of thinkers who were close to Heidegger, or who were influenced by Heidegger, like Jacques Derrida, were asked to respond to the question of Heidegger's Nazism. And there was lots of heated discussion in the newspapers and magazines suggesting that, if Derridean deconstruction as a philosophical procedure owes its existence to Heidegger, and Heidegger was a big-time

13. Victor Farías, *Heidegger and Nazism*, ed. Joseph Margolis and Tom Rockmore, trans. Paul Burrell, Dominic Di Bernardi, and Gabriel R. Ricci (Temple University Press, 1989). German edition: Victor Farías, *Heidegger und der Nationalsozialismus: Mit einem Vorwort von Jürgen Habermas* (S. Fischer, 1989). A more recent entry into this tradition comes in Guillaume Payen, *Martin Heidegger's Changing Destinies: Catholicism, Revolution, Nazism*, trans. Jane Marie Todd and Steven Rendall (Yale University Press, 2023). French edition: Guillaume Payen, *Martin Heidegger: Catholicisme, révolution, nazisme* (Perrin, 2016).

Nazi, then deconstruction is Nazi too. (You know, the usual insane logic of equivalence.[14])

It was a big intellectual scandal in the way that the French used to do with great aplomb.[15] Anyway, into this context appears Emmanuel Levinas, my philosophical hero about whom I have written a lot. Levinas, respected for bringing Heidegger into the Francophone world (Sartre learned his Heidegger from him), was asked to give a response to the Heidegger affair in the form of a talk organized by the Collège International de Philosophie, which had only been founded a few years earlier, at the invitation of the important, too-little-read philosopher Miguel Abensour, who was president of the college at the time. This was in March 1987, when I was a graduate student, actually. It was an important moment for me. And the talk that Levinas gave is called "Dying For . . . ," or "To Die For." In French: "Mourir pour."[16] Levinas had been imprisoned in POW camps in Germany throughout most of the Second World War, for five years, during which time his entire birth family in Lithuania was murdered. Only his wife and daughter were saved in hiding with Catholic nuns in the French provinces. Because Levinas was an officer in the French army, he was not murdered. But unlike non-Jewish officers, he was forced to work in the woods and forests close to his stalag. The war was nightmarish for Levinas. Utterly horrible.

14. Jordan Peterson brings this logic back when he argues that Foucault and Derrida have led to the decay of masculinity and Western culture.

15. Thomas Sheehan made important interventions into this debate at the time, in *The New York Review of Books*. See Thomas Sheehan, "Heidegger and the Nazis," *The New York Review of Books*, June 16, 1988.

16. Emmanuel Levinas, "Dying for . . . ," in *Entre Nous: On Thinking-of-the-Other*, trans. Michael B. Smith and Barbara Harshav (Columbia University Press, 1998), 207–18.

But when asked to respond to the question of Heidegger's Nazism—how much of a Nazi was he? A little bit Nazi, or a whole lot Nazi? Does Nazism contaminate his entire philosophy? —Levinas simply doesn't respond to that question directly. He's more discreet (as he is about the Holocaust too).[17] He just reads this passage and looks at the issue that it raises. I want to read it to you, but if you think of it coming out of the mouth of Levinas, it might be more helpful and more theatrical. But it is already rather dramatic. Heidegger says:

> However, this possibility of representing breaks down completely if the issue is one of representing that possibility-of-Being which makes up Dasein's coming to an end, and which, as such, gives to it its wholeness. *No one can take the other's dying away from him.* Of course someone can "go to his death for another." But that always means to sacrifice oneself for the Other "*in some definite affair.*" Such "dying for" can never signify that the Other has thus had his death taken away in even the slightest degree. Dying is something that every Dasein itself must take upon itself at the time. By its very essence, death is in every case mine, in so far as it "is" at all. And indeed death signifies a peculiar possibility-of-Being in which the very Being of one's own Dasein is an issue. In dying, it is shown that

17. This discretion should not be read as favorable to the question of how Nazi he believes Heidegger was: Levinas wrote in 1968, "One can forgive many Germans, but there are some Germans it is difficult to forgive. It is difficult to forgive Heidegger." See Emmanuel Levinas, *Nine Talmudic Readings*, trans. Annette Aronowicz (Indiana University Press, 2019), 25. Levinas was writing a book on Heidegger from 1930 to 1933, which would have been the first such book in any language, but he stopped this project when Heidegger publicly committed to National Socialism.

> mineness and existence are ontologically constitutive for death. Dying is not an event; it is a phenomenon to be understood existentially; and it is to be understood in a distinctive sense which must be still more closely delimited. (SZ 240 / BT 284)

There's a lot I could say about that passage. But let me come at it in a slightly different way. The phrase I want to focus on, the phrase that Levinas just leaves hanging in the air, is that to die for another would just mean "to sacrifice oneself for the Other." The verb in German, to sacrifice oneself, is *sich opfern*. It's a reflexive verb. An *Opfer* in German is a sacrifice. Another word for sacrifice, an older word for sacrifice, is *holocaustos*, holocaust. To make oneself a sacrifice is in a sense, to make oneself a holocaust. That, for Heidegger, would be secondary to my death. Levinas knows exactly what he is doing here. He is showing how Heidegger, in this philosophical decision, foreshadows the Holocaust (there's much debate about whether this term should be used) or the Shoah by making the death of the other secondary to my own and turning the other into a sacrifice on the altar of Dasein's mineness.

In closing, let me come at this question of sacrifice and the other's dying in a different way. I'm going to change the tone and the approach but come at the same matter from a slightly different angle. There are four criteria in Heidegger's full existential-ontological conception of death. It's nonrelational, certain, indefinite, and not to be outstripped. It's only the first of these criteria that I want to take issue with. The other three are true, if banal. First, it's certain that we're going to die. Second, the instant of our death is indefinite: we don't know when it's going to happen. And third, it's pretty damn important, not to be outstripped. But if the first of these criteria falls, or the first of these criteria can be placed in question, then the whole picture changes. Heidegger insists in the passage I just read on the nonrelational character of death. Because for him, death is ontologically constituted through mineness and existence. Therefore, dying for another would simply be to sacrifice oneself for another or to substitute myself for another. Thus the fundamental experience of finitude is nonrelational, and all relationality is rendered secondary because of the primacy of mineness.

Remember, mineness, *Jemeinigkeit*, is the first concept we saw in *Being and Time*, all the way back to the first page of the existential analytic. All roads lead back to mineness. Now I want to oppose the nonrelational character of being-towards-death with the thought of the fundamentally relational character of finitude—namely, that death is first and foremost experienced in a relation to the death or dying of the other and others, in being with the dying in a caring way and in grieving after they are dead. And such a relation to the other is not a relation of understanding. The other's dying is not like placing an intuition under a concept, in Kant's epistemology.

In other words, the experience of finitude opens up in relationship to a brute fact that escapes my understanding, escapes the reach of my criteria, as someone like Stanley Cavell would put it.[18] The aspiration to wholeness, to totality, can be questioned. Perhaps the aspiration to totality should be given up. Our excursus on Levinas has shown that for him death is primordially relational. In Levinas's *Totality and Infinity*, he argued, following Franz Rosenzweig, that

18. See, e.g., Stanley Cavell, *The Claim of Reason: Wittgenstein, Skepticism, Morality, and Tragedy* (Oxford University Press, 1979), esp. 65–85 and 440–50.

Heidegger and the entire history of philosophy is dominated by the concept of totality. Levinas opposes the concept of totality with the experience of the infinite, the uncontainable otherness of the other. The infinite responsibility of the relation to the other is ethics.[19]

Deliberately twisting the intent of the example that Heidegger gives in paragraph 47, I would say—and this is me speaking, this isn't Heidegger—that the fundamental experience of finitude is rather like being a student of pathological anatomy. In Heidegger's example, the "student of pathological anatomy" sees the dead other as "a *lifeless* material Thing" (SZ 238 / BT 282). With all the terrible lucidity of grief, one watches the person one loves—parent, partner, child, stranger—dying and becoming a "*lifeless* material Thing." That is, there is a thing, a corpse, at the heart of the experience of finitude. And this is why I mourn. Antigone understood this well, it seems to me, staring at the lifeless material thing of her dead brother, Polynices, and demanding justice or burial. No one else could die Antigone's death for her. Yet she dies for another. She substitutes, or sacrifices, herself.

Authentic Dasein—and this is the problem with authentic Dasein—does not mourn. The logic of authenticity makes the act of mourning secondary to my mineness. Because of the primacy of Dasein's mineness, it cannot experience the radicality of loss. And this is why Heidegger can write, as he does, "The dying of Others is not something which we experience in a genuine sense; at most we are always just 'there alongside'" (SZ 239 / BT 282). But as I suggested by appealing to Levinas's work, there is a different way than

19. See Emmanuel Levinas, *Totality and Infinity*.

Heidegger's to think about finitude and death. In Antigone, but even more so in Abraham, we see acts of sacrifice and heroism that suggest a relational finitude. Judith Butler explored a similar relationship, between mourning and solidarity, in the post-9/11 book *Precarious Life*, which borrows the concept of precarity from Levinas.[20] There are four characteristics of death in *Being and Time*: nonrelationality, that it is not to be outstripped, certainty, and indefiniteness. All those are self-evident except the first—namely, its supposed nonrelationality. And it is from this idea of relational finitude that one can launch a devastating critique of Heidegger.

If we challenge Heidegger on this point, as I would like to do, motivated or inspired by Levinas, we come to the view that death and finitude are fundamentally relational. They're constituted in a relationship to someone whom I love, some*thing*, a corpse, some*one* that's become some*thing*, that casts a long, mournful shadow across the self and that undoes the self's authenticity. But if that's the case, if that thought can at least be floated, that there is a problem with Heidegger's analysis of being-towards-death, then this would also lead me to question a distinction that's fundamental to Heidegger's analysis of death, which we haven't talked about, a threefold distinction. How does Heidegger make this threefold distinction? First, he talks about dying, what he calls *Sterben*, which is proper to Dasein, the very mark of Dasein's ownness. Dying is what Daseins do. Second, what he calls "perishing," which is confined to plants and animals. And third, what he calls "demise," which Heidegger calls a

20. Judith Butler, *Precarious Life: The Powers of Mourning and Violence* (Verso, 2004).

Zwischenphänomen, "between phenomenon," between these two extremes of dying and perishing.

Now, one could question Heidegger on this point: Is it the case that plants and animals simply perish? I have my doubts, particularly when it comes to the higher mammals or indeed my beloved cephalopods like octopuses and cuttlefish. But I think we have to leave open the possibility that certain animals die, they undergo dying and not just perishing, as Heidegger insists. I also doubt whether human beings are incapable of perishing. Is it not true that human beings can perish? Or, "dying like a dog," as Kafka's fiction and the facts of war, famine, and global poverty insistently remind us? And what of those persons who die at the end of a mentally debilitating disease or who die whilst being in what is termed a "permanent vegetative state"? Do they cease to be Dasein? I think they do cease to be human on Heidegger's account, and that's a big problem. But even more important than that: If finitude is fundamentally relational—that is, if it is by definition a relation to the fact of the other who exceeds my powers of projection—then the only authentic death is inauthentic. That is, on my account—again, this is me speaking, not Heidegger—an authentic relation to death is not constituted through mineness but rather through otherness. Death enters the world, not through my own fear of death, but rather through my relation to the other's dying, perhaps even through my relation to the other's fear, which I try to assuage as best I can.

This line of thinking is what leads some important thinkers to invert Heidegger's phrasing. When Heidegger talks about death as the possibility of impossibility, Emmanuel Levinas and Maurice

Blanchot strongly disagree: *Death is the impossibility of possibility.*[21] Death is that which shows something over which I have no power. I have no power over the other's death, and I have no power over my own. Death is not a possibility of Dasein but rather describes an empirical limit and a normative limit to all possibility and to all my fateful powers of projection. My relation to finitude limits my potentiality, my potency.

We might say, in closing, that the relational character of finitude, the fact that my relationship to the other is primary in my view, *impotentializes* the self and disables the healthy virility of authentic Dasein. This is important to me. There is a huge problem in Heidegger's analysis that begins from his insistence on the nonrelational character of being-towards-death. I think Heidegger's *Being and Time* is too much of a hymn to potentiality, possibility, potency, and the rest. But I also think there are ways in which Heidegger's thinking can be redescribed that would make a space for weakness, for vulnerability, for dependency, for illness, for sickness. Admittedly, this does require some extra work.

21. Emmanuel Levinas, "Time and the Other," in *The Levinas Reader*, ed. Seán Hand (Blackwell, 1989), 54n2. See also Simon Critchley, *Very Little . . . Almost Nothing* (Routledge, 1997), 68. See also Maurice Blanchot, *Thomas the Obscure*, trans. Robert Lamberton (Station Hill Press, 1973), 13–16.

12. Conscience

Exoteric and esoteric (Leo Strauss) — The ten-minute exoteric version of conscience — The much longer esoteric version: watching the ice floe of Being and Time *break up — The null basis-being of a nullity (the shape of the argument) — The peculiar character of the call of conscience — It says nothing (Cordelia in* King Lear*) — It is silent — Me, myself, and I — The call comes from me and yet beyond me — Stranger voice (Nietzsche) — The nothing of the world — The self divided between two nothings — The three R's: repetition, repetition, repetition (Mark E. Smith) — Guilty! — Heidegger's genealogy of morals — Mortgage, death-pledge, money — Dasein is a nullity that does not have power over itself — Double impotence — Lagging behind one's possibilities — An antiheroic Heidegger (Beckett) — Dasein is a double nullity — The experience of freedom — Guilt is the pre-moral source for any morality (Kafka) — Double zero — Quasi-transcendental philosophy (Derrida) — Weakness is the brand (Maria Bamford) — Wanting to have a conscience — Action! — Resoluteness — Stillness (Stars of the Lid) — Situation as the seizing hold of space — Resoluteness as both taking action and the passivity of resistance.*

YOU MIGHT OR MIGHT NOT KNOW of the philosopher and political theorist Leo Strauss. And it doesn't matter whether you know his work or not. He taught at The New School for a long time. Longer than Hannah Arendt, actually. But we choose not to remember that episode because of his links to conservative politics during his many years at the University of Chicago and influential students like Allan Bloom. Leo Strauss makes a distinction, or a distinction is operative in his work, between the exoteric and the esoteric. The exoteric is for everyone, and the esoteric for the few. He thinks that every major philosophical text from Plato's *Republic* onwards has an exoteric and an esoteric dimension—what you think it's about, and what it's actually about—and should be interpreted accordingly.[1] Regardless of the truth of that distinction, I'm going to try to look at conscience in *Being and Time* in an exoteric manner and then in an esoteric manner. You get to choose either the exoteric version, which will be a consoling plainness, or the esoteric version, which will try to convey the utter strangeness of these pages.

Let's begin with the exoteric version of Heidegger on conscience. After the existential drama of Heidegger's notion of being-towards-death, why do we need a discussion of conscience? As is so often the case in *Being and Time*, Heidegger insists that although his description of being-towards-death is formally or ontologically correct, it needs more compelling content at what he calls the "ontic" level, the level of "empirical experience." Finitude gets a grip on—is attested by—the self through the experience of conscience. For me, the discussion of conscience contains the most exciting and challenging pages in *Being and Time*.

Let me try and sketch the argument as simply as possible. Conscience is a call. It is something that calls one away from one's inauthentic immersion in the homely familiarity of everyday life. It is, Heidegger writes, an uncanny experience of something like an

1. Leo Strauss, "Exoteric Teaching," ed. Hannes Kerber, in *Reorientation: Leo Strauss in the 1930s*, ed. Martin D. Yaffe and Richard S. Ruderman (Palgrave, 2014), 275–92.

external voice in one's head that pulls one out of the hubbub, chatter, and twitter of life in the world and arrests our ceaseless busyness (SZ 277 / BT 322). This idea of conscience is very close to the Christian experience of conscience, *syneidesis*, that one finds in Saint Paul, Saint Augustine, or indeed perhaps most strongly in Martin Luther. In book 8 of the *Confessions*, Augustine describes the entire drama of conversion in terms of hearing an external voice, "as of a child," Augustine says, that leads him to take up the Bible and eventually turn away from paganism and Manicheanism and towards Christ.[2] Luther describes conscience as the work of God in the mind of man.[3] Heidegger is pulling and pushing against Luther in this chapter, but the force of Heidegger's appeal to the conversion or turning around of the self is powerfully Lutheran. It invites atheism, but a Lutheran atheism.

For Heidegger, by contrast, conscience is not God talking to me, but me talking to myself. The uncanny call of conscience—the pang and pain of its sudden appearance—feels like an *alien voice*, it feels strange and uncanny. But it is, Heidegger insists, Dasein calling to itself. I am called back from inauthentic life in the world, complete with what Jean-Paul Sartre would call its "counterfeit immortality"—I'm called back from that, through my mortality, towards myself.[4] Furthermore, that self is, as we saw in the last chapter, defined in

2. Augustine, *Confessions*, book 8, chapter 12, section 29.

3. Martin Luther, "Romans Twelve," in *Commentary on Romans*, trans. J. Theodore Mueller (Kregel, 1976), 165–69.

4. Jean-Paul Sartre, *The Wall (Intimacy) and Other Stories*, trans. Lloyd Alexander (New Directions, 1948), 11. "I wanted to tell myself, this is a beautiful life. But I couldn't pass judgment on it; it was only a sketch; I had spent my time counterfeiting eternity, I had understood nothing. I missed nothing."

terms of being-towards-death. Conscience is the experience of the human being calling itself back to its mortality. A little like Hamlet in the grave with Yorick's skull (HAM 5.1.188ff.). Or like Luther's "Diet of Worms," his summons to appear before the Holy Roman Emperor in the city of Worms to recant his views or be charged with heresy. Luther refuses to bend and continues to stand because of the call of conscience: "*Hier stehe ich. Ich kann nicht anders*" (Here I stand. I can't do otherwise).[5]

What gets said in the call of conscience? Heidegger is crystal clear: like Cordelia in *King Lear*, nothing is said. The call of conscience is silent. "The discourse of the conscience never comes to utterance. Only in keeping silent does the conscience call" (SZ 296 / BT 342–43). It contains no instructions or advice. In order to understand this, it is important to grasp that, for Heidegger, inauthentic life is characterized by chatter, idle talk—for example, the ever-ambiguous hubbub of social media. Conscience calls Dasein back from this chatter, from this hubbub, silently. This call has the character of what Heidegger calls "reticence," which is the privileged mode of language in *Being and Time*, as we saw in chapter 5, division 1. The call of conscience is a silent call that hushes the chatter of the world and brings me back to myself. But what does this uncanny call of conscience give one to understand? Conscience's call can be reduced to one word: Guilty. Or, to say it louder: *Guilty!*

But what does Dasein's guilt really mean? It means that because, as we saw in earlier chapters, the human being is defined in terms of thrown projection, which we already know, the human being always

5. Martin Luther, "Luther at the Diet of Worms," trans. Roger A. Hornsby, in *Career of the Reformer II*, ed. George W. Forell, vol. 32 of *Luther's Works*, ed. Helmut T. Lehmann (Muhlenberg Press, 1958), 113.

has its being to be. That is, human existence is a lack, it is something due to Dasein, a debt that it strives to make up or repay. This is the ontological meaning of guilt as *Schuld*, which can also mean debt. (The call of conscience is like a being-towards-debt.) As Heidegger, perhaps surprisingly, writes—although it should be recalled that he was also living through a time of troubled economic strife in 1920s Weimar Germany—"'Life' is a 'business,' whether or not it covers its costs" (SZ 289 / BT 336). Debt is a way of being. I owe therefore I am. That's really Heidegger's thought in this chapter. *I owe therefore I am*; I'm indebted, I am guilty. I am lacking with regard to myself.

Heidegger goes on to show that this ontological meaning of guilt as indebtedness is the basis for any traditional moral understanding of guilt. In his phenomenology of guilt, he is close to Nietzsche, who digs his claws into conscience in the *Genealogy of Morals*, in particular the stunning second essay of the polemic. Heidegger's relationship with Nietzsche's philosophy could be seen as a subtext for his discussion of conscience here. His phenomenology of guilt claims to uncover the deep structure of ethical selfhood that cannot be defined by morality, since morality already presupposes it.[6] Rejecting any Christian notion of evil as the privation of good (*privatio boni*), Heidegger's claim is that guilt is the pre-moral source for any morality. Listen to that: the pre-moral source for morality is the experience of guilt. As such, it is "beyond good and evil." So, is guilt bad? No. But neither is it good. It is simply what we are. For Heidegger, we *are* guilty. We are lacking, indebted to a lack that we are.

Heidegger insists that Dasein does not load guilt onto itself, like

6. Friedrich Nietzsche, *On the Genealogy of Morals*, "Second Essay: 'Guilt,' 'Bad Conscience,' and the Like," 57–96.

a bad or hostile superego (SZ 283 / BT 328). It simply *is* guilty, always already, as Heidegger liked to say. What changes in being authentic is that the human being understands the call of conscience and takes it into itself. Authentic Dasein comes to understand itself as guilty. And in doing this, Dasein has chosen itself, as Heidegger writes. It's chosen itself (SZ 287 / BT 334). This is very interesting. What is chosen is not *having* a conscience, which Dasein already has because of its ontological want, its ontological indebtedness. What is chosen is, as Heidegger puts it rather awkwardly, "wanting to have a conscience" (SZ 288 / BT 334). This is what he calls *Gewissen-haben-wollen*. That is, if you like, a second-order wanting: I choose to want the want that I am. Only in this way, Heidegger adds, can the human being be answerable or responsible. The German here is *verantwortlich*. Responsibility, or *Verantwortlichkeit*, is the key to any conception of ethics in relation to Heidegger's work. Ethics has to begin in understanding the call, in wanting to have a conscience. To make this choice, Heidegger insists, to choose to have a conscience, to *want* to have a conscience, is to become *resolute*. Resoluteness is a vitally important concept in this chapter and in *Being and Time* as a whole (SZ 296–97 / BT 343).

•

That is the exoteric version of what's going on in this chapter, the version of conscience for the hoi polloi, for the unwashed masses, for the they-self. At this point, you could stop reading if you want. Less flippantly, the argument I've just presented is quite enough to be going on with. But if you want to get the esoteric doctrine in Heidegger, the really weird stuff, then keep on reading. I'm now going to repeat (and repetition will become a key theme) what I said just now about conscience, in a philosophically deeper, textually more rigorous way. Here we go: conscience, the esoteric version.

At times, reading a classical philosophical text is like watching an ice floe break up during global warming. All major philosophical texts are like ice floes, breaking up; things begin to complicate and crack. The compacted, cold assurance of a coherent system begins to become liquid, and great conceptual pieces break off before your eyes and begin to float free on the sea. To be a reader is to try to either keep one's footing as the ice breaks or fall in the icy water and drown. This feeling of the text breaking up is perhaps true of every page of Heidegger's *Being and Time*. But it is nowhere truer than in this discussion of conscience in division 2, which, to my mind, is the most interesting moment in the book. I want to try to show where the ice floe of fundamental ontology begins to crack. At stake here will be trying to bring the human being, us, Dasein, face-to-face with its uncanniness, with the utter strangeness of being human.

This is the thought, the esoteric force of this chapter: *We are the null basis-being of a nullity* (SZ 287 / BT 333). A double zero, suspended between two nothings. I warned you that it was esoteric. As everyone who's read *Being and Time* is aware, what Heidegger is seeking to do in division 2 is to find an authentic potentiality for being a whole. And this turns on the question of the self. If Dasein's inauthentic selfhood is defined in terms of *das Man*, the They—and this is something over which I exert no choice because I'm always falling into it, falling to it—then what Heidegger is after in chapter 2, division 2, is a notion of authentic selfhood defined in terms of choice.

I either choose to choose myself as authentic, or I am lost in the choiceless publicness of *das Man*. Heidegger's claim is that this potentiality for being a whole, for being authentic, is attested in the voice of conscience. Ontologically, conscience discloses something. It discloses Dasein to itself. He writes: "If we analyse conscience more penetratingly, it's revealed as a *call* [*Ruf*]. Calling is a mode of *discourse*. The call of conscience has the character of an *appeal* to Dasein by calling it to its ownmost potentiality-for-Being-its-Self; and this is done by way of *summoning* it to its ownmost Being-guilty" (SZ 269 / BT 314). Conscience is a call, a *Ruf*. And this call is a mode of *Rede*, a mode of discourse, a silent call, as we will see. The call has the character of an *Anruf*, an appeal—that is, a summons or a convocation, *Aufruf*, of Dasein to its utmost being-guilty. Heidegger is playing here with the different modalities, different prefixes of the word *Ruf*. We will see in a little while what Heidegger means by guilt, which is something closer to "lack" than any idea of moral guilt or culpability. And if we wanted to, we could make sense of the idea of guilt as lack in Lacanian terms because it is one of Lacan's many thefts from Heidegger, although I'm not going to insist on that.

Heidegger insists that our understanding of this call, hearing this call, unveils itself as wanting to have a conscience. *Gewissen-haben-wollen*. Adopting this stance, making this choice, this choosing to choose, is the meaning of resoluteness, or decidedness. The word in German is *Entschlossenheit*, which has within it the idea of decision, *Entscheidung*, being determined. Resoluteness is being determined, being decided, having a fixity of purpose.

That's the basic shape of the argument in division 2, chapter 2, and the terminology that Heidegger employs. Heidegger insists that the call of conscience calls one away from one's listening to the they-self, which is always described as listening away to the hubbub of ambiguity, idle talk. Instead, one listens to the call. The call that pulls one away from this hubbub to the silent and strange certainty of conscience. He says, "The call is from afar unto afar. It reaches him who wants to be brought back" (SZ 271 / BT 316). Perhaps it's a little like Socrates's *daimon*, the voice that calls Socrates back at different points in the Platonic dialogues when he's given a bad speech or said something he regrets. Like in the *Phaedrus*, his *daimon* says, "Go back and do better, it wasn't good enough." It's a little bit like that: the call calls to us, silently—calls to Socrates—from afar.[7]

To what is one called in being appealed to in conscience? One is called to one's own self. Conscience calls Dasein to itself in the call. What gets said in the call of conscience? Heidegger is crystal clear. Again, as I said before, in the exoteric version, like Cordelia in *King Lear*, nothing is said. Just stay with that thought for a tiny bit. Remember the scene at the beginning of play, when King Lear's three daughters—Goneril, Regan, and Cordelia—all have to pledge their love to the king. Lear has suddenly decided to stop being king and wants to wander around from castle to castle with this retinue of a hundred, partying basically. He asks Goneril and Regan how much they love him. And they say they love him a lot. They give long, florid speeches. Then he asks Cordelia, whom he loves most, How much do you love me? What do you say, Cordelia? And she says, "Nothing, my Lord." And Lear says, Nothing? Nothing comes

7. "Just as I was about to cross the river, the familiar divine sign came to me which, whenever it occurs, holds me back from something I am about to do. I thought I heard a voice coming from this very spot, forbidding me to leave until I made atonement for some offense" (Plato, *Phaedrus*, trans. Alexander Nehamas and Paul Woodruff, in *Complete Works*, 242B–C).

of nothing. Therefore, speak. And she says, Nothing. My Lord, "I cannot heave / my heart into my mouth."[8]

What does the call of conscience say? Nothing. Heidegger writes:

> But how are we to determine *what is said in the talk* that belongs to this kind of discourse? *What* does the conscience call to him to whom it appeals? Taken strictly, nothing. The call asserts nothing, gives no information about world-events, has nothing to tell. Least of all does it try to set going a "soliloquy" in the Self to which it has appealed. "Nothing" gets called *to* this Self, but it has been *summoned* to itself—that is, to its ownmost potentiality-for-Being. (SZ 273 / BT 318)

The call contains no information, nor is it a soliloquy, like the ever-indecisive Danish prince Hamlet, endlessly and self-piteously soliloquizing, word whistling (oh Lord, shut up), from the first act until the end of the play. It is not that. It is the summoning of Dasein to itself that occurs silently. And this picks up on a remark Heidegger makes: "Vocal utterance . . . is not essential for discourse." We've had that thought before. Vocal utterance is not essential for discourse. "And therefore not for the call either; this must not be overlooked" (SZ 271 / BT 316). Conscience discourses in the mode of silence. It discourses in and as reticence. And this idea of reticence, as we've seen, is given a privilege in the discussion of discourse in paragraph 34 of *Being and Time*. Reticence is the highest form of discourse. One says the most in saying the least.

8. Shakespeare, *King Lear*, in *Shakespeare: Complete Works*, ed. W. J. Craig (Oxford University Press, 1966), act 1, scene 1, lines 78–95.

Pushing on a little bit into paragraph 57, we now get into the nature of the call that says nothing. What is the logic of the call? It is a paradoxical logic. On the one hand, the call of conscience that pulls Dasein out of its immersion and groundless floating in the They is nothing else than Dasein calling to itself, and calling to itself by saying nothing. It is not God calling me. It is not my genes calling to me. It's not my nature calling to me. It's not my superego calling to me. It's not my parents in my head calling to me. To quote De La Soul's 1989 hip-hop classic, it is "Me Myself and I." As we will see, this logic becomes more complex. Heidegger writes:

> But is it at all necessary to keep raising explicitly the question of who does the calling? Is this not answered for Dasein just as unequivocally as the question of to whom the call makes its appeal? *In conscience Dasein calls itself.* This understanding of the caller may be more or less awake [*mehr oder minder wach*[9]] in the factical hearing of the call. Ontologically, however, it is not enough to answer that Dasein is at the same time both the caller and the one to whom the appeal is made. When Dasein is appealed to, is it not "there" in a different way from that in which it does the calling? Shall we say that its ownmost potentiality-for-Being-its-Self functions as the caller?
>
> Indeed the call is precisely something which we ourselves have neither planned nor prepared for nor voluntarily performed, nor have we ever done so. "It" calls, against our expectations and even

9. This word, *wach*, is related to the German *der Wächter*, meaning "guardian" or, more literally, "watcher." The guardian, the person awake to the call of conscience, is also the carer of Being in Heidegger's "Letter on Humanism" (BH 343 / LH 245–46).

> against our will. On the other hand, the call undoubtedly does not come from someone else who is with me in the world. The call comes *from* me and yet *from beyond me and over me.* (SZ 275 / BT 320)

This is a very interesting passage. In German it is "*Der Ruf kommt aus mir und doch über mich*" (The call comes *out of* me and *over* me). And it's this idea of the call coming from beyond me that is so uncanny. It happens against my will, and it's something that I do not voluntarily perform. Dasein is both the caller and the called, and there is no immediate identity between these two sides or faces of the call. The logic of the call is paradoxical. It's me calling to myself, yet it seems to come from beyond me, and there's no identity between these two aspects of the call. How do we explain this? How do we explain this division at the heart of the call of conscience? Heidegger also says that this is a call of conscience that "everyone agrees that he hears" (SZ 281 / BT 326). Does everyone agree? Does everybody hear the call of conscience? I think that's another discussion. Do psychopaths hear the call? I'm not sure.

Let's look at this division within the call. In order to explain this division, Heidegger folds the analysis of the call structure back into the care structure. Of course that's what he's going to do. The call structure mirrors the care structure, which is why that structure is so important to memorize. This is why paragraph 57 is called "Conscience as the Call of Care" (SZ 274 / BT 319). The situation of Dasein being both the caller and the called corresponds to the structure of Dasein as both authentic and inauthentic, the care/call structure, as anxious potentiality for being, or freedom, and thrown lostness in the They. That is, Dasein is both in the truth and in the untruth. We've seen this pattern again and again in *Being and Time.*

Dasein is both disclosed and covered over, existential and factical, projecting and thrown.

This takes Heidegger back into a fascinating discussion of uncanniness that first turned up in the discussion of anxiety in paragraph 40. Heidegger asks here, What if Dasein finds itself in the depths of its uncanniness? What is strange in the call of conscience is that it's us calling to ourselves as if it were an alien voice. We become uncanny to ourselves. "What if this Dasein that finds itself in the very depths of this uncanniness should be the caller of the call of conscience?" (SZ 276 / BT 321). And this leads us to the idea of what Heidegger calls the stranger voice or the alien voice, *die fremde Stimme.* And there's a lot we could say about that. Again, it would be another topic. But this idea of the stranger voice is a topic that I think of in relationship to Nietzsche, when he is thinking about his own work in the 1886 prefaces (which are the very summit of philosophical honesty in my view), where he basically tears into his early writing, *The Birth of Tragedy*, saying that it's terrible because it's Wagnerian and Schopenhauerian and metaphysical. It's an embarrassing artist's metaphysics. But, Nietzsche adds, there is an alien voice, a strange voice that somehow speaks in that work. And when he's reading himself, he can find himself in his own uncanniness, where that voice is both his voice and Zarathustra's voice.[10] Anyway, Nietzsche is a shadow in this analysis, as I said.

Going back to *Being and Time*, Heidegger says, "In its 'who,' the caller is definable in a 'worldly' way by *nothing* at all. The caller is Dasein in its uncanniness: primordial, thrown Being-in-the-world

10. Friedrich Nietzsche, "An Attempt at Self-Criticism," in *The Birth of Tragedy and Other Writings*, trans. Ronald Speirs, ed. Raymond Geuss and Ronald Speirs (Cambridge University Press, 1999), 3–12.

as the 'not-at-home'—the bare 'that-it-is' in the 'nothing' of the world" (SZ 276–77 / BT 321). And in the German, "nothing" does not have quotation marks around it, as it does in Macquarrie and Robinson. The nothing is naked. He goes on: "The caller is unfamiliar to the everyday they-self; it is something like an *alien* voice. What could be more alien to the 'they,' lost in the manifold 'world' of its concern than the Self, which has been individualized down to itself in uncanniness and been thrown into the 'nothing'?" (SZ 277 / BT 321–22). And again, "nothing" in the translation has got quotation marks, but they're not there in the German.

What might be noted here is the repeated emphasis upon the word "nothing" and the general strangeness of the claim that Heidegger makes. The call of conscience is the anxious uncanniness of not being at home in the homeliness or familiarity of the at-home quality of the world. But then this not-at-home is claimed to be the *nothing* of the world. Do you hear that? "The nothing of the world." The self is thrown into the nothing of the world, and into that nothing I hear the silent call that strikes me as alien.

We begin to get the sense here that Dasein's being-in-the-world, which we saw over and over again in the early chapters of division 1—this world full of significance, this world that hangs together, the world we are *in* and that is *there*, the world of concern and solicitude, suddenly becomes nothing. The self is thrown into the nothing of the world, and into that nothing I hear the silent call that strikes me as alien. And that call is also nothing.

Strictly speaking, and this is the thought, this is the esoteric thought: The self is divided between two nothings. On the one hand, the nothing of the world, and on the other, the nothingness of pure possibility revealed in being-towards-death. The nothing of the world and the nothing of pure potentiality. The self is nothing but the movement between two nothings. The nothing of thrownness, and the nothing of projection, which is to say that the uncanniness of being human, being a stranger to oneself, consists in *a double impotentialization*. This thought is much more me speaking rather than Heidegger. I'll try and elucidate that in a little while.

Just stay with this thought: the call reveals the nothing of the world, the nullity of our thrown being-in-the-world. In relation to the nothing of the call, and between these two nothings, we hover. In this hovering, we are pulled out of our being-in-the-world, we've unleashed that radical possibility, the nothingness of pure possibility, and that's being attested in the structure of the self. Heidegger writes:

> The call does not report events; it calls without uttering anything. The call discourses in the uncanny mode of *keeping silent*. And it does this only because, in calling the one to whom the appeal is made, it does not call him into the public idle talk of the "they," but *calls* him *back* from this *into the reticence of his existent* potentiality-for-Being. When the caller reaches him to whom the appeal is made, it does so with a cold assurance which is uncanny but by no means obvious. (SZ 277 / BT 322)

Note the "cold assurance" of the appeal here. I rather like that. The *kalte Sicherheit*, the cool certainty, of the uncanny call. Uncanniness pursues Dasein down into the lostness of its life in the They. A life where it has forgotten itself and it tries to arrest this lostness in a movement that Heidegger will call in the next chapter of *Being and*

Time "repetition" (what I will call "fetching back"). And we'll come back to that idea of repetition, *Wiederholung*. The thought here, which you can just absorb at this point—but we're going to have to live it in the next chapters—is that it's only in the self's repetition to itself of itself that it can momentarily pull clear of the downward plunge of the they-self. It's only in repetition that this can happen. When the self ceases to repeat itself, it forgets and ceases to be itself. We are repetition. The three *R*'s, the three *R*'s, as the legendary Mark E. Smith said, of the mighty band The Fall: repetition, repetition, repetition. I told you this was going to be weird. But please listen to The Fall's "Repetition"; it will help you live your life. Heidegger completes this run of argument in the following way:

> The proposition that Dasein is at the same time both the caller and the one to whom the appeal is made, has now lost its empty formal character and its obviousness. *Conscience manifests itself as the call of care*: the caller is Dasein, which, in its thrownness (in its Being-already-in), is anxious about its potentiality-for-Being. The one to whom the appeal is made is this very same Dasein, summoned to its ownmost potentiality-for-Being (ahead of itself . . .). Dasein is falling into the "they" (in Being-already-alongside the world of its concern), and it is summoned out of this falling by the appeal. The call of conscience—that is, conscience itself—has its ontological possibility in the fact that Dasein, in the very basis of its Being, is care. (SZ 277–78 / BT 322–23)

Let that passage resonate a bit. The call of conscience is entirely intelligible in terms of the care structure. That is, thrown projection, falling, factical existence. Remember the three elements of care? Facticity or thrownness, existence or projection, and falling—idle talk, ambiguity, curiosity. We don't need any other powers to explain conscience. We don't need to resort to God as in Saint Paul or in Luther. We don't need to think about public conscience or world conscience. Heidegger deals with that and dismisses it in the final pages of paragraph 57 (SZ 278–80 / BT 323–24) and again in paragraph 59. What does the uncanny call give one to understand? I formulated this question exactly like that in the exoteric interpretation. What does conscience give us to think about? Conscience's call can be produced in one word, and that word is *guilty*. What does Dasein's guilt really mean? It means that because Dasein's being is thrown projection, it always has its being to be. That is, Dasein's being is a lack. Being is a lack, it's something due to Dasein, a debt that it strives to make up or repay. This is the ontological meaning of guilt, as *Schuld*, which I said before means "guilt," "wrong," even "sin." But it can also just mean "debt." To be *schuldig* is to be guilty or blameworthy, but it also means to give someone their due, to be owing, to be in someone's debt.

Why I'm emphasizing this line of thought is because I'm thinking here of Nietzsche's second essay of the *Genealogy of Morals*. What Nietzsche does so powerfully is to give a genealogy of guilt. For him, this is the Christian idea of guilt, but for us, it is the secular (but actually Protestant) idea of guilt that utterly saturates the world that we live in. Nietzsche traces back the idea of guilt genealogically to *Schulden*, to debts. And these *Schulden* have a material origin, he argues. And if you get interested in this topic, I have written about it elsewhere, with my close friend Tom McCarthy, in relation to

Shakespeare's *Merchant of Venice*.[11] This was a long time ago, but it's work I'm still proud of. The thought here is that life is a series of repayments on a loan that you didn't agree to, with ever-increasing interest, and that will cost you your life. Life is a death pledge: a mortgage. This is what a mortgage means. Debt is a way of being. Arguably it is *the* way of being. And this is why credit and the credence in credit, the belief structure, is so important. Tom and I trace this in relation to money. The key concept in relation to money is credibility, credit, belief, which is always thought in relation to debt. A monetary debt.

Pulling out from that line of thought a little, Heidegger runs through various meanings of guilt, understood as having debts, or being responsible for or owing something or another. Although this would require a longer discussion, it's fascinating to watch Heidegger try to separate his conception of guilt from the usual concept of guilt as a responsibility to others and from any idea of guilt understood in relation to law and in relation to the ought. The idea of a moral ought. The Kantian ought that someone like Hegel relentlessly criticizes and whom Heidegger implicitly follows here. If we reduce morality to the ought and an ought that takes place in conscience, which resembles a law court, a juridical assizes in your head, then we're in a pretty terrible place. We're playing what Bernard Williams

11. Simon Critchley and Tom McCarthy, "Universal Shylockery: Money and Morality in *The Merchant of Venice*," *Diacritics* 34, no. 1 (2004): 3–17. Shylock sees morality in terms of money and unjustly not repaying a debt, and Christians are always banging on about mercy. What's due, a debt, is twisted up with guilt.

would call "the morality game." Sadly, we love playing this game too much because we love hating ourselves. We love guilt, the *jouissance* of guilt. We use it as resentment, to hit people, as with a stick. Heidegger, of course, is trying to get at an ontological meaning of guilt and avoid the usual legal or moralistic connotations of the word. What he is aiming for, in my view, is a pre-ethical, pre-moral understanding of guilt or perhaps what we could call an originary ethical understanding of guilt. Not the standard ethical, moral understanding of it with which we torture and vivisect ourselves.

Can Heidegger pull this off? Let's follow him a little further into some of the weirdest passages in *Being and Time,* perhaps *the* weirdest. So, here we go. Full esoteric. Paragraph 58, which is very strange. The bit I want to draw your attention to seems obscure, but it's actually just really deep. Heidegger says that Dasein is its own "thrown basis" (SZ 284 / BT 330). The word in German is *ein geworfene Grund.* And *Grund*, ground, has a wide range of meanings in German. It also means "reason" (in the sense of a "grounds"). But here it means something like a basis. And Heidegger uses this word to say that Dasein is a *thrown* basis. Dasein projects forth on the basis of the possibilities into which it has been thrown. This is also to say, as we will now see, that Dasein is a null basis (SZ 285 / BT 331). We might well ask ourselves, What is a "null basis"? The German is dense and difficult to render, but the idea is really strange, rich, and powerful:

> In being a basis—that is, in existing as thrown—Dasein constantly lags behind its possibilities. It is never existent *before* its basis, but only *from it* and *as this basis*. Thus "Being-a-basis" means *never* to have power over one's ownmost Being from the ground up. This

> "*not*" belongs to the existential meaning of "thrownness." It itself, being a basis, is a nullity of itself. "Nullity" does not signify anything like not-Being-present-at-hand or not-subsisting; what one has in view here is rather a "not" which is constitutive for this Being of Dasein—its thrownness. The character of this "not" as a "not" may be defined existentially: in being its *Self*, Dasein is, *as* a Self, the entity that has been thrown. It has been *released* from its basis, *not through* itself but to itself, so as to be *as this basis*. Dasein is not itself the basis of its Being, inasmuch as this basis first arises from its own projection; rather, as Being-its-Self, it is the *Being* of its basis. (SZ 284–85 / BT 330)

This is fascinating. The claim is that Dasein is a nullity of itself. Dasein understood as being a basis means that it does not have power over itself, it does not have that power. Dasein is the experience of nullity with regard to itself. The potentiality for being a whole that defines Dasein's power of projection is revealed to be an *impotentialization*, a limit against which it runs and over which it has no power. It's the impotence of Dasein that interests me here. And, as we're going to see, it's a double impotence. What I'm suggesting here is the way in which Heidegger is normally discussed is in terms of potentiality or possibility. Dasein is the virile being that *can do*, and so on. But here we see potentiality constantly bumping up against impotence.

As a thrown basis, Dasein constantly lags behind its possibilities. As he writes in that passage I just read, in being a basis—that is to say, in existing as thrown—Dasein constantly lags behind its possibilities. Do you get that thought? The experience of guilt reveals the being of being human as a lack, as something wanting. As a lag. The self is not just the ecstasy of some heroic leap towards authenticity, energized by the experience of anxiety and being-towards-death. Such would be the heroic reading of the existential analytic, and that possibly was Heidegger's intention (it was certainly his influence in so-called existentialism). On that heroic reading, the goal of the existential analytic would be to achieve a form of self-mastery, a self-sufficiency. What they used to call in Greek *autarkeia*, autarchy. Self-origination. This is what Heidegger calls in paragraph 64 "self-constancy," which we'll get to in the next chapter.

Rather, in my view, the self's fundamental self-relation is to an *unmasterable thrownness.* The burden of a facticity that weighs me down without me ever being able to fully pick it up. This is why I seek to evade myself. I project or throw off a thrownness that catches me in its throw and inverts the movement of possibility by shattering it against impotence. I'm always too late to meet my fate. This is a reading of Heidegger that edges him slightly closer to Samuel Beckett than to the standard heroic, teenage ninja Nietzsche. Heidegger is like a Beckettian Nietzschean. (And here one could contrast the "heroic" Heidegger with the more tragicomic reading that emphasizes the absurd, the unmasterability of existence, and the inescapable nature of inauthenticity.) Think of disability in Beckett. The characters in Beckett are always in bed, or they can't walk, they've lost the use of their legs, or they used to ride bikes but they can't anymore and they're sitting in a pile of dirt or shoved like a bunch of weeds into a jar in the shambles in the bad side of town outside an abattoir trying to concentrate on the rump of a horse. But I digress.

To return to my main thought, Dasein is a being suspended between two nothings: the nullity of thrownness and the nullity of projection. This is where the text gets really, really radical. Hang on tight:

> Not only is the projection, as one that has been thrown, determined by the nullity of Being-a-basis; *as projection* it is itself essentially *null*. This does not mean that it has the ontical property of "inconsequentiality" or "worthlessness"; what we have here is rather something existentially constitutive for the structure of the Being of projection. The nullity we have in mind belongs to Dasein's Being-free for its existentiell possibilities. Freedom, however, is only in the choice of one possibility that is, in tolerating one's not having chosen the others and one's not being able to choose them.
>
> In the structure of thrownness, as in that of projection, there lies essentially a nullity. This nullity is the basis for the possibility of inauthentic Dasein in its falling; and as falling, every inauthentic Dasein factically is. *Care itself, in its very essence, is permeated with nullity through and through*. Thus "care"—Dasein's Being—means, as thrown projection, Being-the-basis of a nullity (and this Being-the-basis is itself null). This means that *Dasein as such is guilty*, if our formally existential definition of "guilt" as "Being-the-basis of a nullity" is indeed correct. (SZ 285 / BT 331)

This quotation is one of the wildest moments in *Being and Time*. It is dizzyingly odd. What does it tell us? I think it tells us the following. *Dasein is a double nullity.* It is simultaneously constituted and divided around this double nullity. And this is the structure of thrown projection and the ontological meaning of guilt. That is, Dasein is guilty. It is indebted doubly. It is null at the heart of its being. Dasein is essentially doubly lacking. What I've emphasized throughout these chapters is the idea of Dasein as thrown projection. But what does thrown projection mean? What he tells us in this passage is that thrown projection is the null basis-being of a nullity,

"*Das (nichtige) Grund-sein einer Nichtigkeit*" (SZ 285 / BT 331).[12] And this is nothing less than the experience of freedom. As Heidegger writes, freedom is the choice of one possibility of being in choosing one's self and not the others. But what one is choosing in such a choice is the nullity of a projection that projects on the nullity of a thrown basis, over which one has no power. If we take this thought seriously, we begin to move away from the tragic-heroic pathos of authenticity. The core of the existential analytic is rather an experience of weakness, powerlessness.

Try to get your head around this thought: *Dasein is the nullity of a projection, projecting onto the nullity of a thrown basis over which one has no power, behind which one is always lagging.* Freedom is the assumption of one's ontological guilt of the double nullity that one is. Heidegger goes on to show that this existential ontological meaning of guilt is the basis for any traditional moral understanding of guilt. Heidegger's phenomenology of guilt, like Nietzsche's in the *Genealogy of Morals*, claims to uncover the deep structure of ethical subjectivity that cannot be defined by morality, as I said earlier, because morality already presupposes it. Heidegger rejects any notion of evil as the privation of good. Heidegger's claim is that *guilt is the pre-moral source for any morality.* Guilt, as he understands it, this double nullity, this double nothingness of thrownness and projection, is the experience of freedom, and it beyond good and evil. Is guilt bad? No. But neither is it good. It's simply what we are. We are guilty. Such is Kafka's share of eternal truth.

12. See Simon Critchley, "The Null Basis-Being of a Nullity, or Between Two Nothings: Heidegger's Uncanniness," in *Interpreting Heidegger: Critical Essays*, ed. Daniel O. Dahlstrom (Cambridge University Press, 2011), 69–78.

Heidegger then brings together a large number of themes in an enormously powerful way and we come back again to the question of uncanniness:

> The call is the call of care. Being-guilty constitutes the Being to which we give the name of "care." In uncanniness Dasein stands together with itself primordially. Uncanniness brings this entity face to face with its undisguised nullity, which belongs to the possibility of its ownmost potentiality-for-Being. To the extent that for Dasein, as care, its Being is an issue, it summons itself as a "they" which is factically falling, and summons itself from its uncanniness towards its potentiality-for-Being. The appeal calls back by calling forth: it calls Dasein *forth* to the possibility of taking over, in existing, even that thrown entity which it is; it calls Dasein *back* to its thrownness so as to understand this thrownness as the null basis which it has to take up into existence. This calling back in which conscience calls forth, gives Dasein to understand that Dasein itself—the null basis for its null projection, standing in the possibility of its Being—is to bring itself back to itself from its lostness in the "they"; and this means that it *is guilty*. (SZ 286–87 / BT 332–33)

This is strong stuff. There's an awful lot going on here. Guilt has been shown to be the innermost meaning of care: the very movement of care, its *kinesis*. Here and indeed elsewhere in his work, Heidegger is simply trying to think movement, *kinesis*, as the rhythm of existence and ultimately the rhythm of being itself. Being itself has a rhythm, a rhythm of a bivalence of giving and withholding. Appropriation and expropriation (*Ereignis*/*Enteignis*). This movement—which is the movement of thrown projection or what I prefer to call thrown throwing off—this movement is the structure of the call that calls back by calling forth. It calls Dasein forth to take over its potentiality for being by taking it back to its thrownness and taking it over. Dasein is this movement back and forth. Almost like a rocking movement. A movement that's based on nothing, a double nothing. No metaphysics, no ground, no basis. Just . . . us . . . doing nothing and leaving nothing undone.

Look closely at Heidegger's words. Dasein is the null basis for its null projection. Dasein is a double nothing, a double zero. And this is the meaning of thrown projection. Guilt is the movement, the *kinesis*, of this nullity, a movement back and forth, a movement to and fro, as Beckett would say. A lot of Beckett's late plays are about simple movement back and forth, like someone in a rocking chair, rocking back and forth in a dramaticule like *Rockaby*, or walking up and down.[13] Such is the strangeness of what it means to be human: the uncanniness of being brought face-to-face with ourselves. As Heidegger writes in the *Introduction to Metaphysics*, "Dasein is the happening of strangeness."[14] Isn't that amazing? Dasein is the *happening* of strangeness, where happening is *Geschehen*, which is the

13. Samuel Beckett, "Rockaby," in *Collected Shorter Plays* (Grove Weidenfeld, 1984), 271–82.

14. Martin Heidegger, *Einführung in die Metaphysik*, ed. Petra Jaeger, vol. 40 of *Gesamtausgabe*, ed. Friedrich-Wilhelm von Herrmann (Klostermann, 1983), 167: "So ist das Da-sein die geschehende Unheimlichkeit selbst. (Die geschehende Unheimlichkeit mußfür uns anfänglich als Da-sein gegründet werden.)"; *Introduction to Metaphysics*, 2nd ed., trans. Gregory Fried and Richard Polt (Yale University Press, 2014), 176: "Thus, Being-here [Fried and Polt's translation of Heidegger's hyphenation of Da-sein] is the happening of un-canniness itself. (The happening of uncanniness must for us be grounded inceptively as Being-here [Da-sein].)" Hereafter cited as EM/IM followed by page numbers.

core meaning of historicity later in *Being and Time*. The human being is the utter strangeness of action between two nothings. The self is a potentiality for being whose sole basis, limit, and condition of possibility is a double impotentialization, which, of course, is to say that it is also a condition of impossibility, an existential quasi-transcendental claim. What Heidegger is describing here are the conditions of possibility for being a self. "Conditions of possibility" talk is transcendental argumentation in the Kantian style. *Being and Time* is transcendental philosophy, identifying the a priori structures of existence—because our being is existence. But those conditions of possibility bump up against impossibility, impotentialization, at both ends of existence, death and birth. Conditions of possibility are also conditions of impossibility, and this is the modification of the concept of the transcendental from the straightforward Kantian idea that philosophy is about identifying the conditions of possibility for X, to the more Derridean idea that deconstruction is about identifying conditions of possibility for X and conditions of the *im*possibility of X at the same time. And we have to learn to stay with that double movement.

Where this takes us—and I think it takes us to an interesting place—is to an idea of weakness. Rather than seeing weaknesses as a signal of something poor or something opposed to strength, I think what Heidegger is pushing towards here is an idea of impotence or an idea of weakness that's more important than the potency of heroic Dasein. As the great Maria Bamford, my favorite female comedian, would say, "Weakness is the brand." "I may be mental," she says, "but at least I'm a millionaire." I digress again. But please watch some stand-up by Bamford, if for some crazy reason you haven't already, and read her wonderful book, *Sure, I'll Join Your Cult.*[15] Heidegger insists that Dasein does not load guilt onto itself. It *is* in its being already guilty. Dasein is guilty, always already. What changes in being authentic is that Dasein understands the call or appeal of conscience and takes it into itself. Authentic Dasein comes to understand itself as guilty, which means that Dasein as potent comes to understand itself as impotent. In doing this, Dasein has somehow chosen itself, as Heidegger says. *Er hat sich selbst gewählt.* This is very interesting. What is chosen is not *having* a conscience. We have conscience by virtue of being Dasein. What is chosen is *wanting* to have a conscience. This is a second-order wanting: *to want the want that one is.* Heidegger says, rounding out this extraordinary line of thought in paragraph 58, "*Wanting to have a conscience is rather the most primordial existentiell presupposition for the possibility of factically coming to owe something.* In understanding the call, Dasein lets its ownmost Self *take action in itself* in terms of that potentiality-for-Being which it has chosen. Only so can it *be* answerable [*verantwortlich*]" (SZ 288 / BT 334). Answerability or responsibility should be the key to any ethics, any morality. But here we're dealing with a pre-moral morality, a pre-ethical ethics. Answerability, responsibility, consists in understanding the call, in wanting to have a conscience. This choice, Dasein's choice of itself, in Heidegger's strange phrasing, is taking action in itself. Taking action in itself. Heidegger

15. Maria Bamford, *Sure, I'll Join Your Cult: A Memoir of Mental Illness and the Quest to Belong Anywhere* (Gallery Books, 2023). Bamford has a tendency, in her stand-up comedy, to give voice to something like an internal voice that takes a form like the disturbed thoughts of her audience. She delivers this in a raspy, worried whisper, often as a rejoinder to whatever she's just said.

reminds us much later on—in the opening of the "Letter on Humanism" that he wrote after the Second World War, at the invitation of a loyal and long-standing French reader and translator of his work, Jean Beaufret—that, "We are still far from pondering the essence of action decisively" (BH 313 / LH 217). The word "action" is one that Heidegger both uses in *Being and Time* and continually reminds us that he wants to avoid, like "spirit" or "consciousness." Such is the logic of Heidegger's avoidances. We have to watch what he's doing when he says he's going to avoid a word and then sometime later in the book we find him using it. But what might action mean, conceived in relation to the double nullity we have described? What might potentiality-for-being mean when its condition of possibility and impossibility is a double impotentialization?

This conception of action might be called tragic. But it might also be called tragicomic. It's why Beckett calls *Waiting for Godot* a tragicomedy. And as one of Beckett's gallery of moribunds, Molloy, asks himself, tongue deep in his cheek, "From where did I get this excess of vigor? From my weakness, perhaps."[16] In Beckett, and much less overtly in Heidegger, we see weakness folded into heroism.

•

That concludes the esoteric part. I've got two more things to do before we finish, but this discussion of conscience never ceases to move me. There are two concepts we haven't discussed that come up at

16. Samuel Beckett, *Molloy*, trans. Patrick Bowles in collaboration with the author (Grove Press, 1955), 114.

the end of the chapter in paragraph 60: *resoluteness* and *situation*. Resoluteness is the authentic structure of Dasein, the truth of Dasein, the truth of existence. Heidegger writes:

> The disclosedness of Dasein in wanting to have a conscience, is thus constituted by anxiety as state-of-mind, by understanding as a projection of oneself upon one's ownmost Being-guilty, and by discourse as reticence. This distinctive and authentic disclosedness, which is attested in Dasein itself by its conscience—*this reticent self-projection upon one's ownmost Being-guilty, in which one is ready for* anxiety—we call "*resoluteness*." (SZ 296–97 / BT 343)

Resoluteness, or *Entschlossenheit*, means, as I said much earlier, decidedness, having fixity of purpose, becoming resolved, determined, where *entschliesen* means to make up one's mind. Resolute Dasein understandingly projects onto its possibilities in the mood of anxiety. And note here that Heidegger reintroduces or keeps going back the concept of anxiety. And he'll speak here of the anxiety of conscience—*Gewissensangst*—where "wanting-to-have-a-conscience becomes a readiness for anxiety." But we have seen that anxiety is not anguished. Rather, it is articulated discursively as reticence. That is, where I pull away from the loud chatter of the they-self, and I move into silence. Anxiety is the mark or proof of the calling, the call of conscience.

I could say a lot more about this in relation to the experience of faith, as I understand it. I begin from someone like Saint Paul here rather than from some warm, fuzzy feelings of faith, some sort of subjective belief or certainty in the existence of a transcendent being. This is not what faith is about. Rather, faith is anxiety. It's the experiential mark or proof in the self of the calling. The way the calling, faith, is felt is as anxiety, and the proper disposition for Saint Paul of being in Christ is to remain in an anguished waiting. And the anguished waiting is itself the induction or inculcation of a stillness in the self. Not a restful stillness, but a resolute stillness, perhaps a cold certainty. In Heidegger's words:

> Only in keeping silent does the conscience call; that is to say, the call comes from the soundlessness of uncanniness, and the Dasein which it summons is called back into the stillness of itself, and called back as something that is to become still. Only in reticence, therefore, is this silent discourse understood appropriately in wanting to have a conscience. It takes the words away from the common-sense idle talk of the "they." (SZ 296 / BT 343)

Conscience calls in silence, and this summons or calls one back to stillness, becoming still. It's a theme that you could plot in Heidegger's later work, when he talks in *On the Way to Language* of "the ringing stillness" (*Geläut der Stille*).[17] For me, it's the sort of stillness that you can find in some musical forms. I think about that soundless sense of uncanniness in relation to the stillness of, say, certain forms of ambient music: Harold Budd, Gavin Bryars, or Brian Eno or a favorite band of mine from Austin, Texas, called Stars of the Lid, who have made amazingly beautiful music. Anyway, to move on: Resoluteness is that becoming stillness in oneself, taking into oneself

17. Martin Heidegger, "Der Weg zur Sprache" (1959), in *Unterwegs zur Sprache*, ed. Friedrich-Wilhelm von Herrmann, vol. 12 of *Gesamtausgabe*, ed. Friedrich-Wilhelm von Herrmann (Klostermann, 1985), esp. 240–42; "The Way to Language," in *On the Way to Language*, trans. Peter D. Hertz (Harper & Row, 1971), esp. 121–22.

decidedly the experience of conscience. To let the call resound is an anxious stillness, focused, hovering between being thrown and projecting.

Let's finish with the concept of the situation. This concept is introduced very quickly and hurriedly by Heidegger. But it's a very important concept that you can find picked up in Sartre and a whole host of other thinkers. I'll mention one more in a second. Resoluteness is always resoluteness at some particular time. It is not a question of eternity here. We're dealing with a temporal being that is in time and for its time. Upon what does Dasein resolve? For what is Dasein authentic? This partially answers an old objection that can be made against existential philosophy, whose originator is Heidegger. Resoluteness is closely linked to the concept of commitment or engagement that comes up more in a more popular way in Sartre. The existentialist pledges themselves to the idea of commitment, but what exactly are they committed to? Commitment or resoluteness risks becoming an abstraction. Heidegger is partially preemptively answering that at the end of this chapter on conscience.

Dasein resolves upon what is factically possible at its time. Otherwise said, in a fantastic phrase, "Resoluteness appropriates untruth authentically" (SZ 299 / BT 345). That is, resoluteness is in the irresoluteness of the They. But it does not acquiesce in that untruth; rather, it appropriates it authentically. It makes their decision my decision. The time and place that resolute Dasein seizes hold of is the situation: "The existential attributes of any possible resolute Dasein include the items constitutive for an existential phenomenon which we call a '*Situation*' and which we have hitherto passed over" (SZ 299 / BT 346).

Situation, as he notes immediately, is a spatial term. It's an idea of a situation as a space, as a location. The idea here, which is an important idea, is rather than emptily drift in the spatiality of being-in-the-world—average everydayness—I seize hold of that space authentically in a situation and make it into my space. This is an important thought. We have an idea of space as a spatiality or aroundness of average everyday being-in-the-world, which is characterized by de-severance and directionality, all that good stuff that we saw in division 1, chapter 3. But here in division 2 we get the idea of the situation as the claim that we could authentically seize hold of space and turn it into something that is genuinely ours. I think this is the way Sartre will pick up the idea. It is a concept with very direct moral, social, and political implications.

This is also the way in which Alain Badiou will adapt the idea of situation into his concept of the *event*.[18] An event is something where one comes to possess a place where one lives, where one loves, where one works, where one *is*. And rather than just experience that space as just where one happens to have ended up or where one happens to be factically thrown, one can seize hold of that place as a situation, as an event. This is the logic of occupation as a political strategy. By contrast the they-self has no situation, the they-self is just adrift in the aroundness of average everydayness. And this leads to an important and interesting conclusion—namely, that resoluteness is not some empty truth of the "transcendental subject." Resoluteness is the name for a specific engagement in a concrete and specific situation. Resoluteness becomes real in a concrete situation. Look at

18. See Alain Badiou, *Being and Event*, trans. Oliver Feltham (Continuum, 2005).

this last quote that I'll give you, it's a really good one: "This phenomenon which we have exhibited as 'resoluteness' can hardly be confused with an empty '*habitus*'"—a mode of life—"or an indefinite 'velleity'" (SZ 300 / BT 347). Velleity means a wish, desire, or inclination. Resoluteness must not be confused with an empty mode of life or an indefinite wish. He goes on:

> Resoluteness does not first take cognizance of a Situation and put that Situation before itself; it has put itself into that Situation already. As resolute, Dasein is already *taking action*. The term "take action" is one which we are purposely avoiding. For in the first place this term must be taken so broadly that "activity" will also embrace the passivity of resistance. In the second place, it suggests a misunderstanding in the ontology of Dasein, as if resoluteness were a special way of behaviour belonging to the practical faculty as contrasted with one that is theoretical. (SZ 300 / BT 347)

Dasein does not present itself with a situation and then act within it. Dasein is already taking action. Although again note the logic of Heidegger's avoidances. He wants to avoid the notion of activity. Also notice that fascinatingly teasing remark about the passivity of resistance. Of course, resistance is not just passive. To resist is to resist, even the way Herman Melville's Bartleby the Scrivener resists when he quietly insists, "I would prefer not to."[19] There can be passive resistance, but it's still a resistance. Bartleby is still doing something by preferring not to do anything. But Heidegger doesn't want

19. Herman Melville, *Bartleby the Scrivener: A Story of Wall Street* (Melville House, 2004).

to insert that idea of situation into a simple activity/passivity distinction or a theory/practice distinction.

It seems to me that the concept of situation, which is left undeveloped by Heidegger, can be thought of in a number of different ways.

1. It could be compared to prudence, practical wisdom, or *phronesis* in Aristotle—namely, the capacity for discrimination, moral judgment, or doing the right thing in the right way at the right time and place. The person who knows what to do, called the *phronimos* by Aristotle.

2. It could be thought of, with Hubert Dreyfus, as pragmatic expertise, possessing expert judgment, like being a highly skilled driver or being able to participate expertly in a specific practice, like playing baseball. (Dreyfusians like John Haugeland are very fond of baseball analogies.)[20]

3. It could be conceived as radical political action, in the sense of the situation as something decided upon by the general or collective will. This is the political sense of the event in Badiou (e.g., when students blocked police from entering the Sorbonne in May 1968), which follows from Sartre's idea, in the *Critique of Dialectical Reason*, of "the fused group" as something that takes shape in a pledge or oath, an act of shared resolution (such as in the aftermath of the French Revolution), which is opposed by

20. See Hubert L. Dreyfus and Stuart E. Dreyfus, *Mind over Machine: The Power of Human Intuition and Expertise in the Era of the Computer*, with Tom Athanasiou (Free Press, 1986).

Sartre to seriality, a line of people, a bus queue, say—which is closer to Heidegger's idea of the they-self.[21]

4. It could also be thought about in relation to the Situationism of Guy Debord and Raoul Vaneigem,[22] where *das Man*, or the They, is the society of the spectacle, the flat audience of anonymous consumers that can be transformed by the seizing hold of the spectacle in political action through strategies of what Debord called *détournement*, rerouting, disfiguring, hijacking, hacking, culture-jamming or whatever. These days, hacking is even sort of twee. As the Situationists used to say, "Under the paving stones, it's a beach."

But let's leave the idea of the situation hanging in the air, because its meaning is uncertain, and we are at the end of the chapter.

We have achieved, first, an understanding of totality or Dasein as being a whole that's answered with the analysis of being-towards-death. And we've also answered the question of authenticity that is given to us in conscience. What we have to do now is to bring together those two concepts. To cut to the chase: The analysis of being-towards-death led to the concept of anticipation. The analysis

21. Jean-Paul Sartre, *Theory of Practical Ensembles*, vol. 1 of *Critique of Dialectical Reason*, trans. Alan Sheridan-Smith, ed. Jonathan Rée (Verso, 1960), book 2, chapter 1, pp. 345ff.

22. See Raoul Vaneigem, *Treatise on Living for the Use of the Young Generation*, trans. Donald Nicholson-Smith (Practical Paradise Publications, 1972); and Guy Debord, *The Society of the Spectacle*, trans. Ken Knabb (Bureau of Public Secrets, 2014).

of conscience leads to the idea of resoluteness. What Heidegger is going to do in division 2, chapter 3, is to bring together anticipation and resoluteness into an idea of *anticipatory resoluteness*, which is going to define the authentic being of Dasein. After that, there are some very interesting remarks on method and hermeneutic violence, before Heidegger gives his full statement of the nature of the self by contrasting it to Kant. Then, in a separate discussion, we will come to the high point of the Dasein-analytic, and in many ways the culmination of the book: *ecstatic temporality*. Our destination is ecstasy.

13. Anticipatory Resoluteness

What do death and the concrete situation of action have in common? — A puzzling footnote on sin — Holding-for-true as the certainty of conscience — Resoluteness is repetition — How death has power over Dasein's illusions — Sober anxiety — The presence of Freud(e) in Being and Time *— Joy — Discourse on the method — How we must do violence to ourselves in order to become authentic — Virtuous circularity — How do we conceive the unity of the self? — Heidegger's relation to Kant's "I think" — How Kant is both right and wrong — Kant's response to Hume on the self —* Cogito *without an* ergo sum *— Self as activity (Fichte) — Heidegger on the constancy of the self — Dasein is itself in the silent resoluteness of action in a situation — The constancy of the self consists in repetition — Performance — Descartes's dilemma with the* cogito *and Heidegger's attempt to think the self without ground — We are our acts, nothing more.*

HEIDEGGER NOW BRINGS TOGETHER chapters 1 and 2 of division 2, the discussion of being-towards-death and the discussion of conscience. He brings those two concepts together in the notion of "anticipatory resoluteness." We will first explain the meaning of anticipatory resoluteness, and then there's a short, but quite fascinating, digression on method in paragraph 63. This is followed by a very interesting discussion of the question of the self that brings out Heidegger's relation to Kant's conception of subjectivity in paragraph 64. It is the most far-reaching account of the self that Heidegger provides in *Being and Time*. That's what we're going to cover here. Looking ahead, once we finish chapter 3, division 2, then we have really completed the Dasein-analytic in the mode of authenticity. But things are far from over. Once the analysis of Dasein's authentic being-a-whole is completed, then the task is one of *repeating* the analytic of everydayness from the perspective of temporality. This is the task of chapter 4, division 2. And don't worry, this is much more interesting than it sounds. We'll also ask the question, Why is this repetition necessary, and what is it exactly?

Paragraph 61 lays out the basic problem of the chapter, which is how to bring together the concepts of anticipation and resoluteness. The way Heidegger puts this is, "What can death and the 'concrete situation' of taking action have in common?" (SZ 302 / BT 349). Death, or being-towards-death, or anticipation, and the concrete situation of taking action, or resoluteness. How are those two things linked? The task of paragraph 62 is the putting together of the outcomes of chapters 1 and 2. Now remember that the two problems that were driving the analysis in division 2, the two problems left over from division 1 that Heidegger initiates division 2 with, are authenticity (which can be thought in connection with resoluteness) and totality (which can be thought with anticipation). How do we get Dasein into our grasp as a whole, and how do we do so authentically? This is achieved with the notion of anticipatory resoluteness. To use a key concept in *Being and Time*, namely *modalization*, being-towards-death and conscience modalize each other. They are modalities of the same basic movement. The modality of being-towards-death, which is ontological and existential, and the modality of conscience, which is ontic and existentiell. But the point is, we need both modalities.

Let's look at this from the two sides of the two concepts. Resoluteness is only possible if it is thought right through to the end, Heidegger says. If resoluteness is the manner in which Dasein takes into itself its authentic potentiality for being, then this potentiality has to be thought through in relation to death as the outermost limit of potentialization, the possibility of impossibility. Heidegger says, "Resoluteness does not just 'have' a connection with anticipation, as with something other than itself. *It harbours in itself authentic Being-towards-death, as the possible existentiell modality of its own authenticity.* This 'connection' must be elucidated phenomenally" (SZ 305 / BT 353). It's therefore only in anticipation that resoluteness becomes fully resolute because only in relation to being-towards-death does resoluteness get a grip on potentiality for being as a whole. Authentic resoluteness—and this is the point—has to be anticipatory resoluteness, it has to be resoluteness all the way to the end.

As Heidegger says, "Only anticipatory resoluteness understands the potentiality-for-being-guilty *authentically and wholly*—that is to say, *primordially*" (SZ 306 / BT 354). Note here footnote 2, which is one of those moments when the ice floe of the book breaks up. As we saw last time in the discussion of conscience, I rather like these moments. Heidegger says, "The Being-guilty which belongs primordially to Dasein's state of Being, must be distinguished from the *status corruptionis* as understood in theology." And then, a few lines further on: "The existential analysis of Being-guilty, proves nothing either *for* or *against* the possibility of sin. Taken strictly, it cannot even be said that the ontology of Dasein *of itself* leaves this possibility open; for this ontology, as a philosophical inquiry, 'knows' in principle nothing about sin" (SZ 306 / BT 496). It's a strange footnote. What's it doing there? It seems to be in the wrong chapter. This should have been in the previous chapter—division 2, chapter 2—or in the 1924 protocol on Luther.[1] Heidegger says that the analysis of being guilty says nothing either for or against the possibility of sin. Then he adds, taken strictly ("*streng genommen*," as they say in German—and you know, Husserlian phenomenology claimed to be a rigorous science, a "*strenge Wissenschaft*"), the ontology of Dasein as philosophical inquiry knows nothing about sin. This is a very peculiar pronouncement. Sin is not a rigorously philosophical concept, Heidegger seems to be saying. But if that's the case, if sin is not a rigorously philosophical concept, then how can we say that falling or guilt or anxiety *are* rigorous philosophical concepts? I think it's an open question, and, although this would be a separate discussion, which I have pursued elsewhere in a book I wrote called *The Faith of the Faithless*,[2] Heidegger could be said to be offering a description of original sin in his analysis of guilt, where guilt is the ontological lack that defines what it means to be human.

We're still in paragraph 62, so let's pick it up from there: resoluteness discloses "the primordial truth of existence" (SZ 307 / BT 355). To such primordial truth, there must correspond a "holding-for-true." And this holding-for-true constitutes a "*Being*-certain," Heidegger says (ibid.). Being certain in German is "*gewiß sein*." And Heidegger

1. Martin Heidegger, "The Problem of Sin in Luther (1924)."

2. Simon Critchley, *The Faith of the Faithless: Experiments in Political Theology* (Verso, 2014).

here is running together a number of things. He's running together the link between conscience as *Gewissen*, and certainty as *Gewissheit* because death is something of which Dasein is certain. This means that resoluteness must be constantly certain of death. There's an appeal to certainty here, which is linked at the level of language to conscience: *Gewissheit*, *Gewissen*.

Heidegger writes, "Since resoluteness is constantly certain of death —in other words, since it *anticipates* it—resoluteness thus attains a certainty which is authentic and whole" (SZ 308 / BT 356). What this means is that the certainty of resoluteness consists in the constant reiteration of its being-towards-death, of its finitude. And this is what Heidegger seems to mean in this passage by the allusion to repetition. Repetition, which in Heidegger's German means something like "fetching back" or retrieval, *Wiederholung*. The linking between conscience's certainty of death with repetition calls to mind the *Ayenbite of Inwyt* of Middle English, meaning literally the "again-biting of inner wit" or "re-morse of conscience."[3] Authentic resoluteness is the repetition of the certainty to itself of death. Heidegger writes, "On the contrary, this holding-for-true [*Für-wahr-halten*], as a resolute holding-oneself-free for taking back, is *authentic resoluteness which resolves to keep repeating itself*" (SZ 308 / BT 355). Note that: "resolves to keep repeating itself." "Thus in an existentiell manner, one's very lostness in irresoluteness gets undermined" (SZ 308 / BT 355–56). This is the point. If authentic resoluteness is the repetition of the certainty to itself of death, then the way in which Dasein com-

3. Dan Michel of Northgate, *Ayenbite of Inwyt*, ed. Rev. Joseph Stevenson (London: J. B. Nichols and Sons, 1855).

ports itself is through repetition and as repetition. Repetition is absolutely central here. For Heidegger, the way in which we maintain our resoluteness up to being-towards-death is through the repetition of ourselves to ourselves. I want to come back to that thought later on.

But that's just looking at things from the side of resoluteness. Resoluteness has to be certain, take conscience into itself, and repeat. What about the side of anticipation? From anticipation, it looks a little different. This takes us back to the idea of being-towards-death as ontologically or existentially correct but lacking any ontic, existentiell facticity. Heidegger tries to answer this question when he says, "We have now shown that anticipation is not just a fictitious possibility which we have forced upon Dasein; it is a mode of an existentiell potentiality-for-Being that is attested in Dasein" (SZ 309 / BT 357). "Attested," that key word in the previous chapter. Conscience is the attestation, the witness, the testimony, of being-towards-death. Anticipation is

> a mode which Dasein exacts of itself, if indeed it authentically understands itself as resolute. Anticipation "is" not some kind of free-floating behaviour, but must be conceived as *the possibility of the authenticity of that resoluteness which has been attested in an existentiell way—a possibility hidden in such resoluteness, and thus attested therewith*. Authentic "thinking about death" is a wanting-to-have-a-conscience, which has become transparent to itself in an existentiell manner. (SZ 309 / BT 357)

He makes a similar point a couple of lines further down, where he says, "*The question of the potentiality-for-Being-a-whole is one which is factical and existentiell. It is answered by Dasein as resolute*" (ibid.). Anticipation, being-towards-death, is ontologically correct. We're going to die. Resoluteness is the way in which we can find testimony for that anticipation in the form of conscience. Then there's a very powerful paragraph I want to read to you:

> Anticipatory resoluteness is not a way of escape, fabricated for the "overcoming" of death; it is rather that understanding which follows the call of conscience and which frees for death the possibility of acquiring *power* over Dasein's *existence* and of basically dispersing all fugitive Self-concealments. Nor does wanting-to-have-a-conscience, which has been made determinate as Being-towards-death, signify a kind of seclusion in which one flees the world; rather, it brings one without Illusions into the resoluteness of "taking action." Neither does anticipatory resoluteness stem from "idealistic" exactions soaring above existence and its possibilities; it springs from a sober understanding of what are factically the basic possibilities for Dasein. Along with the sober anxiety which brings us face to face with our individualized potentiality-for-Being, there goes an unshakable joy in this possibility. In it Dasein becomes free from the entertaining "incidentals" with which busy curiosity keeps providing itself—primarily from the events of the world. But the analysis of these basic moods would transgress the limits which we have drawn for the present Interpretation by aiming towards fundamental ontology. (SZ 310 / BT 357–58)

Anticipatory resoluteness is not an overcoming of death. It's rather the way in which death has a power over us, over Dasein. Death disperses all the fugitive self-concealment of the They and the falling-into what They do. Note here again the paradox of power that runs through these chapters of *Being and Time.* Anticipatory resoluteness is the way in which we can become powerful, gain power over ourselves. And that power shatters itself against its impotentialization in being-towards-death, a power in vulnerability, in memento mori. It's the possibility of impossibility, as Heidegger said two chapters back.

Anticipatory resoluteness does not signify a flight from death or a flight from the world. On the contrary, it brings one into the resoluteness of "taking action," in those inevitable quotation marks. Taking action, along with sober anxiety. And notice the emphasis upon the sobriety of anxiety. It's not drunken anxiety; it's sober, clear-eyed. Anxiety has a Protestant, abstinent temperament. Along with that, there is an unshakable joy. You heard that right: *joy.* This could be the basis for a joke. I won't make it. But, of course, I will. *Joy* in German is *Freude.* If you're interested in the relationship between Heidegger's existential analytic and psychoanalysis, you could ask the question, Well, where is Freud in *Being and Time*? Well, the name Freud means *joy*, which itself is a joke.[4] The joy here is the cheer of sober anxiety. It's the joy of assuming one's finitude. With references like this to joy, Heidegger teases us, we could say, with the possibility of other basic moods; namely, it's not just anxiety

4. On the subject of names: the name Bernstein, of which we lost Richard in 2022, means "amber." In the Netherlands, in the early nineteenth century, the French occupiers required a registry of family names for the purpose of taxing the Dutch. Perhaps as an act of passive-aggressive protest, many made up joke names that the French didn't understand. For example: Naaktgeboren (which means "born naked") and Gekkehuis (meaning "madhouse").

that's going to do the disclosive work in the existential analytic. Perhaps there's a space for joy. Perhaps, as Heidegger says in the 1929 lecture course, *Fundamental Concepts of Metaphysics*, deep boredom can do that work of disclosure, an open daylight to the taxonomy of fundamental moods.[5] Joy will come up again in the next chapter, chapter 4, just to look ahead. This rounds out paragraph 62. What we've done here is knitted together anticipation and resoluteness into this concept of anticipatory resoluteness. In doing that, we've brought together the two analyses of the first two chapters of division 2. In a sense, we've come to a halting point, a standstill in the kinetic forward drive of fundamental ontology. The momentum suddenly stops.

This brings us into paragraph 63. I'm going to skip over some of what's happening here because there's a lot to get through in this chapter. But let me make a few points. What Heidegger is really concerned with in this pause in the analysis is the question of method. And he raises the question of violence. Not physical violence but hermeneutic violence. The claim is this: that which we ourselves *are* is ontologically furthest from us, and therefore we need to wrest our authentic being from the tendency to falling that defines our everyday being-in-the-world (SZ 311 / BT 359). This wresting, this pulling of ourselves from our tendency towards falling, requires a counterpull, what is called here in German a *Gegenzug*, a counter-

5. Heidegger describes three forms of boredom: becoming bored with something, the feeling of being in limbo in regards to the time of being bored with something, and what he calls "profound boredom," which he thinks is the "fundamental attunement" of "contemporary Dasein." Boredom, far from something to be avoided, has the power to disclose metaphysical questions to us by revealing time as a problem for us. See esp. GM 199–249 / FCM 132–68.

tug, a pull or even a railway train (*Zug*) that travels in the opposite direction. And this is what Heidegger means by violence. In order to lay bare who we are authentically, we have to violently push against our tendency towards falling that defines our being. This is what's meant by doing violence, doing hermeneutical violence to ourselves, by pushing ourselves against our tendency towards falling. We could connect Heidegger's thoughts on violence here to other discussions of violence, which are a little bit earlier, in someone like Georges Sorel, *Reflections on Violence*, from 1908.[6] And then famously and more or less contemporary with Heidegger's thinking that emerges in *Being and Time*, Walter Benjamin's "Critique of Violence."[7] We could, if we wanted to, connect what Heidegger is saying about violence to those discussions of violence. But let's put that to one side. It's a big topic, a volatile topic, but the point is that authenticity requires a violent operation. It requires that we exert a violence with regard to ourselves. We can think about other contexts in Heidegger's work where he talks about violence. One context in particular is his 1935 lecture course, *Introduction to Metaphysics*, which is very important for all sorts of reasons, notably its discussion of the *Antigone*. In

6. Georges Sorel, *Reflections on Violence*, trans. T. E. Hulme and J. Roth (Dover Publications, 2012).

7. Walter Benjamin, "Critique of Violence," in *Selected Writings, 1913–1926*, ed. Marcus Bullock and Michael W. Jennings, vol. 1 of *Selected Writings*, edited by Michael W. Jennings, Howard Eiland, and Gary Smith (Harvard University Press, 1996), 236–52. Benjamin's analyses are often close to Saint Paul. Benjamin articulates in 1940 a "weak messianic power": "Like every generation that preceded us, we have been endowed with a weak messianic power, a power on which the past has a claim. Such a claim cannot be settled cheaply" (Walter Benjamin, "On the Concept of History," in *Selected Writings, 1938–1940*, ed. Marcus Bullock and Michael W. Jennings, vol. 4 of *Selected Writings*, edited by Michael W. Jennings, Howard Eiland, and Gary Smith [Harvard University Press, 2006], 390).

Introduction to Metaphysics, Dasein as the human being is understood as the violent one, as the one who wrenches free from the They, wrenches free from the rabble, the one who becomes uncanny, who pushes out into nothingness (EM 158–60 / IM 166–68). And that requires a violence with regard to ourselves. That's the first point I want to make in relation to paragraph 63.

The second point is that if fundamental ontology is violent, then, Heidegger asks, is the interpretation of Dasein arbitrary? If Heidegger's analysis is violent, well, then, does it mean it's just random or hit-and-miss and that Heidegger is just forcing this on us? Obviously the answer to the question is no. But what happens is that Heidegger backs up into a useful and slightly defensive summary of *Being and Time*. If you're feeling a bit lost, not in the they-self but in the forest of Heidegger's words, then these passages (e.g., SZ 313–14 / BT 361–62) may help. The fact that Heidegger's analysis is necessarily violent doesn't mean it's arbitrary.

The third point I want to make is about a discussion in paragraph 63, an elaboration of the problem of presuppositions in fundamental ontology and whether Heidegger is presupposing what he is analyzing, and whether he is therefore moving in a circle. Is Heidegger's pattern of reasoning in *Being and Time* circular? The answer is, of course, yes. The issue for Heidegger, as ever, is not one of avoiding circularity. The question is rather how we enter into circularity, and whether we see circularity as virtuous or vicious. You might remember that the way in which this came up earlier in *Being and Time* was through a discussion of understanding and the relationship between understanding and interpretation—namely, that our understanding is laid out through acts of interpretation. Heidegger now reminds us that this structure of interpretation, understanding, is itself circular: "In analysing the structure of understanding in general, . . . what gets censured inappropriately as a 'circle,' belongs to the essence and to the distinctive character of understanding as such" (SZ 314 / BT 362). The charge of circularity that Heidegger confronts here is based on the argument that if we presuppose "the idea of existence and of Being in general," we will end up interpreting Dasein based upon this presupposed idea, and "the idea of Being may be obtained from it," allowing us to "posit some proposition from which we deduce further propositions about the Being of Dasein." But, Heidegger counters, the "understanding projection" that this "presupposition" really amounts to does not simply allow us to deduce propositions but allows what is interpreted to

> put itself into words for the very first time, so that it may decide of its own accord whether, as the entity which it is, it has that state of Being for which it has been disclosed in the projection with regard to its formal aspects. . . . What common sense wishes to eliminate in avoiding the 'circle,' . . . is nothing less than the basic structure of care. Because it is primordially constituted by care, any Dasein is already ahead of itself. (SZ 314–15 / BT 362–63)

Circularity isn't something to be avoided, it is something to be accepted as part of who we are and entered into in the right way. Thus, Heidegger says, "We cannot ever 'avoid' a 'circular' proof in the existential analytic, because such an analytic does not do *any* proving *at all* by the rules of the 'logic of consistency'" (SZ 315 / BT 363). The question is one of approaching circularity in the right way.

Thrown projection, our pre-understood interpretation of our world, is indeed circular. Dasein as the being that possesses an understanding of its own being—this is also a circle. But Heidegger's thinking is *virtuously* circular. It allows what is understood to show itself as what it takes itself to be and decide whether it has that state of being.

Let's move on to paragraph 64, "Care and Selfhood," in which we get the most detailed discussion of the concept of the self in *Being and Time*. You might remember that I discussed this in chapter 4 when we looked at being-with, where Dasein is not just being-in-the-world but also essentially being-with-others. I said then that the subject or the self is an internal limit or frontier in *Being and Time*, which is continually crossed. Namely, the subject, the self, is a concept that Heidegger wants to avoid because it always entails some form of metaphysical thinking. But we have seen that the self is a concept that he also reinstates. The concept of the subject is something both avoided and reinstated. It is like an internal frontier in the book, a barrier that is repeatedly breached. This is the issue in paragraph 64. The unity of the care structure lies in the unity of the self. A self that is first and foremost characterized by mineness. Yet how do we conceive this unity of the self? Well, the obvious answer is, existentially. Our unity, such as it is, must be conceived of existentially. But there's a problem here, which we'll get to in a moment.

In paragraph 64 Heidegger's constant and compelling point of contrast is with Kant in the *Critique of Pure Reason*. Now, Heidegger admits, as Kant admits, as Descartes admits, that the starting point for any articulation of the self lies in saying "I." In saying "I think." The way Heidegger argues this, as we saw at the beginning of the existential analytic, is through the concept of mineness, the first concept introduced in the existential analytic. So. Kant's notion of the self says, "I think." And the "I think" is that which must accompany all my representations. The argument that Heidegger makes with regard to Kant is very simple. On the one hand, Heidegger says, Kant is true to the "phenomenal content of the I" (SZ 318 / BT 366). The "I think" is not claimed to be any substance of the self. Or, as Heidegger says, Kant's analysis arises "from a genuine pre-phenomenological experience" (ibid.).

Yet, on the other hand, Kant interprets this content, this pre-phenomenological experience of mineness, in relation to the wrong ontology. Kant has the right intuitions with the wrong ontology. Of course, the wrong ontology for Heidegger is the ontology of the present-at-hand. (Heidegger did philosophy in the formative years of his work in a context utterly dominated by neo-Kantianism.) As such, Kant almost pulls clear of the tradition. This is the way Heidegger always reads Kant: he almost pulls clear of the tradition but falls back into it nonetheless. It's a bit like the famous line from *The Godfather* spoken by Michael Corleone on leaving the mob: "Just when I thought I was out, they pull me back in!" This is Kant's relation to the philosophical tradition: he tries to leave, but he keeps getting pulled back in.

But let's look at the details a little bit more closely. Heidegger writes, "Kant makes a more rigorous attempt than his predecessors to keep hold of the phenomenal content of saying 'I'"—so far, so good—"yet even though in theory he has denied that the ontical foundations of the ontology of the substantial apply to the 'I,' he still slips back into *this same* inappropriate ontology" (SZ 318–19 / BT 366). Keep that quote in one half of your mind, and then put the next quote in the other half of your mind.

> Kant grasps the phenomenal content of the "I" correctly in the expression, "I think," or—if one also pays heed to including the practical person when one speaks of intelligence—in the expression "I take action." In Kant's sense we must take saying "I" as saying "I think." Kant tries to establish the phenomenal content of the "I" as *res cogitans* [as thinking substance, a thinking thing]. If in doing so he calls this "I" a "logical subject," that does not mean that the "I" in general is a concept obtained merely by way of logic. The "I" is rather the subject of logical behaviour, of binding together. "I think" means "I bind together." All binding together is an "I bind together." (SZ 319 / BT 367)

Let's keep both those quotes in mind. Obviously, I don't know the reader's familiarity with Kant, especially with the concept of the transcendental unity of apperception. But it's a much simpler thought than it sounds. As we saw in the discussion of reality in chapter 9, the transcendental unity of apperception is something that is logically deduced by Kant from the fact of experience. The problem here is set by David Hume, Edinburgh's favorite son. Hume sets Kant a problem in relation to what we now call "personal identity." Namely, that our personal identity, Hume says, consists merely in bundles of perceptions that do not seem to be gathered together into any unity, into any substantial idea of the self.[8] Now, Kant accepts Hume's challenge and makes the argument that in order to explain the fact that experience is bound together, that experience hangs together, our experiences are *our* experiences—in order to explain that, we have to *logically* presuppose that these experiences

8. David Hume, *A Treatise of Human Nature*, 251–62.

are happening to a self. This is what Kant calls the transcendental unity of apperception.[9] But we cannot *know* that self in the way in which a philosopher like Descartes might claim indubitable and transparent self-knowledge.

Let's look at the outcome of this analysis. Heidegger writes:

> Kant's analysis has two positive aspects. For one thing, he sees the impossibility of ontically reducing the "I" to a substance; for another thing, he holds fast to the "I" as "I think." Nevertheless, he takes this "I" as subject again, and he does so in a sense which is ontologically inappropriate. For the ontological concept of the subject *characterizes not the Self-hood of the "I" qua Self, but the selfsameness and steadiness of something that is always present-at-hand.* To define the "I" ontologically as "*subject*" means to regard it as something always present-at-hand. The Being of the "I" is understood as the Reality of the *res cogitans.* (SZ 319–20 / BT 367)

Conscience gives Dasein its self-constancy. I become Nietzsche's "animal that can keep promises."[10] Without self-constancy, I could be anyone. We can give one cheer or maybe two cheers for Kant, but decidedly not three. In a fascinating, long footnote, Kant is said to hang on to the phenomenal, the pre-phenomenological, experience of the "I think" (SZ 320 / BT 496–97). The great virtue of Kant's approach is that for Kant there is a *cogito* without an *ergo sum*, there is an "I think" without a "therefore I am." Namely, the fact that the "I think" does not entail for Kant any claim to the

9. Immanuel Kant, *Critique of Pure Reason*, A106–8, A117.

10. See Friedrich Nietzsche, *On the Genealogy of Morals*, esp. 57–60.

substantiality or reality of the self as a metaphysical entity.[11] And that claim about the "I think" in Kant, as a subject that is defined through the activity of thought, is the concept that Kant's most brilliant and influential successor, Johann Gottlieb Fichte, will develop in his *Science of Knowledge*, the *Wissenschaftslehre*, when he talks about the subject as an activity.[12] The "'I' is 'I'" in Fichte, the first principle of philosophy, is an idea of thought as activity, which is the way in which Fichte tries to unify theoretical and practical reason and restore the idea of intellectual intuition as action. That's a separate topic.

Kant's virtue lies in positing the "I think" as a *cogito* without an *ergo sum*—there's no substantiality to the self. Yet, negatively, Kant still moves within an ontology of the present-at-hand. Which means that he slips back into the position of an isolated subject. This is what happens when you work with the ontology of the present-at-hand. You slide back willy-nilly into subjectivism and you miss the *phenomenon* of the world, the way the world shows itself as ready-to-hand. Heidegger makes this point: "Kant did not see the phenomenon of the world, and was consistent enough to keep the 'representations' apart from the a priori content of the 'I think.' But as a consequence the 'I' was again forced back to an isolated subject, accompanying representations in a way which is ontologically quite indefinite" (SZ 321 / BT 368). There are ways of challenging this approach to what Heidegger is doing here as a reading of Kant. In

11. For Descartes, I can doubt everything except that I am doubting. What about when I am sleeping? Then God watches over me, continuously creating me while I sleep. According to Berkeley, to be is to be perceived. But for Heidegger, it is possible for us to stop being ourselves.

12. Johann Gottlieb Fichte, *The Science of Knowledge*, trans. and ed. Peter Heath and John Lachs (Cambridge University Press, 1982).

particular, the way in which the subject is described by Kant in relation to practical reason, the subject of the kingdom of ends, the moral subject that is defined by respect for persons: this could lead one to read an intersubjective subject in Kant. This would not be a wild and crazy thing to say, that Kant could be redescribed as a thinker of intersubjectivity, of being-with. But Heidegger's critique is that although Kant gets it right pre-phenomenologically, he falls back into an isolated subject because he has the wrong ontology of the present-at-hand.

If we assume that this is right about Kant, then how are things different for Heidegger's approach? Well, this is where we encounter a problem. Dasein is defined as being-in-the-world, as I've said a thousand times. But this means that Dasein understands itself in terms of the world in such a radical fashion that Dasein falls into the world and it becomes a they-self. I is an other. *Je est un autre*. In the world of the They, in the world of *das Man*, the one, I am them, I am at one with them. "I think" what they think. All of that stuff that we've seen before.

How does this play out? Heidegger says: "What is the motive for this 'fugitive' way of saying 'I'? It is motivated by Dasein's falling; for as falling, it *flees* in the face of itself into the 'they.' When the 'I' talks in the 'natural' manner, this is performed by the they-self. What expresses itself in the 'I' is that Self which, proximally and for the most part, I am *not* authentically" (SZ 322 / BT 368). Dasein's selfhood is only conceivable existentially in the mode of authenticity. Most of the time, we lose ourselves in the world. When we pull ourselves back from the world, authentically, we can achieve something like a self-constancy. Let me offer a long, but important quote here:

> Selfhood is to be discerned existentially only in one's authentic potentiality-for-Being-one's-Self—that is to say, in the authenticity of Dasein's Being *as care*. In terms of care the *constancy of the Self*, as the supposed persistence of the *subjectum*, gets clarified. But the phenomenon of this authentic potentiality-for-Being also opens our eyes for the constancy of the Self in the sense of its having achieved some sort of position. *The constancy of the Self*, in the double sense of steadiness and steadfastness, is the *authentic* counter-possibility to the non-Self-constancy which is characteristic of irresolute falling. Existentially, "*Self-constancy*" signifies nothing other than anticipatory resoluteness. The ontological structure of such resoluteness reveals the existentiality of the Self's Selfhood.
>
> Dasein is *authentically itself* in the primordial individualization of the reticent resoluteness which exacts anxiety of itself. *As something that keeps silent*, authentic *Being*-one's-Self is just the sort of thing that does not keep on saying "I"; but in its reticence it "*is*" that thrown entity as which it can authentically be. (SZ 322–23 / BT 369–70)

The persistence of the self does not refer to any notion of the subject, the subject as *hypokeimenon*, as *subjectum*, as ground, as foundation, as metaphysical first principle. The persistence of the self is simply the experience of Dasein's anticipatory resoluteness. The self is constant insofar as it manifests itself as the counterpossibility, even the counterthrust, against irresolute falling. This goes back to the idea of this countertug that we saw in the discussion of method. The constancy of the self manifests itself in a *violent* counterthrust against irresolute falling, in wrenching ourselves away from the They. Dasein is constantly itself in the reticent resoluteness that is its anxiety as something or someone silent, that does not just keep on saying "I." Rather, the constancy of the self does not consist in endless stories about me, myself, and I. It consists in the silent resoluteness of action in a situation. That's the claim that Heidegger is making. A claim for the constancy of the self as a countermovement against the falling of the self into the They. And this is the problem: this constancy is not constant. This constancy perdures only as long as Dasein authentically is. Which is not for all time, but for a while, for a moment, even if that might be a long moment. For the duration of what Heidegger will soon call the "moment of vision."

Maybe you can begin to see the problem that is appearing at this point in our analysis of Heidegger: the cost of giving up any substantial notion of the subject and the metaphysical concepts of the subject is that the self becomes defined in terms of the unity and constancy that it ever only fleetingly is. The rest of the time it is not itself, but falls. Dasein is inconstant, and I fall to the world. I am, proximally and for the most part, an inconstant self that becomes constant only when I become authentic.

Think about this, because it's an interesting philosophical thought and a compelling existential and personal thought. In what does the unity or constancy of the self consist? Think about the idea that Kant was true to the phenomenal content of the "I," the appearance of the "I think" to itself, but false in attributing that content to a self defined in terms of the ontology of the present-at-hand. For Heidegger, we must think the unity of the self existentially, without reference to any substantialist ontology or any reference to God. The self consists in nothing else but this *performance* of itself, its self-affirmation. This is a little like the subject in Descartes, which is also a performance, the performance of Descartes's thinking in the *Meditations* and the *Discourse on the Method*. But that is done in the mode of the present-at-hand, the mode of *Vorhandenheit*. Heidegger is trying to present, to show, the consistency of the self through the anxious, resolute certainty of itself in conscience. Of course, to solve the problem of self-constancy, Descartes had to introduce the idea of God.

For Descartes, everything can be doubted. In the *Meditations*, he declares that "I shall proceed towards the destruction of all my opinions."[13] But I cannot doubt the existence of the doubter who doubts. And doubting is thinking. A doubting being is a thinking being. And that thinking is what my existence consists in. That's the proof of the *cogito*. But what happens when the *cogito* sleeps? Well, when I'm sleeping and I'm not thinking, am I therefore nonexistent? Well, this is one of the reasons why Descartes makes the appeal to God. God is that being who underwrites the constancy of myself. My *cogito*, my "I think," is underwritten by the constant presence of God. This is the doctrine of *continuous creation*, that God is continuously creating and re-creating the universe and therefore underpins us. For Descartes, this is the thought: Our self-constancy doesn't just consist in the performance of my saying "I think therefore I am" but lies in the fact that that "I think therefore I am" rests upon the power of God. God is the guarantor for the *cogito*.

Heidegger is trying to solve the problem of self-constancy without reference to God. Which means that he is in Descartes's dilemma

13. René Descartes, *Meditations on First Philosophy*, in vol. 2 of *The Philosophical Writings of Descartes*, trans. John Cottingham, Robert Stoothoff, and Dugald Murdoch (Cambridge University Press, 1984), 50–62.

in the *Meditations*. Namely, that the only unity of the self consists in the act whereby it asserts or performs that unity, it performs that constancy. (For Heidegger, however, there is no reference to God. The self is just the performance of itself; it has no guarantor, it only exists in vouching for itself.) For the rest of the time, the self is inconstant and is lost in the they-self.

In other words, there is no deus ex machina. There's no God we can pull out for Heidegger like there is for Descartes. God is not a rabbit in a hat. Let's assume that Heidegger is right. We are selves in some way. Most of the time we fall to the world. We're distracted, we're asleep. We're watching television, we're drunk, or whatever. And our self-constancy consists in our resoluteness for Heidegger, resoluteness thought all the way to the end, in relation to our being-towards-death. Once we lose the idea of God or the idea of any substantial underpinning of the self, then we're left in the following position: The constancy of the self consists in nothing more than the act whereby the self affirms itself in anticipatory resoluteness and in the repetition of that act.

The self is only the performance of itself. The self is a performative self, we're only authentic, only constant, for the time we decide to be resolute. For the rest of the time we fall, and the self is not itself. It is them. It is anyone. As we will see in the next chapter of *Being and Time*, "Temporality and Everydayness," the self that does not repeat itself but forgets itself, is no longer an authentic self. The self is constant only in and through repetition. That's why Heidegger writes, "*Authentic resoluteness* . . . resolves to keep repeating itself" (SZ 308 / BT 355).

Heidegger's position on the self is that we are selves, we have the possibility of a constancy of the self insofar as we anticipate our being-towards-death, and we internalize that through our resoluteness. But once that self-constancy is released through falling to the world, then that self-constancy dissolves and we are nothing more than that. We're nothing more than the acts, the performances that we undergo. There is nothing outside the performance of ourselves and the assurance that we give ourselves when we're being ourselves, that we are those selves. The rest of the time we fall, we flee, we disperse. In a deep sense, we *are* not.

That will do for anticipatory resoluteness. I'm going to come back and complete division 2, chapter 3, with a discussion of ecstatic temporality, which in many ways is the high point of division 2 and arguably of *Being and Time* as a whole.

14. Ecstatic Temporality

Unwavering discipline — Authentic Dasein is the unity of the three ecstases *of time — The three-step rhythm of temporality — The future as coming towards itself — What comes out of the future is my having-been, the past — The past is null, but not annulled — The curtains lift on action — The future opens up the past and releases itself into the present — The moment of vision, the glance of the eye —* Kairos *as the fulfillment of time (without Christ) — How can time be said without metaphysics? — Verbalizing nouns, speaking in tautology — Temporality temporalizes itself — Dasein is the ecstatic one (Taminiaux's story) — The meaning of finitude — End-ly time — Towards oneself! Existing as the nullity that I am — Sketch of a critique of Heidegger on time: the primacy of the past over the future.*

THE PAGES OF PARAGRAPH 65 are written with what Heidegger calls "*unwavering discipline*" (SZ 321 / BT 370). Interesting wording: unwavering discipline. Heidegger's concept of "ecstatic temporality" is the apex of the existential analytic and the culmination of the book. Let's focus on the key thought in ecstatic temporality, that authentic Dasein is the movement of temporality, the unity of the three *ecstases*. It is this lived stretch of past, present, and future; of having-been, waiting towards, and the to-come. Although for Heidegger, as we'll see, the primary phenomenon of authentic time is the future. The unity of the three *ecstases* is viscerally felt through the compact power of these pages. There is a concentration of thought here that is amazing. This is a schema (fig. 3) to accompany our discussion of paragraph 65.

The Structure of Ecstatic Temporality (fig. 3)

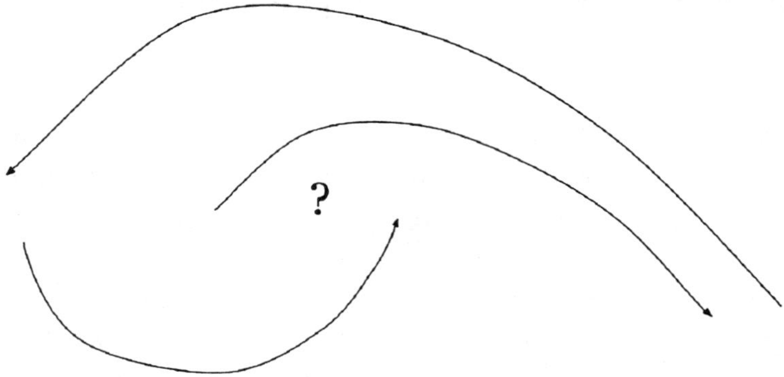

Past	**Present**	**Future** *(Zu-kunft)*
Having-been-ness *(Gewesenheit)*	Waiting-towards *(Gegenwart)*	To-come *(Zu-kommen)*
Guilt—Resoluteness	Moment of vision *(Augenblick)*	Death—Anticipation
	Situation—"Taking Action"	

This structure is a unity, the unity of three *ecstases*, and it temporalizes itself. Dasein "is" the movement of this temporalization. This movement is the experience of finitude.

As we've gone over previously, care is the being of Dasein. We know that. But the *meaning* of care is temporality. And "meaning" is understood here as it was in division 1, in the discussion of understanding and interpretation, namely as that upon which care is based. The question is, What does care mean? Care is the being of Dasein, but what is it that makes possible the unity of the articulated structure of care? Remember, care is composed of three elements: Being-ahead-of-itself, or projection. Being-already-in, or facticity. And being-alongside-a-world, or falling. And the unity of those three elements is temporality, Heidegger claims.

Let's dig in here, follow the text closely, and see how Heidegger unpacks this claim. And the text has a rhythm. It's a one-step, two-step, three-step choreography. Future, past, present. The three dimensions, the three steps of time. It is in temporality that we can think of the movement, the *kinesis*, of Heidegger's analysis. There will be a few long quotes here, but we need them. We need to see exactly what Heidegger says:

> That which was projected in the primordial existential projection of existence has revealed itself as anticipatory resoluteness. What makes this authentic Being-a-whole of Dasein possible with regard to the unity of its articulated structural whole? Anticipatory resoluteness, when taken formally and existentially, without our constantly designating its full structural content, is *Being towards* one's ownmost, distinctive potentiality-for-Being. This sort of thing is possible only in that Dasein *can, indeed,* come towards itself in its ownmost possibility, and that it can put up with this possibility as a possibility in thus letting itself come towards itself—in other words, that it exists. This letting-itself-*come-towards*-itself in that distinctive possibility which it puts up with, is the primordial phenomenon of the *future as coming towards*. If either authentic or inauthentic Being-towards-death belongs to Dasein's Being, then such Being-towards-death is possible only as something futural, in the sense which we have now indicated, and which we have still to define more closely. By the term "futural," we do not here have in view a "now" which has not yet become "actual" and which sometime will be for the first time. We have in view the coming [*Kunft*] in which Dasein, in its ownmost potentiality-for-Being, comes towards itself. (SZ 325 / BT 372–73)

In authenticity, in its potentiality for being a whole, in anticipatory resoluteness, Dasein comes towards itself. This coming towards itself is the primary meaning of the future, where the future is not understood as the now that is not yet but as the coming, the *Kunft*, the futurity that Dasein is always already moving towards, coming towards. Notice here—and again, Heidegger is playing with the possibilities of his language, German (and we could play with the possibilities of English)—the link he's making between *Zukunft*, the to-come of the future, and *zukommen*, or coming-towards. *Zukunft* and *zukommen*. The future is something we come towards. But what Dasein takes over in the future is its being guilty. What it takes over is its primary ontological lack, what was called in the last chapter, chapter 2 of division 2, the "nullity of its thrown basis." Dasein comes towards itself in the future. What it's taking over in this coming towards is its being guilty, the nullity of its thrown basis, which it already was. I come towards my nullity. Heidegger writes:

> Anticipatory resoluteness understands Dasein in its own essential Being-guilty. This understanding means that in existing, one takes over Being-guilty; it means *being* the thrown basis of nullity. But taking over thrownness signifies *being* Dasein authentically *as it already was*. Taking over thrownness, however, is possible only in such a way that the futural Dasein can *be* its ownmost "as-it-already-was" —that is to say, its "been" [*sein "Gewesen"*]. (SZ 325–26 / BT 373)

Dasein can be its *been*. Heidegger is aware of the grammatical awkwardness of this way of expressing things. But this is the thought.

> Only in so far as Dasein is as an "I-*am*-as-having-been," can Dasein come towards itself futurally in such a way that it comes *back*. As authentically futural, Dasein is authentically as "*having been.*" Anticipation of one's uttermost and ownmost possibility is coming back understandingly to one's ownmost "been." Only so far as it is futural can Dasein *be* authentically as having been. The character of "having been" arises, in a certain way, from the future. (SZ 326 / BT 373)

Listen to that. It's a strange thought. "The character of 'having been' arises, in a certain way, from the future." What is it that comes out of the future, that we can come towards? What comes out of the future is myself in my having-been-ness. My *bin*—my *ich bin*, my "I am"—is *been*. I am my having-been. There's a tricky thought at work here. I come towards myself futurally in anticipation. But I come towards myself in a way that I come back to what I have been. In other words, projection throws forth from its thrown basis. *My pastness comes to me from out of the future.* I do not, as is clear from the analysis of chapter 2 on conscience, have power over this null thrown basis. The past is not annulled. It is null, but it is not annulled. All that changes is the way in which I seize hold of the past. What comes out of the future is my having-been-ness, which I can take over, take into myself. In conscience we have a nullity of projection and also a null basis of our thrownness; we hover between these two nullities. And now we can see how this dual nullity has a temporal dimension through and through.

That's the future and past. Which leaves us with the present. Here is the key quotation:

> Anticipatory resoluteness discloses the current Situation of the "there" in such a way that existence, in taking action, is circumspectively concerned with what is factically ready-to-hand environmentally. Resolute Being-alongside what is ready-to-hand in the Situation—that is to say, taking action in such a way as to let one encounter what *has presence* environmentally—is possible only by *making* such an entity *present*. Only as the *Present* [*Gegenwart*] in the sense of making present, can resoluteness be what it is: namely, letting itself be encountered undisguisedly by that which it seizes upon in taking action. (SZ 326 / BT 373–74)

The first thing to note here is the way in which Heidegger usually puts quotation marks around the word "action." Here in this passage, you see those quotation marks lift like the curtains in a theater, and we're presented with simply taking action. One could track this process in Heidegger. He will use a word that he was employing under erasure (*sous rature*, as Jacques Derrida would say[1]) or using cautiously, and suddenly that caution begins to evaporate as the momentum of the book accelerates, as it achieves maximum flight velocity. Heidegger is writing here without a draft, and you can feel the force of the thinking that's happening.

What anticipatory resoluteness discloses is Dasein in the situation of its there, which it has seized hold of in the form of taking action. That is, Dasein is being-alongside. It is being-alongside not in the inauthentic mode of just tarrying along thoughtlessly. It is alongside in the mode of seizing hold and making entities present.

1. Jacques Derrida, *Of Spirit: Heidegger and the Question*, trans. Geoffrey Bennington and Rachel Bowlby (University of Chicago Press, 1989).

This making-entities-present is *Gegenwärtigen*, which is linked to the idea of *Gegenwart*, the present, so that means making things present in the present. Look at the schema carefully and you see the future, the to-come, which opens up onto having-been-ness and releases into the present.

Heidegger brings these three dimensions of time together, these three tenses, with his usual philosophical formality (i.e., ugly language). But there is theater here. There is a beautiful mise-en-scène in these pages of *Being and Time*. Now we're going to pull the three *ecstases* of temporality together. In the next paragraph, we read the following sentence: "Coming back to itself futurally, resoluteness brings itself into the Situation by making present" (SZ 326 / BT 374). Coming back to itself futurally: the to-come of the future. Resoluteness brings itself into the situation: the past. And Dasein does this by making present. The three dimensions of temporality are operative in that sentence.

> The character of "having been" arises from the future, and in such a way that the future which "has been" (or better, which "is in the process of having been") releases from itself the Present. This phenomenon has the unity of a future which makes present in the process of having been; we designate it as "temporality." Only insofar as Dasein has the definite character of temporality, is the authentic potentiality-for-Being-a-whole of anticipatory resoluteness, as we have described it, made possible for Dasein itself. *Temporality reveals itself as the meaning of authentic care.* (ibid.)

"Authentic care"—we've not seen that expression before. Care has just been described as the being of Dasein. Neither authentic nor inauthentic. But here we have an idea of authentic care.

Now, what is fascinating here in these passages is the thought that what comes out of the future is our having-been. What the future opens is our past, which releases itself in the present and as the present. The having-been that comes out of the future makes present. And it's this threefold structure that Heidegger calls temporality. This is the meaning of authentic care. Heidegger then makes some remarks distinguishing authentic from inauthentic time, and that will become clearer in the next chapter of *Being and Time*, "Temporality and Everydayness."

Heidegger folds the structure of care into the structure of temporality. I think this is one of the beautiful things that he does in this book: namely, that we are presented with these structures, we isolate these formal analytical matrices, and then they're mobilized and combined, almost musically, symphonically, with increasing acceleration and momentum. Care is being-ahead-of-itself, being-already-in, and being-alongside, as we have seen. That being-ahead-of-itself is the future, is projection, is existentiality. Being-already-in is having-been, is thrownness, is facticity. And being-alongside is the present as authentic life in the world. Care is temporalized.

You can see how the threefold structure of care is now the threefold structure of temporality. And those three *ecstases* are a unity, but the primary meaning of temporality, authentic temporality, is the future.

It is at this point that Heidegger introduces the notion of the moment of vision: the *Augenblick*, the glance of the eye or the twinkling of the eye. What is Heidegger doing with this concept? He takes it from Kierkegaard. In Danish, it is *øjeblik.* The term may be originally taken from Luther in his translation of Saint Paul. It's a term that translates the idea of *kairos* in Greek, which can be thought of as the right moment, the opportune moment, the passing instant that is seized hold of (*le bon moment* in French). *Kairos* was particularly important in Saint Paul and in early Christianity, and then again in Protestantism, which is essentially a revival of Paul. The *kairos* was the moment of the appearance of Christ, and the *kairos* —the moment of vision—was the fulfillment or the redemption of time in the Gospels.

Remember that, for a Christian, the Messiah is the fulfillment of time and the redemption of the world from sin and death. The way in which time as the time of the past, as the time of sin, can be fulfilled and redeemed is through the *kairos* in which Christ appears. It is the possibility of life, of salvation, and of healing. For Heidegger (and this can be seen in the schema provided at the beginning of this discussion), the future opens up—and, in a sense, creates—the past. As will not be surprising, instead of Christ appearing, what shows itself as the fulfillment of time, in Heidegger, is ecstatic temporality. Namely, it's *me* as the fulfillment of time. It's Dasein as the fulfillment of time. This is what Heidegger will call, in the next chapter, in another powerful word, the experience of *rapture*. He writes:

> When resolute, Dasein has brought itself back from falling, and has done so precisely in order to be more authentically "there" in the "moment of *vision*" as regards the Situation which has been disclosed.
>
> Temporality makes possible the unity of existence, facticity, and falling, and in this way constitutes primordially the totality of the structure of care. The items of care have not been pieced together

> cumulatively any more than temporality itself has been put together "in the course of time" ["*mit der Zeit*"] out of the future, the having been, and the Present. Temporality "is" not an *entity* at all. It is not, but it *temporalizes* itself. Nevertheless, we cannot avoid saying, "Temporality 'is'... the meaning of care," "Temporality 'is'... defined in such and such a way"; the reason for this can be made intelligible only when we have clarified the idea of Being and that of the "is" in general. Temporality temporalizes, and indeed it temporalizes possible ways of itself. These make possible the multiplicity of Dasein's modes of Being, and especially the basic possibility of authentic or inauthentic existence. (SZ 328 / BT 376–77)

Taken strictly, and again, note the logic of avoidance, we cannot avoid saying that time is this or that. The problem is that when we do this we are reifying time. But time by definition is movement and is *only* as movement. If we say time is X or Y, we're substantializing time. We're already providing an answer to the being of time and falling into a metaphysics of time, which is precisely the thing that Heidegger wants to avoid. Heidegger says, in German, "*die Zeitlichkeit zeitigt sich*" (temporality temporalizes itself) (SZ 331 / BT 380). And it temporalizes possible ways for it to be. Time thought as temporality is its auto-temporalization, where time produces itself from itself in a sort of immaculate conception, we might say. The idea that time temporalizes itself means that there is no being outside of time, determining the meaning of time. That was traditionally the case within theology, in the way in which time was understood in relation to eternity, which was prior to and more important than time.

Heidegger's thought here is that temporality is this movement of the unfolding of the three *ecstases* of time. How does one freeze that motion in words? How does one express movement in words? That's a problem that Heidegger wrestles with here and indeed throughout his long philosophical itinerary. One way in which he addresses it is by verbalizing nouns. If I say "temporality is X or Y," then the copula suggests a substance. Time becomes reified. One way around that problem is to verbalize the noun. To say that "time temporalizes itself" rather than "time is temporality." For Heidegger, that which is, being, is verbal. It's movement, it's temporal. But how does one say that? It's a problem that will beset Heidegger's work from beginning to end and that he expressed in different ways. Here he says, "*die Zeitlichkeit zeitigt sich*" (temporality temporalizes itself) in the three *ecstases* of time. In his later work, he writes, "*die Welt weltet*" (the world worlds; UK 30–31 / OWA 43), or "*das Ereignis ereignet*" (the appropriative event appropriates itself; ZS 29 / TB 24[2]). And so on, so forth. You get the idea.

This is a way of speaking tautologically, which is really quite interesting if you think it through. I raised the topic in one of the earlier chapters. There is something about a tautology, which through its repetition, through its verbalization of a noun, allows us to push through to some different experience of language. Phenomenology might best proceed through tautologies (an idea explored in Heidegger's work by Jean-François Courtine[3]). Heidegger wants to make the nouns move, even dance and groove.

2. Stambaugh's translation of this sentence reads "Appropriation appropriates."

3. Jean-François Courtine, "Phenomenology and/or Tautology," in *Reading Heidegger: Commemorations*, ed. John Sallis (Indiana University Press, 1993), 241–57.

Now we've arrived at the thought of ecstatic temporality. Let's read the key passage.

> The future, the character of having been, and the Present, show the phenomenal characteristics of the "towards-oneself," the "back to," and the "letting-oneself-be-encountered-by." The phenomena of the "towards . . . ," the "to . . . ," and the "alongside . . . ," make temporality manifest as the ἐκστατικόν [*ekstatikon*] pure and simple. *Temporality is the primordial "outside-of-itself" in and for itself.* We therefore call the phenomena of the future, the character of having been, and the Present, the "ecstases" of temporality. Temporality is not, prior to this, an entity which first emerges from itself; its essence is a process of temporalizing in the unity of the ecstases. What is characteristic of the "time" which is accessible to the ordinary understanding, consists, among other things, precisely in the fact that it is a pure sequence of "nows," without beginning and without end, in which the ecstatical character of primordial temporality has been levelled off. (SZ 328–29 / BT 377)

Temporality is the *ekstatikon* itself, the outside of itself. And it is composed of these three *ecstases*. Let me tell you a story that I haven't told you before. I was very influenced by a Belgian philosopher called Jacques Taminiaux, who used to teach at Leuven in Belgium. I met him when I was a graduate student, and he made a big impression on me because he wrote clearly and deeply about difficult thinkers, and he told terrific stories. He talked to me once about being in a seminar with Heidegger in the late 1960s with a bunch of French interpreters of Heidegger's work and the poet René Char at the latter's house at Le Thor, in Provence. They were there to discuss

Heidegger's project in *Being and Time*. According to Jacques (he always came out well in the stories he told), the seminar began with Heidegger going around the table—there were only about six or seven people there, with participants including my teacher, Dominique Janicaud, and others like Barbara Cassin, Giorgio Agamben, Kostas Axelos, and Jean Beaufret. Heidegger asked each of them what they thought *Being and Time* was about. And you can imagine that this puts you on the spot. You might reply, Well, Herr Heidegger, it's about being. But Heidegger says, No! And someone else says, It's about Dasein. No!, shouts Heidegger. It's about being-in-the-world. No! *Nein!* And he went round the table like this until everybody had given the wrong answer. And then Heidegger said, *Dasein ist das Ekstatische*. Dasein is the ecstatic one. That's the message of *Being and Time*. And where that ecstatic character is most manifest is in ecstatic temporality. This is the kernel, the apex, the bee's knees, of *Being and Time*.

What is time? Time is nothing prior to the process of its own emergence. Again, the standard way of dealing with the question of time is by distinguishing time from eternity. This is what Saint Augustine does in the *Confessions*, in books 11 and 12.[4] We understand time with reference to something that is other than time, that precedes time: eternity is timelessness. For Heidegger, that's a nonstarter. Rather, temporality is something that is its own process of emergence. Its essence consists in the process of temporalizing the unity of its *ecstases*. This ecstatic dimension of temporality gets leveled off in our vulgar or ordinary experience of time into a series of now-points.

We're going to come back to that ordinary experience of time

4. Augustine, *Confessions*, book 11, chapters 11–16; book 12, chapter 11, section 13.

when we read the final chapter of *Being and Time*. The ordinary experience of time would be to say, there's the present, which is now; there's the past, which is the no-longer-now; and there's the future, which is the not-yet-now. Now, no-longer-now, and not-yet-now. Present, past, future. They're distinguishable, and they're organized as if they were on a line. In this linear idea of time, there's a now that is here, a now that is going to arrive, and a now that has elapsed. In such a conception, time is a linear and infinite series of now-points. But with the idea of ecstatic temporality, Heidegger is describing a more dynamic, almost cyclical vision of time as the unity of the three *ecstases*: where time comes out of the future, which unleashes the past and releases itself in the present.

This takes us to the idea of finitude. Ecstatic temporality is finite. It is *endlich*, endly. It comes to an end in death, and its fundamental character is shaped by the anticipation of being-towards-death. And that language of finitude, which we haven't seen very much of in *Being and Time*, really begins to pepper the text at this point. If you read pages 378 through to 380, you'll see that the word "finitude" pops up a number of times. Given that time is existentially *our* time, a time that is defined by death, time is *endlich*, it has a finite character. That's what has to be absolutely kept in mind. Heidegger writes:

> Its finitude does not amount primarily to a stopping, but is a characteristic of temporalization itself. The primordial and authentic future is the "towards-oneself" (to *oneself*!), existing as the possibility of nullity, the possibility which is not to be outstripped.... Primordial and authentic coming-towards-oneself is the meaning of existing in one's ownmost nullity. (SZ 330 / BT 378–79)

Feel the radicality of that language: the primordial and authentic coming-towards-oneself is the meaning of existing in one's ownmost nullity. Heidegger goes on, his language ever more unbound: "In our thesis that temporality is primordially finite, we are not disputing that 'time goes on'; we are simply holding fast to the phenomenal character of primordial temporality—a character which shows itself in what is projected in Dasein's primordial existential projecting" (SZ 330 / BT 379). Primordial time is finite. It's the very movement of Dasein's finitude. And finitude is the coming-towards-oneself, where the oneself that one comes towards is the nullity of a thrown basis. The finitude that one comes towards is the oneself that comes from the future. And what is taken over in that future is the nullity of one's thrown basis. Finitude is *coming back towards the nullity that I am.* That's why conscience is so important. The double nullity of thrownness and projection.

Please absorb that thought, take your time with it, keep it in your mouth for a while and savor it. Finitude is coming towards the double nullity of thrownness and projection. That's who we are. That's what it means to be human, mortal, in motion, and to exist without ground. The question now, which is the question of the remaining chapters of *Being and Time*, is, How does inauthentic time arise out of authentic, finite time? And more particularly, how does inauthentic time produce an endless infinite now-time out of itself? With this question, we will start to see the emergence of Heidegger's conception of history.

The point about the vulgar and ordinary experience of time is really important because we have an idea of time as a line, and we have an idea of time as an infinite series, an infinite possibility of ever-new nows pouring out of the future through the present, into the past. How does that inauthentic idea of time get a grip on us? Well, that's going to lead us back to everydayness, back to inauthenticity. It's going to lead us back to repeating that analysis of average everydayness from the standpoint of temporality. From there we move into another key moment, the final crescendo of *Being and Time*, which is a discussion of historical time, historicity, and then time-reckoning, or time-measuring, in the final two chapters of the book.

We're not at the end yet, but we're within sight of it. You know, we can see the shoreline: *Land ahoy!*[5] Or something like that. We're not that far out at sea. Heidegger summarizes his conclusions very crisply and simply:

> Our analysis of primordial temporality up to this point may be summarized in the following theses. Time is primordial as the temporalizing of temporality, and as such it makes possible the Constitution of the structure of care. Temporality is essentially ecstatical. Temporality temporalizes itself primordially out of the future. Primordial time is finite. (SZ 331 / BT 380)

I could insert some points of—I don't want to use the word "critique" because it's so vulgar—dissent or difference of emphasis that I would make from what Heidegger is doing. But I think I'll leave

5. Hegel says "Land ahoy!" in regards to Descartes's *ego cogito*, as though it represents, in Heidegger's words, "the firm ground upon which philosophy can truly and completely settle." See G. W. F. Hegel, *Lectures on the History of Philosophy*, vol. 3, trans. Elizabeth S. Haldane and Frances H. Simson (London, 1895), 217; and Martin Heidegger, "Hegel and the Greeks," trans. Robert Metcalf, in *Pathmarks*, ed. William McNeill (Cambridge University Press, 1998), 325.

that to one side because I don't want to bore you to death. But in some published work that I've done, in a paper I wrote quite a long time ago now called "Originary Inauthenticity," there's a critique of Heidegger's analysis of time, in particular his insistence on the primacy of the future. I try to articulate a view of time that gives primacy to the past over the future, which would give us a different vision of temporality.[6]

I very much agree with Heidegger that we are temporal beings and that temporality temporalizes itself in these three dimensions. That's what it means to live without ground, finitely. The question is where the emphasis or privilege should be placed in the analysis. Whether we agree with Heidegger and give primacy to the future or whether we can think about temporality in relation to the primacy of the past. In defense of the primacy of inauthenticity in the context of time, we might consider the example of Heidegger's son and the overseer of his *Gesamtausgabe* (*Complete Works*), Hermann Heidegger, who revealed in 2005, at age eighty-four, that he was actually not Martin's biological son.[7] Did being "illegitimate" make him any less his father's child? But I'm just going to leave that thought in your mind. We can take some moments to catch our breath and come back fresh, ecstatic, and invigorated for our discussion of the next chapter of *Being and Time*, "Temporality and Everydayness."

6. Simon Critchley, "Originary Inauthenticity—on Heidegger's *Sein und Zeit*," in Simon Critchley and Reiner Schürmann, *On Heidegger's* Being and Time, ed. Steven Levine (Routledge, 2008), 132–51.

7. Hermann Heidegger, Afterword, *Letters to His Wife: 1915–1970*, by Martin Heidegger, ed. Gertrud Heidegger, trans. R. D. V. Glasgow (Polity, 2008).

15. The Time of the Everyday World

The repetition of average everydayness from the standpoint of authentic temporality — Awaiting — Rapture — More unfairness to Kierkegaard — Making present — Forgetting — Repetition and memory — Bewildered fear, a burning house — Beckett — Resolute Dasein has no fear — Anxiety in the face of the nothing of the world — A taxonomy of moods — Hope comes in to tickle your feet — Equanimity — Being dragged along in thrownness — The authentic and inauthentic ecstases *do not come in succession — The future is not later than the past — The genesis of the theoretical view of the world — An existential conception of science — Lifeworld practices and science — Phenomenology as pre-science — Deliberation is not absent from Heidegger (Ernst Tugendhat) — Different ways of philosophizing with a hammer — The importance of place — How nature is seen by modern science — Mathematical projection, data, modeling — Dasein is its world existingly — How the transcendence of being-in-the-world gets covered over in subject-object dualism — Horizonal schemas defined (with Kant) — The time of spatiality — Taking space in, literally — Leeway, play-space — Back to the everyday — The relation between authenticity and the everyday — There is no breakout from the cave — The enigma of everydayness — Philosophy and the enigmatic a priori.*

THE CHAPTER that is under our eyes is division 2, chapter 4, titled "Temporality and Everydayness." What's going on in this chapter? To hearken back once again to Mark E. Smith of The Fall: "The three

R's: repetition, repetition, repetition." What we get in this chapter is the repetition of the analytic of everydayness from the perspective of temporality. We will take the analyses that we've seen in division 1—of understanding, state-of-mind, falling, discourse, spatiality, and the rest—and repeat them from the standpoint of temporality. We've got to the level of authentic existence with the discussion of ecstatic temporality, so the framing question here is, How does average everydayness look from the vantage point of authentic existence? This is a long and messy chapter, a hodgepodge of different topics, and it lacks the condensed drama of the chapters that preceded it.

Let's dig in at paragraph 68(a): "The Temporality of Understanding." Having given the structure of ecstatic temporality as authentic care in paragraph 65, here we get the ecstatic temporality of *in*authenticity. We've had an account of ecstatic temporality as that of authentic existence. But how do the three *ecstases* of time appear in their inauthentic modes? The first thing to note is that inauthentic Dasein is still ecstatic. It is always already outside, alongside things. Even as inauthentic, Dasein is still ecstatic, just differently so. Inauthentic life has chosen the They, not itself, as its hero. What does that look like, to have an ecstatic immersion in a they-self? To have chosen to be immersed in the They, while still outside, alongside the world? And, more to the point, what does this ecstatic inauthenticity look like in terms of *time*? Given that proximally and for the most part—words that, after much repetition, are finally going to be glossed at the end of this chapter—Dasein is irresolute and inconstant, how do we characterize this inauthentic temporality? What does ecstatic temporality look like when it's couched in the proximal irresolute inconstancy of everyday Dasein?

Heidegger analyzes inauthentic temporality in terms of the threefold structure of future, past, and present. As we've seen so often in *Being and Time*, he is obliquely illuminating. He often illuminates the thing that he's discussed previously by discussing a new thing. For example, the discussion of the world becomes clearer in comparing it with Descartes's conception of the world. Likewise, when Heidegger discusses inauthentic temporality, it sheds light on what he means by authentic time. As we think through Heidegger's three inauthentic *ecstases* of time, we will start to gain a better understanding of what he means by an authentic temporality.

First, the future. Remember that the authentic future is anticipatory, it is Dasein's constant being-ahead-of-itself, which is the consummation of our being-towards-death. Then how do we characterize the inauthentic future? In contrast to anticipation, it is characterized in terms of *awaiting*: "Dasein does not come towards itself primarily in its ownmost non-relational potentiality-for-Being [namely, ecstatically], but it *awaits this* concernfully *in terms of that which yields or denies the object of its concern*. Dasein comes towards itself from that with which it concerns itself. The inauthentic future has the character of *awaiting*" (SZ 337 / BT 386). Try to get a feel for the difference, which we thought about previously (see ch. 11, "Death"), between awaiting and anticipating. In German, *awaiting* is *Gewärtigen*. Heidegger links it to the ideas of waiting (*Warten*) and expectation (*Erwarten*). Anticipation (*Vorlaufen*) means running ahead. Conceptually, it should be easy to see how anticipation does some work of bringing forth rather than simply waiting for. To run ahead of is to already do some futural work of preparing for. Meanwhile *Warten* (the root of *Gewärtigen*), which is etymologically related to the work that a warden does—which is essentially to watch something as it's happening, to watch over or look at—feels less futural. Another phrase, "to look after," can also characterize *Warten*, but only in the sense that when looking after, we fail to look before. That is to say, we remain behind, rather than ahead of, that which we look after.

Second, the inauthentic present is explained in contrast with the idea of being carried away in the rapture of ecstatic temporality: "Corresponding to the inauthentic future (awaiting), there is a special way of Being-alongside the things with which one concerns oneself. This way of Being-alongside is the Present—the 'waiting-towards'" (SZ 337–38 / BT 387). Heidegger is deliberately hyphenating and playing with the German here, "waiting-towards" is *Gegen-wart*. He's breaking up the term *present* in everyday German into its two component parts: waiting-towards.

> This way of Being-alongside is the Present—the "waiting-towards"; this ecstatical mode reveals itself if we adduce for comparison this very same ecstasis, but in the mode of authentic temporality. To the anticipation which goes with resoluteness, there belongs a Present in accordance with which a resolution discloses the Situation. In resoluteness, the Present is not only brought back from distraction with the objects of one's closest concern, but it gets held in the future and in having been. That Present which is held in authentic temporality and which thus is authentic itself, we call the "moment of vision." This term must be understood in the active sense as an ecstasis. It means the resolute rapture with which Dasein is carried away to whatever possibilities and circumstances are encountered in the Situation

> as possible objects of concern, but a rapture which is *held* in resoluteness. The moment of vision is a phenomenon which *in principle* can *not* be clarified in terms of the "*now*." (SZ 337–38 / BT 387–88)

The quote goes on:

> The "now" is a temporal phenomenon which belongs to time as within-time-ness: the "now" "in which" something arises, passes away, or is present-at-hand. "In the moment of vision" nothing can occur; but as an authentic Present or waiting-towards, the moment of vision permits us to encounter for the first time what can be "in a time" as ready-to-hand or present-at-hand. (SZ 338 / BT 388)

And then Heidegger opens a footnote on Kierkegaard that we'll come back to in an instant: "In contradistinction to the moment of vision as the authentic Present, we call the inauthentic Present 'making present.' Formally understood, every Present is one which makes present, but not every Present has the character of a 'moment of vision'" (SZ 338 / BT 497). It's a long quote. Forgive me. The inauthentic future is described as waiting towards the dispersed being-alongside things in the mode of concern. And to this Heidegger contrasts the authentic present of the moment of vision, what he calls (in a very compressed German phrase) the "in-resoluteness-held-rapture" (ibid., trans. mod.). What a phrase this is: "*In der Entschlossenheit gehaltene Entrückung*." We then get a fascinating little adumbration of this moment of vision as a present that is distinct from the vulgar now. Then we get this little footnote on Kierkegaard. We've seen this logic before, in the references to Kierkegaard, which are grotesquely unfair. Kierkegaard sees the moment of vision, the *Augenblick*. But, Heidegger insists, Kierkegaard clings to the vulgar concept of time and therefore never knows the moment of vision properly. Heidegger again acknowledges that he borrows something from Kierkegaard and then disavows it because he is allegedly still trapped within the traditional way of looking at things: existentiell and ontic rather than existential and ontological.

In contrast to the moment of vision as the authentic present, the inauthentic present is called making-present (*Gegenwärtigen*), which is the way in which we make things present in an everyday manner. Heidegger again is playing with the German here. What you have to think of is that we have the authentic present, which is the moment of vision, and then the inauthentic present, which is just the way things show up in average everyday life. That's the future and the present. Which leaves us with the past. If the authentic past is having-been, having-been-ness whereby I come back to my ownmost self in the act of repetition, then the inauthentic past is Dasein's forgetting of itself. Look at the passage:

> In anticipating, Dasein *brings* itself *again forth* into its ownmost potentiality-for-Being. If Being-as-having-been is authentic, we call it "*repetition*." But when one projects oneself inauthentically towards those possibilities which have been drawn from the object of concern in making it present, this is possible only because Dasein has *forgotten* itself in its ownmost *thrown* potentiality-for-Being. (SZ 339 / BT 388)

We here get the idea of forgetting, as a backing away from one's ownmost potentiality for being that Dasein closes off from itself.

When I do not repeat who I am in the certainty of self-constancy, when I'm not repeating myself, ensuring the constancy of the self in the unity of the three *ecstases* of time, I forget myself and I slip into self-oblivion. The self is constant only for as long as it repeats itself. When it doesn't, it forgets and becomes inconstant.

We might notice here that this idea of repetition that we saw in the last chapter is a little bit like memory. Heidegger doesn't use the term, but it is suggested by the contrast with forgetting. If forgetting is the way in which I close myself off from myself, then my having-been, being myself authentically, is an act of memory. Repetition is like memory or, better, *re-memory.*

This is the future, present, and past, inauthentically conceived. What is the structure of inauthentic ecstatic temporality? Heidegger gives us the answer to that question in the last lines of the paragraph: "The *awaiting which forgets and makes present* is an ecstatical unity in its own right, in accordance with which inauthentic understanding temporalizes itself with regard to its temporality. The unity of these ecstases closes off one's authentic potentiality-for-Being, and is thus the existential condition for the possibility of irresoluteness" (SZ 339 / BT 389). Inauthentic ecstatic temporality is an awaiting that forgets and makes present. This is a rather odd formulation but worth pondering. It is then linked to the concept of fear, which appears in the next paragraph, where Heidegger discusses the temporality of fear as a forgetting that awaits and makes present. What he also calls a form of "*bewildered making-present.*" He uses an example (and thank God he does use some examples). He talks about the way in which fear can lead to a bewilderedness with regard to time and experience. He says, "It is well known, for instance, that the inhabitants of a burning house will often 'save' the most indifferent things that are most closely ready-to-hand" (SZ 342 / BT 392). When we are scared, we do strange things. We don't operate resolutely, we forget stuff and we pick up the oddest things when our house is burning down.

We might think about this further in relation to what this inauthentic experience of time suggests. As you will know by now, I'm an enthusiastic reader of Beckett. I think about a play like *Waiting for Godot*. Beckett's *Godot* is an example of inauthentic temporality, where the constant awaiting, the waiting for Godot, is an awaiting that continually forgets. People in *Godot*, Vladimir, Estragon, or Pozzo, are constantly forgetting things, which makes possible a present that is nonrapturous and where nothing really happens.[1] The characters in Beckett do not rapturously anticipate their finitude. They await forgetfully. Think of someone like Krapp in Beckett's *Krapp's Last Tape*, who's trying to recover the elements of his memory that are recorded on tape.[2] And that he has forgotten.

Let's continue into paragraph 68(b). It deals with the temporality of state-of-mind and therefore with mood, and with the question: What is the temporality of mood? Of course, mood is the way in which state-of-mind is disclosed. This analysis is fascinating, especially because of what it says about fear and the contrast between the time of fear and the time of anxiety. But let me skip to the end of these pages, where we get a rearview illumination of something that has come up before: the precise nature of anxiety. In a suggestive remark, Heidegger writes, "But even though the Present of anxiety

1. Samuel Beckett, *Waiting for Godot: A Tragicomedy in Two Acts* (Grove, 1954).
2. Samuel Beckett, *Krapp's Last Tape*, in *The Collected Shorter Plays*, 47–58.

is *held on to*, it does not as yet have the character of the moment of vision, which temporalizes itself in a resolution. Anxiety merely brings one into the mood for a *possible* resolution. The Present of anxiety holds the moment of vision *at the ready* [*auf dem Sprung*]; as such a moment it itself, and only itself, is possible" (SZ 344 / BT 394). This throws some light on anxiety. Anxiety is the condition of possibility for the moment of vision. It's what makes the moment of vision possible. But it is not yet the moment of vision. It is the mood for a possible resolution that holds it at the ready, *auf dem Sprung*, like a sprinter waiting on the blocks or a plane on the runway. Anxiety is at the ready in a "ready, set, go" sort of way. Anxiety brings Dasein to the ready. But Dasein has to get going. And when it takes off, that is the moment of vision. This also illuminates the distinction between anxiety and fear, which we've returned to a couple of times: "Fear is occasioned by entities with which we concern ourselves environmentally. Anxiety, however, springs from Dasein itself. When fear assails us, it does so from what is within-the-world. Anxiety arises out of Being-in-the-world as thrown Being-towards-death" (SZ 344 / BT 395). Fear is fear of things in the world. Fear of determinate, specific entities that threaten us. As you know, I like examples of bears and badgers. Anxiety, by contrast, is something that springs from Dasein itself. Therefore, resolute Dasein has no fear. Fear is transformed into anxiety. Anxiety is courage. It is bravery. Heidegger goes on: "But anxiety can mount authentically only in the Dasein, which is resolute. He who is resolute, knows no fear, but he understands the possibility of anxiety as the possibility of the very mood which neither inhibits nor bewilders him. Anxiety liberates him *from* possibilities which 'count for nothing.' And lets him become free *for* those which are authentic" (ibid.). Anxiety allows Dasein to become free for its possibilities. Not those that count for nothing, which are those of the world. Those (no)things of the world bewilder us, scare us. They inhibit us. That's not what we're concerned with. To become free is to become free *from* those things of the world. In relation to the last quotation, if we flip back to page 393, there are two telling remarks. First, Heidegger says, "Anxiety is anxious in the face of the 'nothing' of the world" (SZ 343 / BT 393). Again, note the striking contrast with the idea of the world in division 1, which was this world of fullness, meaning, and significance. It is now revealed to be nothing. Second, Heidegger says, "Anxiety discloses an insignificance of the world" (ibid.). The world that is meaningful has become a world that is meaning*less*. This is one of the places where we can think about what Hans Jonas called the nihilism of *Being and Time*,[3] whether we think of nihilism critically or not.

We then get a taxonomy of moods, which is suggestive, and much more could be said. If the moods of fear and anxiety are linked to our inauthentic and authentic existence, and which are both grounded in having-been, then what about other moods? And what about *hope* in particular? Hope, which does not relate to the past but relates surely to the future. A future good. A *bonum futurum*. Look at the quotation here: "How is a temporal meaning to be found in the pallid lack of mood which dominates the 'grey everyday' through and through?" Do you see how gray the everyday has now become? Heidegger goes on:

3. See Hans Jonas, "Gnosticism and Modern Nihilism," *Social Research* 19, no. 4 (1952): 430–52, esp. 447–52.

> And how about the temporality of such moods and affects as hope, joy, enthusiasm, gaiety? Not only fear and anxiety, but other moods, are founded existentially upon one's having been; this becomes plain if we merely mention such phenomena as satiety, sadness, melancholy, and desperation. Of course these must be Interpreted on the broader basis of an existential analytic . . . that has been well worked out. But even a phenomenon like hope, which seems to be founded wholly upon the future, must be analysed in much the same way as fear. (SZ 345 / BT 395)

What can we say about this? Well, first, let's notice the moods that are being discussed here. We've seen a few moods, but primarily anxiety and fear. Here we've got joy, enthusiasm, and cheerfulness. We have satiety, sadness, melancholy (the black bile), and even desperation. But the analysis of hope is peculiar here because it is entirely self-directed. Heidegger goes on to say that "here [hope's] character as a mood lies primarily in hoping as *hoping for something for oneself*" (SZ 345 / BT 396). Hope is not hope for a better future but hoping for something for myself. As such, even hope is related to our burdens, our thrown ground, our mineness. That raises the question, at least in my mind, about whether on a Heideggerian basis we can think about hope for something other than what I hope for myself. Can I hope for a future that is not my future, say? Can I hope for another's future? For Heidegger, that would still have to be referred back to the idea of hoping for something for myself.

We could take this into a much longer discussion of hope, but that would be another discussion for a separate occasion. But let's just say that I have doubts about the concept of hope. I think about an extraordinary moment in Anne Carson's *Antigonick*. Not the

Antigone, but the *Antigonick*, where Nick is this character who is silent, who just draws shapes on the stage throughout the performance, doesn't say a word. And there's this line about hope. Oh, look, here comes hope to tickle your feet. Without realizing the soles of those feet are on fire.[4] Heidegger goes from that discussion of hope to a discussion of indifference, which is inauthentic, and distinguished from equanimity. Heidegger thinks that indifference is a pallid lack of mood. Whereas equanimity flows from resoluteness. Equanimity is a form of equalness or balance of courage. Being equal to oneself in one's mood (*Gleichmut*). It is similar to the idea of *ataraxia* (ἀταραξία) in Epicurus, an equanimity or tranquility that, for Epicurus, flows from the pondering of the cosmos and not fearing death.[5]

This whole chapter is like a minestrone of different ingredients, little moments of teasing where we can find concepts that we already know in the book that are being elaborated in suggestive ways. For example, there is a remark about the being of animals, the being of that which merely has life. What does it mean to say that an animal merely has life? Heidegger writes, "It remains a problem in itself to define ontologically the way in which the senses can be *stimulated* or *touched* in something that merely has life, and how and where, the Being of animals, for instance, is constituted by some kind of 'time'" (SZ 346 / BT 396). One can see here Heidegger's implicit and classical humanism in *Being and Time*. We introduced the idea earlier to the effect that Dasein is a human being, but perhaps not

4. Anne Carson, *Antigonick* (New Directions, 2012), 23.

5. Diogenes Laertius, *Lives of Eminent Philosophers*, trans. and ed. Stephen White (Cambridge University Press, 2021), book 10, Epicurus, 80.850 (Letter to Herodotus).

just human. Perhaps nonhumans can be thought of as Daseins, like the proverbial octopus. I have no problem with that. But in passages like the one just quoted, Heidegger's humanism is clear as day: Animals are poor in world. Stones are without world. We have a world. Such is the chain of being in Heidegger's early work.

Let's turn to paragraph 68(c), "The Temporality of Falling." I just want to note a key remark on page 400. There is also a discussion of curiosity in these pages, which is well worth looking at. Curiosity, concupiscence, lust of the eyes, which Heidegger talks about in chapter 5, division 1, as one of the features of inauthentic life, and curiosity is defined as a "craving for the new." In German, it's *Neugier*. It is the craving for the new that takes us away in curiosity. But let's turn directly to the key remark in this paragraph:

> Proximally, the "throw" of Dasein's Being-thrown into the world is one that does not authentically get "caught." The "movement" which such a "throw" implies does not come to "a stop" because Dasein now "is there." Dasein gets dragged along in thrownness; that is to say, as something which has been thrown into the world, it loses itself in the "world" in its factical submission to that with which it is to concern itself. (SZ 348 / BT 400)

What's interesting about this remark is that Dasein's being-in-the-world, the throw of Dasein's facticity, does not get caught. It does not come to a stop. Rather, Dasein gets dragged along in thrownness, we get dragged along by our throw. It's a fascinating image, like someone being pulled behind a runaway car or dragged along in a balloon or on a horse that has lost control or something like that.

The point is that for as long as Dasein is, it is in the throw. We are always in the throes of our thrownness, and that thrownness is always dragging us along, pulling us from a past that we can never fully grasp towards a future that is beyond our reach.

Turning to paragraph 68(d), "The Temporality of Discourse," Heidegger makes a wonderful summarizing remark about the structure of temporality, which brings together the analyses of authentic and inauthentic temporality we have been following. It's a long quote, but I think it's a useful and illuminating one: "Understanding is grounded primarily in the future (whether in anticipation or in awaiting). States-of-mind temporalize themselves primarily in having been (whether in repetition or in having forgotten). Falling has its temporal roots primarily in the Present (whether in making-present or in the moment of vision)" (SZ 350 / BT 401). You can see that in this little description we get the analyses of temporality authentically and inauthentically combined. Ecstatic temporality: anticipation, having-been, and the moment of vision; and inauthentic temporality: awaiting, forgetting, and making present. Continuing the quote:

> All the same, understanding is in every case a Present which "is in the process of having been." All the same, one's state-of-mind temporalizes itself as a future which is "making present." And all the same, the Present "leaps away" from a future that is in the process of having been, or else it is held on to by such a future. Thus we can see that *in every ecstasis, temporality temporalizes itself as a whole; and this means that in the ecstatical unity with which temporality has fully temporalized itself currently, is grounded the totality of the structural*

> *whole of existence, facticity, and falling—that is, the unity of the care-structure.* (ibid.)

Note these next words: "Temporalizing does not signify that ecstases come in a 'succession.' The future is *not later* than having been, and having been is *not earlier* than the Present. Temporality temporalizes itself as a future which makes present in the process of having been" (ibid.). Understanding, state-of-mind, and falling, these three key structural moments of care, are experientially indistinguishable. It's not that there is understanding, and then there is state-of-mind, followed by falling. Rather those three moments come together in the unity of care. So too with the *ecstases* of temporality, whether authentic or inauthentic. In every *ecstasis*, in every dimension of time, temporality temporalizes itself as a whole. That is, crucially, the *ecstases* do not come in succession. The future is not later than having been, but rather temporalizes itself as the "having-been-making-present-future." In other words, temporality temporalizes itself as a future that makes present in the process of having been. Time past, time present, and time future are all contained together as structural moments that do not come in a linear order but as a piece, and a piece, moreover, that has two modes: the authentic and the inauthentic.

The idea of succession in relationship to the thinking of time is something that is going to become clearer when we get to the last chapter of *Being and Time*, where Heidegger focuses on the idea of time-reckoning and the vulgar experience of time, which has, since Aristotle and through to Kant, viewed time in relation to the concept of succession: not-yet-now = the future; now = the present; no-longer-now = the past. Heidegger's prime target in *Being and*

Time and his later work is the tyranny of linear time. This is what has to be destroyed.

Let's move on to paragraph 69. Now this is another interesting ragbag of a paragraph, and it's rather long-winded. Heidegger is trying to deal with the question that we thought we'd already dealt with and was done and dusted. The question is: In what way is anything like a world possible at all? And particularly, in what sense can we speak of the transcendence of the world with respect to Dasein? I'm going to be a little brutal with this paragraph, otherwise we'll never get done. I'm going to skip part A because it's fairly straightforward. That's called "The Temporality of Circumspective Concern." Catchy title, no?

The themes it raises get focused more clearly in paragraph 69(b), which is particularly compelling because the issue that Heidegger is dealing with here is one that has been suggested throughout *Being and Time*, but not really directly discussed, which concerns the genesis of the theoretical attitude. As we know, Heidegger is deeply critical of the ontology of the present-at-hand: when we view things theoretically as objects that are present to a subject. But he is concerned here with the genesis of that theoretical attitude, which is the attitude that's consummated in philosophy as epistemology and in the natural-scientific conception of the world. For Heidegger, the theoretical attitude is generated out of the world of circumspective concern: the inauthentic, average, everyday world. Theory has its genesis in praxis.

We know that already. But there is a phrase here that's always intrigued me and that suggests something new: "an *existential conception of science*" (SZ 357 / BT 408). What is an "existential conception of science"? What is Heidegger really up to with regard to science and the scientific view of the world? In order to get a sense of what Heidegger is about in these pages—in a wider, broader focus—I think we have to get clear on the issue of scientism. And look at the way phenomenology criticizes scientism.

What is scientism? This is really what Heidegger is criticizing throughout *Being and Time*. Scientism rests on the fallacious claim that the theoretical or natural-scientific way of viewing things, what Heidegger calls the present-at-hand, provides the primary and most significant access to ourselves and to our world. Scientism is the claim that the methodology of the natural sciences is not just true for the natural sciences but also provides the best form of explanation for all phenomena in general. Heidegger, in these pages and elsewhere, tries to show how the scientific conception of the world is derivative or parasitic upon a prior practical understanding of the world as ready-to-hand. The latter view of the world as ready-to-hand is what he means by the environing world, the world of the *Umwelt* that is closest, most familiar, and most meaningful to us. The world, that is, to speak in a more conventional way, that is already colored by our values, our cognitive, aesthetic, and ethical values. Heidegger won't speak in those terms, but let's just do that in order to make matters a little easier and more accessible.

What Heidegger is up to in these pages and throughout *Being and Time* is adapting and extending a claim that was being made by his teacher, Edmund Husserl. It's implied throughout Husserl's early work and developed more polemically in his later work, *The Crisis of European Sciences*. Now the issue for Husserl is what he calls "objectivism." Scientism is objectivism. This is the idea that we can treat everything that is, all entities, as objects that can be treated in scientific terms. Husserl's critique of scientism or objectivism is that it overlooks the *lifeworld*. What he calls the *Lebenswelt* is for him the enabling condition for scientific practice. And Husserl, in *The Crisis of European Sciences*, describes the lifeworld as the condition of possibility, out of which the scientific conception of the world arises.[6]

This leads to the following point, which really explains what Heidegger's up to in these pages. The critique of scientism, at least within phenomenology, does not seek to refute or negate the results of scientific research in the name of some mythic unity of man and nature. Which is a risk in some areas of phenomenology, perhaps some of the slightly ecstatical pronouncements of the later Merleau-Ponty on the flesh of the world. Rather, Heidegger's view, following Husserl, is that science does not provide the primary or most significant access to a sense of ourselves and the world. Anti-scientism does not at all entail an anti-scientific attitude. Being opposed to scientism doesn't mean that we're anti-science. It rather means that we have to link the scientific conception of the world to the conditions of possibility that make that conception possible, which are found in the lifeworld or, in Heidegger's terms, our everyday being-in-the-world.

What's required, I think, and this is what returns us to Heidegger's phrase, is an existential conception of science that would show how the practices of the natural sciences arise out of lifeworld practices. The lifeworld precedes the world of the natural sciences. Another way of putting this point about an existential conception of science would be to refer it to a concept that Heidegger introduces

6. Edmund Husserl, *The Crisis of European Sciences and Transcendental Phenomenology*, trans. David Carr (Northwestern University Press, 1970). See, e.g., 51–52.

in his 1924 lecture, "The Concept of Time," which is the lecture that Hans-Georg Gadamer famously described as the "primal form," the "Ur-Form," of *Being and Time*.[7] In this lecture, three years prior to *Being and Time*, Heidegger describes what he is doing as a "pre-science" that comes before scientific research and that has its genesis in lifeworld practices.

This has an important consequence. For Heidegger, and indeed Husserl, philosophy is not the "under-laborer" to science. This is the famous description that Locke uses in the *Essay*.[8] Rather, philosophy is something that, as Heidegger says, leaps ahead of the sciences by showing their basis in a fundamental ontology of persons, things, and world—what we might call the pre-theoretical layer of experience. The task is not seeing philosophy as an under-laborer to science that tidies up the furniture in the great mansion of science and cleans the floors and makes sure everything is in its right place.

Rather, what philosophy does is that it provides a phenomenology of persons, things, and world, which shows how the theoretical attitude of the sciences finds its condition of possibility in our lifeworld practices. And those everyday practices have their own form of validity. But it is not the validity of the causal remarks that we attribute to natural science. It's rather something else, what we might

7. The twenty-page lecture, distinct from the 1924 article of the same name and year, is published in a bilingual German-English edition. See Martin Heidegger, *Der Begriff der Zeit*, ed. Friedrich-Wilhelm von Herrmann, vol. 64 of *Gesamtausgabe*, ed. Friedrich-Wilhelm von Herrmann (Klostermann, 2004); *The Concept of Time*, trans. William McNeill (Blackwell, 1992). Gadamer makes the remark in 1964 in "Heidegger and Marburg Theology," in *Philosophical Hermeneutics*, trans. and ed. David E. Linge (University of California Press, 1976), 198–212, esp. 199.

8. John Locke, *An Essay Concerning Human Understanding* (London, 1836), ix.

call a redescription, elucidation, or hermeneutical clarification, what Heidegger calls repeatedly in *Being and Time* "formal indication." This is a form of validity or truth claim that is not reducible to the empirical truth claims of natural science. Again, we're not being remotely anti-scientific. These formal indications are what precede science. They are pre-science.

In my view, this is what is going on in these pages of *Being and Time*. The claim for an existential conception of science is suggestive. Sadly, Heidegger doesn't really develop the thought in his work, and it remains unclear in *Being and Time* where this insight is going. I think it's a possible avenue to be taken in saving Heidegger from the claim that he is an anti-scientific thinker, or indeed saving Heidegger from himself when he says things in his later work like "Science does not think" (*Die Wissenschaft denkt nicht*), which is not a particularly helpful thought. I like to think that what Heidegger is indicating here is the possibility of an alternative praxis of science that would also have to acknowledge that science itself is an activity. Science is not just a theoretical attitude towards the world but also a praxis that emerges out of a complex lifeworld and that is highly context-, history-, and time-dependent.

There is a lot going on in these pages. For example, there is a passing discussion of deliberation (SZ 359 / BT 410–11). There is a line of critique of Heidegger that can be found in major philosophers like Ernst Tugendhat, who wrote a very important book on the concept of truth in Husserl and Heidegger.[9] This critique is developed more

9. Ernst Tugendhat, *Der Wahrheitsbegriff bei Husserl und Heidegger* (De Gruyter, 1967).

influentially by Habermas and the Habermasians.[10] They will often claim that there is no space for deliberation in Heidegger. Heidegger is a decisionist, in philosophy and politics, and the absence of deliberation, the very stuff of democratic life, is revealing. Well, if you look at these passages of *Being and Time,* you will see that Heidegger indeed has a discussion of deliberation. It's an essential part of inauthentic life. It is subordinate to the decision to be resolute, but it is hardly absent.

Let's return more closely to the text and see where Heidegger is philosophizing with a hammer (and philosophizing about a hammer). We met the hammer in division 1, chapter 3. The issue he is dealing with here is the sentence "The hammer is heavy." If someone says, "The hammer is heavy," then these words can be understood in two ways. We can understand the words as a ready-to-hand remark about the fact that the hammer is hard to handle. Too heavy! It's hard to handle! *Or* I can understand the remark as a proposition about the physical property of heaviness, which the hammer has as something I regard as present-at-hand. We're already familiar with this distinction between the hammer as it is used, ready-to-hand in banging nails into something, and the hammer as something that is viewed objectively as having physical properties. What changes in this shift from the ready-to-hand to the present-at-hand is explained in terms of the categorial distinction that we are familiar with but that is here given an interesting elucidation:

10. See, e.g., Jürgen Habermas, *The Philosophical Discourse of Modernity: Twelve Lectures*, trans. Frederick G. Lawrence (MIT Press, 1990).

> Why is it that what we are talking about—the heavy hammer—shows itself differently when our way of talking is thus modified? Not because we are keeping our distance from manipulation, nor because we are just looking *away* from the equipmental character of this entity, but rather because we are looking *at* the ready-to-hand thing which we encounter, and looking at it "in a new way" as something present-at-hand. *The understanding of Being* by which our concernful dealings with entities within-the-world have been guided *has changed over*. (SZ 361 / BT 412)

Changed over. What happens in the categorial shift from the ready-to-hand to the present-at-hand is a changeover in our understanding of being. If we relate that back to the existential conception of science, we can see that an existential conception of science would look at things in their ready-to-handedness as that which makes possible their present-at-handness, the way in which they're consummated as theoretical objects of science.

The point is we need both ways of seeing the hammer. The present-at-hand needs to be not just supplemented by the ready-to-hand but undergirded by it. If we lose sight of the ready-to-hand, then what we lose sight of is the *place* of the thing, how a piece of equipment is found spatially in a *Platz*, in a way that directly recalls the discussion of space in chapter 3. Science is not done by Martians. It's not done by aliens. Rather, science is a radically context-dependent praxis. A lifeworld practice with an origin that can be reactivated, as Husserl would have said. We need to reactivate what is sedimented in science and see it in a different way. The point that Heidegger is making here is very simple: The theoretical view of the world, which is consummated in scientism, loses the experience of place, the experience of lived space, the existential context for things. We always have to go back to context, because that is where we live.

> In the "physical" assertion that "the hammer is heavy" we *overlook* not only the tool-character of the entity we encounter, but also something that belongs to any ready-to-hand equipment: its place. Its place becomes a matter of indifference. This does not mean that what is present-at-hand loses its "location" altogether. But its place becomes a spatio-temporal position, a "world-point." (SZ 361–62 / BT 413)

That's the point. Heidegger then talks very suggestively about the way in which nature is viewed in modern science. It anticipates his later analyses of modern science, which are known most famously in his 1953 essay on technology, "The Question Concerning Technology." What he's talking about here is the way in which nature is understood in mathematical physics (SZ 362 / BT 413). Specifically, the way physics is related to what he calls the "mathematical projection of Nature" (SZ 362 / BT 414). Heidegger has in mind thinkers like Galileo, Francis Bacon, and Descartes. This idea of nature in modern science as something mathematically projected can be linked to the idea of a crisis in the sciences in Husserl, which is implicit throughout Heidegger's work. What is crucial and decisive in the rise of the mathematical view of nature is the fact that it derives from an a priori understanding of being that needs to be deconstructed. We are fully in the grips of a mathematical projection of nature right now with our obsession with data, statistics, modeling, and the rest. Heidegger is not saying this is wrong but that it is derived and secondary to the lived context of the world:

> In the mathematical projection of Nature, moreover, what is decisive is not primarily the mathematical as such; what is decisive is that this projection *discloses something that is a priori.* Thus the paradigmatic character of mathematical natural science does not lie in its exactitude or in the fact that it is binding for "Everyman"; it consists rather in the fact that the entities which it takes as its theme are discovered in it in the only way in which entities can be discovered—by the prior projection of their state of Being. (SZ 362 / BT 414)

What Heidegger is saying here is that the mathematical projection of nature that is characteristic of modern science is the ontological framework by which we grasp things. It's not simply an ontic or an empirical matter. Another way of putting this is that modern science is the deployment of a distinct ontological framework. A distinct understanding of being. The way Heidegger will understand that understanding of nature in modern science in his later work is what he will call the enframing, *das Gestell*, which reduces everything to a standing reserve or a resource, like the ghastliness of human resources in institutions, where every feature of human life can be reduced to data points that are measured, modeled, projected, and captured. There's a lot more to say about this paragraph, but we'll have to make do with a final remark:

> We shall not trace further how science has its source in authentic existence. It is enough now if we understand that the thematizing of entities within-the-world presupposes Being-in-the-world as the basic state of Dasein, and if we understand how it does so.
>
> If the thematizing of the present-at hand—the scientific projection of Nature—is to become possible, *Dasein must transcend* the

> entities thematized. Transcendence does not consist in Objectifying, but is presupposed by it. (SZ 363 / BT 415)

The question of an existential conception of science, and more particularly of a science that flows from authentic and inauthentic existence, is fascinating in this period of Heidegger's work, but it's left undeveloped in *Being and Time*. The point is that science has to offer an account of and take into account our lived being-in-the-world. The point of this point is that the objectification of science, the way in which science operates through projecting everything as an object that can be viewed theoretically and measured and projected, presupposes Dasein's transcendence towards the world. Dasein's transcendence towards the world is the condition of possibility for the scientific explanation of nature. The mathematical projection of nature has its basis in life as it is lived.

With that mention of transcendence, we move towards paragraph 69(c): "The Temporal Problem of the Transcendence of the World." Let me summarize here. How is there a world for Dasein? In what way must the world be if Dasein is being-in-the-world? These are questions that we've come back to in this book, that we've asked before. How is there a world for Dasein? The answer is that *existingly* Dasein is its world. I think that's a very nice way, a very crisp way, of putting what Heidegger is up to in this book. "Dasein *is* its world existingly" (SZ 364 / BT 416). Dasein is the worldhood of the world. Dasein is the existential condition of possibility for there being a world. "There never was a world for her / Except the one she sang and, singing, made." "She was the single artificer of the world / In which she sang"—think of those lines from Wallace Stevens. And recall here the discussions of idealism and realism in paragraph 43, the discussion of reality.

Here Heidegger says, "If no *Dasein* exists, no world is 'there' either" (SZ 365 / BT 417). The world is Dasein-dependent. If we already know that, then what's new here is that Dasein is now conceived in terms of ecstatic temporality. The world is not distinct from Dasein's activity of temporalizing. If Dasein is being-in-the-world, then the world *is* the temporalizing of temporality:

> In so far as Dasein temporalizes itself, a world *is* too. In temporalizing itself with regard to its Being as temporality, Dasein *is* essentially "in a world," by reason of the ecstatico-horizonal constitution of that temporality. The world is neither present-at-hand nor ready-to-hand, but temporalizes itself in temporality. It "is," with the "outside-of-itself" of the ecstases, "there." If no *Dasein* exists, no world is "there" either. (SZ 365 / BT 417)

We cannot make a distinction between Dasein and its world. A world *is* insofar as Dasein temporalizes itself. Dasein is temporal, and therefore the world is temporal. Does this mean that the world is subjective? I think you're going to know the answer to this one. No, it is not subjective. Heidegger continues:

> Thus the significance-relationships which determine the structure of the world are not a network of forms which a worldless subject has laid over some kind of material. What is rather the case is that factical Dasein, understanding itself and its world ecstatically in the unity of the "there," comes back from these horizons to the entities encountered within them. (SZ 366 / BT 417)

We can link this to a rhetorical move that we've seen elsewhere in *Being and Time*, which Heidegger recalls here and which we'll come back to shortly: "If the 'subject' gets conceived ontologically as an existing Dasein whose Being is grounded in temporality, then one must say that the world is 'subjective.' But in that case, this 'subjective' world, as one that is temporally transcendent, is 'more Objective' than any possible 'Object'" (SZ 366 / BT 418). If we concede that the world is "subjective," then it is also more "objective" than any object. Being-in-the-world is the condition of possibility for subject-object dualism. In the earlier discussion of idealism and realism Heidegger will concede, Well, Dasein's being-in-the-world is idealist, but it's also realist. And it's more realistic than any realism and more effective than any idealism. Heidegger tends to spin these terms around rhetorically. The problem of transcendence is not the question of how a subject comes out of itself towards an object or into an objective world. The question is, rather, how does it happen that the essential transcendence of Dasein, its essence as temporalized being-in-the-world, gets translated into a subject-object dualism?

The world is neither subjective nor objective. It is prior to both concepts. Subject-object dualism and the entire epistemological construal of the world presuppose the transcendence of being-in-the-world. If the standard problem of philosophy is conceived of epistemologically (i.e., the question of how we can know an objective world or how we can come out of "ourselves" [a subjective "cabinet of consciousness"] and know objects), then everything will get conceived of as a relationship of subjects to objects. That is the outcome of the epistemological construal of the world. But Heidegger's point is that the subject-object dualism presupposes Dasein as being-in-the-world and forgets that Dasein is the transcendence of being-in-the-world.

One thing I just passed over—and let me look at it quickly because it's a little bit of terminology quoted here that has not been explained—is the idea of *horizonal schemas*, which pops up in these paragraphs and we've not discussed. We know that on the first page of *Being and Time*, the untitled preface to the book, Heidegger talks about time as the horizon for any understanding of being. There is always this constant background reference to the horizon. But look at what he says while explaining *ecstases*. What are the *ecstases* of time?

> Ecstases are not simply raptures in which one gets carried away. Rather, there belongs to each ecstasis a "whither" to which one is carried away. This "whither" of the ecstasis we call the "horizonal schema." In each of the three ecstases the ecstatical horizon is different. The schema in which Dasein comes towards itself *futurally*, whether authentically or inauthentically, is the "*for-the-sake-of-itself*." The schema in which Dasein is disclosed to itself in a state-of-mind as thrown, is to be taken as that *in the face of which* it has been thrown and that *to which* it has been abandoned. This characterizes the horizonal schema of *what has been*. In existing for the sake of itself in abandonment to itself as something that has been thrown, Dasein, as Being-alongside, is at the same time making present. The horizonal schema for the *Present* is defined by the "*in-order-to*." (SZ 365 / BT 416)

Okay, take a deep breath. Time is understood in terms of temporality. Temporality is the horizon for any understanding of being. The word "schema" is borrowed from Kant, where he uses it in discussing the schematism of the imagination in the foreword to the Analytic of Principles in the Transcendental Analytic in the *Critique of Pure Reason*. If you want to go there, it's fascinating stuff, and Heidegger's got a lot to say about it in *Kant and the Problem of Metaphysics*, a couple of years after *Being and Time*. What is a schema? A schema is a form. More precisely, a schema is a form or a rule of the productive imagination through which the understanding is able to apply its categories to the manifold of intuition as a way of explaining how we know the world. For Kant, the schematism is the way in which the imagination works and which bridges sensibility and the intellect. Behind the understanding is the productive imagination (and this is where Romanticism found its philosophical motor and motivation), and that imagination is what's generating the framework, the categories by which we're able to make sense of the perceptual field that is the world, the manifold of intuition, as Kant calls it. Schematism is a key concept, what Kant famously calls "a hidden art in the depths of the human soul."[11]

If we replace the language of categories with that of the existentials, and if we replace the subjectivist language of the transcendental imagination with Dasein as being-in-the-world, then the *ecstases* of temporality are that by which we can understand Dasein as being-in-the-world in an existential and temporal manner. Once again, Kant comes awfully close to a key insight into the ontological problematic, but according to Heidegger, always shrinks back. Kant had the right intuitions but the wrong ontology. Because of that, he always falls back into the tradition that needs to be destroyed.

We're moving towards the end of the chapter, in paragraph 70, which is the relation between temporality and space. Heidegger gives a

11. Immanuel Kant, *Critique of Pure Reason*, A141/B180.

very good recap of the argument about space and spatiality that we saw in division 1, chapter 3. Dasein's being-in-the-world has a distinctive spatiality. This spatiality is not the placelessness of world-space, but it's a lived place that is constituted, Heidegger says, by directionality and de-severance. The way space occurs is by being directed towards things in the manner of bringing them close. The way Heidegger puts that is that Dasein has a tendency towards the near, a drive to bring things close. That's our fundamental operation with regard to space. This existential spatiality, which is prior to objective space, is defined in terms of region or neighborhood, what he calls, in German, *Gegend*.

So far, so good. Now, what is the time of this space? What is the temporality of this lived neighborhood? Well, Heidegger begins by claiming that Dasein's spatiality is rooted in temporality. Space is subordinate to time. Now this relation of subordination will get more complicated in the later Heidegger, where he will talk about an idea of time-space, *Zeit-Raum*, in the lecture "Time and Being," from 1962, which is a mirror text from the end of his career, the speculum to *Being and Time* (ZS 18–24 / TB 14–21). But that's another story. The first claim is that spatiality is rooted in temporality. He then makes a very suggestive remark:

> We must first remember in what way Dasein is spatial. Dasein can *be* spatial only as care, in the sense of existing as factically falling. Negatively this means that Dasein is never present-at-hand in space, not even proximally. Dasein does not fill up a bit of space as a Real Thing or item of equipment would, so that the boundaries dividing it from the surrounding space would themselves just define that

> space spatially. Dasein takes space in; this is to be understood literally. (SZ 367–68 / BT 419)

And look at the German text here: "*Das Dasein nimmt—im wortlichen Verstande—Raum ein.*" Dasein literally takes space in. "It is by no means just present-at-hand in a bit of space which its body fills up. In existing, it has already made room for its own leeway. It determines its own location in such a manner that it comes back from the space it has made room for to the 'place' which it has reserved" (SZ 368 / BT 419). Note that term, "leeway." What's fascinating here is the idea that Dasein takes space in. Dasein is not a thing occupying space with its body, but as existing it has already made room for its leeway. It's a lovely translation by Macquarrie and Robinson, because even if they sometimes make questionable decisions, they're always explained and glossed. In the translators' footnote on page 419, we find the German sentence "*Existierend hat es sich je schon einen Spielraum eingeräumt.*" *Spielraum*. That's the German word being translated as "leeway." Literally it means something like "play-space." Existing Dasein has made room for its play-space. That word *Spielraum* has different connotations in German: we could call it "scope" or "elbow room" or "clearance" or something like that.

Also, suggestively, Heidegger doesn't say this, but we could imagine a link to the idea in Friedrich Schiller, of the *Spieltrieb*, the "play drive."[12] Dasein is driven by the desire to play. The point being that Dasein takes space in as play-space, as an existential playing with spatiality, as an opening up of leeway, an opening up of place. I find

12. Friedrich Schiller, *On the Aesthetic Education of Man*, trans. Reginald Snell (Dover, 2004).

that really rather compelling, that spatiality is rooted in temporality that has now been described as ecstatico-horizonal. As Heidegger writes, "Only on [that] basis... is it possible to break into space" (SZ 369 / BT 421).

Once again, the view of things that Heidegger is opposing here is the idea that space has an objectivity in that there is a third-person view of space, a view-from-nowhere, and we occupy a neutral point in world-space. Rather, for Heidegger, we take space in, we are there as that leeway of space, play-space. We are there with the things of the world, the things of our concern, our fascination. What's been shown in paragraph 70 is how what undergirds the whole structure of the spatiality of Dasein, of being directed towards things and bringing them close, is time. Temporality is the ecstatical play-space for the understanding of being.

Let's turn to the final paragraph of this long ragbag of a chapter. Paragraph 71 is called "The Temporal Meaning of Dasein's Everydayness," and it gives us a couple of real insights and hints into what's going on in *Being and Time*, into its *enigma*. Heidegger goes back to the question of the everyday, a question that I think has been on our minds. How does the everyday look from the standpoint of authenticity? The first question is, What is everydayness? This isn't exactly a new question. We've been kicking it around for a while. But Heidegger gives us a lovely description of it here:

> "Everydayness" manifestly stands for that way of existing in which Dasein maintains itself "every day." And yet this "every day" does not signify the sum of those "days" which have been allotted to Dasein in its "lifetime." Though this "every day" is not to be understood

> calendrically, there is still an overtone of some such temporal character in the signification of the "everyday." But what we have primarily in mind in the expression "everydayness" is a definite "how" of existence by which Dasein is dominated through and through "for life." In our analyses we have often used the expression "proximally and for the most part." "Proximally" signifies the way in which Dasein is "manifest" in the "with-one-another" of publicness, even if "at bottom" everydayness is precisely something which, in an existentiell manner, it has "surmounted." "For the most part" signifies the way in which Dasein shows itself for Everyman, not always, but "as a rule." (SZ 370 / BT 422)

Everydayness is a "how" of Dasein's existing, arguably *the* how. We also get an elaboration of the phrase that he's used endlessly in the book, "proximally and for the most part," or the way I prefer to render that, "most closely and mostly." "Proximally" means life with one another in the public realm; "for the most part" means not always, but as a rule. So, most closely and mostly we are always there in everydayness. Heidegger elaborates: "In everydayness everything is all one and the same, but whatever the day may bring is taken as diversification. Everydayness is determinative for Dasein even when it has not chosen the 'they' for its 'hero'" (SZ 371 / BT 422). Even when it has not chosen the They for its hero, Dasein still lives with the They, lives with *das Man*, lives in accordance with what they say. Even when Dasein has chosen to reject the They, when Dasein has chosen to be authentic, it is still determined by the everyday. Even when Dasein chooses itself, it is still utterly shot through with everydayness. Heidegger will repeat the expression "for its hero" in the next chapter, on historicity, so keep that idea of choosing a hero in mind. But what is the relation between the everyday and the authentic? Heidegger addresses this further down the same page: "In everydayness Dasein can undergo dull 'suffering,' sink away in the dullness of it, and evade it by seeking new ways in which its dispersion in its affairs may be further dispersed. In the moment of vision, indeed, and often just 'for that moment,' existence can even gain mastery over the 'everyday'; but it can never extinguish it" (ibid.). We can sink into the dull suffering and dispersions of the everyday. That's inauthentic life. *Or*, if we choose to be authentic, we can master the everyday in the moment of vision. But even in that mastery of the everyday, we can never extinguish it. Everydayness remains determinative for Dasein even when it has not chosen the They as its hero. Continuing this idea, we confront the enigma at the heart of the existential analytic: "That which is *ontically* so familiar in the way Dasein has been factically interpreted that we never pay any heed to it, hides enigma after enigma existential-ontologically. The 'natural' horizon for starting the existential analytic of Dasein is *only seemingly self-evident*" (SZ 371 / BT 423). What does this mean? The existential analytic renders enigmatic the everyday ontic fundament of life, what Husserl calls the natural attitude, what Plato calls the realm of opinion, of *doxa*, the cave. But—and this is crucial—Heidegger does not say that the existential analytic overcomes or permanently brackets out the natural attitude of ontic life. We do not achieve some permanent breakout from the Platonic cave. Rather, everydayness is determinative for Dasein, even when it has not chosen the They for its hero. Even when I choose to become authentically who I am—resolute, self-certain, constant—the everyday is not extinguished. It is, rather, rendered enigmatic. Or "uncanny," to use another word that Heidegger's deployed in this book. That which is ontically so familiar hides enigma after enigma ontologically. In the words of the opening paragraph of the existential analytic, "The ontically nearest and familiar is the ontologically furthest." That which is ontically nearest is ontologically furthest. That which is closest to us, so close that it's under our noses, is furthest from us and most enigmatic: the everyday. The existential analytic of Dasein seems to return ceaselessly to the enigma from which it begins, an enigma that, in Heidegger's words, shatters the seeming self-evidence of any natural attitude from which phenomenology might begin, in order to force the philosopher to formulate anew the question of being-in-the-world.

That is, Heidegger transforms the beginning point of phenomenology. We move from what in Husserl's work is the self-evidence of the natural attitude to the enigma of facticity, the wondrous fact that one *is*. That's what we live. That's what we always return to. That's what we can achieve momentary mastery over. But that's what we always fall back into. The enigma of everyday life.

Authenticity is not one and done, it is not a once and for all time permanent condition. It's transitory. And it overlays a fundamental opacity, what I've called, in "Originary Inauthenticity," an enigmatic a priori that both seems to resist phenomenological description and is that in relationship to which the phenomenologist describes.[13] Philosophy is an activity that attempts to be equal to the enigma of our being-in-the-world whilst knowing all the time that it cannot. Philosophy is this complex back-and-forth or to-and-fro with every-

13. Simon Critchley, "Originary Authenticity," 135–40.

dayness and our determination by everydayness. We leave the cave, we return to it.

That will do for this long, messy but quite fascinating chapter, which completes and extends a whole number of elements of the analysis that we've seen in *Being and Time*. You might be thinking, well, how many more surprises can there be in this wonderful book? Well, ladies and gentlemen, there's one or two more, and I think there is a big one in the next chapter on historicity. The latter is a compelling theme, and it's also the moment when the issue of politics is going to come to the surface, very much to the fore. It's a question of the following thought: Heidegger seems to be obsessed with death, being-towards-death, mortality, finitude . . . the end, the end, the end. But what about the beginning? What about the start of existence? What about birth? Well, Heidegger's going to deal with that.

Remember—and this can serve as a tantalizing clue—that when, in 1936, Heidegger went to Rome, he met a former student of his, Karl Löwith, a very important philosopher in his own right. Löwith, who had been forced out of Germany because of who he was, his Jewish heritage, meets his former teacher in Rome. Heidegger was apparently wearing the National Socialist insignia on his lapel, a little like American politicians with their stars and stripes. Löwith asked Heidegger, "What's the relationship between what you said in *Being and Time*, between your philosophy, and your attitude towards politics?" Heidegger's one-word answer was "Historicity." Next chapter, we will try to figure out what that answer means.

16. Historicity and Politics

Death-obsessed male philosophers and the question of birth (Arendt) — The power of beginning — Stretchedness, persistence, movedness — Historizing as happening — History is rare — World history (a Greek temple) — The past lies ahead of us — Natural history — The ring — The loyalty of existence — Inauthentic history and the craving for the new — Recurrence and the love of fate — How to approach the science of history on the basis of existence — Monumental, antiquarian, and critical (Nietzsche on history) — Heidegger's debt to Dilthey — Philosophy as a manifestation of life — Fatigue — Why paragraph 74 of Being and Time *is so important — Heidegger's reply to Löwith — "See you later, alligator": excursus on Philippe Lacoue-Labarthe — Heidegger's revolutionary conception of history — Fate as individual, destiny as collective — My struggle — The People (Volk) — "The inner truth and greatness of National Socialism" — Young Habermas — Who are the Germans? — Heidegger's thinking of the One (henology) — Against authenticity — Defusing the link between philosophy and politics — Radicalizing facticity — Tragic and comic Heidegger — Powerlessness — Heritage as the counter-concept to tradition — History as repetition — A weird thought: we repeat from out of the future — History is not the past — History comes to be from out of the future (Sandinistas).*

WE'RE NOW LOOKING at the penultimate chapter of *Being and Time*, "Temporality and Historicality" (SZ 372–404 / BT 424–55). I'd like to discuss these pages in a particular way. The main focus for the discussion is paragraph 74, titled "The Basic Constitution of Historicality" (SZ 382–87 / BT 434–39). This is where I want to make the connection to Heidegger's politics. The path from *Being and Time* to Heidegger's political commitment to National Socialism in 1933 leads through paragraph 74, which brings up the awkward question of what authentic collective Dasein might look like. But I'll save that for the climax of the discussion. I'm going to organize the other paragraphs around 74 and do those first. So let's begin with paragraph 72, where he lays out the plan for the chapter.

Heidegger opens this chapter with a familiar question, by asking whether we have truly brought Dasein into view as a whole. He's used that phrase throughout *Being and Time*, but here it opens up a new possibility for his analysis: "Have we indeed brought the whole of Dasein, as regards its authentically *Being*-a-whole, into the fore-having of our existential analysis?" (SZ 372 / BT 424). This introduces a fascinating question that Heidegger hasn't discussed yet—namely, that of *birth*, the question of Dasein's beginning, not just its end in death. Heidegger writes:

> It may be that as regards *Being-towards-the-end* the question itself may even have found its answer. But death is only the "end" of Dasein; and, taken formally, it is just *one* of the ends by which Dasein's totality is closed round. The other "end," however, is the "beginning," the "birth." Only that entity which is "between" birth and death presents the whole which we have been seeking. (SZ 373 / BT 425)

We've heard a lot about being-towards-death, but what about being-towards-birth? Introducing this concept places in question the *doxa* that Heidegger, like most male philosophers, is a death-obsessed chap. "Isn't the question of natality absent in *Being and Time*?" People often say this from a perspective influenced by Hannah Arendt. She raises the issue of natality in her wonderful book *The Life of the Mind*.[1] She says that there is an obsession with the question of death amongst philosophers. True. What about birth and beginning? But the idea that this is absent from Heidegger's work—that Heidegger is your usual death-bound, male philosopher—is not true. Here at the beginning of this chapter, we find him talking about birth. The other theme that arises in relation to this question of birth is a way of thinking about beginning in Heidegger's work.

Now, in relation to Hannah Arendt, this question of beginning is the way she thinks about revolution, the revolutions that she endorses, like the American Revolution of 1776, and the revolutions that she disparages, like the French Revolution of 1789. Such a judgment is based on the distinction between the political, exemplified in the American Revolution, and the social, exemplified in the French Revolution—an extremely questionable distinction in my view. Why would awful social conditions like poverty *not* be the legitimate basis for a demand for the transformation of the state?[2] But where, for Arendt, the American Revolution is an exemplification of the power of beginning, for Heidegger there is also a revolution, but that revolution, at least briefly glimpsed in 1933, is the National Socialist revolution. Fascinatingly, in Heidegger's National Socialist texts, where he's writing as the rector, the provost of Freiburg University, you find the concept of beginning mentioned

1. Hannah Arendt, *The Life of the Mind* (Harvest, 1978), 109–10, 212–17.

2. Hannah Arendt, *On Revolution* (Viking, 1963).

on a number of occasions. It is central to the infamous rectoral address that he delivered in May 1933, "The Self-Assertion of the German University."[3] (When the fascists come to power, the first thing to be sorted out is the universities—sound familiar?) The question of beginning doesn't go away in Heidegger's work. It's there in the work of the later 1930s, when he begins to talk more gnomically about "the other beginning."[4] This other beginning would be the epoch that would succeed the time of the completion of metaphysics and global nihilism, which is where Heidegger comes to think that we live, move, and have our being.[5] I have strong doubts about the veracity of such talk. But the question of the beginning, of the *Anfang*, is a key issue in Heidegger's work, and it has different valences. Let's just put it neutrally in that way.

So we need to consider Dasein's beginning and not just its end. And we need to try to account for the idea of Dasein "stretching itself along between birth and death" (SZ 373 / BT 425). If we are not

3. Martin Heidegger, "Die Selbstbehauptung der deutschen Universität," in *Reden und andere Zeugnisse eines Lebensweges*, ed. Hermann Heidegger, in vol. 16 of *Gesamtausgabe*, ed. Friedrich-Wilhelm von Herrmann (Klostermann, 2000), 107–17; "The Self-Assertion of the German University," trans. Karsten Harries, *Review of Metaphysics* 38, no. 3 (1985): 467–502.

4. Martin Heidegger, *Der Wille zur Macht als Kunst*, in *Nietzsche*, ed. Brigitte Schillbach, vol. 6.1 of *Gesamtausgabe*, ed. Friedrich-Wilhelm von Herrmann (Klostermann, 1996), esp. 211–12; *The Will to Power as Art*, trans. David Farrell Krell, vol. 1 of *Nietzsche* (HarperCollins, 1979), esp. 208.

5. See Martin Heidegger, *Die Geschichte des Seyns*, ed. Peter Trawny, vol. 69 of *Gesamtausgabe*, ed. Friedrich-Wilhelm von Herrmann (Klostermann, 1998); *The History of Beyng: 1. The History of Beyng, 1938–40; 2. Κοινόν, Out of the History of Beyng, 1939–40*, trans. Jeffrey Powell and William McNeill (Indiana University Press, 2015).

just beings who are bound to die but are between birth and death, and being between birth and death is the connectedness of a life, then how do we account for such connectedness? Dasein is the being of the between. In Heidegger's words, "Dasein *is* the 'between'" of birth and death (SZ 374 / BT 427). We are the stretch between a beginning and an end. Birth and death are connected through this stretch, in some way we haven't quite figured out yet.

Another way Heidegger puts this is that Dasein "persists" (SZ 373, 375 / BT 425, 427). We could think about that persistence in relationship to a thinker like Spinoza—who's not on Heidegger's mind in *Being and Time* or elsewhere, which is decidedly odd for such a specialist in the history of philosophy—namely, in the *conatus essendi*, persistence in being, that characterizes us for Spinoza.[6] And also in relation to movement, *kinesis*, which I've mentioned already, that really what Heidegger is trying to do in *Being and Time* can be thought very simply as thinking and describing the human being in movement and as movement. Here, in this chapter, movement is importantly described as Dasein's *historizing*. A better way of thinking about it is as "happening." The word in German is *Geschehen*, and Heidegger is playing with a number of terms in this chapter that are linked to the idea of history (*Geschichte*). This idea of historizing, *Geschehen*, the happening of Dasein, the taking place of us, is the key. When you see "historizing" in the text, and you think, What does that mean? Well, it's not a bad translation of *Geschehen*. It's easy to criticize translations but much more difficult

6. Benedict de Spinoza, *A Spinoza Reader: The* Ethics *and Other Works*, trans. and ed. Edwin Curley (Princeton University Press, 1994), 159.

to translate, especially with a thinker like Heidegger who works through the building of sonic connotations. His thinking is like a sound bath.

Think of Dasein as the existential core of history. Heidegger's claim is that this historizing, this happening, understood existentially and temporally as the *movedness* of existence, is the basis for historicality, or historicity. Happening is the basis for any understanding of history. Again, Heidegger is forming these poetic, almost rhyming word sequences. Historicity is *Geschichtlichkeit*, and, as Heidegger puts it, the essence of history lies in Dasein's historizing (SZ 378 / BT 429–30). Heidegger wants to trace the idea of history, *Geschichte*, back to Dasein's historizing, *Geschehen*. And then he wants to link that idea of historizing as happening back to the notion of temporality. Temporality, which is the being of Dasein, is the condition of possibility for historicality. That's the basic point in this chapter.

Throughout these pages, Heidegger is making a distinction between history in his sense, the German term *Geschichte*, and the Greco-Latin derivations of *Historie*, which is translated in this chapter as "historiology." It's not confusing if you get with the rhythm of the translation. We have a basic distinction between historicality (*Geschichtlichkeit*) and the scientific study of history (*Historie*). Heidegger wants to trace the science of history back to the happening of Dasein. Then he wants to think about that happening as the eruption of the new, as what he calls, in the rectoral address, "the power of beginning," and later, "the other beginning." What Heidegger is pushing towards in these pages, and it's important, is an idea of history not as a continuum, but as a *happening* that occurs at certain key, decisive points. It's not happening all the time. History, as Heidegger will say, is seldom. "History is rare."[7] Other thinkers, such as Walter Benjamin, were saying similar things.[8] History is not a continuous succession of events, one damn thing after another. It is the rarity of a happening. Such history can seize hold of a people or be seized by a people in moments of crisis and decision. Decision is the key to Heidegger's understanding of politics and history. The only question is whether a Dasein or the Dasein of a people is prepared to make the decision to become truly historical. Can a people make a decision to begin, to become something? True politics is about such decisions. Here, Heidegger's work intersects with that of another Nazi fellow traveler: Carl Schmitt.[9]

So much by way of preface. In paragraph 73, Heidegger makes the following remark:

> We contend that what is *primarily* historical is Dasein. That which is *secondarily* historical, however, is what we encounter within-the-world—not only equipment ready-to-hand, in the widest sense, but also the environing Nature as "the very soil of history." Entities other than Dasein which are historical by reason of belonging to the world,

7. Martin Heidegger, *Erläuterungen zu Hölderlins Dichtung*, ed. Friedrich-Wilhelm von Herrmann, vol. 4 of *Gesamtausgabe*, ed. Friedrich-Wilhelm von Herrmann (Klostermann, 1981), 76; *Elucidations of Hölderlin's Poetry*, trans. Keith Hoeller (Humanity Books, 2000), 97.

8. Walter Benjamin, "On the Concept of History," 397.

9. See Martin Heidegger, "Heidegger and Schmitt: The Bottom Line," special issue, *Telos* 72 (1987): 132.

> are what we call "world-historical." It can be shown that the ordinary conception of "world history" arises precisely from our orientation to what is thus secondarily historical. (SZ 381 / BT 433)

What is primarily historical is Dasein, us. We are the essence of history. What is secondarily historical is what we encounter in the world: equipment, the stuff that surrounds us and makes up our environing world. But, Heidegger concedes, nature is also a basis for history, as the material out of which history is made. And you could even link that to an idea of natural history. At this point in Heidegger's work, "world-historical" denotes the history of entities, of things, that are other than Dasein. An example that he gives of something other than Dasein is a Greek temple, where "a 'bit of the past' is still 'in the present'" (SZ 378 / BT 430). That Greek temple will come back in an important way in later texts from the mid-1930s, like "The Origin of the Work of Art," when Heidegger talks about the temple as the setting up of a world and the interruption of that world by the presence of the earth, which is another way of thinking about nature (UK 25–29 / OWA 39–42).

What Heidegger is thinking about here is that history should not be understood solely or primarily in relation to the past. If we begin an understanding of history from the ontology of the present-at-hand, then we're here in the now, understood theoretically or objectively, the future is what is to come, and history is what is no longer. If we approach history as a linear, objective flow of pasts that were once present, then we're misunderstanding the point of discussing history. It is more accurate to say that the past for Heidegger—as what he's been calling *Gewesen* or "having-been"—does not really lie behind us. Rather, history, as Heidegger says in paragraph 74, is what lies ahead of us (SZ 386 / BT 438). *History has its roots in the future.* This is what makes history so important for his project. Its roots are in the future rather than the past. Furthermore, if we see time ecstatically, then the past is not the past. The past, having-been, is what is *to come*, and we as Daseins are always ahead of ourselves. It can come to pass either authentically or inauthentically. In either case, it is lived.

We're going to come back to paragraph 74. That's the main course for today's dinner. Let me move ahead now to paragraph 75, "Dasein's Historicality, and World-History" (SZ 387–92 / BT 439–44). In this paragraph, Heidegger continues the discussion of world history and mentions equipment, books, and institutions as having a history, and he even talks in passing about natural history in relation to the countryside. There's a fascinating quote where he puts "natural history" in quotation marks. "But nature is historical as a countryside, as an area that has been colonized [a loaded word!] or exploited, as a battlefield, or as the site of a cult" (SZ 388–89 / BT 440). World history is something that also has to be understood in relation to the shaping of nature historically: a countryside, a battlefield, the site of a cult, the remains of a Greek temple or church, whatever it might be.

In these pages, there is another example of what he's thinking about in relation to world history, with respect to equipment or things. He gives the curious example of a ring. What is involved in the handing over of a ring from one person to another? When we give a ring and that ring is an act of inheritance, what is entailed by that? This is also an aspect of historicizing. The change when we give a ring from one person to another person isn't just a change of location but a movement of historicizing. He says:

> When, for instance, a ring gets "handed over" to someone and "worn," this is a kind of Being in which it does not simply suffer changes of location. The movement of historizing in which something "happens to something" is not to be grasped in terms of motion as change of location. This holds for all world-historical "processes" and events, and even, in a certain manner, for "natural catastrophes." Quite apart from the fact that if we were to follow up the problem of the ontological structure of world-historical historizing, we would necessarily be transgressing the limits of our theme, we can refrain from this all the more because the very aim of this exposition is to lead us face to face with the ontological enigma of the movement of historizing in general. (SZ 389 / BT 441)

We go from the exchange of a ring to the ontological movement of historizing in general. It's an interesting thought. Think about that in relation to inheritance. We all have examples like this, I'm sure: I have a ring on my right hand, on the third finger. It was my grandfather's ring from Liverpool. My grandmother kept it for me from when he died (in 1968) until about a year before she died (in 2001). When she handed over the ring to me, she said, "This is now yours, Si." And I wear it to this day. And in that ring, to get a little bit sentimental, I can almost feel my grandmother's hand pressing on my hand. I can feel her touch. And my grandfather, too, whom I didn't know very well. But my grandmother, or "Nan," I knew very well indeed and loved her very much.

Much less sentimentally, think about that scene in Tarantino's

Pulp Fiction with Bruce Willis's character, Butch. The Christopher Walken character, who is an army vet, comes in and gives the kid Butch a watch and says, Here's this watch, kid. It was your father's watch. I had it up my ass for two years after your father died. Your father had it up his ass for another three years in the prisoner-of-war camp. So it's been up my ass, or your father's ass, for five years, and here it is. I'm giving it to you. It's yours. That's another sort of handing down. This experience of inheritance, this movement of one thing to another, is also a movement of historizing. And there is an enigma to this movement. It can't just be understood in terms of a change of location. The point is that the ring that's on my finger or the watch that was in, as it were, Christopher Walken's ass, can't just be understood in terms of a change of location from one place to another. It has to be understood in a more existential sense.

Heidegger then adumbrates the idea of the connectedness of life, as Dasein pulling itself together out of the dispersion and disconnectedness of inauthentic life, and introduces the idea of what he calls the "*steadiness which has been stretched along*" (SZ 390 / BT 442). Again, the question that's been announced at the beginning of this chapter is, How do we account for the connectedness of life? That connectedness is a stretched-along steadiness. The German there is rather nice: *die erstreckte Stätigkeit*, where we pull together the disconnectedness of our life into a unified narrative. Such a unified narrative is also something that we can use to articulate resoluteness understood as the *loyalty* of existence to itself (SZ 390–91 / BT 443). This is a way we can think about the movement of historizing, the connectedness of life in terms of a unification, a steady stretching along that is a fidelity to existence itself: "Resoluteness constitutes the *loyalty* of existence to its own Self" (SZ 391 / BT 443). The German word for loyalty is *die Treue*, which is also one of the roots of the idea of truth—namely, that truth is a kind of troth, a fidelity, a loyalty. This is a concept that the philosopher Alain Badiou has incorporated into his idea of the event. The event is something that happens, that erupts, and that is something we have to be faithful to, loyal to, in an experience of decision. That's authentic connectedness.

What does inauthentic historicity look like? Heidegger writes, and this is pretty interesting:

> In inauthentic historicality, on the other hand, the way in which fate has been primordially stretched along has been hidden. With the inconstancy of the they-self Dasein makes present its "today." In awaiting the next new thing, it has already forgotten the old one. The "they" evades choice. Blind for possibilities, it cannot repeat what has been, but only retains and receives the "actual" that is left over, the world-historical that has been, the leavings, and the information about them that is present-at-hand. Lost in the making present of the "today," it understands the "past" in terms of the "Present." On the other hand, the temporality of authentic historicality, as the moment of vision of anticipatory repetition, deprives the "today" of its character as present, and weans one from the conventionalities of the "they." When, however, one's existence is inauthentically historical, it is loaded down with the legacy of a "past" which has become unrecognizable, and it seeks the modern. But when historicality is authentic, it understands history as the "recurrence" of the possible, and knows that a possibility will recur only if existence

> is open for it fatefully, in a moment of vision, in resolute repetition. (SZ 391–92 / BT 443–44)

It's a very rich paragraph: inauthentic historicity is dominated by the present, the present of today and the bored awaiting of the next new thing, which is combined with the forgetfulness of the past. A past that weighs us down inauthentically. Inauthentic historicality is experienced as a flow of inconsequential events. We live in a superficial present, we await the next new thing, and we're weighed down with the past that we immediately forget. Heidegger is building this inauthentic historicity into his analysis of inauthentic ecstatic temporality understood in terms of forgetting, or an awaiting that forgets (SZ 339 / BT 389).

By contrast, authentic historicity deprives the today of its character as presence and understands history as the recurrence of the possible. Listen to that: "*the 'recurrence' of the possible*" (SZ 391–92 / BT 444). The word for "recurrence" is similar to the idea of repetition. Think of repetition and recurrence together. Repetition is *Wiederholung*, fetching back, as I said last time. Recurrence here is *Wiederkehr*, a re-turn not in the sense of a U-turn or turning back but an occurring again. Authentic Dasein knows that a possibility will only recur if existence is open to it and to itself fatefully in the moment of vision. Authentic historicity is the affirmation of the recurrence of history.

This could obviously be linked to an idea that is on Heidegger's mind but that is not explicit in the text and that returns in the years that follow, in particular in the mid-1930s when he's giving his four years of lectures on Nietzsche, from 1936 to 1940. Fateful years! This idea of recurrence can be connected to *eternal recurrence* in Nietzsche and linked to the idea of *amor fati*, love of fate. Heidegger doesn't talk about love, of course (except once in passing, in a footnote about fear and anxiety [SZ 190 / BT 492]), but we could think about authentic historicity as love of fate, as love of the possible.

Inauthentic historicity is mired in the past that has become unrecognizable, and what it craves is the modern, the new, the novel (SZ 391 / BT 444). What is the new thing? What are the new trends? What ten new things are happening in the world next week? This longing for the modern, the new, the novel, is rooted for Heidegger in the inauthentic disavowal of historicity. It is the work of ideology.

Heidegger is indeed a deeply anti-modernist philosopher. But he's got a point. In fact, he's got a bloody strong point. Bourgeois society is addicted to the new, to the next big thing. Bourgeois society, modern society, can be defined as curiosity, as endless craving for the novel, new trends, new commodities. We covered Heidegger's concept of curiosity in paragraph 36 in the discussion of inauthentic life. The German word for curiosity is *Neugier*, which has within it this reference to the new, the *Neu*, a craving, *Gier*, for the new. Curiosity becomes the craving for new gear. Can we get some new gear? We lust after the new, we crave after the new. That is the function of inauthentic time, because we seemingly cannot abide the present.

The final words of paragraph 75 are fascinating. If we had the time and shared linguistic know-how, then we'd need to look at the German text in order to get a feel for the full range of metaphors in this paragraph, metaphors of light and shadow that Heidegger's been using throughout this book and that are concentrated in certain moments like the following:

> The existential Interpretation of Dasein's historicality is constantly getting eclipsed unawares. The obscurities are all the harder to dispel when we have not disentangled the possible dimensions of the appropriate inquiry, and when everything is haunted by the *enigma of Being*, and, as has now been made plain, by that of *motion*. Nevertheless, we may venture a projection of the ontological genesis of historiology as a science in terms of Dasein's historicality. This projection will serve to prepare us for the clarification of the task of destroying the history of philosophy historiologically—a clarification which is to be accomplished in what follows. (SZ 392 / BT 444)

The existential interpretation of historicity is constantly and in unseen ways moving into the shadows. These are the shadows or obscurities or, better, *darknesses*. The German here is *Dunkelheiten*. These darknesses are hard to dispel when everything is haunted by the enigma of being and the enigma of being's movement, the enigma of motion, or *kinesis*. Two things to note here. First, this idea of enigma, which I have pointed out a number of times, is front and center in the chapter on historicity, and the enigma of being is linked to the enigma of motion. And this movement, this motion, is the essence of being. What Heidegger is trying to describe in *Being and Time* and throughout his entire work is movement. Essence as movement. Essence as something that essences, something that happens. It is not something static.

The great error of philosophy and religion and much else besides (think of gender or race in this regard) is to think essence in static terms, to freeze it into some being like God and then to say that all things exist by virtue of this essence as a ground. Heidegger's trying to think of essence as groundless movement. This is the enigma. Also notice in the paragraph I just quoted how Heidegger's language is progressively loosening up. He's not mentioned destroying the history of philosophy before. The claim he makes in the introduction to *Being and Time* was about the destruction of the history of ontology. That's what he says in paragraph 6, and it's come up a number of times since. But here the deconstruction is really the destruction of the history of philosophy understood historiologically. The way in which philosophy has conducted itself, in particular, the way in which philosophy has thought the notion of history, has to be deconstructed. That deconstruction is something that is going to reveal the enigma of history. Making sense of that is the task that Heidegger reserves for his future work, much of which is on the history of philosophy and the way in which the history of philosophy has produced an ever-increasing forgetfulness or oblivion of being.

Let me talk just briefly about paragraph 76. It is pretty self-explanatory. It just does what it says on the tin. It is a question of the existential origin of *Historie*—the science of history, what's being translated here as historiology—out of Dasein's historicity. The issue here is how we explain the emergence of history as a science out of existential analysis. There are interesting moments in this paragraph, with mentions of remains, monuments, ruins, and records, which are all things that greatly interest me. But what is most interesting is the way in which Heidegger assimilates an authentic historiology, or the study of the science of history, to the three dimensions of ecstatic temporality—the to-come, the having-been, and the moment of vision.

He assimilates those three dimensions of ecstatic temporality into Nietzsche's tripartite distinction of history, "The Uses and Abuses of History for Life." That threefold division is between the monumental, the antiquarian, and the critical conceptions of history.[10] It's very suggestive because it's the point in *Being and Time* where you can really see Heidegger trying to fold his analysis into a set of Nietzschean categories. And the result of that is quite simple. The monumental is tied to the idea of the future. The antiquarian is linked to having-been. And the present, or moment of vision, is linked to the idea of critique. In this way, an authentic science of history becomes a way in which the "today" is deprived of the present, and we can detach ourselves from inauthentic fallen publicness (SZ 396–97 / BT 448–49).

The existential conception of historicity is based on this idea of the monumental, the antiquarian, and the critical; the future or the to-come (the monumental), the having-been (the antiquarian), and presence as the moment of vision (the critical). If we can assume that idea of historicity, then we can pull away from fallen publicness. Such a pulling away is going to be painful, Heidegger says. Painful but essential. So, "Authentic historiology" (namely, an authentic science of history)

> becomes a way in which the "today" gets deprived of its character as present; in other words, it becomes a way of painfully detaching oneself from the falling publicness of the "today." As authentic, the historiology which is both monumental and antiquarian is necessarily a critique of the "Present." Authentic historicality is the foundation for the possibility of uniting these three ways of historiology. But

10. See Martin Heidegger, *Unzeitgemäßer Betrachtung*, ed. Hans-Joachim Friedrich, vol. 46 of *Gesamtausgabe*, ed. Friedrich-Wilhelm von Herrmann (Klostermann, 2003); *Interpretation of Nietzsche's Second Untimely Meditation*, trans. Ullrich Haase and Mark Sinclair (Indiana University Press, 2016).

> the ground on which authentic historiology is founded is *temporality* as the existential meaning of the Being of care. (SZ 397 / BT 449)

A critique of the present—even a history of the present, as Foucault will pick up this thought—requires a basis in Dasein's historicity. The three dimensions of history—what has been, what's happening now, what is to come; antiquarian, critical, and monumental—have their roots in the existential temporality of care as the basis for Dasein's happening, Dasein's historizing. That's the basic idea. Perhaps we can interpret the primacy of the future in Heidegger in relation to Nietzsche's praise of the monumental, heroic idea of history, which can be thought of as the power of beginning.

Paragraph 77 is somewhat odd, but it's also pretty self-explanatory. Heidegger is trying to substantiate the claim that he made in the opening pages of chapter 5, where he talks about the research of Wilhelm Dilthey, who's very important for the project of *Being and Time*, not someone who's read very widely, but from whom we get the idea of a philosophy of life and hermeneutics. Those ideas, along with the idea of factical life, are picked up from Dilthey and developed by Heidegger. Indeed, Heidegger gave a ten-part series of lectures on Dilthey in 1925 called "Wilhelm Dilthey's Research and the Struggle for a Historical Worldview."[11] That is to say, just before

11. Martin Heidegger, "Wilhelm Diltheys Forschungsarbeit und der gegen wärtige Kampf um eine historische Weltanschauung," in *Vorträge, 1915–1932*, ed. Günther Neumann, vol. 80.1 of *Gesamtausgabe*, ed. Friedrich-Wilhelm von Herrmann (Klostermann, 2016), 103–58; "Wilhelm Dilthey's Research and the Struggle for a Historical Worldview," trans. Charles Bambach, in *Supplements*, 147–76.

the writing of *Being and Time*. Heidegger is a Diltheyan. The chapter on historicity that we're reading is very clearly the extension of Dilthey's thought that philosophy is a manifestation of life and that life is a historical phenomenon. Therefore, we have to approach historical inquiry from the idea of existential historicity.

What is essential if we're thinking historically—and we *have* to think historically—is that there is no way of doing philosophy outside of history, beginning with some a priori, ahistorical set of eternal ideas or any sort of formalism. Nonhistorical philosophy is a nonsense nonstarter for Dilthey and for Heidegger (and indeed for me too). In paragraph 77, a defense of Dilthey is offered by way of a lengthy paraphrase of a letter exchange. It's rather sweet, actually. It's a correspondence between Dilthey and Count Paul Yorck von Wartenburg, who's a little-known figure but who lives on through the way he is picked up in these pages by Heidegger. Yorck's views are consistent with Heidegger's views on hermeneutics, with what Heidegger would call productive logic—he talks about that in the introduction to *Being and Time* (SZ 10 / BT 30–31).

Between you and me—and don't let this go any further—I have no idea why Heidegger included these pages in *Being and Time*. They really have the character of notes. That's a good thing to realize about Heidegger, or indeed any philosopher. If Heidegger is a major thinker (and hopefully we all think he is), then how does he develop his work? Well, he developed his work by reading and making notes. My advice for philosophical success is simple: *write notes*. Develop the neglected art of the précis. And what we see in paragraph 77 are really a bundle of notes that might have been developed into something more focused and systematic if Heidegger had the time. But he's writing division 2 of *Being and Time* very fast, incredibly quickly, in six to eight weeks, because he is up against the Berlin tenure clock.

At the end of this chapter, you can feel Heidegger's fatigue. He's exhausted, the book begins to run out of steam. I find that interesting. I find exhaustion interesting. When we feel exhausted, that's not necessarily a bad thing for thinking. People complain, "Oh, I'm tired all the time. It's terrible." Yes, of course. But isn't fatigue phenomenologically interesting? Doesn't it mean that some other set of experiences becomes possible? Isn't fatigue, deep weariness, suggestive of different ways of experiencing and thinking about human existence and specifically the nature of facticity? Emmanuel Levinas certainly thought so, and he gives a wonderful analysis of the movement between fatigue and effort in section 2 of chapter 1 of his *Existence and Existents*.[12]

Being and Time is beginning to exhaust itself. Heidegger is exhausted writing it. And here we are with this odd little discussion of Dilthey and Yorck before we turn to the final chapter of *Being and Time,* which will actually deal with . . . drumroll . . . *time*!

•

But we've forgotten something. We've forgotten to serve the main course in the dinner on historicity. All that I've said so far is just the canapés and the hors d'oeuvres and maybe a little bit of pasta to stop

12. Emmanuel Levinas, *Existence and Existents*, trans. Alphonso Lingis (Duquesne University Press, 1978), 18–26.

you feeling hungry. Now we're going to get stuck into the meat and mashed potatoes of the chapter on historicity: paragraph 74.

Let me first remind you of why paragraph 74 is so important by linking it to the argument I've been trying to hone in my reading of *Being and Time* and that I've tried to develop in some of the writing I've done on Heidegger over the years. Throughout this book, I've kept repeating that there are two formulas that provide a clue to understanding what takes place in *Being and Time*: Dasein exists factically, and Dasein is thrown projection. Ultimately, I would like to modify the way we hear the formulations "thrown projection" or "factical existing" by placing the emphasis on the *thrown* and the *factical* rather than on projection and existence. That is, my interpretation of *Being and Time*, which has been quietly driving much of what I've been saying, is that Dasein is fundamentally a *thrown* throwing off, a *factical* existing. Rather than a thrown *projection*, a factical *existence*, I want to place the emphasis upon thrownness and facticity.

What's continually appealed to in Heidegger, in *Being and Time*, and even more so in the later work, is a change in our capacity for hearing. That is, whether we listen away (*hinhören auf*) to the They and get caught up in their chatter or hear the appeal that Dasein makes to itself—namely, if we hear the appeal that we make, where the "we" is understood as being-in-the-world. It's my hope that a change in the way we hear these key formulas in *Being and Time* will produce what we might call, in a Wittgensteinian spirit, an aspect change in the way we understand the project of fundamental ontology. That's my overriding ambition.

I'm a Heideggerian. But I'm a rather worried Heideggerian,

because I'm troubled by all sorts of aspects of his work. How might we rethink his concepts in ways that avoid or make more complicated certain outcomes in Heidegger's own itinerary, particularly with respect to ethics and politics? I'm asking for an aspect change. A shift in our understanding.

But why is that necessary? Why is that aspect change necessary? It's necessary, I think, in order to move our understanding of *Being and Time* away from the heroic political pathos of authenticity, which is consummated in the discussions of fate and destiny in paragraph 74 of *Being and Time*. Remember the story I told you at the end of our last discussion, about Karl Löwith's meeting with Heidegger in Rome, when Heidegger visited the Eternal City as part of a German cultural delegation in 1936. That itself was a terrible error of judgment. Heidegger and Löwith met in Rome and in Frascati, and when he asked Heidegger about the relation between his philosophy and his political commitment to National Socialism, the one-word answer was "Historicity." Historicity is the connecting tissue.[13]

Now, what does that mean? Well, that is what's going on in paragraph 74. I'm building off the work of a number of philosophers in making the argument I'm going to propose, but I won't bore you with listing all those scholars. But I am standing on some broad shoulders. One philosopher who did important work on this issue is Philippe Lacoue-Labarthe, who was always known as "Lacoue." He wrote an important text called "Transcendence Ends in Politics,"

13. Karl Löwith, "My Last Meeting with Heidegger in Rome, 1936," trans. Richard Wolin, special issue, *New German Critique* 45 (1988): 115–16.

which is contained in a book called *Typography*, published in English in 1989.[14] And in that essay, he shows the direct influence of paragraph 74 on Heidegger's political texts from 1933 to 1934. I'm secretly borrowing that argument. Well, not so secretly now.

Philippe Lacoue-Labarthe is a very interesting figure. He is indicative of the way things go with philosophy, the way certain people get remembered and certain people don't. Philippe Lacoue-Labarthe and Jean-Luc Nancy worked together for twenty or thirty years, lived in the same house, and actually exchanged wives at a certain point. Or the wives exchanged husbands—depends how you see things. They both taught at the University of Strasbourg in the northeast of France and made that place a really vibrant intellectual center for a good long time. Nancy is well remembered, but Lacoue-Labarthe not as much. Lacoue died younger, in 2007, because he smoked like a sailor. He somehow taught in the United States, at Berkeley for a number of years, without ever speaking a single word of English (a remarkable feat). I met him a number of times, and the only time I ever heard him speak English was one night when he was leaving a group of late-night carousers in a cafe in Urbino, Italy. We'd all been talking French and, as he was depart-

14. Philippe Lacoue-Labarthe, "Transcendence Ends in Politics," trans. Peter Caws, in *Typography*, 267–300. Readers interested in the relation between Heidegger and politics, particularly in relation to the so-called Black Notebooks—Heidegger's journals written between 1931 and 1970 (thirty-four notebooks totaling about 3,384 pages), first published in the mid-2010s—should read the excellent work of Peter Trawny, *Heidegger and the Myth of a Jewish World Conspiracy*, trans. Andrew J. Mitchell (University of Chicago Press, 2015). See also Richard Polt, "Inception, Downfall, and the Broken World: Heidegger Above the Sea of Fog," in *Heidegger's "Black Notebooks": Responses to Anti-Semitism*, ed. Andrew J. Mitchell and Peter Trawny (Columbia University Press, 2017).

ing, he said in English, "See you later, alligator." Anyway, Lacoue-Labarthe was a *reader* of texts. He wasn't doing first philosophy, he wasn't doing metaphysics in the way that Nancy went on to develop in an important series of books. Lacoue wasn't trying to write some new ontology. He was simply commenting on texts. His ambitions were more modest, and he was a fascinating figure. Anyways, he's in the background. Enough about that.

Let me briefly restate the argument about the relationship between historicity and politics, because it will be far from an obvious argument why and how those two things are connected. Dasein's authentic anticipation of its death is called fate (*Schicksal*). And this is designated as the originary historicizing—*Geschehen*—the original happening of Dasein. Heidegger's claim in division 2, chapter 5, is that the condition of possibility for any authentic understanding of history or any science of history lies in Dasein's historicity, in its historizing. Which means the self-understanding of the temporal character of being human, the condition of possibility for any historical understanding, lies in finitude. So, to repeat, the meaning of the being of being human, the being of Dasein, is temporality. The meaning of temporality is finitude. We saw that in detail in chapter 14, "Ecstatic Temporality." Dasein's authentic self-understanding of finitude is fate, and this originary historicizing is the condition of possibility for any authentic relation to history. And any authentic relation to history is what Heidegger means by world-historical historicizing, which he discusses in paragraph 75, which we just discussed.

It's clear that political events such as revolutions, the founding of a state, or general social transformations would qualify as world-historical events for Heidegger. The essence of history is Dasein's historizing, and history is rare. Heidegger's is a transformative theory of history. It has similarities with revolutionary theories of history. Again, not unlike Walter Benjamin. He writes:

> Only by the anticipation of death is every accidental and "provisional" possibility driven out. Only Being-free *for* death, gives Dasein its goal outright and pushes existence into its finitude. Once one has grasped the finitude of one's existence, it snatches one back from the endless multiplicity of possibilities which offer themselves as closest to one—those of comfortableness, shirking, and taking things lightly—and brings Dasein into the simplicity of its *fate* [*Schicksals*]. This is how we designate Dasein's primordial historizing, which lies in authentic resoluteness and in which Dasein *hands* itself *down* to itself, free for death, in a possibility which it has inherited and yet has chosen. (SZ 384 / BT 435)

Heidegger is saying that history is something we choose, something we rise up into when we can move away from the usual shirking and taking things lightly that defines our everyday existence. When we rise up into history, we rise up into the simplicity of our fate.

Now, it was established in division 1, chapter 4 that Dasein is always being-with, Dasein is *Mitsein*. Being-there is being-with. The a priori condition of being-in-the-world is being together with others in that world. We've known that for a long time. As is well known, the everyday ontic actuality of this a priori condition of being-with is what Heidegger calls the They, the one, anyone, *das Man*. This is determined as inauthentic because in such everyday experience Dasein is not truly itself but is lived through by the customs and conventions of the existing social world. Think back to that remark in division

1, chapter 4: in the They, "everyone is the other, and no one is himself" (*Jeder ist der Andere und Keiner er selbst*; SZ 128 / BT 165).

Returning more properly to the arguments of paragraph 74, if fateful authentic Dasein is always already *Mitsein*, then such historicizing has to be what Heidegger calls co-historicizing. Here is the key claim: Historicizing isn't just my individual historicity, it is the co-historicizing of my being-with-others. An authentic individual life is not one lived in isolation and in opposition to the shared life of the community, understood as "them." The question therefore arises: What is the authentic mode of being together with others that escapes or masters the inauthenticity of the They? I've signaled this question a number of times in this book. Heidegger writes, "But if fateful Dasein, as Being-in-the-world, exists essentially in Being-with Others, its historizing is a co-historizing and is determinative for it as *destiny* [*Geschick*]" (SZ 384 / BT 436).

The way I rise up into the acceptance of my fate (*Schicksal*) is with resoluteness. I decidedly anticipate my being-towards-death, I become ecstatically temporal. Destiny, or *Geschick*, is a common fate I share with others. Destiny is the authentic happening that I experience with others insofar as my individual fate is always already bound up with the collective destiny of the community to which I belong. Heidegger goes on: "Our fates [a more accurate translation would be "the fates"] have already been guided in advance, in our Being with one another in the same world and in our resoluteness for definite possibilities. Only in communicating and in struggling does the power of destiny become free" (ibid.). The fates are guided from the front in being with one another in the same world. Resoluteness, my individual fate, is always already guided by the fate that I share with others. This is the power of destiny that becomes free in communication and struggle, *Mitteilung und Kampf.* The mention of "struggle" is repeated in these lines. The fates of authentic individual Daseins are guided from the front by the destiny of the collective, a destiny that first becomes free for itself or self-conscious—not a term that Heidegger would use—in communication and struggle. Now, obviously, the word struggle, *Kampf,* has acquired some dubious political connotations between the period that saw the publication of *Being and Time* and the present. Heidegger reads Hitler's *Mein Kampf,* or *My Struggle,* in 1930 and is impressed. Let's just say that the notion of struggle is a little bit compromised and leave it at that.

This is where this argument is taking us. Fateful Dasein is with others. That co-historizing with others is destiny. Such destiny is expressed in communication and struggle. He then says, a few lines further up, "This is how we designate the historizing of the community, of a people" (*Damit bezeichnen wir das Geschehen der Gemeinschaft, des Volkes*; ibid.). The concept here is *Volk,* people. Strong concept. Problematic concept. But maybe an essential concept. The argument is that the authentic communal mode of being-with, which masters the inauthenticity of the They, is the people. The question that's been on our minds in reading *Being and Time* is, If the They is our inauthentic average everydayness, our being-with-others defined by curiosity, ambiguity, and idle talk, then what is the counter-concept to the They? The counter-concept, it would appear, is *the people.* And in my view, it's the possible political realization of a resolute and authentic people in opposition to the inauthentic nihilism or bourgeois liberalism of modernity, with its craving for the modern and the new that we just saw in paragraph 75. This possible realization of a people, in opposition to empty modern individualism—particularly as this gets understood by Heidegger in relationship to technology from the mid-1930s onwards—is what he would call in the *Introduction to Metaphysics* in 1935, a year after resigning as rector of his home university of Freiburg, "the inner truth and greatness of [National Socialism]" (EM 208 / IM 222).

What does "the inner truth and greatness of National Socialism" mean? I think it's the idea of the formation of a people. That would be a way of opposing a modernity defined by technology, by individualized alienation, reification, and capitalism. What's interesting is that Heidegger used those words, "the inner truth and greatness of National Socialism," in 1935. But he only published *Introduction to Metaphysics* in 1953. And in those years—from 1935 to 1953—quite a lot happened in Germany and in the world. I realize that this is a considerable and rather stupid understatement. In the preface to the 1953 German edition, he says, "I have made no change in the content" (EM 1 / IM xlv). Does it really ring true that Heidegger changed nothing in the text from the lecture notes to the published book? The worst of it is that it is clear that he did alter some passages from *Introduction to Metaphysics* when they were published. For example, there is a discussion of Rudolf Carnap's lacerating critique of Heidegger that the latter edited out of the book. So Heidegger's lying, and he knows he's lying. But while he did alter the 1953 edition, he did not change the remark about the inner truth and greatness of National Socialism, because he didn't feel he had to. He still believed it to be true.

Imagine that you're a young German. An educated, intelligent, young German after the Second World War. And you're trying to make sense of what happened in the National Socialist catastrophe. You find out that you've actually been taught by people at university that have been through the war, have tacitly collaborated with the National Socialist regime, and haven't really changed the way they're teaching after the war. They just go on working as professors of philosophy regardless. This was the situation of a young philosopher, a young German philosopher, whom I will name in a moment. The first thing this young German philosopher publishes is a review of *Introduction to Metaphysics* from 1953. It's a review published in a newspaper in Frankfurt, the *Frankfurter Allgemeine Zeitung*, a very important newspaper. The review is called, in German, "*Mit Heidegger gegen Heidegger denken*" ("Thinking with Heidegger against Heidegger").

That's how I would describe the effort that I've been trying to make throughout this book. To think with Heidegger and against Heidegger at the same time. Going with the grain and against the grain. The piece in 1953 was published by none other than Jürgen Habermas. It was his first publication.[15] It's a significant moment in his thinking about the following question: How, given the horror that had happened in German history, can you leave unchanged that remark about the inner truth and greatness of National Socialism? Heidegger's deafness to what had happened in the Second World War is deafening, as is his silence about any culpability on his part until his death in 1976.

In my interpretation, there is a systematic philosophical basis to Heidegger's political commitment, which is due to the specific way

15. Jürgen Habermas, "Mit Heidegger gegen Heidegger denken: zur Veröffentlichung von Vorlesungen aus dem Jahre 1935," *Frankfurter Allgemeine Zeitung*, 1953.

in which Heidegger develops the concept of authenticity in division 2 of *Being and Time*. This culminates in the concept of the people. This concept isn't accidental or peripheral; it's central. That is, the only way in which Heidegger can conceive of an authentic mode of human being-together or community is in terms of the unity of a specific people, a particular nation. It is the political expression of that possibility that Heidegger saw in National Socialism in 1933. And the name of that people is the Germans, *Die Deutschen*. An interesting name for a people because the Germans as a people were not unified by a state until 1871. Unlike Britain or France, the German state was formed very late as a consequence of Prussian military, political, and territorial hegemony. Before that, Germany didn't exist. Germany was a set of regions, free cities, principalities, electorates, dukedoms, and so on and so forth, in the loose-fitting robes of the remains of the Holy Roman Empire. What unified the German-speaking lands was language. That language had spread all across northern Europe in different dialects, all the way from Dutch to Yiddish. Heidegger often quotes Friedrich Hölderlin when he's thinking about the Germans. But for Hölderlin, writing in the late eighteenth and very early nineteenth centuries, the Germans are a language group spread over a vast geographic area, a series of lands. He's not thinking about a state. In many ways, Heidegger conflates those two ideas: language and the state.

There's another point that can be made here, which is that Heidegger is a thinker of *the One*. What I mean is that when Heidegger's thinking about who we are, we begin from an idea of mineness. Mineness, which is me. That's where we begin *Being and Time*. That's one. I'm one. Then there is the idea of social life in inauthentic, average everydayness as the One (anyone), as the they-self. And as we've just seen, the counter-concept to the They is the One of the people. Heidegger's thinking is a series of interrelated moves across different conceptions of the One. To that extent, we can say there's a henological logic in Heidegger's work that could be challenged. We could think about beginning with the Two or the Three, and that would be a way of breaking with this henology. There are thinkers who try to think in terms of the Two, or more. These are thinkers of multiplicity like Levinas, Deleuze, and Badiou. Another philosopher who thinks along those lines is Hannah Arendt, who begins from a strong concept of plurality—namely, that we have to commence our political thinking from an idea of plurality as something positive. For Heidegger, plurality is a negative concept, or it has to be captured within the horizon of the One, the They, anyone. There can be no positive idea of plurality in Heidegger's work.

Where does this argument take us? In my view, the urgent task of Heidegger interpretation, provided, of course, that one is not a Nazi—and by the way, I'm not one—and provided that one is still in the business of thinking that Heidegger is a great philosopher, as I do, is to try and defuse the systematic link between Heidegger's philosophy and his politics. How do we do that? The key concept for establishing the link between philosophy and politics in Heidegger is authenticity. I think it's the dream of authenticity that's driving the logic of politics in *Being and Time*.

Therefore, if we want to criticize Heidegger's politics, we have to question the primacy of authenticity. This is why the idea of inauthenticity, and what I call *originary inauthenticity*, is important.[16] There are different ways of making this argument. It's surprising, actually, that the concept of authenticity often survives intact in work that's influenced by Heidegger, but resolutely opposed to his political views. There are many moral and political thinkers who, unsurprisingly, will be opposed to Nazism but in favor of authenticity, usually a normative idea of authenticity. But I don't think it's so simple. I think we have to rethink the concept of authenticity. There are philosophers who will say if we want to hang on to an idea of personal responsibility and individual autonomy that's influenced by Heidegger, then we have to hang on to a normative idea of authenticity. I'm not so sure. There's a lovely remark that Adorno makes in the *Minima Moralia* about the need to take on board the stigma of the inauthentic, and not to endorse some crazy, secularized, but essentially religious notion of authentic personhood.[17]

I've tried to do this in some of my work by developing an idea of originary inauthenticity and a concept of what I call *hetero-affectivity*.[18] We need to rethink concepts like autonomy and what I call the "autonomy orthodoxy" that dominates many ways of thinking about

16. Simon Critchley, "Originary Inauthenticity—on Heidegger's *Sein und Zeit*," 132–51.

17. Theodor Adorno, *Minima Moralia: Reflections from Damaged Life*, trans. E. F. N. Jephcott (NLB, 2005), 154: "Genuineness is nothing other than a defiant and obstinate insistence on the monadological form which social oppression imposes on man. Anything that does not wish to wither should rather take on itself the stigma of the inauthentic. For it lives on the mimetic heritage. The human is indissolubly linked with imitation: a human being only becomes human at all by imitating other human beings."

18. Simon Critchley, *Infinitely Demanding*, esp. 56–61.

moral and political philosophy. The idea behind originary inauthenticity is that human existence is fundamentally shaped in relationship to a brute facticity, or thrownness, that cannot be mastered through any existential projection. Authenticity always slips back into a prior inauthenticity from which it cannot escape but that it would like to evade. It's this movement of evasion, of the self's turning away from itself, where Dasein's embeddedness in factical existence is disclosed. From the perspective of an originary inauthenticity, human existence is something that is first and foremost experienced as a burden or weight, something to which I'm riveted without being able to know why or know further. Inauthentic existence has the character of an irreducible and intractable thatness, what Heidegger calls in chapter 5, division 1, "*das Daß seines Da*" (the that of its there; SZ 136 / BT 175). I feel myself bound to the that of my there, the sheer fact of my facticity that invites some sort of response.

This is where my proposed aspect change begins to kick in. The nature of the response to our facticity will not and does not necessarily have to be as it is in division 2 of *Being and Time*—namely, the authentic and heroic decision of existence that comes into the simplicity of its fate by shattering itself against death. That phrase "*shattering [yourself] against death*" is a wording Heidegger uses in italics, shouting it out loud (SZ 385 / BT 437). No, the response to our facticity does not need to be the heroic mastery of the everyday in the authentic presence of the moment of vision that produces an ecstasy or rapture. On the contrary, the response to the fact of my finitude can be a more passive and less heroic decision, made in the face of a facticity whose demand can never be mastered, and which faces me like a riddle I cannot solve. This situation might be said to call for comic acknowledgment, humorous self-defeat, rather than ecstatic tragic affirmation. The key to human existence in my view is finding yourself ridiculous, which is something I have perfected over the years.

The problem of Heidegger's politics is its tragic-heroic character. I'm more compelled by the comic, by the humor of being human. I think my reading of Heidegger can be developed on the basis of the text itself. And maybe now you will begin to get a sense of why I place the emphasis upon certain concepts in *Being and Time*, especially on the phrasing of those concepts. The whole discussion, a few chapters back, of the double nullity of Dasein in the experience of conscience—this double impotentialization at the heart of Dasein's movement, its happening, its historizing—gives us another way of thinking about authenticity. Everything turns here on questions of power and powerlessness. Heidegger writes:

> If Dasein, by anticipation, lets death become powerful in itself, then, as free for death, Dasein understands itself in its own *superior power*, the power of its finite freedom, so that in this freedom, which "is" only in its having chosen to make such a choice, it can take over the *powerlessness* of abandonment to its having done so, and can thus come to have a clear vision for the accidents of the Situation that has been disclosed. (SZ 384 / BT 436)

The issue is whether Dasein, in letting death become powerful in itself, can take over the powerlessness of abandonment and overcome it, or whether power always has to be measured by the limitation of power, by an essential impotence. What has to be worked out here is Heidegger's phrase "Fate is that powerless superior power" (SZ 385 / BT 436). *Die ohnmächtige Übermacht*: the impotent superpower. (It sounds a bit like the United States.) How might we think about the relationship between power and powerlessness? How do we map that onto questions of authenticity and inauthenticity? And what are the implications of that for us, in thinking about history and politics? These are the kinds of things that are on my mind.

•

There's a lot more to say, but I've already said far too much. Let me conclude by mentioning two more concepts from this chapter, concepts that we need to understand to get a sense of what is going on in this rich and interesting discussion. Those two concepts are *heritage* and *repetition*. And they're linked. We're still in paragraph 74, but a little further on. What is heritage? Heritage is the way in which we take over our factical thrownness authentically and resolutely. When we take it over, we hand it down to ourselves, Heidegger says (SZ 385 / BT 437). Heidegger is using grammatical and lexical elements that are more readily possible in German to make his point. "Tradition" can be thought of as an idea of handing down, like the handing down of a ring. *Überliefern* means to hand something down; *Überlieferung* is tradition. This idea of tradition as handing something down is a way of thinking about the past that wouldn't just be some stale or dead inheritance. Here's the headline: the counterconcept that opposes the idea of tradition as dead past is *heritage* (*Erbe*). Heritage is an active, living handing down. And I've got a lot to say about the idea of heritage. It's a key concept in Derrida's work, by the by. He very often, especially in the later work, described deconstruction as no more but no less than a thinking of heritage. But in these pages, it is linked by Heidegger to the idea of repetition: "In repetition, fateful destiny can be disclosed explicitly as bound up with the heritage which has come down to us. By repetition, Dasein first has its own history made manifest" (SZ 386 / BT 438). History is made manifest in and as repetition. There's a link between heritage and repetition. Heritage is the active, living, existential way of taking something over, which is opposed to a dead tradition, and which is enacted as repetition, fetching back.

The intrication of the concepts of heritage and repetition is developed in the following passage, which is the last thing I'm going to quote here. It's a long quote. But it's a good one.

> The resoluteness which comes back to itself and hands itself down, then becomes the *repetition* of a possibility of existence that has come down to us. *Repeating is handing down explicitly*—that is to say, going back into the possibilities of the Dasein that has-been-there. The authentic repetition of a possibility of existence that has been—the possibility that Dasein may choose its hero—is grounded existentially in anticipatory resoluteness; for it is in resoluteness that one first chooses the choice which makes one free for the struggle of loyally following in the footsteps of that which can be repeated. But when one has, by repetition, handed down to oneself a possibility that has been, the Dasein that has-been-there is not disclosed in order to be actualized over again. The repeating of that which is possible does not bring again something that is "past," nor does it bind the "Present" back to that which has already been "outstripped." Arising, as it does, from a resolute projection of oneself, repetition

> does not let itself be persuaded of something by what is "past," just in order that this, as something which was formerly actual, may recur. Rather, the repetition makes a *reciprocative rejoinder* to the possibility of that existence which has-been-there. But when such a rejoinder is made to this possibility in a resolution, it is made *in a moment of vision; and as such* it is at the same time a *disavowal* [*Widerruf* (i.e., a call against and a call towards)] of that which in the "today," is working itself out as the "past." Repetition does not abandon itself to that which is past, nor does it aim at progress. In the moment of vision authentic existence is indifferent to both these alternatives.
>
> We characterize repetition as a mode of that resoluteness which hands itself down—the mode by which Dasein exists explicitly as fate. But if fate constitutes the primordial historicality of Dasein, then history has its essential importance neither in what is past nor in the "today" and its "connection" with what is past, but in that authentic historizing of existence which arises from Dasein's *future*. As a way of Being for Dasein, history has its roots so essentially in the future that death, as that possibility of Dasein which we have already characterized, throws anticipatory existence back upon its *factical* thrownness, and so for the first time imparts to *having been* its peculiarly privileged position in the historical. *Authentic Being-towards-death—that is to say, the finitude of temporality—is the hidden basis of Dasein's historicality*. Dasein does not first become historical in repetition; but because it is historical as temporal, it can take itself over in its history by repeating. For this, no historiology is as yet needed. (SZ 385–86 / BT 437–38)

This is a brilliant passage. Complex, dense, difficult. Allow me a few thoughts. Repetition is that which underwrites and sustains the constancy of the self. Self-constancy consists in the performative repetition of the three *ecstases* of temporality: the to-come, the having-been, and the present in the moment of vision. The three at once, in one time, as one finite stretch of existence. Repetition at the level of Dasein's historicity, of our happening, is the handing-over, the explicit taking over of possibilities that have come down to us from the past. Our happening as a being, our happening as a living existence, is the taking over of possibilities that have come to us from the past. That's repetition. How does one characterize this repetition? It is not, Heidegger says, a bringing back. It is not a bringing back to itself of something that is past, something that's passed. It's rather, he says, the "reciprocative rejoinder" to a possibility of existence that has-been-there. This word "rejoinder," *Erwiderung*, is a response to the possibilities of the past.

The point that Heidegger is making is that repetition has a very peculiar character. Make whatever sense you can of this. Repetition is not the reiteration of something past but the repetition of the possible that is orientated towards the future. *One repeats from out of the future.* History has its existential roots in the future of death that throws Dasein back onto its factical thrownness. The concealed ground of Dasein's historicity, of its happening, is authentic being-towards-death. That is to say, historicity leans towards anticipation, towards possibility. It is because Dasein is temporal, in terms of the three *ecstases* of time, that it's historical, and then it can take itself over in repetition and become history.

History is not something that *is* because of the fact that it *was*. History is not the past. History is not reducible to the events of the past, to what is scattered around in a present-at-hand way that we can pull together into a history book or a popular history podcast.

History is something that comes to be from out of the future. That is to say—and this is the point—*history is something that has the power of a beginning.* History has the power of what starts, what commences. This is why history is seldom, why history is rare. History is an event, a transformative event, like a revolution, like an utter personal and collective transformation.

What I'm asking you to think about is this idea of history as a transformative event that has the power of a beginning. We can think about history as one damn thing after another, as a series of events within a present-at-hand linear sequence of time: now this, now that. This is the time of the news, the ever-new now that is just like the past now. This would be an idea of history as a flat continuity, a dead chronicle. This flatness and deadness gives us stability and preserves the status quo. *Or* we can think about history as the possibility of emergence. The possibility of beginning. Which we can initiate collectively. And this idea of history is something you can find in Heidegger, in Benjamin, and in other revolutionary thinkers.

I remember when I was a student in the 1980s at the University of Essex in England, and we were very fond of the Sandinistas in Nicaragua, a revolutionary socialist movement that seized power in 1979. They were the heroes of people like me: stupid, pasty-faced, English leftists. But there was this Sandinista poster that said in Spanish something like "The future is ours because we possess our history." This is a revolutionary idea of history where we can possess the future, make it our own, because we possess our history, where our history comes towards us from out of the future. The final question would be, Is that good? Is that bad? What do we think of that?

That comes down to what we think about the idea of revolution and the uses and abuses of that idea in politics. But that is a separate topic for another occasion.

So much for historicity and politics. If you made it all the way to the end of this chapter, then I salute you. It has been a long and breathless ride. All that remains is the final chapter of *Being and Time*, which will be precisely on time.

17. Existential Time and Vulgar Time

Three conceptions of time: primordial time, world-time, and vulgar time — The genesis of the tyranny of the present — Time-reckoning — Datability — Duration or the span of world-time — Inauthentic Dasein has no time — Authentic Dasein is its time and always has time — Take your time — Time first shows itself in the sky — Brightness, night — Tide — Heidegger is not a thinker of the sea — "Primitive" Dasein, the peasant's clock, shadow, and the sundial — "Advanced" Dasein and electricity — Footnote on the theory of relativity — Time and physis *— The genesis of linear now-time — Time in Aristotle's* Physics *— Counted nows, the flow, the stream of time — The error of eternity — Primordial time is finite, reversible, ecstatic, and episodic — How time becomes leveled off — Time and the soul — Spirit falling into time (Hegel) — Why Hegel is wrong — The long, final footnote to* Being and Time *(Derrida) — The endurance of Aristotle's conception of time in Bergson — Time-consciousness in Husserl.*

We're approaching the end of *Being and Time*. In the closing pages of the book, Heidegger is going to question whether time is really the meaning of being, a potentially infuriating undersell given that he's just spent 437 pages arguing this point. But we'll get to that. First I want to talk about paragraphs 78–82, then give you a bonus discussion of paragraph 83, the final paragraph of *Being and Time*. I want to take in the broader picture of where we are in *Being and Time*, what we've learned, and what's going to happen next. This is going to be called "Opening and Presence." There's a late essay where Heidegger says that *Being and Time* could be retitled *Opening and Presence*.[1] We will try and figure out what that means. But we are here, in the final chapter of *Being and Time*, and there is finally an account of the concept of time that Heidegger is trying to criticize in *Being and Time*: the ordinary or vulgar conception of time that's been alluded to, been at work in the whole book, but not really been made explicit. The main claim is to establish the relation of priority between three conceptions of time.

First, there is primordial time. This first layer of time, which we've already seen in the previous chapters of division 2, is the conception of "ecstatic temporality" Heidegger described in paragraph 65. Second, there is what Heidegger calls here the "within-time-ness," *Innerzeitigkeit*, of "world-time." World-time is understood as a ready-to-hand conception of time—at the end of the book, we get a retrieval of the analysis of the *zuhanden*, the handy. So, the first layer of time: primordial time, ecstatic time. Second order of time: world-time, the time of the ready-to-hand world. The third conception of time is vulgar time, to make it sound slightly pejorative. The word *vulgär* in German means "ordinary." *Vulgär*—the common or garden present-at-hand notion of time—is what Heidegger calls now-time. This is a notion of time understood as an infinite succession of

1. Martin Heidegger, "Das Ende der Philosophie und die Aufgabe des Denkens" (1964), in *Zur Sache des Denkens*, ed. Friedrich-Wilhelm von Herrmann, vol. 14 of *Gesamtausgabe*, ed. Friedrich-Wilhelm von Herrmann (Klostermann, 2007), 90; "The End of Philosophy and the Task of Thinking," in *On Time and Being*, 73.

uniform nows: the now of the present, the not-yet-now of the future, and the no-longer-now of the past.

Heidegger's broad claim—which he is now making good on explaining to us—is that this conception of time as a series of nows has its origin in Aristotle's *Physics*, and it influences the entire philosophical and theological thinking of time up to Hegel, who is discussed in paragraph 82, and even Henri Bergson, whose thinking of time is a highly influential precursor (by a few decades) and competitor to what Heidegger is up to in *Being and Time*. Bergson has been barely mentioned in this book but was the most famous philosopher in the world in the first decades of the twentieth century and won the Nobel Prize in Literature in 1927, the same year *Being and Time* was published.[2] (As brilliant as Bergson's thought was, he was unfortunately French, and for Heidegger one cannot really think in French. You can order dinner or maybe sing a cheery *chanson* in French, but for Heidegger true thinking takes place in German and Greek—this is obviously a ridiculous prejudice.) There is a long, important footnote at the end of this chapter where Bergson is mentioned and summarily dismissed. The main claim here is that we have three conceptions of time—primordial time, world-time, and the ordinary conception of time. The ordinary or vulgar conception

2. Bergson's first lecture in the US is widely credited as the source of the first traffic jam on New York's Broadway. As Henry May describes it in his 1959 book, *The End of American Innocence*, "When Bergson finally came to America to lecture at Columbia in February 1913, a line of automobiles (still the vehicles of the well-to-do) clogged Broadway, one lady fainted in the crush at the lecture-room door, and regular students were crowded out of their seats by well-dressed auditors" (Henry F. May, *The End of American Innocence: A Study of the First Years of Our Own Time, 1912–1917* [Knopf, 1959], 228).

of time arises as what Heidegger calls the "leveling off" of primordial time and world-time.

This final chapter of *Being and Time* documents the genesis of the tyranny of the now. The genesis of a certain *metaphysics of presence*. Heidegger doesn't use this expression; it's an expression that's associated with the work of Derrida. And we'll come back to Derrida very briefly at the end of this chapter. The domination of the thinking of time by an idea of the present and a metaphysics of presence is really what's being documented in this chapter. I'm going to work through these pages in a fairly straightforward way. Then we'll draw some more dramatic conclusions or wider inferences in a coda on opening and presence, where we will see what we've learned in the time we've spent thinking together through this long book.

In paragraph 78, Heidegger begins by acknowledging an oversight in his previous analysis. There's always an oversight in Heidegger, isn't there? There's always something left over. We might think we've got Dasein into our grasp as a whole, but we haven't until this, that, or the other has been analyzed. Heidegger keeps making that move, so here we have another oversight—namely, that the approach to primordial time that we've had explained in the previous chapters does not take account of the way in which Dasein reckons with time in its everyday life. "Reckoning with time" is a good way of translating the German verb *rechnen*: counting or reckoning. You can say in colloquial English, "I *reckon* it's about twenty minutes since I left home." "I reckon it's about a day since that happened." Or in response to a question, "Is the railway station close to here?," "I reckon." Reckoning is an everyday way of talking about the counting and accounting of time in an approximative and everyday way, as a form of estimation. And that's what Heidegger wants to focus on in this chapter. In English, *reckoning* also has a sense of past events being "reckoned with," a historical "reckoning," for example.[3] But this is not the sense of *reckoning* that Heidegger has in mind. Historical "reckoning with" in this sense would be closer to what Heidegger described in the last chapter as fateful historical decisiveness, a facing of historical recurrence.

The first task in relation to time-reckoning is elucidating the relation between Dasein's everyday concern, its everyday dealing with time, and ecstatic temporality. This is the work of paragraph 79, which we're going to turn to now. What is the relationship between primor-

3. This sense of reckoning, as a settling of accounts, is captured in phrases like "chickens coming home to roost"—which has its roots in Geoffrey Chaucer paraphrasing Paul in "The Parson's Tale"—and Paul's phrase "reap what you sow" in Galatians 6:7–8. See Meeks, *The Writings of St. Paul*, 20; and Geoffrey Chaucer, *The Parson's Tale*, trans. Mary Farrell Pomerleau (Tumblar House, 2011), 81: "Cursing separates man from the reign of God, as Saint Paul says. And oftentimes, the cursing returns to him who cursed, as a bird returns again to his own nest." Chaucer may be alluding to Ephesians 4:30: "Do not grieve the Holy Spirit of God, with whom you were sealed for the day of redemption" (Meeks, *The Writings of St. Paul*, 120). But the point comes back to remaining in the condition in which you were called (see ch. 8, "Anxiety and Care"). Your actions get away from you because "You are not your own" (1 Corinthians 6:19 [Meeks, *The Writings of St. Paul*, 30]). Thus the modern sense of "historical reckoning" is closer (minus the social justice overtones) to Heidegger's conception of historical conscience we looked at in "Historicity and Politics": letting history act through you (historical decisiveness replacing God's will). So Heidegger's thought has a place for that idea, but when he uses the word itself, *rechnen*, he means something entirely different, the ordinary sense of simply counting or approximating the passage of time.

dial time and world-time, or authentic time and inauthentic time? Obviously, the relation between these two conceptions of time is one of derivation. World-time, like the vulgar conception of time, is also dominated by a notion of the now. Now this, then that, and now the other. World-time is also dominated by an idea of the now, but it's an everyday now, not a scientific now. Not a present-at-hand now but a ready-to-hand now. The "nows" that we experience in day-to-day activity are datable, Heidegger says. And the structure of world-time has the structure of what Heidegger calls *datability* (*Datierbarkeit*; SZ 407 / BT 459). Now, he's not referring to dating apps. Heidegger's not an OkCupid or Tinder kind of chap. He's talking about datability in terms of the marking of the nows in the ordering of time. Heidegger claims that inauthentic world-time is derived from ecstatic temporality. What you see here is Heidegger folding in his analysis of the structure of inauthentic time as a making-present that awaits and forgets. Remember, that's what happened two chapters back in the relation between temporality and everydayness (division 2, chapter 4). He's going to fold this idea of a making present that awaits and forgets into his notion of time-reckoning.

Heidegger says, "The making present which awaits and retains, interprets *itself*" (SZ 408 / BT 460). He's looking at the way in which we move from ecstatic time to everyday time. What Heidegger's saying here, in relation to what he's calling datability, the marking of different nows of time, is incredibly simple in German. He's using a very colloquial, everyday language. He's talking about "now," "then," and "on a former occasion." He talks about a *jetzt* (now), *dann* (then), and *damals* (on a former occasion). You can say "*Damals habe ich*..." or "At that time, I have..." or whatever. Heidegger says: "The horizons which belong to the 'now,' the 'then,' and the 'on that former occasion' [the *Jetzt, dann*, and *damals*] all have their source in *ecstatical* temporality; by reason of this, these horizons too have the character of datability as 'today, when . . . ,' 'later on, when . . . ,' and 'earlier, when . . .'" (SZ 408–9 / BT 461). The very colloquialism of the German makes it harder to understand in English. But the claim here is very simple. It's that our everyday way of dealing with time, what we call "time" in our day-to-day activity, is rooted in the ecstatical unity of temporality, even when this might be unknown to us. What we normally think of as time, our everyday lived time, is rooted in ecstatic temporality. But that primordial experience of time is for the most part concealed. And rightly so. We couldn't live constantly in the moment of vision. It would be exhausting.

Heidegger then draws a contrast between duration, the "during" of our everyday concern with time, and the ecstatic stretch of temporality. If the ecstatic stretch of temporality is one where the three *ecstases* are happening at the same time—out of the future comes my having-been, which is affirmed in the moment of vision—then in everyday world-time, that becomes a duration, or what Heidegger calls "a span" (*Spanne*), "a span of time" (SZ 409 / BT 462). Heidegger writes, "When Dasein is 'living along' in an everyday concernful manner, it just never understands itself as running along in a Continuously enduring sequence of pure 'nows'" (ibid.). This is a thought that's going to be worked out in subsequent paragraphs—namely, the distinction between this everyday, ready-to-hand idea of world-time and the vulgar, or ordinary, experience of time.

In the next move, we get a distinction between authentic and inauthentic temporalizing. This is discussed in an illuminating way.

Heidegger says that the irresolute person, the person that lacks serious authentic resolve, loses their time by living in a present that awaits the next thing that befalls them and then forgets what happened. Another way of thinking about inauthentic time is in terms of losing time. The inauthentic person always says, "I have no time." If an inauthentic Dasein loses time, then an authentic Dasein always has time. Insofar as it exists in the situation of the moment of vision, Dasein makes the time in which and through which it moves. Authentic Dasein is its time and always *takes* its time. Inauthentic Dasein has none, it never possesses time. It lives in a constant state of incipient forgetfulness. Taken by time, but having none. Making haste, but losing speed. The larger point in *Being and Time* is that we are both authentic and inauthentic. We both *have* time and *lose* time. Time can slide by, slipping through our fingers. Heidegger then describes the nature of inauthentic Dasein's relation to time:

> Hence his characteristic way of talking—"I have no time." But just as he who exists inauthentically is constantly losing time and never "has" any, the temporality of authentic existence remains distinctive in that such existence, in its resoluteness, never loses time and "always has time."... When such a moment makes the Situation authentically present, this making-present does not itself take the lead, but is *held* in that future which is in the process of having-been. (SZ 410 / BT 463)

That is the structure of ecstatic temporality. "One's existence in the moment of vision temporalizes itself as something that has been stretched along in a way which is fatefully whole in the sense of the authentic historical *constancy* of the Self" (SZ 410 / BT 463). The constancy of the self that was discussed in division 2, chapter 3, is now deepened, or filled out with an idea of a *historical* constancy. The essence of history is us. History begins with our happening: "*The 'there' is disclosed in a way which is grounded in Dasein's own temporality as ecstatically stretched along, and with this disclosure a 'time' is allotted to Dasein; only because of this can Dasein, as factically thrown, 'take' its time and lose it*" (ibid.). The takeaway here is that authentic Dasein takes its time. Inauthentic Dasein has no time. Remember what Wittgenstein says: "What does one philosopher say to another philosopher? Take your time."[4] Doing philosophy is taking your time. The sophist, the lawyer, the inauthentic professor, never has any time. They're too busy with their pressing appointments and with their alleged "labor."

Let's turn to paragraph 80. The concern here is indeed with our old friend from division 1: concern. Heidegger calls the time of concern "within-time-ness." This refers to the time of entities other than us, entities other than Dasein. This is world-time. The time of concern, the time of our everyday circumspective concern in the world. In this paragraph we get a fascinating little history of time-reckoning from the "primitive," natural time of Dasein, which is the sun and one's shadow—and we'll come back to "sun and shadow" as the primal time of Dasein—to advanced, alienated Dasein, who measures time with clocks. Heidegger emphasizes time-reckoning in order to emphasize that the way in which everyday Dasein reckons with time does not necessarily imply what he calls the application

4. Ludwig Wittgenstein, *Culture and Value*, trans. Peter Winch, ed. G. H. von Wright in collab. with Heikki Nyman (Blackwell, 1980), 80.

of precise "numerical procedures in dating" (SZ 412 / BT 465). The time of the everyday world is *estimated*. It is not an exact measure. That exactitude happens at a third stage.

Primordial time, world-time, ordinary or vulgar time. There's a lot of good material in this paragraph. Particularly the line that I'll come back to in a few moments, where Heidegger says, "'Time' first shows itself in the sky" (SZ 419 / BT 471). Remember that showing itself is the definition of "phenomenon" for Heidegger—meaning that time is first revealed in the sky, as the sky, in the space of the world as a lived time-space unity. This recalls what we have already said about "time-space," and we might link it to the concept of play-space we briefly discussed (see ch. 15, "The Time of the Everyday World").

But let's back up a tiny bit. What about this everyday way of reckoning with time? What are we talking about here? The first way of reckoning with time, Heidegger insists, is natural. It is the distinction between day and night. That is to say, the first clock is the sun. And we begin—and I find this very compelling and rather poetic—with brightness. *Helle* in German. Heidegger writes:

> Everyday circum*spective* Being-in-the-world needs the possibility of sight (and this means that it needs brightness) if it is to deal concernfully with what is ready-to-hand within the present-at-hand. With the factical disclosedness of Dasein's world, Nature has been uncovered for Dasein. In its thrownness Dasein has been surrendered to the changes of day and night. Day with its brightness gives it the possibility of sight; night takes this away. (SZ 412 / BT 465)

The most natural measure of time is the day. The movement between day and night. Time is first marked by brightness. I wake up in the morning and it's bright; I know it is daytime. This is what Heidegger means by saying that time first shows itself in the sky. Time first shows itself in the heavens. The word in German is *Himmel*, heavens or sky. This is fascinating. The movement between day and night is the first way of reckoning with time and reckoning with it publicly in a way that is shared. What we have in the first clock, the natural clock, the clock of the movement between day and night, is a shared, public, common clock.

Heidegger then talks about the journeying sun, the time of the sky, and heavenly bodies. He says, "This public dating, in which everyone assigns himself his time, is one which everyone can 'reckon' on simultaneously" (SZ 413 / BT 466). Again, this natural clock of the day, movement between the day and the night, is held in common. It's a common reckoning here.

> It uses a publicly available *measure*. This dating reckons with time in the sense of a *measuring of time*; and such measuring requires something by which time is to be measured—namely, a clock. This implies that *along with the temporality of Dasein as thrown, abandoned to the "world," and giving itself time, something like a "clock" is also discovered—that is, something ready-to-hand which in its regular recurrence has become accessible in one's making present awaitingly*. (ibid.)

Notice here the phrase "making present awaitingly." He's building the analysis of inauthentic time of the everyday world back into this analysis here in a very skillful way. The everyday measure of time is the sky, the movement between day and night.

Let me add a little sidebar. Another way of thinking about the measure of time would be not just in relationship to the sun but also in relationship to water, something I mentioned way back in the discussion of anxiety and care in chapter 8, the link between time and the movement of the tide, the rising and lowering of the sea. In some languages—in many of the languages with which I am more and less familiar (say, some of the northern European and Nordic languages, Swedish, Danish, Norwegian, Dutch, Flemish) and in English, or at least in older forms of English—*tide is time and time is tide*. The natural clock is the rhythm of the sea. Heidegger doesn't talk about this. Heidegger is a thinker of rivers, a thinker of hills, and a thinker of mountains, woods, and forests.

This is perfectly fine if you live in a natural world of rivers, hills, mountains, and woods, which is very much how things would have appeared in the southwestern German landscape that Heidegger was intimately familiar with, especially the Black Forest. But if you grow up near the sea or on an island or on the coast, where the primary element is the rising and falling of seawater through the movement of tide, then one has a different idea of the "natural clock." In order to make sense of that, Heidegger would have to become a thinker of the sea, which is something that he's very much not. He even mentions climbing in this chapter. Heidegger liked to climb. He liked to hike. He liked to ski. That's great if it's your bag. But if you're from an island, the estuaries and coastal plains where the natural clock is the movement of tide and indeed where the clock is measured not just in terms of time but also in terms of the perceived salt quantity in the air and the gravitational pull of the moon—people that live on the coasts can smell and feel the tide changing. The natural clock can even be an olfactory clock.

Heidegger is a thinker of rivers, and this comes out particularly in his discussions of the great poet Hölderlin, many of whose long, hymnal poems are about rivers: the Rhine, the Danube, and so on. But he is not a poet of the sea like Melville. For me, the sea is primal. It's why I love my part of Brooklyn, because it is close to the harbor and the vast dark shoulder of the Atlantic. I could never live in the middle of the country. I don't understand life so far away from the sea.

The first clock arises out of Dasein's being-in-the-world insofar as our thrown abandonment is to a world regulated by the movement between day and night. The oldest artificial clocks, sundials for instance, are rooted in the natural clock. Sundials are manufactured, they are works of artifice. But the way they read time is rooted in the movement of the sun. Artificial clocks are rooted in this natural clock and thereby in Dasein's temporality. Heidegger writes:

> Dasein has its basis in temporality, and the "natural" clock which has already been discovered along with Dasein's factical thrownness furnishes the first motivation for the production and use of clocks which will be somewhat more handy; it also makes this possible. Indeed it does this in such a manner that these "artificial" clocks must be "adjusted" to that "natural" one if the time which is primarily discoverable in the natural clock is to be made accessible in its turn. (SZ 413–14 / BT 466–67)

This is the meaning of world-time. World-time does not refer to a present-at-hand conception of the world but to the temporality of the everyday ready-to-hand relationship to the world:

> Hence the time which makes itself public in the temporalizing of temporality is what we designate as "*world-time*." And we designate it thus not because it is *present-at-hand* as an entity *within-the-world* (which it can never be), but because it belongs *to the world* in the sense which we have Interpreted existential-ontologically. In the following pages we must show how the essential relations of the world-structure (the "*in-order-to*," for example) are connected with public time (the "then, when . . . ," for example) by reason of the ecstatico-horizonal constitution of temporality. Only now, in any case, can the time with which we concern ourselves be completely characterized as to its structure: it is datable, spanned, and public; and as having this structure, it belongs to the world itself. (SZ 414 / BT 467)

The time of concern, world-time, or the time of the ready-to-hand is *datable*, *spanned*, and *public*. We then get some fascinating remarks about different forms of clocks. Heidegger makes a distinction between "primitive" and "advanced" Dasein. We've seen primitive Dasein before on a number of occasions, all the way back to the beginning of the existential analytic in paragraph 11. I do not believe that "advanced Dasein" has been mentioned before, except once to note that the structure of falling is not to be compared negatively against "more advanced stages of human culture" (SZ 176 / BT 220). But here in the final chapter, Heidegger says, "Comparison shows that for the 'advanced' Dasein the day and the presence of sunlight no longer have such a special function as they have for the 'primitive' Dasein on which our analysis of 'natural' time-reckoning has been based" (SZ 415 / BT 468). What he is suggesting is that the way in which the world-time of the ready-to-hand being-in-the-world has been described is more consistent with "primitive" Dasein.

How stable is that distinction between primitive and advanced? Continuing the quote:

> For the "advanced" Dasein has the "advantage" of even being able to turn night into day. Similarly we no longer need to glance explicitly and immediately at the sun and its position to ascertain the time. The manufacture and use of measuring-equipment of one's own permits one to read off the time directly by a clock produced especially for this purpose. The "what o'clock is it?" (ibid.)

So, advanced Dasein can read "what o'clock" it is using its time-measuring equipment, as well as turn night into day. Electricity! You can do this for yourselves at home. Turn on the light. Turn off the light. You can entirely disregard the sun for telling the time. You can just look at your phone alone in the dark. Heidegger clearly has a preference for what he calls the "peasant's clock" (SZ 416 / BT 468). The peasant's clock is the shadow, the way in which I can measure time in relation to the way it shows itself in the sky. I can measure it by the length of my shadow. I can become a sundial. I can figure out whether it's noon or evening or morning or whatever. The other example he gives is the public sundial (SZ 416 / BT 469), where time is given by the shadow of the pointer. Out of these natural, inexact, and locationally relative clocks, obviously the sun will appear in different ways. Time will be different in different places, whether you are living in Mexico City or Nova Scotia, Marrakesh or Melbourne, Rochester or Riyadh, Quito or Queens. The way in which the time shows up naturally will be different depending on one's position on the earth. I spent a few June days in Iceland in 2002. I remember on one occasion getting up, showered, and dressed and bursting hungrily into the breakfast room of the small hotel I was staying at, only to discover it was three a.m. The clock depends on one's geolocation. Heidegger says:

> Thus when *time* is *measured*, it is *made public* in such a way that it is encountered on each occasion and at any time for everyone as "now and now and now." This time which is "universally" accessible in clocks is something that we come across as a *present-at-hand multiplicity of "nows,"* so to speak, though the measuring of time is not directed thematically towards time as such. (SZ 417 / BT 470)

What we're moving towards now is the way in which, out of this natural idea of time (the peasant's clock and so on), we develop the present-at-hand idea of time. This is the third level of time as a multiplicity of nows, which he's going to develop in paragraph 81.

There is also a footnote reference to relativity theory: "Here we shall not go into the problem of the measurement of time as treated in the theory of relativity" (SZ 417n1 / BT 499n4). It's good that Heidegger was aware of the theory of relativity. In many ways, the universe that Heidegger is assuming, as the universe in relation to which there will be the present-at-hand conception of time, is a Newtonian universe rather than an Einsteinian one. We might think about whether all these questions of time and the relation of time to space would be made more complicated by a fuller understanding of the implications of the theory of relativity and the emergence of the quantum world. Might the weird quantum world actually be closer to ecstatic temporality? What might be the temporality of the quantum world? Deep physical reality is totally and truly strange.

To close out this paragraph, we find a rhetorical trick, familiar to us from previous gestures Heidegger has made in *Being and Time*, especially in the discussion of reality. He says that world-time is neither subjective nor objective. But it's more objective than any object, and more subjective than any subject. He says, "*World-time is 'more Objective' than any possible Object*" (SZ 419 / BT 471). This rhetorical move is also a characteristic argumentative move when Heidegger asks, Well, does my account of the primacy of the temporality of Dasein mean that I'm making everything subjective? He will reply: No, to presuppose an idea of the subject would be to presuppose a metaphysical ground, a substratum or basis (whether ideal or material, Hegelian or Marxist) to experience. To contrast that with objectivity is to beg the question about the legitimacy and priority of the ontology of the present-at-hand and the theoreticism and intellectualism that accompany that conception of the world.

World-time as the time of the ready-to-hand is more subjective than any subject and more objective than any object. Another interesting side thought here is that although time is not subjective, there is a way of thinking about the connection in the history of philosophy between time and the soul, or the spirit, or *psyche* (ψυχή). Heidegger's going to come back to that a little bit later on in this chapter in a discussion of Hegel and Aristotle. But *is* time just psychical or soulful? Might we also think about time as physical or natural?

This is something we haven't really thought about much in *Being and Time*. The way Heidegger will develop the relationship between time and nature in his later work, from "The Origin of the Work of

Art" onwards, implicitly or more explicitly, is in relationship to the early Greek theme of nature, or *physis* (φύσις), which is the basis for Heidegger's concept of earth (*Erde*). What is the time of nature, the time of *physis*, given that *physis* as nature, as earth, is a donation of time, and also a shaping of time, a framing of time? Heidegger thinks this in relation to Heraclitus's enigmatic fragment 123, φύσις κρύπτεσθαι φιλεῖ (Nature likes to hide).[5] There is both a dimension of the physical framing of time and the hiddenness or withdrawal of *physis*, in the experience of time. Earth is that out of which a world is formed, but which hides or withdraws from the world.

When Heidegger says time first shows itself in the sky, he suggests a physical understanding of time, provided that "physical" is not understood in a present-at-hand sense within a naturalist ontology but in a more lived, experiential way. This is just a speculative thought, but the point is that time is not just the form of inner sense, as Kant would say, it's not just "in my head" or even "in my head" conceived of Dasein-ly as being-in-the-world. Time is also something that emerges in relationship to nature. There's a physical, earthly dimension of time here that needs to be kept in mind. This would be the possible link to the concept of *Gaia* as both nonsubjective and nonobjective that I mentioned in chapter 9, "Reality."

In paragraph 81, Heidegger turns to the genesis of the vulgar or ordinary conception of time. The problem here is how world-time, as he's just described it, the time of the ready-to-hand world, the peasant's clock, is covered up in what Heidegger calls now-time. In now-time, time is construed as a constant presence-at-hand. That is, as an infinite

5. Heraclitus, *The Art and Thought of Heraclitus*, fr. 123.

succession of nows flowing from the future to the past through the present. This is the idea of time as a line of uniform nows: the now that is the present; the no-longer-now that is the past; and the not-yet-now that is the future. These nows are all identical: *now* is the same as *now*, which is the same as *now*. These nows are infinite: they have no beginning and no end. And they come in *succession*: one after another.

The key philosophical source for the vulgar conception of time is Aristotle's *Physics*, book 4. Aristotle is trying to understand the nature of time and sees it in terms of the idea of counting, of time as what is counted in the movement from the earlier to the later.[6] (Good old sensible, plain-minded Aristotle.) Heidegger's reading of Aristotle is at times bizarre, to be sure, but it is also precise. It's actually right there in the text of the *Physics*. Time, for Aristotle, is counted nows, a linear series that comes in endless succession. When we count a now, we make it present. This is what Heidegger calls *Jetzt-Zeit*, now-time. It is a conception of time that gives priority to presence. Heidegger introduces it in the following way: "'Now here, now here, and so on.' The 'nows' are what get counted. And these show themselves 'in every "now"' as 'nows' which will 'forthwith be no-longer-now' and 'nows' which have 'just been not-yet-now.' The world-time which is 'sighted' in this manner in the use of clocks, we call the 'now-time'" (SZ 421 / BT 474). Time is a sequence of nows simultaneously passing away and coming along endlessly, like a flowing stream, an unending flow of nows. Time on this conception is endless, infinite, and has no gaps. Time *can* have no gaps. It is an infinite fullness. Heidegger writes that "Even Plato" (given that Heidegger thinks Plato is

6. Aristotle, *Physics*, trans. R. P. Hardie and R. K. Gaye, in *The Basic Works of Aristotle*, ed. Richard McKeon (Modern Library, 2001), 219a–220a.

quite good) in the *Timaeus* calls time the "moving image of eternity" (SZ 423 / BT 475).[7] Eternity, one imagines, is static, constant, and unchanging. The point is that now-time is an ever-changing, never-ending movement from the future to the past through the present. Now-time is a mobile mirror of the eternal Now.

Heidegger will question all of these assumptions about time. For Heidegger, time is finite. We have considered that in detail. Furthermore, it's not just finite, it's *reversible*. (Now-time, the unending, infinite succession full of nows, can't be reversed.) Ecstatic temporality is a reversal in the order of time. Out of the future, the to-come, emerges the past, the having-been, which is released into the present of the moment of vision. Also, time is not constant. It's not a constantly flowing stream. Time is intermittent. It has gaps and moves at different speeds at different moments and in various moods, when I'm anxious or bored, excited or indifferent. When I'm bored, time drags, it spreads out. It becomes long, unfolded and spilled out before me, like something I wish to erase or contain. When I'm excited, it quickens, accelerates, even spikes in short, quick bursts.

Time is not a flow, it's more of a series of episodic moments, blips, glitches, moments of vision, and periods of forgetting.[8] In a footnote, Heidegger talks about eternity in relationship to the *nunc stans*, the

7. Plato, *Timaeus*, trans. Donald J. Zeyl, in *Complete Works*, 37D.

8. There's much more to say about this. I am against narrative ideas of the self. See Galen Strawson, "Against Narrativity," *Ratio* 17, no. 4 (2004): 428–52. If you live the lie that your life is somehow a biographical narrative story, you're probably faking it. (It's also the pathology of narcissism.) The episodic idea of the self is liberatory. You don't need to constrain your existence into a linear story. To hell with narrative ideas of the self.

standing now. Heidegger thinks that this idea of eternity has been derived from the vulgar concept of time in Aristotle, which attempts to underpin that conception with an idea of God's eternity as a constant presence-at-hand. Heidegger writes, "If God's eternity can be 'construed' philosophically, then it may be understood only as a more primordial temporality which is 'infinite.'" He concludes the footnote by saying, "Whether the way afforded by the *via negationis et eminentiae* is a possible one, remains to be seen" (SZ 427n1 / BT 499n13).

Heidegger is alluding to the idea of eternity as the standing now. And he's also showing the way in which there is, in philosophy and theology, a divinization of the vulgar concept of time as a way of thinking about the constant presence-at-hand of God to himself and to us. The idea is that we have, on the one hand, time as a flowing stream of nows, as a moving image of eternity, and then, on the other hand, an appeal to eternity as the being of God. (Eternity becomes temporal in the person of Christ.) What is happening here is that the vulgar conception of time is being elevated into a metaphysics, an onto-theology of constant presence. The reference to "the way of negation" and "the way of eminence" is an allusion to different ways of predicating the being of God, ascribing properties to the divine that can be found in thinkers like Duns Scotus and Aquinas. For the way of negation, God cannot be stated positively, and for the way of eminence God's perfection surpasses all attempts to describe it. We have to remember that Heidegger wrote his *Habilitation* thesis on this material and he's no slouch on this topic.[9]

9. Martin Heidegger wrote his *Habilitationsschrift* in 1915 on Duns Scotus. See *Die Kategorien- und Bedeutungslehre des Duns Scotus* (Mohr, 1916); *Duns Scotus's Doctrine of Categories and Meaning*, trans. Joydeep Bagchee and Jeffrey D. Gower (Indiana University Press, 2022).

I think these references to God show Heidegger being polite and circumspect to an audience that is still overwhelmingly or at least overtly Christian. He doesn't want to get into any trouble with the clerics and their minions in the universities. But there is no disguising the fact that he's not at all persuaded that we need an idea of God in order to understand time. On the contrary. We need to get rid of the idea of eternity and get rid of the idea of God.

Being and Time is an atheist book. But he's not going to shout this from the rooftops. This atheism is not a constant feature of Heidegger's work, but at this time in the 1920s it comes out of his residual Lutheranism. (Protestantism is the gateway drug to atheism.) Whether this atheism continues in Heidegger's later work is a moot point that can be discussed and debated. I think Heidegger's view, echoing that of Hölderlin, is that we are too late for the Gods but too early for Being.[10]

Why do we say that time cannot be reversed? The present-at-hand, vulgar conception of time is an irreversible stream that flows from the future to the past through the present. But if we conceive of time ecstatically, then it is reversible. At the very least, authentic time is back to front. The present, as the moment of vision, comes out of the having-been, which comes out of the future. This is the structure of historicity that we saw in the last chapter: historicity is the repetition of the past that comes out of the future, in the event of the new.

Heidegger wants to argue that primordial time is reversible and

10. Martin Heidegger, "Aus der Erfahrung des Denkens," 76; "The Thinker as Poet," 4; Friedrich Hölderlin, "Bread and Wine," in *Poems of Friedrich Hölderlin: The Fire of the Gods Drives Us to Set Forth by Day and by Night*, trans. James Mitchell (Ithuriel's Spear, 2007), 19.

that vulgar time's irreversibility is simply a secondary, derivative way of thinking. Heidegger's claim is not that the vulgar conception of time is wrong but that it's secondary. It is founded and not founding. The way he puts that thought in these pages is that the vulgar conception of time is a *leveling off*. Indeed, it is a double leveling off. It is both a leveling off of world-time or the within-time-ness of the ready-to-hand, and a leveling off of ecstatic temporality. In adopting the vulgar conception of time, we level off and lose the true significance of time. Heidegger writes, "In the ordinary interpretations of time as a sequence of 'nows,' both datability and significance are *missing*" (SZ 422 / BT 474). Datability is the way of marking the everyday, ready-to-hand experience of time. Significance, remember, is the nature of worldhood (i.e., the way we find the world meaningful). In the vulgar, ordinary conception of time (in philosophy and in Newtonian conceptions of time), datability and significance are lost. "These two structures are *not* permitted to 'come to the fore' when time is characterized as a pure succession. The ordinary interpretation of time *covers them up*. When these are covered up, the ecstatico-horizonal constitution of temporality, in which the datability and the significance of the 'now' are grounded, gets *levelled off*" (SZ 422 / BT 474). How does this leveling off happen? How do we slip from authentic to inauthentic time? And how do we slip from inauthentic time into vulgar time? If Heidegger is right, how does it happen that our experience is dominated by this vulgar conception of time? The reason lies, once again, in Dasein's *falling*. Insofar as Dasein cannot bear to look its finitude in the face, it looks away from itself towards the world at which it falls, and it falls away from an ordinary conception of world-time to a more scientific conception of time. The reason why there was a leveling off in our experience of time is explained by our propensity to fall to the world.

Just after asking the question, Why cannot time be reversed?, Heidegger says, "The ordinary way of characterizing time as an endless, irreversible sequence of 'nows' which passes away, arises from the temporality of falling Dasein" (SZ 426 / BT 478). However, such an evasion of our primordial experience of temporality must still possess awareness of finitude. It must still be there in what Heidegger calls in these pages a "fugitive" way. This is intriguing. We must still know time as it truly is primordially, although in a hidden manner, in a way that is fleeting and evanescent. True time is still there, but it's hidden away. We refuse to see it and we fall. Heidegger talks about this in relation to a theme that we've kept coming back to in our discussion of *Being and Time*, the theme of *enigma*. This is the last enigma in *Being and Time*, the last reference to *Rätsel*, to a riddle. He writes, "But just as he who flees in the face of death is pursued by it even as he evades it, and just as in turning away from it he must see it none the less, even the innocuous infinite sequence of 'nows' which simply runs its course, imposes itself 'on' Dasein in a remarkably enigmatical way" (SZ 425 / BT 477). The thought here is that we flee from the enigma. We run away from the riddle of ourselves. We cover it up. It's still there, in the same way as being-towards-death is still there, quietly throbbing away inside our being. But we can forget about it, cover it up, conceal it.

Heidegger draws out the contrast between vulgar and ecstatic temporality in terms of the contrast between the primacy of the present versus the primacy of the future.

> Ecstatico-horizonal temporality temporalizes itself [the same verbalizing of nouns we saw in paragraph 65] *primarily* in terms of the *future*. In the way time is ordinarily understood, however, the basic phenomenon of time is seen in the "*now*," and indeed in that pure "now" which has been shorn in its full structure—that which they call the "Present." One can gather from this that there is in principle no prospect that *in terms of this kind of "now"* one can clarify the ecstatico-horizonal phenomenon of the *moment of vision* which belongs to temporality, or even that one can derive it thus. (SZ 426–27 / BT 479)

What Heidegger is describing here is the emergence of the metaphysics of presence, which misconceives time in order to disregard or level off the experience of finitude and radically mortal time. The futurity that opens to having-been, that explodes in the present of the moment of vision: that ecstatical experience of temporality gets misconceived, misdescribed, leveled off, lost, forgotten. Dumbed down.

Let's turn to paragraph 82. Heidegger pulls on two quotations, one in Greek, one in Latin, from Aristotle and Augustine. The quotation from Aristotle is very roughly that, if there were no soul or *psyche*, time would be impossible.[11] And with Augustine, the perplexity is with the idea of time as extended (SZ 427 /BT 479–80). Augustine says, "It seemed to me that time is nothing else than extendedness. But of what sort of thing it is an extendedness, I do not know; and it would be surprising if it were not an extendedness of the soul

11. Aristotle, *Physics*, in *The Basic Works of Aristotle*, 223a25–30.

itself."[12] Puzzled by the enigma of time, both Aristotle and Augustine think that there is a connection between time and the soul.

If such a connection between time and the soul makes sense, then how might it be thought about? What suggests itself to Heidegger at the very end of this long book is the connection between time and spirit. This is what leads him to ponder Hegel's idea of Spirit falling into time. What, we might ask, is the point of this little discussion of Hegel, sort of shoved in at the end of *Being and Time*? Hegel has barely come up in the book up to this point. To be clear, it is not in order to criticize Hegel, because that is considered pointless, as Heidegger says.

> We shall make no claim to give even a relatively full treatment of the allied problems in Hegel, especially since "criticizing" him will not help us. Because Hegel's conception of time presents the most radical way in which the ordinary understanding of time has been given form conceptually, and one which has received too little attention, a comparison of this conception with the idea of temporality which we have expounded is one that especially suggests itself. (SZ 428 / BT 480)

Hegel is the most radical and representative way in which the vulgar conception of time has come down to us philosophically. Interestingly, Heidegger acknowledges, in another footnote, that Kant has a more radical understanding of time than Hegel: "On the other hand, the extent to which an even more radical understanding of time than Hegel's makes itself evident in Kant will be shown in the first division of the second part of this treatise" (SZ 427n4 / BT 499n16). We're going to come back to that thought. Hegel is a great

12. Augustine, *Confessions*, book 11, chapter 26, section 33.

philosophical illustration of the vulgar, dominant understanding of time in the philosophical tradition. Hegel gives us an illustration of the privileging of the now in the metaphysics of presence, where one can say that only the present is: the future is the not-yet present, the past is the no-longer present. All of the three dimensions of time have to be subsumed in this ordinary conception of time consummated in Hegel in terms of the domination of the present.

For Heidegger, spirit does not fall into time. Rather, "spirit," in quotation marks—and remember how these quotation marks have been dancing around the entirety of *Being and Time*—exists as the temporalizing of temporality, for Heidegger (SZ 427 / BT 479). This is the thought: If spirit is the temporalizing of temporality, then Dasein can be thought of as spirit. Dasein is time, finite time. The falling of Dasein is not into time from eternity, but out of primordial time into world-time and from there into vulgar now-time. You can see the point that Heidegger's making. Spirit does not fall into time, as Hegel says, rather Dasein is time, but falls out of primordial time into world-time and from thence into the vulgar conception of time. We fall into progressively more leveled-off experiences of temporality. Heidegger concludes this little discussion of Hegel thusly:

> Our existential analytic of Dasein, on the contrary, starts with the "concretion" of factically thrown existence itself in order to unveil temporality as that which primordially makes such existence possible. "Spirit" does not first fall into time, but it *exists as* the primordial *temporalizing* of temporality. Temporality temporalizes world-time, within the horizon of which "history" can "appear" as historizing within-time. "Spirit" does not fall *into* time; but factical existence "falls" as falling *from* primordial, authentic temporality. This "falling,"

> however, has itself its existential possibility in a mode of its temporalizing—a mode which belongs to temporality. (SZ 435–36 / BT 486)

This is not very elegantly stated, I'll grant you that. The book is running out of steam. Yet the point is philosophically clear. Spirit does not fall into time. Dasein is time but falls out of itself towards world-time and into the vulgar experience of time as an infinite and irreversible series of now-points, which is the origin of the (mistaken, in Heidegger's view) idea of eternity.

We are now at the point where we can read the final paragraph of *Being and Time.* We are tantalizingly close to the end. But I want to frustrate you a little bit and linger on a moment at the end of paragraph 82, which will hopefully raise some interesting questions and perhaps even doubts. I want to just draw your attention to a dense and very detailed footnote. In 1968, Jacques Derrida wrote an essay called "*Ousia* and *Grammē*: Note on a Note from *Being and Time*."[13] It was published in an interesting *Festschrift* for Heidegger's most loyal and well-known French translator, Jean Beaufret.[14] Derrida's essay unravels the entirety of *Being and Time* from the perspective of this footnote.[15] If you want to know how to read a philosophical text, then read Derrida. He uses this footnote as a lever to open up and question the whole book.

13. Jacques Derrida, "*Ousia* and *Grammē*: Note on a Note from *Being and Time*," in *Margins of Philosophy*, trans. Alan Bass (University of Chicago Press, 1982), 29–67.

14. René Char, *L'Endurance de la pensée: Pour saluer Jean Beaufret* (Plon, 1968).

15. Find the loose end and pull it apart. People don't read enough Derrida anymore. They'd rather listen to Žižek talk about movies.

The basic point that Heidegger makes in the footnote, as Derrida's brilliant essay exposes, is that the origin of the vulgar conception of time in Hegel has its roots in Aristotle, and these roots continue to spread and grow everywhere philosophers try to think of time. Here, right at the end of the book, in the penultimate paragraph of *Being and Time,* Heidegger reveals what he has been assuming all the way through the book. Not that he's been hiding it, but he's not exactly been as forthcoming as one might have wished. Heidegger's basic insight running throughout his analysis has been that Aristotle sees the essence of time in the *now* (i.e., the *nŷn* or *vũv*).

> Aristotle sees the essence of time in the *vũv*, Hegel in the "now." Aristotle takes the *vũv* as ὅρος [*horos*: horizon]; Hegel takes the "now" as a "boundary" [*péras*, πέρας]. Aristotle understands the *vũv* as στιγμή [*stigmē*, number]; Hegel interprets the "now" as a point [*stigmē* also means *point*]. Aristotle describes the *vũv* as τόδε τι [*tode ti*, "a this," "some this," or "a something"]; Hegel calls the "now" the "absolute this." Aristotle follows tradition in connecting χρόνος [*chrónos*] with the σφαῖρα [*sphaîra*]; Hegel stresses the "circular course" of time. [*Chrónos* is *time* in Greek; *sphaîra* is a rounded sphere] To be sure, Hegel escapes the central tendency of the Aristotelian analysis—the tendency to expose a foundational connection ἀκολουθεῖν [*akolouthein* also means to follow along] between the *vũv*, the *horos*, the *stigmē*, and the τόδε τι [*tode ti*]. (SZ 432–33n1 / BT 500n30)

The footnote is really rather confusing and packed with a lot of Greek. But the philosophical point that Heidegger is making is simple: that Hegel's privileging of the "now" is the way in which he inherits Aristotle's conception of time based upon the primacy of the now, the primacy of the point, of the "absolute this," as well as even the idea of time as a sphere or well-rounded circle, which we can find in Parmenides in his celebrated "Poem": "Being is a well-rounded circle."[16]

Going back to the words of the footnote, look at what happens next: "In its results, Bergson's view is in accord with Hegel's thesis that space 'is' time, in spite of the very different reasons they have given. Bergson merely says the reverse: that time (*temps*) is space" (ibid.). Henri Bergson is significant competition for Heidegger because he is someone who has thought about lived, subjective time in terms of what Bergson calls duration, *la durée*, in distinction from the objective time of the clock. Heidegger's claim here is that Bergson simply reverses Aristotle. But by reversing him, he simply affirms the traditional conception of time. Bergson is as much a traditional thinker of time as Hegel. They are both indebted to, and indeed draw directly from, this Aristotelian idea of time. So much for Bergson.[17] Heidegger is never exactly fair with his close precursors and competitors. But I rather like him in apex-predator mode, killing his competition.

The last thing I want to pick out from this footnote is very nerdy: "So far as anything essential has been achieved in today's analyses,

16. Parmenides of Elea, *Fragments*, fr. 8.42–9.

17. Heidegger cites both Bergson's "Essai sur les données immédiates de la conscience" and his "Quid Aristoteles de loco senserit." See Henri Bergson, *Time and Free Will: An Essay on the Immediate Data of Consciousness*, trans. F. L. Pogson (Dover Publications, 1960), and Henri Bergson, "Aristotle's Concept of Place," trans. John K. Ryan, in *Ancients and Moderns*, ed. John K. Ryan (Catholic University of America Press, 2018), 13–72.

which will take us beyond Aristotle and Kant, it pertains more to the way in which time is grasped and to our 'consciousness of time.' We shall come back to this in the first and third divisions of Part Two." And the text of the English edition includes the following translators' note: "The preceding sentence has been deleted in the later editions" (ibid.). Those parts of *Being and Time* that Heidegger is referring to were not written (or, at the very least, were not published) during his lifetime. What Heidegger is alluding to here is the idea of time-consciousness, *Zeitbewusstsein*, which Edmund Husserl was thinking about in the years before, during, and after *Being and Time*. Together with Edith Stein,[18] Heidegger edited and introduced Husserl's lectures on "internal time-consciousness" in 1928.[19] These matters were clearly on Heidegger's mind. The point that Heidegger is making elliptically and not as clearly as one would like is that all philosophical conceptions of time derive from Aristotle, Hegel and Bergson included. If there's a way of stepping beyond that, then we have to look at time-consciousness. In other words, Husserl is onto something.

The problem with time-consciousness is, of course, the concept of "consciousness." What we've seen in *Being and Time* is a persistent refusal to talk about the human being in terms of consciousness. Deeper than *Bewusstsein* lies *Dasein*. But isn't there also something faintly patronizing in what Heidegger is suggesting here? Husserl's

18. Stein, a German Jewish philosopher who converted to Catholicism and became a nun, was Husserl's first assistant. She was murdered in Auschwitz in 1942 and beatified as Sister Teresa of the Carmelite Order by Pope John Paul II in 1987.

19. Edmund Husserl, *The Phenomenology of Internal Time-Consciousness*, trans. James S. Churchill, ed. Martin Heidegger (Indiana University Press, 1964).

idea of time-consciousness prefigures what Heidegger achieves in *Being and Time*. This is both a compliment to his teacher and a not-so-subtle slap in his face, showing where and how the student has (or imagines himself to have) gone beyond the teacher.

Philosophy is nothing if not parricide. Recall the analysis of heritage in the previous chapter, in particular the role of disavowal. Perhaps Heidegger sees himself as offering a repudiation or revision of his own philosophical "past." Philosophy is this endless rebuke to its history. There is always more to say that hasn't quite been said. *Being and Time* is technically an unfinished work, but we will come back for a coda that will kick off from the final two pages of the book in a different key. It will be called "Opening and Presence."

18. Coda
Opening and Presence

The relation between construction and destruction in Being and Time *— How the incompletion of Heidegger's book has been exaggerated — The fetishization of failure and the provisional — Heidegger's marginalia on his own book — What happens when we destroy the tradition — Presencing out of the source — The end of philosophy and the task for thinking — Being as opening, time as presence — The last words of* Being and Time*; or, WTF, Martin? — The risk of self-paralysis in false philosophical modesty — Against towardsism — Having the courage of your convictions — The massive amount that we have learned from* Being and Time *— Polemic: why this is the right path to be taken in philosophy — Finitude restated — Naming the experience of the appropriative event — There is no turning in Heidegger's work, just a deepening — The importance of building castles in the air (Gadamer) — All that there is, is air — How* Being and Time *changes one's life and approach to philosophy — Leaving the climate of Heidegger's thinking (Levinas).*

At the end of the last chapter, we looked at a long footnote, very close to the end of *Being and Time*, where Heidegger traces the vulgar conception of time that he is opposing back to Aristotle (in the *Physics*) and shows how, in figures like Hegel and indeed Heidegger's near contemporary Bergson, their thinking of time still derives from this Aristotelian conception of time as an infinite succession of counted nows. The conception of time that Heidegger is against finally comes out in the last chapter. And this move, at the end, brings us back to the beginning of *Being and Time*. In the end is our beginning, as T. S. Eliot would say.[1] I want to begin this ending by thinking about the relationship between the phenomenological and deconstructive aspects of this huge book that we're tantalizingly close to finishing.

What is the relationship between the constructive project of *Being and Time*, its extraordinary philosophical conceptual innovations, and the work of destruction, what we're trying to dismantle and get away from? This question leads us back to the way in which Heidegger designs the treatise itself. At the very end of the introduction, in a brief paragraph called "Design of the Treatise," you will see that *Being and Time* was originally envisaged to be in two parts. The first part is called "the Interpretation of Dasein in terms of temporality, and the explication of time as the transcendental horizon for the question of Being." Part 2 is called "basic features of a phenomenological destruction of the history of ontology, with the problematic of Temporality as our clue" (SZ 39 / BT 63–64). Those two parts break down into three divisions each (see fig. 4). What we've done in this book is to read as closely as we can part 1, divisions 1 and 2, which constitute the published version of *Being and Time*. Division 3, "Time and Being," was never published, but scholars

1. T. S. Eliot, *Four Quartets* (Harcourt, 1943), 17.

have attempted to reconstruct it from the 1927 lecture course *The Basic Problems of Phenomenology*.[2]

Here I just want to remind you that *Being and Time* was projected to be in these two parts, the first part of which we've covered, divisions 1 and 2. Likewise, part 2 has three divisions. As Heidegger tirelessly insisted, the whole project of the published portion of *Being and Time* is provisional and conditional upon the deconstruction of the history of ontology.[3] As he says at the end of paragraph 6 of the introduction, "The question of Being does not achieve its true concreteness until we have carried through the process of destroying the ontological tradition" (SZ 26 / BT 49). As I've said a number of times, the whole premise of the phenomenology of Dasein as a preparation for the question of being as such is conditional upon part 2: the destruction of the history of ontology, none of which was published.

As is perhaps too well known, Heidegger didn't complete his task. He didn't finish the book called *Being and Time*. And therefore the published volume remains a fragment or a torso. Many writers on Heidegger—John Sallis, Otto Pöggeler, Thomas Sheehan—have made this point well over the years. And it's undeniably true. We have just looked at part of the much larger picture, the full scale of which can only be imagined and pieced together largely from posthumously published texts and lecture courses.

2. Martin Heidegger, *Die Grundprobleme der Phänomenologie*, ed. Friedrich-Wilhelm von Herrmann, vol. 24 of *Gesamtausgabe*, ed. Friedrich-Wilhelm von Herrmann (Klostermann, 1975); *The Basic Problems of Phenomenology*, trans. Albert Hofstadter, rev. ed. (Indiana University Press, 1982).

3. For an examination of the hidden assumptions of such a project, see Dennis Schmidt, *The Ubiquity of the Finite: Hegel, Heidegger, and the Entitlements of Philosophy* (MIT Press, 1988).

The Design of *Being and Time* (fig. 4)

	Preface	
	Introduction	Exposition of the question of the meaning of Being
	PART ONE	THE INTERPRETATION OF DASEIN IN TERMS OF TEMPORALITY, AND THE EXPLICATION OF TIME AS THE TRANSCENDENTAL HORIZON FOR THE QUESTION OF BEING
	Division 1	The preparatory fundamental analysis of Dasein
	Division 2	Dasein and temporality
UNPUBLISHED	Division 3	Time and Being
UNPUBLISHED	PART TWO	BASIC FEATURES OF A PHENOMENOLOGICAL DESTRUCTION OF THE HISTORY OF ONTOLOGY, WITH THE PROBLEMATIC OF TEMPORALITY AS OUR CLUE
UNPUBLISHED	Division 1	Kant's doctrine of schematism and time, as a preliminary stage in a problematic of temporality
UNPUBLISHED	Division 2	The ontological foundation of Descartes's "*cogito sum*," and how the medieval ontology has been taken over into the problematic of the "*res cogitans*"
UNPUBLISHED	Division 3	Aristotle's essay on time, as providing a way of discriminating the phenomenal basis and the limits of ancient ontology

But—and I cannot overstate this point—there's a tendency to exaggerate the fragmentary nature of *Being and Time*. There's a predilection in Heidegger himself and amongst Heideggerians towards what I would call a *pathological provisionality*, where they continually emphasize the preparatory or provisional character of Heidegger's thinking and overstate the questioning or even questing character of what's being articulated. We're often told by Heidegger and Heideggerians that we're just on the way to something else. We're never quite there, we're always already *Unterwegs*, underway.

Indeed, Heidegger's collected works, the *Gesamtausgabe*, which are in many, many, many volumes, currently 102 unsearchable tomes without any useful critical apparatus, has a motto. This was Heidegger's own plan and certainly his own motto: "*Wege, nicht Werke.*" Ways and not works. And Heidegger was very fond of this idea of the way, of the *Feldweg*, of the *Holzweg*, of the way in the field, the wood, the country path. This is linked to the idea of thinking as a way, a path, even an errant wandering. One of the collections of essays published in his lifetime is called *Unterwegs zur Sprache* (*On the Way to Language*).

For Heidegger, we're always on the way. His thinking isn't a work, a system, a completed edifice. It is something we are moving towards. Fine, one might say. But what this leads to is the risk of a fetishization of preparatory thinking, a fetishization of the provisional. It also leads to a fake humility in Heidegger's work and (even worse) in the work of Heideggerians, which undersells the philosophical achievement of *Being and Time*.

If we go back to paragraph 8, "The Design of the Treatise," then we know all about part 1, divisions 1 and 2. And if we look at part 2 of the imagined treatise, it's in three parts. The first part is on Kant, the second on Descartes, and the third on Aristotle. We can pretty much imagine what's going to take place in divisions 1 through 3 of part 2. Everything turns back to Aristotle. We've seen the basic shape of this claim about Aristotle in the very long footnote at the end of the last chapter of *Being and Time* and from the entire last chapter of division 2. The argument about Kant was published as *Kant and the Problem of Metaphysics*, a lecture series that appeared in 1929.

The focus of the discussion of Kant was the schematism of time and the abyss of temporality from which Kant allegedly shrinks back, which we have touched on.[4] What Heidegger says about Kant has been thoroughly sketched in different moments in *Being and Time*, particularly paragraph 64 in the discussion of care and selfhood, and we have covered that ground previously (see ch. 9, "Reality," and ch. 13, "Anticipatory Resoluteness"). We can imagine or reconstruct pretty clearly what Heidegger is going to say about Kant. As for the discussion of Descartes, we can very much imagine that, too, from the various critiques of world, extension, and spatiality, thinking all the way back to division 1, chapter 3. In our discussion of space in chapter 3, we saw how Heidegger contrasts his discussion of being-in-the-world with the Cartesian picture. It would not take Einstein to give a thumbnail sketch of the contents of part 2, divisions 1 through 3.

We need to lose a little bit of the talk of provisionality in Heidegger studies, and the whole fetish of the "preparatory," the "towards," and the "on the way to." When we've done that, we can face up to what we've actually achieved in our reading of Heidegger and how much we have learned.

What, then, is missing from *Being and Time*? What is absent is part 1, division 3: "Time and Being."[5] Heidegger himself explicitly returned to this topic thirty-five years later. Indeed, he used the title "Time and Being" in a brilliant but extremely difficult lecture that

4. Martin Heidegger, *Kant und das Problem der Metaphysik*, 167–68; *Kant and the Problem of Metaphysics*, 117–18.

5. See the outline of *Being and Time* at the end of its introduction. (SZ 39–40 / BT 63–64)

he gave in January 1962.[6] If I have the good fortune, I hope to follow up this book with a careful discussion of that lecture. As I mentioned, there is an important lecture course that Heidegger gives in summer semester 1927, right after the publication of *Being and Time*, called *The Basic Problems of Phenomenology*, where he tries to see through the analysis of the temporality of being in a way that is not only compelling but also slightly disappointing. Long story short, I think that Heidegger's provisionality is somewhat overdone.

Where can we take that thought? For some insight here, we might look at the comments that Heidegger made on his own copy of *Sein und Zeit*. These can be found at the end of the German edition, on pages marked "marginal comments from the author's copy of the book."[7] These are fascinating because you can see Heidegger reading Heidegger, which is a strange and complicated thing to do when you've written your own books. There's a play of blindness and insight that characterizes reading one's own writing, how you read your own books and what you say about them.[8] Heidegger comments on his

6. See p. 59, n. 3.

7. Speaking of marginal notes, Husserl's commentary on *Being and Time* has been translated into English and published. He was at times extremely critical. One can only imagine what was going through his mind as he read this eccentric work by his greatest pupil. See Edmund Husserl, "Husserl's Marginal Notes in Martin Heidegger, *Being and Time*," in *Psychological and Transcendental Phenomenology and the Confrontation with Heidegger, 1927–1931*, trans. and ed. Thomas Sheehan and Richard E. Palmer (Springer, 1997), 258–422.

8. Paul de Man, *Blindness and Insight: Essays in the Rhetoric of Contemporary Criticism*, 2nd ed. rev. (Routledge, 1983). De Man would say that you can't read your own writing. Therefore, Heidegger has no particular insight into his own work.

own copy of *Being and Time* just where he's laid out the design of the treatise, which we just discussed. He says the following, first in German: "*Die transcendendenzhafte Differenz. Die Überwindung des Horizonts als Solchen. Die Umkehr in die Herkunft. Das Anwesen aus dieser Herkunft.*" And now in my translation: "The transcendentality of difference. The overcoming of the horizon as such. The turning back into the source. And the presencing from out of this source."[9]

What does that mean? Why does he write these words when he is commenting on his own copy and his reimagining of the design of his own unfinished book? I think he means the following. The destruction of the history of ontology sketched as part 2 of *Being and Time* is the path of a turning back into the source of time and being: time as being and being as time. What we receive when we turn back, when we've destroyed the tradition, is the *presencing* from out of this source: the original donation of being. The other thing that needs to be mentioned is that Heidegger says that the overcoming of the horizon as such has to be linked to his claim that time is the horizon for the understanding of the meaning of being, first mentioned on the untitled first page of the book, where Heidegger quotes from Plato's *Sophist*. But the concept of horizon is a troubled term for Heidegger because it's too spatial. The horizon is a form of limit, a boundary or frontier by which we understand the operation of time. Horizon presupposes that we're understanding time in terms of spatial metaphorics, and that is something that the later Heidegger thinks needs to be called into question. The concept of

9. Martin Heidegger, *Sein und Zeit*, vol. 2 of *Gesamtausgabe*, ed. Friedrich-Wilhelm von Herrmann (Klostermann, 1977), 53.

horizon blocks the turning back into the source of time and being. Once you turn back into the source, once you've destroyed the tradition, once you have overcome the concept of horizon, then there can be a presencing, an *Anwesen*, out of this source, an original gift of being. This presencing or *Anwesenheit* will assume great importance in Heidegger's later work, but it is also mentioned earlier in the introduction to *Being and Time* in connection with the Greek meaning of being (SZ 25 / BT 46–47).

I hope all this doesn't sound too mysterious. I don't want it to sound too gnomic. But, to make matters probably even worse, there is a very late text by Heidegger called "The End of Philosophy and the Task of Thinking." It's a lecture from 1964. Most of Heidegger's work after *Being and Time* took the form of lectures and essays. These are the last words of the lecture:

> Does not the title for the task of thinking then read instead of *Being and Time*: Opening and Presence?
>
> But where does the opening come from and how is it given? What speaks in the "There is/It gives" [*es gibt*]?
>
> The task of thinking would then be the surrender of previous thinking to the determination of the matter for thinking.[10]

As often in Heidegger, there is illumination and there is a little obscurity. But what I want you to ponder here is that what *Being and Time* is trying to get us to think about and to really *do* in our

10. Martin Heidegger, "Das Ende der Philosophie und die Aufgabe des Denkens," 90; "The End of Philosophy and the Task of Thinking," 73 (trans. mod.).

thinking can be thought under the title of Being as *Opening*. We can think about Dasein, the human being, as opening, as a cleared opening, as a thrown clearedness. Dasein is the opening, it is ecstatic. And this ecstatic opening that is ours and that is us as being-in-the-world is an opening to presencing. This isn't the metaphysics of presence—that conception is frozen in the idea of now-time—but rather a fuller and more compelling, more lived and more rounded idea of presencing as the ecstatic temporality into which we can enter and of which we're a part. This lived presencing of the source is possible when we temporally stand outside ourselves, releasing our having-been into the moment of vision, the historicizing vision, by projecting our thrownness-into-the-world into our future—that is, when we throw our temporal opening, like throwing light through time. We are opening and presencing as temporal, as these three *ecstases* at once.

Think about *Being and Time* as opening and presence. The condition of possibility for the concrete experience of opening and presence is the deconstruction of the ontological tradition. And once that turning back into the source and that deconstruction of the tradition have been achieved, then there can be an opening, an experience of releasement or even relaxation. (This is more me than Heidegger.) This released existence would be a presencing from out of the source that is not hemmed in by the tyranny of the vulgar metaphysics of time. Heidegger picks up the concept of releasement, *Gelassenheit*, from the truly radical and wonderful medieval German theologian Meister Eckhart. The idea of releasement here has its roots in the Christian mystical conception of a soul released from its constriction, becoming a "Free Spirit" to move closer to God.

God is relaxed. And we have to be released and relaxed. The letting go of itself, and our letting go of ourselves, releases us from constrictive guilt and sin. Heidegger's thought here is, as elsewhere, deeply influenced by what I've argued is a conversion narrative, in this case a liberation theology.

•

With those slightly weird thoughts in mind, let's turn back to the very end of *Being and Time*, in paragraph 83. I'm going to read you the final words of the book and then make some comments. I will read four quotes. Heidegger opens with the following statement: "In our considerations hitherto, our task has been to Interpret the *primordial whole* of factical Dasein with regard to its possibilities of authentic and inauthentic existing, and to do so in an existential-ontological manner in terms of its very basis. *Temporality* has manifested itself as this basis and accordingly as the meaning of the Being of care" (SZ 436 / BT 486). We know this from chapter 3 of division 2, and we went over it in chapter 14, "Ecstatic Temporality." Temporality is the meaning of the being of care. Heidegger continues here with a lovely summary of the argument of *Being and Time*. Here is the second quote, and wait for the hesitation:

> So that which our *preparatory* existential analytic of Dasein contributed *before* temporality was laid bare, has now been *taken back* into temporality as the primordial structure of Dasein's totality of Being. In terms of the possible ways in which primordial time can temporalize itself, we have provided the "grounds" for those structures which

> were just "pointed out" in our earlier treatment. Nevertheless [here comes the hesitation], our way of exhibiting the constitution of Dasein's Being remains only *one way* which we may take. Our *aim* is to work out the question of Being in general. The *thematic* analytic of existence, however, first needs the light of the idea of Being in general, which must be clarified beforehand. (SZ 436 / BT 486–87)

The first two quotes are joined together in the text. The third quote is about a page further on: "One must seek a *way* of casting light on the fundamental question of ontology, and this is the way one must *go*. Whether this is the *only* way or even the right one at all can be decided only *after one has gone along it*" (SZ 437 / BT 487). Heidegger is giving us a summary of what he's done magnificently in this book, and then he seems to be pulling it away from us. This is only *one way* of casting light on the fundamental question of ontology, only *one way* and not obviously the right way. Our aim was to work out the question of being, but still we haven't thought about what "the idea of Being in general" really *means*. Here we are, in the final moments of the book, and if we don't know by now whether it was the right way, then is that a problem? I think so. It's as though Heidegger is threatening to take back the whole brilliant argument that he's just summarized for us.

Then he says, continuing the third quote, "The conflict as to the Interpretation of Being cannot be allayed, *because it has not yet been enkindled*" (ibid.). That reference to conflict is an allusion to the "conflict of the giants" in Plato's *Sophist*, the *gigantomachia peri tes ousias* (γιγαντομαχία περὶ τῆς οὐσίας), the Battle of the Giants around being, which Heidegger alludes to at the very beginning of the introduction to *Being and Time* (SZ 2 / BT 21).[11] That conflict, that gigantic struggle, hasn't even begun, and it can only be decided whether it was the right way after one has gone along it.

Let's look at how the book finishes, with the fourth quotation. Heidegger says, "Something like 'Being' has been disclosed in the understanding-of-Being which belongs to existent Dasein as a way in which it understands. Being has been disclosed in a preliminary way, though non-conceptually." Notice this emphasis on the preliminary, on the provisionality of what Heidegger is up to here. He goes on:

> ... and this makes it possible for Dasein as existent Being-in-the-world to comport itself *towards entities*—towards those which it encounters within-the-world as well as towards itself as existent. *How is this disclosive understanding of Being at all possible for Dasein?* Can this question be answered by going back to the *primordial constitution-of-Being* of that Dasein by which Being is understood? The existential-ontological constitution of Dasein's totality is grounded in temporality. Hence the ecstatical projection of Being must be made possible by some primordial way in which ecstatical temporality temporalizes. How is this mode of the temporalizing of temporality to be Interpreted? Is there a way which leads from primordial *time* to the meaning of *Being*? Does *time* itself manifest itself as the horizon of *Being*? (SZ 437 / BT 488)

With those words, the book ends. At which point, I am inclined to say (and excuse my French): *What the fuck, Martin?* You said on the

11. Plato, *Sophist*, 246A.

untitled first page of *Being and Time* that time is the horizon for the understanding of being. And now you ask at the very end: *Does* time itself manifest itself as the horizon of being? Why are you pulling everything away?

I want to finish this book, Heidegger's and my own, by casting some doubt on this rhetorical move—indeed a series of rhetorical moves—at the end of *Being and Time*, which is something that is indicative of his style and the style of thinkers who claim to follow him. It is indeed true that the hermeneutic of Dasein that we've followed in *Being and Time* remains only one way in which the question of being might be pursued, and the Dasein-analytic is only a point of departure for universal phenomenological ontology, as he says on the book's final page. It is also true that the question as to whether this is the only way that can or should be pursued is something that can only be decided after one has gone along it —and as such the conflict, the *gigantomachia* with regard to the interpretation of being, has not even begun, has not even been enkindled. And yet, despite the fact that this investigation is only (and this is such a key Heideggerian word) *underway*, there is absolutely no question that Heidegger has given us so much more than a clue as to how the closing questions of *Being and Time* are to be answered.

Let's go back to those final words. I fear that Heidegger loses us with these rhetorical questions. It's as if the whole ladder that we have climbed is being kicked away, and we have learned nothing. By contrast, I think we've learned a lot more than nothing. He says, in the final two sentences of *Being and Time*, "Is there a way which leads from primordial *time* to the meaning of *Being*?" To which the answer is "Yes, there is undoubtedly a path from the primordial temporalizing of temporality—ecstatic Dasein has been described beautifully and powerfully in *Being and Time*—to the meaning of being." And, in the last sentence of the book, Heidegger asks, "Does *time* itself manifest itself as the horizon of *Being*?" Again, a resounding "Yes, time understood in this sense *does* manifest itself as the horizon of being, even if the concept of horizon must ultimately be overcome." Going back to that marginal comment that Heidegger makes, even if we have to find or develop a better concept than "horizon," there's no question that time is the way in which being as such, its meaning or truth, is to be understood.

This is what I want to emphasize at the end of our long journey. I worry about the false modesty of Heidegger's potentially paralyzing rhetorical questions at the end of *Being and Time*. Both the false modesty and the willful self-paralyzing of readers of Heidegger. I could turn this into a more general remark upon my hatred for book titles and paper titles that include the word "Towards," like "Towards a Theory of X" or, even worse, "Notes Towards a Theory of Y." I've seen this so many times with academics and their students. They'll write a paper usually with three abstract nouns: "knowledge," "possibility," "truth," or "reference," "meaning," "necessity," or even worse, "universality," "recognition," "justice" [*colon*] "Notes towards a theory of" blah, blah, blah, blah. Please don't include that in any of your writing and thinking. Never use the word "towards." Never say that your words are "notes." Either write something or don't write something. Have the courage of your convictions. This is just the failure of *towardsism*. And I really don't like that.

In relation to the enormous ground that we have covered in these readings, this long journey through the entirety of *Being and Time*, we've learned a ton of stuff, a massive amount. We need to acknowledge that, absorb it, and then use it to transform our living understanding of everything: ourselves, others, world, space, time, and history. You name it. I do believe that *Being and Time* is a book that changes one's life. It doesn't mean that everything Heidegger says is right. There are lots of criticisms I have of *Being and Time*, as has become clear in the course of this book. But I passionately believe that this is the right orientation. This is the right path to be followed in philosophy. Even if some of the turnings that Heidegger took on that path are doubtful, even malign, and other ways could be chosen. Existential phenomenology is the right method for thinking.

At the end of *Being and Time*, in the final two pages, Heidegger looks back on another quote from the introduction and directly cites himself: "Philosophy 'is universal phenomenological ontology, and takes its departure from the hermeneutic of Dasein, which, as an analytic of *existence*, has made fast the guiding-line for all philosophical inquiry at the point where it *arises* and to which it *returns*'" (SZ 38, 436 / BT 62, 487). Having returned to this concern for a "universal phenomenological ontology" from the introduction, Heidegger now asks the following: "Can one provide *ontological* grounds for ontology, or does it also require an *ontical* foundation? And *which* entity must take over the function of providing this foundation?" The very distinction on which this question is based—

the distinction between "the Being of existing Dasein and the Being of entities" that are not Dasein—this distinction is only "the *point of departure*" for the Dasein-analytic (SZ 436–37 / BT 487).

Let's try to clarify this. Yes, the Dasein-analytic is a point of departure. Yes, it's also something to which a universal phenomenological ontology must return. That's correct. There's a tendency in Heidegger and in discussions of his work to say something like the following: The early Heidegger was on the right track, but he was much too concerned with the anthropological perspective of the human being. Our focus needs to be on being as such. I dispute that claim. The thought that Dasein is being-in-the-world is a much broader claim than any form of humanism, and we have all the elements in place in *Being and Time* for an understanding of being, of the meaning of being, of the truth of being. It is self-evident that we have to continually return to the standpoint of us, of Dasein and our mineness. That's where we begin and end. The Dasein-analytic, the analysis of us, is ineliminable, even if its character must change, and its *form* must change (although its method does not need to change—from beginning to end, Heidegger is a phenomenologist). Indeed, the forms of Heidegger's thinking do change in his later work.

Furthermore, we've been told what constitutes the meaning of being human. The meaning of being human is finitude. That's it. The clues for unraveling the experience of finitude are the two statements I have persistently come back to: Dasein is thrown projection, and Dasein exists factically. What that means is that we are finite, we are mortal. And these are clues that we have followed assiduously in our thinking with and against Heidegger on death, conscience, temporality, and historicity, each of which deepens and thickens the insight into thrown projection, into the being of what Heidegger calls "care." Our view of life has changed, and I want that to stay with us, to resonate with us, and to abide with us.

But, and here's a closing thought: If the being of being human is finite, then can the meaning of being as such be anything other than finite? No. The way Heidegger throws up his arms at the end of this book is perfectly pointless. We are finite. The being of being human is finite. The being of beings as such is also finite. As Heidegger will say in his later work: Being is epochal. It's dispensed in finite historical epochs.[12] Being cannot be thought independently from the experience of thrown projection, from factical existence and the Daseinness of our experience. This is something that I would like to use and to lean upon as a clue in reading Heidegger's later work. Heidegger does not give us a metaphysical determination of being. He does not give us a new thesis on being. Rather, he names an experience. Heidegger is trying to name an experience that corresponds to the lack of a word for being that characterizes our age. Every period has a word for being: for Plato, it was the forms; for Aristotle, substance; for Aquinas, the self-caused cause. For Nietzsche, it is will to power and eternal return. Heidegger refuses to add a word to that list. What characterizes our age is the loss of

12. Martin Heidegger, "Protokoll zu einem Seminar über den Vortrag 'Zeit und Sein'" (1962), in *Zur Sache des Denkens*, 61–62; "Summary of a Seminar on the Lecture 'Time and Being,'" in *On Time and Being*, 51–52.

the word for being.[13] But that loss is also a possibility. Indeed, it is the experience of possibility as such. It is a dizzying freedom.

This is where Heidegger is going to go after *Being and Time.* The master word of Heidegger's later work is *das Ereignis*, the appropriative event, the clearing or the opening. So, the question of being, the question of the meaning of being, is ultimately a question about *meaning*, about what shows itself and what can be brought to language. Heidegger is not trying to get us to think about being as if it was something hidden beneath the surfaces of things, as if there was some sort of occluded obscurity that somehow evades our grasp and that is ineffable and untouchable. Rather, meaning is what lies under our noses. It is the blindingly banal and everyday experience of existence; it's something we do every day in an ordinary way. That ordinariness is the mystery. In Heidegger's later work, the word "being" is even crossed out repeatedly. Heidegger uses "being" under erasure.[14] The opening event of appropriation is another way of thinking about thrown projection. It's another way of thinking about the enigmatic depth or difficulty of what it means to be—to be human, and perhaps something more than human in the usual sense of that word associated with humanism.

13. Martin Heidegger, "Die Onto-theo-logische Verfassung der Metaphysik" (1956/57), in *Identität und Differenz*, 51–80; "The Onto-theo-logical Constitution of Metaphysics," in *Identity and Difference*, 42–76.

14. Martin Heidegger, "Zur Seinsfrage," in *Wegmarken*, ed. Friedrich-Wilhelm von Herrmann, vol. 9 of *Gesamtausgabe*, ed. Friedrich-Wilhelm von Herrmann (Klostermann, 1976), 385–426; Martin Heidegger, *The Question of Being*, trans. William Kluback and Jean T. Wilde, bilingual ed. (Twayne Publishers, 1958), 33–109.

I don't see a fundamental shift or turning (the so-called *Kehre*) in Heidegger's thought at all. I think that Heidegger has different ways in his work of approaching the same matter, the matter of thinking, which is the question of being. And that is something that has to pass through an analysis of who we are. We can't do that by bracketing ourselves out of the picture. (With Heideggerians, there's sometimes a tendency to talk like there's something big "up there," doing all the work: a capital-*B* "Being" pulling all the strings.)

The answer to who we are in *Being and Time* is thrown projection. The answer to who we are in the later work is the appropriative event. Those two ways of thinking are, I think, consistent with and echo each other. But it would take another book on the later Heidegger to establish that claim. In my fancy at least, I imagine it as a sequel to this book.

Where does that leave us in closing? *Being and Time* is essential to the entirety of Heidegger's project. The later work is not a break with the early in the sense of a rupture or a turning away. Rather, I think there is a deepening of the matter of thinking. It's not that nothing changes in the later work. I think (and indeed I could show) how Heidegger's relation to metaphysics undergoes a profound transformation that turns on his complex understanding of the overcoming of metaphysics. Again, he does not want to overcome metaphysics but rather deepen the roots of its questioning. Also, the focus of Heidegger's later work is very often historical. He reflects in extraordinary detail upon key figures from the history of philosophy, like his yearslong confrontation with Nietzsche. But Heidegger is always trying to dig into the same matter with the same phenomenological method. From way before *Being and Time* in 1927 to "Time and Being" in 1962 and until his death, there is a continuity. There is a deepening. There is an ever-returning questioning and experimenting with the forms of thinking. That's my view. There's not a Heidegger 1 and a Heidegger 2. There's continuity throughout his career. He gives some things up, and emphasizes others, but it's recognizably the same project, the same effort.

•

Let me end with a story. In July 1986, I traveled from Essex to Italy for the first time in my life in order to attend a monthlong phenomenology summer school on Heidegger's Marburg lectures just outside Perugia in Umbria. My friend Will and I arrived after a very long and difficult train and ferry journey and, being English, instantly got drunk on local red wine and went to sleep in the impossible heat of the Umbrian night. I woke early, badly hungover and bitten by mosquitoes, and went downstairs to find breakfast, which was extremely meager and disappointing. I was sitting on my own in the corner trying to avoid any human contact, eat what I could find, and gather my wits.

At which point, Hans-Georg Gadamer, Heidegger's great—possibly greatest—student, appeared in the dining room. Gadamer had a walking stick (he'd been sporty as a youth but contracted polio in 1922, the aftereffects of which stayed with him the rest of his life), and moved rather slowly, but deliberately. I then realized he was coming towards my table. He sat down. I thought, "Oh no, it's bloody Gadamer. What am I going to talk to him about?" But he just sat down with his assistant, placed his walking stick beside his chair, and started to talk. In fact, there was no way of stopping him. I thought he'd never shut up. By this point, very late in his career—he was in his eighties, and he still loved drinking red wine and telling stories (he said to a group of us one evening, "For Plato, wine moves the blood")—Gadamer had decided that he was going to serve as a messenger for his teacher. He thought of Heidegger as a greater philosopher than himself, whether rightly or wrongly. The Marburg lectures of 1923–28, which Gadamer had attended as a student and which had opened the doors of philosophy to him, began to appear in German in the late 1970s and in English in the early to mid-1980s. Gadamer had decided to travel to talk about them.

So he sat there at the breakfast table in Perugia and was talking to me. I was no one to him and no one in general. But he took the time. And he said to me, as I'm sure he had said to many others, that Heidegger wrote just one book, *Being and Time*. Everything else he published was collections of essays, or lecture courses, or talks and fragments that were put together into book form. And that's true. But *Being and Time* was his *book*. And it's a failure in the sense in which Heidegger abandons the method of classical transcendental philosophy as a way of access to the question of being. Heidegger comes to think that there are other ways of approaching this question. In his later work, these approaches might be historical, reflective, poetic, dialogical, political, or indeed polemical. Gadamer said that Heidegger spent his career building castles in the air. This sounds like criticism. But it is not.

The activity of thinking is building castles in the air. All that there is, is air. This is the air I'm speaking to you through or the air between this page you're reading and your eyes. Language is just sounds moving in the air. Hopefully it's not just hot air. But we build castles and that's just fine. And then we discover those castles are not perfect and we construct other castles and we go on building. I think that *Being and Time*, unlike any other book of philosophy that I know, can fundamentally shift the way in which one approaches oneself, others, and the world. Heidegger's thinking—phenomenology—is a way of focusing on the simplicity, obviousness, and delightful banality of everyday existence, and a way of finding concepts that are the equivalent to our lived experience.

Heidegger gives us a language for making sense of the ways in which we make sense. And it gives us a language for what hovers at the edges of sense: anxiety, mortality, responsibility, ecstasy, the rush of lived time, and the happening of history. *Being and Time* makes other, more traditional ways of doing philosophy redundant and gives us a language of thinking that is simply closer to life as it is lived. Heidegger's work opens up a whole series of questions, and a powerful tradition of thinking is continued in figures like Sartre, De Beauvoir, Merleau-Ponty, Arendt, Jonas, Strauss, Levinas, and Derrida and so many others, who are all thinkers who raise questions to Heidegger, but from within the framework that Heidegger has established.

I'll finish with a remark that Levinas makes at the beginning of a book published in 1947 called *De l'existence à l'existant* (published in English as *Existence and Existents*, though a more literal translation would read *From Existence to the Existent*), which itself is a provocation. Heidegger is trying to take us from beings to the question of Being, and Levinas is reversing that direction, taking us from the question of existence in general to the existent, back to subjectivity, even to consciousness. Levinas says at the beginning of that book that we must leave the climate of Heidegger's philosophy for obvious reasons, reasons that are at the very least moral and political but also ontological and conceptual. But, Levinas goes on, we cannot leave that climate for a philosophy that would be pre-Heideggerian.[15] Namely, what happens in *Being and Time* is a paradigm shift in the way in which we see the basic questions of philosophy and thinking. Once we accept that paradigm shift, then we move on and develop those questions in freer, deeper, less tradition-bound and more innovative and independent ways. That's the task that awaits each of us in the thinking that we will go on to develop in our lives, with the air in our lungs and the words on our lips. I hope that the way in which you continue to think is influenced by Heidegger's *Being and Time*. If you've got to the end of this marathon, then I admire your stamina and I thank you deeply for your attention. I hope to continue this adventure with you again before too long. As long as we have the time.

15. Emmanuel Levinas, *Existence and Existents*, 4.

Thanks

I WOULD LIKE TO THANK Ben Olson for his splendid and exhaustive work on this book. It simply would not have happened without him. I would like to thank Edwin Frank, Sara Kramer, Linda Hollick, Nick During, and everyone at NYRB for their belief in my work and ongoing trust. I would like to thank my agent and dear friend, Nemonie Craven. I would like to thank Susan Foote for her friendship and support, and the Onassis Foundation, especially Anthony Papadimitriou. I would like to thank my wife, Sophia Critchley, for her love and light in the darkness. This book is dedicated to the generations of students to whom I have taught Heidegger's *Being and Time* for close to forty years and from whom I have learned so much. Thank you for putting up with me.

Bibliography

Works by Martin Heidegger in German

Works outside the *Gesamtausgabe*

"Das Problem der Sünde bei Luther." In *Sachgemäße Exegese: Die Protokolle aus Rudolf Bultmanns Neutestamentlichen Seminaren 1921–1951*, edited by Bernd Jaspert. N. G. Elwert, 1996. [student transcript]

"Heidegger and Schmitt: The Bottom Line." *Carl Schmitt: Enemy or Foe?*, Special issue of *Telos* 72 (1987): 132. [English/German bilingual translation of Heidegger's letter to Carl Schmitt of 22 August 1933]

Sein und Zeit. Max Niemeyer, 1979.

Volumes of the *Gesamtausgabe*

1. *Frühe Schriften*. Edited by Friedrich-Wilhelm von Herrmann. Vol. 1 of *Gesamtausgabe*, edited by Friedrich-Wilhelm von Herrmann. Klostermann, 1978.

2. *Sein und Zeit*. Edited by Friedrich-Wilhelm von Herrmann. Vol. 2 of *Gesamtausgabe*, edited by Friedrich-Wilhelm von Herrmann. Klostermann, 1977.

3. *Kant und das Problem der Metaphysik*. Edited by Friedrich-Wilhelm von Herrmann. Vol. 3 of *Gesamtausgabe*, edited by Friedrich-Wilhelm von Herrmann. Klostermann, 1991.

4. *Erläuterungen zu Hölderlins Dichtung*. Edited by Friedrich-Wilhelm von Herrmann. Vol. 4 of *Gesamtausgabe*, edited by Friedrich-Wilhelm von Herrmann. Klostermann, 1981.

6.1. *Nietzsche*. Edited by Brigitte Schillbach. Vol. 6.1 of *Gesamtausgabe*, edited by Friedrich-Wilhelm von Herrmann. Klostermann, 1996.

6.2. *Nietzsche*. Edited by Brigitte Schillbach. Vol. 6.2 of *Gesamtausgabe*, edited by Friedrich-Wilhelm von Herrmann. Klostermann, 1997.

7. *Vorträge und Aufsätze*. Edited by Friedrich-Wilhelm von Herrmann. Vol. 7 of *Gesamtausgabe*, edited by Friedrich-Wilhelm von Herrmann. Klostermann, 2000.

9. *Wegmarken*. Edited by Friedrich-Wilhelm von Herrmann. Vol. 9 of *Gesamtausgabe*, edited by Friedrich-Wilhelm von Herrmann. Klostermann, 1976.

11. *Identität und Differenz*. Edited by Friedrich-Wilhelm von Herrmann. Vol. 11 of *Gesamtausgabe*, edited by Friedrich-Wilhelm von Herrmann. Klostermann, 2006.

12. *Unterwegs zur Sprache*. Edited by Friedrich-Wilhelm von Herrmann. Vol. 12 of *Gesamtausgabe*, edited by Friedrich-Wilhelm von Herrmann. Klostermann, 1985.

13. *Aus der Erfahrung des Denkens*. Edited by Hermann Heidegger. Vol. 13 of *Gesamtausgabe*, edited by Friedrich-Wilhelm von Herrmann. Klostermann, 1983.

14. *Zur Sache des Denkens*. Edited by Friedrich-Wilhelm von Herrmann. Vol. 14 of *Gesamtausgabe*, edited by Friedrich-Wilhelm von Herrmann. Klostermann, 2007.

15. *Seminare*. Edited by Curd Ochwadt. Vol. 15 of *Gesamtausgabe*, edited by Friedrich-Wilhelm von Herrmann. Klostermann, 1986.

16. *Reden und andere Zeugnisse eines Lebensweges*. Edited by Hermann Heidegger. Vol. 16 of *Gesamtausgabe*, edited by Friedrich-Wilhelm von Herrmann. Klostermann, 2000.

24. *Die Grundprobleme der Phänomenologie*. Edited by Friedrich-Wilhelm von Herrmann. Vol. 24 of *Gesamtausgabe*, edited by Friedrich-Wilhelm von Herrmann. Klostermann, 1975.

26. *Metaphysische Anfangsgründe der Logik im Ausgang von Leibniz*. Edited by Klaus Held. Vol. 26 of *Gesamtausgabe*, edited by Friedrich-Wilhelm von Herrmann. Klostermann, 1978.

29/30. *Die Grundbegriffe der Metaphysik: Welt—Endlichkeit—Einsamkeit*. Edited by Friedrich-Wilhelm von Herrmann. Vols. 29/30 of *Gesamtausgabe*, edited by Friedrich-Wilhelm von Herrmann. Klostermann, 1992.

40. *Einführung in die Metaphysik*. Edited by Petra Jaeger. Vol. 40 of *Gesamtausgabe*, edited Friedrich-Wilhelm von Herrmann. Klostermann, 1983.

46. *Unzeitgemäßer Betrachtung*. Edited by Hans-Joachim Friedrich. Vol. 46 of *Gesamtausgabe*, edited by Friedrich-Wilhelm von Herrmann. Klostermann, 2003.

47. *Nietzsches Lehre vom Willen zur Macht als Erkenntnis*. Edited by Eberhard Hanser. Vol. 47 of *Gesamtausgabe*, edited by Friedrich-Wilhelm von Herrmann. Klostermann, 1989.

64. *Der Begriff der Zeit*. Edited by Friedrich-Wilhelm von Herrmann. Vol. 64 of *Gesamtausgabe*, edited by Friedrich-Wilhelm von Herrmann. Klostermann, 2004.
69. *Die Geschichte des Seyns*. Edited by Peter Trawny. Vol. 69 of *Gesamtausgabe*, edited by Friedrich-Wilhelm von Herrmann. Klostermann, 1998.
75. *Zu Hölderlin—Griechenlandreisen*. Edited by Curd Ochwadt. Vol. 75 of *Gesamtausgabe*, edited by Friedrich-Wilhelm von Herrmann. Klostermann, 2000.
80.1. *Vorträge, 1915–1932*. Edited by Günther Neumann. Vol. 80.1 of *Gesamtausgabe*, edited by Friedrich-Wilhelm von Herrmann. Klostermann, 2016.
89. *Zollikoner Seminare*. Edited by Peter Trawny. Vol. 89 of *Gesamtausgabe*, edited Friedrich-Wilhelm von Herrmann. Klostermann, 2017.

Works by Martin Heidegger in English

Shorter Works

"Heidegger and Schmitt: The Bottom Line." *Carl Schmitt: Enemy or Foe?*, Special issue of *Telos* 72 (1987): 132 [English/German bilingual translation of Heidegger's letter to Carl Schmitt of 22 August 1933]
"The Self-Assertion of the German University." Translated by Karsten Harries. *Review of Metaphysics* 38, no. 3 (1985): 467–502.

Volumes and Primary Texts

Basic Writings: From Being and Time *(1927) to* The Task of Thinking *(1964)*. Rev. ed. Edited by David Farrell Krell. HarperCollins, 1993.
Being and Time. Translated by John Macquarrie and Edward Robinson. Blackwell; Harper & Row, 1962.
Duns Scotus's Doctrine of Categories and Meaning. Translated by Joydeep Bagchee and Jeffrey D. Gower. Indiana University Press, 2022.
Elucidations of Hölderlin's Poetry. Translated by Keith Hoeller. Humanity Books, 2000.
Four Seminars. Translated by Andrew Mitchell and François Raffoul. Indiana University Press, 2003.
Identity and Difference. Translated by Joan Stambaugh. Harper & Row, 1969.
Interpretation of Nietzsche's Second Untimely Meditation. Translated by Ullrich Haase and Mark Sinclair. Indiana University Press, 2016.
Introduction to Metaphysics. 2nd ed. Translated by Gregory Fried and Richard Polt. Yale University Press, 2014.

Kant and the Problem of Metaphysics. 5th ed. Translated by Richard Taft. Indiana University Press, 1997.

The Metaphysical Foundations of Logic. Translated by Michael Heim. Indiana University Press, 1984.

Nietzsche: Volumes 1 and 2: The Will to Power as Art and *The Eternal Recurrence of the Same*. Edited by David Farrell Krell. HarperCollins, 1991.

Nietzsche: Volumes 3 and 4: The Will to Power as Knowledge and as Metaphysics and *Nihilism*. Translated by David Farrell Krell. HarperCollins, 1991.

On Time and Being. Translated by Joan Stambaugh. Harper & Row, 1972.

On the Way to Language. Translated by Peter D. Hertz. Harper & Row, 1971.

Pathmarks. Edited by William McNeill. Cambridge University Press, 1998.

Poetry, Language, Thought. Translated by Albert Hofstadter. Harper & Row, 1971.

Sojourns: The Journey to Greece. Translated by John Panteleimon Manoussakis. State University of New York Press, 2005.

Supplements: From the Earliest Essays to Being and Time *and Beyond*. Edited by John van Buren. State University of New York Press, 2002.

The Basic Problems of Phenomenology. Rev. ed. Translated by Albert Hofstadter. Indiana University Press, 1982.

The Concept of Time. Translated by Ingo Farin and Alex Skinner. Continuum, 2011.

The Concept of Time. Translated by William McNeill. Blackwell, 1992.

The Fundamental Concepts of Metaphysics: World, Finitude, Solitude. Translated by William McNeill and Nicholas Walker. Indiana University Press, 1995.

The Heidegger Reader. Translated by Jerome Veith. Edited by Günter Figal. Indiana University Press, 2009.

The History of Beyng: 1. The History of Beyng, 1938–40; 2. Κοινόν, Out of the History of Beyng, 1939–40. Translated by Jeffrey Powell and William McNeill. Indiana University Press, 2015.

The Question of Being. Bilingual ed. Translated by William Kluback and Jean T. Wilde. Twayne Publishers, 1958.

The Question Concerning Technology and Other Essays. Translated by William Lovitt. Garland Publishing, 1977.

Zollikon Seminars: Protocols—Conversations—Letters. Edited by Medard Boss. Translated by Franz Mayr and Richard Askay. Northwestern University Press, 2001.

Works by Simon Critchley

ABC of Impossibility. Univocal Publishing, 2015.
Continental Philosophy: A Very Short Introduction. Oxford University Press, 2001.
Infinitely Demanding: Ethics of Commitment, Politics of Resistance. Verso, 2007.
The Faith of the Faithless: Experiments in Political Theology. Verso, 2012.
Very Little . . . Almost Nothing: Death, Philosophy, Literature. Routledge, 1997.
"Life Doesn't Need a Narrative Arc." *The New York Times*, June 12, 2023.
"With Being-With? Notes on Jean-Luc Nancy's Rewriting of *Being and Time*." *Phänomenologische Forschungen* 3, no. 2 (1998): 198–210.
with Reiner Schürmann. *On Heidegger's* Being and Time. Edited by Steven Levine. Routledge, 2008.
with Tom McCarthy, "Universal Shylockery: Money and Morality in *The Merchant of Venice*." *Diacritics* 34, no. 1 (2004): 3–17.

Works by Other Authors

Adorno, Theodor W. *The Jargon of Authenticity*. Translated by Knut Tarnowski and Frederic Will. Northwestern University Press, 1973.
——. *Minima Moralia: Reflections from Damaged Life*. Translated by E. F. N. Jephcott. NLB, 2005.
Thomas Aquinas. *The Disputed Questions on Truth*. Translated by Robert W. Mulligan. H. Regnery, 1952.
Arendt, Hannah. *On Revolution*. Viking, 1963.
——. *The Origins of Totalitarianism*. Harcourt, 1968.
——. *The Life of the Mind*. Harvest, 1978.
Aristotle. *The Basic Works of Aristotle*. Edited by Richard McKeon. Modern Library, 2001.
——. *Complete Works of Aristotle: The Revised Oxford Translation*. Vol. 1. Edited by Jonathan Barnes. Princeton University Press, 1983.
——. *Complete Works of Aristotle: The Revised Oxford Translation*. Vol. 2. Edited by Jonathan Barnes. Princeton University Press, 1984.
——. *Metaphysics*. Translated by C. D. C. Reeve. Hackett Publishing, 2016.
Augustine. *Confessions*. 2nd ed. Translated by F. J. Sheed. Edited by Michael P. Foley. Hackett Publishing, 2006.
Badiou, Alain. *Being and Event*. Translated by Oliver Feltham. Continuum, 2005.
Bamford, Maria. *Sure, I'll Join Your Cult: A Memoir of Mental Illness and the Quest to Belong Anywhere*. Gallery Books, 2023.

Beaufret, Jean. "A propos de l'existentialisme." *Confluences* Nos. 2–6 (1945).
Becker, Ernest. *The Denial of Death*. Free Press, 1973.
Beckett, Samuel. *The Collected Shorter Plays*. Grove Weidenfeld, 1984.
——. *Molloy*. Translated by Patrick Bowles in collaboration with author. Grove Press, 1955.
——. *Waiting for Godot: A Tragicomedy in Two Acts*. Grove Press, 1954.
Benjamin, Walter. *Selected Writings, 1913–1926*. Edited by Marcus Bullock and Michael W. Jennings. Vol. 1 of *Selected Writings*, edited by Michael W. Jennings. Harvard University Press, 1996.
——. *Selected Writings, 1938–1940*. Edited by Howard Eiland and Michael W. Jennings. Vol. 4 of *Selected Writings*, edited by Michael W. Jennings. Harvard University Press, 2006.
Benner, Patricia. *From Novice to Expert: Excellence and Power in Clinical Nursing Practice*. Prentice Hall Health, 2001.
Bergson, Henri. *Time and Free Will: An Essay on the Immediate Data of Consciousness*. Translated by F. L. Pogson. Dover Publications, 1960.
Berkeley, George. *Philosophical Writings*. Edited by Desmond Clarke. Cambridge University Press, 2012.
Bernasconi, Robert. *Heidegger in Question: The Art of Existing*. Humanities Press, 1993.
Blake, William. *Collected Poems*. Edited by W. B. Yeats. Routledge, 2002.
——. *The Marriage of Heaven and Hell*. University of Miami Press, 1965.
Blanchot, Maurice. *Thomas the Obscure*. Translated by Robert Lamberton. Station Hill Press, 1973.
Boswell, James. *Life of Samuel Johnson*. Edited by Herbert Vaughan Abbott. Scott, Foresman, 1923.
Burdach, Konrad. "Faust und die Sorge." *Deutsche Vierteljahrsschrift für Literaturwissenschaft und Geistesgeschichte* 1 (1923): 1–60.
Butler, Judith. *Precarious Life: The Powers of Mourning and Violence*. Verso, 2004.
Camus, Albert. *The Stranger*. Translated by Matthew Ward. Vintage International, 1989.
Canetti, Elias. *Crowds and Power*. Translated by Carol Stewart. Viking Press, 1962.
Carson, Anne. *Antigonick*. New Directions, 2012.
Cavell, Stanley. *The Claim of Reason: Wittgenstein, Skepticism, Morality, and Tragedy*. Oxford University Press, 1979.
Char, René. *L'Endurance de la pensée: Pour saluer Jean Beaufret*. Plon, 1968.
Chaucer, Geoffrey. *The Parson's Tale*. Translated by Mary Farrell Pomerleau. Tumblar House, 2011.
Dahlstrom, Daniel O., ed. *Interpreting Heidegger: Critical Essays*. Cambridge University Press, 2011.

de Man, Paul. *Blindness and Insight: Essays in the Rhetoric of Contemporary Criticism*. 2nd rev. ed. Routledge, 1983.

Debord, Guy. *The Society of the Spectacle*. Translated by Ken Knabb. Bureau of Public Secrets, 2014.

Defoe, Daniel. *A Journal of the Plague Year*. Lerner Publishing, 2022.

——. *Robinson Crusoe*. Edited by Thomas Keymer. Oxford University Press, 2007.

Derrida, Jacques. *Margins of Philosophy*. Translated by Alan Bass. University of Chicago Press, 1982.

——. *Of Spirit: Heidegger and the Question*. Translated by Geoffrey Bennington and Rachel Bowlby. University of Chicago Press, 1989.

Descartes, René. *The Philosophical Writings of Descartes*. Vol. 1. Translated by John Cottingham, Robert Stoothoff, and Dugald Murdoch. Cambridge University Press, 1985.

——. *The Philosophical Writings of Descartes*. Vol. 2. Translated by John Cottingham, Robert Stoothoff, and Dugald Murdoch. Cambridge University Press, 1984.

Dickinson, Emily. *Poems, Second Series*. Edited by Mabel Loomis Todd and T. W. Higginson. Roberts Brothers, 1891.

Dodd, James. *The Heresies of Jan Patočka: Phenomenology, History, and Politics*. Northwestern University Press, 2023.

Dreyfus, Hubert L. *Being-in-the-World: A Commentary on Heidegger's* Being and Time, *Division I*. MIT Press, 1991.

Dreyfus, Hubert L., and Stuart E. Dreyfus. *Mind over Machine: The Power of Human Intuition and Expertise in the Era of the Computer*. With Tom Athanasiou. Free Press, 1986.

Durkheim, Émile. *The Division of Labour in Society*. Translated by W. D. Halls. Macmillan, 1984.

——. *Suicide: A Study in Sociology*. Translated by John A. Spaulding and George Simpson. Edited by George Simpson. Routledge, 1952.

Eliot, T. S. *Four Quartets*. Harcourt, 1943.

Farías, Victor. *Heidegger und der Nationalsozialismus: Mit einem Vorwort von Jürgen Habermas*. S. Fischer, 1989.

——. *Heidegger and Nazism*. Edited by Joseph Margolis and Tom Rockmore. Translated by Paul Burrell, Dominic Di Bernardi, and Gabriel R. Ricci. Temple University Press, 1989.

Fichte, Johann Gottlieb. *The Science of Knowledge*. Translated and edited by Peter Heath and John Lachs. Cambridge University Press, 1982.

Flusser, Vilém, and Louis Bec. *Vampyroteuthis Infernalis: A Treatise, with a Report by the Institut Scientifique de Recherche Paranaturaliste*. Translated by Valentine A. Pakis. University of Minnesota Press, 2012.

Foucault, Michel. *The Order of Things: An Archaeology of the Human Sciences*. Routledge, 1989.

Friedman, Michael. *A Parting of the Ways: Carnap, Cassirer, and Heidegger*. Open Court, 2000.

Fynsk, Christopher. *Heidegger: Thought and Historicity*. Expanded ed. Cornell University Press, 1986.

Gadamer, Hans-Georg. *Philosophical Hermeneutics*. Translated and edited by David E. Linge. University of California Press, 1976.

——. *Truth and Method*. 2nd rev. ed. Translated by Joel Weinsheimer and Donald G. Marshall. Continuum, 1996.

Godfrey-Smith, Peter. *Other Minds: The Octopus, The Sea, and the Deep Origins of Consciousness*. Farrar, Straus and Giroux, 2016.

Habermas, Jürgen. "Mit Heidegger gegen Heidegger denken: zur Veröffentlichung von Vorlesungen aus dem Jahre 1935." *Frankfurter Allgemeine Zeitung*, July 25, 1953.

——. *The Philosophical Discourse of Modernity: Twelve Lectures*. Translated by Frederick G. Lawrence. MIT Press, 1990.

von Harnack, Adolf. *Marcion: The Gospel of the Alien God*. Translated by John E. Steely and Lyle D. Bierma. Labyrinth, 1990.

——. *History of Dogma*. Translated by Neil Buchanan. 7 vols. Dover, 1961.

Haugeland, John. *Dasein Disclosed: John Haugeland's Heidegger*. Edited by Joseph Rouse. Harvard University Press, 2013.

Hegel, G. W. F. *Lectures on the History of Philosophy*. Vol. 3. Translated by Elizabeth S. Haldane and Frances H. Simson. Kegan Paul, 1896.

——. *Phenomenology of Spirit*. Translated by A. V. Miller. Oxford University Press, 1977.

——. *Philosophy of Right*. Translated by T. M. Knox. Oxford University Press, 1952.

Heraclitus. *The Art and Thought of Heraclitus: An Edition of the Fragments with Translation and Commentary*. Translated and edited by Charles H. Kahn. Cambridge University Press, 1979.

Hölderlin, Friedrich. *Poems of Friedrich Hölderlin: The Fire of the Gods Drives Us to Set Forth by Day and by Night*. Translated by James Mitchell. Ithuriel's Spear, 2007.

Hume, David. *A Treatise of Human Nature*. Edited by L. A. Selby-Bigge. Clarendon Press, 1888.

Husserl, Edmund. *The Crisis of European Sciences and Transcendental Phenomenology: An Introduction to Phenomenological Philosophy*. Translated by David Carr. Northwestern University Press, 1970.

——. *The Phenomenology of Internal Time-Consciousness*. Translated by James S. Churchill. Edited by Martin Heidegger. Indiana University Press, 1964.

——. *Psychological and Transcendental Phenomenology and the Confrontation with Heidegger, 1927–1931*. Translated and edited by Thomas Sheehan and Richard E. Palmer. Springer, 1997.

——. *Shorter Works*. Edited by Peter McCormick and Frederick A. Elliston. University of Notre Dame Press, 1981.

Ibn Sina. *Ibn Sina's Remarks and Admonitions: Physics and Metaphysics: An Analysis and Annotated Translation*. Translated by Shams Inati. Columbia University Press, 2014.

Ibsen, Henrik. *Hedda Gabler*. Translated by Edmund Gosse and William Archer. Lerner Publishing Group, 2015.

——. *Little Eyolf*. Translated by Henry L. Mencken. John W. Luce, 1909.

Israeli, Isaac. *Isaac Israeli: A Neoplatonic Philosopher of the Early Tenth Century: His Works Translated with Comments and an Outline of his Philosophy*. Translated by Alexander Altmann and Samuel M. Stern. University of Chicago Press, 2009.

James, William. *Essays in Radical Empiricism*. Edited by H. G. Callaway. Lexington Books, 2022.

Jones, James. *Whistle*. Delacorte Press, 1978.

Kafka, Franz. *The Trial*. Translated by Mike Mitchell. Oxford University Press, 2009.

Kant, Immanuel. *Critique of Pure Reason*. Translated by Paul Guyer and Allen W. Wood. Cambridge University Press, 1998.

——. *Theoretical Philosophy, 1755–1770*. Translated and edited by David Walford and Ralf Meerbote. Cambridge University Press, 1992.

Lacan, Jacques. *Écrits: The First Complete Edition in English*. Translated by Bruce Fink. W. W. Norton, 2006.

Lacoue-Labarthe, Philippe. *Typography: Mimesis, Philosophy, Politics*. Edited by Christopher Fynsk. Stanford University Press, 1998.

Laertius, Diogenes. *Lives of Eminent Philosophers*. Translated and edited by Stephen White. Cambridge University Press, 2020.

Latour, Bruno. *Facing Gaia: Eight Lectures on the New Climatic Regime*. Translated by Catherine Porter. Polity, 2017.

——. *We Have Never Been Modern*. Translated by Catherine Porter. Harvard University Press, 1993.

Levinas, Emmanuel. *Basic Philosophical Writings*. Edited by Adriaan T. Peperzak, Simon Critchley, and Robert Bernasconi. Indiana University Press, 1996.

——. *Entre Nous: On Thinking-of-the-Other*. Translated by Michael B. Smith and Barbara Harshav. Columbia University Press, 1998.

——. *Existence and Existents*. Translated by Alphonso Lingis. Duquesne University Press, 1978.

——. *Difficult Freedom: Essays on Judaism*. Translated by Seán Hand. Athlone, 1991.

——. *The Levinas Reader*. Edited by Seán Hand. Blackwell, 1989.

——. *Nine Talmudic Readings*. Translated by Annette Aronowicz. Indiana University Press, 2019.

——. *Totality and Infinity: An Essay on Exteriority*. Translated by Alphonso Lingis. Duquesne University Press, 1969.

Livergood, Norman D. *Activity in Marx's Philosophy*. Martinus Nijhoff, 1968.

Locke, John. *An Essay Concerning Human Understanding*. T. Tegg, 1836.

Lovelock, James. *Gaia: A New Look at Life on Earth*. Oxford University Press, 1979.

Löwith, Karl. "My Last Meeting with Heidegger in Rome, 1936." Translated by Richard Wolin. Special Issue on Bloch and Heidegger, *New German Critique* 45 (1988): 115–16

Luckhardt, C. Grant, ed. *Wittgenstein: Sources and Perspectives*. Cornell University Press, 1979.

Lukács, Georg. *The Destruction of Reason*. Translated by Peter Palmer. Humanities Press, 1981.

——. *The Theory of the Novel: A Historico-Philosophical Essay on the Forms of Great Epic Literature*. Translated by Anna Bostock. MIT Press, 1971.

Luther, Martin. "Romans Twelve." In *Commentary on Romans*, translated by J. Theodore Mueller. Kregel, 1976.

Lynch, David. *Catching the Big Fish: Meditation, Consciousness, and Creativity*. Tarcher, 2006.

May, Henry F. *The End of American Innocence: A Study of the First Years of Our Own Time, 1912–1917*. Knopf, 1959.

McCarthy, Tom. *Remainder*. Vintage, 2005.

Meeks, Wayne A., and John T. Fitzgerald, eds. *The Writings of St. Paul*. W. W. Norton, 2007.

Meillassoux, Quentin. *After Finitude: An Essay on the Necessity of Contingency*. Translated by Ray Brassier. Continuum, 2008.

Melville, Herman. *Bartleby the Scrivener: A Story of Wall Street*. Melville House, 2004.

Merleau-Ponty, Maurice. *Phenomenology of Perception*. Translated by Colin Smith. Routledge, 1962.

Dan Michel of Northgate. *Ayenbite of Inwyt*. Edited by Rev. Joseph Stevenson. London, 1855.

Milton, John. *Paradise Lost*. Routledge, 1905.

Monk, Ray. *How to Read Wittgenstein*. Granta, 2005.

Montaigne, Michel de. *The Complete Essays*. Translated by M. A. Screech. Penguin Books, 1987.

Nancy, Jean-Luc. *Being Singular Plural*. Translated by Robert Richardson and Anne O'Byrne. Stanford University Press, 2000.

Nietzsche, Friedrich. *The Birth of Tragedy and Other Writings*. Translated by Ronald Speirs. Edited by Raymond Geuss and Ronald Speirs. Cambridge University Press, 1999.

——. *Ecce Homo: Nietzsche's Autobiography*. Translated by Anthony M. Ludovici. Vol. 17 of *The Complete Works of Friedrich Nietzsche*, edited by Oscar Levy. Macmillan, 1911.

——. *On the Genealogy of Morals*. Translated and edited by Walter Kaufmann. Random House, 1967.

——. *The Will to Power*. Translated by Walter Kaufmann and R. J. Hollingdale. Edited by Walter Kaufmann. Random House, 1967.

Parmenides of Elea. *Fragments: A Text and Translation with an Introduction*. Translated by David Gallop. University of Toronto Press, 1984.

Payen, Guillaume. *Martin Heidegger: Catholicisme, révolution, nazisme*. Perrin, 2016.

——. *Martin Heidegger's Changing Destinies: Catholicism, Revolution, Nazism*. Translated by Jane Marie Todd and Steven Rendall. Yale University Press, 2023.

Pessoa, Fernando. *Fernando Pessoa & Co.: Selected Poems*. Translated and edited by Richard Zenith. Grove Press, 1998.

Plato. *Plato: Complete Works*. Edited by John M. Cooper. Hackett Publishing, 1997.

——. *The Republic of Plato*. Translated by Allan Bloom. Harper Collins, 1968.

Pöggeler, Otto. *Martin Heidegger's Path of Thinking*. Translated by Daniel Magurshak and Sigmund Barber. Humanities Press International, 1987.

Polt, Richard. "Inception, Downfall, and the Broken World: Heidegger Above the Sea of Fog." In *Heidegger's "Black Notebooks": Responses to Anti-Semitism*, edited by Andrew J. Mitchell and Peter Trawny. Columbia University Press, 2017.

Quine, W. V. *Ontological Relativity and Other Essays*. Columbia University Press, 1969.

Rayner, Claire. *Claire Rayner's Lifeguide: A Commonsense Approach to Modern Living*. New English Library, 1980.

Richardson, William J. *Heidegger: Through Phenomenology to Thought*. 4th ed. Fordham University Press, 2003.

Rieff, Philip. *The Triumph of the Therapeutic: Uses of Faith after Freud*. University of Chicago Press, 1987.

Riesman, David, Nathan Glazer, and Reuel Denney. *The Lonely Crowd: A Study of the Changing American Character*. Abr. and rev. ed. Yale University Press, 2020.

Rimbaud, Arthur. *Complete Works, Selected Letters: A Bilingual Edition*. Translated by Wallace Fowlie. Revised by Seth Whidden. University of Chicago Press, 2005.

——. *Illuminations*. Translated by John Ashbery. W. W. Norton, 2012.

Rorty, Richard. *Philosophy and the Mirror of Nature*. Princeton University Press, 1979.

Ryan, John K., ed. *Ancients and Moderns*. Catholic University of America Press, 2018.

Ryle, Gilbert. "Review of Martin Heidegger, *Sein und Zeit*." *Mind* 38, no. 151 (1929): 355–70

Sallis, John, ed. *Reading Heidegger: Commemorations*. Indiana University Press, 1993.

Sartre, Jean-Paul. *Being and Nothingness: A Phenomenological Essay on Ontology*. Translated by Hazel E. Barnes. Washington Square Press, 1956.

——. *The Condemned of Altona: A Play in Five Acts*. Translated by Sylvia and George Leeson. Vintage, 1961.

——. *Theory of Practical Ensembles*. Vol. 1 of *Critique of Dialectical Reason*, translated by Alan Sheridan-Smith, edited by Jonathan Rée. Verso, 1960.

——. *Existentialism and Human Emotions*. Philosophical Library, 1957.

——. *Nausea*. Translated by Lloyd Alexander. New Directions, 2013.

——. *The Wall (Intimacy) and Other Stories*. Translated by Lloyd Alexander. New Directions, 1948.

Schiller, Friedrich. *On the Aesthetic Education of Man*. Translated by Reginald Snell. Dover, 2004.

Schmidt, Dennis J. *The Ubiquity of the Finite: Hegel, Heidegger, and the Entitlements of Philosophy*. MIT Press, 1988.

Schmitt, Carl. *Dictatorship: From the Origin of the Modern Concept of Sovereignty to the Proletarian Class Struggle*. Translated by Michael Hoelzl and Graham Ward. Polity, 2014.

——. *Political Theology: Four Chapters on the Concept of Sovereignty*. Translated by George Schwab. University of Chicago Press, 1985.

Sellars, Wilfrid. *Empiricism and the Philosophy of Mind*. Harvard University Press, 1997.

Shakespeare, William. *Hamlet*. Edited by Burton Raffel. Yale University Press, 2003.

——. *Shakespeare: Complete Works*. Edited by W. J. Craig. Oxford University Press, 1966.

Sheehan, Thomas, ed. *Heidegger: The Man and the Thinker*. Precedent Publishing, 1981.

——. *Making Sense of Heidegger: A Paradigm Shift*. Rowman and Littlefield, 2015.

——. "Heidegger and the Nazis." *The New York Review of Books*, June 16, 1988.

Sloterdijk, Peter. *Critique of Cynical Reason*. Translated by Michael Eldred. University of Minnesota Press, 1988.

Sorel, Georges. *Reflections on Violence*. Translated by T. E. Hulme and J. Roth. Dover Publications, 2012.

Spinoza, Benedict de. *A Spinoza Reader:* The Ethics *and Other Works*. Translated and edited by Edwin Curley. Princeton University Press, 1994.

Stevens, Wallace. *The Collected Poems of Wallace Stevens*. Knopf, 1954.

Strauss, Leo. "Exoteric Teaching." Edited by Kenneth Hart Green. *Interpretation* 14, no. 1 (1986): 51–59.

——. "On a Forgotten Kind of Writing." *Chicago Review* 8, no. 1 (1954): 64–75.

——. "Persecution and the Art of Writing." *Social Research* 8, no. 4 (1941): 488–504.

Strawson, Galen. "Against Narrativity." *Ratio* 17, no. 4 (2004): 428–52.

Taylor, Charles. *Philosophical Arguments*. Harvard University Press, 1995.

Tolstoy, Leo. *The Death of Ivan Ilyich and Other Stories*. Translated by Rosemary Edmonds. Penguin Books, 1995.

Trawny, Peter. *Heidegger and the Myth of a Jewish World Conspiracy*. Translated by Andrew J. Mitchell. University of Chicago Press, 2015.

Tugendhat, Ernst. *Der Wahrheitsbegriff bei Husserl und Heidegger*. De Gruyter, 1967.

Vladislav, Jan, ed. *Václav Havel: Living in Truth*. Faber and Faber, 1989.

Vaneigem, Raoul. *Treatise on Living for the Use of the Young Generation*. Translated by Donald Nicholson-Smith. Practical Paradise Publications, 1972.

Watt, Ian. *The Rise of the Novel: Studies in Defoe, Richardson and Fielding*. University of California Press, 1957.

Wittgenstein, Ludwig. *Culture and Value*. Translated by Peter Winch. Edited by G. H. von Wright in collaboration with Heikki Nyman. Blackwell, 1980.

——. *Philosophical Investigations*. 2nd ed. Translated by G. E. M. Anscombe. Blackwell, 1958.

——. *Tractatus Logico-Philosophicus*. Translated by D. F. Pears and B. F. McGuinness. Routledge, 2014.

Index

A

ABC of Impossibility, by Simon Critchley 184n
Abensour, Miguel 266
"action," loss of quotation marks 338
action, tragicomic 301
actuality and possibility 251–2
adequation of thoughts to things 221–2, 224
Adorno, Theodor 11, 21, 229n, 417
After Finitude, by Quentin Meillassoux 201
Alētheia (truth) 106, 219, 227–8, 232
alienation 158–60
alien voice, of conscience 276, 286–7
the "always already" 13, 23, 28, 33–5, 46–7, 59, 81–2, 97, 102, 104
ambient music 303
ambiguity, and fallen language 155–6
ambiguity, and sociality 156–7, 181
animals, non-human 143, 272, 361–2
anthropology
  anticipation *(das Vorlaufen)* 251, 253
  a priori 168
  Christian 197–8
  distinction from awaiting 251, 352
  as excessive in division I 199
anticipatory resoluteness 242, 309–13, 315, 318–19, 328, 331–2, 335–6, 338–9, 421
Antigone 270–1, 319
the Antigonick 360–1
anxiety *(Angst)*
  and anticipatory resoluteness 317–18
  anxiety of conscience 302
  author's youthful anxiety 176
  and being-towards-death 257–8
  Christianity and 192–4
  distinction between fear and 119, 185, 189, 358
  and the emergent self 82, 180, 182–4
  Kierkegaard on 243
  as nothing and nowhere 189–90
  the Present of 357–8
  an understanding state-of-mind 166
  usefulness 177, 192

apophantical statements 132–3, 142
Apply-Degger podcast 4, 239
the appropriative event *(das Ereignis)* 476–7
architecture 12, 69
architecture, linguistic 88, 90
Arendt, Hannah 11, 79, 97, 274, 388, 416, 480
Aristotle 24, 55, 109, 134, 141, 197–8, 213–14, 219, 220, 222, 225, 364, 442, 446, 451, 454–6, 462
  On Interpretation 128–9, 222n
  The Metaphysics 16, 151
  ostensible subject of *Being and Time* 128, 197
  phronesis 307
  Physics 428, 444, 450, 458
  Politics 114
  Rhetoric 118, 185
"Art and Space" *(Die Kunst und der Raum)* 61
aspect changes 406–7, 418
atheism,
  Heidegger's fusion as 197–8
  a Lutheran atheism 276, 447
Augustine, Saint 14, 24, 151–2, 192, 197, 276, 345, 450–1
authenticity
  and being towards death 240, 250, 254
  choosing or loosing ourselves 19–21
  and falling 116–17
  by modification of our understanding 99–100
  and totality 311
"average everydayness" 21–2, 26, 36, 38, 55, 76–7, 80, 82, 92, 105, 158–9, 161, 163, 240, 246–8, 254, 258, 305, 348, 351, 412, 416
Avicenna (Ibn Sina) 222

B

Badiou, Alain 105, 305, 307, 397, 416
Bamford, Maria 299, 300n
bear example 118
Beaufret, Jean 35, 301, 345, 453
Becker, Ernest 247
Beckett, Samuel 294, 298, 301, 357
being
  Being in general 169, 321, 469
  meaning of 17, 105
  modes of 20
  as superficial 123
  words for 475–6
being-ahead-of-itself/oneself 170, 249, 256, 340, 352
being-alongside 338, 340, 353, 377
being and thinking, as the same 222–3
Being and Time (Sein und Zeit)
  alternative title 427
  claimed provisionality 463, 470
  Heidegger's own comments 464–5
  as a paradigm shift 481
  structure and scope 1–2, 459–60, 462
  summary 320
"being-a-self" 82
being-a-whole 241, 243, 265, 311, 315, 335, 339, 387
"being-in-the-world"
  Dasein as 13, 23, 26
  truth of 128–9
Being-its-Self, potential 281, 284, 293

the being of care 242, 403, 468
"being-towards-birth" 387
"being-towards-death"
  and anticipation 308
  and anxiety 257–8, 358
  attestation 315
  basis of Dasein's historicality 422–3
  characterized 258
  contrasted with everyday death 246
  and the emergence of the self 83, 331–2
  exclusions 262
  four criteria 248, 254, 268
  full existential conception 241
  as futural 335
  introduced 14
"being-with-others" *(Mitsein)* 26, 74, 75, 84, 95, 99, 101, 136–7, 178, 185–6, 250, 257, 259, 322, 410–12
  and being alone 87–8
Benjamin, Walter 319, 392, 410, 424
Benner, Patricia 89
Bergson, Henri 428, 455–6, 458
Berkeley, Bishop George 200–1, 213, 326n
Bernasconi, Robert 76, 92n, 247
"bewildered making-present" 356
birth 387
Blake, William 243, 248
Blanchot, Maurice 272–3
The Bonzo Dog Doo-Dah Band 108
The Book of Definitions, by Isaac Israeli 222
book titles, forms of 472–3
boredom 318
Boss, Medard 12, 172n
Bücheler, Franz 195
Bultmann, Rudolf 193
Burdach, Konrad 195, 196n
burning house example 356
Butler, Judith 271

C

"cabinet of consciousness" 29, 376
Caeiro, Alberto (Fernando Pessoa) 69–74
Camus, Albert 179
Canetti, Elias 80n
care *(Sorge)* 87, 88, 166
  authentic care 237–8, 339–40
  the being of care 242, 403, 468
  call of, and conscience 289, 297
  disclosed in anxiety 175
  meaning of care 334
  myth of care 195–6
  Reality dependent on 215
  structure of 167–71, 175, 212, 285, 289, 321–2, 340–1, 348
  and temporality 339–40
  three elements of 334, 340
car indicator example 45
Carnap, Rudolf 11, 149, 413
Carson, Anne 360–1
Cartesian worldview
  inherited by Kant 57
  unlearning 71
  see also Descartes, René
Cassirer, Ernst 148–9
"castles in the air" 479–80
Catching the Big Fish: Meditation, Consciousness, and Creativity, by David Lynch 187
categories
  distinguished from existentials 24
  ready-to-hand and present-at-hand 24–5

"categories of the understanding" 203
cave allegory (Plato) 22, 34, 259, 383, 444
Cavell, Stanley 269
cephalopods 144–6, 272, 362
Char, René 344, 453
Chaucer, Geoffrey 430n
chimpanzees 143, 145
Christian anthropology 197–8
Christian idea of guilt 290
Christianity
  and anxiety 192–4
  and conscience 276
  and the fulfillment of time 341
Christian notion of evil 278
churches and graves, orientation 63–4
Cioffi, Frank 221
circular reasoning 320–2
circumspection *(Umsicht)* 39, 41, 44, 68, 90, 126, 131–4, 338, 365, 434–5, 447
clocks 435–41, 443–4
co-historicizing 411
communal ethical life *(Sittlichkeit)* 40
"concern" *(Besorgen)* 27–30, 37–8, 44, 88, 100
concupiscence 152, 192, 362
The Condemned of Altona (play), by Jean-Paul Sartre 190n
"conditions of possibility" 47, 64, 113, 125, 213, 216–17, 252, 299, 301, 358, 367–8, 374, 376, 391, 409–10, 467
Confessions, by Saint Augustine 14, 24, 151–2, 192, 276, 345, 450–1
connectedness of life 390, 396–7
conscience 83, 241–2, 261, 308, 419
  call of 83, 261, 276–9, 281–7, 289, 297, 302–3, 316
  esoteric version 279–92, 300
  exoteric version 275–9, 282
  wanting to have a conscience 279, 281, 300, 316
conscience-anxiety *(Gewissensangst)* 258, 302
consciousness, seat of 58
considerateness *(Rücksicht)* 90
constancy of the self 84, 294, 325, 327–32, 356, 423, 433–4
"Continental philosophy" 11, 148, 149n
Continental Philosophy: A Very Short Introduction, by Simon Critchley 149n
conversion stories 262
Copernican turn (Kant) 201, 202, 208
correspondence, truth as 221, 223–4
"counterfeit immortality" 276
courage 186, 189, 191, 358, 361
Courtine, Jean-François 343
COVID pandemic 3, 26, 239
The Crisis of European Sciences, by Edmund Husserl 223, 366–7
Critchley, Simon
  ABC of Impossibility 184n
  Continental Philosophy: A Very Short Introduction 149n
  The Faith of the Faithless 313
  Infinitely Demanding: Ethics of Commitment, Politics of Resistance 261, 417
  Life Doesn't Need a Narrative Arc 206n
  "Originary Inauthenticity—on Heidegger's *Sein und Zeit*" 349, 384n, 416, 417n

Critique of Dialectical Reason, by Jean-Paul Sartre 307–8
Critique of Pure Reason, by Immanuel Kant 200n, 202–3, 206n, 208n, 209, 322, 325n, 377, 378n
Critique of Violence, by Walter Benjamin 319
curiosity *(Neugier)* 115, 125–6, 148, 151–6, 158, 163, 168, 362, 399

D

darkness and anxiety 191
Dasein
  an a priori structure 33
  as being-alongside 338, 377
  as "being-in-the-world" 13, 23, 26, 102–3, 210
  as "being-with" *(Mitsein)* 77–9
  as the between of birth and death 389–90
  as care 102, 171
  as the clearing 104
  described in spatial terms 59
  existence, truth and 233
  forgetting itself 355–6
  as the fulfillment of time 341
  as futural 336–7
  happening of 390
  as the happening of strangeness 298
  individuation 83, 256
  as "in the truth" 230
  as *l'ouverture* 35
  meaning and translation 13
  meaning of the world and 52
  potentiality for being 335–6
  primitive and advanced 439–40
  as thrown projection 107, 474–5
the Dasein-analytic 237, 244, 309, 311, 471, 474
datability 431, 448
death
  certain but indefinite 248–9, 314
  as Dasein's end 178, 244
  death of others 245, 247
  dying, perishing, and demise 271–2
  evasion of 246, 247, 250
  existential-ontological structure 244, 248, 268
  as non-relational 244–5, 248–9, 255–6, 269, 271, 352
  "not to be outstripped" 244–5, 248–9, 256, 268–9, 271, 346, 421
  as a null basis 292
  as the possibility of the impossible 244, 251–3, 272
  projection of, as possibility 254
  see also being-towards-death
The Death of Ivan Ilyich, by Leo Tolstoy 246
Debord, Guy 308
debt, guilt, and indebtedness 278, 291
deconstruction
  of the Cartesian worldview 33, 53–5, 71
  Derrida and 299
  of the ontological tradition 57, 280, 401, 460, 467
  truth as correspondence 224
  whether Nazi 266
Defoe, Daniel 179, 180n
Deleuze, Gilles 12, 105, 416
deliberation 369–70
democracy 98
Derrida, Jacques 11, 186, 221n, 224–5, 265, 266n, 274, 299, 338, 420, 426, 429, 453–4, 480

Descartes, René
coverage in missing sections 463
Meditations on First Philosophy 81, 330
reliance on mathematics 55–6
terminology rejected 12
transcendental realism 200
worldview deconstruction 33, 53–5
destruction *(Destruktion)*
of philosophical tradition 12
the devil/Satan 116, 158, 243
Dickinson, Emily 77
dictatorship of the They 95–6, 98
Dilthey, Wilhelm 122, 214, 403–5
disavowal *(Widerruf)* 133–4, 399, 422, 457
disclosure/disclosedness 104–5, 231
discourse *(Rede)*
and the call of conscience 283
and intelligibility 135–6
as one of three existentials 105–6, 113–15, 147–8, 168–9, 171
as thrown projection 168
Discourse on the Method, by René Descartes 330
dispersal amongst the world 27–8, 99
Dreyfus, Hubert 29, 84, 89, 186, 197, 307
Dreyfus, Stuart 89n, 307n
Duchamp, Marcel 246
Duns Scotus 446
duration 432, 455
"dying for" another 266–7

E

earth, relationship with the world 42, 443
Echo and the Bunnymen 176, 177n
ecstases of time 352
unity of 346, 356
ecstatic temporality 15, 174, 242, 309, 332–4, 341, 344–6, 351, 353, 356, 363, 375, 398, 401, 409, 427, 430–3, 442, 445, 448–9, 467–8
of inauthenticity 351, 356
primacy of the future 449
ecstatico-horizonal temporality 375, 381, 439, 448, 450
Eisenman, Peter 12
"elementary propositions" 130
Eliot, T. S. 459
empirical idealism 200, 203
empirical realism 200–1, 204, 215
enigma of being and of motion 400
enigma of everyday life 384
enigma of humanity 168, 170, 384, 400, 476
enigma of time 450–1
entanglement 160–2
the episodic self (Hume) 206
"epistemological construal" of the world 32, 76, 205, 211, 212, 376
epistemology
of Kant 202
and lived experience 30, 103
equanimity and indifference 361
"errance" 231
essence as movement 400
estimation, measurement as 66–7, 430, 435
eternity 174, 276, 304, 342, 345, 445–6, 452–3
ethical experience, theory of 261
an "event" 305, 307
everydayness
analytic of 351
and authenticity 117–18, 381
and falling 158

"Everyone is the other, and no one is himself" 77, 79, 98, 411
"Every secret loses its force" 96–7
existence, human, defined by time 18–19
the "existential analytic" 18, 84, 91, 100, 240, 269, 294, 296, 317–18, 321–3, 333, 340, 360, 383–4, 439, 452, 468
  rhythm of 100
the existential and the existentiell 260
existential conception of science 365–7, 369, 371, 374
existential concepts 105, 113, 115, 171–2
"existentials" 23–5, 39, 56–7, 105, 115, 148, 168–9, 171
  categories distinguished from 24
  the They *(das Man)* 94
existential spatiality 68–9, 74, 81, 379
exoteric and esoteric dimensions 275
experience, preceded by care 173
extension in space 54
  derived from Aristotle 55
extremes of solicitude 89–90, 120

F

Facing Gaia: Eight Lectures on the New Climatic Regime, by Bruno Latour 217
factical concepts 106
factical existing 406
"facticity" 27, 106, 115, 166–7, 170–1, 173, 182, 290, 294, 315, 334, 340–1, 362, 384, 405–6, 418
faith as anxiety 302
The Faith of the Faithless, by Simon Critchley 313
The Fall (band) 289, 350
falling *(das Verfallen)*
  and being-towards-death 247, 254
  as an existential concept 115
  and the experience of anxiety 183–4
  five characteristics of 158, 231
  inauthentic concepts 105–6, 148
  three models of 115–16
Farias, Victor 265
"fascination" *(Benommenheit)* 29–30, 38, 76, 172, 381
fate, the powerless superior power 420
fatigue as productive 405
fear
  of death 260
  distinguished from anxiety 119, 185, 189, 358
  opposed to confidence 118
  and the pandemic 239–40
Fernandez, Ramon 50–2
Fichte, Johann Gottlieb 326
Ficker, Ludwig von 130
finitude
  and being human 474
  demands of 257
  ecstatic temporality and 346
  fundamentally relational 269, 271–3
  making meaning from 15
Finland 140–1
flight from flight, anxiety as 183–4
flirtation 154, 180
forbearance *(Nachsicht)* 90
Foucault, Michel 11, 175, 221n, 266n, 403

"freedom"
first felt in anxiety 180
and the nullity of projection 295–6
from things of the world 359
through acting in the world and being-towards-death 111, 258–60
through being futural 170
"freeing" our relationship to the world 68–9
Friedman, Michael 148, 149n
Fundamental Concepts of Metaphysics (Die Grundbegriffe der Metaphysik) 143
the future, inauthenticity 352
the future, and the monumental 402
the future, as primordial 333, 335, 346, 349
futurity of death 249
Fynsk, Christopher 81

G

Gadamer, Hans-Georg 11, 122, 368, 478–9
Gaia 217, 443
On the Genealogy of Morals, by Friedrich Nietzsche 160, 278, 290, 296, 325n
the geocentric universe 63
German history 415
God
and eternity 446–7
and finitude 261–2
and self-constancy 330–1
the self without reference to 14–15
speculation about 73
Godfrey-Smith, Peter 144
gods, clearing away 61
guilt 290–1, 296–7, 300
Guilty! (as the call of conscience) 277–8

H

Habermas, Jürgen 11, 265, 370, 414
Hamlet 17, 19n 40, 113, 131, 148, 277, 283
hammer illustration 38, 41, 43, 74, 127, 132, 142, 370–2
handy things *(das Zeug)* as stuff 37–8, 41, 43, 45–6, 51, 58, 60, 62, 76
Hartmann, Nicolai 250
Haugeland, John 93, 117, 307
Havel, Vaclav 230
having-been, arising from the future 337
Heat (movie, 1995) 192
Hedda Gabler, by Henrik Ibsen 153–4
Hegel, G. W. F. 40, 78, 81, 193, 207, 243, 251, 291, 348, 428, 442, 451–2, 454–6, 458
Heidegger, Hermann (M. H.'s son) 61n, 349, 389n
Heidegger, Martin
Art and Space (Die Kunst und der Raum) 61
The Basic Problems of Phenomenology 460, 464
The End of Philosophy and the Task of Thinking (Das Ende der Philosophie und die Aufgabe des Denkens) 427n, 466
influence beyond philosophy 12
Introduction to Metaphysics (Einfuhrung in die Metaphysik) 298, 319–20, 413–14

Kant and the Problem of Metaphysics (Kant und das Problem der Metaphysik) 205n, 378, 462, 463n
"Letter on Humanism" ("Brief über den 'Humanismus'") 35, 114, 284n, 301
Origin of the Work of Art (Der Ursprung der Kunstwerkes) 42, 216–17, 393, 442–3
The Question Concerning Technology (Die Frage nach der Technik) 18, 372
"The Thinker as Poet" *(Aus der Erfahrung des Denkens)* 173
On the Way to Language (Unterwegs zur Sprache) 303, 462
What Is Metaphysics? (Was ist Metaphysik?) 119, 175, 182, 190
Heidegger: Through Phenomenology to Thought, by William J. Richardson 108
Heidegger and Nazism, by Victor Farias 265
Heidegger Thinking
basis in lectures 2–3
successor to a podcast 4
Heisenberg, Werner 234
Heraclitus 219, 227, 443
heritage and repetition 421
heritage and tradition 420–1
the hermeneutical circle 127
hermeneutical violence 224, 318–19
hermeneutics
of Dasein 471
of facticity 27
invention of 122
preceding logic 133
"hetero-affectivity" 417
Hetherington, Fiona 28
historical constancy 434
historicality 386–7, 391, 394, 397–8, 400, 402, 422
historicity 242, 257, 348, 382, 385, 391, 397–407, 411, 423–5, 447
inauthentic 397–8
and politics 409
historicizing 394–5, 409, 411, 467
historiology 391, 400–3, 422
authentic historiology 402
ontological genesis 400
historizing 390, 392, 395–6, 403, 409–12, 419, 422, 452
history
nature of 423–4
roots in the future 394
three dimensions of 403
Hitler, Adolf 412
Hölderlin, Friedrich 415, 438, 447
the Holocaust 267–8
hope 359–60
the horizon, overcoming 465–6
horizonal schemas 377
humanism 35, 143, 301, 361–2, 474, 476
humanity, enigma of 168, 170, 384, 400, 476
humans as rational animals 114, 141
Hume, David, and Kant 206, 235, 324
humor of being human 419
Husserl, Edmund 63n, 85, 190, 223, 230, 261, 313, 366–9, 371–2, 383–4, 456, 464n
hypokeimenon 31, 318

I

ice floe analogy 280, 312
idealism vs realism, both wrong 213–14

idealism vs realism, four theses 200
The Idea of Order at Key West, by Wallace Stevens 48–50
identity, distinguished from sameness 223
idle talk, defined 149–50
illustrations, use of outmoded 43–4, 65
impotence/impotentialization 246, 273, 288, 293, 299, 317, 419
inauthentic life, not a negative 94
indifference and equanimity 361
individuation 83, 180, 255, 257
Infinitely Demanding: Ethics of Commitment, Politics of Resistance, by Simon Critchley 261, 417
"intellection" of Descartes 54
interpretation *(Auslegung)*, and understanding 121–2, 126–9
interpretative tendencies 36
intersubjectivity problem 81, 111, 136, 327
Introduction to Metaphysics (Einfuhrung in die Metaphysik) 298, 319–20, 413–14
involvement *(Bewandtnis)* 46–7, 52, 68, 103, 121, 126, 189
Irigaray, Luce 246
Israeli, Isaac 222
Izzard, Eddie 34

J

James, William 37, 161
Janicaud, Dominique 35, 186, 345
Johnson, Samuel 201
Jonas, Hans 359, 480
Jones, James 187, 188n
joy, and anticipatory resoluteness 317–18

K

Kafka, Franz 83, 272, 296
Kant, Immanuel
- Cartesian worldview inherited 57
- concept of the self 322
- Copernican turn 201, 202, 208
- and David Hume 206, 235, 324
- epistemology 202, 269
- moral "ought" 291
- presuppositions questioned 207
- problem of orientation 67–8
- rejection of idealism 201, 203, 208
- on time 364, 451
- transcendental unity of apperception 206, 324–5
- worldview deconstruction 53–4

Kant and the Problem of Metaphysics (Kant und das Problem der Metaphysik) 205n, 378, 462, 463n
Kaurismaki, Aki 141
Kierkegaard, Søren 14, 186, 193, 197, 242–3, 340, 354–5
King Lear 277, 282, 283n
Knausgård, Karl Ove 97
knowing as being-in-the-world 204–5

L

Lacan, Jacques 161n, 216, 281
Lacoue-Labarthe, Philippe 84n, 407–9
language
- and being human 114, 141–2
- as ecstatic 137
- form prior to propositions 131
- highest form in propositions 128–9

as the house of being 114
inherited terminology 24–5
not inside the head 23
and phenomenology 23
prejudice 428
Rede and *Sprache* 134
as the uncovering of things 227
as unifying Germany 415
see also discourse
Latour, Bruno 56, 217
leaping ahead of the other *(vorausspringen)* 89–90, 120
Lectures on the History of Philosophy, by G. W. F. Hegel 348
leeway *(Spielraum)* 380–1
"Letter on Humanism" ("Brief über den 'Humanismus'") 35, 114, 284n, 301
"leveling down" 97
leveling off of time 448, 452
Levinas, Emmanuel 91–2, 111n, 246, 266–73, 405, 416, 458, 480–1
Life Doesn't Need a Narrative Arc, by Simon Critchley 206n
The Life of the Mind, by Hannah Arendt 388
"Life" (poem) 78
lifeworld 367–9, 371
Lingis, Alphonso 111n, 186, 405
the "linguistic turn" 113
liveliness 153–4
"living in the truth" (Havel) 230
Locke, John 368
The Lonely Crowd: A Study of the Changing American Character, by David Riesman et al 184–5
Lovelock, James 217
Löwith, Karl 385, 407
Lukács, Georg 229
Luther, Martin 14, 192, 193n, 197, 261, 276–7, 290, 313, 340
Lutheran(ism) 256, 262, 276, 447
Lynch, David 187

M

Macquarrie, John 11n, 13, 46, 117n, 287, 380
Madsbjerg, Christian 4
Magnolia (film) 110
making-present *(Gegenwärtigen)* 355–6, 364, 431, 433
Malick, Terrence 88, 186–8
manifold of intuition (Kant) 206, 378
the Marburg lectures 192, 378–9
Marko, Zenon 3, 239
Marx, Karl 201, 442
mathematization of nature 55–6, 372–4
matter, status of 216
McCarthy, Tom 179, 290
meaning as an existential of Dasein 122
meaning of care 174
Meditations on First Philosophy, by René Descartes 81, 330
Meillassoux, Quentin 201
Mein Kampf (My Struggle), by Adolf Hitler 412
Meister Eckhart 467
Melville, Herman 83, 306, 438
memory and repetition 355–6
The Merchant of Venice 291
Merleau-Ponty, Maurice 11, 47, 367, 480
metaconcepts (Aristotle) 24
metaphors, anachronistic 43–4, 65
metaphors, maritime 83, 178, 181

metaphysical terms rejected 25, 105
metaphysics, definition 220
metaphysics, overcoming 477
metaphysics, poem 72–3
mineness *(Jemeinigkeit)* 18–19, 21, 95, 268–70, 272, 322–3, 360, 415, 474
modalization 311
modernity in philosophy 55–6, 265, 399
the moment of vision *(Augenblick)* 329, 333–4, 340–1, 353–5, 358, 363, 383, 397–8, 401–2, 418, 422–3, 432–3, 445, 447, 450, 467
Monk, Ray 129
Montaigne, Michel de 259–60
mood as one of three existentials 107–8, 110–11, 167, 177–8
  see also anxiety
moods, and rate of time passing 445
moods, taxonomy of 359
morality, pre-moral source 278
Morrison, Jim 110
mortality and meaninglessness 85
"most closely and mostly" *(zunachst und zumeist)* 21, 24, 28, 36, 64, 75, 109, 115, 131, 204, 382
mourning 264, 270–1
music, experience of 106
mysteries as delusional 73–4
"the Myth of the Given" (Sellars) 203

N

Nancy, Jean-Luc 84n, 408–9
Naturalism 35–6
"natural light" in man 104, 151–5
Nazi (National Socialist) Party 84n, 91, 94, 96, 234, 265–7, 385, 387–8, 392, 407, 413–17
The New School, New York 5, 95, 274
Newton, Isaac 60–1, 68, 233, 441, 448
Nietzsche, Friedrich 15, 160, 251, 256–7, 278, 286, 290, 294, 296, 325, 389n, 398, 401–3, 475, 477
nihilism 160n, 257n, 359, 389, 412
"nobodaddy" (Blake) 248
non-human creatures 143, 272, 361–2
note-taking, importance of 404
the not-yet 249, 253, 452
the not-yet-now 346, 428, 444
novels, anxiety and 179–80
the "now," privileging 454
now-time *(Jetzt-Zeit)* 42, 242, 427, 435, 443–4, 448
"The Null Basis-Being of a Nullity...," by Simon Critchley 296n
nullity of projection 294–5, 298
nullity of thrownness 294, 336–7
nursing theory 88–9

O

objectivism 366
octopus 144–6, 272, 362
Of Spirit: Heidegger and the Question, by Jacques Derrida 338
Olson, Ben 5
Onassis Foundation 4
"On the Essence of Truth" (*Vom Wesen der Wahrheit*, 1930 lecture) 220, 231, 303
On the Way to Language (Unterwegs zur Sprache) 303, 462

the ontic and the ontological 83, 241, 275, 323, 355, 473
ontological tradition, deconstruction 57, 280, 401, 460, 467
ontology
  of the present-at-hand 33, 54, 207–10, 213, 226, 325–6, 329, 365, 393, 442
  presuppositions in 320
"opening and presence" 427, 429, 466–7
orientation, problem of 67–8
Origin of the Work of Art (Der Ursprung der Kunstwerkes) 42, 216–17, 393, 442–3
"Originary Inauthenticity—on Heidegger's *Sein und Zeit*," by Simon Critchley 349, 384n, 416, 417n
Other Minds: The Octopus, the Sea, and the Deep Origins of Consciousness, by Peter Godfrey-Smith 144
others
  death of others 245, 247, 263–4
  detectable via things 87
  keeping a distance 94–5
  sanctity of 92

P

Parmenides of Elea 222–3, 232, 455
parties 181
A Parting of the Ways: Carnap, Cassirer, and Heidegger, by Michael Friedman 148–9
passions and reason 109
Patočka, Jan 230
Paul, Saint 14, 194, 197, 276, 290, 302–3, 319n, 340–1, 430n
the people *(Volk)* 412, 415–16
perception, problem of 32
persistence of the self 328
Pessoa, Fernando (poems) 69–74, 123
Peterson, Jordan 266n
phenomenal content of the dead 264
phenomenology 22–3, 56–7, 114, 135, 142, 147, 170, 214, 261, 278, 296, 313, 343, 366–8, 384, 460, 464, 473, 478, 480
  beginning point 384
  phenomenology and scientism 366–7
  phenomenology of direction 68
  phenomenology of existence 56, 109
  phenomenology of flirtation 154
  phenomenology of guilt 278
  phenomenology of hearing 137–8
  phenomenology of moods 110
  phenomenology of truth 217
  pre-phenomenological experience 323, 325, 327
  universal phenomenological ontology 471, 473–4
Phenomenology and/or Tautology, by Jean-François Courtine 343
Phenomenology of Perception, by Maurice Merleau-Ponty 47n
Phenomenology of Spirit, by G. W. F. Hegel 6, 207
phenomenon of the world 38
  avoiding splitting 103–4, 165, 218
  ignored by Descartes 54–5

philosophy
- critical turn in 202
- as drama 190
- inherited terminology 24–5
- knowing how to die 259
- modern, as epistemology 212
- as parricide 457
- "scandal of" 209–10
- sunlight, sickness and 72–4, 123
- and taking time 2, 434

philosophy, existential 304
physics, truths of 233–4
The Pixies 81
Plato 105, 128, 200–2, 235, 475, 479
- *Phaedrus* 282n
- *The Republic* 22, 34, 259, 275
- *The Sophist* 16, 465, 469–70
- *The Timaeus* 444–5

plurality, whether positive or negative 416
poetry
- "The Idea of Order at Key West," by Wallace Stevens 48–50
- "Life" 78
- as pre-propositional 137
- "To not think of anything is metaphysics enough" 72–3
- "What we see of things are the things" 70–1, 123

Pöggeler, Otto 169, 460
political moments 84–5
politics of ambiguity 157
possibility and actuality 251–2
potentiality-for-being 117, 138, 160, 248, 253–6, 281, 283–4, 289, 297, 301, 312, 315–16, 328, 335, 339, 352, 355–6
Precarious Life, by Judith Butler 271n
"pre-science" 368
presence, metaphysics of 429
presencing *(Anwesenheit)* 465–7
present-at-hand *(vorhanden)*
- distinguished from ready-at-hand 39
- ontology of 33, 54, 207–10, 213, 226, 323, 326–7, 329, 365, 393, 442
- science and 36
- the world seen as 32, 54

professorship at Marburg 197, 250
projection *(Entwurf)* 85, 111–12, 125, 166, 168–71, 173, 231, 251, 253, 272–3, 288, 290, 293–8, 302, 321, 324, 337, 340, 347, 421
- ecstatic projection 470
- existential projection 335, 418
- understanding linked to 168

propositions
- not unique truths 128
- in Wittgenstein's *Tractatus* 129–30

Protestant theology 116, 255, 290, 317, 341, 447
- Calvinism in the USA 262n
- idea of guilt 290
- *see also* Lutheranism

"proximally and for the most part" *(zunächst und zumeist)* 21, 36, 39, 63, 75, 78, 116, 131, 204, 207, 327, 329, 351, 382
psychoanalysis 108, 111, 161, 176, 317
publicness 78, 97, 148, 151, 158, 281, 382, 402
public transport, distancing 95
Pulp Fiction (movie) 395–6

Q

quantum mechanics 234, 441–2
quasi-transcendental methods 224–5, 299

the question of totality 237–8
Quine, W. V. 212

R

rabbits, giant 138
rabbits in hats 123–4
radio metaphor 65
rapture 341, 350, 353–4, 377, 418
Rayner, Claire 183
the ready-to-hand *(Zuhandenheit)* 24–5, 37
  announcing itself 42
  associated with a "what-for" 41–2
  car indicator example 45
  distinguished from present-at-hand 39, 370
  features of the space of 60, 62
the Real *(das Reales)*, dependent on Dasein 214–15
realism vs idealism, both wrong 213–14
realism vs idealism, four theses 200
reality and the real 215–17
reciprocative rejoinders 422–3
"recurrence of the possible" 397–8
referential structure/totality *(Verweisungsganzheit)* 38, 45–6, 58, 60, 62, 76, 78, 126
Reflections on Violence, by Georges Sorel 319
relativity theory 441
releasement 467
repetition *(Wiederholung)* 289, 314–15, 331, 351, 356, 398, 420
resolution/resoluteness 83–4, 279–81, 302–6, 308, 315
  anticipatory resoluteness 242, 309–13, 315, 318–19, 328, 331–2, 335–6, 338–9, 421
  which resolves to keep repeating itself 314, 422
reticence (and a reticence bar) 140–1, 151, 154–5, 256, 277, 283
revolutions 388, 424–5
rhythm
  of being itself 297
  of disclosure and concealment 106
  of the "existential analytic" 100
Richardson, Father William 108
Rieff, Philip 161
Riesman, David 184–5
Rimbaud, Arthur 79n, 80n
ring, gift of 394–5
Robinson, Edward 11n, 13, 46, 117n, 287, 380
Rorty, Richard 131
Rosenzweig, Franz 269
Ryle, Gilbert 186

S

the sacred, and sacred spaces 62
Sallis, John 220n, 343n, 460
sameness, distinguished from identity 223
Sartre, Jean-Paul 11, 47n, 179, 185, 190, 266, 276, 304–5, 307–8
Satan/the devil 116, 158, 243
Scheler, Max 214
Schelling, F. W. J. 265
Schiller, Friedrich 380
Schmitt, Carl 96, 98, 392
science, existential conception of 365–7, 369, 371, 374
"Science does not think" 369
scientific conception of nature 57–8
scientific conception of space 60–1, 68–9

scientism 366, 371–2
seeing without thinking 70–1
the self
  authenticity and choice 280
  finitude and conscience 275
  impotentialized 273
  narrative idea 445n
  nonmetaphysical idea of 85
  not ontically obvious 86
  persistence of 328–9
  precipitation of 178, 182
  presupposing a worldless self 211
  unity of 322
  without reference to God 14–15
self-constancy 84, 294, 325, 327–32, 356, 423, 433–4
Sellars, Wilfrid 203
Shakespeare, William 113, 258, 283n, 291
  see also Hamlet
Sheehan, Thomas 15, 250, 266n, 460, 464n
sickness and philosophy 73–4
sight
  and our relation to things 126
  and solicitude 90
  taking delight in 151–2
  and time reckoning 435
"significance" 46–8, 52, 68, 142, 189, 448
signs, as indicators 43–5, 127
silence, call of conscience as 277, 303
sin, possibility of 312
the "situation" 304–8, 338, 341, 353
Situationism 308
skepticism 229, 234–6
sleep, and self-constancy 330
Sloterdijk, Peter 93–4, 98
smartphones, scrolling 95
Smith, Mark E. 289, 350
social media 65–6, 80, 97, 99, 155, 277
social ontology 79
Socrates, trial and execution 259
Socrates's *daimon* 282
solicitude *(fursorge)* 88–90, 100, 120, 173, 180, 246, 259, 287
  vorausspringen 89, 120
  see also care
solipsism 120, 184
Sorel, Georges 319
space
  bringing things closer *(Entfernung)* 60, 64–6
  conceived of as roominess 58
  existential spatiality 61–2, 68–9, 74, 81
  genesis of modern view 60–1, 68–9
  no third-person view 381
Spinoza, Baruch 390
"spirit" and temporality 452–3
splitting the phenomenon 103–4, 165, 218, 226
Stanshall, Vivian 108
state-of-mind
  co-state-of-mind 120
  as one of three existentials 105, 107–9, 119, 147–8, 168–9
status corruptionis 312
Stein, Edith 456
Stevens, Wallace 48, 76, 374
stigma of the inauthentic 417
Strauss, Leo 274–5, 480
struggle *(Kampf)* 97–8
the subject as always present-at-hand 325–6
subject-object dualism 32, 165, 205, 225, 376, 442

subway travel 95
succession, time and 364
sundials 438, 440
Sure, I'll Join Your Cult, by Maria Bamford 300

T

taking-as as interpretation 126, 132
Taminiaux, Jacques 344
Tarantino, Quentin 395–6
tautology 130, 333, 343
Taylor, Charles 32
temporality
  autotemporalization 342–3
  dimensions 334
  inauthentic 352
  and the meaning of care 196, 334, 346
  temporality of discourse 363
  temporality of mood 357
  temporalizing of 174, 348, 375, 439, 452, 470, 472
  see also ecstatic temporality
theoretical attitudes, genesis of 365
theoreticism 30, 76, 442
theory-practice distinction 38–9
therapy, culture of 161
the They *(das Man)* 77, 92–3, 99, 246, 280–1, 308, 327, 382, 410
  as an existential 94, 99
  dictatorship of 95–6, 98–9
  as fun 181
  immersion in 351
  in relation to death 246–7, 254
the they-self 82, 85, 94–5, 97–9, 158, 178, 185, 246, 248, 258–9, 279, 282, 287, 289, 302, 305, 308, 320, 327, 331, 351, 397, 416
"The Thinker as Poet" *(Aus der Erfahrung des Denkens)* 173
thinking things and extended things 33, 54, 205, 209
The Thin Red Line (movie) 88, 187, 188n
Thomas Aquinas, Saint 221–2, 446, 475
"thrownness" *(Geworfenheit)* 19, 26, 100, 106–7, 110–13, 115, 125, 157, 166–8, 170–1, 173, 231, 234, 288–90, 293–8, 336–7, 340, 347, 362–3, 406, 435, 467
  structural elements 162
"thrown projection" 26, 101, 107, 112–13, 115, 125, 151, 162, 168–71, 173–4, 185, 229–32, 259, 277, 289–90, 295, 297–8, 322, 406, 474–7
  and the appropriative event 477
  articulated thrown projection 169–70
time
  Christian idea of fulfillment 341
  defining human existence 18–19
  enigma of 451
  as finite 14
  as the horizon of being 470–1
  "making present awaitingly" 436
  philosophy as taking time 434
  primordial time 426–9, 435, 447, 452, 468, 470, 472
  reckoning with time 429, 434–5
  reifying 342–3
  three conceptions 427
  three dimensions 339
  and tides 181–2, 437
  whether reversible 445, 447, 449
  see also temporality
"Time and Being" (lecture) 59
time-consciousness 456–7
time measurement 438–41

"time-space" *(Zeit-Raum)* 59, 379, 435
Tolstoy, Leo 246
"To not think of anything is metaphysics enough" (poem) 72–3
Totality and Infinity, by Emmanuel Levinas 111n, 269, 270n
tourism 183–4n
towards-oneself, the future as 346
Tractatus Logico-Philosophicus, by Ludwig Wittgenstein 129–30
tranquillity/tranquilization 159, 182, 256–7, 361
transcendence 23, 59, 167, 173, 365, 374, 376
  the transcendental ego 85
  transcendental idealism 200, 201–2, 204, 207–8, 216
  transcendental ontology 260
  transcendental realism 200–1, 203, 208, 299
transcendental unity of apperception (Kant) 206, 324–5
transference 176
the trivial, unintelligibility of 139–40
truth
  as correspondence 221, 223–4
  as disclosure 224, 230
  as movement 106
  not eternal 233, 235
  three theses 225–6
truth and propositions 220–1
Tugendhat, Ernst 369

U

the uncanny, experiencing 164, 177, 180, 190–1, 275–7, 280, 285–8, 290, 297
unconcealment and concealment 228–9
uncoveredness, being-uncovered *(Entdeckend-sein)* 226–7, 230
understanding
  as one of three existentials 105–6, 113–15, 147–8, 166, 168–9, 171, 351, 364
  in relationship with others 90, 99
"underway" 461, 471
unlearning, lessons in 70–1

V

value, adding to material nature 57–8
Vaneigem, Raoul 308
velleity 306
verbalizing nouns 343, 350, 450
virility and impotence 246, 273
vulgar time (now-time) 42, 242, 427, 435, 443–4, 448, 451

W

Waiting for Godot, by Samuel Beckett 301, 357
the "waiting-towards 333–4, 353–4
Watt, Ian 180
Weimar Germany 94, 278
Western Buddhism 160
What Is Metaphysics? (Was ist Metaphysik?) 119, 175, 182, 190
"What we see of things are the things" (poem) 70–1
Why is there something rather than nothing? 17
Williams, Bernard 291–2
within-timeness 354, 427, 434, 452
Wittgenstein, Ludwig 74, 113, 129–31, 137, 186, 188, 251, 254, 406, 434

Tractatus Logico-Philosophicus 129–30
words, elemental 229, 236
workshop example 86
the world
announcing itself 34–5, 40–1, 43
falling towards 116
humans as already in 13, 28
meaning dependence on the singer 51
of non-human animals 143, 145
as nothing and nowhere 189
phenomenological description 38
proof of its existence 209, 211
relationship with earth 42
whether subjective 375–6, 442
world history as secondary 392–5, 397
"worldhood" 26, 33, 52, 64, 68, 144, 189, 213, 216, 374, 448
world-time 427–9, 431–2, 434–5, 438–40, 442–4, 448, 452–3

Y

Yorck von Wartenburg, Count Paul 404–5

SIMON CRITCHLEY has written more than twenty books, including studies of Greek tragedy, David Bowie, football, suicide, Shakespeare, and how philosophers die, as well as a novella. He is the Hans Jonas Professor of Philosophy at the New School for Social Research in New York. In 2024, New York Review Books published his nonfiction work *Mysticism*.

BEN OLSON works on Heidegger and historicity, specifically applying Heidegger's thought to the question of family. He is currently an Onassis Fellow at the New School for Social Research and teaches as an adjunct lecturer in philosophy at New York City College of Technology.